Chiral Amine Synthesis

Edited by
Thomas C. Nugent

Related Titles

Blaser, H.-U., Federsel, H.-J. (eds.)

Asymmetric Catalysis on Industrial Scale

Challenges, Approaches and Solutions

Second Edition

2010

ISBN: 978-3-527-32489-7

Dunn, P., Wells, A., Williams, M. T. (eds.)

Green Chemistry in the Pharmaceutical Industry

2010

ISBN: 978-3-527-32418-7

Otera, J., Nishikido, J.

Esterification

Methods, Reactions, and Applications

Second Edition

2010

ISBN: 978-3-527-32289-3

Kollár, L (ed.)

Modern Carbonylation Methods

2008

ISBN: 978-3-527-31896-4

Börner, A. (ed.)

Phosphorus Ligands in Asymmetric Catalysis

Synthesis and Applications

2008

ISBN: 978-3-527-31746-2

Yamamoto, H., Ishihara, K. (eds.)

Acid Catalysis in Modern Organic Synthesis

2008

ISBN: 978-3-527-31724-0

Ricci, A. (ed.)

Amino Group Chemistry

From Synthesis to the Life Sciences

2008

ISBN: 978-3-527-31741-7

Christmann, M., Bräse, S. (eds.)

Asymmetric Synthesis - The Essentials

2008

ISBN: 978-3-527-32093-6

Hudlicky, T., Reed, J. W.

The Way of Synthesis

Evolution of Design and Methods for Natural Products

2007

ISBN: 978-3-527-32077-6

Chiral Amine Synthesis

Methods, Developments and Applications

Edited by
Thomas C. Nugent

WILEY-VCH Verlag GmbH & Co. KGaA

The Editor

Prof. Thomas C. Nugent
Jacobs University Bremen
Department of Chemistry
Campus Ring 1
28759 Bremen
Germany

The cover design was created by Michael Holt (Bremen, Germany).

All books published by **Wiley-VCH** are carefully produced. Nevertheless, authors, editors, and publisher do not warrant the information contained in these books, including this book, to be free of errors. Readers are advised to keep in mind that statements, data, illustrations, procedural details or other items may inadvertently be inaccurate.

Library of Congress Card No.: applied for

British Library Cataloguing-in-Publication Data
A catalogue record for this book is available from the British Library.

Bibliographic information published by the Deutsche Nationalbibliothek
The Deutsche Nationalbibliothek lists this publication in the Deutsche Nationalbibliografie; detailed bibliographic data are available on the Internet at http://dnb.d-nb.de.

© 2010 WILEY-VCH Verlag GmbH & Co. KGaA, Weinheim

All rights reserved (including those of translation into other languages). No part of this book may be reproduced in any form – by photoprinting, microfilm, or any other means – nor transmitted or translated into a machine language without written permission from the publishers. Registered names, trademarks, etc. used in this book, even when not specifically marked as such, are not to be considered unprotected by law.

Cover   Formgeber, Eppelheim
Typesetting   Thomson Digital, Noida, India
Printing and Binding   T. J. International Ltd., Padstow, Cornwall

Printed in Great Britain
Printed on acid-free paper

ISBN: 978-3-527-32509-2

*This book is dedicated to my twins, Kian and Helen,
and their wonderful mother Elisa.*

Foreword

A vast majority of drugs are amines or contain functional groups derived from amines, and an increasing number of these molecules are chiral and nonracemic. Thus, the development of many biologically active molecules, whether derived from natural or synthetic sources, relies on the development of general and efficient methods to prepare chiral amines. With this need in mind, many researchers made the enantioselective synthesis of amines a focus of their research during the past decade. During the early 1990s, several groups illustrated that the asymmetric hydrogenation of imines was possible and that the asymmetric hydrogenation of enamides without the accompanying ester functionality present in the reagent for the classic hydrogenation to form L-DOPA could be achieved. However, this work also revealed many of the challenges faced by these types of hydrogenation. Further work on hydrogenation of heteroarenes illustrated both progress and additional challenges. Finally, methods to form chiral amines by C–C and C–N bond formation began to emerge. Again, progress was made, but this chemistry confronted many challenges.

The asymmetric synthesis of amines is a difficult problem for many reasons described in this monograph. Although a comprehensive list of challenges cannot be provided here, I outline a few of them. First, acyclic imines can adopt E and Z isomers, and these isomers often lead to opposite enantiomers of the addition product, thereby reducing enanioselectivity. Second, both imines and amines can be ligands for a transition metal, and amines or amides can form metal–amide or metal–amidate complexes in which the product becomes a part of the structure of the active – or deactivated – catalyst. Third, nitrogen protective groups are a particularly thorny component of any synthetic endeavor, and the most useful protective groups for a synthetic sequence, due to their ability to be removed under mild conditions, need not be the substituents on nitrogen that lead to highly enantioselective chemistry to form amine derivatives.

However, the creativity of a large number of synthetic chemists has led to tremendous progress toward addressing these challenges and the development of truly practical methods to prepare chiral amines. Perhaps the best illustration of how an open mind can lead to significant advances toward the asymmetric synthesis of amines is the development of the process to manufacture the blockbuster diabetes drug Januvia by the asymmetric hydrogenation of an unstabilized N–H imine.

Chiral Amine Synthesis: Methods, Developments and Applications. Edited by Thomas C. Nugent
Copyright © 2010 WILEY-VCH Verlag GmbH & Co. KGaA, Weinheim
ISBN: 978-3-527-32509-2

This book describes many valuable approaches to the enantioselective preparation of chiral amines using both chemical and biological catalysts. As is the case for the synthesis of any substructure or set of molecules containing common functionality, synthetic chemists need a box filled with diverse tools to ensure that he or she can construct compounds of different sizes, shapes, and architecture. Thus, the synthesis of amines by reduction, C–C bond formation, or C–N bond formation are all useful and together provide a means to prepare some of the most challenging molecules – those containing basic nitrogens. An old adage states that each nitrogen in a molecule increases a graduate student's career by at least 1 year; this statement is likely to need editing as a result of the creative research described in this book.

Urbana, Illinois
December 2009

J. Hartwig

Contents

Foreword *VII*
Preface *XVII*
List of Contributors *XXIII*

1	**Stereoselective Synthesis of α-Branched Amines by Nucleophilic Addition of Unstabilized Carbanions to Imines** *1*	
	André B. Charette	
1.1	Introduction *1*	
1.2	Overview of the Methods for the Preparation of Imines *3*	
1.2.1	N-Aryl and N-Alkyl Imines and Hydrazones *3*	
1.2.2	N-Sulfinyl Imines *3*	
1.2.3	N-Sulfonyl Imines *4*	
1.2.4	N-Phosphinoyl Imines *5*	
1.2.5	N-Acyl and N-Carbamoyl Imines *6*	
1.3	Chiral Auxiliary-Based Approaches *6*	
1.3.1	Imines Derived from Chiral Aldehydes *7*	
1.3.2	Imines Bearing a Chiral Protecting/Activating Group *9*	
1.4	Catalytic Asymmetric Nucleophilic Addition to Achiral Imines *15*	
1.4.1	Catalytic Asymmetric Addition of sp^3 Hybridized Carbanions *16*	
1.4.1.1	Copper-Catalyzed Dialkylzinc Additions *16*	
1.4.1.2	Zinc Alkoxide-Catalyzed Dialkylzinc Additions *20*	
1.4.1.3	Early Transition Metal (Zr, Hf)-Catalyzed Dialkylzinc Additions *20*	
1.4.1.4	Rhodium-Catalyzed Dialkylzinc Addition Reactions *23*	
1.4.2	Catalytic Asymmetric Allylation of Imines *24*	
1.4.3	Catalytic Asymmetric Addition of sp^2 Hybridized Carbanions *29*	
1.4.3.1	Catalytic Asymmetric Vinylation *31*	
1.4.3.2	Catalytic Asymmetric Arylation *32*	
1.4.3.2.1	Amino Alcohol-Catalyzed Addition of Organozinc Reagents *32*	
1.4.3.2.2	Rhodium Phosphine-Catalyzed Arylation of Imines *34*	
1.4.3.2.3	Rhodium Diene-Catalyzed Arylation of Imines *38*	
1.4.4	Catalytic Asymmetric Addition of sp Hybridized Carbanions *39*

Chiral Amine Synthesis: Methods, Developments and Applications. Edited by Thomas C. Nugent
Copyright © 2010 WILEY-VCH Verlag GmbH & Co. KGaA, Weinheim
ISBN: 978-3-527-32509-2

1.5	Conclusion 42
	References 44

2	**Asymmetric Methods for Radical Addition to Imino Compounds** 51
	Gregory K. Friestad
2.1	Background and Introduction 51
2.2	Intermolecular Radical Addition Chiral N-Acylhydrazones 52
2.2.1	Design of Chiral N-Acylhydrazones 52
2.2.2	Preparation of Chiral N-Acylhydrazones 54
2.2.3	Tin-Mediated Addition of Secondary and Tertiary Radicals 55
2.2.4	Tin-Free Radical Addition 58
2.2.5	Manganese-Mediated Radical Addition 59
2.2.6	Manganese-Mediated Coupling with Multifunctional Precursors 60
2.2.6.1	Hybrid Radical–Ionic Annulation 60
2.2.6.2	Precursors Containing Hydroxyl or Protected Hydroxyl Groups 62
2.2.6.3	Ester-Containing N-Acylhydrazones 64
2.2.6.4	Additions to Ketone Hydrazones 65
2.3	Asymmetric Catalysis of Radical Addition 67
2.4	Closing Remarks 68
	References 69

3	**Enantioselective Synthesis of Amines by Chiral Brønsted Acid Catalysts** 75
	Masahiro Terada and Norie Momiyama
3.1	Introduction 75
3.2	Carbon–Carbon Bond Forming Reactions 76
3.2.1	Mannich and Related Reactions 76
3.2.1.1	Mannich Reaction 76
3.2.1.2	Nucleophilic Addition of Diazoacetates to Aldimine 81
3.2.1.3	Vinylogous Mannich Reaction 83
3.2.1.4	Aza-Petasis–Ferrier Rearrangement 84
3.2.2	One-Carbon Homologation Reactions 85
3.2.2.1	Strecker Reaction 85
3.2.2.2	Aza-Henry Reaction 86
3.2.2.3	Imino-Azaenamine Reaction 87
3.2.3	Friedel–Crafts and Related Reactions 87
3.2.3.1	Friedel–Crafts Reaction via Activation of Aldimines 87
3.2.3.2	Friedel–Crafts Reaction via Activation of Electron-Rich Alkenes 91
3.2.3.3	Pictet–Spengler Reaction 93
3.2.4	Cycloaddition Reactions 94
3.2.4.1	Hetero-Diels–Alder Reaction of Aldimines with Siloxydienes 94
3.2.4.2	Direct Cycloaddition Reaction of Aldimines with Cyclohexenone 95
3.2.4.3	Inverse Electron-Demand Aza-Diels–Alder Reaction (Povarov Reaction) 96

3.2.4.4	1,3-Dipolar Cycloaddition Reaction 97	
3.2.5	Aza–Ene-Type Reactions 99	
3.2.5.1	Aza–Ene-Type Reaction of Aldimines with Enecarbamates 99	
3.2.5.2	Cascade Transformations Based on Tandem Aza–Ene-Type Reaction 99	
3.2.5.3	Two-Carbon Homologation of Hemiaminal Ethers 100	
3.2.5.4	Homocoupling Reaction of Enecarbamates 102	
3.2.6	Miscellaneous Reactions 104	
3.2.6.1	Aza-Cope Rearrangement 104	
3.2.6.2	Aldol-Type Reaction of Azlactones with Vinyl Ethers 104	
3.2.6.3	Cooperative Catalysis by Metal Complexes and Chiral Phosphoric Acids 106	
3.3	Carbon–Hydrogen Bond Forming Reactions 108	
3.3.1	Transfer Hydrogenation of Acyclic and Cyclic Imines 109	
3.3.2	Cascade Transfer Hydrogenation of Quinoline and Pyridine Derivatives 113	
3.3.3	Application of Transfer Hydrogenation to Cascade Reaction 116	
3.4	Carbon–Heteroatom Bond Forming Reactions 117	
3.4.1	Hydrophosphonylation (Kabachnik–Fields Reaction) 117	
3.4.2	Formation of (Hemi)Aminals 119	
3.4.3	Nucleophilic Ring Opening of Aziridines and Related Reactions 121	
3.5	Conclusion 123	
	References 125	
4	**Reduction of Imines with Trichlorosilane Catalyzed by Chiral Lewis Bases** 131	
	Pavel Kočovský and Sigitas Stončius	
4.1	Introduction 131	
4.2	Formamides as Lewis-Basic Organocatalysts in Hydrosilylation of Imines 132	
4.3	Other Amides as Organocatalysts in Hydrosilylation of Imines 141	
4.4	Sulfinamides as Organocatalysts in Hydrosilylation of Imines 143	
4.5	Supported Organocatalysts in Hydrosilylation of Imines 144	
4.6	Mechanistic Considerations 147	
4.7	Synthetic Applications 149	
4.8	Conclusions 151	
4.9	Typical Procedures for the Catalytic Hydrosilylation of Imines 152	
4.9.1	Catalytic Hydrosilylation of Simple Imines 152	
4.9.2	Catalytic Hydrosilylation of Enamines 153	
	References 154	
5	**Catalytic, Enantioselective, Vinylogous Mannich Reactions** 157	
	Christoph Schneider and Marcel Sickert	
5.1	Introduction 157	
5.2	Vinylogous Mukaiyama–Mannich Reactions of Silyl Dienolates 159	
5.3	Direct Vinylogous Mannich Reactions of Unmodified Substrates 170	

5.4	Miscellaneous 174
5.5	Conclusion 175
	References 176

6	**Chiral Amines from Transition-Metal-Mediated Hydrogenation and Transfer Hydrogenation** 179
	Tamara L. Church and Pher G. Andersson
6.1	Scope and Related Publications 179
6.2	Chiral Amines with a Disubstituted Nitrogen Atom, HNR^*R^1 179
6.2.1	Direct Asymmetric Hydrogenation of Alkyl- and Aryl-Substituted Imines 179
6.2.1.1	Development 180
6.2.1.1.1	A Representative Synthesis 183
6.2.1.2	Pressure in the Asymmetric Hydrogenation of Alkyl- and Aryl-Substituted Imines 183
6.2.1.3	Reducing the Environmental Impact of the Reaction 185
6.2.2	Direct Asymmetric Hydrogenation of Heteroaromatics 190
6.2.2.1	Quinolines and Isoquinolines 190
6.2.2.1.1	Quinolines – A Representative Synthesis 195
6.2.2.1.2	Isoquinolines – A Representative Synthesis 196
6.2.2.2	Quinoxalines 197
6.2.2.3	Pyridines 198
6.2.3	Direct Asymmetric Hydrogenation of "Activated" Imines 202
6.2.4	Asymmetric Transfer Hydrogenation of Imines 204
6.2.4.1	Reducing the Environmental Impact of the Reaction 207
6.2.4.2	Syntheses Using the Asymmetric Transfer Hydrogenation of Imines as a Key Step 211
6.3	Chiral Amines with Trisubstituted Nitrogen, $NR^*R^1R^2$ 211
6.3.1	Hydrogenation and Transfer Hydrogenation of N,N-Disubstituted Iminiums 211
6.3.2	Hydrogenation and Transfer Hydrogenation of Enamines 213
6.4	Conclusion 216
	References 218

7	**Asymmetric Reductive Amination** 225
	Thomas C. Nugent
7.1	Introduction 225
7.2	Transition Metal-Mediated Homogeneous Reductive Amination 226
7.3	Enantioselective Organocatalytic Reductive Amination 231
7.4	Diastereoselective Reductive Amination 234
7.4.1	Stereoselective Reductive Amination with Chiral Ketones 234
7.4.2	The Phenylethylamine Auxiliary and Stereoselective Reductive Amination 237
7.4.3	The *tert*-Butylsulfinamide Auxiliary and Stereoselective Reductive Amination 240

7.5	Conclusions	243
	References	244

8	**Enantioselective Hydrogenation of Enamines with Monodentate Phosphorus Ligands** *247*	
	Qin-Lin Zhou and Jian-Hua Xie	
8.1	Introduction	247
8.2	Asymmetric Hydrogenation of Enamides	249
8.2.1	Chiral Monodentate Phosphoramidite Ligands	249
8.2.2	Chiral Monodentate Phosphite Ligands	257
8.2.3	Other Chiral Monodentate Phosphorus Ligands	262
8.2.4	Mixed Chiral Monodentate Phosphorus Ligands	263
8.3	Asymmetric Hydrogenation of *N,N-Dialkyl Enamines*	264
8.4	Conclusion and Outlook	269
	References	270

9	**Bidentate Ligands for Enantioselective Enamide Reduction** *273*	
	Xiang-Ping Hu and Zhuo Zheng	
9.1	Introduction	273
9.2	Catalytic Enantioselective Hydrogenation of Enamides	274
9.2.1	Synthesis of Enamides	274
9.2.2	Catalytic Asymmetric Hydrogenation of Acyclic Enamides	276
9.2.2.1	Chiral Phospholane Ligands for Rh-Catalyzed Asymmetric Hydrogenation	276
9.2.2.2	Chiral 1,4-Diphosphine Ligands for Rh-Catalyzed Asymmetric Hydrogenation	278
9.2.2.3	Bisaminophosphine Ligands for Rh-Catalyzed Asymmetric Hydrogenation	283
9.2.2.4	Unsymmetrical Hybrid Phosphorus-Containing Ligands for Rh-Catalyzed Asymmetric Hydrogenation	284
9.2.3	Catalytic Asymmetric Hydrogenation of Cyclic Enamides	289
9.3	Conclusions	296
	References	297

10	**Enantioselective Reduction of Nitrogen-Based Heteroaromatic Compounds** *299*	
	Da-Wei Wang, Yong-Gui Zhou, Qing-An Chen, and Duo-Sheng Wang	
10.1	Asymmetric Hydrogenation of Quinolines	299
10.1.1	Ir- and Ru-Catalyzed Asymmetric Hydrogenation of Quinolines	299
10.1.2	Organocatalyzed Asymmetric Transfer Hydrogenation of Quinolines	318
10.2	Asymmetric Hydrogenation of Isoquinolines	320
10.3	Asymmetric Hydrogenation of Indoles	322
10.4	Asymmetric Hydrogenation of Pyrroles	327
10.5	Asymmetric Hydrogenation of Quinoxalines	329

10.6	Asymmetric Hydrogenation of Pyridine Derivatives *329*
10.7	Summary and Outlook *336*
	References *337*

11	**Asymmetric Hydroamination** *341*
	Alexander L. Reznichenko and Kai C. Hultzsch
11.1	Introduction: Synthesis of Amines via Hydroamination *341*
11.2	Hydroamination of Simple, Nonactivated Alkenes *342*
11.2.1	Intermolecular Hydroamination of Simple Alkenes *342*
11.2.2	Intramolecular Asymmetric Hydroamination of Simple Aminoalkenes *346*
11.2.2.1	Rare Earth Metal-Based Catalysts *346*
11.2.2.2	Alkali Metal-Based Catalysts *353*
11.2.2.3	Group 4 Metal-Based Catalysts *356*
11.2.2.4	Organocatalytic Asymmetric Hydroamination *358*
11.3	Hydroamination of Dienes, Allenes, and Alkynes *360*
11.3.1	Intermolecular Hydroaminations *360*
11.3.2	Intramolecular Reactions *361*
11.4	Hydroamination with Enantiomerical Pure Amines *363*
11.4.1	Hydroaminations Using Achiral Catalysts *363*
11.4.2	Kinetic Resolution of Chiral Aminoalkenes *366*
11.5	Synthesis of Chiral Amines via Tandem Hydroamination/Hydrosilylation *368*
11.6	Conclusions *369*
11.7	Experimental Section *369*
	References *372*

12	**Enantioselective C–H Amination** *377*
	Nadège Boudet and Simon B. Blakey
12.1	Introduction *377*
12.2	Background *378*
12.3	Racemic C–H Amination *379*
12.3.1	Intramolecular C–H Amination *379*
12.3.2	Intermolecular C–H Amination *382*
12.4	Substrate-Controlled Chiral Amine Synthesis via C–H Amination *384*
12.5	Enantioselective C–H Amination of Achiral Substrates *386*
12.5.1	Enantioselective C–H Amination with Rhodium(II) Catalysts *386*
12.5.2	Enantioselective C–H Amination with Ruthenium(II) Catalysts *390*
12.6	Conclusion *392*
	References *394*

13	**Chiral Amines Derived from Asymmetric Aza-Morita–Baylis–Hillman Reaction** *397*
	Lun-Zhi Dai and Min Shi
13.1	Introduction *397*

13.2	Recent Mechanistic Insights 398
13.3	Asymmetric Aza-MBH Reaction 400
13.4	Chiral Auxiliary-Induced Diastereoselective Aza-MBH Reaction 400
13.5	Chiral Tertiary Amine Catalysts 401
13.5.1	Cinchona-Derived Bifunctional Catalysts 401
13.5.2	Chiral Binol-Derived Bifunctional Amine Catalysts 408
13.5.3	Chiral Acid/Achiral Amine 410
13.6	Chiral Phosphine Catalysts 411
13.7	Chiral Bifunctional N-Heterocyclic Carbenes 418
13.8	Chiral Ionic Liquids as Reaction Medium 419
13.9	Aza-MBH-Type Reaction to Obtain Chiral Amines 419
13.10	Strategies for the Removal of Protecting Groups 422
13.11	Selected Typical Experimental Procedures 423
13.11.1	Typical Procedures for 1a-Catalyzed Aza-MBH Reaction of Methyl Acrylate with N-Benzylidene-4-Nitrobenzenesulfonamide 423
13.11.2	Typical Procedures for β-ICD-Catalyzed Aza-MBH Reaction of MVK with N-(p-Ethylbenzenesulfonyl)Benzaldimine 423
13.11.3	Typical Procedures for Chiral Phosphine 23-Catalyzed Aza-MBH Reaction of MVK with N-(Benzylidene)-4-Chlorobenzenesulfonamide 424
13.11.4	General Procedures of Aza-MBH Reactions Involving Aliphatic Imines 424
13.11.5	Typical Procedures for 25a and Benzoic Acid-Catalyzed Aza-MBH Reaction of N-Sulfonated Imine with MVK 424
13.11.6	Typical Procedures for Trifunctional Phosphine 27-Catalyzed Aza-MBH Reaction of N-Tosylimines with MVK 424
13.11.7	General Procedures for the Synthesis of Enantiomerically Enriched Aza-MBH-Type Adducts Catalyzed by Chiral Sulfide 29 425
13.11.8	General Procedures for the Removal of N-p-Toluenesulfinyl Group 425
13.11.9	General Procedures for the Removal of N-Tosyl Group 425
13.11.9.1	Reduction of the Aza-MBH Reaction Product with $LiAlH_4$ 425
13.11.9.2	Boc-Protection 426
13.11.9.3	Detosylation 426
13.12	Summary and Outlook 426
	References 428

14 Biocatalytic Routes to Nonracemic Chiral Amines 431
Nicholas J. Turner and Matthew D. Truppo

14.1	Introduction 431
14.2	Kinetic Resolution of Racemic Amines 432
14.2.1	Hydrolytic Enzymes 432
14.2.2	Transaminases 441
14.2.3	Amine Oxidases 443
14.3	DKR and Deracemization of Amines 444

14.3.1	DKR Using Hydrolytic Enzymes and Racemization Catalysts	*444*
14.3.2	Deracemization Reactions Using Amine Oxidases	*448*
14.4	Asymmetric Synthesis of Amines Using Transaminases	*450*
14.5	Conclusions and Future Perspectives	*455*
	References	*457*

Appendix: Solution *461*

Index *479*

Preface

Chiral amines are powerful pharmacophores for defining new pharmaceutical drugs (Figure P.1) due to their high density of structural information and inherent ability for hydrogen bonding, yet their synthesis remains a challenge. That challenge is further appreciated when framed in the context of the following goal: the introduction of nitrogen into a commodity chemical or an advanced intermediate via an operationally simple procedure, preferably one-step, with complete chemo-, regio-, diastereo-, and enantiocontrol. Ten years ago this was an interesting goal to set and ponder, today it is within our grasp and the diversity of new and optimized methods developed since 2000 has enabled this sea change in technical capability. The book before you is therefore the first attempt to organize the results of this expanding field, by providing the overarching strategies that have elevated chiral amine synthesis from a specialized subtopic into the large field of diverse methods it now is. Perhaps unsurprisingly, the methods and concepts that have allowed this rise in stature are different from those in the field of chiral amino acid synthesis. The collected chapters that follow examine the vast majority of the relevant chiral amine literature (nonamino acids) from the years 2000 to 2009.

The key to assessing any methodology is substrate breadth; with this knowledge an immediate sense of substrate applicability, or lack thereof, is evident. From this vantage point a more restricted list of methods is chosen, with the final decision almost always driven by what is deemed as carrying higher priority: short timelines (number of reaction steps and overall yield of lower importance) or efficiency (number of reaction steps and overall yield of high importance). In general, the top methods can provide both, but this is not always the case and the chapter material has been carefully chosen to meet the demands of both types of research needs, namely, those of medicinal chemists and process research chemists. Academics will additionally enjoy the fact that each main chapter author has contributed three questions: one at the undergraduate level and two at the graduate level. Thus, this book could easily become a template for part of a graduate course.

Our working definition of a "chiral amine," which is better described as an "α-chiral amine," is a nitrogen atom with an adjacent, or α, stereogenic carbon with the exception that α- and β-amino acids have been purposely excluded. In this

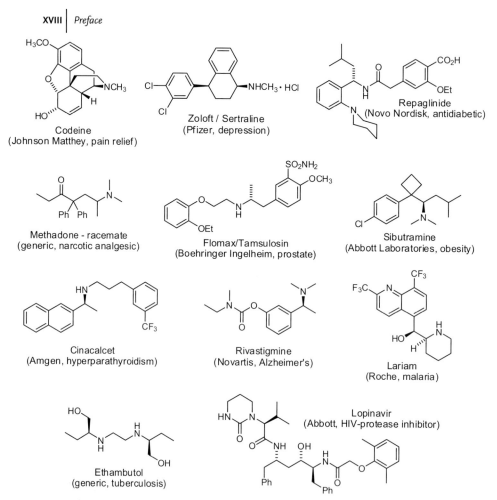

Figure P.1 Examples of chiral amine pharmaceutical drugs.

context, two generic structural features of chiral amines are useful to point out (Figure P.2): first, the nitrogen can be primary, secondary, tertiary, or even quaternary (ammonium salt), and second, the α-stereogenic carbon, by necessity, can only be secondary or tertiary. Methods leading to the vast majority of all permeations of these basic structural features are addressed in this book. Of those, chiral primary amine building blocks are often sought after because of the flexibility they lend to the synthetic design of pharmaceutical drugs and alkaloid natural products. These important compounds overwhelmingly contain chiral amines as secondary or tertiary amines, but are sometimes masked as secondary or tertiary amides (Figure P.2).

The authoritative summaries and insights found in the chapters that follow will allow the reader to assess and define the benefits and limitations of each strategy. As a consequence, the readers should also be able to define the methodology(ies) that will lay the foundation for their future research project. Each chapter is authored by

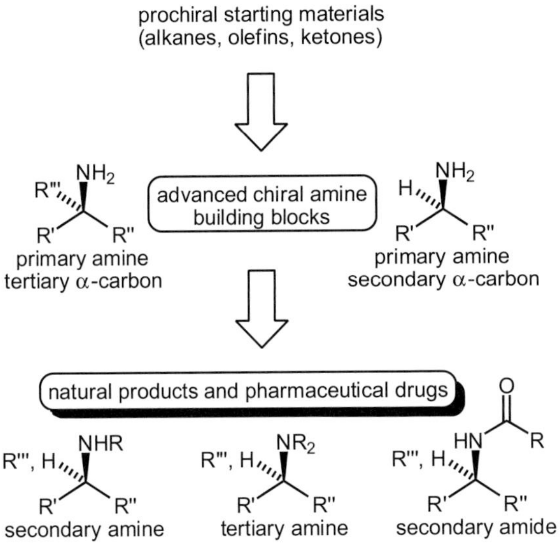

Figure P.2 General utility of chiral primary amines.

an expert in the respective field that is summarized, and all the methods rely on alkanes, alkenes, or ketones in combination with a source of nitrogen.

A traditional approach to chiral amine synthesis calls for the addition of a "carbanion" to an imine, but as Charette demonstrates in the opening chapter, this field has been revolutionized by the development of mild prochiral carbanions that can be enantioselectively added to prochiral imines. Friestad follows in Chapter 2 with methods using the same disconnection approach, albeit via radical additions. An important theme for the reader to be aware of, regarding these two opening chapters, is the greatly broadened functional group compatibility now possible. Later, when comparing carbanion methods with hydrogen (or hydride)-based methods, a global trend is clearly established: where the one fails, the other usually excels; in simple terms, they are currently complementary.

The use and development of organocatalytic methods continues on a log rhythmic trajectory of discovery, and carbon–nitrogen bond formation is no exception as Terada eloquently describes for a plethora of different reaction types in Chapter 3. The methods rely on a chiral Brønsted acid catalyst, generally a BINOL-monophosphoric acid template, and have ushered in a new age of chiral amine forming reactions. Kočovský complements this chapter with a focused summary of prochiral N-aryl imine reduction via chiral organocatalysts based on trichlorosilane (Chapter 4). Schneider follows with a wonderful transition chapter (Chapter 5) demonstrating how to take advantage of both organocatalytic and transition metal-mediated methods for enantioselective vinylogous Mannich reactions; the chiral amine products thereof will truly be considered as advanced building blocks due to the useful coexisting functionality. Chapter 6 follows with an examination of transition metal-based enantioselective imine reduction methods by Andersson. Chapter 7 is the last chapter to discuss methods requiring the use of imines, but this time they are only intermediates as elaborated on by Nugent in his summary of all known asymmetric reductive amination methods. Reductive amination is of interest because the requirement for imine formation and isolation is removed.

The enantioselective reduction of *N*-acetyl enamides has a rich history and several substrate classes of *N*-acetyl enamides can now be reliably converted to enantiopure amines (>99% ee) using rhodium-based catalysts. The field has bifurcated over the past 10 years with the utility of monodentate versus bidentate ligands at the fore of the discussion. In Chapter 8, Zhou lays a strong foundation for the monodentate phosphorus ligand accomplishments and challenges. Chapter 9 provides a wonderful complement by Zheng, who fully summarizes the bidentate phosphorus ligand literature over the past 10 years.

Starting from Chapter 10, less traditional methods of chiral amine syntheses are examined, but the reader needs to be aware that they are at the forefront of chiral amine synthesis because of the great strides achieved in each of the respective areas. For example, Zhou starts by skillfully showing how the enantioselective reduction of nitrogen-containing heteroaromatic compounds can be taken advantage of for cyclic chiral amine synthesis. In Chapter 11, Hultzsch provides a strong theoretical background and summary of the advances made in the field of enantioselective hydroamination. This is followed by Blakey's description of carbon–hydrogen amination in Chapter 12. Both of these methods, hydroamination and carbon–hydrogen amination, are of particular interest when considering reaction step efficiency and therefore hold tremendous potential. The last chemical method examined is the Aza-Morita–Baylis–Hillman reaction, with Shi providing a comprehensive account of the most up-to-date methods in Chapter 13. Here, the products would be considered as high-value building blocks because of the great number of diverse products they can be converted into.

Finally, but certainly no less important, is an evaluation of biocatalytic/enzymatic routes allowing chiral amine synthesis. Traditional organic chemists may tend to avoid this material because of terminology or lack of basic understanding, but Turner and Truppo have written Chapter 14 from a perspective that an organic chemist will appreciate. The importance of this approach cannot be understated and will continue to challenge the chemical methods. For example, a formal reductive amination approach using ammonia and a cheap reductant (glucose or formate) has recently been demonstrated and appears promising [1]. The challenge for this field of endeavor will likely be the identification of new or modified enzymes capable of greater substrate acceptability; here, omega-transaminases might be expected to have a major impact (personal communication with Wolfgang Kroutil, University of Graz, Austria).

Emerging at a fast rate, but not examined in this book, are the syntheses of allylic chiral amines [2] and propargylic chiral amines [3]. In addition, the very recent development of efficient catalytic hydroaminoalkylation, albeit racemic, holds great promise [4].

Finally, I would like to gratefully acknowledge all chapter authors, without this truly international mix of top scientists making a full commitment to this project, the book simply would not have materialized. Furthermore, it was a true pleasure to work with Wiley-VCH representatives Dr. Elke Maase and Lesley Belfit, who always provided timely and excellent ideas, guidance, and feedback at all stages of the preparation of this book. I would also like to personally thank Michael Holt (Bremen, Germany) for his generous and professional rendering of the graphic art that enabled the book cover to become a reality.

References

1 Koszelewski, D., Lavandera, I., Clay, D., Guebitz, G.M., Rozzell, D., and Kroutil, W. (2008) *Angew. Chem., Int. Ed.*, **47**, 9337–9340.
2 (a) Leitner, A., Shekhar, S., Pouy, M.J., and Hartwig, J.F. (2005) *J. Am. Chem. Soc.*, **127**, 15506–15514; (b) Pouy, M.J., Leitner, A., Weix, D.J., Ueno, S., and Hartwig, J.F. (2007) *Org. Lett.*, **9**, 3949–3952; (c) Johns, A.M., Liu, Z., and Hartwig, J.F. (2007) *Angew. Chem., Int. Ed.*, **46**, 7259–7261.
3 Gommermann, N., Koradin, C., Polborn, K., and Knochel, P. (2003) *Angew. Chem., Int. Ed.*, **42**, 5763–5766.
4 (a) Herzon, S.B., and Hartwig, J.F. (2008) *J. Am. Chem. Soc.*, **130**, 14940–14941; (b) Herzon, S.B. and Hartwig, J.F. (2007) *J. Am. Chem. Soc.* **129**, 6690–6691.

Thomas C. Nugent
Jacobs University Bremen,
Bremen, Germany

List of Contributors

Pher G. Andersson
Uppsala University
Department of Biochemistry
and Organic Chemistry
Box 576, Husargatan 3
S751 23 Uppsala
Sweden

Simon B. Blakey
Emory University
Department of Chemistry
1515 Dickey Drive
Atlanta, GA 30322
USA

Nadège Boudet
Emory University
Department of Chemistry
1515 Dickey Drive
Atlanta, GA 30322
USA

André B. Charette
Université de Montréal
Département de Chimie
P.O. Box 6128, Station Downtown
Montréal, Québec
Canada H3C 3J7

Qing-An Chen
Chinese Academy of the Sciences
Dalian Institute of Chemical Physics
State Key Laboratory of Catalysis
457 Zhongshan Road
Dalian 116023
China

Tamara L. Church
Uppsala University
Department of Biochemistry and
Organic Chemistry
Box 576, Husargatan 3
S751 23 Uppsala
Sweden

Lun-Zhi Dai
Chinese Academy of Sciences
Shanghai Institute of Organic
Chemistry
State Key Laboratory of Organometallic
Chemistry
354 Fenglin Lu
Shanghai 200032
China

Gregory K. Friestad
University of Iowa
Department of Chemistry
Iowa City, IA 52242
USA

Xiang-Ping Hu
Chinese Academy of Sciences
Dalian Institute of Chemical Physics
457 Zhongshan Road
Dalian 116023
China

Kai C. Hultzsch
Rutgers, The State University
of New Jersey
Department of Chemistry and
Chemical Biology
610 Taylor Road
Piscataway, NJ 08854-8087
USA

Pavel Kočovský
University of Glasgow
Department of Chemistry
Glasgow G12 8QQ
UK

Norie Momiyama
Tohoku University
Graduate School of Science
Department of Chemistry
6-3 Aza-Aoba, Aramaki, Aoba-ku
Sendai 980-8578
Japan

Thomas C. Nugent
Jacobs University Bremen
School of Engineering and Science
Department of Chemistry
Campus Ring 1
28759 Bremen
Germany

Alexander L. Reznichenko
Rutgers, The State University
of New Jersey
Department of Chemistry and
Chemical Biology
610 Taylor Road
Piscataway, NJ 08854-8087
USA

Christoph Schneider
Universität Leipzig
Institut für Organische Chemie
Johannisallee 29
04103 Leipzig
Germany

Min Shi
Chinese Academy of Sciences
Shanghai Institute of Organic
Chemistry
State Key Laboratory of Organometallic
Chemistry
354 Fenglin Lu
Shanghai 200032
China

Marcel Sickert
Universität Leipzig
Institut für Organische Chemie
Johannisallee 29
04103 Leipzig
Germany

Sigitas Stončius
Vilnius University
Department of Organic Chemistry
Naugarduko 24
03225 Vilnius
Lithuania

Masahiro Terada
Tohoku University
Graduate School of Science
Department of Chemistry
6-3 Aza-Aoba, Aramaki, Aoba-ku
Sendai 980-8578
Japan

Matthew D. Truppo
University of Manchester
School of Chemistry
Manchester Interdisciplinary Biocentre
131 Princess Street
Manchester M1 7DN
UK

Nicholas J. Turner
University of Manchester
School of Chemistry
Manchester Interdisciplinary Biocentre
131 Princess Street
Manchester M1 7DN
UK

Da-Wei Wang
Chinese Academy of the Sciences
Dalian Institute of Chemical Physics
State Key Laboratory of Catalysis
457 Zhongshan Road
Dalian 116023
China

Duo-Sheng Wang
Chinese Academy of the Sciences
Dalian Institute of Chemical Physics
State Key Laboratory of Catalysis
457 Zhongshan Road
Dalian 116023
China

Jian-Hua Xie
Nankai University
Institute of Elemento-Organic
Chemistry
94 Weijin Road
Tianjin 300071
China

Zhuo Zheng
Chinese Academy of Sciences
Dalian Institute of Chemical Physics
457 Zhongshan Road
Dalian 116023
China

Qin-Lin Zhou
Nankai University
Institute of Elemento-Organic
Chemistry
94 Weijin Road
Tianjin 300071
China

Yong-Gui Zhou
Chinese Academy of the Sciences
Dalian Institute of Chemical Physics
State Key Laboratory of Catalysis
457 Zhongshan Road
Dalian 116023
China

1
Stereoselective Synthesis of α-Branched Amines by Nucleophilic Addition of Unstabilized Carbanions to Imines

André B. Charette

1.1
Introduction

Amines bearing a stereocenter α to the nitrogen atom are among the most commonly found subunits in bioactive molecules, natural products, and chiral ligands. As such, there has been much interest in developing methods for their synthesis. Of these methods, the addition of nonstabilized carbanions to activated imines constitutes one of the most reliable ways for synthesizing this important class of compounds [1, 2].

The preparation of α-branched amines from the corresponding carbonyl derivatives typically requires three steps: (1) imine formation; (2) nucleophilic addition; and (3) protecting or activating group (PG) cleavage (Scheme 1.1). Over the past 25 years, numerous variants of this process allowing the preparation of a large variety of substituted α-branched amines have been developed.

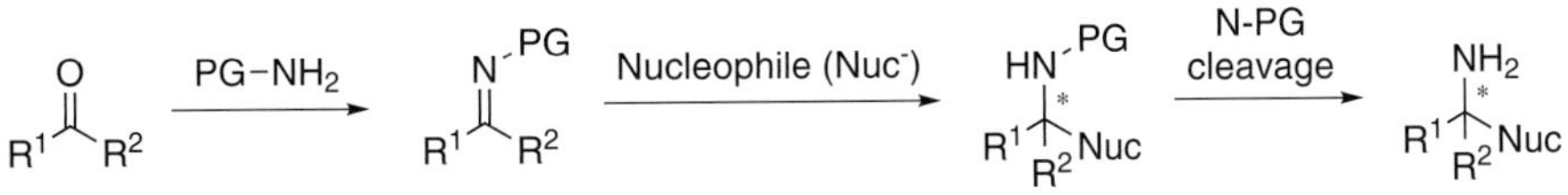

Scheme 1.1 Preparation of chiral amines from carbonyl derivatives.

As oxygen is more electronegative than nitrogen, N-alkyl-substituted imines are less electrophilic but more Lewis basic than their carbonyl counterparts. For this reason, only very reactive nucleophiles, such as organolithium [3], Grignard [4], organocuprate/BF$_3$ [3], and organocerium reagents [3], have been reported to react with N-alkyl imines (Scheme 1.2). Conversely, N-alkyl-substituted imines are usually the candidates of choice for Lewis acid-catalyzed processes, such as the Strecker reaction [5], Mannich condensation [6], and Ugi condensation [7].

Chiral Amine Synthesis: Methods, Developments and Applications. Edited by Thomas C. Nugent
Copyright © 2010 WILEY-VCH Verlag GmbH & Co. KGaA, Weinheim
ISBN: 978-3-527-32509-2

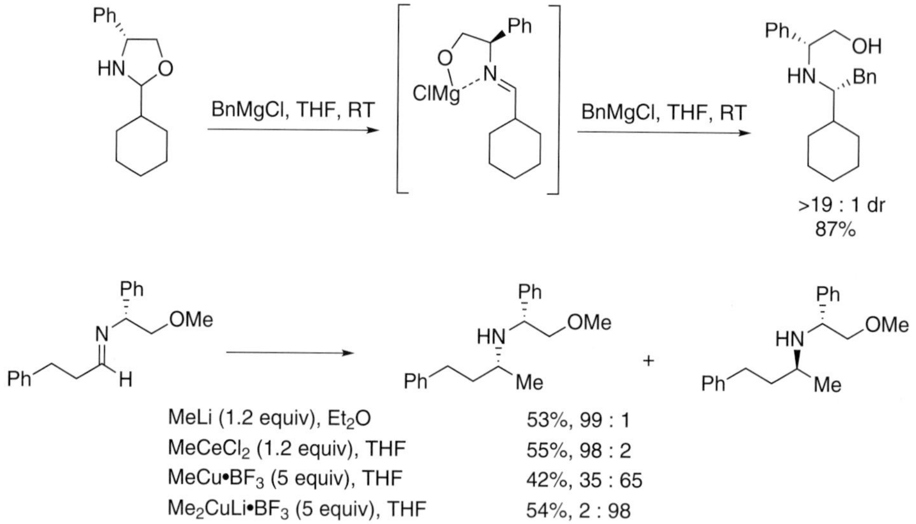

Scheme 1.2 Types of nucleophiles that add to N-alkyl-substituted imines.

The electrophilicity of these imines can be increased via the selection of an appropriate electron-withdrawing N-substituent, a strategy that has been quite important in increasing the scope of potential nucleophiles suitable for this chemistry. With the appropriate activating group (phosphinoyl, sulfonyl, etc.), Grignard reagents add smoothly in high yields to various imines. This method of activation is especially important in the catalytic asymmetric addition of nonstabilized carbanion reagents, where several activated imines have been reported. This modification opened the door to copper-, rhodium-, iridium-, and palladium-catalyzed nucleophilic addition chemistry. The increase in electrophilicity as a function of the electron-withdrawing ability of the activating group can be appreciated by comparing the *ab initio* calculations of the LUMO energy of various electrophiles to that of the imine derived from benzaldehyde and N-methylamine (Figure 1.1) [8]. These results suggest that N-acyl imines are substantially more electrophilic than the other types of imines bearing alkyl or aryl groups.

This chapter will discuss methods for the preparation of these electrophiles as well as their applications in various stereoselective transformations.

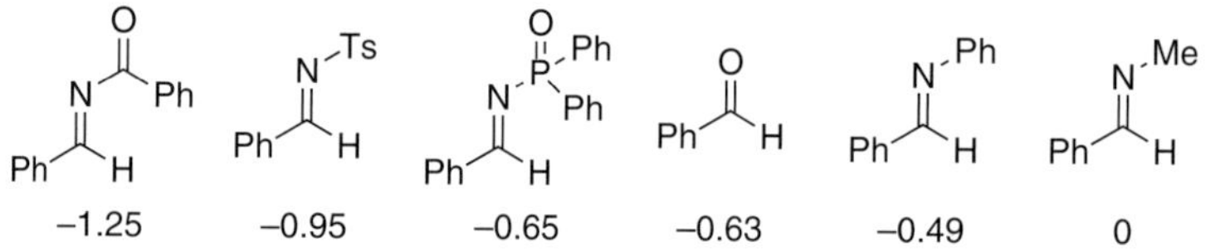

Figure 1.1 Calculations of the LUMO energy in eV (geometry optimized at B3LYP/6-31G(d)) relative to N-methyl imine derived from benzaldehyde.

1.2
Overview of the Methods for the Preparation of Imines

1.2.1
N-Aryl and N-Alkyl Imines and Hydrazones

The method most commonly used for the preparation of imines derived from aldehydes and anilines [9], primary amines, or hydrazines [10] consists in mixing a nucleophilic amine with the corresponding carbonyl derivative under Dean–Stark conditions or in the presence of a dehydrating agent, such as magnesium sulfate or molecular sieves. For example, the preparation of the imine derived from benzaldehyde and o-anisidine is accomplished by stirring both reagents in benzene in the presence of molecular sieves (Scheme 1.3). Conversely, the imine derived from pivalaldehyde requires heating the reagents in benzene in the presence of a catalytic amount of p-TsOH. Alternatively, it is also possible to use N-sulfinyl amines as precursors to generate N-aryl imines [11].

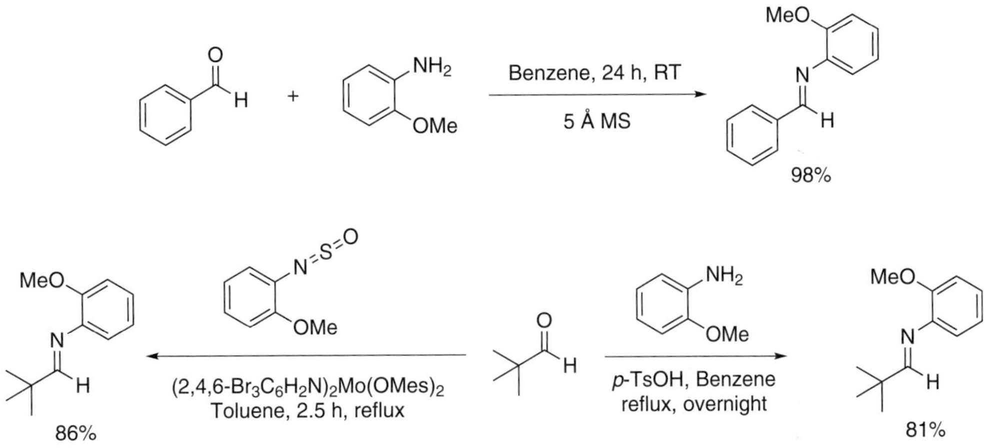

Scheme 1.3 Preparation of N-aryl imines.

1.2.2
N-Sulfinyl Imines

Although N-sulfinyl imines can be prepared by several other methods (oxidation of sulfenimines [12] and iminolysis of sulfinates [13]), they are most conveniently synthesized by condensing aldehydes with tert-butanesulfinamide in the presence of copper(II) sulfate or magnesium sulfate/pyridinium p-toluenesulfonate as a Lewis acid catalyst and water scavenger (Scheme 1.4) [14]. When ketones, more sterically hindered aldehydes or even electronically deactivated aldehydes are used, titanium tetraethoxide is the preferred Lewis acid and water scavenger. These methods are convenient to synthesize both alkyl- and aryl-substituted imines.

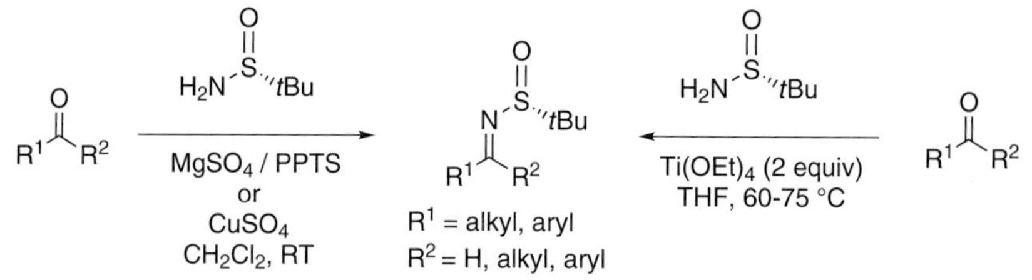

Scheme 1.4 Preparation of N-sulfinyl imines.

1.2.3
N-Sulfonyl Imines

A plethora of methods are available for the preparation of N-sulfonyl imines (Scheme 1.5). However, some are only compatible when R^1 is an aryl group or when the starting carbonyl compound is nonenolizable. Trost reported that aldehydes could be treated with the reagent obtained by mixing chloramine-T and tellurium to generate N-tosyl imines in high yields (method A) [15]. Alternatively, the direct condensation of an aldehyde and p-toluenesulfonamide is greatly facilitated by the addition of a strong Lewis acid, such as $TiCl_4$ [16] or $Si(OEt)_4$ [17] (method B). N-Sulfonylation of N-(trimethylsilyl) imines conveniently leads to N-tosyl imines when aryl-substituted aldehydes are used (method C) [18]. One of the most general methods compatible with both nonenolizable and enolizable aldehydes involves the initial formation of the sulfinic acid adduct of the imine from an aldehyde and p-toluenesulfinic acid formed *in situ* from formic acid and sodium benzene sulfinate

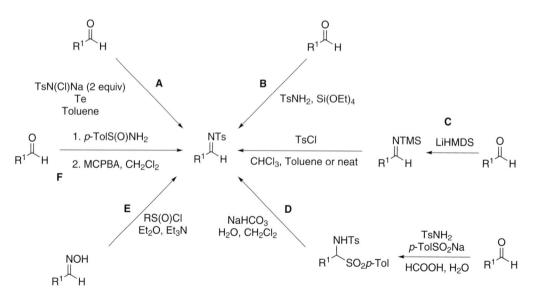

Scheme 1.5 Methods for the preparation of N-sulfonyl imines.

(method D) [19]. A subsequent base treatment generates the N-tosyl imines. Oximes can be treated with a sulfinyl chloride and undergo the Hudson reaction to generate the N-tosyl imine (method E) [20]. Finally, it is also possible to initially prepare the N-sulfinyl imine from the more nucleophilic sulfinamide and oxidize it with MCPBA to generate the N-tosyl imine (method F) [21]. Along method D, this route is the most effective in generating N-tosyl imines derived from either enolizable or nonenolizable aldehydes.

1.2.4
N-Phosphinoyl Imines

Although N-phosphinoyl imines can be prepared by most of the methods applied in the synthesis of N-tosyl imines, the most commonly used methodologies are illustrated in Scheme 1.6. The direct condensation of diphenylphosphonamide with an aldehyde is efficient in the presence of a titanium Lewis acid that also acts as a water scavenger (method A) [16]. The carbonyl moiety can also first be treated with hydroxylamine to generate the corresponding oxime, which is then treated with chlorodiphenylphosphine. The O-phosphinoyl oxime undergoes a spontaneous N−O bond homolytic cleavage followed by a radical pair recombination to yield the corresponding N-phosphinoyl aldimine (method B) [22]. One of the most reliable procedures for generating N-diphenylphosphinoyl imines from enolizable and nonenolizable aldehydes is a two-step sequence involving the initial formation of the p-toluenesulfinic acid adduct of the imine (method C) [23, 24], and subsequent base treatment leads to the readily isolable imine. Depending on the structure of the imine and the reaction in which it is used, it is sometimes possible to use the sulfinic acid adduct directly in the nucleophilic addition reaction.

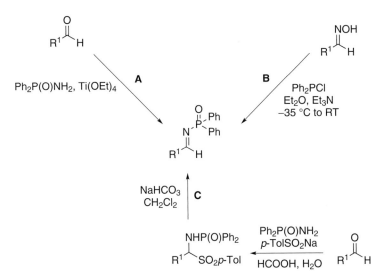

Scheme 1.6 Most widely described methods for the preparation of N-phosphinoyl imines.

The Kresze reaction has also been used in the synthesis of N-phosphinoyl imines derived from nonenolizable aldehydes, although usually in modest yields [25, 26]. The N-phosphinoyl imine derived from trifluoromethyl ketones can be prepared via the ethanolate intermediate (Scheme 1.7) [27]. This adduct is directly used in the subsequent nucleophilic addition reaction.

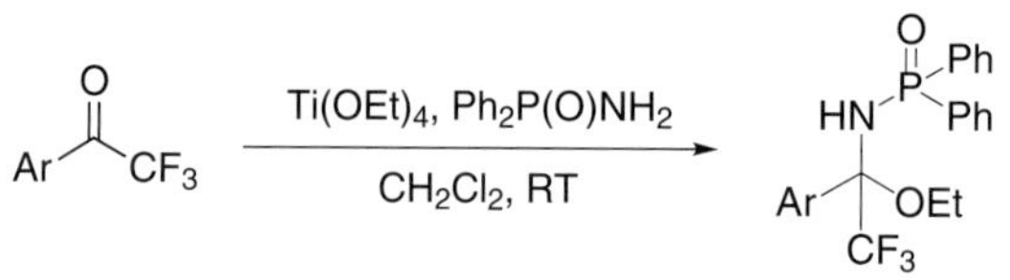

Scheme 1.7 Preparation of trifluoromethyl-substituted ketimines via the ethanolate adduct.

1.2.5
N-Acyl and N-Carbamoyl Imines

N-Acyl imines can be achieved from nonenolizable aldehydes by the formation of N-silyl imines obtained through treatment of the carbonyl derivative with lithium hexamethyldisilazide [28]. The resulting N-silyl imines can be reacted with an acylchloride to generate the corresponding N-acyl imines. Due to the instability of the latter imines, they must be reacted directly with a nucleophile without further purification (Scheme 1.8). In addition, it is also possible to prepare N-acyl and N-carbamoyl imines via the following two-step sequence [29]. The treatment of an aldehyde with the corresponding amide in the presence of an arenesulfinic acid and formic acid in water generates the corresponding α-acyl [30] or α-carbamoyl sulfone [31]. Subsequent basic treatment leads to N-acyl or N-carbamoyl imines.

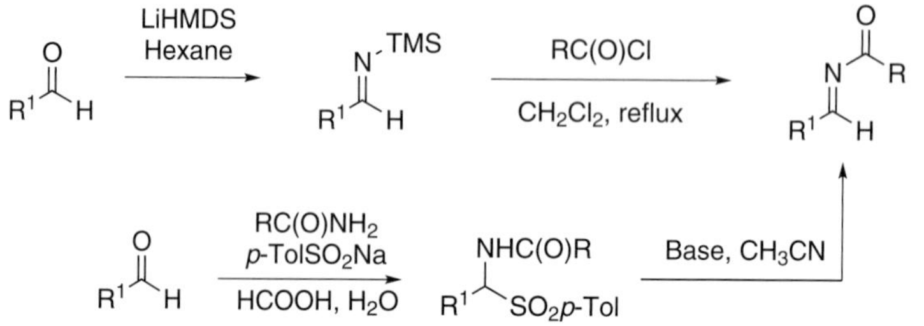

Scheme 1.8 Preparation of N-acyl imines from nonenolizable aldehydes.

1.3
Chiral Auxiliary-Based Approaches

The nucleophilic addition to imines bearing a chiral auxiliary is an extremely reliable, efficient, and powerful strategy to generate α-branched amines [32]. Imines are structurally unique, as the chirality can reside on either terminus of the imine double bond (Figure 1.2). Both strategies are complementary and will be reviewed separately in the following two sections.

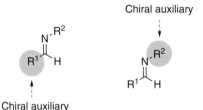

Figure 1.2 Complementary approaches for the incorporation of a chiral auxiliary.

1.3.1
Imines Derived from Chiral Aldehydes

Masked forms of the monoimine from glyoxal have been extensively used as chiral precursors to α-amino acids and α-amino aldehydes (Figure 1.3) [33]. For example, glyoxal can be converted into aminal **6** in 90% yield upon treatment with diamine **5** (Scheme 1.9) [33b]. Imine formation, followed by Grignard addition, leads to amine **7** in 86% yield as a single diastereomer. Acid hydrolysis affords the α-amino aldehyde **9** in good yields.

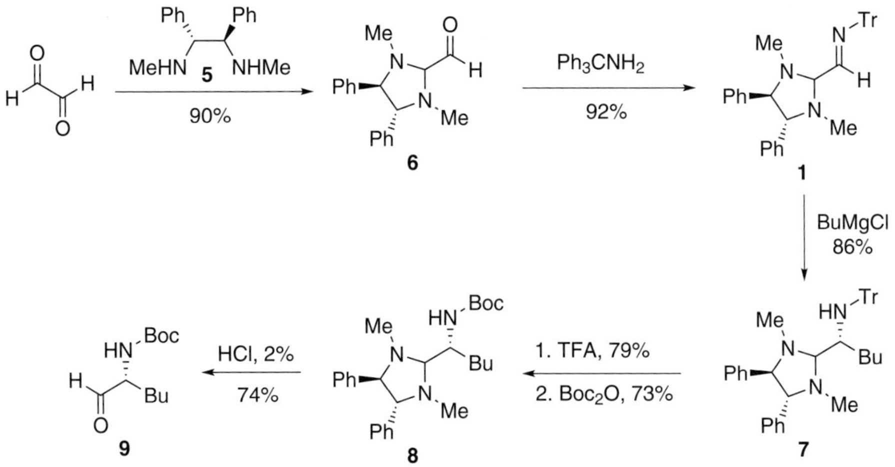

Scheme 1.9 Preparation of α-amino aldehydes from glyoxal.

An alternative strategy to the recoverable chiral auxiliary utilizes chiral α-hydroxy aldehydes as starting materials. For example, 2,3-di-O-benzyl-D-glyceraldehyde (**10**) is converted to the imine **11** (Scheme 1.10) [33e] and treatment with phenylmagnesium bromide leads to amine **12** as a single diastereomer. Phenylglycine **14** is obtained following protecting group cleavage and oxidation of the resulting diol.

Reports of diastereoselective additions to chiral ketimines are much more rare. One example describes the sequential addition of two different nucleophiles (Grignard and organocerium reagents) to the tartaric acid-derived nitrile **15** (Scheme 1.11) [34]. The second addition proceeds with good yields and excellent

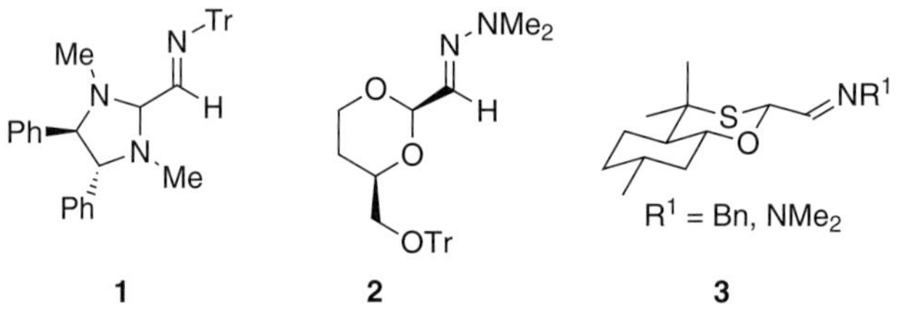

Figure 1.3 Selected chiral imines derived from glyoxal (**1–3**) and chiral α-hydroxyaldehydes **4**.

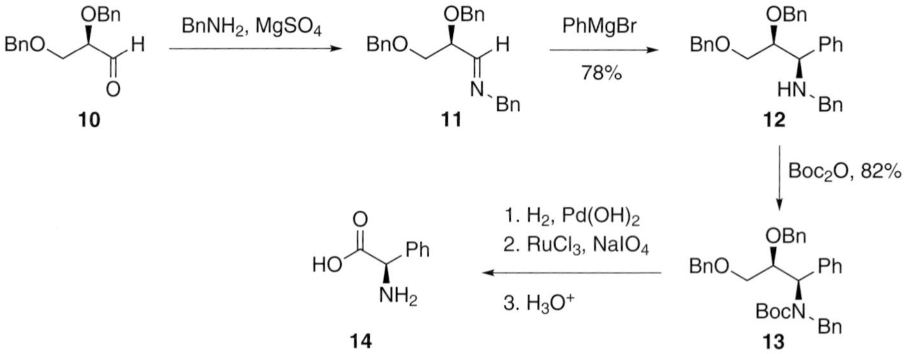

Scheme 1.10 Preparation of α-amino acids from protected glyceraldehyde.

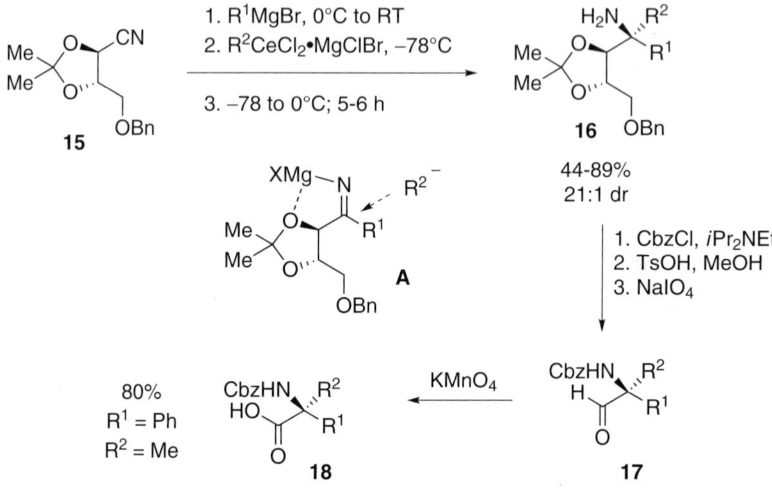

Scheme 1.11 Sequential diastereoselective addition to chiral nitriles.

diastereocontrol that can be explained through nucleophilic addition from the least hindered face of the postulated magnesium chelate **A** (Scheme 1.11). The reaction products are converted into α,α-disubstituted α-amino aldehydes and α,α-disubstituted α-amino acids [35].

1.3.2
Imines Bearing a Chiral Protecting/Activating Group

One of the most practical methodologies used to generate α-branched amines relies on the stoichiometric use of covalently bonded chiral auxiliaries on the nitrogen substituent. As such, a large number of chiral amines have been applied toward the generation of chiral imines suitable for nucleophilic addition chemistry (Figure 1.4).

The imine **19** derived from α-phenylethylamine has been reported to react with various nucleophiles, such as organolithium reagents and cuprates, to generate the diastereoenriched amine following hydrogenolysis (H$_2$/Pd) of the chiral auxiliary. However, the diastereoselectivities of these reactions are typically modest [36–38]. The seminal works of Takahashi [39] and later Pridgen [40] demonstrated that higher

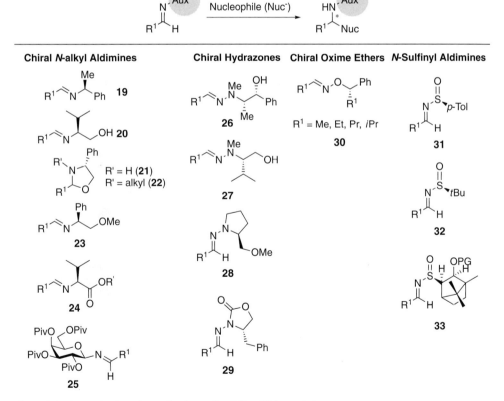

Figure 1.4 Some chiral auxiliaries for the nucleophilic addition to imines.

diastereoselectivities are observed when Grignard, organolithium, allylzinc halides, or organocerium reagents are added to the imine **20** or the oxazolidine **21** bearing a chiral auxiliary derived from either valine or phenylglycine (Scheme 1.12). The imine derived from phenylglycinol spontaneously cyclizes to the oxazolidine **21** upon formation. Further treatment of **21** with excess Grignard or organolithium reagent induces the ring opening to generate a chelated imine–metal alkoxide intermediate (**A**, Scheme 1.12) that undergoes the nucleophilic addition to yield the desired product. In addition, high diastereocontrol is observed with N-alkylated oxazolidines (**22**). The cleavage of the auxiliary is achieved through either hydrogenolysis conditions (phenylglycinol auxiliary) or oxidative conditions (valinol or phenylglycinol auxiliaries). Other chiral moieties, such as the imine **23** bearing an amino ether chiral auxiliary, have been developed that behave in a similar fashion [3]. Several other chiral auxiliaries bearing ester groups (**24** [41] and **25** [42]) have been reported, although the scope is typically limited to allylation reactions.

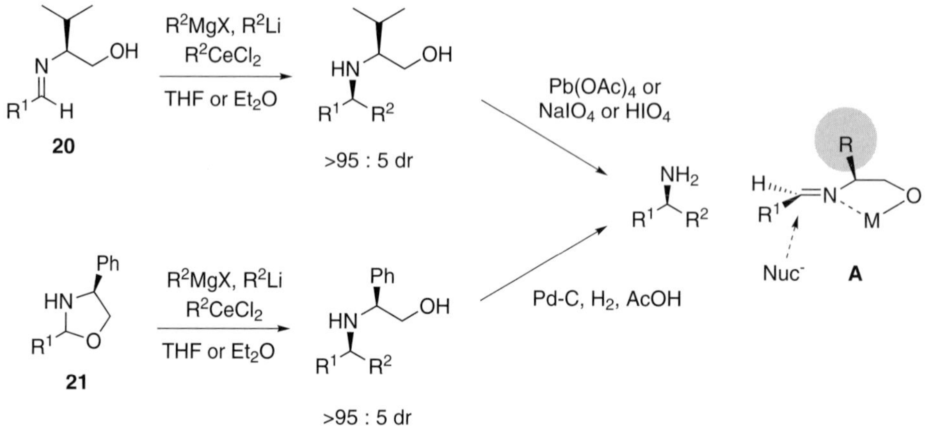

Scheme 1.12 Takahashi's valinol and phenylglycinol chiral auxiliaries.

Hydrazones derived from chiral hydrazines have also been applied as auxiliaries for the nucleophilic addition chemistry. Takahashi [43] reported that chiral hydrazones prepared from L-ephedrine (**26**) and L-valinol (**27**) react with Grignard reagents with good to excellent diastereoselectivities. Following the addition, the N−N bond cleavage can be achieved by hydrogenolysis (H$_2$, Pd). Both RAMP and SAMP hydrazines [44] have been developed and used by the Enders [45] and Denmark [46] groups as hydrazone **28** precursors. These groups have demonstrated that both organolithium reagents and organocerium reagents (prepared from Grignard or organolithium reagents) can be added to the hydrazone in high yields and excellent diastereocontrol. The cleavage of the N−N bond is accomplished either with hydrogenolysis, with H$_2$/Raney Ni, or with radical cleavage by lithium in ammonia.

Friestad's auxiliary **29** has been applied in diastereoselective indium(III) triflate-mediated allylation reactions with tetraallylsilane (Table 1.1) [47, 48]. The reaction proceeds with high diastereocontrol with imines derived from aryl-substituted

Table 1.1 Diastereoselective allylation of hydrazone **29**.

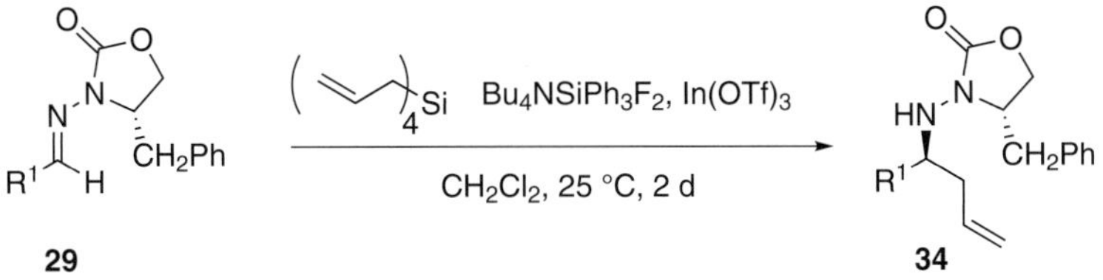

R^1	Yield (%)	dr
p-Tolyl	94	98:2
m-Nitrophenyl	71	>99:1
2-Naphthyl	82	98:2
2-Furyl	58	96:4
Phenyl	78	>99:1
(E)-PhCH=CH	60	95:5
Et	51	82:18

aldehydes. The reductive cleavage of the N−N bond is accomplished upon treating the resulting product with samarium diiodide.

Several oxime ethers (i.e., **30**) have been used as chiral auxiliaries; however, the diastereoselectivities for the nucleophilic addition reactions are usually modest [49].

Chiral N-sulfinyl imines **31–33** are among the most popular and widely used chiral auxiliaries in the preparation of α-branched amines. Although the p-tolylsulfinyl group **31** has been sporadically used for the addition of Grignard reagents, the transformation generally suffers from an important side reaction resulting from an undesired attack of the nucleophile on the electrophilic sulfur atom [50] (Scheme 1.13).

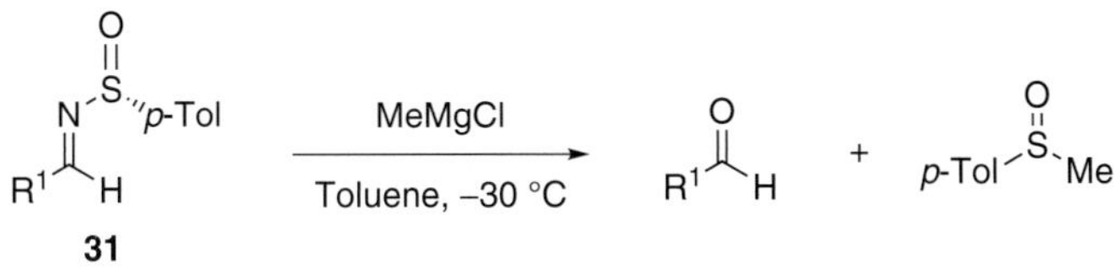

Scheme 1.13 Addition of methylmagnesium chloride to N-sulfinyl imines.

As seen in Scheme 1.14, exceptions to this are when benzyl- [50], allylmagnesium chloride [51], vinylaluminum [52, 53], or vinylcopper reagents are used as nucleophiles [54].

The N-tert-butanesulfinyl chiral auxiliary **32** has become the reagent of choice for the preparation of α-branched amines due to its excellent level of diastereocontrol and the fact that the steric hindrance caused by the tert-butyl group aids in the minimization of side reactions [32d]. The preparation of this chiral auxiliary in enantiomerically pure form is easily achieved in two steps and 65% overall yield starting from di-tert-butyl disulfide (Scheme 1.15) [55]. The nucleophilic addition of

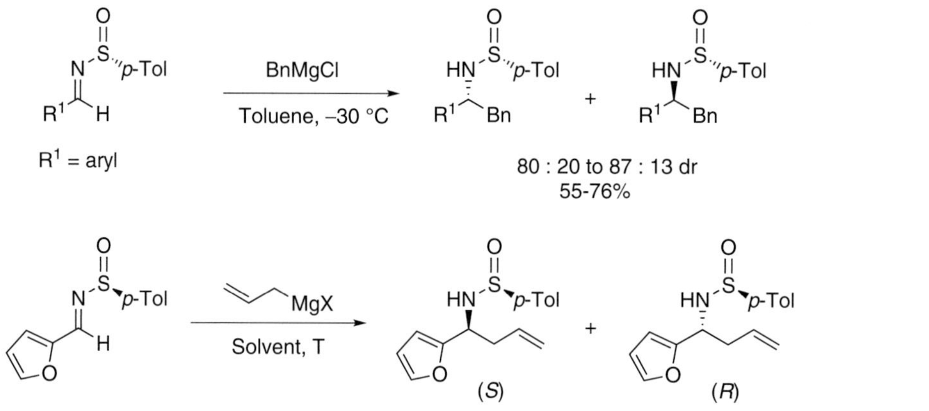

1 Stereoselective Synthesis of α-Branched Amines by Nucleophilic Addition

X	Additive	Solvent	T	Yield [%]	dr [(S) : (R)]
Br	—	CH_2Cl_2	−78 °C	98	9 : 91
Cl	$BF_3 \cdot OEt_2$	Et_2O	−78 to −40 °C	83	>99 : 1

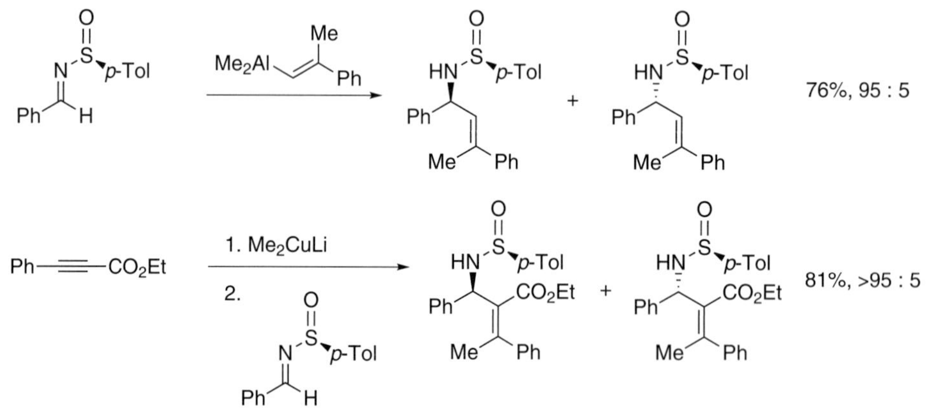

Scheme 1.14 Allylation, benzylation, and alkenylation of N-sulfinyl imines.

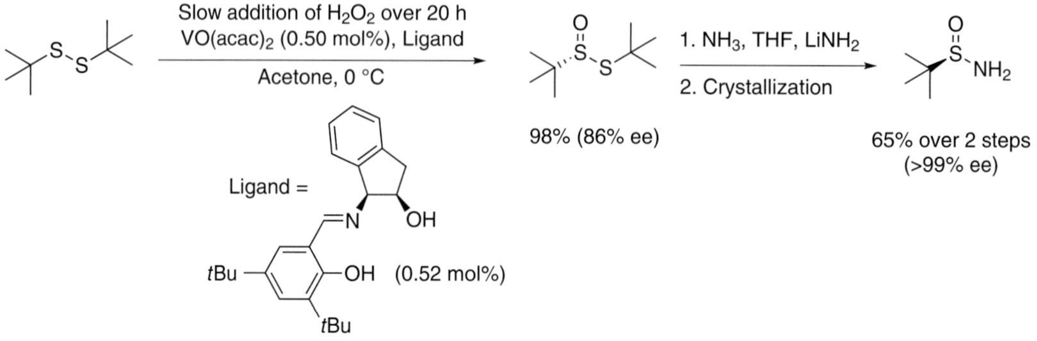

Scheme 1.15 Preparation of tert-butanesulfinamide.

Table 1.2 Diastereoselective addition to N-tert-butanesulfinyl imines derived from aldehydes.

Entry	R¹	Nucleophile	Yield 35 (%)	dr	Yield 36 (%)
1	Et	MeMgBr	96	93:7	97
2	Et	iPrMgBr	97	98:2	92
3	Et	PhMgBr	>98	96:4	90
4	i-Pr	MeMgBr	97	98:2	97
5	i-Pr	EtMgBr	>98	97:3	93
6	i-Pr	PhMgBr	98	89:11	91
7	i-Pr	VinylMgBr	90	88:12	78
8	Ph	MeMgBr	96	97:3	88
9	Ph	EtMgBr	98	92:8	94
10	Ph	iPrMgBr	29	—	—
11	Ph	VinylMgBr	79	94:6	93
12	Bn	MeMgBr	89	95:5	95
13	Bn	EtMgBr	85	92:8	98
14	Bn	VinylMgBr	81	91:9	97
15	Bn	PhMgBr	81	95:5	99
16	p-MeOPh	EtMgBr	88	99:1	>98
17	p-MeOPh	PhCCLi	93	98:2	—
18	p-MeOPh	TMSCCLi	75	98:2	—
19	p-MeOPh	nBuCCLi	88	94:6	—
20	n-Bu	PhCCLi	76	97:3	—

Grignard reagents to these imines proceeds very efficiently and generally with high diastereocontrol [56]. The reaction is thought to proceed via the chair transition-state intermediate depicted in Table 1.2, in which the Lewis acidic magnesium atom is believed to activate the imine. In addition, cleavage of the chiral auxiliary occurs under mild acidic conditions to generate α-branched amines **36** [57]. It is also possible to add sp^2 hybridized carbanions (see Table 1.2, entries 3, 6, 7, 11, 14, and 15) [58] as well as substituted lithium acetylides to these chiral electrophiles with high diastereocontrol (Table 1.2, entries 17–20) [59, 60].

Imines derived from N-tert-butanesulfinamide also undergo diastereoselective rhodium(I)-catalyzed addition of arylboronic acids [61] (Scheme 1.16), a reaction that has been primarily developed for catalytic asymmetric processes (see below).

The hydrogenation of 1,3-enynes or 1,3-diynes in the presence of N-sulfinyl-iminoesters enables highly regio- and stereoselective reductive coupling to afford diene- and enyne-containing α-amino acid esters [62]. The reaction has been extended to include ketimines used in the synthesis of tertiary carbinamines [63] with excellent diastereoselection (Table 1.3) [64, 65]. As with the aldimines, one key

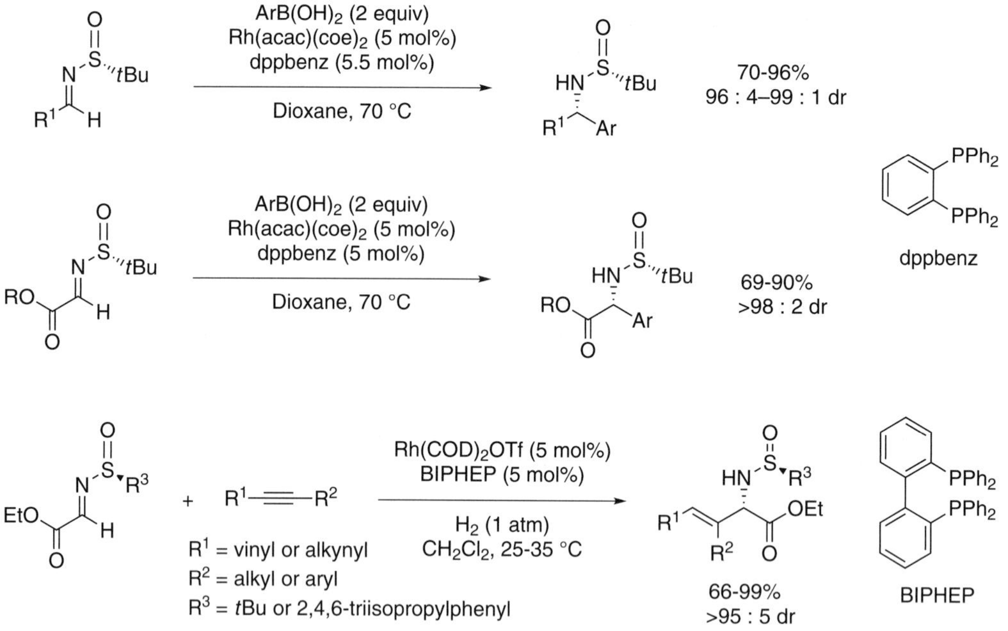

Scheme 1.16 Rhodium(I)-catalyzed arylboronic acid diastereoselective addition.

Table 1.3 Diastereoselective addition to N-sulfinyl ketimines.

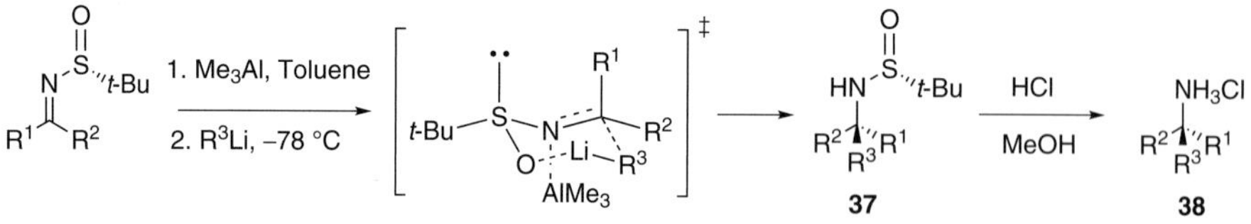

Entry	R¹	R²	R³Li	Yield 37 (%)	dr
1	Me	i-Pr	BuLi	61	99 : 1
2	Me	i-Pr	PhLi	93	97 : 3
3	Bu	i-Pr	MeLi	82	91 : 9
4	Bu	i-Pr	PhLi	82	91 : 9
5	Me	Ph	BuLi	86	98 : 2
6	Me	Bu	PhLi	93	89 : 11
7	Me	i-Bu	PhLi	62	85 : 15
8	Me	2-Naphthyl	PhLi	62	99 : 1
9	Bu	Ph	MeLi	>98	99 : 1
10	Me	Ph	AllylMgBr (no AlMe₃)	85	>99 : 1
11	Me	i-Pr	AllylMgBr (no AlMe₃)	93	>95 : 5
12	Me	i-Pr	n-PrCCLi	71	>99 : 1
13	Me	t-Bu	TMSCCLi	81	>99 : 1

feature of this methodology is the ability to generate the starting imine as a major diastereomer. The reaction can also be applied to the preparation of α,α-dibranched propargylamines (Table 1.3, entries 12 and 13) [66].

1.4
Catalytic Asymmetric Nucleophilic Addition to Achiral Imines

The first catalytic asymmetric addition of organolithium reagents to imines in the presence of a chiral amino alcohol was reported by Tomioka in 1991 (Scheme 1.17) [67]. This report was quickly followed by Soai [68], who showed that a chiral amino alcohol can also be used in the nucleophilic addition of diorganozinc to *N*-diphenylphosphinoyl imines. Four years after Tomioka's seminal report, Denmark disclosed that sparteine [69] and bisoxazolines [70] can be used in catalytic amounts in the transfer of organolithium reagents to *N*-aryl imines. All three approaches are founded on the concept that the chiral ligand RMet complex should be more reactive than the corresponding uncomplexed organometallic reagent. These initial results provided the foundation for the development of new, effective, and more practical methods for the transition metal-catalyzed asymmetric alkylation of imines, which will be discussed in the following sections.

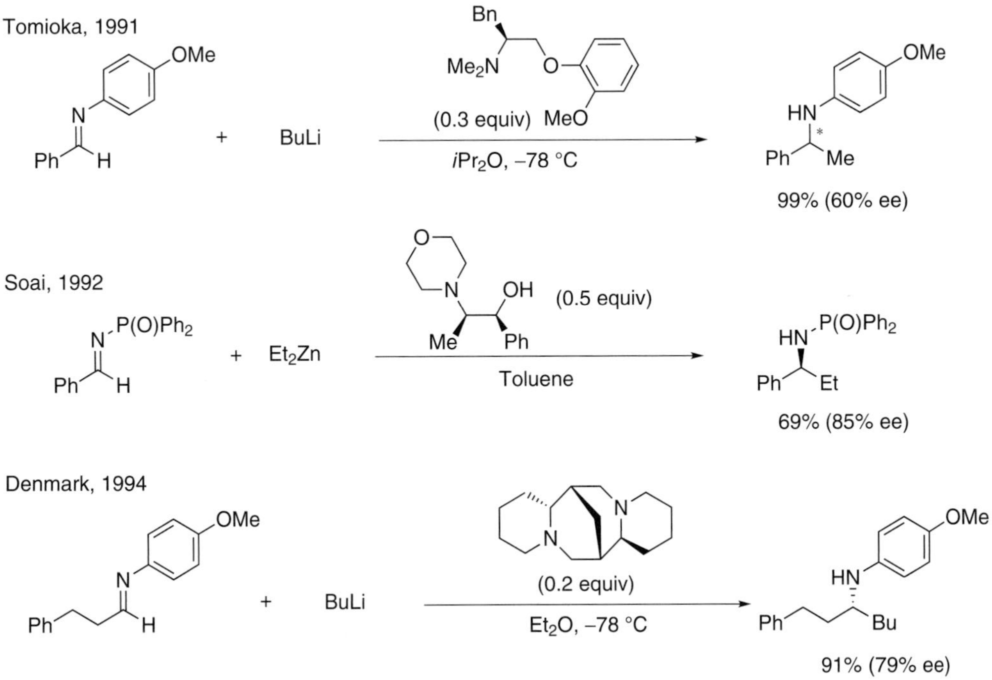

Scheme 1.17 Initial reports of the catalytic asymmetric addition of organometallic reagents to achiral aldimines.

1.4.1
Catalytic Asymmetric Addition of sp³ Hybridized Carbanions

Several transition metal-based catalysts have been developed for the addition of sp^3 hybridized carbanions to imines (Cu, Zr/Hf, Ti, Zn, and Rh). Since most of these reactions are mechanistically different, they will be addressed separately.

1.4.1.1 Copper-Catalyzed Dialkylzinc Additions

The breakthrough for the efficient development of a catalytic asymmetric addition of nucleophiles to imines was first made by Tomioka [71], when he reported that the conditions developed by Feringa and Alexakis for conjugate addition chemistry [72] were very effective in minimizing background reactions observed with uncatalyzed nucleophilic additions to the imine (Scheme 1.18). Treatment of **39** with diethylzinc in toluene leads to a mixture of the starting material (50%), the ethyl addition product **40**, and the reduction product **41**. The addition of copper(II) trifluoromethanesulfonate triphenylphosphine complex improves the yield of **40**, while minimizing the yield of the reduced imine **41**. The substitution of triphenylphosphine with chiral ligands **42–44** provided excellent yields and selectivities for the addition of diethylzinc to N-tosyl imines (Table 1.4).

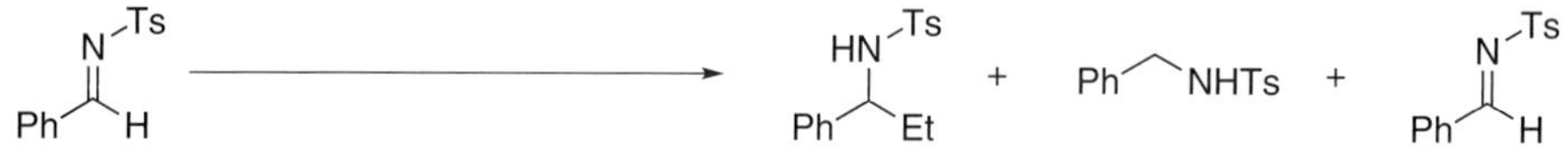

	40	**41**	**39**
Et$_2$Zn, Toluene, 0 °C, 4 h	10%	40%	50%
Et$_2$Zn, Cu(OTf)$_2$, Toluene, RT, 12 h	46%	32%	18%
Et$_2$Zn, Cu(OTf)$_2$•2 PPh$_3$ Toluene, 0 °C, 4 h	57%	15%	22%

Scheme 1.18 Conditions to minimize the uncatalyzed background reaction.

The addition of dimethyl- or diisopropylzinc proceeds with much lower efficiency, requiring higher catalyst loading and producing lower enantioselectivities (Table 1.4, entries 3 and 4). The ethyl transfer process is very efficient with aryl- and alkyl-substituted N-mesyl (Ms), N-trimethylsilylethanesulfonyl (SES), or N-tosyl imines (Ts). The cleavage of the sulfonyl group is accomplished with standard conditions; however, some racemization is observed when the methanesulfonyl group is cleaved using RedAl (Table 1.5). Although extensive studies to elucidate the reaction mechanism have not yet been completed, it is likely that the reaction proceeds via a pathway similar to that postulated for conjugate additions. The reaction of a dialkylzinc with copper(I) or copper(II) triflate affords an alkylcopper(I) species that is complexed to the chiral ligand. This species then undergoes an enantioselective nucleophilic attack to the imine. A final copper–zinc transmetallation completes the catalytic cycle and regenerates the nucleophile (Figure 1.5).

1.4 Catalytic Asymmetric Nucleophilic Addition to Achiral Imines

Table 1.4 Catalytic asymmetric addition to N-tosyl imines.

Entry	R¹	R²	R³	Ligand	Loading x (mol%)	Yield (%)	ee (%)
1	Ph	Et	Tol	42	8	98	93
2	Ph	Et	Tol	43	5	97	96
3	Ph	Me	Tol	43	19.5	39	82
4	Ph	iPr	Tol	44	19.5	92	78
5	Ph	Et	Ms	42	1	97	94
6	Ph	Et	SES	42	8	98	90
7	1-Naphthyl	Et	Ms	42	5	79	92
8	2-Naphthyl	Et	Ms	42	1	94	93
9	4-(MeO)C_6H_4	Et	Ms	42	5	83	92
10	4-ClC_6H_4	Et	Ms	42	1	95	94
11	2-Furyl	Et	Ms	42	1	98	93
12	c-C_6H_{11}	Et	Ts	43	5	84	96
13	Ph(CH$_2$)$_2$	Et	Ts	43	5	69	93

Table 1.5 Cleavage of the N-sulfonyl group.

Entry	R³	Conditions	Yield (%)	ee (%) (starting material)	ee (%) (product)
1	Ts	SmI$_2$, THF-HMPA, reflux, 5 h	84	89	89
2	SES	CsF, DMF, 95 °C, 40 h	84	86	86
3	Ms	RedAl, C$_6$H$_6$, reflux, 12 h	87	94	72

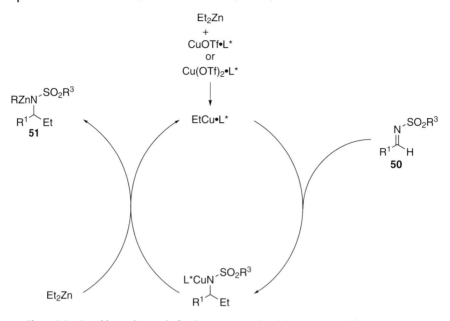

Figure 1.5 Possible catalytic cycle for the copper-catalyzed diorganozinc addition to imines.

Related chiral ligands have been developed for the copper-catalyzed addition to N-sulfonyl imines, but the enantioselectivities are usually lower and/or the scope is more limited (Figure 1.6) [73].

A copper-catalyzed catalytic asymmetric addition of diorganozinc reagents to N-phosphinoyl imines was developed by Charette [74, 75]. The most effective ligand was determined to be the Me-DuPHOS monoxide, a hemilabile, bidentate ligand prepared in two steps from Me-DuPHOS. The scope of the reaction is quite large, and this method offers the advantage of facile cleavage of the N-phosphinoyl group under mild acidic conditions (HCl, MeOH) (Table 1.6). In addition, Charette also reported that diorganozinc reagents prepared from a reaction between Grignard reagents and zinc methoxide produce chiral amines with high yields and enantiomeric excesses [76].

Since the stability of N-phosphinoyl imines derived from enolizable aldehydes can be problematic, an *in situ* formation of the N-phosphinoyl imine from its

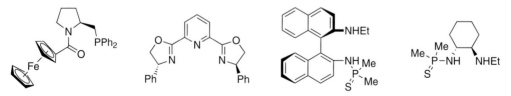

Figure 1.6 Other chiral ligands tested in the copper-catalyzed diorganozinc addition to N-sulfonyl imines.

Table 1.6 Selected examples of catalytic asymmetric addition of diorganozinc reagents to N-phosphinoyl imines.

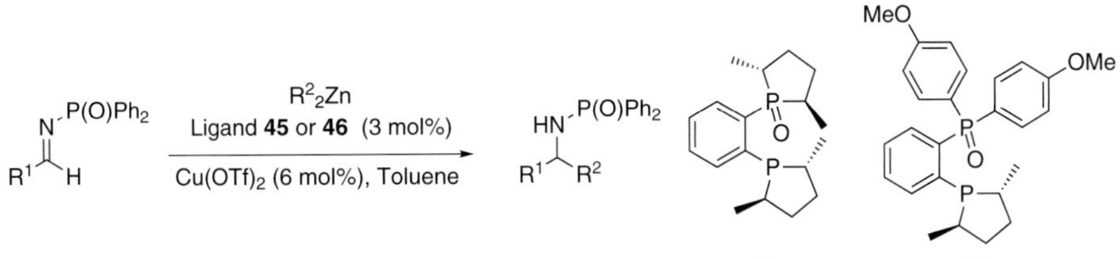

Entry	R¹	R²	Ligand	Loading (mol%)	Yield (%)	ee (%)
1	Ph	Et	45	3	96	98
2	Ph	Et	46	5	84	94
3	4-MeC$_6$H$_4$	Et	45	3	94	98
4	3-MeC$_6$H$_4$	Et	45	3	96	97
5	4-(MeO)C$_6$H$_4$	Et	45	3	91	98
6	1-Naphthyl	Et	45	3	93	97
7	2-Naphthyl	Et	45	3	96	97
8	2-Furyl	Et	45	3	97	96
9	Cyclopropyl	Et	45	3	95	94
10	2-(MeO)C$_6$H$_4$	Et	45	3	98	98
11	Ph	Me	45	5	87	97
12	Ph	nBu	45	5	92	96
13	Ph	iPr	45	5	84	95
14	Ph	nC$_{10}$H$_{21}$	45	10	73	97

corresponding *p*-toluenesulfonic acid adduct has been developed (Table 1.7), again obtaining the products in high yields and enantiomeric excesses [23, 74d]. A procedure for the one-pot preparation of α-branched amines from aldehydes and the corresponding amine has also been developed using the stable precatalyst CuOTf (Me-DuPHOS(MO))$_2$ [74c]. In addition, this catalytic system is effective for the preparation of chiral α,α,α-trifluoromethylamines (Scheme 1.19) [27].

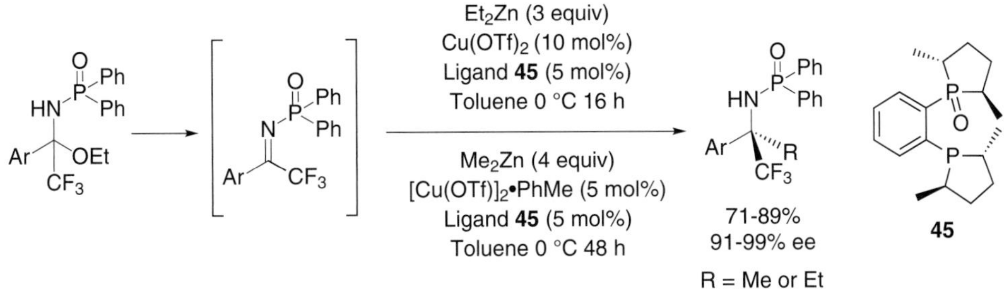

Scheme 1.19 Preparation of α,α,α-trifluoromethylamines.

Table 1.7 Enantioselective addition to alkyl-substituted N-phosphinoyl imines.

Entry	R¹	Yield (%)	ee (%)
1	Ph	87	97
2	cC_5H_{11}	92	95
3	$PhCH_2CH_2$	98	96
4	cC_6H_{11}	89	96
5	C_6H_{13}	98	95
6	iPr	86	96
7	Me	97	90
8	iBu	97	96
9	$TrOCH_2$	84	97

Feringa has reported a Cu phosphoramidite-catalyzed addition of diorganozinc and organoaluminum reagents to N-acyl imines [77]. The imines are generated in situ from aromatic and aliphatic α-amidosulfones. The reaction proceeds with high yield and a high degree of enantiocontrol (Table 1.8). However, the scope of the reaction is limited to aryl-substituted imines.

1.4.1.2 Zinc Alkoxide-Catalyzed Dialkylzinc Additions

Since the first report by Soai, which suggested that zinc alkoxides derived from amino alcohols could be used as substoichiometric additives to induce the nucleophilic attack of diorganozinc reagents to N-phosphinoyl imines, limited success has been achieved in developing a catalytic process using these electrophiles [78]. A breakthrough in this area was reported by Bräse in 2002 when it was found that substitution of the electrophile for a N-formyl imine led to a very efficient catalytic reaction when the chiral paracyclophane 48 [79] was employed as a catalyst [80]. Although the stability of N-acyl imine precluded their use as starting materials, they could be readily generated in situ from their corresponding sulfinic acid adducts (Table 1.9). Furthermore, hydrolysis of the N-formyl amide to the free α-branched amine is accomplished under mild acidic conditions (HCl, MeOH, 50 °C, 1.5 h).

1.4.1.3 Early Transition Metal (Zr, Hf)-Catalyzed Dialkylzinc Additions

Hoveyda and Snapper reported an alternative approach to the preparation of α-branched amines. Through the screening of parallel libraries of peptide-based ligands, they discovered that Schiff base dipeptide ligands were quite effective in mediating zirconium-catalyzed dialkylzinc addition to N-aryl imines (Table 1.10) [81]. The chiral catalyst complex is believed to increase the electrophilicity of the imine

Table 1.8 Copper-catalyzed diethylzinc addition to N-formyl imines.

Entry	R¹	Yield (%)	ee (%)
1	Ph	99	96
2	4-ClC$_6$H$_4$	94	97
3	4-BrC$_6$H$_4$	94	99
4	3-MeC$_6$H$_4$	99	95
5	3-(MeO)C$_6$H$_4$	96	95
6	2-(MeO)C$_6$H$_4$	99	47
7	cHexyl	99	45
8	nHexyl	99	70

through a Zr imine complex and to increase the nucleophilicity of the dialkylzinc through complexation (Figure 1.7). When dimethyl zinc is employed as the nucleophile, a large excess (15 equiv) is required for the reaction to proceed in good to excellent yields (Table 1.10, entries 8 and 9).

To extend this methodology to N-aryl imines derived from alkyl-substituted aldehydes, they developed a three-component approach involving *in situ* imine formation from an appropriate aldehyde and o-anisidine followed by the alkylation

Table 1.9 Diethylzinc addition to N-formyl imines.

Entry	Ar	Loading (mol%)	Yield (%)	ee (%)
1	Ph	2	>99	95
2	4-ClC$_6$H$_4$	2	>99	89
3	4-ClC$_6$H$_4$	5	97	90
4	4-(MeO)C$_6$H$_4$	2	97	95
5	4-(MeOOC)C$_6$H$_4$	2	90	94
6	2,6-Cl$_2$C$_6$H$_3$	2	98	95
7	4-tBuC$_6$H$_4$	2	>99	75
8	4-MeC$_6$H$_4$	2	>99	95
9	3-ClC$_6$H$_4$	5	99	93

Table 1.10 Zirconium-catalyzed alkylation of N-aryl imines.

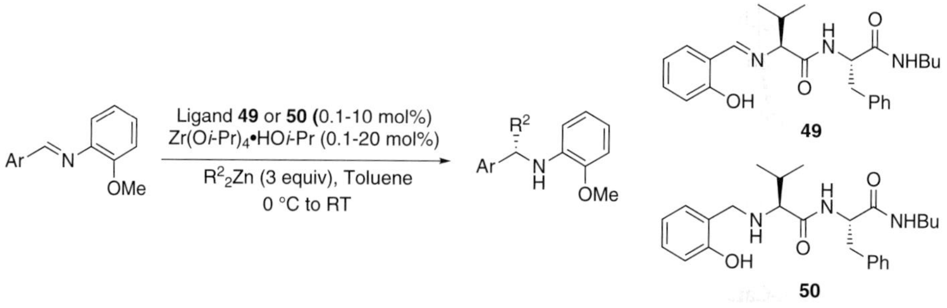

Entry	Ar	R²	Ligand	Yield (%)	ee (%)
1	Ph	Et	49	82	93
2	1-Furyl	Et	49	87	97
3	2-Naphthyl	Et	49	67	92
4	1-Naphthyl	Et	49	86	91
5	4-(CF$_3$)C$_6$H$_4$	Et	50	87	88
6	2-BrC$_6$H$_4$	Et	50	62	90
7	4-(MeO)C$_6$H$_4$	Et	50	71	91
8	Ph	Me	50	79	88
9	1-Furyl	Me	50	98	84

reaction (Table 1.11) [82]. While the more Lewis acidic zirconium alkoxide afforded lower yields, higher yields with equally good or slightly lower enantioselectivities were obtained using hafnium tetraisopropoxide [83].

These conditions using the same family of chiral ligands have been recently applied to the enantioselective synthesis of N-substituted quaternary carbon centers derived from ketoimines (Scheme 1.20) [84, 85].

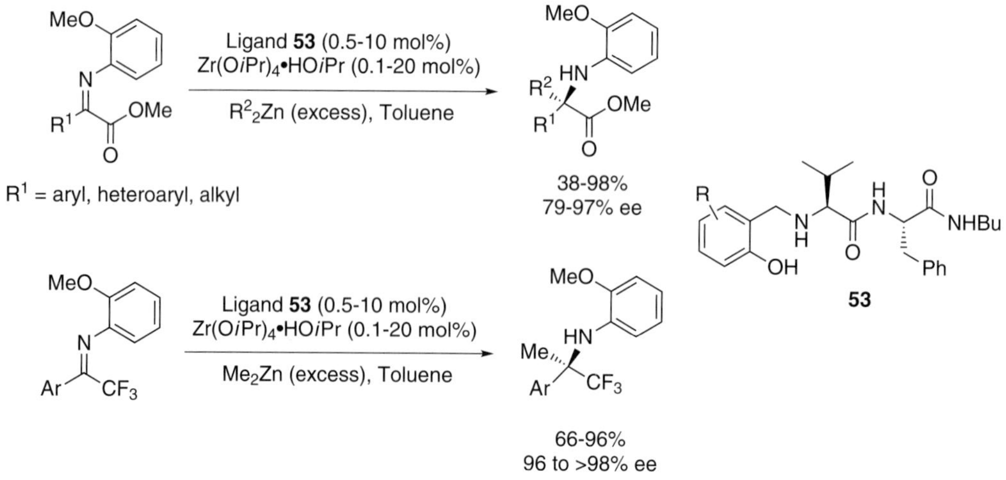

R¹ = aryl, heteroaryl, alkyl

38–98%
79–97% ee

66–96%
96 to >98% ee

Scheme 1.20 Zirconium-catalyzed addition to N-aryl imines.

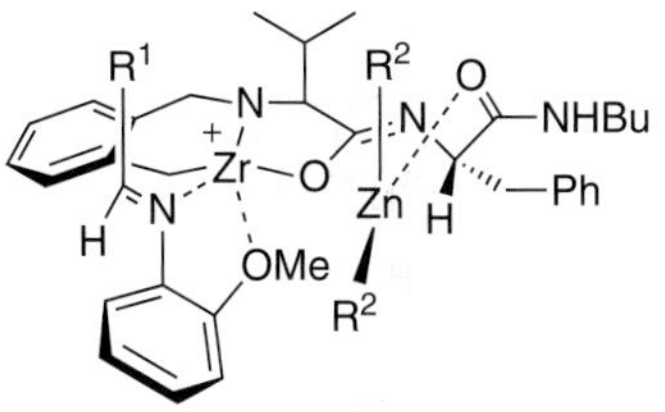

Figure 1.7 Model for the zirconium-catalyzed diorganozinc addition to N-aryl imines.

1.4.1.4 Rhodium-Catalyzed Dialkylzinc Addition Reactions

As mentioned, in most copper- and zirconium-catalyzed alkylations of imines, the addition of less reactive dimethylzinc (relative to diethylzinc) usually requires a large excess of the organometallic reagent. Hayashi reported that the methylation of N-tosyl imines could be achieved using only 1.5 equiv of dimethylzinc in the presence of a chiral rhodium complex that is coordinated with chiral diene **54** [86]

Table 1.11 Three-component synthesis of α-branched amines.

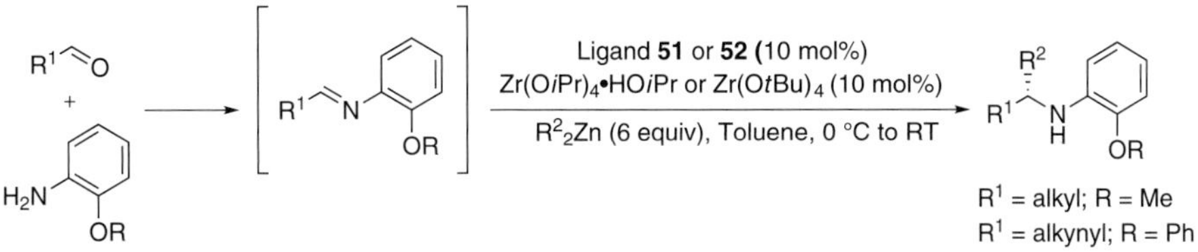

R^1 = alkyl; R = Me
R^1 = alkynyl; R = Ph

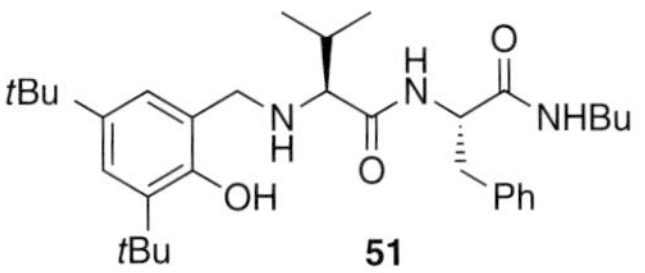

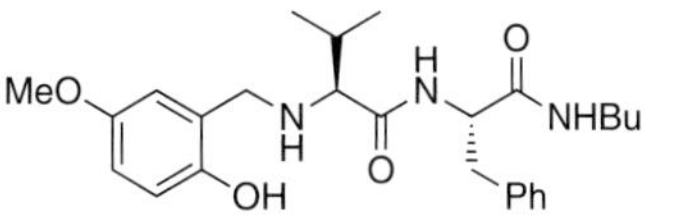

Entry	R^1	R^2	Ligand	Yield (%)	ee (%)
1	C_4H_9	Et	52	69	97
2	C_6H_{13}	Et	52	62	97
3	$PhCH_2CH_2$	Et	52	>98	94
4	(E)-$C_5H_{11}CH=CHCH_2CH_2$	Et	20	60	>98
5	$iPrCH_2$	Et	52	58	95
6	cC_3H_5	Et	52	83	98
7	$Br(CH_2)_5$	Et	52	57	95
8	$nC_5H_{11}CC$	Me	51	87	82
9	$TBSOCH_2CC$	Me	51	75	92
10	$TBSOCH_2CC$	Et	51	70	>98
11	PhCC	Me	51	84	80
12	PhCC	Et	51	85	>98
13	$TMSOC(Me_2)CC$	Et	52	86	84

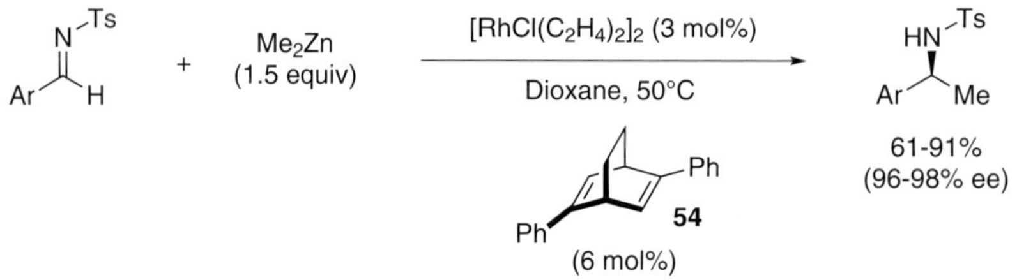

Scheme 1.21 Rhodium-catalyzed dimethylzinc addition to N-tosyl imines.

(Scheme 1.21). This catalyst provides the addition product derived from the N-tosyl imines of aryl-substituted aldehydes in high yield and enantiocontrol. It is believed that the addition proceeds via the formation of a methylrhodium species obtained by the reaction between the rhodium catalyst and dimethylzinc.

1.4.2
Catalytic Asymmetric Allylation of Imines

The enantioselective asymmetric allylation of imines has been a synthetic challenge, the initial solutions of which required stoichiometric amounts of chiral allylboron [87], allylsilane [88], allylzinc [89], or allylindium reagents [90]. Itsuno showed that a chiral B-allyloxazaborolidine derived from norephedrine could add to the N-trimethylsilyl imine prepared from benzaldehyde in high yield and enantiomeric excess (Scheme 1.22) [91]. Brown later reported that B-allyldiisopinocamphenylborane is also very effective for the allylation of the same electrophiles, but the addition of a molar amount of water is necessary to obtain high yields [92]. The diastereo- and

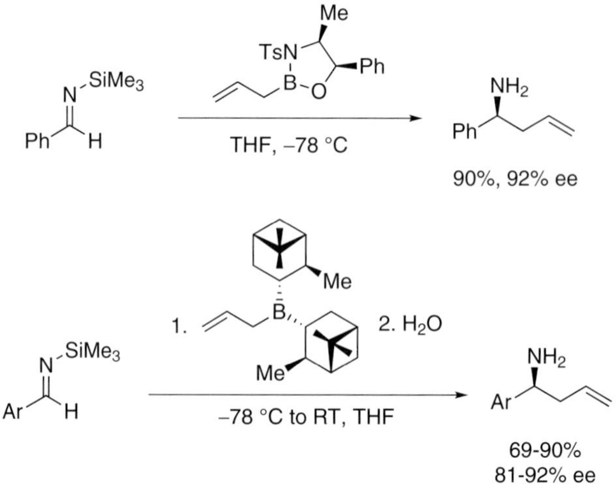

Scheme 1.22 Allylation of N-silyl imines.

enantioselective crotylation of *N*-alumino imines (obtained by the DIBAL-H reduction of the nitrile) using the related *E*- or *Z*-crotylboranes has also been reported [93].

An interesting methodology employing the related allylsilane chemistry has been reported by Leighton [88c,d]. Chiral allylsilanes derived from pseudoephedrine are reacted with aldehyde- and ketone-derived hydrazones to generate the homoallylic hydrazide in good to excellent enantioselectivities and yields (Scheme 1.23). *cis*-Crotylsilanes and *trans*-crotylsilanes react with the hydrazone to afford the *anti*-hydrazide and the *cis*-hydrazide, respectively, with high enantio- and diastereocontrol. The stereochemistry is the opposite to that observed for the addition of crotylboranes to aldehydes. That is why a two-point binding/double activation has been proposed to account for these observations. Reductive cleavage of the N−N bond is accomplished in high yield using samarium diiodide.

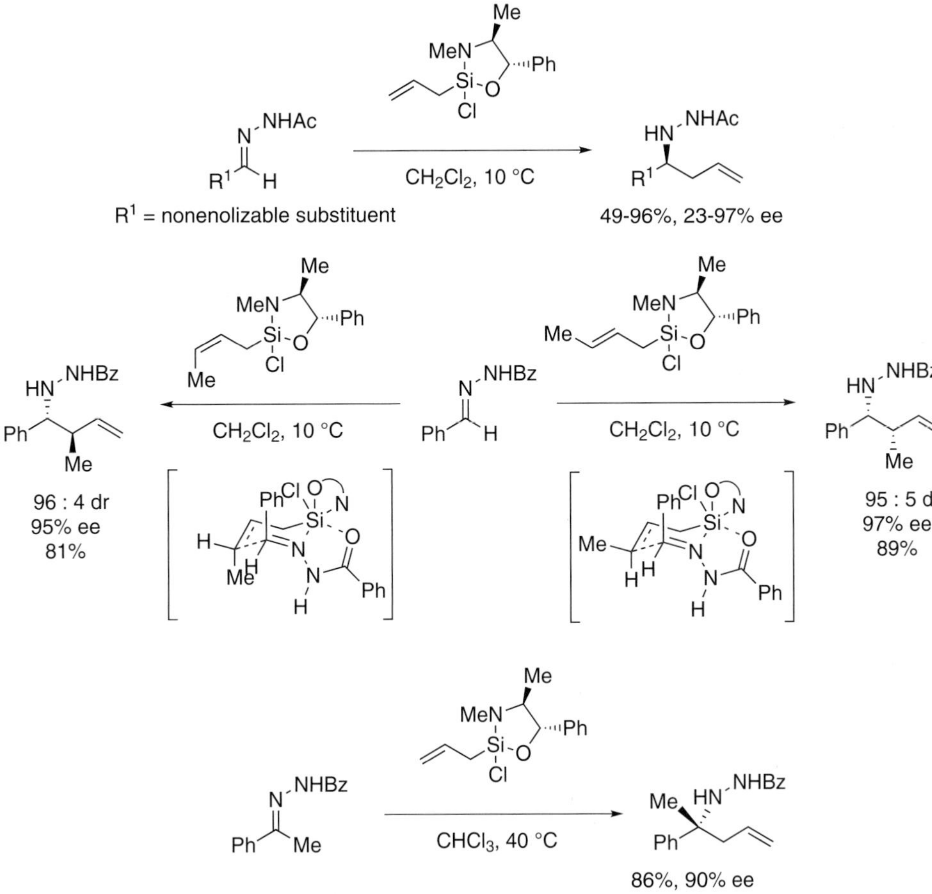

Scheme 1.23 Allylation of hydrazones.

Yamamoto reported the first palladium-catalyzed addition of allyltributyltin or allyltrimethylsilane to N-benzyl imines [94]. Higher enantioselectivities are obtained if the reaction is performed in the presence of 1 equiv of water (Scheme 1.24).

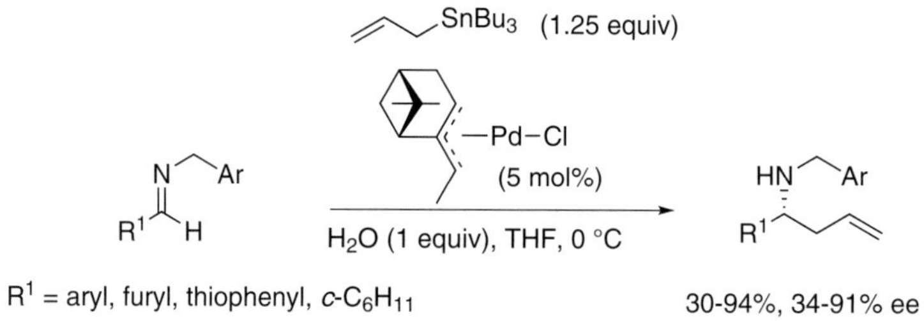

Scheme 1.24 Palladium-catalyzed allylation of N-benzyl imines.

The first step of the catalytic cycle reaction is presumably an initial tin to palladium transmetallation to generate an η^3 allylpalladium species that adds on the imine. The catalyst is then regenerated following a second N-Pd to N-Sn transmetallation.

Kobayashi has demonstrated that a chiral zirconium catalyst derived from 3,3′-dibromo- or 3,3′-dichloro-BINOL can catalyze the addition of substituted allylstannanes to the imines derived from 2-hydroxyaniline and aryl- or hetereoarylsubstituted aldehydes (Scheme 1.25) [95]. The active catalyst is possibly generated by the bonding of the alcohol functionalities of the imine and of allylstannane to the zirconium center.

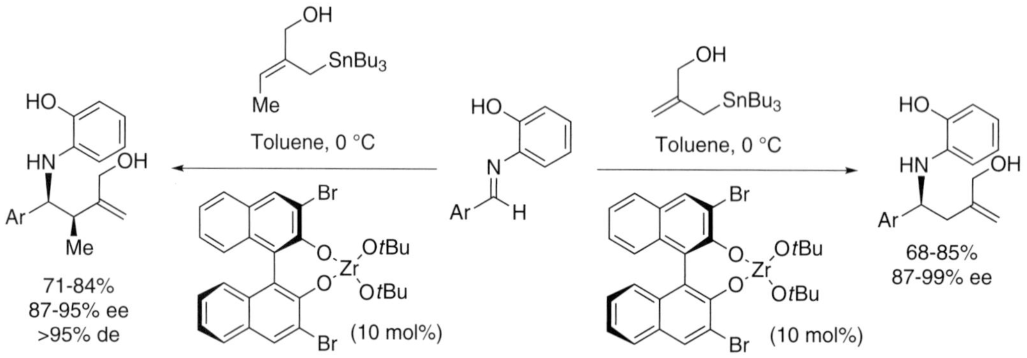

Scheme 1.25 Zirconium-catalyzed allylation of N-aryl imines.

Cook reported that a 3,3′-bis(trifluoromethyl)-BINOL-catalyzed asymmetric addition of allylindium to hydrazones proceeds in modest to good enantioselectivities (10–92% ee) [90]. The stoichiometric version of this reaction yields much higher enantioselectivities (84–97% ee). Jacobsen later found that a chiral urea catalyst is effective in catalyzing a similar transformation [96]. The bifunctional catalyst **55** bearing a hydrogen-bond donor and a Lewis base that are properly

Table 1.12 Allylation of hydrazones.

Entry	R	Yield (%)	ee (%)
1	C_6H_5	87	92
2	$pClC_6H_4$	83	92
3	2-Furyl	90	87
4	2-Thienyl	82	93
5	$p(MeOOC)C_6H_4$	92	76
6	$oBrC_6H_4$	78	93
7	$oMeC_6H_4$	89	95
8	1Naphthyl	89	95
9	$p(MeO)C_6H_4$	79	93

positioned is very effective in catalyzing the allylindium addition to N-acylhydrazones (Table 1.12).

The last class of allylation reactions that are amenable to asymmetric catalysis employs allylboronate derivatives. Schaus reported that several chiral BINOL derivatives catalyze the enantioselective asymmetric allylboration of acyl imines [97]. This reaction is most effective when 3,3′-diphenyl-BINOL acts as the catalyst and allyldiisopropoxyborane is the nucleophile. The allylation products are obtained in good yields (75–94%) and excellent enantiomeric excesses (>90% ee) for both aromatic and aliphatic imines (Table 1.13).

The reactions of both isomers of the corresponding crotyldiisopropoxyboranes are stereoconvergent, giving the *anti*-isomer with excellent diastereocontrol (Scheme 1.26). Mechanistic studies suggest that a facile exchange between an isopropoxy group of the nucleophile and a BINOL unit takes place to give the active allylation reagent. Two transition states have been proposed for the reaction of E-crotyl- and Z-crotylboronates (Figure 1.8).

Shibasaki also reported related allylation chemistry using allylboronic acid pinacol ester [98], where an allylation procedure of ketoimines using 10 mol% of CuF cyclopentyl-DuPHOS, 30 mol% of lithium isopropoxide, and *t*-butanol was developed. High enantiocontrol is observed with a variety of aryl methyl ketoimines (Table 1.14). The enantioselectivity decreases with dialkyl-substituted ketoimines. Mechanistic studies suggest that lithium isopropoxide accelerates the reaction rate by increasing the concentration of the allylcopper species.

Table 1.13 Allylation of N-benzoyl imines.

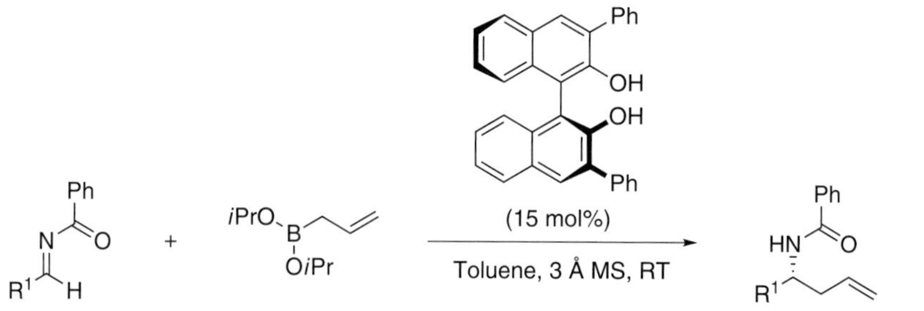

Entry	R^1	Yield (%)	ee (%)
1	C_6H_5	87	98
2	$pMeC_6H_4$	83	96
3	$pBrC_6H_4$	86	95
4	$p(MeO)C_6H_4$	85	90
5	pFC_6H_4	94	96
6	oFC_6H_4	91	91
7	$m(CF_3)C_6H_4$	89	95
8	2-Furyl	83	92
9	2-Thienyl	81	90
10	2-Naphthyl	88	92
11	(E)-PhCH=CH	82	91
12	$PhCH_2CH_2$	83	99
13	cC_6H_{11}	80	96
14	tBu	81	99
15	$BnOCH_2$	84	93
16	(Z)-EtCH=CH(CH$_2$)$_2$	82	91

Several catalytic asymmetric allylation reactions have been developed specifically for iminoester derivatives, but the level of enantioselection remains typically only in the low to high 80s (Table 1.15). Allylstannanes [99], allylsilanes [100], allyltrimethoxysilanes [101], and allylboronates [102] are suitable nucleophiles in these copper- or zinc-catalyzed processes.

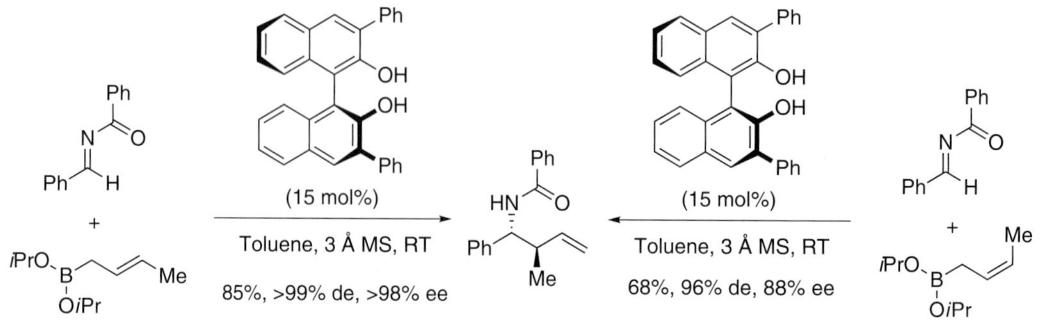

Scheme 1.26 Crotylation of N-acyl imines.

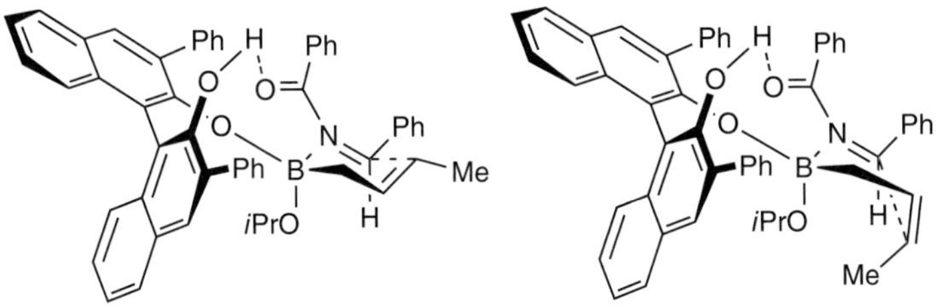

Figure 1.8 Transition-state models for the crotylation of N-acyl imines.

1.4.3
Catalytic Asymmetric Addition of sp² Hybridized Carbanions

The various possibilities for the preparation of chiral allylic amines or α-aryl-substituted amines are outlined in Figure 1.9. Although the addition reaction of a carbon nucleophile to an imine derived from an aryl-substituted aldehyde is very efficient (**B**), the related addition to an α,β-unsaturated imine (**A**) can sometimes proceed via a 1,4-addition pathway. Similarly, the asymmetric C=N reduction reaction (**C** and **D**) is sometimes hampered by the possibility of either obtaining conjugate reduction (in the case of **C**) or low enantioselectivities (in **D** when R = aryl). The addition of sp² hybridized carbanions to imines (**E**) is a particularly effective

Table 1.14 Allylation of N-benzyl ketimines.

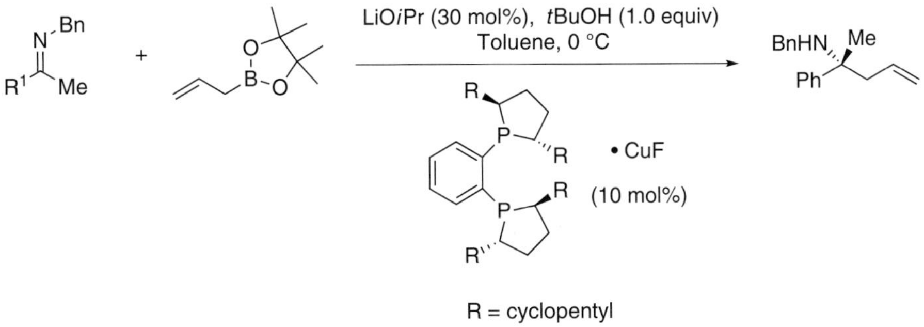

Entry	R¹	Yield (%)	ee (%)
1	Ph	92	89
2	3-MeC$_6$H$_4$	96	91
3	3-(MeO)C$_6$H$_4$	97	93
4	3-FC$_6$H$_4$	89	87
5	4-(MeO)C$_6$H$_4$	76	85
6	4-ClC$_6$H$_4$	82	81
7	2-Naphthyl	88	92
8	PhCH$_2$CH$_2$	98	23

1 Stereoselective Synthesis of α-Branched Amines by Nucleophilic Addition

Table 1.15 Selected catalytic systems for the allylation of imines.

Electrophile	Nucleophile	Catalyst	Product	Author, year
Ts-N=CH-C(O)OEt	allyl-SnBu$_3$	(R)-Tol-BINAP · CuPF$_6$ (5 mol%)	NHTs-CH(allyl)-C(O)OEt, 91%, 83% ee	Jørgensen, 1999
HN(Ts)-CH(OH)-C(O)OEt	Ph-C(=CH$_2$)-CH$_2$-SiMe$_3$	(R)-Tol-BINAP · CuClO$_4$ (6 mol%)	TsHN-CH(CH$_2$C(Ph)=CH$_2$)-C(O)OEt, 85%, 90% ee	Lectka, 1999
4-MeO-C$_6$H$_4$-C(O)NH-N=CH-C(O)OR^1	R^2-C(=CH$_2$)-CH$_2$-Si(OMe)$_3$	bis(2-methoxybenzyl) diamine ligand, ZnF$_2$ (10 mol%, 20 mol%)	4-MeO-C$_6$H$_4$-C(O)NH-NH-CH(CH$_2$C(R^2)=CH$_2$)-C(O)OR^1, 61–92%, 78–83% ee	Kobayashi, 2003
4-NMe$_2$-C$_6$H$_4$-C(O)NH-N=CH-C(O)OMe	allyl-Bpin (R-substituted)	bis(3-methoxy-2-naphthylmethyl) diamine ligand, ZnF$_2$ (12 mol%, 5 mol%)	4-NMe$_2$-C$_6$H$_4$-C(O)NH-NH-CH(CH(R)-CH=CH$_2$)-C(O)OMe, 65 to >99%, >99:1 dr, 82–88% ee	Kobayashi, 2008

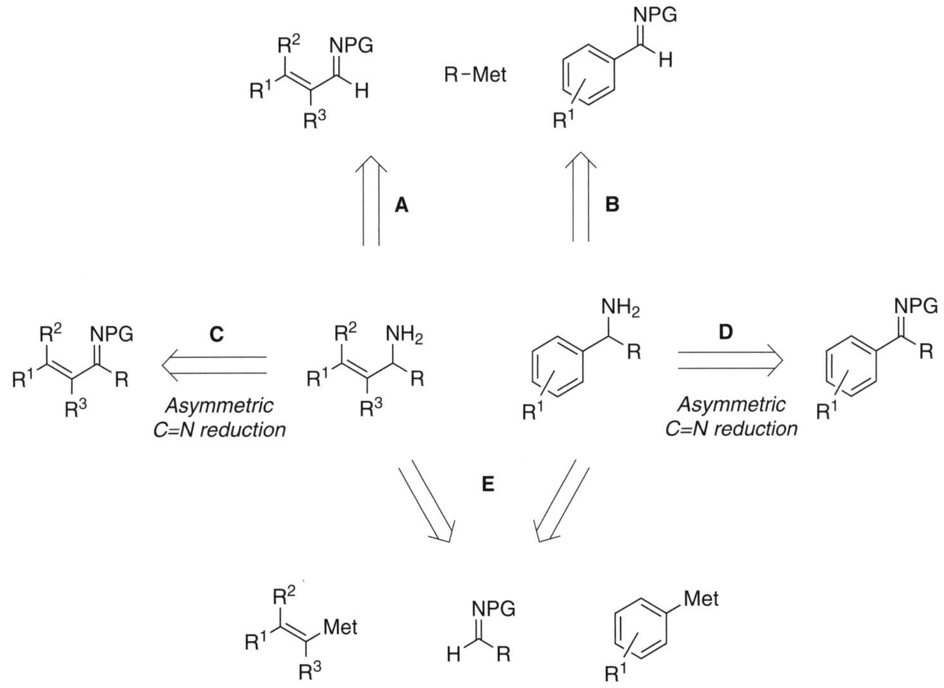

Figure 1.9 Various approaches to allylic and benzylic amines.

method to generate chiral allylic amines or diarylcarbinamines when the other strategies outlined in Figure 1.9 are not appropriate.

1.4.3.1 Catalytic Asymmetric Vinylation

The catalytic asymmetric addition of an unsubstituted vinylorganometallic reagent to an imine remains an important synthetic challenge and, to date, only limited success has been reported with appropriately substituted vinyl units. Jamison has reported a nickel-catalyzed three-component coupling of alkynes, imines, and organoboron reagents to afford tetrasubstituted allylic amines in a single step as well as high reaction yields and good enantioselectivities (Table 1.16) [103]. When unsymmetrical alkynes are used, a regioisomeric mixture of allylic amines is usually obtained unless both substituents on the alkyne are sterically or electronically significantly different (Table 1.16, entry 10 versus entry 11). The removal of the (*tert*-butyldimethylsilyloxy)ethyl protecting group is achieved using a two-step process (Bu$_4$NF then H$_5$IO$_6$).

An enantioselective iridium-catalyzed hydrogenative coupling of alkynes to aromatic and aliphatic *N*-arylsulfonyl imines has been reported by Krische (Table 1.17) [104]. The reaction is very efficient with a wide range of imines and alkynes. When unsymmetrical alkynes are used, the carbon–carbon bond forming process occurs at the more substituted position (Table 1.7, entries 17–21).

Table 1.16 Vinylation of imines.

Entry	R¹	R²	R³	Yield (%)	ee (%)
1	Ph	nPr	nPr	85	89
2	Ph	nBu	nBu	83	89
3	Ph	Et	Et	89	83
4	oMeC$_6$H$_4$	nPr	nPr	74	85
5	p(MeO)C$_6$H$_4$	nPr	nPr	75	82
6	p(CF$_3$)C$_6$H$_4$	nPr	nPr	91	85
7	2-Naphthyl	nPr	nPr	90	73
8	2,3-(OCH$_2$O)C$_6$H$_3$	nPr	nPr	95	73
9	cC$_6$H$_{11}$	Et	Et	53	51
10	Ph	Ph	Et	62	71
11	Ph	2-Naphthyl	Me	42 (85:15 rr)	70

The reaction is believed to proceed via the formation of an aza-iridacyclopentene that undergoes protolytic cleavage of the nitrogen–iridium bond (Scheme 1.27). Subsequent reductive elimination leads to the product.

A related rhodium-catalyzed enantioselective reductive coupling of acetylene to N-arylsulfonyl imines leads to the formation of (Z)-dienyl allylic amines (Scheme 1.28) [105]. The scope of the reaction is comparable to that demonstrated for the analogous iridium-catalyzed process. The reaction between the acetylene and rhodium leads to the oxidative dimerization of acetylene to form a cationic rhodacyclopentadiene that then reacts with the imine to generate the product after the protolytic cleavage and reductive elimination.

The Petasis reaction is a mild multicomponent reaction that allows the condensation of a boronic acid, an amine, and a carbonyl derivative to generate an allylic amine. Although several diastereoselective Petasis reactions have been reported [106], the first catalytic asymmetric reaction was described in 2008 (Scheme 1.29) [107]. It was shown that the condensation proceeds in high yields and enantiomeric excesses, affording the corresponding protected α-vinylglycine derivatives.

1.4.3.2 Catalytic Asymmetric Arylation

1.4.3.2.1 Amino Alcohol-Catalyzed Addition of Organozinc Reagents
Despite the relatively large number of efficient catalytic asymmetric methods available for alkyl group transfer from diorganozinc reagents to imines, the analogous reaction employing a diarylzinc reagent is much more difficult to achieve. A major obstacle

Table 1.17 Iridium-catalyzed vinylation of N-sulfonyl imines.

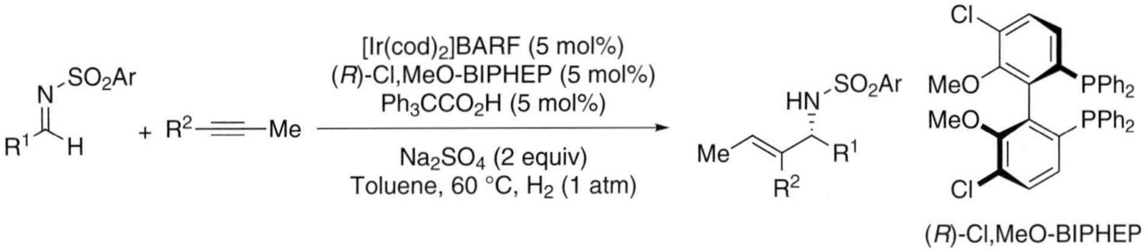

Entry	R^1	Ar	R^2	Yield (%)	ee (%)
1	C$_6$H$_5$	Ph	Me	70	98
2	4-(MeO)C$_6$H$_4$	Ph	Me	64	99
3	3-(MeO)C$_6$H$_4$	Ph	Me	72	98
4	4-ClC$_6$H$_4$	4-Tol	Me	67	97
5	4-(MeOOC)C$_6$H$_4$	4-Tol	Me	72	97
6	2-Naphthyl	Ph	Me	64	94
7	(E)-PhCH=CH	Ph	Me	76	99
8	1-Furyl	Ph	Me	80	97
9	1-Thienyl	Ph	Me	81	98
10	cC$_6$H$_{11}$	4-Tol	Me	74	98
11	cC$_5$H$_9$	Ph	Me	66	99
12	cC$_3$H$_5$	Ph	Me	80	92
13	Me	4-Tol	Me	67	97
14	nPr	Ph	Me	70	97
15	iPr	Ph	Me	69	94
16	iBu	4-Tol	Me	70	99
17	1-Furyl	Ph	nPr	78	97
18	cC$_3$H$_5$	Ph	nPr	65	99
19	1-Furyl	Ph	iPr	80	97
20	cC$_3$H$_5$	Ph	iPr	66	94
21	1-Furyl	Ph	CH$_2$CH$_2$OTBS	69	98

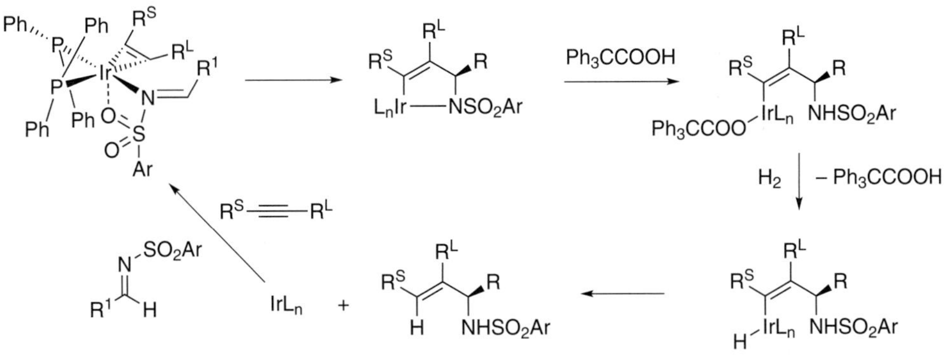

Scheme 1.27 Proposed mechanism for the iridium-catalyzed vinylation.

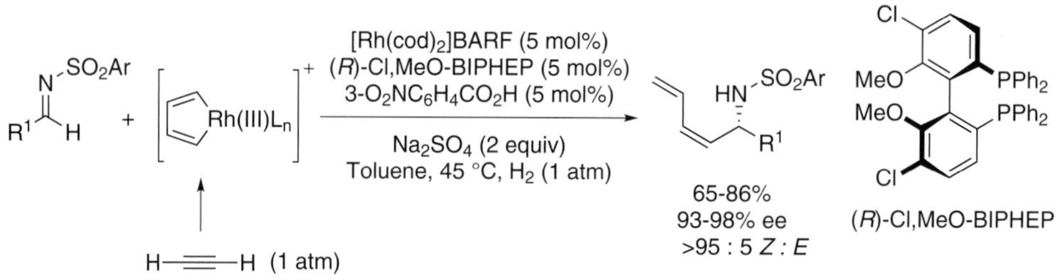

Scheme 1.28 Rhodium-catalyzed vinylation of N-arenesulfonyl imines.

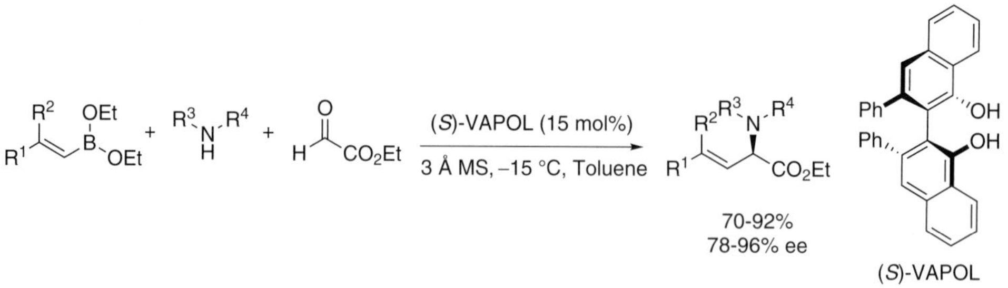

Scheme 1.29 Asymmetric Petasis reaction.

resides in the fact that the aryl group transfer from a diarylzinc reagent to an imine is a more facile process, thus rendering the uncatalyzed aryl group transfer competitive with the catalyzed asymmetric process. A solution to overcome this problem is to apply mixed diorganozinc reagents that possess lower levels of reactivity [108]. Bräse and Bolm have reported that the *in situ* preparation of a mixed diorganozinc leads to increased enantioselectivities for the phenyl group transfer to N-formyl imines catalyzed by cyclophane **56** [109] (Scheme 1.30). Thus far, the reaction has not been applied to the transfer of other aryl groups.

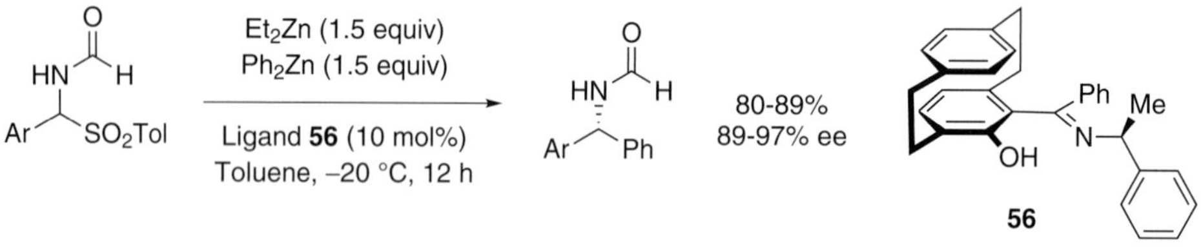

Scheme 1.30 Arylation of N-formyl imines.

1.4.3.2.2 Rhodium Phosphine-Catalyzed Arylation of Imines

The asymmetric rhodium-catalyzed arylation of imines is among the most efficient methods for the preparation of α-aryl chiral amines. Hayashi first reported that arylstannane [110] and

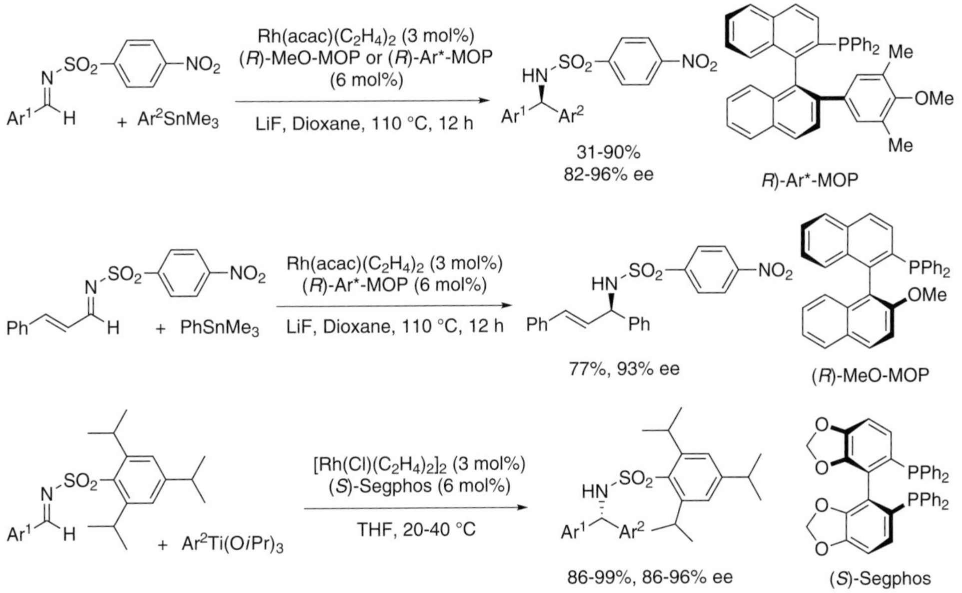

Scheme 1.31 Rhodium-catalyzed arylation of N-arenesulfonyl imines.

aryltitanium [111] reagents can be used as stoichiometric nucleophiles in the rhodium-catalyzed addition to N-sulfonyl imines (Scheme 1.31). Both reactions presumably proceed via the formation of an arylrhodium species that is complexed to the chiral ligand. Several aryl sulfonyl groups were screened, and the p-nitro derivatives gave not only the highest ee's when (R)-Ar*-MOP was used as the ligand but also the highest yields when 4 equiv of various arylstannanes were employed. The cleavage of the sulfonyl group to liberate the free amine is accomplished in high yields upon treating the product with thiophenol and potassium carbonate. Interestingly, the arylation of an α,β-unsaturated imine leads to the 1,2-addition product in high ee's under these reaction conditions. The arylation of N-sulfonyl imines with 2 equiv of an aryltitanium reagent leads to the product in good yields and ee's. SEGPHOS was determined to be the optimal ligand with this nucleophile, but the sterically hindered 2,4,6-triisopropylbenzenesulfonyl group afforded the highest enantioselectivities. The cleavage of the bulky group could be accomplished with no epimerization with samarium diiodide in HMPA, lithium in ammonia, or with RedAl.

Building upon Miyaura's pioneering work on the rhodium-catalyzed addition of organoboron reagents to N-tosyl imines [112], Tomioka reported [113] the first catalytic enantioselective version of the reaction involving several chiral phosphines (Scheme 1.32). The enantiomeric excesses are usually >90% when Ar^1 is an o-trimethylsilylphenyl. Although the enantioselectivities were good, several more effective phosphorus-based ligands were since developed to give high enantioselectivities and yields. Zhou later found that a rhodium complex bearing a chiral monophosphite ligand gave good enantioselectivities [114].

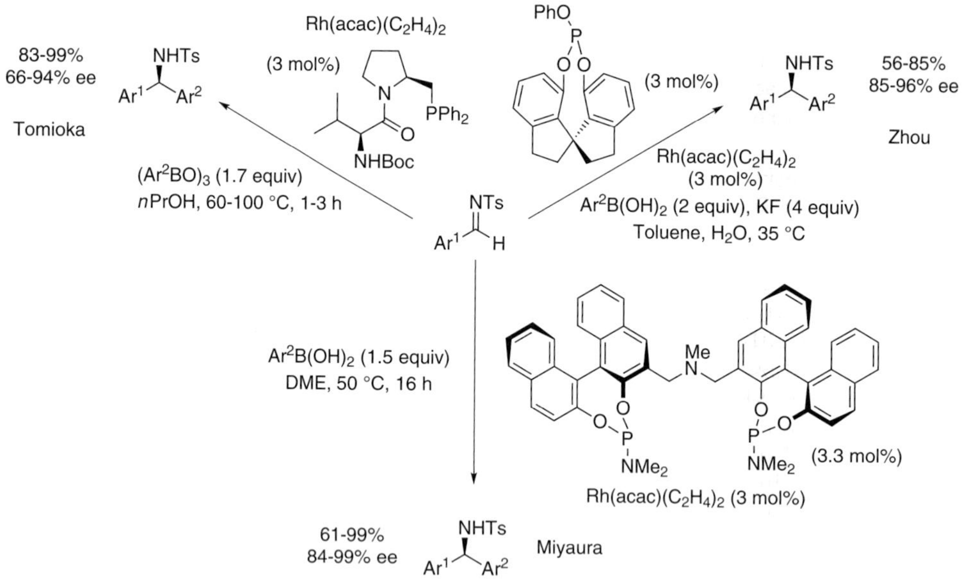

Scheme 1.32 Arylation of N-tosyl imines.

Finally, a chiral N-linked C_2-symmetric bidentate phosphoramidite was developed by Miyaura for the arylation of N-sulfonyl imines [115]. Over 38 different chiral N-tosyl amines were prepared in high yields and enantiomeric excesses using this novel ligand. These conditions were also quite effective (72–99% yield and 93–98% ee) for the arylation of N-p-nitrobenzenesulfonyl imines, which can be cleaved under mild conditions to liberate the free amine.

The catalytic enantioselective addition of arylboronic acids to aliphatic imines proceeds with broad substrate scope, high yields, and enantiocontrol (Table 1.18) [116]. In this case, DeguPHOS (57) was found to be the optimal chiral ligand.

The nature of the activating group on the imine can also be modified to facilitate its cleavage after the addition. Again, DeguPHOS (57) was found to be equally effective for the nucleophilic addition to both aryl- [61a] and alkyl-substituted N-phosphinoyl imines [116]. The diphenylphosphinoyl group was easily cleaved using HCl/MeOH to afford high yield of the pure amine hydrochloride (Scheme 1.33).

Alternatively, it is also possible to use *in situ* generated N-Boc imines as electrophiles [117]. When α-carbamoyl sulfones are treated under the rhodium-catalyzed addition of arylboronic acids, the imine is formed *in situ*, and the nucleophilic addition proceeds smoothly to generate the N-Boc-protected amine (Scheme 1.34).

The catalytic asymmetric synthesis of diarylmethylamines by a rhodium/phosphoramidite-catalyzed addition of arylboronic acids to N,N-dimethylsulfamoyl-protected aldimines has been reported by de Vries and Feringa [118]. The reaction produces very high yields and high enantioselectivities of the protected amine. Deprotection of the amine is achieved without any racemization upon heating the product in the microwave with 1,3-diaminopropane (Scheme 1.35).

1.4 Catalytic Asymmetric Nucleophilic Addition to Achiral Imines

Table 1.18 Addition of arylboronic acids to N-tosyl imines.

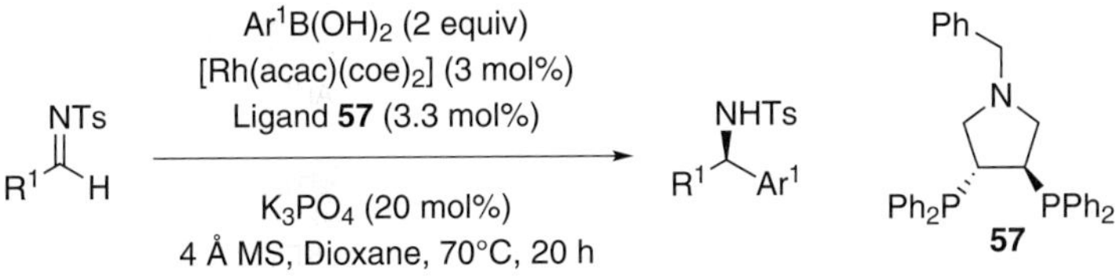

Entry	R^1	Ar^1	Yield (%)	ee (%)
1	PhCH$_2$CH$_2$	4-ClC$_6$H$_4$	94	95
2	CH$_3$CH$_2$CH$_2$	4-ClC$_6$H$_4$	89	93
3	(CH$_3$)$_2$CHCH$_2$	4-ClC$_6$H$_4$	96	91
4	CH$_2$=CHCH$_2$CH$_2$	4-ClC$_6$H$_4$	87	98
5	cC$_6$H$_{11}$	4-ClC$_6$H$_4$	80	96
6	cC$_6$H$_{11}$	4-MeC$_6$H$_4$	71	96
7	cC$_6$H$_{11}$	4-MeOC$_6$H$_4$	74	90
8	cC$_6$H$_{11}$	4-CF$_3$C$_6$H$_4$	89	91
9	cC$_6$H$_{11}$	3-ClC$_6$H$_4$	75	90
10	cC$_6$H$_{11}$	3-AcC$_6$H$_4$	81	89

The mechanism of the rhodium-catalyzed arylation of imines presumably occurs via an initial transmetallation to produce the arylrhodium species **59** (Figure 1.10). Aryl transfer followed by a second rhodium–boron transmetallation (or protonation by water) completes the catalytic cycle.

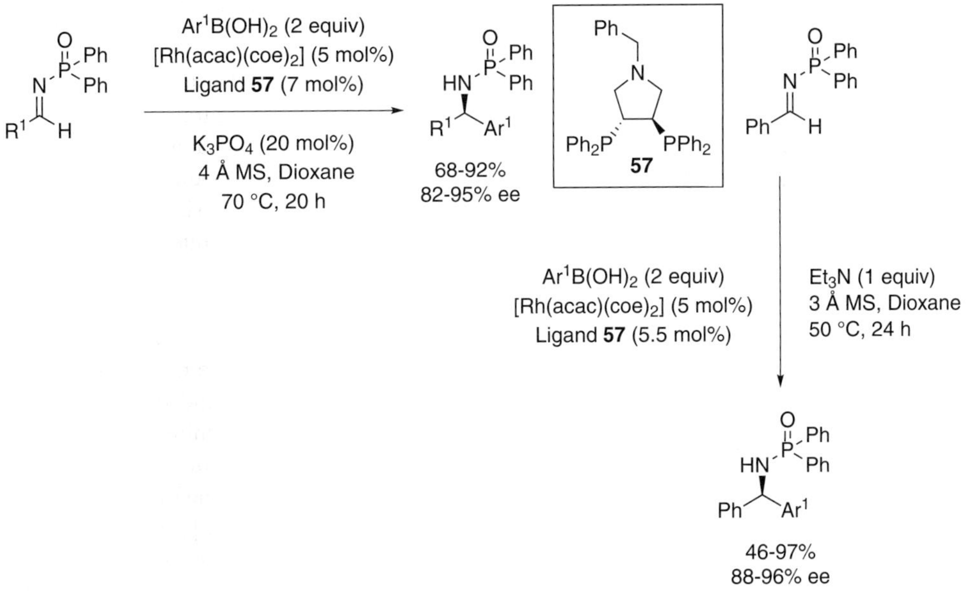

Scheme 1.33 Addition of arylboronic acid to N-phosphinoyl imines.

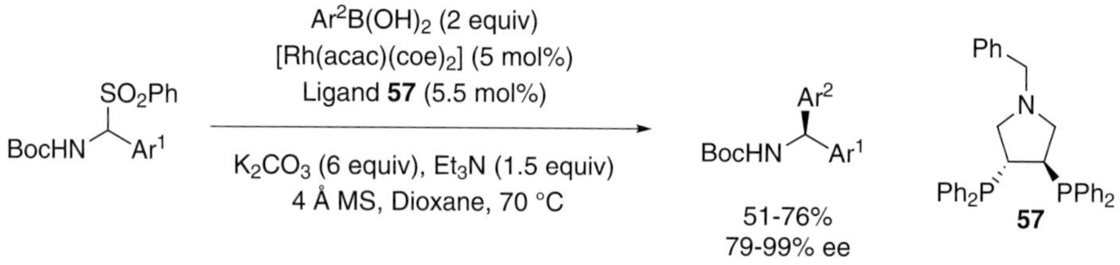

Scheme 1.34 Arylation of N-Boc imines.

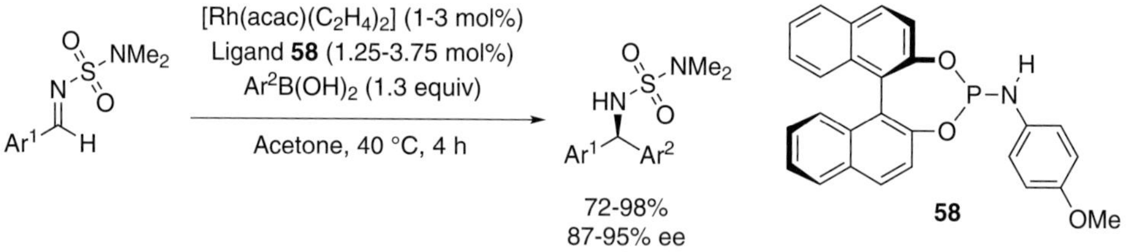

Scheme 1.35 Arylation of N-sulfamoyl imines.

1.4.3.2.3 Rhodium Diene-Catalyzed Arylation of Imines

Hayashi has shown that the asymmetric synthesis of diarylmethylamines could be realized with high enantiocontrol by the rhodium-catalyzed arylboronic acid addition to N-tosyl imines [119]. The rhodium catalyst bears the C_2-symmetrical bicyclo[2.2.1]heptadiene ligand **54**.

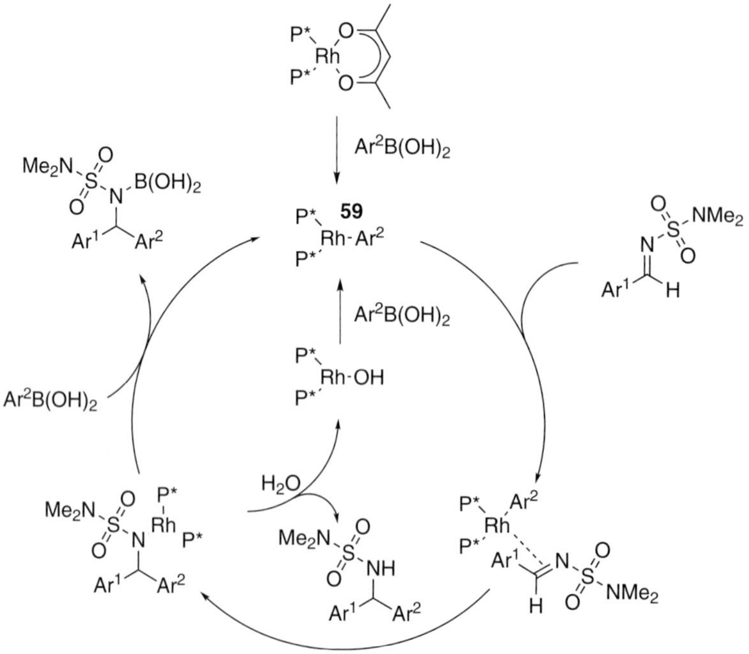

Figure 1.10 Proposed mechanism for the rhodium-catalyzed arylation of imines.

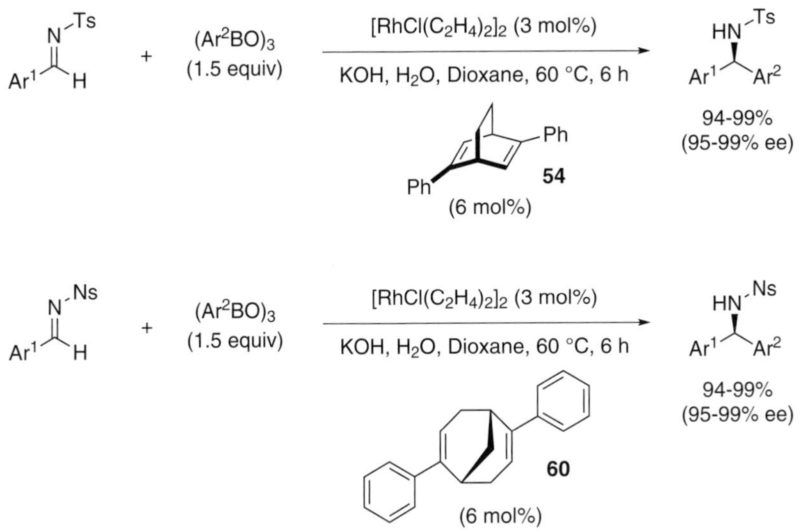

Scheme 1.36 Rhodium-catalyzed arylation of N-arenesulfonyl imines.

This chiral diene ligand displays a clear superiority over chiral phosphorus ligands in both enantioselectivity and catalytic activity. A second-generation catalytic system was later reported for the highly enantioselective addition to N-nosyl imines, for which the product is more easily converted into the free amine than that bearing a N-tosyl group [120]. The more effective chiral ligand **60** had to be used (Scheme 1.36).

1.4.4
Catalytic Asymmetric Addition of sp Hybridized Carbanions

The first examples of enantioselective alkynylation of imines relied on the use of stoichiometric chiral additives [121] or chiral reagents [122]. It was only in 2002 that the first reports of catalytic asymmetric copper-catalyzed terminal alkyne addition to imines appeared. Li reported that a three-component approach, in which N-aryl imines are generated *in situ* from an aldehyde and aniline, leads to propargylamines in excellent yields and enantioselectivities (Table 1.19) [123, 124]. Quite remarkably, the reaction can be carried out either in toluene or in water, with the yields and enantioselectivities being slightly higher in toluene. However, the scope is limited to aryl-substituted aldehydes. The reaction presumably involves the *in situ* formation of an alkynylcopper nucleophile that adds to the imine. Several similar chiral bidentate ligands were developed, but they offer little advantages over the original pybox ligand derived from phenylglycinol [125].

An enantioselective copper-catalyzed addition of alkynes to enamines was reported by Knochel [126]. The reaction was later extended to an enantioselective three-component reaction for the synthesis of N,N-disubstituted propargylamines (Scheme 1.37) [127].

Table 1.19 Copper-catalyzed alkynylation of N-aryl imines.

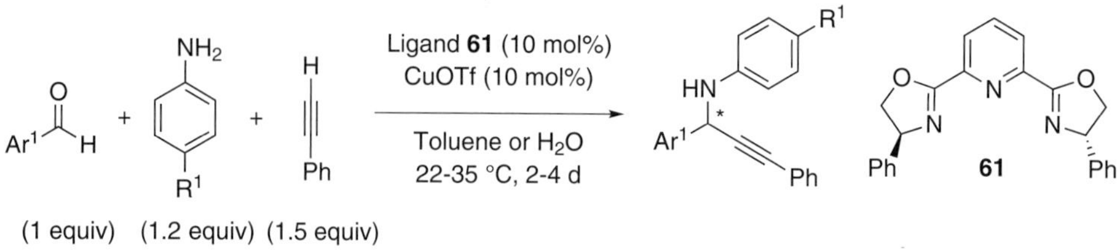

Entry	Ar¹	R¹	Toluene		H₂O	
			Yield (%)	ee (%)	Yield (%)	ee (%)
1	Ph	H	78	96	71	84
2	4-MeC$_6$H$_4$	H	85	92	86	81
3	4-ClC$_6$H$_4$	H	85	94	70	87
4	4-PhC$_6$H$_4$	H	81	94	48	84
5	2-Naphthyl	H	63	88	57	86
6	4-(CF$_3$)C$_6$H$_4$	H	71	93	56	87
7	Ph	Br	93	91	82	83
8	Ph	Cl	92	91	77	84
9	Ph	Me	93	94	68	91

Although the level of enantioselectivity depends upon the dialkylamine used for the enamine precursor, the reaction is highly divergent, as four different substituents can be modified in one step. Although the use of the dibenzylamine generally provides the highest enantioselectivities, its removal under hydrogenolysis conditions also leads to the reduction of the alkyne. The deprotection of the protected propargylamine is possible using palladium(0) when diallylamine or bis(phenallyl)amine is used (Scheme 1.38) [128]. However, the enantioselectivity of the conden-

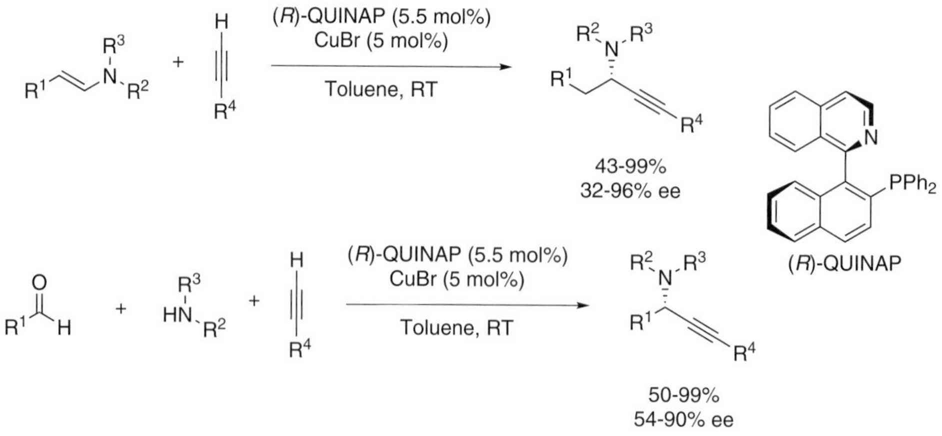

Scheme 1.37 Alkynylation of enamines.

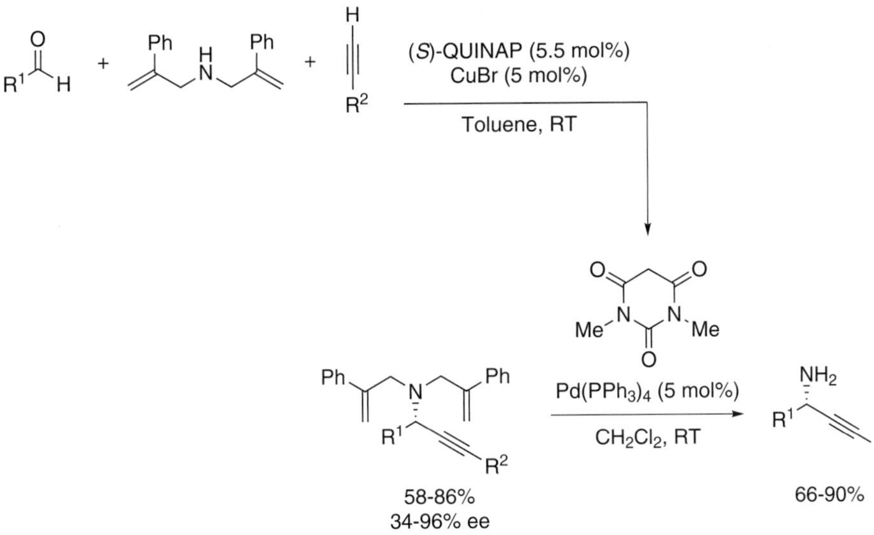

1.4 Catalytic Asymmetric Nucleophilic Addition to Achiral Imines | 41

Scheme 1.38 Three-component synthesis of propargylamines.

sation reaction is usually higher with bis(phenallyl)amine than with diallylamine or bis(methallyl)amine.

A variant of this reaction has been reported by Carreira using a novel PINAP ligand [129]. The use of 4-piperidone hydrochloride hydrate affords the tertiary amine in good yield and enantioselectivity (Scheme 1.39). The deprotection of the amine can

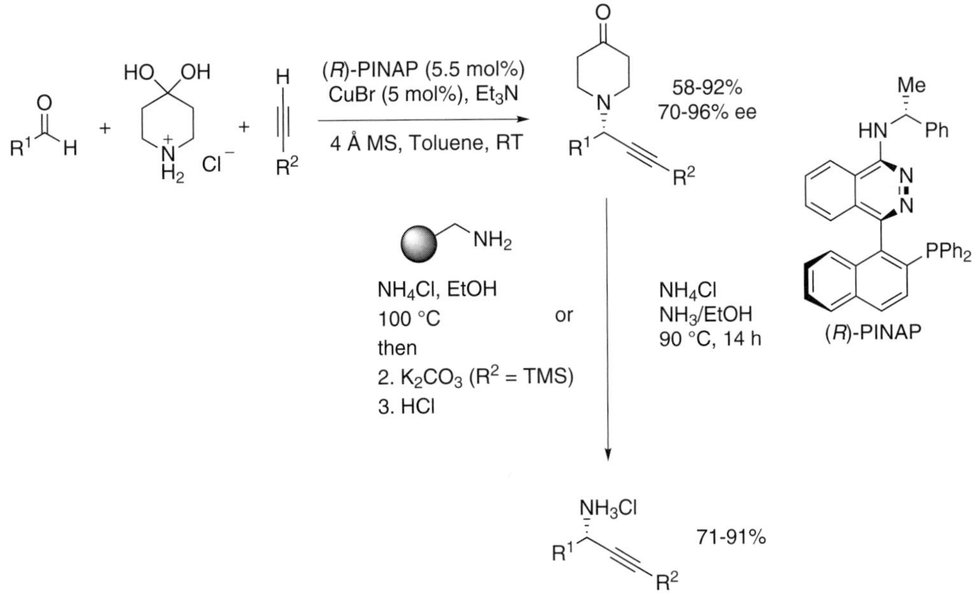

Scheme 1.39 Preparation of unprotected propargylamines.

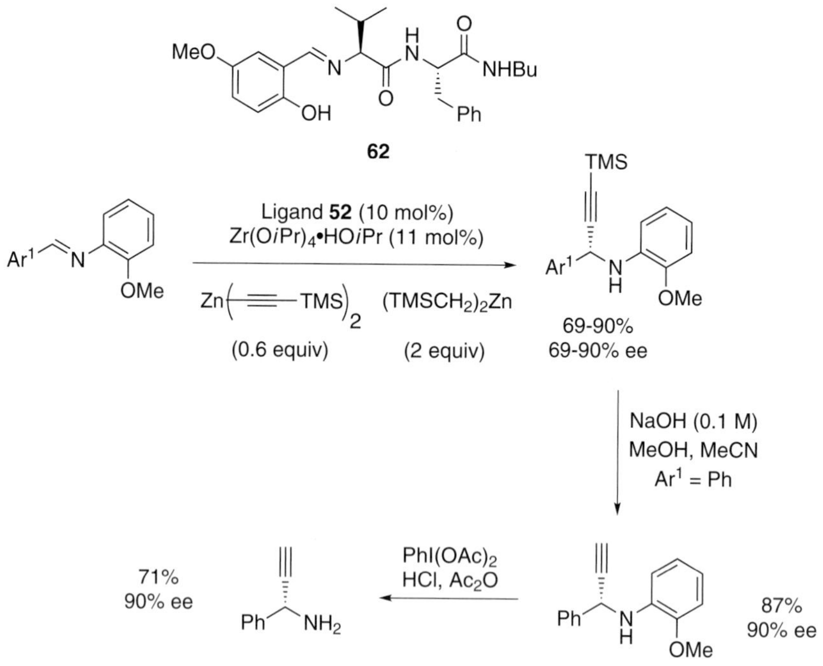

Scheme 1.40 Zirconium-catalyzed alkynylation of N-aryl imines.

be accomplished in high yield upon treatment with ammonia or, more conveniently, a polystyrene-supported scavenger amine.

Hoveyda and Snapper reported that the addition of mixed alkynylzinc reagents to various aryl imines can be catalyzed by Schiff base ligand **52** and zirconium tetraisopropoxide to afford the protected propargylamines in good yields and enantioselectivities [130]. The oxidative removal of the o-anisidyl group affords the propargylamine without any racemization (Scheme 1.40).

1.5
Conclusion

The importance of α-branched chiral amines in nature and as substructures in biologically active unnatural products led to the rapid development of an impressive number of methods to synthesize this class of compounds in enantiomerically enriched forms. It is quite spectacular to see how many very efficient methods have appeared since the early twenty-first century. Furthermore, catalytic asymmetric methods now exist to transfer sp^3, sp^2, or sp nonstabilized carbanion nucleophiles to a wide range of activated imines. Even if the past 10 years have witnessed phenomenal advances in this area, Ellman's chiral auxiliary method is still very attractive. Indeed, it is the only one that allows the addition of the three kinds of nucleophiles using the same class of reagents and it also possesses numerous other advantages (easy hydrolysis of the auxiliary and commercial availability of numerous Grignard reagents and of *tert*-butanesulfonamide's chiral auxiliary). These features make this process even more attractive when large amounts of an α-branched chiral amines are needed.

Questions

1.1. The cryptostylines and their analogues have been used as pharmacological probes, such as the D_1 dopamine receptor, an antagonist of substance P, and a peptide neurotransmitter. Propose a catalytic asymmetric synthesis of (S)-(+)-cryptostyline II starting from methyl 2-(2-formyl-4,5-dimethoxyphenyl)acetate and 4-bromo-1,2-dimethoxybenzene.

1.2. Propose a mechanism for the following formation of the aza-Morita–Baylis–Hillman reaction product that is obtained from an α-hydroxypropargylsilane (1) and the *N-tert*-butanesulfinyl imine. Provide the structure of intermediate **A** obtained upon slow addition of *n*-BuLi to (±)-1.

1.3. Propose a catalytic asymmetric synthesis of N-acetylcolchinol starting from 3-hydroxybenzaldehyde and 3,4,5-trimethoxybenzaldehyde.

References

1. For reviews on this topic, see (a) Yamada, K. and Tomioka, K. (2008) *Chem. Rev.*, **108**, 2874; (b) Friestad, G.K. and Mathies, A.K. (2007) *Tetrahedron*, **63**, 2541; (c) Ferraris, D. (2007) *Tetrahedron*, **63**, 9581; (d) Vilaivan, T., Bhanthumnavin, W., and Sritana-Anant, Y. (2005) *Curr. Org. Chem.*, **9**, 1315; (e) Kobayashi, S. and Ishitani, H. (1999) *Chem. Rev.*, **99**, 1069; (f) Bloch, R. (1998) *Chem. Rev.*, **98**, 1407; (g) Enders, D. and Reinhold, U. (1997) *Tetrahedron: Asymmetry*, **8**, 1895.
2. This chapter will not include imine reduction chemistry. For excellent reviews on this topic, see (a) Deutsch, C., Krause, N., and Lipshutz, B.H., (2008) *Chem. Rev.*, **108**, 2916; (b) Gladiali, S. and Alberico, E. (2006) *Chem. Soc. Rev.*, **35**, 226; (c) Blaser, H.U., Malan, C., Pugin, B., Spindler, F., Steiner, H., and Studer, M. (2003) *Adv. Synth. Catal.*, **345**, 103.
3. Ukaji, Y., Watai, T., Sumi, T., and Fujisawa, T. (1991) *Chem. Lett.*, 1555.
4. Takahashi, H., Chida, Y., Yoshii, T., Suzuki, T., and Yanaura, S. (1986) *Chem. Pharm. Bull.*, **34**, 2071.
5. (a) Merino, P., Marqués-López, E., Tejero, T., and Herrera, R.P. (2009) *Tetrahedron*, **65**, 1219; (b) Gröger, H. (2003) *Chem. Rev.*, **103**, 2795.
6. (a) Marques, M.M.B. (2006) *Angew. Chem., Int. Ed.*, **45**, 348; (b) Mukherjee, S., Yang, J.W., Hoffmann, S., and List, B. (2007) *Chem. Rev.*, **107**, 5471; (c) Verkade, J.M.M., van Hemert, L.J.C., Quaedflieg, P.J.L.M., and Rutjes, F.P.J.T. (2008) *Chem. Soc. Rev.*, **37**, 29.
7. (a) Wessjohann, L.A., Rivera, D.G., and Vercillo, O.E. (2009) *Chem. Rev.*, **109**, 796; (b) Domling, A. (2006) *Chem. Rev.*, **106**, 17; (c) An, H.Y. and Cook, P.D. (2000) *Chem. Rev.*, **100**, 3311.
8. Charette, A.B., Boezio, A.A., Côté, A., Moreau, E., Pytkowicz, J., Desrosiers, J.-N., and Legault, C. (2005) *Pure Appl. Chem.*, **77**, 1259.
9. (a) Colebourne, N., Foster, R.G., and Robson, E. (1967) *J. Chem. Soc. C*, 685; (b) Mangeney, P., Tejero, T., Alexakis, A., Grosjean, F., and Normant, J. (1988) *Synthesis*, 255; (c) Bigelow, L.A. and Eatough, H. (1941) *Org. Synth.*, **1**, 80; (d) Taguchi, K. and Westheimer, F.H. (1971) *J. Org. Chem.*, **36**, 1570.
10. Enders, D. and Eichenauer, H. (1979) *Chem. Ber.*, **112**, 2933.
11. Zhizhin, A.A., Zarubin, D.N., and Ustynyuk, N.A. (2008) *Tetrahedron Lett.*, **49**, 699.
12. Davis, F.A., Reddy, R.T., and Reddy, R.E. (1992) *J. Org. Chem.*, **57**, 6387.
13. Annunziata, R., Cinquini, M., and Cozzi, F. (1982) *J. Chem. Soc., Perkin Trans. 1*, 339.
14. Liu, G.C., Cogan, D.A., Owens, T.D., Tang, T.P., and Ellman, J.A. (1999) *J. Org. Chem.*, **64**, 1278.
15. Trost, B.M. and Marrs, C. (1991) *J. Org. Chem.*, **56**, 6468.
16. Jennings, W.B. and Lovely, C.J. (1991) *Tetrahedron*, **47**, 5561.
17. (a) Wynne, J.H., Price, S.E., Rorer, J.R., and Stalick, W.M. (2003) *Synth. Commun.*, **33**, 341; (b) Love, B.E., Raje, P.S., and Williams, T.C. II (1994) *Synlett*, 493.
18. Georg, G.I., Harriman, G.C.B., and Peterson, S.A. (1995) *J. Org. Chem.*, **60**, 7366.
19. Chemla, F., Hebbe, V., and Normant, J.-F. (2000) *Synthesis*, 75.
20. (a) Artman, G.D., Bartolozzi, A., Franck, R.W., and Weinreb, S.M. (2001) *Synlett*, 232; (b) Boger, D.L. and Corbett, W.L. (1992) *J. Org. Chem.*, **57**, 4777.
21. Ruano, J.L.G., Alemán, J., Cid, M.B., and Parra, A. (2005) *Org. Lett.*, **7**, 179.
22. (a) Brown, C., Hudson, R.F., Maron, A., and Record, K.A. (1976) *Chem. Commun.*, 663; (b) Krzyzanowska, B. and Stec, W.J. (1978) *Synthesis*, 521.
23. Côté, A., Boezio, A.A., and Charette, A.B. (2004) *Proc. Natl. Acad. Sci. USA*, **101**, 5405.
24. Yamaguchi, A., Matsunaga, S., and Shibasaki, M. (2006) *Tetrahedron Lett.*, **47**, 3985.
25. Kim, Y.H. and Shin, J.M. (1985) *Tetrahedron Lett.*, **26**, 3821.
26. Lauzon, C., Desrosiers, J.-N., and Charette, A.B. (2005) *J. Org. Chem.*, **70**, 10579.

27 Lauzon, C. and Charette, A.B. (2006) *Org. Lett.*, **8**, 2743.
28 (a) Hart, D.J., Kanai, K., Thomas, D.G., and Yang, T.K. (1983) *J. Org. Chem.*, **48**, 289; (b) Vidal, J., Damestoy, S., Guy, L., Hannachi, J.C., Aubry, A., and Collet, A. (1997) *Chem. Eur. J.*, **3**, 1691; (c) Kupfer, R., Meier, S., and Würthwein, E.-U. (1984) *Synthesis*, 688.
29 Petrini, M. (2005) *Chem. Rev.*, **105**, 3949.
30 (a) Ballini, R. and Petrini, M. (1999) *Tetrahedron Lett.*, **40**, 4449; (b) Olijnsma, T., Engberts, J.B.F.N., and Strating, J. (1967) *Recl. Trav. Chim. Pays-Bas*, **86**, 463.
31 Kanazawa, A.M., Denis, J.N., and Greene, A.E. (1994) *J. Org. Chem.*, **59**, 1238.
32 For reviews of chiral auxiliary-based methods for the nucleophilic addition to imines, see (a) Enders, D. and Reinhold, U. (1997) *Tetrahedron: Asymmetry*, **8**, 1895; (b) Davis, F.A., Zhou, P., and Chen, B.C. (1998) *Chem. Soc. Rev.*, **27**, 13; (c) Alvaro, G. and Savoia, D. (2002) *Synlett*, 651; (d) Ellman, J.A., Owens, T.D., and Tang, T.P. (2002) *Acc. Chem. Res.*, **35**, 984; (e) Zhou, P., Chen, B.C., and Davis, F.A. (2004) *Tetrahedron*, **60**, 8003; (f) Friestad, G.K. (2005) *Eur. J. Org. Chem.*, 3157.
33 (a) Alexakis, A., Lensen, N., and Mangeney, P. (1991) *Tetrahedron Lett.*, **32**, 1171; (b) Alexakis, A., Tranchier, J.P., Lensen, N., and Mangeney, P. (1995) *J. Am. Chem. Soc.*, **117**, 10767; (c) Thiam, M. and Chastrette, F. (1992) *Bull. Chem. Soc. Fr.*, **129**, 161; (d) Matsubara, S., Ukita, H., Kodama, T., and Utimoto, K. (1994) *Chem. Lett.*, 831; (e) Badorrey, R., Cativiela, C., Diaz-de-Villegas, M.D., and Galvez, J.A. (1997) *Tetrahedron*, **53**, 1411.
34 Charette, A.B. and Mellon, C. (1998) *Tetrahedron*, **54**, 10525.
35 For recent reviews describing various approaches to α,α-disubstituted α-amino acids, see (a) Vogt, H. and Bräse, S. (2007) *Org. Biomol. Chem.*, **5**, 406; (b) Cativiela, C. and Díaz-de-Villegas, M.D. (2007) *Tetrahedron: Asymmetry*, **18**, 569.
36 Alvaro, G., Savoia, D., and Valentinetti, M.R. (1996) *Tetrahedron*, **52**, 12571.
37 (a) Beuchet, P., Lemarrec, N., and Mosset, P. (1992) *Tetrahedron Lett.*, **33**, 5959; (b) Wang, D.K., Dai, L.X., and Hou, X.L. (1995) *Tetrahedron Lett.*, **36**, 8649; (c) Wang, D.K., Dai, L.X., Hou, X.L., and Zhang, Y. (1996) *Tetrahedron Lett.*, **37**, 4187; (d) Gao, Y.A. and Sato, F. (1995) *J. Org. Chem.*, **60**, 8136.
38 Wakchaure, V.N., Mohanty, R.r., Shaikh, A.J., and Nugent, T.C. (2007) *Eur. J. Org. Chem.*, 959.
39 (a) Takahashi, H., Chida, Y., Higashiyama, K., and Onishi, H. (1985) *Chem. Pharm. Bull.*, **33**, 4662; (b) Takahashi, H., Chida, Y., Suzuki, T., Onishi, H., and Yanaura, S. (1984) *Chem. Pharm. Bull.*, **32**, 2714; (c) Takahashi, H., Suzuki, Y., and Hori, T. (1983) *Chem. Pharm. Bull.*, **31**, 2183; (d) Takahashi, H., Chida, Y., Suzuki, T., Yanaura, S., Suzuki, Y., and Masuda, C. (1983) *Chem. Pharm. Bull.*, **31**, 1659; (e) Suzuki, Y. and Takahashi, H. (1983) *Chem. Pharm. Bull.*, **31**, 2895; (f) Suzuki, Y. and Takahashi, H. (1983) *Chem. Pharm. Bull.*, **31**, 31; (g) Takahashi, H., Suzuki, Y., and Inagaki, H. (1982) *Chem. Pharm. Bull.*, **30**, 3160.
40 (a) Muralidharan, K.R., Mokhallalati, M.K., and Pridgen, L.N. (1994) *Tetrahedron Lett.*, **35**, 7489; (b) Mokhallalati, M.K. and Pridgen, L.N. (1993) *Synth. Commun.*, **23**, 2055; (c) Pridgen, L.N., Mokhallalati, M.K., and Wu, M.J. (1992) *J. Org. Chem.*, **57**, 1237; (d) Wu, M.J. and Pridgen, L.N. (1991) *J. Org. Chem.*, **56**, 1340; (e) Wu, M.J. and Pridgen, L.N. (1990) *Synlett*, 636.
41 Tanaka, H., Inoue, K., Pokorski, U., Taniguchi, M., and Torii, S. (1990) *Tetrahedron Lett.*, **31**, 3023.
42 Laschat, S. and Kunz, H. (1990) *Synlett*, 51.
43 (a) Takahashi, H., Tomita, K., and Otomasu, H. (1979) *J. Chem. Soc., Chem. Commun.*, 668; (b) Takahashi, H., Tomita, K., and Noguchi, H. (1981) *Chem. Pharm. Bull.*, **29**, 3387; (c) Takahashi, H. and Inagaki, H. (1982) *Chem. Pharm. Bull.*, **30**, 922; (d) Takahashi, H. and Suzuki, Y. (1983) *Chem. Pharm. Bull.*, **31**, 4295.
44 For a review of the chemistry of SAMP and RAMP, see Job, A., Janeck, C.F., Bettray, W., Peters, R., and Enders, D. (2002) *Tetrahedron*, **58**, 2253.

45 Enders, D., Schubert, H., and Nubling, C. (1986) *Angew. Chem., Int. Ed.*, **25**, 1109.

46 (a) Denmark, S.E., Weber, T., and Piotrowski, D.W. (1987) *J. Am. Chem. Soc.*, **109**, 2224; (b) Denmark, S.E., Edwards, J.P., and Nicaise, O. (1993) *J. Org. Chem.*, **58**, 569.

47 Friestad, G.K. and Ding, H. (2001) *Angew. Chem., Int. Ed.*, **40**, 4491.

48 For a review on acyl hydrazone chemistry, see Sugiura, M. and Kobayashi, S. (2005) *Angew. Chem., Int. Ed.*, **44**, 5176.

49 (a) Brown, D.S., Gallagher, P.T., Lightfoot, A.P., Moody, C.J., Slawin, A.M.Z., and Swann, E. (1995) *Tetrahedron*, **51**, 11473; (b) Gallagher, P.T., Lightfoot, A.P., Moody, C.J., and Slawin, A.M.Z. (1995) *Synlett*, 445.

50 Moreau, P., Essiz, M., Merour, J.Y., and Bouzard, D. (1997) *Tetrahedron: Asymmetry*, **8**, 591.

51 Koriyama, Y., Nozawa, A., Hayakawa, R., and Shimizu, M. (2002) *Tetrahedron*, **58**, 9621.

52 Wipf, P., Nunes, R.L., and Ribe, S. (2002) *Helv. Chim. Acta*, **85**, 3478.

53 Li, G.G., Wei, H.X., Whittlesey, B.R., and Batrice, N.N. (1999) *J. Org. Chem.*, **64**, 1061.

54 For other examples of allylation and crotylation using trifluoroborate reagents, see Li, S.W. and Batey, R.A. (2004) *Chem. Commun.*, 1382.

55 Weix, D.J. and Ellman, J.A. (2003) *Org. Lett.*, **5**, 1317.

56 (a) Liu, G.C., Cogan, D.A., and Ellman, J.A. (1997) *J. Am. Chem. Soc.*, **119**, 9913; (b) Cogan, D.A., Liu, G.C., and Ellman, J. (1999) *Tetrahedron*, **55**, 8883.

57 For the preparation of a library of α-branched amines, see Mukade, T., Dragoli, D.R., and Ellman, J.A. (2003) *J. Comb. Chem.*, **5**, 590.

58 For the preparation of aza-Morita–Baylis–Hillman, see Reynolds, T.E., Binkley, M.S., and Scheidt, K.A. (2008) *Org. Lett.*, **10**, 5227.

59 Ding, C.H., Chen, D.D., Luo, Z.B., Dai, L.X., and Hou, X.L. (2006) *Synlett*, 1272.

60 For diastereoselective addition reactions of trialkoxysilylalkynes to *N-tert*-butanesulfinyl imines, see Lettan, R.B. and Scheidt, K.A. (2005) *Org. Lett.*, **7**, 3227.

61 (a) Weix, D.J., Shi, Y.L., and Ellman, J.A. (2005) *J. Am. Chem. Soc.*, **127**, 1092; for addition to imino esters, see (b) Beenen, M.A., Weix, D.J., and Ellman, J.A. (2006) *J. Am. Chem. Soc.*, **128**, 6304; (c) Bolshan, Y. and Batey, R.A. (2005) *Org. Lett.*, **7**, 1481.

62 Kong, J.R., Cho, C.W., and Krische, M.J. (2005) *J. Am. Chem. Soc.*, **127**, 11269.

63 For an alternative approach to this class of compounds, see (a) Steinig, A.G. and Spero, D.M. (1999) *J. Org. Chem.*, **64**, 2406; (b) Spero, D.M. and Kapadia, S.R. (1997) *J. Org. Chem.*, **62**, 5537.

64 Cogan, D.A. and Ellman, J.A. (1999) *J. Am. Chem. Soc.*, **121**, 268.

65 For an application of this methodology, see Shaw, A.W. and deSolms, S.J. (2001) *Tetrahedron Lett.*, **42**, 7173.

66 Patterson, A.W. and Ellman, J.A. (2006) *J. Org. Chem.*, **71**, 7110.

67 (a) Tomioka, K., Inoue, I., Shindo, M., and Koga, K. (1991) *Tetrahedron Lett.*, **32**, 3095; (b) Tomioka, K., Inoue, I., Shindo, M., and Koga, K. (1990) *Tetrahedron Lett.*, **31**, 6681.

68 Soai, K., Hatanaka, T., and Miyazawa, T. (1992) *Chem. Commun.*, 1097.

69 (a) Denmark, S.E., Nakajima, N., and Nicaise, O.J.C. (1994) *J. Am. Chem. Soc.*, **116**, 8797; (b) Denmark, S.E. and Nicaise, O.J.C. (1996) *Chem. Commun.*, 999; (c) Denmark, S.E., Nakajima, N., Stiff, C.M., Nicaise, O.J.C., and Kranz, M. (2008) *Adv. Synth. Catal.*, **350**, 1023.

70 (a) Denmark, S.E. and Stiff, C.M. (2000) *J. Org. Chem.*, **65**, 5875; (b) Denmark, S.E., Nakajima, N., Nicaise, O.J.C., Faucher, A.M., and Edwards, J.P. (1995) *J. Org. Chem.*, **60**, 4884.

71 (a) Fujihara, H., Nagai, K., and Tomioka, K. (2000) *J. Am. Chem. Soc.*, **122**, 12055; (b) Nagai, K., Fujihara, H., Kuriyama, M., Yamada, K., and Tomioka, K. (2002) *Chem. Lett.*, 8; (c) Soeta, T., Nagai, K., Fujihara, H., Kuriyama, M., and Tomioka, K. (2003) *J. Org. Chem.*, **68**, 9723.

72 (a) de Vries, A.H.M., Meetsma, A., and Feringa, B.L. (1996) *Angew. Chem., Int. Ed.*, **35**, 2374; (b) Alexakis, A., Vastra, J.,

and Mangeney, P. (1997) *Tetrahedron Lett.*, **38**, 7745.
73 Figure 1.6: (a) Wang, M.C., Xu, C.L., Zou, Y.X., Liu, H.M., and Wang, D.K. (2005) *Tetrahedron Lett.*, **46**, 5413; (b) Li, X., Cun, L.F., Gong, L.Z., Mi, A.Q., and Jiang, Y.Z. (2003) *Tetrahedron: Asymmetry*, **14**, 3819; (c) Wang, C.J. and Shi, M. (2003) *J. Org. Chem.*, **68**, 6229; (d) Shi, M. and Zhang, W. (2003) *Tetrahedron: Asymmetry*, **14**, 3407.
74 (a) Boezio, A.A. and Charette, A.B. (2003) *J. Am. Chem. Soc.*, **125**, 1692; (b) Boezio, A.A., Pytkowicz, J., Côté, A., and Charette, A.B. (2003) *J. Am. Chem. Soc.*, **125**, 14260; (c) Côté, A. and Charette, A.B. (2005) *J. Org. Chem.*, **70**, 10864; (d) Desrosiers, J.-N., Côté, A., and Charette, A.B. (2005) *Tetrahedron*, **61**, 6186; (e) Charette, A.B., Côté, A., Desrosiers, J.-N., Bonnaventure, I., Lindsay, V.N.G., Lauzon, C., Tannous, J., and Boezio, A.A. (2008) *Pure Appl. Chem.*, **80**, 881; (f) Bonnaventure, I. and Charette, A.B. (2008) *J. Org. Chem.*, **73**, 6330.
75 For other less effective chiral ligands, see (a) Shi, M. and Wang, C.J. (2003) *Adv. Synth. Catal.*, **345**, 971; (b) Wang, M.C., Liu, L.T., Hua, Y.Z., Zhang, J.S., Shi, Y.Y., and Wang, D.K. (2005) *Tetrahedron: Asymmetry*, **16**, 2531; (c) Kim, B.S., Kang, S.W., Kim, K.H., Ko, D.H., Chung, Y., and Ha, D.C. (2005) *Bull. Korean Chem. Soc.*, **26**, 1501; (d) Chen, J.M., Li, D., Ma, H.F., Cun, L.F., Zhu, J., Deng, J.G., and Liao, J. (2008) *Tetrahedron Lett.*, **49**, 6921.
76 Côté, A. and Charette, A.B. (2008) *J. Am. Chem. Soc.*, **130**, 2771.
77 Pizzuti, M.G., Minnaard, A.J., and Feringa, B.L. (2008) *J. Org. Chem.*, **73**, 940.
78 (a) Andersson, P.G., Guijarro, D., and Tanner, D. (1997) *J. Org. Chem.*, **62**, 7364; (b) Brandt, P., Hedberg, C., Lawonn, K., Pinho, P., and Andersson, P.G. (1999) *Chem. Eur. J.*, **5**, 1692.
79 For the preparation of this ligand, see (a) Rozenberg, V., Danilova, T., Sergeeva, E., Vorontsov, E., Starikova, Z., Lysenko, K., and Belokon, Y. (2000) *Eur. J. Org. Chem.*, 3295; (b) Cipiciani, A., Fringuelli, F., Mancini, V., Piermatti, O., Scappini, A.M., and Ruzziconi, R. (1997) *Tetrahedron*, **53**, 11853; (c) Cipiciani, A., Fringuelli, F., Mancini, V., Piermatti, O., Pizzo, F., and Ruzziconi, R. (1997) *J. Org. Chem.*, **62**, 3744.
80 Dahmen, S. and Bräse, S. (2002) *J. Am. Chem. Soc.*, **124**, 5940.
81 Porter, J.R., Traverse, J.F., Hoveyda, A.H., and Snapper, M.L. (2001) *J. Am. Chem. Soc.*, **123**, 984.
82 (a) Porter, J.R., Traverse, J.F., Hoveyda, A.H., and Snapper, M.L. (2001) *J. Am. Chem. Soc.*, **123**, 10409; (b) Akullian, L.C., Snapper, M.L., and Hoveyda, A.H. (2003) *Angew. Chem., Int. Ed.*, **42**, 4244.
83 Akullian, L.C., Porter, J.R., Traverse, J.E., Snapper, M.L., and Hoveyda, A.H. (2005) *Adv. Synth. Catal.*, **347**, 417.
84 Fu, P., Snapper, M.L., and Hoveyda, A.H. (2008) *J. Am. Chem. Soc.*, **130**, 5530.
85 A titanium-catalyzed asymmetric addition of diorganozinc to α-aldiminoesters was shown to proceed with moderate level of enantioselectivities: Basra, S., Fennie, M.W., and Kozlowski, M.C., (2006) *Org. Lett.*, **8**, 2659.
86 Nishimura, T., Yasuhara, Y., and Hayashi, T. (2006) *Org. Lett.*, **8**, 979.
87 (a) Chataigner, I., Zammattio, F., Lebreton, J., and Villieras, J. (1998) *Synlett*, 275; (b) Itsuno, S., Yokoi, A., and Kuroda, S. (1999) *Synlett*, 1987; (c) Sugiura, M., Hirano, K., and Kobayashi, S. (2004) *J. Am. Chem. Soc.*, **126**, 7182; (d) Wu, T.R. and Chong, J.M. (2006) *J. Am. Chem. Soc.*, **128**, 9646; (e) Canales, E., Hernandez, E., and Soderquist, J.A. (2006) *J. Am. Chem. Soc.*, **128**, 8712; (f) Chataigner, I., Zammattio, F., Lebreton, J., and Villiéras, J. (2008) *Tetrahedron*, **64**, 2441.
88 (a) Panek, J.S. and Jain, N.F. (1994) *J. Org. Chem.*, **59**, 2674; (b) Schaus, J.V., Jain, N., and Panek, J.S. (2000) *Tetrahedron*, **56**, 10263; (c) Berger, R., Rabbat, P.M.A., and Leighton, J.L. (2003) *J. Am. Chem. Soc.*, **125**, 9596; (d) Berger, R., Duff, K., and Leighton, J.L. (2004) *J. Am. Chem. Soc.*, **126**, 5686.
89 (a) Hanessian, S. and Yang, R.Y. (1996) *Tetrahedron Lett.*, **37**, 8997; (b) Nakamura, M., Hirai, A., and Nakamura, E. (1996) *J. Am. Chem. Soc.*, **118**, 8489.

90 Cook, G.R., Kargbo, R., and Maity, B. (2005) *Org. Lett.*, **7**, 2767.
91 Itsuno, S., Watanabe, K., Ito, K., El-Shehawy, A.A., and Sarhan, A.A. (1997) *Angew. Chem., Int. Ed.*, **36**, 109.
92 Chen, G.M., Ramachandran, P.V., and Brown, H.C. (1999) *Angew. Chem., Int. Ed.*, **38**, 825.
93 Ramachandran, P.V. and Burghardt, T.E. (2005) *Chem. Eur. J.*, **11**, 4387.
94 (a) Nakamura, H., Nakamura, K., and Yamamoto, Y. (1998) *J. Am. Chem. Soc.*, **120**, 4242; (b) Nakamura, K., Nakamura, H., and Yamamoto, Y. (1999) *J. Org. Chem.*, **64**, 2614; (c) Fernandes, R.A., Stimac, A., and Yamamoto, Y. (2003) *J. Am. Chem. Soc.*, **125**, 14133; (d) Fernandes, R.A. and Yamamoto, Y. (2004) *J. Org. Chem.*, **69**, 3562; (e) Fernandes, R.A. and Yamamoto, Y. (2004) *J. Org. Chem.*, **69**, 735.
95 Gastner, T., Ishitani, H., Akiyama, R., and Kobayashi, S. (2001) *Angew. Chem., Int. Ed.*, **40**, 1896.
96 Tan, K.L. and Jacobsen, E.N. (2007) *Angew. Chem., Int. Ed.*, **46**, 1315.
97 Lou, S., Moquist, P.N., and Schaus, S.E. (2007) *J. Am. Chem. Soc.*, **129**, 15398.
98 Wada, R., Shibuguchi, T., Makino, S., Oisaki, K., Kanai, M., and Shibasaki, M. (2006) *J. Am. Chem. Soc.*, **128**, 7687.
99 Fang, X.M., Johannsen, M., Yao, S.L., Gathergood, N., Hazell, R.G., and Jørgensen, K.A. (1999) *J. Org. Chem.*, **64**, 4844.
100 Ferraris, D., Dudding, T., Young, B., Drury, W.J., and Lectka, T. (1999) *J. Org. Chem.*, **64**, 2168.
101 Hamada, T., Manabe, K., and Kobayashi, S. (2003) *Angew. Chem., Int. Ed.*, **42**, 3927.
102 Schneider, U., Chen, I.H., and Kobayashi, S. (2008) *Org. Lett.*, **10**, 737.
103 (a) Patel, S.J. and Jamison, T.F. (2004) *Angew. Chem., Int. Ed.*, **43**, 3941; (b) Patel, S.J. and Jamison, T.F. (2003) *Angew. Chem., Int. Ed.*, **42**, 1364.
104 Ngai, M.Y., Barchuk, A., and Krische, M.J. (2007) *J. Am. Chem. Soc.*, **129**, 12644.
105 Skucas, E., Kong, J.R., and Krische, M.J. (2007) *J. Am. Chem. Soc.*, **129**, 7242.
106 (a) Petasis, N.A. and Zavialov, I.A. (1997) *J. Am. Chem. Soc.*, **119**, 445; (b) Harwood, L.M., Currie, G.S., Drew, M.G.B., and Luke, R.W.A. (1996) *Chem. Commun.*, 1953; (c) Koolmeister, T., Sodergren, M., and Scobie, M. (2002) *Tetrahedron Lett.*, **43**, 5969; (d) Nanda, K.K. and Trotter, B.W. (2005) *Tetrahedron Lett.*, **46**, 2025; (e) Southwood, T.J., Curry, M.C., and Hutton, C.A. (2006) *Tetrahedron*, **62**, 236.
107 Lou, S. and Schaus, S.E. (2008) *J. Am. Chem. Soc.*, **130**, 6922.
108 (a) Nehl, H. and Scheidt, W.R. (1985) *J. Organomet. Chem.*, **289**, 1; (b) Oppolzer, W. and Radinov, R.N. (1992) *Helv. Chim. Acta*, **75**, 170; (c) Lutz, C. and Knochel, P. (1997) *J. Org. Chem.*, **62**, 7895.
109 Hermanns, N., Dahmen, S., Bolm, C., and Bräse, S. (2002) *Angew. Chem., Int. Ed.*, **41**, 3692.
110 (a) Hayashi, T. and Ishigedani, M. (2001) *Tetrahedron*, **57**, 2589; (b) Hayashi, T. and Ishigedani, M. (2000) *J. Am. Chem. Soc.*, **122**, 976.
111 Hayashi, T., Kawai, M., and Tokunaga, N. (2004) *Angew. Chem., Int. Ed.*, **43**, 6125.
112 (a) Ueda, M., Saito, A., and Miyaura, N. (2000) *Synlett*, 1637; (b) Ueda, M. and Miyaura, N. (2000) *J. Organomet. Chem.*, **595**, 31.
113 Kuriyama, M., Soeta, T., Hao, X.Y., Chen, O., and Tomioka, K. (2004) *J. Am. Chem. Soc.*, **126**, 8128.
114 Duan, H.-F., Jia, Y.-X., Wang, L.-X., and Zhou, Q.-L. (2006) *Org. Lett.*, **8**, 2567.
115 Kurihara, K., Yamamoto, Y., and Miyaura, N. (2009) *Adv. Synth. Catal.*, **351**, 260.
116 Trincado, M. and Ellman, J.A. (2008) *Angew. Chem., Int. Ed.*, **47**, 5623.
117 Nakagawa, H., Rech, J.C., Sindelar, R.W., and Ellman, J.A. (2007) *Org. Lett.*, **9**, 5155.
118 Jagt, R.B.C., Toullec, P.Y., Geerdink, D., de Vries, J.G., Feringa, B.L., and Minnaard, A.D.J. (2006) *Angew. Chem., Int. Ed.*, **45**, 2789.
119 Tokunaga, N., Otomaru, Y., Okamoto, K., Ueyama, K., Shintani, R., and Hayashi, T. (2004) *J. Am. Chem. Soc.*, **126**, 13584.
120 Otomaru, Y., Tokunaga, N., Shintani, R., and Hayashi, T. (2005) *Org. Lett.*, **7**, 307.
121 (a) Huffman, M.A., Yasuda, N., Decamp, A.E., and Grabowski, E.J.J. (1995) *J. Org. Chem.*, **60**, 1590; (b) Kauffman, G.S., Harris, G.D., Dorow, R.L., Stone, B.R.P., Parsons, R.L., Pesti, J.A., Magnus, N.A., Fortunak, J.M., Confalone, P.N., and

Nugent, W.A. (2000) *Org. Lett.*, **2**, 3119; (c) Frantz, D.E., Fassler, R., and Carreira, E.M. (1999) *J. Am. Chem. Soc.*, **121**, 11245.

122 For example, see Wu, T.R. and Chong, J.M. (2006) *Org. Lett.*, **8**, 15.

123 (a) Wei, C.M., Mague, J.T., and Li, C.J. (2004) *Proc. Natl. Acad. Sci. USA*, **101**, 5749; (b) Wei, C.M. and Li, C.J. (2002) *J. Am. Chem. Soc.*, **124**, 5638.

124 For a related silver-catalyzed reaction that proceeds in lower ee's, see Ji, J.X., Wu, J., and Chan, A.S.C. (2005) *Proc. Natl. Acad. Sci. USA*, **102**, 11196.

125 (a) Colombo, F., Benaglia, M., Orlandi, S., Usuelli, F., and Celentano, G. (2006) *J. Org. Chem.*, **71**, 2064; (b) Orlandi, S., Colombo, F., and Benaglia, M. (2005) *Synthesis*, 1689; (c) Benaglia, M., Negri, D., and Dell'Anna, G. (2004) *Tetrahedron Lett.*, **45**, 8705; (d) Bisai, A. and Singh, V.K. (2006) *Org. Lett.*, **8**, 2405; (e) Liu, B., Huang, L., Liu, J.T., Zhong, Y., Li, X.S., and Chan, A.S.C. (2007) *Tetrahedron: Asymmetry*, **18**, 2901; (f) Hatano, M., Asai, T., and Ishihara, K. (2008) *Tetrahedron Lett.*, **49**, 379; (g) Irmak, M. and Boysen, M.M.K. (2008) *Adv. Synth. Catal.*, **350**, 403.

126 (a) Koradin, C., Polborn, K., and Knochel, P. (2002) *Angew. Chem., Int. Ed.*, **41**, 2535; (b) Koradin, C., Gommermann, N., Polborn, K., and Knochel, P. (2003) *Chem. Eur. J.*, **9**, 2797; (c) Gommermann, N. and Knochel, P. (2004) *Chem. Commun.*, 2324; (d) Gommermann, N. and Knochel, P. (2005) *Synlett*, 2799; (e) Gommermann, N. and Knochel, P. (2006) *Chem. Eur. J.*, **12**, 4380.

127 Gommermann, N., Koradin, X., Polborn, K., and Knochel, P. (2003) *Angew. Chem., Int. Ed.*, **42**, 5763.

128 Gommermann, N. and Knochel, P. (2005) *Chem. Commun.*, 4175.

129 (a) Knöpfel, T.F., Aschwanden, P., Ichikawa, T., Watanabe, T., and Carreira, E.M. (2004) *Angew. Chem., Int. Ed.*, **43**, 5971; (b) Aschwanden, P., Stephenson, C.R.J., and Carreira, E.M. (2006) *Org. Lett.*, **8**, 2437.

130 Traverse, J.F., Hoveyda, A.H., and Snapper, M.L. (2003) *Org. Lett.*, **5**, 3273.

2
Asymmetric Methods for Radical Addition to Imino Compounds

Gregory K. Friestad

2.1
Background and Introduction

A wide range of bioactive synthetic targets contain the chiral α-branched amine functionality, and consequently a variety of methods have been developed for direct asymmetric amine synthesis by addition of carbonyl imino derivatives to the C=N bond (Figure 2.1) [1]. Typical objectives for introducing new methods in this area are to improve stereocontrol in the carbon–carbon bond construction and enhance compatibility with various functional groups commonly encountered in the synthesis of natural products. Nucleophilic additions to C=N bonds, according to Figure 2.1, may be compromised by the basicity of many organometallic reagents, leading to functional group incompatibilities or competing aza-enolization [2]. Milder Strecker [3] and Mannich [4] addition reactions allow a wider range of functionalities in the imino acceptor, but these are quite limited with respect to the incoming nucleophile.

In comparison to organometallic nucleophiles, free radicals offer complementary functional group compatibility, so the use of these neutral reactive intermediates could allow the presence of diverse functionality in both precursors, including functional groups that might not survive strong nucleophilic conditions. Thus, in search of more versatile methods for highly stereocontrolled addition to C=N bonds, we have pursued the development of free radical addition to imino compounds [5]. In the course of these investigations, we introduced two new modes of stereocontrol for intermolecular radical additions to N-acylhydrazones (1, Figure 2.1). This chapter focuses on the design and implementation of these methods.

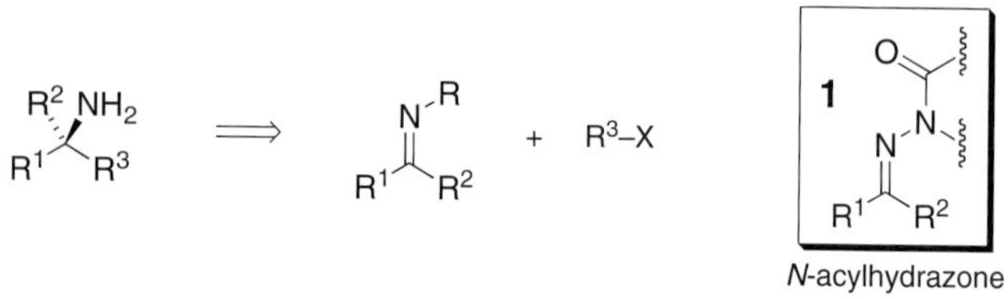

Figure 2.1 The C—C bond disconnection approach to asymmetric amine synthesis and a generalized structure of N-acylhydrazones (**1**).

2.2
Intermolecular Radical Addition Chiral N-Acylhydrazones

2.2.1
Design of Chiral N-Acylhydrazones

The use of simple imines in intermolecular radical additions has been rarely exploited, in part, because of modest reactivity of the C=N bond toward nucleophilic radicals; such acceptors may benefit from the presence of electron-withdrawing groups at the imine carbon. Thus, Bertrand found these reactions to be particularly successful with α-iminoesters **2** (Figure 2.2a) [6]. More reactive oxime ethers have been extensively exploited for radical addition, mainly through the longstanding efforts of Naito [7]. Stereocontrol has been imparted through the substituent on the imino carbon, as exemplified by camphorsultam **4** (Figure 2.2b). Attempts to use chiral O-substituents on oximes for stereocontrol have been ineffective, presumably due to poor rotamer control [8, 9].

We regarded hydrazones as potentially superior radical acceptors; they offer reactivity enhancement similar to that of oxime ethers, and the additional valence of the nitrogen (versus the oxygen of an oxime) could facilitate the conformational constraints that had been missing from oxime ethers. While E/Z mixtures complicate the use of oxime ethers, aldehyde hydrazones generally adopt C=N E-geometry. If

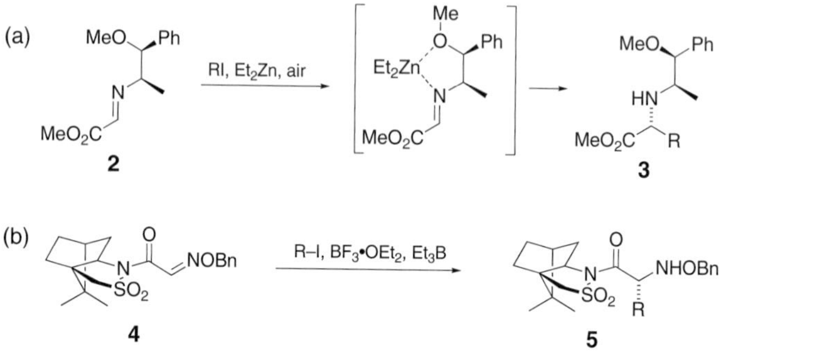

Figure 2.2 Exploitation of α-iminoglyoxylate derivatives for asymmetric induction in radical addition reactions by (a) Bertrand and (b) Naito.

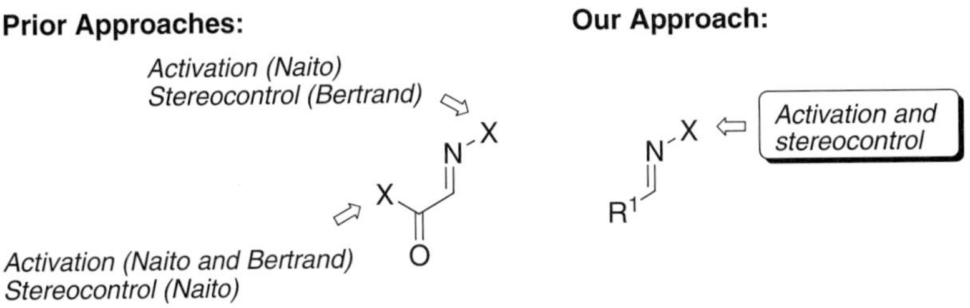

2.2 Intermolecular Radical Addition Chiral N-Acylhydrazones | 53

Prior Approaches:

Activation (Naito)
Stereocontrol (Bertrand)

Our Approach:

Activation and stereocontrol

Activation (Naito and Bertrand)
Stereocontrol (Naito)

Figure 2.3 Approaches to stereocontrol for radical addition to C=N bonds. Earlier approaches require both C- and N-substituents of the C=N bond to play separate roles for activation and stereocontrol. In our approach, the C-substituent R^1 has no role in the reaction, improving versatility.

both activation and stereocontrol could be achieved via a removable N-substituent on the imine, successful addition would then be independent of the identity of substituents on the imine carbon, which would potentially broaden the scope of the reaction.

Chiral N,N-dialkylhydrazones figured prominently in the early development of asymmetric α-alkylation of carbonyl compounds [10] and have also been applied to addition of organometallic reagents to the C=N bond [11–13]. We chose to develop a new type of chiral hydrazone, tailored for use in free radical addition reactions. Toward this end, we conceived a nitrogen-linked chiral auxiliary approach incorporating Lewis acid (LA) activation [14] and stereocontrol enhancement through restriction of rotamer populations [15]. Our approach is distinguished from prior work (Figure 2.3) by combining both activation and stereocontrol into one N-linked substituent, freeing the imino carbon substituent R^1 from any controlling role in the process and thereby enabling greater versatility.

The assumption of an early transition state for radical addition enables the ground-state structure to serve as a reasonable approximation of the transition-state geometry. With this in mind, we hypothesized that steric blocking of one of the enantiotopic approach trajectories by a substituent above or below the plane of hydrazone C=N bond should lead to an enantioselective process.

Our design addressed two specific features desirable for selective blocking of one face of the C=N bond, *restricted rotamer populations* and *Lewis acid activation*. N,N-Dialkylhydrazone **A** (Figure 2.4) offers little restriction to the rotation of C–N and

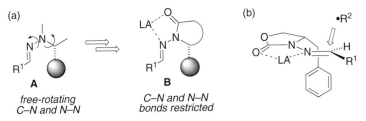

Figure 2.4 (a) Design of a hypothetical N-linked auxiliary approach for stereocontrolled addition to C=N bonds, with Lewis acid (LA) chelation inducing a rigid, electron-deficient imino acceptor. (b) Implementation with N-acylhydrazones derived from 4-benzyl-2-oxazolidinone.

N−N bonds, while N-acylhydrazone **B** constrains the C−N bond within a ring and includes a carbonyl group for two-point binding of a Lewis acid in a rigid chelate structure. The Lewis acid would also increase reactivity toward nucleophilic alkyl radicals by lowering the LUMO energy of the C=N bond, ensuring that the chelated structure would react preferentially. Finally, we noted solid precedent for reductive cleavage of N−N bonds [16], whereby an N-linked auxiliary could be released for reuse after stereoisomer purification. For the first test of our design, N-amino derivatives of oxazolidinones [17, 18] were chosen as precursors for condensation with aldehydes and ketones. The derived chiral N-acylhydrazones possess both the carbonyl for two-point binding and a ring-templated stereocontrol element expected to hinder approach of an alkyl radical to one face of the C=N bond (Figure 2.4b). Prior to our work, N-amino derivatives of oxazolidinones had appeared in the literature only rarely [19] and had never been used for asymmetric synthesis.

2.2.2
Preparation of Chiral N-Acylhydrazones

Electrophilic amination at nitrogen was required to convert the commercial oxazolidinones to their N-amino derivatives. Upon deprotonation of the oxazolidinones 6a–6e (Scheme 2.1) and introduction of various NH_2^+ equivalents, N-aminooxazolidinones 7a–7e could be obtained in good yield. Although O-(mesitylenesulfonyl)hydroxylamine ($MtsONH_2$) was initially employed as the aminating reagent, its use is discouraged due to occasional uncontrolled exothermic decomposition during preparation and storage. Preferred reagents are O-(p-nitrobenzoyl)hydroxylamine ($NbzONH_2$), O-(diphenylphosphinyl)-hydroxylamine ($Ph_2P(=O)ONH_2$), and monochloramine (NH_2Cl) [20]. Our optimized procedure for N-amination with $NbzONH_2$ entailed deprotonation with NaH (or KH) in hot dioxane, followed by introduction of $NbzONH_2$ as a solid at ambient temperature [21].

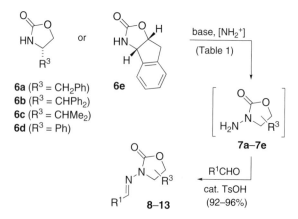

Scheme 2.1

Table 2.1 Amination of oxazolidinones and condensation with aldehydes (Scheme 2.1).

Entry	Oxazolidinone	R[1]	Method[a]	Product[b]
1	6a	Et	A	8a, 81%
2	6a	iPr	A	9, 70%
3	6a	tBu	A	10, 72%
4	6a	cC_6H_{11}	A	11, 65%
5	6a	Ph	A	12, 67%
6	6a	Ph	B[c]	12, 80% (74%)[d]
7	6a	CO_2Me	A	13a, 71%
8	6b	Et	B	8b, 75%
10	6c	Et	B	8c, 49%
12	6d	Et	B[c]	8d, 65%
14	6e	Et	B	8e, 87%

a) Method A: i. nBuLi, THF, −78 °C, 40 min; ii. MtsONH$_2$, −78 °C → rt; iii. R^{1}CHO, rt. Method B: (1) i. KH, dioxane, 60 °C, 1 h; ii. NBzONH$_2$, rt. (2) R^{1}CHO, cat. pTsOH, toluene, rt.
b) Isolated yield except where noted.
c) NaH was used in place of KH.
d) Yield on 10 g scale, employing crystallization.

Condensation with aldehydes proceeds reliably, with or without isolation of the intermediate N-aminooxazolidinone, to afford a wide range of chiral hydrazones 8–13 in good overall yields (Table 2.1). Small amounts of 6a that may remain after amination do not interfere with the condensation of 7a with aldehydes, so it is generally convenient to directly use unpurified 7a in the condensation step. Chiral N-acylhydrazones bearing different substituents on the oxazolidinone unit were prepared in a similar fashion (Scheme 2.1), and good yields of chiral N-acylhydrazones may be obtained on multigram scale. For example, benzaldehyde hydrazone 12 (9.4 g) was obtained in 74% yield from 6a using recrystallization. One can prepare a large quantity of a crystalline, indefinitely stable hydrazone for long-term storage and then exchange the carbonyl component as desired [21].

Chiral N-acylhydrazone derivatives may also be prepared from various ketones by condensation with N-aminooxazolidinone 7a (Table 2.2) [22]. Mixtures of E/Z isomers were usually obtained, although ketone N-acylhydrazones 14d and 15d, with highly branched tertiary butyl (tBu) substituents, were formed as single isomers. Others have recently used the amination and condensation procedures to prepare very similar chiral N-acylhydrazones from ketones with excellent results [23].

2.2.3
Tin-Mediated Addition of Secondary and Tertiary Radicals

The first test of the chiral N-acylhydrazones was in tin-mediated radical addition [24]. Addition of isopropyl iodide to propionaldehyde hydrazone 8a was chosen for initial screening (Scheme 2.2). Using the tin hydride method with triethylborane initiation [25] (Bu$_3$SnH, Et$_3$B/O$_2$), with InCl$_3$ and ZnCl$_2$ as Lewis acid additives, desired

Table 2.2 Preparation of ketone hydrazones.

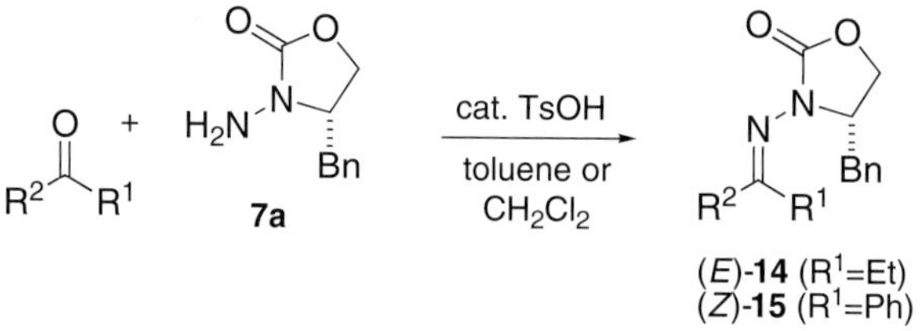

Entry	R¹	R²	Hydrazone (yield[a])	E/Z Ratio[b]
1	Et	iPr	14a (84%)	77:23
2	Et	cC_5H_9	14b (49%)	79:21
3	Et	cC_6H_{11}	14c (64%)	85:15
4	Et	tBu	14d (67%)	>98:2
5	Ph	iPr	15a (55%)	19:81
6	Ph	cC_5H_9	15b (59%)	26:74
7	Ph	cC_6H_{11}	15c (55%)	13:87
8	Ph	tBu	15d (73%)	<2:98

a) Isolated yield.
b) Ratio measured by integration of ^{1}H NMR spectra.

adduct **18** was obtained with high diastereoselectivity. In contrast, **18** was produced with poor selectivity (dr 2:1) in the absence of Lewis acid.

The scope of the reaction was evaluated by variations in both the radical and the radical acceptor. In the presence of $ZnCl_2$, the propionaldehyde N-acylhydrazone **8a** was subjected to radical additions of various organic iodides (Table 2.3, entries 1–4). Reaction conditions entailed addition of Bu_3SnH (5 equiv) and O_2 (7 ml/mmol) by syringe pump to a mixture of iPrI (10 equiv), Et_3B (10 equiv), and Lewis acid (2 equiv) in 2:1 CH_2Cl_2/ether at −78 °C, gradually bringing the mixture to ambient temperature after the addition. Under these conditions, ethyl radical (from the triethylborane) can compete for the radical acceptor, and as a result, the separable ethyl radical adduct **17** (Scheme 2.2) was observed (<10% yield) in all cases. With simple secondary and tertiary alkyl iodides as radical precursors (entries 1–4), additions to **8a** occurred with moderate yields to afford N-acylhydrazines. Radical reactivity is

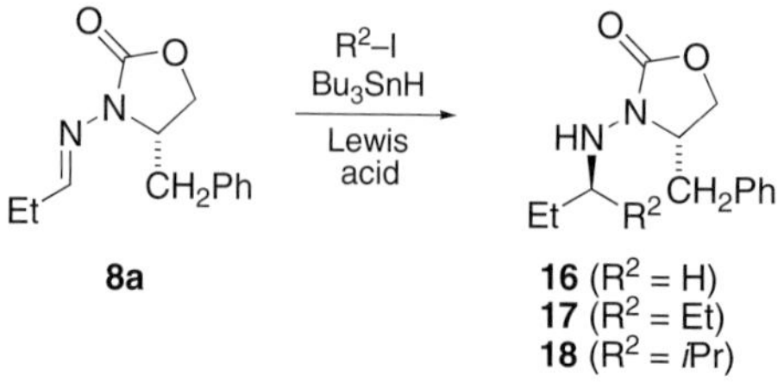

Scheme 2.2

Table 2.3 Reactivity scope of tin-mediated radical addition to N-acylhydrazones ($R^3 = Bn$) in the presence of $ZnCl_2$.

Entry	R^1	R^2	Product, yield[a] (dr)
1	Et (3a)	iPr	18a, 60% (99:1)
2	Et (3a)	cC_5H_9	19, 59% (96:4)
3	Et (3a)	cC_6H_{11}	20, 28% (97:5)
4	Et (3a)	tBu	21, 54% (95:5)
5	Et (3a)	iBu	22, 6%
6	Et (3a)	allyl	23, 7%
7	iPr (4)	Et	epi-18a, 6%
8	iPr (4)	cC_6H_{11}	24, 9%
9[a]	cC_6H_{11} (6)	Et	epi-20, 15%
10	cC_6H_{11} (6)	iPr	epi-24, 9%
11	Ph (7)	iPr	25, 42% (99:1)
12	Ph (7)	cC_5H_9	26, 59% (96:4)
13	Ph (7)	cC_6H_{11}	27, 30% (99:1)
14	Ph (7)	tBu	28, 83% (93:7)
15	CO_2Me (8a)	iPr	29, 0%

a) Isolated yield, %.

important in predicting the success of the addition reactions: primary, aromatic, allylic, and electrophilic radicals were ineffective under these conditions [24b].

A variety of aldehyde hydrazones were screened [24b]. Branching at a saturated α-carbon was detrimental in the tin-mediated radical additions, but an aromatic benzaldehyde hydrazone **12** offered some success, with yields ranging from 30 to 83% (Table 2.3, entries 5–8). With the exception of **8a**, which decomposed under the reaction conditions, the reactions were quite clean. Even in the examples with lower yields, the mass balance after recovery of the hydrazone precursor was generally 80–90%, demonstrating the excellent chemoselectivity of the reactions of radicals with N-acylhydrazones. We were delighted to find that the radical additions had occurred with excellent stereocontrol in all secondary and tertiary radical additions to hydrazones **8** and **12** (Table 2.3), with diastereomer ratios ranging from 93:7 to 99:1 [24].

Next, the effects of varying the stereocontrol elements on the oxazolidinone moiety were assessed, with the main goal to examine the change in diastereoselectivity. Without optimizing for yield, isopropyl radical additions to several N-acylhydrazones **8a–8e** (See Scheme 2.1 for structures) were compared for stereoselectivity (Table 2.4). Although the measurement was not available for **8c**, all auxiliaries gave very high diastereoselectivity in addition of isopropyl radical to propionaldehyde hydrazone [24b].

Table 2.4 Effect of different stereocontrol elements on diastereoselectivity in isopropyl radical addition.

Entry	Hydrazone	R³	dr[a]
1	8a	CH₂Ph	99:1
2	8b	CHPh₂	>98:2
3	8c	iPr	—[b]
4	8d	Ph	94:6
5	8e	Bicyclic[c]	95:5

Reaction conditions: See Table 2.2.
a) Diastereomer ratios by HPLC, GCMS, or ¹H NMR integration versus authentic mixtures.
b) Ratio not available.
c) See Scheme 2.1 for structure.

2.2.4
Tin-Free Radical Addition

Triethylborane or diethylzinc has been proposed to have multiple roles in radical additions to C=N bonds, including both initiation and chain propagation, offering the potential for a radical chain process without tin hydride [6, 7]. Accordingly, we attempted triethylborane-mediated tin-free additions of various halides, simply deleting Bu₃SnH from the conditions outlined in Table 2.3 and using InCl₃ as the Lewis acid (Table 2.5). As in the case of tin-mediated additions, the secondary iodides worked quite well in additions to the propionaldehyde hydrazone (entries 2–4). Although other primary radicals were ineffective, chloroiodomethane did lead to

Table 2.5 Scope of halide in tin-free radical addition to 8a in the presence of InCl₃.

Entry	R²	Product, yield[a]
1	Et	17a, 33%
2	iPr	18a, 75%[b]
3	cC₅H₉	19, 47%
4	cC₆H₁₁	20, 56%
5	CH₂Cl	30, 33%

a) Isolated yield.
b) dr >95:5 (¹H NMR).

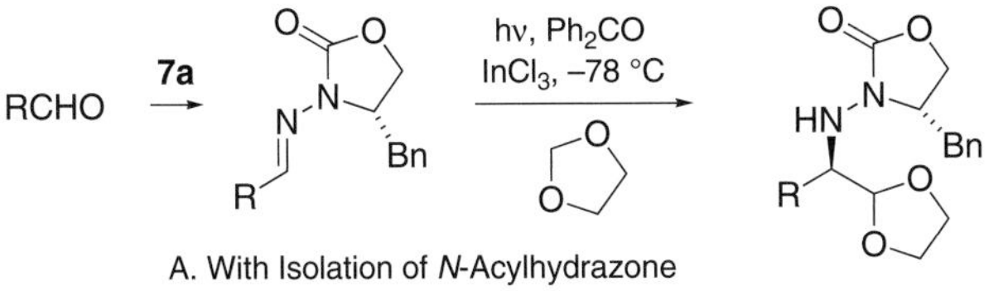

Scheme 2.3

RCHO → **7a** (N-acylhydrazone with oxazolidinone, Bn) — hv, Ph$_2$CO, InCl$_3$, −78 °C, 1,3-dioxolane → product (HN-N, Bn, dioxolane adduct)

A. With Isolation of N-Acylhydrazone
- R = CO$_2$Et — **31**, 68%, dr 80:20
- R = Et — **32**, 87%, dr 91:9

B. One-Pot Protocol
- R = Et — **32**, 96%, dr 98:2
- R = cyclopropyl — **33**, 85%, dr 97:3
- R = dimethoxymethyl — **34**, 75%, dr 95:5
- R = p-methoxyphenyl — **35**, 88%, dr 50:50
- R = 2-furyl — **36**, 93%, dr 80:20

successful addition of the •CH$_2$Cl group (entry 5), retaining the chloride for subsequent functional group manipulations.

Alonso and Fernàndez reported a tin-free method for addition of formyl radical equivalents to these chiral N-acylhydrazones [26]. Photolysis in 1,3-dioxolane in the presence of 1 equiv benzophenone led to H-atom abstraction from the dioxolane followed by intermolecular radical addition to chiral N-acylhydrazones (Scheme 2.3). Consistent with our results, stereocontrol here was low in the absence of Lewis acid but rose to excellent levels in the presence of InCl$_3$. After N—N bond cleavage and oxidation at the formyl carbon, preparation of α-amino acids was achieved with high stereoselectivity. The preferred diastereomer was that suggested by the Lewis acid chelate model, as we had observed in the results described above. In contrast to our findings, the reactions of glyoxylate hydrazone **13b** could be achieved under these conditions, although the stereoselectivity was limited to 4.1 : 1 diastereomer ratio. Interestingly, the addition of Lewis acid did not improve the selectivity in this case. A one-pot protocol was introduced that begins with the aldehyde; the N-acylhydrazone need not be isolated. For a series of aldehydes, the corresponding adducts **31–36** were obtained with yields ranging from 75 to 99%. Stereocontrol was variable, with the more synthetically useful levels observed in aliphatic products **32–34**. Alonso's impressive method provides an independent validation of our stereocontrol hypothesis and renders the radical additions applicable to a broader range of functionalized targets.

The synthetic potential of the intermolecular radical additions would be dramatically enhanced by developing conditions compatible with primary radicals. Less stable 1° radicals (versus 2° or 3°) often suffer premature reduction by hydrogen atom abstraction processes and are impractical under Et$_3$B/O$_2$ initiation conditions due to the competition with ethyl radicals. These considerations led us to seek alternatives to triethylborane.

2.2.5
Manganese-Mediated Radical Addition

Photolytic radical generation with hexamethylditin has shown promise for addition of primary radicals to N-acylhydrazones. Applying Kim's photolysis conditions [27] using hexamethylditin in the presence of InCl$_3$, ethyl and isopropyl additions to hydrazone **8a** occurred in reasonable yield [24b]. Unfortunately, further increase in efficiency was prevented by a carbonyl exchange side reaction [28] with acetone (employed as a sensitizer) to give the acetone hydrazone.

We became interested in manganese carbonyl [$Mn_2(CO)_{10}$] [29], which requires no sensitizer (λ_{max} 340 nm, $\sigma_{Mn-Mn} \rightarrow \sigma^*_{Mn-Mn}$) for homolytic metal–metal bond cleavage. This process had apparently been scarcely recognized by synthetic organic chemists before a series of studies were published by Parsons [30].

The manganese carbonyl reagent proved to be a dramatic improvement [31]. For ethyl iodide addition to **8a** (Table 2.6), irradiation (300 nm) with $Mn_2(CO)_{10}$ using $InCl_3$ as a Lewis acid furnished the ethyl adduct in 85% yield, much higher than with triethylborane or hexamethylditin. Control experiments revealed a requirement for both irradiation and $Mn_2(CO)_{10}$. Without $InCl_3$, the reaction was slow (21% yield after 2 d). Several other halides, including methyl iodide and difunctional halides, were also effective (Table 2.6), with the exception of 2-chloroethyl addition, which gave low yield presumably due to radical fragmentation. Ethyl radical addition to nine additional hydrazones occurred in good yields (entries 12–20). These adducts are epimeric to those derived from hydrazone **8a** with respect to the new stereogenic center, as a result of simply changing the roles of the aldehyde and iodide precursors. Synthetically useful yields and reliably high diastereomer ratios were obtained from these reactions, with some adducts bearing additional functionality for further elaboration.

2.2.6
Manganese-Mediated Coupling with Multifunctional Precursors

2.2.6.1 Hybrid Radical–Ionic Annulation
Interestingly, the 3-chloro-1-iodopropane addition (Table 2.6, entry 9) led exclusively to pyrrolidine **37** (Scheme 2.4); none of the acyclic adducts was found [31]. Presumably, radical addition was followed by *in situ* S_N2-type cyclization. The same type of cyclization, giving the epimeric pyrrolidine (*epi*-**37**), occurs upon ethyl addition to the 3-chlorobutyraldehyde hydrazone (Table 2.6, entry 18). These reactions are hybrid radical–ionic annulations of the C=N bond, a new class of radical–polar crossover reactions [32].

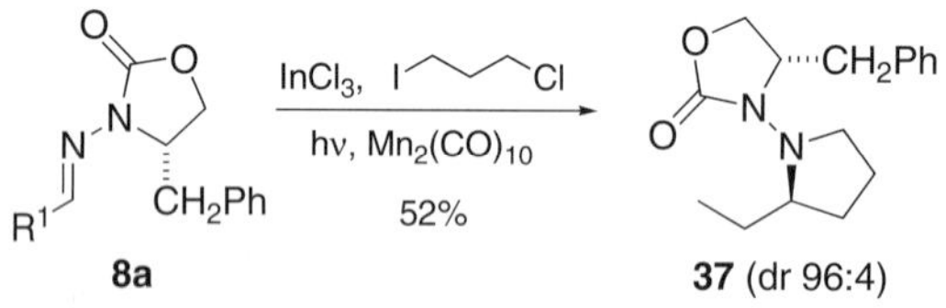

Scheme 2.4

The hybrid radical–ionic annulation was envisioned to be useful for piperidine alkaloid synthesis. The alkaloid coniine [33] (Scheme 2.5) offers a simple test case. Starting with the butyraldehyde hydrazone **38**, Mn-mediated photolysis with 4-chloro-iodobutane afforded the acyclic adduct **39** in 66% yield with a diastereomer ratio of 95 : 5 [31]. In this case, Finkelstein conditions were needed for cyclization,

Table 2.6 Results of metal-mediated radical addition to propionaldehyde hydrazone **8a**.

R¹CHO —**7a**→ [hydrazone with N-N-C(=O)-O ring, CH₂Ph, R¹ on C=N] —R²X, InCl₃, hv, Mn₂(CO)₁₀→ [HN-N-C(=O)-O ring with CH₂Ph, R¹ and R² on carbon]

Entry	Aldehyde (or acetal)	Yield of hydrazone[a]	Halide R²X	Yield, configuration[b]	dr
1	CH₃CH₂CHO	81%	CH₃CH₂I	85%	—
2			CH₃I	48%[c),d], S	95:5[e]
3			⌇I (propyl)	66%, R	94:6[e]
4			⌇I (butyl)	78%, R	95:5[e]
5			⌇I (pentyl)	79%, R	96:4[e]
6			iPr-I	54%[c], R	95:5[f]
7			tBu-I	75%, R	95:5[f]
8			ClCH₂I	63%, R	93:7[e]
9			Cl⌇I	52%, R	96:4[f]
10			Cl⌇⌇I	55%, R	96:4[e]
11			Cl₂CHBr	38%[c),d], R	98:2[f]
12	CH₃CHO	66%	CH₃CH₂I	66%, R	95:5[e]
13	⌇CHO	87%		63%, S	95:5[e]
14	⌇⌇CHO	89%		72%, S	97:3[e]
15	⌇⌇⌇CHO	88%		77%, S	97:3[e]
16	iPrCH₂CHO	85%		65%, S	95:5[f]
17	ClCH₂CH(OMe)₂	85%		57%, S	93:7[e]
18	Cl⌇⌇CHO	95%		60%, S	93:7[f]
19	Cl⌇⌇⌇CHO	89%		62%, S	97:3[e]
20	Cl₂CHCH(OEt)₂	54%		34%[c], S	89:11[f]

Reaction conditions: (1) Aldehyde or acetal (5–10 equiv), **7a**, p-toluenesulfonic acid, CH₂Cl₂, rt. (2) Hydrazone in deoxygenated CH₂Cl₂ (0.1 M), InCl₃ (2.2 equiv), Mn₂(CO)₁₀ (1–2 equiv), R²X (10 equiv), hv (300 nm, pyrex), 1–2 d, about 35 °C.

a) Isolated yield.
b) Isolated yields of purified diastereomer mixtures. R or S denotes the configuration of the new stereogenic center. Addition of methyl iodide gives opposite configurations due to the lower priority of the methyl ligand.
c) 20 equiv of R²X was used.
d) 1,8-Diazabicyclo[5.4.0]undec-7-ene (DBU) was used in the removal of Mn by-products.
e) Ratio by HPLC (Chiralcel OD, 2-PrOH/hexane).
f) Ratio by ¹H NMR.

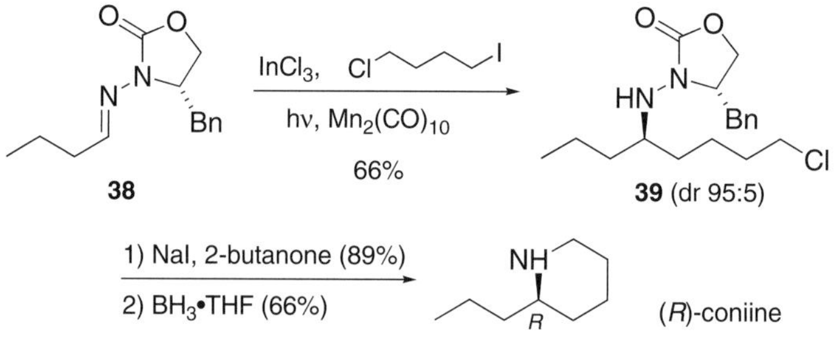

Scheme 2.5

and reductive removal of the auxiliary afforded coniine in 34% overall yield for four steps. An interesting aspect of this reaction sequence is the direct comparison of the efficiency of radical- and carbanion-based syntheses. Using the same retrosynthetic disconnection, a carbanion approach requires 9–10 steps [33b,f]. This illustrates the potential for improved efficiency through novel radical addition strategies.

2.2.6.2 Precursors Containing Hydroxyl or Protected Hydroxyl Groups

The hybrid radical–ionic annulation illustrates that precursors containing additional electrophilic functionality can be used to synthetic advantage. A number of other examples of functional group compatibility have been examined, including oxygen-containing radical precursors and acceptors.

In the key step of an approach to total synthesis of the antimalarial alkaloid quinine, involving coupling of hydrazone **40** with iodide **41** (Scheme 2.6), not only functional group combatibility but also stoichiometry was a concern. In most intermolecular radical additions to imino compounds, large excesses (10–20 equiv or more) of radical precursors are required. Clearly, this would be a prohibitive stoichiometric requirement for an iodide such as **41**, prepared through several synthetic steps. To our great delight, the Mn-mediated coupling of **40** with only 1.25 equiv **41** proceeded in 93% yield in 1 mmol scale, giving **42** as a single diastereomer [34]. The low stoichiometric requirement in the coupling of the multifunctional alkyl group of **41** to an imino compound is quite attractive and should enable broader applications of this Mn-mediated coupling process in complex target synthesis.

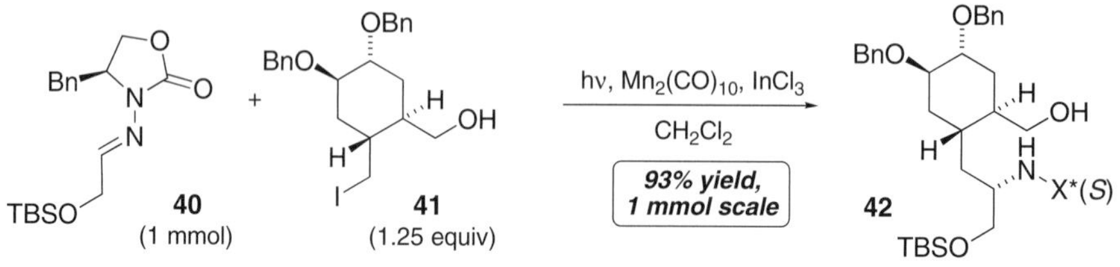

Scheme 2.6

2.2 Intermolecular Radical Addition Chiral N-Acylhydrazones

Although not as thoroughly explored as α- and β-amino acids, γ-amino acids are of significant interest as building blocks for bioorganic and medicinal chemistry [35–38]. Recently, we exploited the Mn-mediated photolysis for a novel synthesis of γ-amino acids **43** and **44** (Figure 2.5) [39]. The C—C bond disconnections shown would require oxygen-containing iodides or hydrazones and therefore constitute an important test of the synthetic versatility of the Mn-mediated coupling reactions.

For synthesis of α-alkoxy-γ-amino acid **43**, nonbasic conditions would be a necessity, considering the potential for β-elimination of the alkoxy group from the hydrazone precursor **45** (Scheme 2.7). Here, the addition of isopropyl iodide under the Mn-mediated photolysis conditions afforded **46** as a single diastereomer in 77% yield, without any evidence of β-elimination.

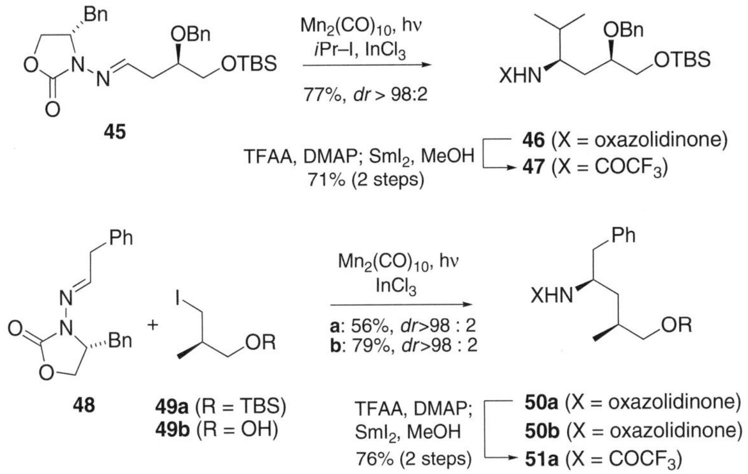

Scheme 2.7

For the γ-amino acid **44** (Figure 2.5), phenylacetaldehyde N-acylhydrazone **48** (Scheme 2.7) was employed as the radical acceptor. Mn-mediated addition of difunctional iodide **49a** (5 equiv) proceeded in 56% yield, affording **50a** as a single diastereomer [39]. Recently, we found that the unprotected alcohol **49b** couples with hydrazone **48** in 79% yield, using a modest excess (3.5 equiv) of the iodide. Considering that typical intermolecular radical additions often require large

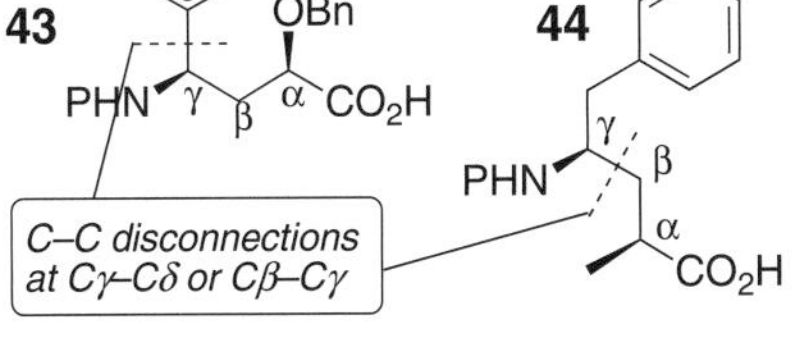

Figure 2.5 Alternative C—C bond disconnections of the tubulysin γ-amino acids at the γ—δ and β—γ bonds.

excesses (10–20 equiv) of the radical precursor, the more manageable stoichiometry in this reaction may invite expanded applications of radical addition chemistry. From adducts **46** and **50a**, cleavage of the N−N bond afforded **47** and **51a** in good yield, and subsequent oxidation of the primary alcohol functionality provided γ-amino acids **43** and **44** bearing synthetically useful trifluoroacetyl amine protection [39].

2.2.6.3 Ester-Containing N-Acylhydrazones

Efforts to develop a γ-amino acid synthesis in which the oxidation state need not be adjusted after coupling led to the hypothesis that γ-hydrazonoesters may be competent radical acceptors. Would the stereocontrol model be applicable in the presence of an additional Lewis basic ester function in the hydrazone? This question was addressed via prototypical Mn-mediated photolytic conditions with $InCl_3$ as the Lewis acid. Successful coupling was achieved between isopropyl iodide and a variety of γ-hydrazonoesters **52a–52d** (Table 2.7, entries 1–4) bearing methyl, dimethyl, and benzyloxy substituents at the position β to the hydrazone (α to the ester). Consistently high diastereoselectivities and excellent yields (91–98%) of the isopropyl adducts **53a–53d** indicated that the substitution patterns examined in this study had little effect on reaction selectivity and efficiency.

A range of secondary and primary iodides were examined (Table 2.7). Secondary iodides coupled with γ-hydrazonoester **52a** in excellent yields and selectivities. When

Table 2.7 Additions to Ester-Containing N-Acylhydrazones.

a: $R^1 = R^2 = H$; b: $R^1 = H$, $R^2 = Me$; c: $R^1 = R^2 = Me$; d: $R^1 = H$, $R^2 = OBn$

Hydrazone	Iodide R^3I	Product, yield	dr
52a	iPrI	53a, >99%	94:6
52b		53b, 98%	95:5
52c		53c, 96%	99:1
52d		53d, 96%	90:10
52a	(iBuI)	54, 82%	—
	(n-pentyl-I)	55, 66%	94:6
	TBSO~~~I	56, 37%	96:4
	HO~~~I	57, 61%	97:3
	HO~~I	58, 33%	96:4
	Cl~~~I	59, 45%	85:15

primary iodides were subjected to coupling with **52a**, the desired adducts were obtained in moderate yields and excellent diastereoselectivity was maintained. Silyl ether and alcohol functionality were accommodated. For **53a**, trifluoroacetylation under microwave irradiation [40] in the presence of Et_3N and DMAP (Scheme 2.8), followed by exposure to SmI_2, smoothly furnished known γ-aminoester **61** and offered proof of absolute configuration [41].

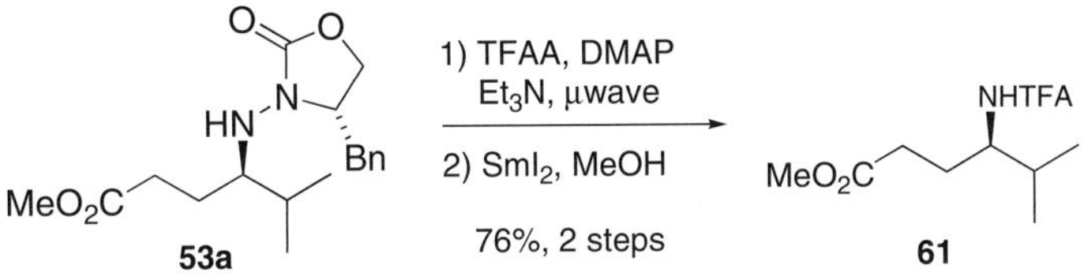

Scheme 2.8

Although the yield of the additions to γ-hydrazonoesters in the absence of $InCl_3$ was modest, surprisingly, the selectivity was only slightly diminished. This prompted a reassessment of the stereocontrol model for these examples, taking into consideration the possibility that the ester might participate in binding Lewis acid, in turn influencing the acceptor geometry and stereoselectivity. However, upon mixing the γ-hydrazonoesters with $InCl_3$, NMR spectra showed characteristic changes in chemical shifts that were indicative of chelation by the imino nitrogen and the oxazolidinone carbonyl in the usual way, without significant interference by the ester function.

2.2.6.4 Additions to Ketone Hydrazones

Despite the aforementioned developments with aldimine-type acceptors, the corresponding additions to ketimine acceptors have not yet reached their synthetic potential [42, 43]. The Mn-mediated coupling of iodides with ketone-derived imino compounds [44] is of particular interest as this strategy may provide a diverse range of *tert*-alkyl amines not conveniently prepared by nucleophilic substitution. A key issue in additions to ketimine derivatives is ensuring the reaction takes place exclusively through one imine geometry. With aldehydes, this issue is of little consequence; the N-acylhydrazone derivatives are exclusively obtained in *E* geometry. On the other hand, ketone hydrazones are generally formed as mixtures of *E* and *Z* isomers.

The first efforts to address this issue involved hydrazone (*E*)-**62** (Scheme 2.9), prepared as a mixture (*E/Z* 92 : 8) and separated via flash chromatography as the pure (*E*)-isomer in 75% yield. In this preparation, the oxazolidinone N-amination was accomplished using a solution of monochloramine in methyl *tert*-butyl ether [45], furnishing a quantitative yield of the N-amino-2-oxazolidinone, which in turn was condensed with methyl pyruvate to afford **62**. Addition of ethyl iodide to (*E*)-**62** using the Mn-mediated photolysis conditions as described above gave 66% yield of the ethyl adduct, with a modest diastereomer ratio of 70 : 30, while the corresponding isopropyl addition was very effective (85% yield, dr 92 : 8). Variation in the stoichiometry indicated that amounts less than 2 equiv of Lewis acid proportionally lowered the diastereoselectivity, suggesting that, in this case, the ester may participate

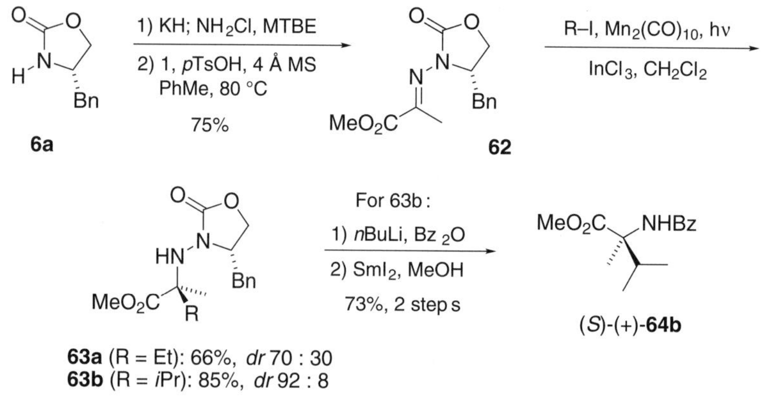

2 Asymmetric Methods for Radical Addition to Imino Compounds

Scheme 2.9

63a (R = Et): 66%, dr 70 : 30
63b (R = iPr): 85%, dr 92 : 8

in Lewis acid coordination along with the N-acylhydrazone. Cleavage of the N—N bond was achieved upon sequential treatment of isopropyl adduct **63b** with n-butyllithium, benzoic anhydride, and SmI$_2$/MeOH (Scheme 2.9). This sequence afforded known benzamide (S)-(+)-**64b** [46] in good yield, confirming the assigned configuration.

The Mn-mediated radical additions offer an inherently flexible carbon–carbon bond construction approach to amine synthesis. Because of the broad functional group compatibility in both the radical precursor and the aldehyde hydrazone acceptor, the roles of these precursors can be switched to result in the construction of either of two C—C bonds at the chiral amine (Scheme 2.10) with excellent stereocontrol. The epimeric configuration can be selected by either (a) employing the enantiomeric auxiliary or (b) interchanging the roles of R^1 and R^2 in the alkyl halide and aldehyde precursors [47]. By combining these two tactics, the optimal roles of R^1 and R^2 with respect to yield and selectivity can be chosen. Such strategic flexibility contributes to the synthetic potential of these radical addition reactions.

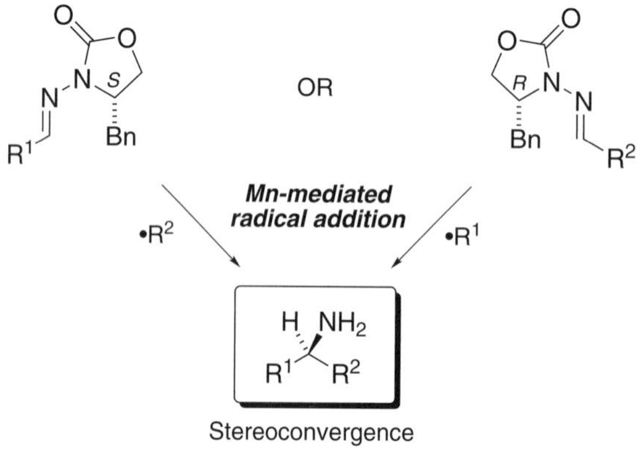

Scheme 2.10

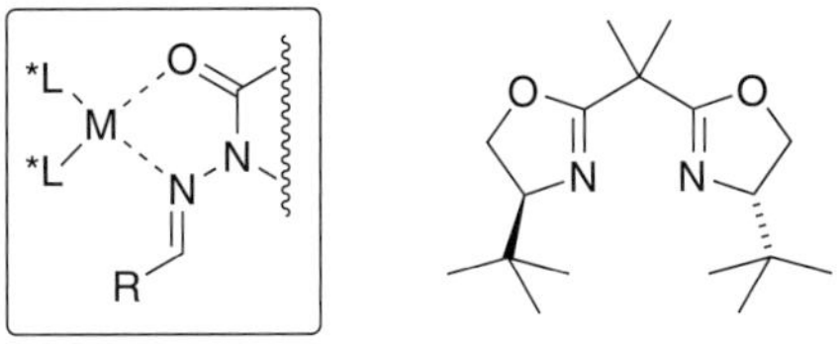

Figure 2.6 Two-point binding of N-acylhydrazones involving Lewis acid and chiral ligand(s).

2.3
Asymmetric Catalysis of Radical Addition

Asymmetric catalysis of radical addition to C=N bonds remains a challenge for further synthetic methodology development [48]. Our effort toward this goal began with the hypothesis that the two-point binding of Lewis acids by N-acylhydrazones (Figure 2.6) would facilitate the development of a versatile means of stereocontrol.

The first successes exploited the tin-free conditions described above, with radical initiation by triethylborane and oxygen. Using valerolactam-derived achiral N-acylhydrazone acceptor **66a** (Table 2.8), we discovered highly enantioselective radical additions of isopropyl iodide, promoted by 1 equiv each of Lewis acid and bisoxazoline ligand **65** (Figure 2.6). InCl$_3$, Mg(ClO$_4$)$_2$, and Cu(OTf)$_2$ offered only modest yields (entries 1–3), but we were gratified to find these initial efforts to have already set

Table 2.8 Studies of Lewis acids, reaction medium, and chiral ligand structure in isopropyl addition to **66a**.[a]

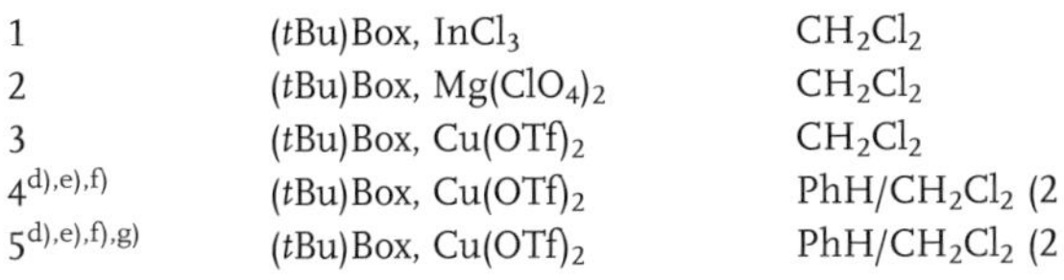

Entry	Chiral ligand, Lewis acid	Solvent	Yield[b]	ee, %[c]
1	(tBu)Box, InCl$_3$	CH$_2$Cl$_2$	33	57 (R)
2	(tBu)Box, Mg(ClO$_4$)$_2$	CH$_2$Cl$_2$	41	66 (R)
3	(tBu)Box, Cu(OTf)$_2$	CH$_2$Cl$_2$	41	59 (R)
4[d],[e],[f]	(tBu)Box, Cu(OTf)$_2$	PhH/CH$_2$Cl$_2$ (2:1)	66	95 (R)
5[d],[e],[f],[g]	(tBu)Box, Cu(OTf)$_2$	PhH/CH$_2$Cl$_2$ (2:1)	94	86 (R)

a) Reaction conditions: Lewis acid (1 equiv), chiral ligand (1 equiv), 2-iodopropane (6 equiv), Et$_3$B/O$_2$ (6 equiv), 25 °C.
b) Isolated yield, %.
c) Enantiomeric excess by HPLC (95:5 hexane:2-propanol, Chiralcel OD or AD).
d) Et$_3$N was added after the reaction to facilitate product isolation.
e) In the presence of powdered 4A molecular sieves.
f) Preformed aquo complex Cu(tBu-Box)(H$_2$O)$_2$(OTf)$_2$ was used.
g) Larger amounts (10 equiv) of 2-iodopropane and Et$_3$B were used.

a new standard for selectivity in radical additions to C=N bonds [49]. Upon changing to benzene/CH$_2$Cl$_2$ (entry 4), the selectivity further increased to 95% ee (66% yield). A less polar medium was assumed to facilitate the assembly of a ternary complex of ligand, Lewis acid, and substrate. The yield improved to 94% with larger amounts of 2-iodopropane and Et$_3$B (entry 5), but this came at the expense of some selectivity.

Various radical precursors and acceptors may be employed with high enantioselectivity (Table 2.9). Isopropyl additions to electron-rich and electron-deficient aromatic hydrazones **66b** and **66c** were all highly enantioselective, as were additions of various radicals, including chloromethyl, to **66a** (entries 1–6). To test the potential for development of asymmetric catalysis, we checked for turnover by lowering the catalyst loading (Table 2.9, entries 7–10). The yield remained high, while enantioselection decreased. With 46% ee and 74% yield at 10 mol% catalyst loading, a catalytic cycle involving **65** is implied – the first evidence of asymmetric catalysis in radical addition to C=N bonds.

2.4
Closing Remarks

The development of radical additions to C=N bonds is a field with great promise for applications in the synthesis of multifunctional compounds; the functional group compatibility of radical chemistry complements that of carbanion reagents to expand

Table 2.9 Scope of radical addition to **66a–66c** promoted by Cu(tBu-Box)(H$_2$O)$_2$(OTf)$_2$ and effects of Cu(II) catalyst loading.[a]

Entry	Halide	Hydrazone (R^1)	Catalyst load	% Yield[b] (% ee[c])
1	iPrI	66b (pMeOC$_6$H$_4$)	1 equiv	46 (90)
2	iPrI	66c (pClC$_6$H$_4$)	1 equiv	53 (81)
3	EtI[d]	66a (Ph)	1 equiv	88 (83)
4	cC$_5$H$_9$I[d]	66a (Ph)	1 equiv	86 (84)
5	cC$_6$H$_{11}$I[d]	66a (Ph)	1 equiv	84 (89)
6	ClCH$_2$I[d]	66a (Ph)	1 equiv	44[e] (95)
7	iPrI	66a (Ph)	1 equiv	66 (95)
8	iPrI	66a (Ph)	0.5 equiv	71 (81)
9	iPrI	66a (Ph)	0.2 equiv	83 (58)
10	iPrI	66a (Ph)	0.1 equiv	74 (46)

a) Reaction conditions: see Table 2.8.
b) Isolated yield, %.
c) Enantiomeric excess, % (hexane:2-propanol, Chiralcel OD or AD).
d) 10 equiv of alkyl halide was used.
e) 56% recovery of unreacted hydrazone.

the versatility of the C—C bond construction approach to chiral amine synthesis. The nonbasic conditions we have developed for Mn-mediated coupling demonstrate compatibility with unprotected hydroxyl groups and other electrophilic functionality in both precursors and are readily applicable to the synthesis of amino acids and alkaloids. High asymmetric induction by chiral Cu(II) catalysts illustrate another promising direction for future development of radical additions to imino compounds.

Acknowledgments

Portions of this chapter appeared in a prior review article (G.K. Friestad, Chiral N-acylhydrazones: versatile imino acceptors for asymmetric amine synthesis. *Eur. J. Org. Chem.* **2005**, 3157–3172. Copyright Wiley-VCH Verlag GmbH & Co. KGaA. Reproduced with permission.). Generous support to our work by NSF (CHE-0096803 and CHE-0749850), NIH (R01-GM67187), Research Corporation, Petroleum Research Fund, and Vermont EPSCoR is deeply appreciated, as are the hard work and insight of numerous students and postdoctoral associates who have participated in our program in chiral amine synthesis.

Questions

2.1. Which of the following pairs of reactants would be expected to give synthetically useful yields in the Sn-mediated addition reaction shown in Table 2.3? Explain.
 (a) *n*-Butyl iodide and hydrazone **8a**
 (b) *sec*-Butyl iodide and hydrazone **8a**
 (c) Iodobenzene and hydrazone **8a**

2.2. For the Mn-mediated coupling of ethyl iodide with N-acylhydrazone **8a** (Table 2.6, entry 1), propose a source of the N—H hydrogen.

2.3. Design a synthesis of halosaline, using a Mn-mediated coupling reaction as a C—C bond construction step.

halosaline

References

1 Reviews: (a) Friestad, G.K. (2009) Addition of carbanions to azomethines, *Science of Synthesis, Vol. 40a: Compounds with One Saturated Carbon–Heteroatom Bond: Amines and Ammonium Salts* (eds D. Enders and E. Shaumann), Thieme, Stuttgart, Germany, pp. 305–342;
(b) Yamada, K.-I. and Tomioka, K. (2008) *Chem. Rev.*, **108**, 2874–2886; (c) Friestad, G.K. and Mathies, A.K. (2007) *Tetrahedron*, **63**, 2541–2569; (d) Ding, H. and Friestad, G.K. (2005) *Synthesis*, 2815–2829; Alvaro, G. and Savoia, D. (2002) *Synlett*, 651–673;

(e) Kobayashi, S. and Ishitani, H. (1999) *Chem. Rev.*, **99**, 1069–1094; (f) Bloch, R. (1998) *Chem. Rev.*, **98**, 1407–1438; (g) Davis, F.A., Zhou, P., and Chen, B.-C. (1998) *Chem. Soc. Rev.*, **27**, 13–18; (h) Enders, D. and Reinhold, U. (1997) *Tetrahedron Asymmetry*, **8**, 1895–1946; (i) Denmark, S.E. and Nicaise, O.J.-C. (1996) *J. Chem. Soc., Chem. Commun.*, 999–1004.

2 (a) Aza-enolization of imines with Grignard reagents: Stork, G. and Dowd, S.R., (1963) *J. Am. Chem. Soc.*, **85**, 2178–2180; Wittig, G., Frommeld, H.D., and Suchanek, P. (1963) *Angew. Chem., Int. Ed. Engl.*, **2**, 683; (b) Deprotonation of iminium ions can be competitive with addition: Guerrier, L., Royer, J., Grierson, D.S., and Husson, H.-P., (1983) *J. Am. Chem. Soc.*, **105**, 7754–7755; (c) Less basic organocerium reagents also exhibit aza-enolization: Enders, D., Diez, E., Fernandez, R., Martin-Zamora, E., Munoz, J.M., Pappalardo, R.R., and Lassaletta, J.M., (1999) *J. Org. Chem.*, **64**, 6329–6336.

3 Reviews: Ohfune, Y. and Shinada, T., (2005) *Eur. J. Org. Chem.*, 5127–5143; Spino, C. (2004) *Angew. Chem., Int. Ed.*, **43**, 1764–1766; Groger, H. (2003) *Chem. Rev.*, **103**, 2795–2827; Yet, L. (2001) *Angew. Chem., Int. Ed.*, **40**, 875–877.

4 Reviews: Arend, M., Westerman, B., and Risch, N., (1998) *Angew. Chem., Int. Ed.*, **37**, 1044–1070; Denmark, S. and Nicaise, O.J.-C. (1999) *Comprehensive Asymmetric Catalysis*, vol. 2 (eds E.N. Jacobsen, A. Pfaltz, and H. Yamomoto), Springer, Berlin, p. 93; Kobayashi, S. and Ishitani, H. (1999) *Chem. Rev.*, **99**, 1069–1094; Liu, M. and Sibi, M.P. (2002) *Tetrahedron*, **58**, 7991–8035; Ma, J.-A. (2003) *Angew. Chem., Int. Ed.*, **42**, 4290–4299; Ueno, M. and Kobayashi, S. (2005) *Enantioselective Synthesis of β-Amino Acids*, 2nd edn (eds E. Juaristi and V.A. Soloshonok), John Wiley & Sons, Inc., New York, pp. 139–157; Marques, M.M.B. (2006) *Angew. Chem., Int. Ed.*, **45**, 348–352.

5 Reviews of radical additions to imines and related acceptors: (a) Friestad, G.K., (2001) *Tetrahedron*, **57**, 5461–5496; (b) Fallis, A.G. and Brinza, I.M. (1997) *Tetrahedron*, **53**, 17543–17594.

6 For a review of the Bertrand group's work in this area, see Bertrand, M., Feray, L., and Gastaldi, S. (2002) *Comptes Rend. Acad. Sci. Paris, Chimie*, **5**, 623–638.

7 For a review of the Naito group's work in this area, see Miyabe, H., Ueda, M., and Naito, T. (2004) *Synlett*, 1140–1157.

8 Booth, S.E., Jenkins, P.R., Swain, C.J., and Sweeney, J.B. (1994) *J. Chem. Soc., Perkin Trans. 1*, 3499–3508.

9 Booth, S.E., Jenkins, P.R., and Swain, C.J. (1998) *J. Braz. Chem. Soc.*, **9**, 389–395.

10 Review: Enders, D. (1984) *Asymmetric Synthesis* (ed. J.D. Morrison), Academic Press, New York, pp. 275–339.

11 Enders, D., Schubert, H., and Nubling, C. (1986) *Angew. Chem., Int. Ed. Engl.*, **25**, 1109–1110.

12 For organometallic additions to other chiral hydrazones, see ephedrine-derived hydrazone: (a) Takahashi, H., Tomita, K., and Otomasu, H. (1979) *J. Chem. Soc., Chem. Commun.*, 668–669; (b) valine-derived hydrazone: Takahashi, H. and Suzuki, Y., (1983) *Chem. Pharm. Bull.*, **31**, 4295–4299; (c) proline-derived hydrazone: Denmark, S.E., Weber, T., and Piotrowski, D.W., (1987) *J. Am. Chem. Soc.*, **109**, 2224–2225.

13 Breuil-Desvergnes, V., Compain, P., Vatèle, J.-M., and Goré, J. (1999) *Tetrahedron Lett.*, **40**, 5009–5012.

14 Protonation of imines leads to improved yields in radical addition. For reviews of Lewis acid effects in radical reactions, see (a) Renaud, P. and Gerster, M., (1998) *Angew. Chem., Int. Ed.*, **37**, 2563–2579; (b) Guerin, B., Ogilvie, W.W., and Guindon, Y. (2001) *Radicals in Organic Synthesis* (eds P. Renaud and M. Sibi), John Wiley & Sons, Inc, New York, pp. 441–460.

15 Friestad, G.K. (2005) *Eur. J. Org. Chem.*, 3157–3172.

16 (a) Burk, M.J. and Feaster, J.E. (1992) *J. Am. Chem. Soc.*, **114**, 6266–6267; (b) Sturino, C.F. and Fallis, A.G. (1994) *J. Am. Chem. Soc.*, **116**, 7447–7448.

17 Evans, D.A. and Kim, A.S. (1995) *Encyclopedia of Reagents for Organic Synthesis*, vol. **1** (ed. L.A. Paquette), John Wiley & Sons, Inc., New York, pp. 345–356.

References

18 Oxazolidinones have been used previously for stereocontrolled radical addition to alkenes. (a) Sibi, M.P., Jasperse, C.P., and Ji, J., (1995) *J. Am. Chem. Soc.*, **117**, 10779–10780; (b) Sibi, M.P., Ji, J., Sausker, J.B., and Jasperse, C.P. (1999) *J. Am. Chem. Soc.*, **121**, 7517–7526.

19 (a) Kim, M. and White, J.D. (1977) *J. Am. Chem. Soc.*, **99**, 1172–1180; (b) Ciufolini, M.A., Shimizu, T., Swaminathan, S., and Xi, N. (1997) *Tetrahedron Lett.*, **38**, 4947–4950; (c) Evans, D.A. and Johnson, D.S. (1999) *Org. Lett.*, **1**, 595–598.

20 Hynes, J. Jr, Doubleday, W.W., Dyckman, A.J., Godfrey, J.D. Jr, Grosso, J.A., Kiau, S., and Leftheris, K. (2004) *J. Org. Chem.*, **69**, 1368–1371.

21 Shen, Y. and Friestad, G.K. (2002) *J. Org. Chem.*, **67**, 6236–6239.

22 Qin, J. and Friestad, G.K. (2003) *Tetrahedron*, **59**, 6393–6402.

23 Lim, D. and Coltart, D.M. (2008) *Angew. Chem., Int. Ed.*, **47**, 5207–5210.

24 (a) Friestad, G.K. and Qin, J. (2000) *J. Am. Chem. Soc.*, **122**, 8329–8330; (b) Friestad, G.K., Draghici, C., Soukri, M., and Qin, J. (2005) *J. Org. Chem.*, **70**, 6330–6338.

25 Nozaki, K., Oshima, K., and Utimoto, K. (1991) *Bull. Chem. Soc. Jpn.*, **64**, 403–409; Brown, H.C. and Midland, M.M. (1972) *Angew. Chem., Int. Ed. Engl.*, **11**, 692–700.

26 Fernández, M. and Alonso, R. (2003) *Org. Lett.*, **5**, 2461–2464.

27 (a) Kim, S., Lee, I.Y., Yoon, J.-Y., and Oh, D.H. (1996) *J. Am. Chem. Soc.*, **118**, 5138–5139; (b) Kim, S. and Yoon, J.-Y. (1997) *J. Am. Chem. Soc.*, **119**, 5982–5983; (c) Ryu, I., Kuriyama, H., Minakata, S., Komatsu, M., Yoon, J.-Y., and Kim, S. (1999) *J. Am. Chem. Soc.*, **121**, 12190–12191; (d) Jeon, G.-H., Yoon, J.-Y., Kim, S., and Kim, S.S. (2000) *Synlett*, 128–130; (e) Kim, S., Kim, N., Yoon, J.-Y., and Oh, D.H. (2000) *Synlett*, 1148–1150; (f) Kim, S. and Kavali, R. (2002) *Tetrahedron Lett.*, **43**, 7189–7191; (g) For a related method using primary alkyl tellurides, see Kim, S., Song, H.-J., Choi, T.-L., and Yoon, J.-Y. (2001) *Angew. Chem., Int. Ed.*, **40**, 2524–2526.

28 The exchange reactions between certain *N*-acylhydrazones and carbonyl compounds constitute a useful preparative method. See Ref. [21].

29 Reviews: (a) Pauson, P.L. (1995) *Encyclopedia of Reagents for Organic Synthesis*, vol. 2 (ed. L.A. Paquette), John Wiley & Sons, Inc., New York, pp. 1471–1474; Meyer, T.J. and Caspar, J.V. (1985) *Chem. Rev.*, **85**, 187; Gilbert, B.C. and Parsons, A.F. (2002) *J. Chem. Soc., Perkin Trans. 2*, 367–387; (b) Halogen atom transfer to •Mn(CO)$_5$: Herrick, R.S., Herrinton, T.R., Walker, H.W., and Brown, T.L., (1985) *Organometallics*, **4**, 42–45.

30 (a) Gilbert, B.C., Kalz, W., Lindsay, C.I., McGrail, P.T., Parsons, A.F., and Whittaker, D.T.E. (1999) *Tetrahedron Lett.*, **40**, 6095–6098; (b) Gilbert, B.C., Lindsay, C.I., McGrail, P.T., Parsons, A.F., and Whittaker, D.T.E. (1999) *Synth. Commun.*, **29**, 2711–2718; (c) Gilbert, B.C., Kalz, W., Lindsay, C.I., McGrail, P.T., Parsons, A.F., and Whittaker, D.T.E. (2000) *J. Chem. Soc. Perkin Trans. 1*, 1187–1194; (d) Huther, N., McGrail, P.T., and Parsons, A.F. (2002) *Tetrahedron Lett.*, **43**, 2535–2538; (e) Gilbert, B.C., Harrison, R.J., Lindsay, C.I., McGrail, P.T., Parsons, A.F., Southward, R., and Irvine, D.J. (2003) *Macromolecules*, **36**, 9020–9023; (f) Huther, N., McGrail, P.T., and Parsons, A.F. (2004) *Eur. J. Org. Chem.*, 1740–1749.

31 Friestad, G.K., Qin, J., Suh, Y., and Marié, J.-C. (2006) *J. Org. Chem.*, **71**, 7016–7027; Friestad, G.K. and Qin, J. (2001) *J. Am. Chem. Soc.*, **123**, 9922–9923.

32 For other recent examples of radical–polar crossover reactions, see Friestad, G.K. and Wu, Y. (2009) *Org. Lett.*, **11**, 819–822; Maruyama, T., Mizuno, Y., Shimizu, I., Suga, S., and Yoshida, J. (2007) *J. Am. Chem. Soc.*, **129**, 1902–1903; Denes, F., Cutri, S., Perez-Luna, A., and Chemla, F. (2006) *Chem. Eur. J.*, **12**, 6506–6513; Ueda, M., Miyabe, H., Sugino, H., Miyata, O., and Naito, T. (2005) *Angew. Chem., Int. Ed.*, **44**, 6190–6193; Jahn, U. and Rudakov, D. (2004) *Synlett*, 1207–1210; Rivkin, A., de Turiso, F.G.L., Nagashima, T., and Curran, D.P. (2004) *J. Org. Chem.*, **69**, 3719–3725; Crich, D., Ranganathan, K., Neelamkavil, S., and Huang, X.H. (2003) *J. Am. Chem.*

Soc., **125**, 7942–7947; Denes, F., Chemla, F., and Normant, J.F. (2003) *Angew. Chem., Int. Ed.*, **42**, 4043–4046; Bazin, S., Feray, L., Siri, D., Naubron, J.V., and Bertrand, M.P. (2002) *J. Chem. Soc., Chem. Commun.*, 2506–2507; Clark, A.J., Dell, C.P., McDonagh, J.M., Geden, J., and Mawdsley, P. (2003) *Org. Lett.*, **5**, 2063–2066; Harrowven, D.C., Lucas, M.C., and Howes, P.D. (2001) *Tetrahedron*, **57**, 791–804; Bashir, N. and Murphy, J.A. (2000) *J. Chem. Soc., Chem. Commun.*, 627–628.

33 For selected asymmetric syntheses of coniine, see (a) Guerrier, L., Royer, J., Grierson, D.S., and Husson, H.-P. (1983) *J. Am. Chem. Soc.*, **105**, 7754–7755; (b) Enders, D. and Tiebes, J. (1993) *Liebigs Ann. Chem.*, 173–177; (c) Yamazaki, N. and Kibayashi, C. (1997) *Tetrahedron Lett.*, **38**, 4623–4626; (d) Reding, M.T. and Buchwald, S.L. (1998) *J. Org. Chem.*, **63**, 6344–6347; (e) Wilkinson, T.J., Stehle, N.W., and Beak, P. (2000) *Org. Lett.*, **2**, 155–158; (f) Kim, Y.H. and Choi, J.Y. (1996) *Tetrahedron Lett.*, **37**, 5543–5546; (g) for reviews of asymmetric syntheses of piperidine alkaloids, see Laschat, S. and Dickner, T. (2000) *Synthesis*, 1781–1813; O'Hagan, D. (2000) *Nat. Prod. Rep.*, **17**, 435–446.

34 Korapala, C.S., Qin, J., and Friestad, G.K. (2007) *Org. Lett.*, **9**, 4246–4249.

35 Reviews: (a) Ordonez, M. and Cativiela, C., (2007) *Tetrahedron: Asymmetry*, **18**, 3–99; (b) Trabocchi, A., Guarna, F., and Guarna, A. (2005) *Curr. Org. Chem.*, **9**, 1127–1153.

36 Examples: (a) Matthew, S., Schupp, P.J., and Leusch, H., (2008) *J. Nat. Prod.*, **71**, 1113–1116; (b) Kunze, B., Bohlendorf, B., Reichenbach, H., and Hofle, G. (2008) *J. Antibiot.*, **61**, 18–26; (c) Oh, D.-C., Strangman, W.K., Kauffman, C.A., Jensen, P.R., and Fenical, W. (2007) *Org. Lett.*, **9**, 1525–1528; (d) Milanowski, D.J., Gustafson, K.R., Rashid, M.A., Pannell, L.K., McMahon, J.B., and Boyd, M.R. (2004) *J. Org. Chem.*, **69**, 3036–3042; (e) Williams, P.G., Luesch, H., Yoshida, W.Y., Moore, R.E., and Paul, V.J. (2003) *J. Nat. Prod.*, **66**, 595–598; (f) Horgen, F.D., Kazmierski, E.B., Westenburg, H.E., Yoshida, W.Y., and Scheuer, P.J. (2002) *J. Nat. Prod.*, **65**, 487–491.

37 (a) Dado, G.P. and Gellman, S.H. (1994) *J. Am. Chem. Soc.*, **116**, 1054–1062; (b) Hanessian, S., Luo, X., Schaum, R., and Michnick, S. (1998) *J. Am. Chem. Soc.*, **120**, 8569–8570; (c) Seebach, D., Brenner, M., Rueping, M., and Jaun, B. (2002) *Chem. Eur. J.*, **8**, 573–584; (d) Sanjayan, G.J., Stewart, A., Hachisu, S., Gonzalez, R., Watterson, M.P., and Fleet, G.W.J. (2003) *Tetrahedron Lett.*, **44**, 5847–5851; (e) Watterson, M.P., Edwards, A.A., Leach, J.A., Smith, M.D., Ichihara, O., and Fleet, G.W.J. (2003) *Tetrahedron Lett.*, **44**, 5853–5857; (f) Baldauf, C., Gunther, R., and Hofmann, H.-J. (2003) *Helv. Chim. Acta*, **86**, 2573–2588; (g) Seebach, D., Schaeffer, L., Brenner, M., and Hoyer, D. (2003) *Angew. Chem., Int. Ed.*, **42**, 776–778; (h) Farrera-Sinfreu, J., Zaccaro, L., Vidal, D., Salvatella, X., Giralt, E., Pons, M., Albericio, F., and Royo, M. (2004) *J. Am. Chem. Soc.*, **126**, 6048–6057; (i) Vasudev, P.G., Shamala, N., Ananda, K., and Balaram, P. (2005) *Angew. Chem., Int. Ed.*, **44**, 4972–4975; (j) Sharma, G.V.M., Jayaprakash, P., Narsimulu, K., Sankar, A.R., Reddy, K.R., Krishna, P.R., and Kunwar, A.C. (2006) *Angew. Chem., Int. Ed.*, **45**, 2944–2947; (k) Vasudev, P.G., Ananda, K., Chatterjee, S., Aravinda, S., Shamala, N., and Balaram, P. (2007) *J. Am. Chem. Soc.*, **129**, 4039–4048.

38 Sasse, F., Steinmetz, H., Heil, J., Höfle, G., and Reichenbach, H. (2000) *J. Antibiot.*, **53**, 879–885; Höfle, G., Glaser, N., Leibold, T., Karama, U., Sasse, F., and Steinmetz, H. (2003) *Pure Appl. Chem.*, **75** (2–3), 167–178.

39 Friestad, G.K., Deveau, A.M., and Marié, J.-C. (2004) *Org. Lett.*, **6**, 3249–3252.

40 (a) Salazar, J., Lopez, S.E., and Rebollo, O. (2003) *J. Fluorine Chem.*, **124**, 111–113; (b) Iranpoor, N. and Zeynizadeh, B. (1999) *J. Chem. Res., Synop.*, 124–125; (c) Prashad, M., Hu, B., Har, D., Repic, O., and Blacklock, T.J. (2000) *Tetrahedron Lett.*, **41**, 9957–9961.

41 Schaum, R. (1998) Allylation of **15** (LDA, allyl bromide, HMPA, −78 °C) afforded a known derivative with optical rotation matching previously reported data. PhD thesis, Université de Montreal, Montreal, Canada.

42 Torrente, S. and Alonso, R. (2001) *Org. Lett.*, **3**, 1985–1987.

43 Miyabe, H., Yamaoka, Y., and Takemoto, Y. (2005) *J. Org. Chem.*, **70**, 3324–3327.

44 (a) Review: Ramon, D.J. and Yus, M., (2004) *Curr. Org. Chem.*, **8**, 149–183; (b) For examples, see Cogan, D.A. and Ellman, J.A. (1999) *J. Am. Chem. Soc.*, **121**, 268–269; Davis, F.A., Lee, S., Zhang, H., and Fanelli, D.L. (2000) *J. Org. Chem.*, **65**, 8704–8708; Ogawa, C., Sugiura, M., and Kobayashi, S. (2002) *J. Org. Chem.*, **67**, 5359–5364; Berger, R., Duff, K., and Leighton, J.L. (2004) *J. Am. Chem. Soc.*, **126**, 5686–5687; Ding, H. and Friestad, G.K. (2004) *Synthesis*, 2216–2221; Dhudshia, B., Tiburcio, J., and Thadani, A.N. (2005) *Chem. Commun.*, 5551–5553; Lauzon, C. and Charette, A.B. (2006) *Org. Lett.*, **8**, 2743–2745; Wada, R., Shibuguchi, T., Makino, S., Oisaki, K., Kanai, M., and Shibasaki, M. (2006) *J. Am. Chem. Soc.*, **128**, 7687–7691; Canales, E., Hernandez, E., and Soderquist, J.A. (2006) *J. Am. Chem. Soc.*, **128**, 8712–8713.

45 Hynes, J. Jr, Doubleday, W.W., Dyckman, A.J., Godfrey, J.D. Jr, Grosso, J.A., Kiau, S., and Leftheris, K. (2004) *J. Org. Chem.*, **69**, 1368–1371.

46 Obrecht, D., Bohdal, U., Broger, C., Bur, D., Lehmann, C., Ruffieux, R., Schönholzer, P., Spiegler, C., and Müller, K. (1995) *Helv. Chim. Acta*, **78**, 563–580.

47 For related examples of this tactic, see (a) Enders, D. (1984) *Asymmetric Synthesis* (ed. J.D. Morrison), Academic Press, New York, pp. 275–339; (b) Husson, H.-P. and Royer, J. (1999) *Chem. Soc. Rev.*, **28**, 383–394.

48 For examples of efforts toward this goal, see Jang, D.O. and Kim, S.Y. (2008) *J. Am. Chem. Soc.*, **130**, 16152–16153; Halland, N. and Jorgensen, K.A. (2001) *J. Chem. Soc., Perkin Trans. 1*, 1290–1295; Miyabe, H., Ushiro, C., Ueda, M., Yamakawa, K., and Naito, T. (2000) *J. Org. Chem.*, **65**, 176–185.

49 Friestad, G.K., Shen, Y., and Ruggles, E.L. (2003) *Angew. Chem., Int. Ed.*, **42**, 5061–5063.

3
Enantioselective Synthesis of Amines by Chiral Brønsted Acid Catalysts
Masahiro Terada and Norie Momiyama

3.1
Introduction

Over the past decade, enantioselective catalysis by a small organic molecule, the so-called organocatalysis, has become a rapidly growing area of research as it offers operational simplicity, uses mild reaction conditions, and is environmentally benign [1]. In recent years, chiral Brønsted acids have emerged as efficient organocatalysts for enantioselective transformations [2]. Among hitherto reported chiral Brønsted acids, chiral phosphoric acids derived from axially chiral biaryls represent an attractive and widely applicable class of enantioselective organocatalysts for a variety of organic transformations [3]. Today, the activation of imines and other nitrogen-containing substrates by chiral phosphoric acid catalysts is one of the most efficient and powerful methods for preparing optically active amines.

The desirable features of phosphoric acids as chiral Brønsted acid catalysts are summarized as follows (Figure 3.1):

1) Phosphoric acids are expected to capture electrophilic components through hydrogen bonding interactions without forming loose ion pairs due to their relatively strong yet appropriate acidity [4].
2) The phosphoryl oxygen would function as a Brønsted basic site and hence it is anticipated that it would convey acid/base *dual function* even to *monofunctional* phosphoric acid catalysts.
3) An acidic functionality is available even with the introduction of a ring system. It is likely that this ring system effectively restricts the conformational flexibility of the chiral backbone.
4) Substituents (G in Figure 3.1) can be introduced to the ring system to provide a chiral environment for enantioselective transformations.

It is anticipated that an efficient substrate recognition site would be constructed around the activation site of the phosphoric acid catalyst, namely, the acidic proton, as

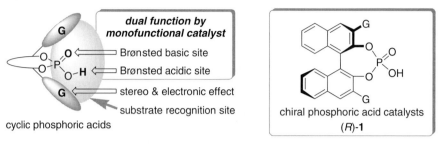

Figure 3.1 Chiral phosphoric acids as chiral Brønsted acid catalysts.

a result of the acid/base dual function and the stereo and/or electronic influence of the substituents (G).

In this chapter, we focus on recent achievements in the enantioselective synthesis of chiral amines using 1,1′-bi-2-naphthol (BINOL)-derived monophosphoric acid (**1**) or related phosphoric acids as chiral Brønsted acid catalysts [2, 3]. The contents are arranged according to the type of bond forming reaction, including carbon–carbon, carbon–hydrogen, and carbon–heteroatom bond forming reactions, followed by specific reaction types.

3.2
Carbon–Carbon Bond Forming Reactions

In 2004, our research group demonstrated a highly enantioselective direct Mannich reaction that used BINOL-derived phosphoric acids (**1**) [5] as chiral Brønsted acid catalysts. In the same year, Akiyama et al. independently reported enantioselective catalysis in the Mukaiyama-type Mannich reaction using similar phosphoric acids [6]. Because of the successful development of these enantioselective Mannich reactions, chiral phosphoric acids have been widely utilized as efficient enantioselective organocatalysts for numerous asymmetric carbon–carbon bond forming reactions. The electrophilic activation of imines by chiral phosphoric acid has been proven to be an attractive and efficient method for preparing a wide range of nitrogen-containing compounds in optically active forms [7].

3.2.1
Mannich and Related Reactions

3.2.1.1 Mannich Reaction
Enantioselective Mannich reactions are widely utilized for the construction of optically active β-amino carbonyl compounds [8] that serve as versatile intermediates for the synthesis of biologically active compounds and drug candidates. Highly enantioselective Mannich reactions have been established using other types of organocatalysts, such as proline and its derivatives, chiral secondary amines [9].

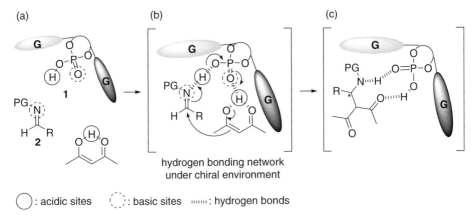

○ : acidic sites   ○ : basic sites   ⋯⋯ : hydrogen bonds

Figure 3.2 Assumed mechanism of enantioselective direct Mannich reaction catalyzed by chiral phosphoric acid.

In 2004, Uraguchi and Terada developed the chiral phosphoric acid (**1**)-catalyzed enantioselective direct Mannich reaction of imine (**2**) with acetylacetone [5]. In the present direct Mannich reaction, it is anticipated that the dual function of the phosphoric acid moiety would smoothly accelerate the reaction (Figure 3.2). The enol proton of the acetylacetone tautomer and the O−H proton of **1** function as acidic sites, while the nitrogen atom of **2** and the phosphoryl oxygen function as basic sites. In the direct Mannich reaction of **2** with acetylacetone, it is considered that **1** would enable the formation of a transient structure through a hydrogen bonding network that connects the acidic and basic sites with each other (Figure 3.2b). Thus, phosphoric acid catalyst **1** electrophilically activates **2** through the acidic proton, and Brønsted basic phosphoryl oxygen interacts with the O−H proton of the enol form of acetylacetone. Subsequent bond recombination results in the formation of the Mannich product and the regeneration of catalyst **1** (Figure 3.2b and c). From this mechanistic assumption, it seems likely that the phosphoric acid catalysts would accelerate the reaction smoothly. More important, it is anticipated that the reaction would proceed under a chiral environment created by the chiral conjugate base of **1**.

Chiral phosphoric acid **1a** exhibited extremely high catalytic activity for the direct Mannich reaction of N-Boc-protected imine with acetylacetone (Scheme 3.1). The resulting β-amino ketone was obtained in an optically active form (12% ee). The beneficial effects of the diaryl substituents at the 3,3'-positions of the binaphthyl backbone are noteworthy in regard to the enantioselectivity. For instance, performing the direct Mannich reaction using 3,3'-phenyl-substituted phosphoric acid **1b** furnished the corresponding product in 56% ee. Interestingly, the simple extension of aromatic substitution to the *para*-position improved the enantioselectivity dramatically. Use of **1d** as a catalyst further increased the enantioselectivity to 95% ee, giving the product in nearly quantitative yield.

The present catalytic reaction was applicable to *ortho*-, *meta*-, and *para*-substituted N-Boc-protected arylimines and the corresponding products were obtained in excellent chemical yields with high enantioselectivities (Scheme 3.2). The reaction

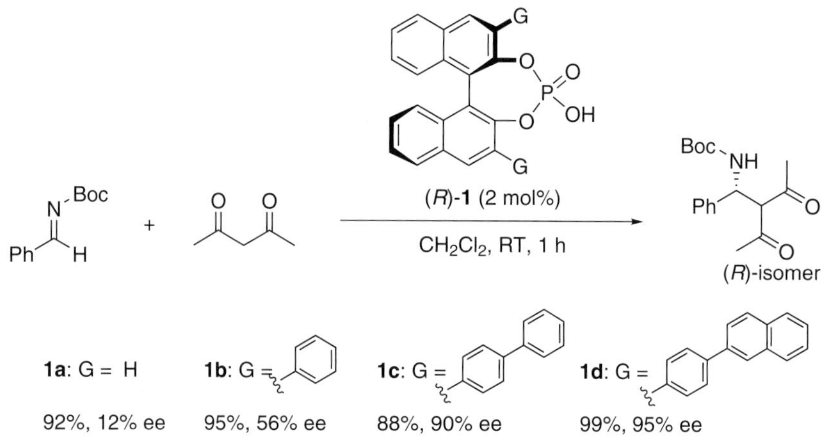

3 Enantioselective Synthesis of Amines by Chiral Brønsted Acid Catalysts

Scheme 3.1 Enantioselective direct Mannich reaction catalyzed by 1.

1a: G = H — 92%, 12% ee
1b: G = phenyl — 95%, 56% ee
1c: G = biphenyl — 88%, 90% ee
1d: G = naphthyl-phenyl — 99%, 95% ee

proceeded smoothly without any detrimental effects even on a gram scale and the catalyst load could be decreased to 1 mol% while maintaining high yield and enantioselectivity. In addition, more than 80% of catalyst **1d** could be recovered.

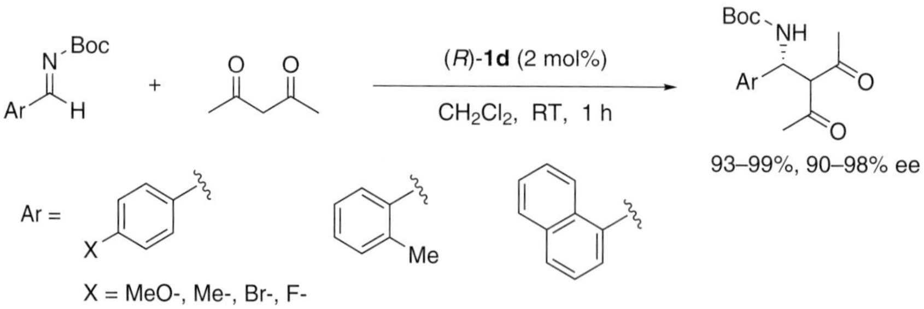

Scheme 3.2 Enantioselective direct Mannich reaction of various imines.

93–99%, 90–98% ee

Ar = 4-X-C$_6$H$_4$ (X = MeO-, Me-, Br-, F-), 2-Me-C$_6$H$_4$, naphthyl

Akiyama et al. independently reported the Mukaiyama-type Mannich reaction of ketene silyl acetals with aldimines derived from 2-hydroxyaniline (Scheme 3.3) [6]. The introduction of 4-nitrophenyl groups at the 3,3′-positions of the catalyst binaphthyl backbone, giving **1e**, yielded the best results in terms of catalytic activity and enantioselectivity. Mannich products were obtained in high *syn*-selectivity and the enantioselectivity of the major *syn*-isomers reached as high as 96% ee. In the present Mannich reaction, an *ortho*-hydroxyl functionality introduced to the N-aryl moiety of the imine is essential to achieving high stereoselectivities. They rationalized the *re*-facial selectivity of the present reaction on the basis of experimental results and density functional calculations [10], in which the phosphoric acid catalyst and the imine form a nine-membered cyclic structure (**A**) through a double hydrogen bonding interaction.

The creation of a new structural motif of chiral phosphoric acids is a challenging task to broaden the scope of chiral Brønsted acid catalysis. Akiyama et al. demon-

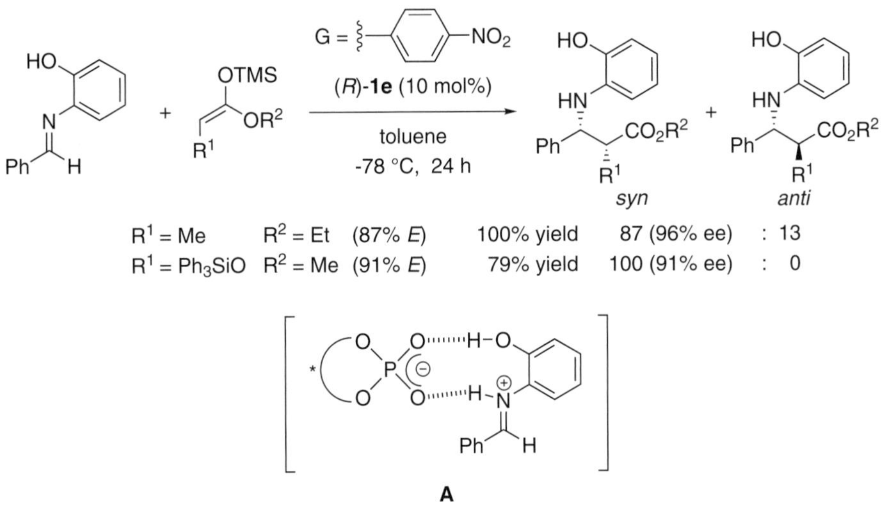

Scheme 3.3 Mukaiyama-type Mannich reaction of imines with ketene silyl acetals.

strated that chiral phosphoric acid **3** derived from TADDOL (*tetraaryl-1,3-dioxolane-4,5-dimethanol*) functioned as an efficient enantioselective catalyst for the Mukaiyama-type Mannich reaction of imines with ketene silyl acetals (Scheme 3.4a) [11]. Terada et al. also developed phosphorodiamidic acid **4** to be used as an efficient Brønsted acid catalyst in the direct Mannich reaction of N-acyl imines with 1,3-dicarbonyl compounds (Scheme 3.4b) [12]. Although the asymmetric induction of the Mannich reaction is still moderate, phosphorodiamidic acid **4** is a viable

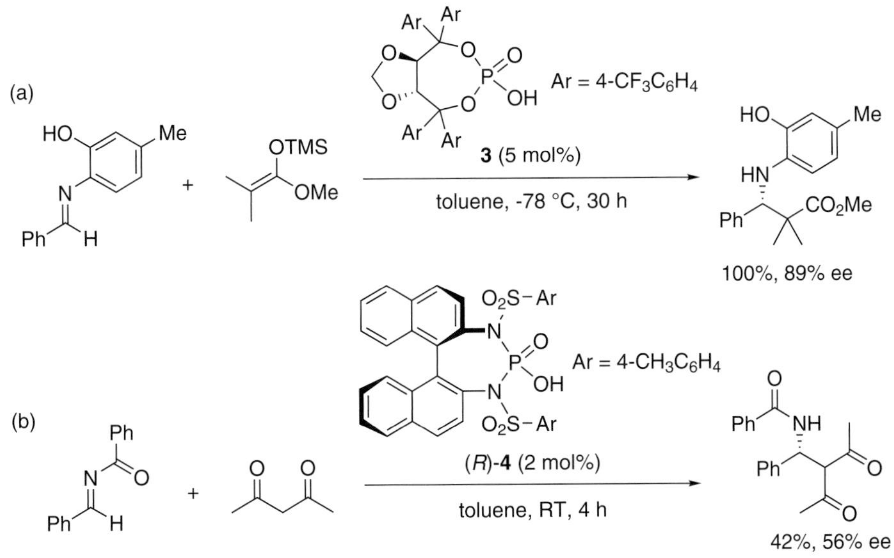

Scheme 3.4 Novel structural motifs of chiral phosphoric acid catalysts.

structural motif of chiral Brønsted acid catalysts. Further modification of the chiral diamine backbone or the substituents on the nitrogen atoms of the catalyst could lead to its becoming an efficient enantioselective catalyst.

Gong and coworkers applied chiral phosphoric acid catalysts to the three-component direct Mannich reaction, realizing a one-pot reaction among aromatic aldehyde, aniline derivatives, and ketones (Scheme 3.5) [13]. The reaction of cyclohexanone as a nucleophilic component proceeded smoothly in the presence of 0.5 mol% **1f** or 2 mol% **5a**, giving the desired product in good yield with high enantio- and *anti*-selectivities. They also extended the coupling method to the reaction of acyclic ketones, such as acetone and acetophenone, which gave products in good yields when the catalyst load was increased to 5 mol% and excess amounts of these ketones (10 equiv) were used, albeit slight decreases in enantioselectivities. Shortly thereafter, Rueping *et al.* reported achiral Brønsted acid-assisted chiral Brønsted acid catalysis in the direct Mannich reaction of acyclic ketones [14]. The reaction of *N*-aryl imines with acetophenone was conducted using a chiral phosphoric acid catalyst in combination with acetic acid as cocatalyst. The corresponding products were obtained in acceptable yields even when the amount of acetophenone employed was decreased (2 equiv).

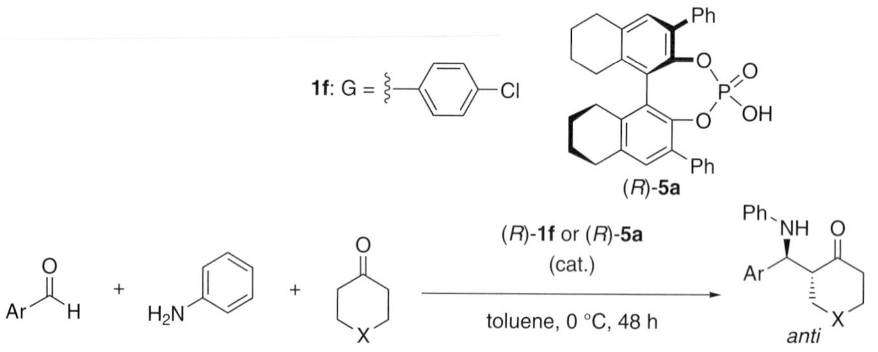

X = CH_2, O, NBoc (*R*)-**1f** (0.5 mol%) 83–99%, 77–92% *anti*, 80–95% ee
X = S (*R*)-**5a** (2 mol%) 74–97% , 89–98% *anti*, 83–98% ee

Scheme 3.5 Three-component direct Mannich reaction.

Gong and coworkers reported the enantioselective Biginelli reaction via three-component coupling (Scheme 3.6) [15]. Utilizing the Mannich reaction as the initial step, the method enabled efficient access to 3,4-dihydropyrimidine-2(1*H*)-one derivatives [16] that possess a wide array of pharmaceutical properties and hence are considered to be medicinally relevant compounds. Catalyst **5a** having the H_8-binaphthyl backbone most effectively furnished the corresponding pyrimidinone derivatives in high yields with high enantioselectivities. A wide variety of substrates, including aldehydes and β-ketoesters, could be employed in the present reaction.

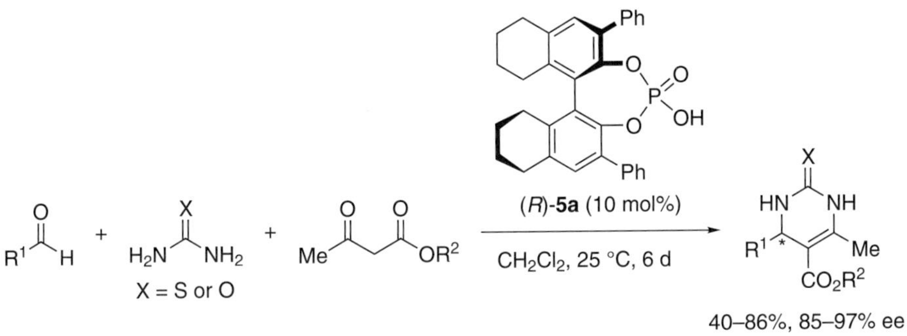

Scheme 3.6 Enantioselective Biginelli reaction via three-component coupling.

3.2.1.2 Nucleophilic Addition of Diazoacetates to Aldimine

α-Diazocarbonyl compounds have been extensively studied and their most important application is for the generation of metal carbene species. α-Diazocarbonyl compounds have an electronically unique sp^2-carbon to which the diazo group is attached. Thus, these compounds have a partially negative charge and function as nucleophiles. In the reactions of imines, α-diazocarbonyl compounds are commonly used in aziridine formation reactions (aza-Darzens reaction) under Lewis [17] and Brønsted [18] acidic conditions. Meanwhile, Terada and coworkers reported enantioselective direct substitution at the α-position of diazoacetate using a chiral phosphoric acid catalyst [19]. Phosphoric acid catalyst **1g** efficiently promoted the substitution reaction of α-diazoacetates with *N*-acyl imines to give Mannich-type products, β-amino esters having a diazo substituent at the α-position, in optically active forms (Scheme 3.7). Interestingly, the electronic properties of the acyl protective group of the imine nitrogen profoundly affected the enantioselectivity. The *para*-substituents of the *N*-acyl aromatic moiety had a marked effect on the enantioselectivity; the introduction of an electron-donating dimethylamino moiety gave the best results.

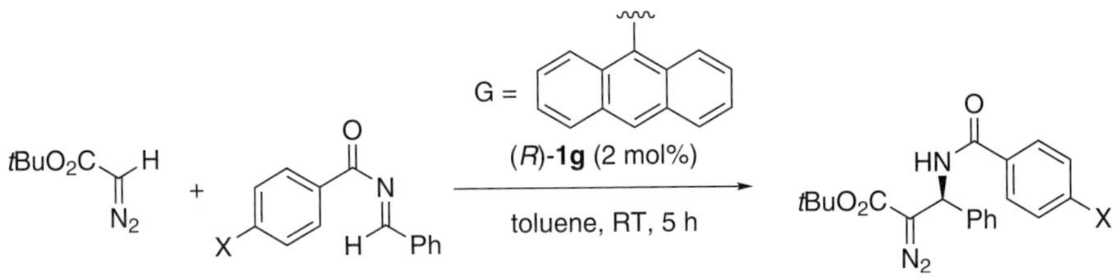

X =	Br	H	Me	MeO	Me2N
	68%	59%	72%	73%	81%
	86% ee	90% ee	91% ee	93% ee	97% ee (for 24 h)

Scheme 3.7 Enantioselective α-substitution reaction of diazoacetate with imine.

A series of aromatic imines are applicable to the present substitution reactions (Scheme 3.8). *para*-Substituted aromatics showed excellent enantioselectivities irrespective of their electronic properties. *Ortho*- and *meta*-substitutions as well as fused-ring systems were also applicable. Thus obtained β-amino-α-diazoester products could be transformed into common synthetic intermediates, that is, β-amino acid derivatives, via simple reduction or oxidation of the diazo moiety.

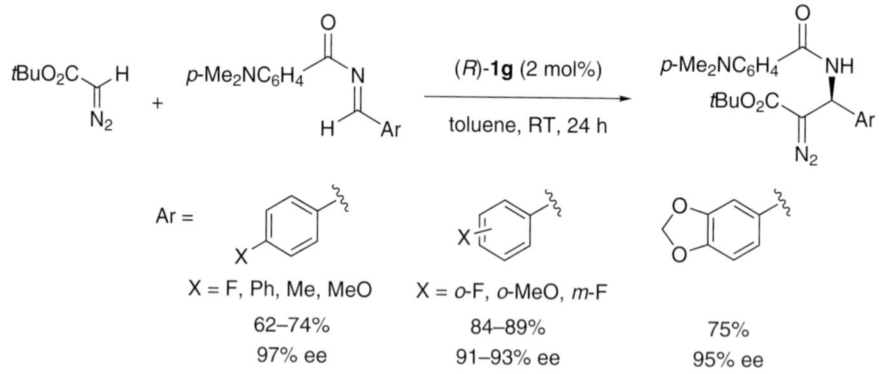

Scheme 3.8 α-Substitution reaction of diazoacetate with various imines.

The phosphoric acid is expected to promote the substitution reaction as a result of its dual function (Figure 3.3a). Intracomplex deprotonation from **B** by the basic phosphoryl oxygen would allow the direct substitution of diazoacetate, giving a Mannich-type product without the formation of aziridine products.

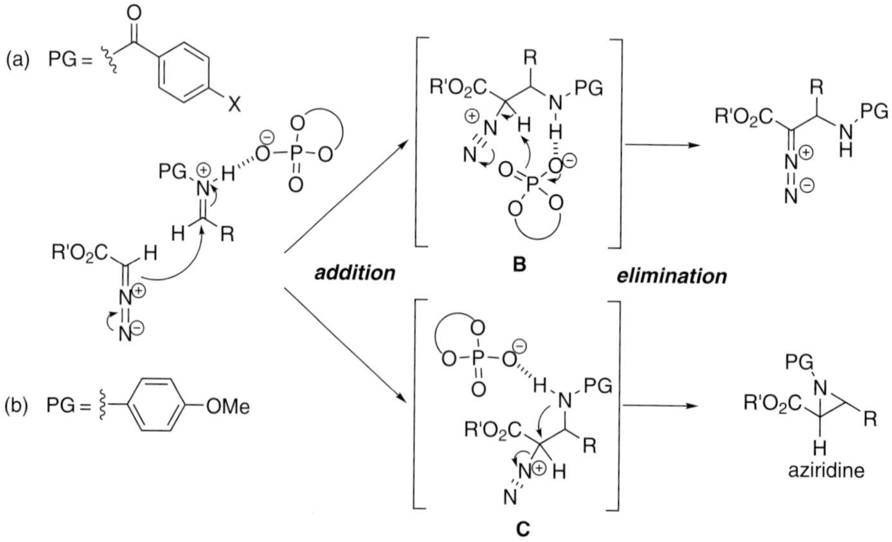

Figure 3.3 Mechanistic proposal for diazoacetate reactions.

Meanwhile, Akiyama et al. reported the aziridine formation reaction (aza-Darzens reaction) using PMP (p-methoxyphenyl)-protected imines and α-diazoacetate (Scheme 3.9) [20]. Chiral phosphoric acid catalyst **1h** gave the corresponding aziridine products with high enantioselectivities. The PMP protective group would preserve the nucleophilicity of the nitrogen atom and hence the intramolecular substitution by the nitrogen atom would proceed exclusively via intermediate **C** (Figure 3.3b). This contrasts the reaction of N-benzoyl imines where Mannich-type products were obtained in high yields. The electron-withdrawing property of the N-benzoyl protective group significantly decreased the electron density of the nitrogen atom, effectively suppressing nucleophilic substitution by the nitrogen atom.

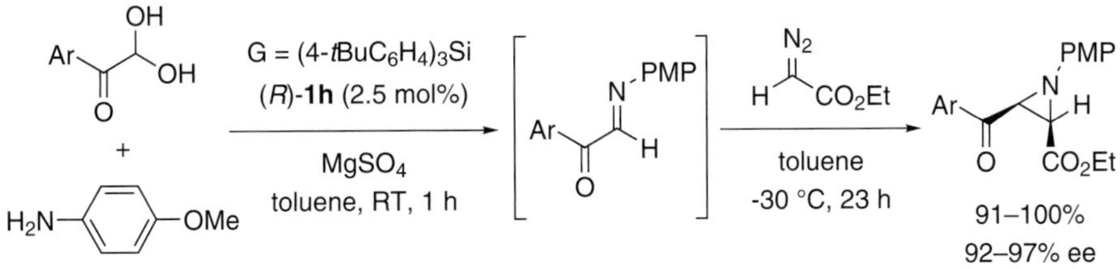

Scheme 3.9 Enantioselective aza-Darzens reaction.

3.2.1.3 Vinylogous Mannich Reaction

The asymmetric Mannich reaction is one of the most ubiquitous carbon–carbon bond forming reactions in organic chemistry. The vinylogous extension of this fundamental carbon–carbon bond forming reaction to nucleophilic components, namely, the vinylogous Mannich reaction, has been little exploited [21], despite its potential to provide efficient access to highly functionalized δ-amino carbonyl compounds bearing a double bond. 2-Siloxyfuran is an attractive vinylogous nucleophile that has been extensively utilized in vinylogous variants of fundamental transformations, such as the aldol reaction [22], because the reactions provide γ-substituted butenolides, an important structural motif existing in naturally occurring products and biologically active compounds. Akiyama et al. successfully developed the enantioselective vinylogous Mannich reaction of 2-trimethylsiloxyfuran with aldimines using chiral phosphoric acid catalyst **1i** bearing iodine groups at 6,6′-positions of the binaphthyl backbone (Scheme 3.10) [23]. A series of aldimines, including aliphatic ones, can be utilized in the present reaction and γ-butenolide products were obtained in good yields with moderate to high diastereo- and enantioselectivities.

Sickert and Schneider demonstrated the first catalytic and enantioselective vinylogous Mannich reaction of an acyclic silyl dienol ether with aldimines using chiral phosphoric acid catalyst **1j** (Scheme 3.11) [24]. The reaction provided highly valuable δ-amino-α,β-unsaturated carboxylic esters in optically active forms with complete regioselectivities. Schneider and coworkers also reported the enantioselective vinylogous Mannich reaction of vinylketene silyl N,O-acetals with aromatic imines [25]. The piperidine-derived vinylketene silyl acetals proved to be more suitable substrates, furnishing the corresponding vinylogous products, δ-amino-α,β-unsaturated amides, in good yields with high enantioselectivities.

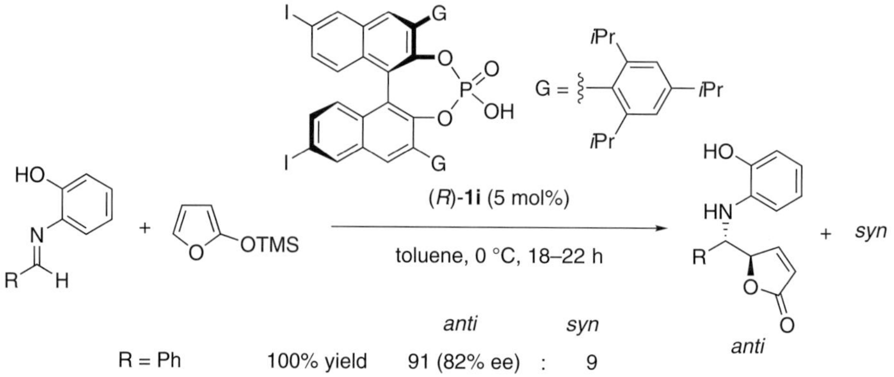

Scheme 3.10 Vinylogous Mannich reaction of 2-siloxyfuran.

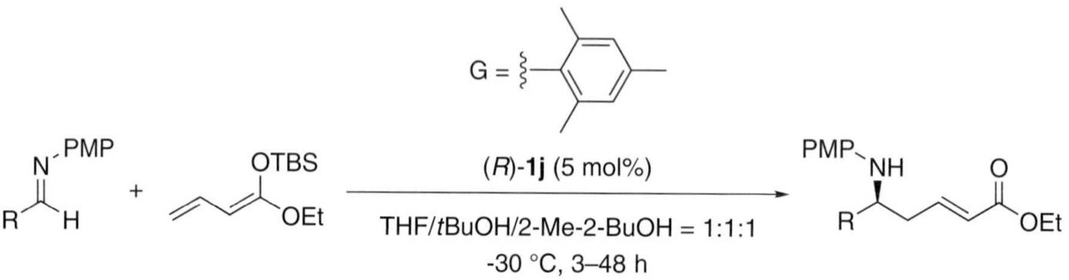

		anti	syn
R = Ph	100% yield	91 (82% ee) :	9
R = p-CF$_3$C$_6$H$_4$	95% yield	69 (99% ee) :	31
R = p-O$_2$NC$_6$H$_4$	85% yield	97 (96% ee) :	3
R = iPr	84% yield	88 (92% ee) :	12

Scheme 3.11 Vinylogous Mannich reaction of silyl dienol ether.

R = Ph   87%, 88% ee
R = p-C$_2$H$_5$C$_6$H$_4$ 88%, 92% ee
R = tBu   83%, 82% ee

3.2.1.4 Aza-Petasis–Ferrier Rearrangement

Organocatalysis in the direct Mannich reactions of aldehydes with aldimines using chiral secondary amine catalysts has emerged as a powerful tool to provide β-amino aldehydes with high diastereo- and enantioselectivities [26]. However, one critical drawback inherent to the methodologies reported to date is that aromatic or glyoxylate-derived aldimines are employed as adaptive substrates in most cases [26c]. The enantioselective direct Mannich reaction of aliphatic aldimines has largely been unexploited. Terada and Toda developed an alternative strategy to furnish optically active β-amino aldehydes having an aliphatic substituent (R) at the β-position by combining two catalytic reactions (Scheme 3.12) [27]. The sequence involves initial metal-catalyzed (Z)-selective isomerization of a double bond, followed by chiral phosphoric acid-catalyzed aza-Petasis–Ferrier rearrangement [28], using readily available hemiaminal allyl ethers as the substrate. The aza-Petasis–Ferrier rearrangement of hemiaminal vinyl ethers proceeded via C−O bond cleavage of the ether moiety by acid catalyst **1k**, generating a reactive iminium intermediate and an enol form of the aldehyde. Subsequent recombination with C–C bond formation resulted in rearranged products, thus providing β-amino aldehydes having not only aliphatic but also aromatic substituents at the β-position with high *anti-* and enantioselectivities.

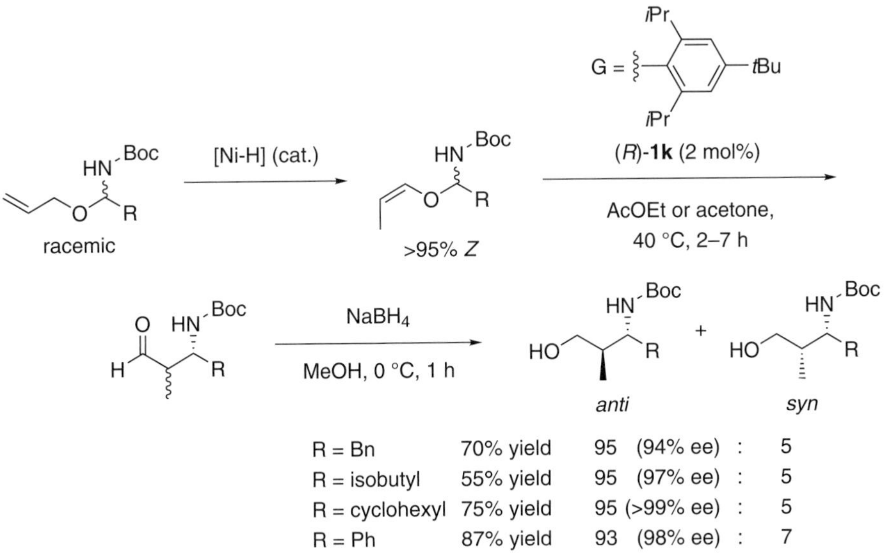

Scheme 3.12 Double bond isomerization/aza-Petasis–Ferrier rearrangement sequence for preparing β-amino aldehydes.

3.2.2
One-Carbon Homologation Reactions

3.2.2.1 Strecker Reaction

The hydrocyanation of imines, namely, the Strecker reaction, offers a practical route to α-amino acids or 1,2-diamines, which can be transformed from the corresponding α-amino nitrile products through simple hydrolysis or reduction of the nitrile moiety. The development of the catalytic enantioselective Strecker reaction is hence a vital step toward the efficient synthesis of these amines in optically active forms. A number of methods for the enantioselective Strecker reaction have been intensively investigated using either chiral metal catalysts or organocatalysts [29]. Chiral phosphoric acid catalysts were also successfully applied to the enantioselective Strecker reaction (Scheme 3.13). Rueping *et al.* were able to attain excellent performance using chiral catalyst **1l**, which has sterically demanding 9-phenanthryl substituents at the 3,3′-positions of the binaphthyl backbone [30]. The method offers efficient access to a broad range of aromatic amino nitriles with high enantioselectivities. They also developed the enantioselective Strecker reaction of ketimines catalyzed by the same chiral phosphoric acid **1l** to give products having a chiral quaternary stereogenic center with moderate to high enantioselectivities [31, 32].

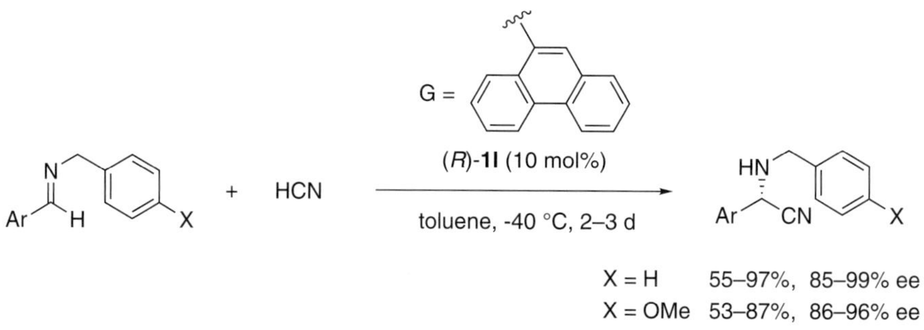

Scheme 3.13 Enantioselective Strecker reaction.

X = H   55–97%, 85–99% ee
X = OMe 53–87%, 86–96% ee

3.2.2.2 Aza-Henry Reaction

The synthesis of β-nitro amines via the addition of nitroalkanes to imines, or the so-called aza-Henry reaction, is an attractive tool to create carbon–carbon bonds. The products obtained can be readily converted into vicinal diamines and α-amino acids by simple reduction and the Nef reaction, respectively, highlighting the several important synthetic applications of these compounds. In this context, enantioselective versions of the aza-Henry reaction have been actively investigated using chiral metal complexes and organocatalysts [33]. Rueping and Antonchick demonstrated the enantioselective direct aza-Henry reaction of α-imino esters with nitroalkanes using chiral phosphoric acid catalysts (Scheme 3.14) [34]. Catalyst **5b** having sterically demanding triphenylsilyl substituents significantly accelerated the reaction to provide β-nitro-α-amino acid esters in good yields. A variety of nitroalkanes were applicable to the present reaction, giving the desired products with high diastereo- and enantioselectivities.

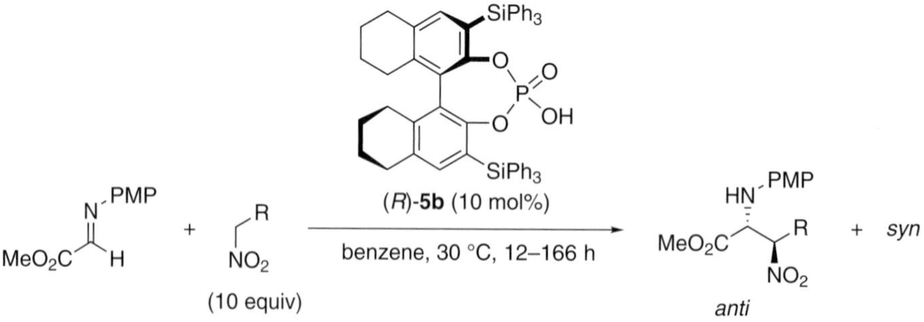

R = Me   61% yield   91 (92% ee) : 9
R = CH$_3$(CH$_2$)$_4$ 65% yield   90 (92% ee) : 10
R = Bn   93% yield   93 (88% ee) : 7

Scheme 3.14 Enantioselective aza-Henry reaction of α-imino ester.

3.2.2.3 Imino-Azaenamine Reaction

Formaldehyde dialkylhydrazones, such as azaenamines (a nitrogen analogue of enamines), possess a carbon atom that shows nucleophilic character and are utilized as formyl anion equivalents [35]. The reaction enables efficient one-carbon homologation of electrophilic substrates. Rueping *et al.* utilized this nucleophilic species in a reaction of imines under the influence of chiral phosphoric acid **5c** (Scheme 3.15) [36]. The reaction of pyrrolidine-derived methylenehydrazine with a series of N-Boc aromatic imines provided α-amino hydrazones with moderate to high enantioselectivities. The corresponding products were proven to be useful synthetic intermediates that could be readily transformed into a diverse array of chiral nitrogen-containing compounds, such as α-amino aldehydes, α-amino nitriles, and 1,2-diamines, without racemization.

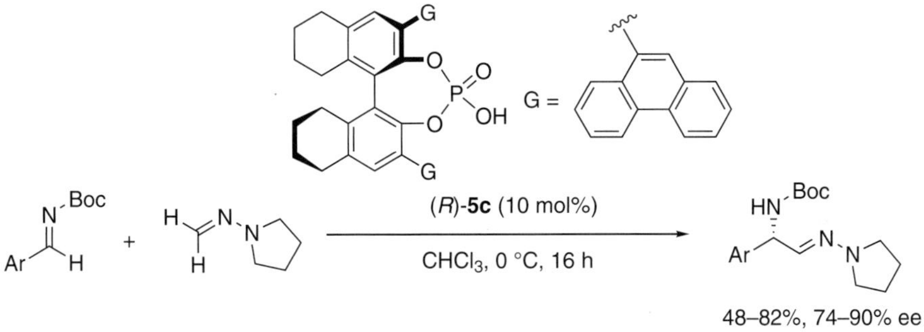

Scheme 3.15 Enantioselective imino-azaenamine reaction of hydrazone.

3.2.3
Friedel–Crafts and Related Reactions

The Friedel–Crafts (F–C) reaction via activation of electrophiles functionalized by a nitrogen atom, such as imines, is undoubtedly the most practical and atom-economical approach to introduce a nitrogen-substituted side chain to aromatic compounds. The enantioselective version of the F–C reaction of nitrogen-substituted substrates, including imines, with electron-rich aromatic compounds enables efficient access to enantioenriched aryl methanamine derivatives [37]. Several excellent approaches to highly enantioselective F–C reactions have been established using chiral phosphoric acid catalysts.

3.2.3.1 Friedel–Crafts Reaction via Activation of Aldimines

Terada and coworkers successfully demonstrated for the first time an enantioselective 1,2-aza-F–C reaction of 2-methoxyfuran with N-Boc aldimines using a catalytic amount of chiral phosphoric acid (Scheme 3.16) [38]. In the presence of 2 mol% **1m** having sterically hindered 3,5-dimesitylphenyl substituents, the corresponding F–C products were obtained in excellent enantioselectivities irrespective of the electronic properties of the aromatic imines employed. Most notable was that the reaction could be performed in the presence of as little as 0.5 mol% **1m** without any detrimental effects even on a gram scale (Ar = Ph: 95%, 97% ee).

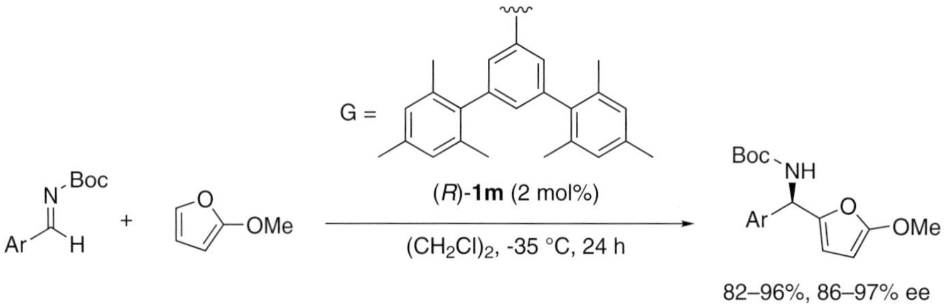

Scheme 3.16 Enantioselective 1,2-aza-F–C reaction of N-Boc imines with 2-methoxyfuran.

The synthetic utility of the present transformation is highlighted by the derivatization of the furyl ring to form γ-butenolide (Scheme 3.17). As the γ-butenolide architecture is a common building block in the synthesis of various natural products, the F–C reaction product represents a new entry to the synthetic precursors of nitrogen-containing molecules. The aza-Achmatowicz reaction, followed by reductive cyclization of the 1,4-dicarbonyl intermediate under Luche conditions, produced the γ-butenolide in good yield.

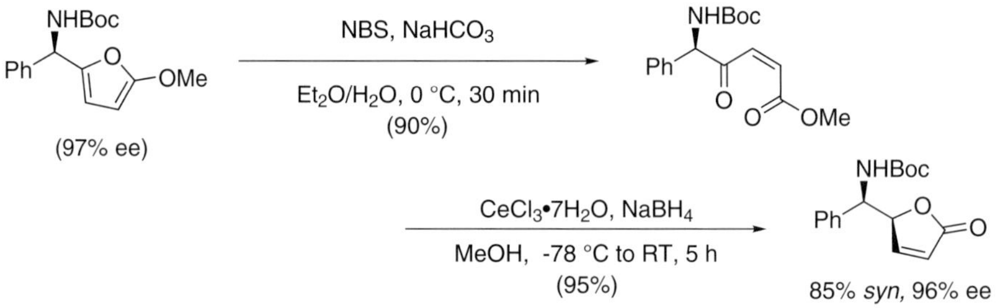

Scheme 3.17 Synthetic utility of furan-2-yl amine products.

Further application of the chiral phosphoric acid-catalyzed 1,2-aza-F–C reaction was investigated by several research groups and the developed methods yielded a diverse array of optically active arylmethaneamine derivatives with high enantioselectivities. In particular, the 1,2-aza-F–C reaction of indoles was intensively investigated because the enantioenriched products, namely, 3-indolyl methanamine derivatives, are widely known as valuable structures among pharmacophores and are present in thousands of natural products and many medicinal agents possessing versatile therapeutic effects [39]. You and coworkers (Scheme 18a) [40], Antilla and coworkers (Scheme 18b) [41], and our group (Scheme 18c) [42] independently developed the highly enantioselective F–C reaction of indoles with aromatic imines. The reaction of

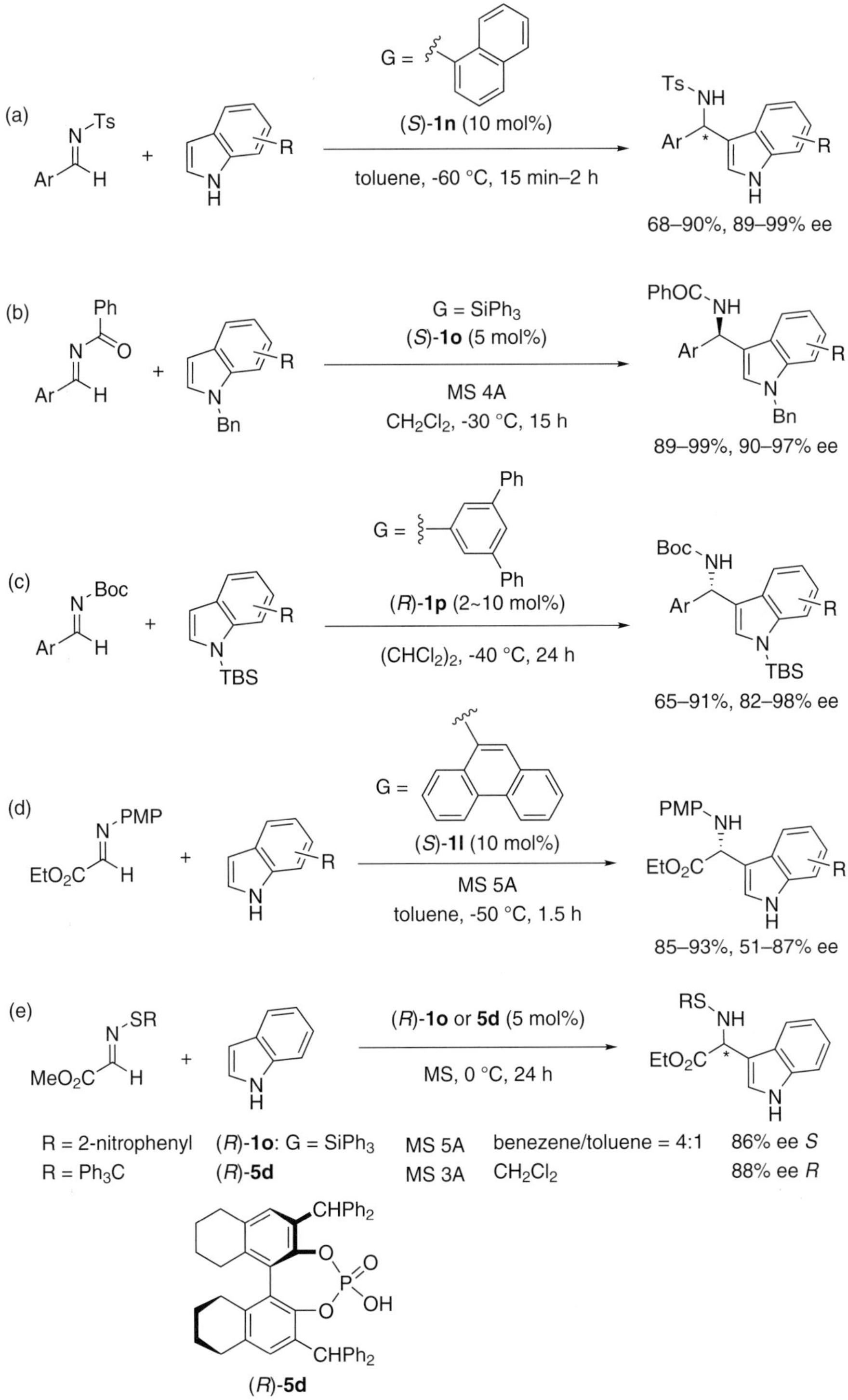

Scheme 3.18 Enantioselective 1,2-aza-F–C reaction of indoles with various imines.

α-imino esters as an electrophilic component was also reported by You and coworkers (Scheme 18d) [43] and Hiemstra and coworkers (Scheme 18e) [44]. In Hiemstra's approach, alkylation products of indole were obtained with opposite absolute configurations depending on the sulfur substituent even when chiral phosphoric acids having axial chirality in the same (R)-configuration were employed.

The enantioselective F–C reactions catalyzed by chiral phosphoric acids were further applied to such electron-rich aromatic compounds as pyrroles (Scheme 3.19a and b) [45, 46] and 4,7-dihydroindoles (Scheme 3.19c) [47]. The reaction of 4,7-dihydroindoles yielded 2-substituted indole derivatives following oxidation of the F–C products. The present approach well complements current studies of enantioselective F–C reactions of parent indoles (see Scheme 3.18), where 3-substituted indole derivatives were obtained in most cases.

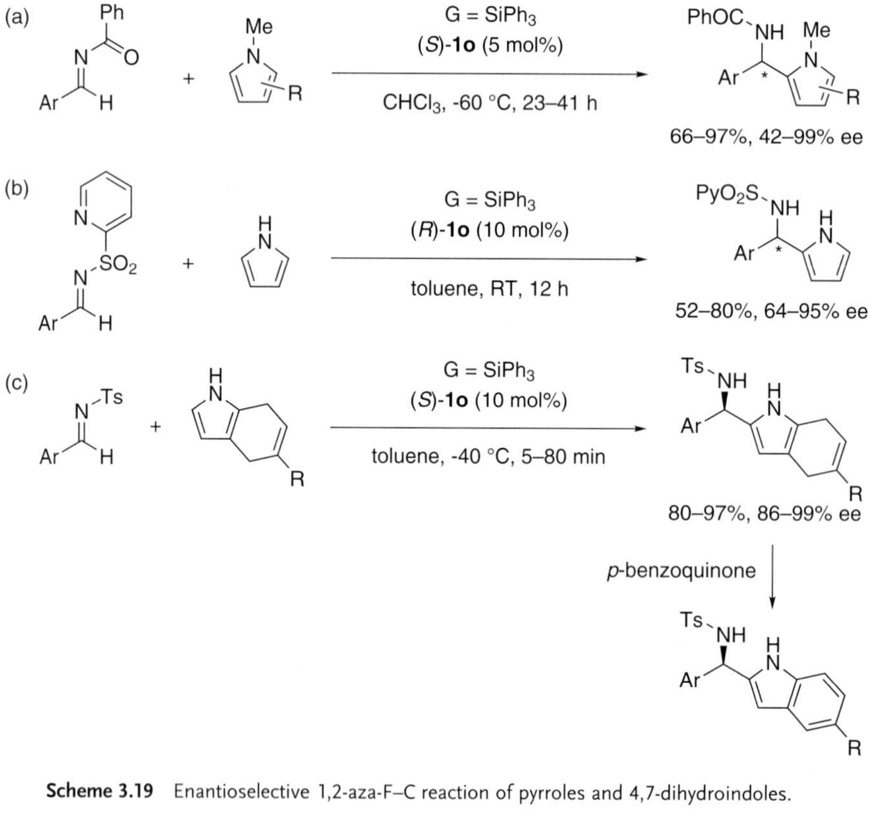

Scheme 3.19 Enantioselective 1,2-aza-F–C reaction of pyrroles and 4,7-dihydroindoles.

Enders et al. demonstrated the one-pot synthesis of enantioenriched isoindolines based on a consecutive transformation involving a chiral Brønsted acid-catalyzed F–C reaction followed by a base-catalyzed aza-Michael addition reaction using indoles and bifunctional ε-iminoenoates as substrates (Scheme 3.20) [48]. In the present one-pot

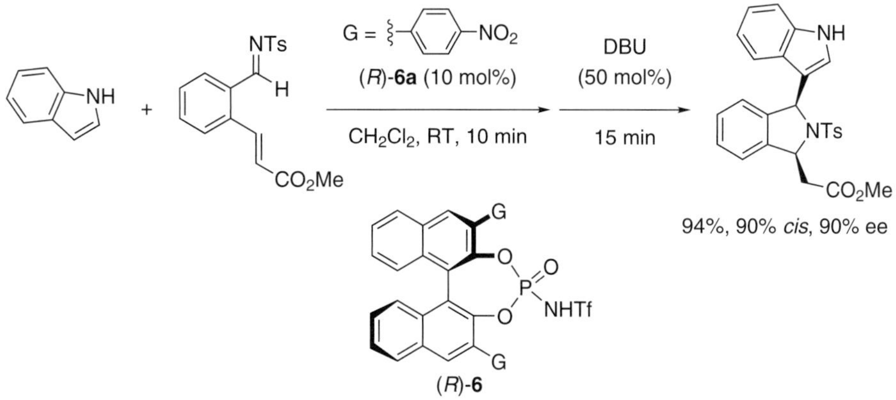

Scheme 3.20 One-pot synthesis of enantioenriched isoindolines based on chiral Brønsted acid catalyzed F–C reaction/base-catalyzed aza-Michael addition reaction sequence.

transformation, the initial F–C reaction catalyzed by chiral BINOL-derived N-triflyl phosphoramide (**6**), which was originally developed by Nakashima and Yamamoto [49], followed by the DBU-promoted intramolecular aza-Michael addition, furnished isoindoline derivatives in good yields with high diastereo- and enantioselectivities.

3.2.3.2 Friedel–Crafts Reaction via Activation of Electron-Rich Alkenes

Enantioselective F–C reactions have been intensively investigated using metal-based chiral catalysts or chiral organocatalysts [37]. These enantioselective catalyses have been accomplished via the activation of electron-deficient multiple bonds, such as C=O, C=NR, and C=C–X (X: electron-withdrawing group). The acid-catalyzed F–C reactions of arenes with electron-rich alkenes are practical and atom-economical methods for the production of alkylated arenes and have been applied to numerous industrial processes. However, there are no previous reports of the enantioselective catalysis of the F–C reaction initiated by the activation of electron-rich multiple bonds, even using chiral metal catalysts. Recently, Terada and Sorimachi successfully developed a highly enantioselective F–C reaction initiated by the activation of electron-rich multiple bonds using a chiral Brønsted acid catalyst. Chiral phosphoric acid **1q** exhibited excellent performance for the activation, utilizing the catalytic reaction of indoles with enecarbamates as electron-rich alkenes (Scheme 3.21) [50]. Uniformly high enantioselectivities and chemical yields were achieved in the reaction of indole with enecarbamates bearing either a linear or branched alkyl group as well as an aromatic substituent. In addition, the enantioselectivities were maintained at an equally high level for a wide variety of indole derivatives, irrespective of their electronic properties. The present approach provides efficient access to enantioenriched 3-indolyl methanamines with a variety of aliphatic substituents and effectively complements previous methods that afforded aromatic group substituted 3-indolyl methanamines via the activation of aromatic imines (see Scheme 3.18).

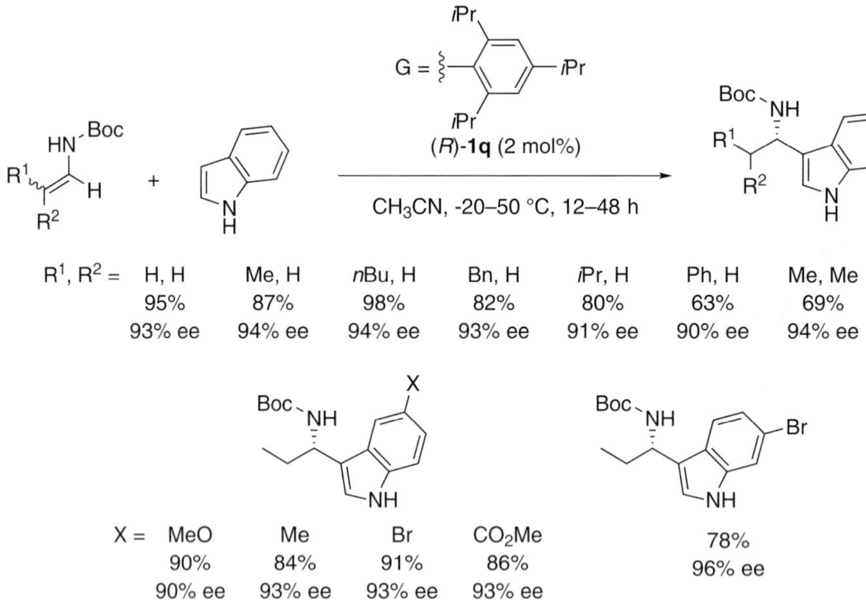

Scheme 3.21 F–C reaction via activation of enecarbamates by phosphoric acid catalyst.

The present F–C reaction proceeded through the *in situ* generation of aliphatic imines that were delivered via the protonation of the enecarbamates by the phosphoric acid catalyst (Figure 3.4). Phosphoric acid functioned as an efficient catalyst for the dual transformation that involved the *in situ* generation of imine and the enantioselective carbon–carbon bond formation with indole. This protocol offers the distinct advantage of generating *in situ* unstable aliphatic imines from storable and thus easily handled enecarbamates, and hence is applicable to other organic transformations. In fact, Terada et al. applied the present method to an enantioselective direct Mannich reaction [51]. The method provides an efficient pathway to β-alkyl-β-aminocarbonyl derivatives in optically active forms.

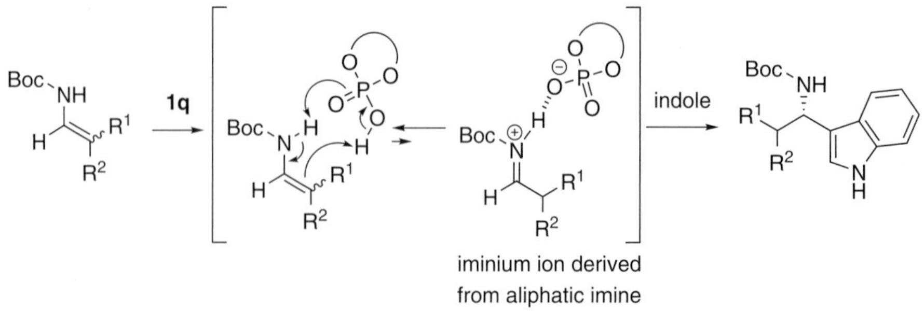

Figure 3.4 Mechanistic considerations of the F–C reaction of enecarbamates with indole.

Shortly thereafter, Zhou and coworkers independently reported the enantioselective F–C reaction of indoles with α-aryl-substituted enamides catalyzed by chiral phosphoric acid catalyst **1q** (Scheme 3.22) [52], in which the quaternary stereogenic center bearing the nitrogen atom was constructed in a highly enantioselective manner.

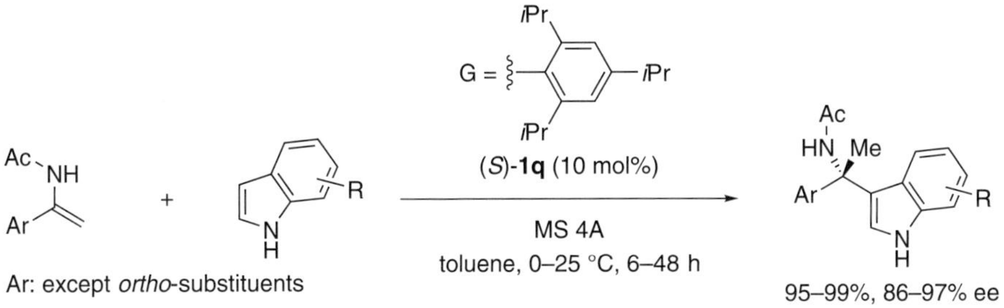

Scheme 3.22 Enantioselective formation of quaternary stereogenic center bearing a nitrogen atom in the F–C reaction.

3.2.3.3 Pictet–Spengler Reaction

The Pictet–Spengler reaction of tryptamine or its substituted analogue with an aldehyde is a powerful and efficient method for preparing tetrahydro-β-carbolines or tetrahydroisoquinolines [53], a structural motif of many alkaloids and related naturally occurring compounds. Enantioselective variants of the Pictet–Spengler reaction are in high demand because of the versatile synthetic applicability of the corresponding products to biologically active molecules. The reaction consisted of a two-step transformation: an initial dehydrative imine formation, followed by an intramolecular F–C reaction. An excess amount of Brønsted acid is usually required to promote the reaction. List and coworkers accomplished the enantioselective Pictet–Spengler reaction by directly using substituted tryptamines and aldehydes in the presence of chiral phosphoric acid catalyst **1q** (Scheme 3.23a) [54]. The method provided efficient access to tetrahydro-β-carbolines with high enantioselectivities, although the introduction of a *gem*-disubstituent adjacent to the reactive imine was required to suppress competing aldol pathways. Hiemstra and coworkers circumvented this intrinsic drawback, the requirement of a *gem*-disubstituent, by taking an iminium ion strategy. They used N-sulfenyliminium ions as intermediates in the Pictet–Spengler reaction and established an efficient method for preparing tetrahydro-β-carbolines in optically active forms using chiral phosphoric acid **1r** (Scheme 3.23b) [55]. In the present method, not only aliphatic but also aromatic substituents can be introduced to the stereogenic center. Recently, Hiemstra and coworkers accomplished the asymmetric total synthesis of the tetracyclic indole alkaloid (−)-arboricine using the enantioselective Pictet–Spengler reaction as a key step (Scheme 3.23c) [56].

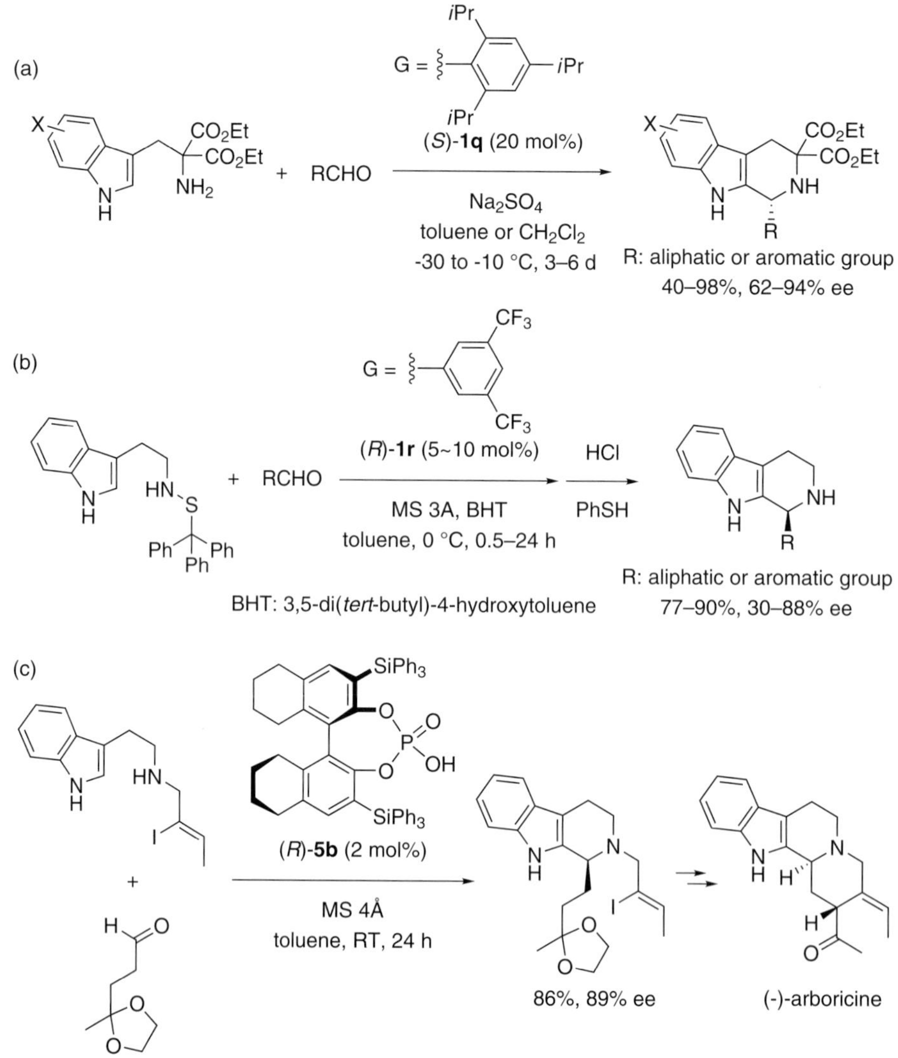

Scheme 3.23 Enantioselective Pictet–Spengler reaction of tryptamine derivatives with aldehydes.

3.2.4
Cycloaddition Reactions

3.2.4.1 Hetero-Diels–Alder Reaction of Aldimines with Siloxydienes

The enantioselective hetero-Diels–Alder (D–A) reaction of siloxydienes, such as Danishefsky's dienes and Brassard's dienes, with imines provides an efficient route for the preparation of functionalized nitrogen heterocycles in optically active forms, which can be utilized as synthetic precursors of biologically interesting alkaloid families and aza sugars. Akiyama et al. developed the enantioselective hetero-D–A reaction of Danishefsky's dienes with aromatic imines using a chiral phosphoric acid catalyst (Scheme 3.24a) [57]. The reaction using catalyst **1q** provided dihydropyridin-4 (1*H*)-one derivatives in good yields with high enantioselectivities in the presence of acetic acid as a stoichiometric additive. Akiyama and coworkers also reported an enantioselective hetero-D–A reaction of Brassard's diene with imines (Scheme 3.24b) [58]. One of the intrinsic problems associated with Brassard's diene lies in its instability under acidic conditions. In order to suppress decomposition of the diene by the phosphoric acid catalysts, they employed pyridinium salts of the catalysts. The pyridinium salts of **1g** displayed excellent performance for the hetero-D–A reaction of Brassard's diene with a broad range of imines, including aromatic and aliphatic ones. Uniformly high enantioselectivities and chemical yields of dihydropyridin-2(1*H*)-one products were noted after treatment with an acid.

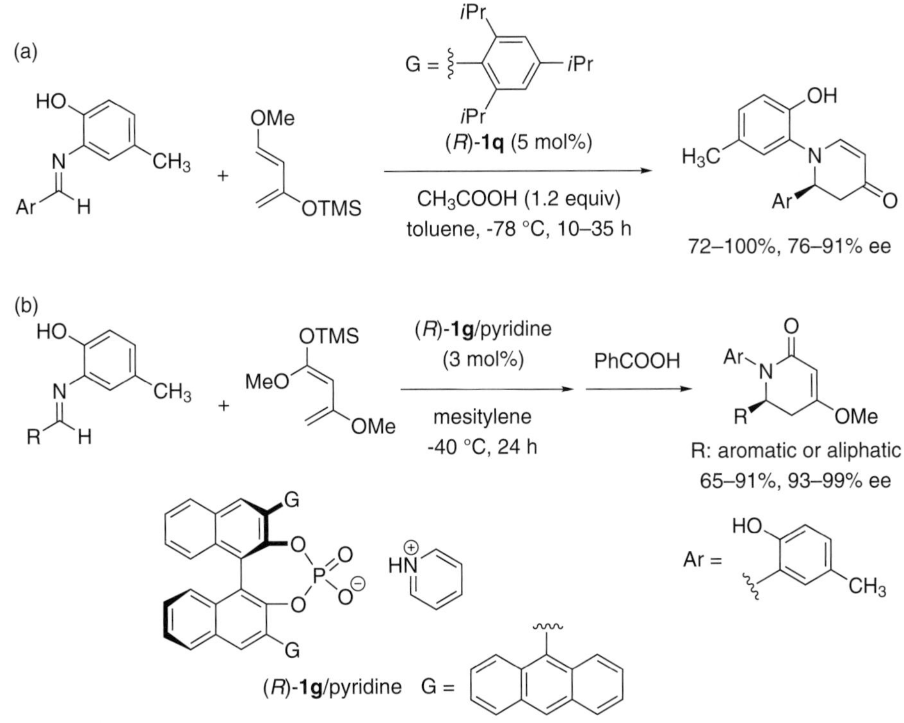

Scheme 3.24 Enantioenriched nitrogen heterocycles by hetero-D–A reaction of siloxydienes.

3.2.4.2 Direct Cycloaddition Reaction of Aldimines with Cyclohexenone

Gong and coworkers [59] and Rueping and Azap [60] independently reported the enantioselective direct cycloaddition reaction of cyclohexenone with aromatic imines. Rueping and Azap employed acetic acid as the cocatalyst to accelerate the tautomerization of cyclohexenone, by which a reactive dienol for the enantioselective cycloaddition was generated *in situ* (Scheme 3.25). They proposed a stepwise mechanism for the present reaction. The initial step is the chiral phosphoric acid-catalyzed Mannich reaction of imines with dienol generated in the presence of acetic acid, giving intermediate **D**. The subsequent intramolecular aza-Michael reaction provides the corresponding cycloaddition products, isoquinuclidine derivatives, in moderate to good yields.

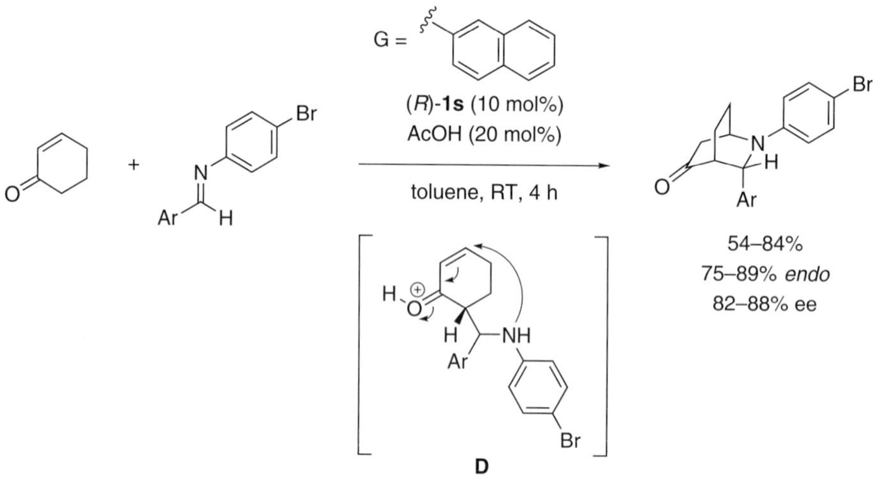

Scheme 3.25 Direct cycloaddition reaction of imines with cyclohexenone.

3.2.4.3 Inverse Electron-Demand Aza-Diels–Alder Reaction (Povarov Reaction)

The Povarov reaction, an inverse electron-demand aza-D–A reaction of 2-azadienes with electron-rich alkenes, enables facile access to tetrahydroquinoline derivatives of pharmaceutical and biological importance. Akiyama et al. developed the highly enantioselective Povarov reaction of N-aryl imines with vinyl ethers using chiral phosphoric acid **1g** (Scheme 3.26a) [61]. The method provides an efficient approach to enantioenriched tetrahydroquinolines having an oxygen functionality at the 4-position. Masson and coworkers applied a three-component coupling process to the enantioselective Povarov reaction using enecarbamates, instead of vinyl ethers, as electron-rich alkenes (Scheme 3.26b) [62]. The three-component approach employing 4-methoxyaniline, N-vinylenecarbamate, and aldehydes in the presence of chiral phosphoric acid **5e** is applicable to a broad range of aromatic aldehydes, giving the corresponding products with excellent enantioselectivities, irrespective of their electronic properties. Aliphatic aldehydes were applicable to the present reaction. Interestingly, the tetrahydroquinolines have different absolute stereochemistries (Scheme 3.26a versus b), although chiral phosphoric acids have the same axial chirality in both cases. Masson and coworkers also demonstrated the power of the present reaction in the enantioselective synthesis of torcetrapib, an inhibitor of cholesteryl ester transfer protein.

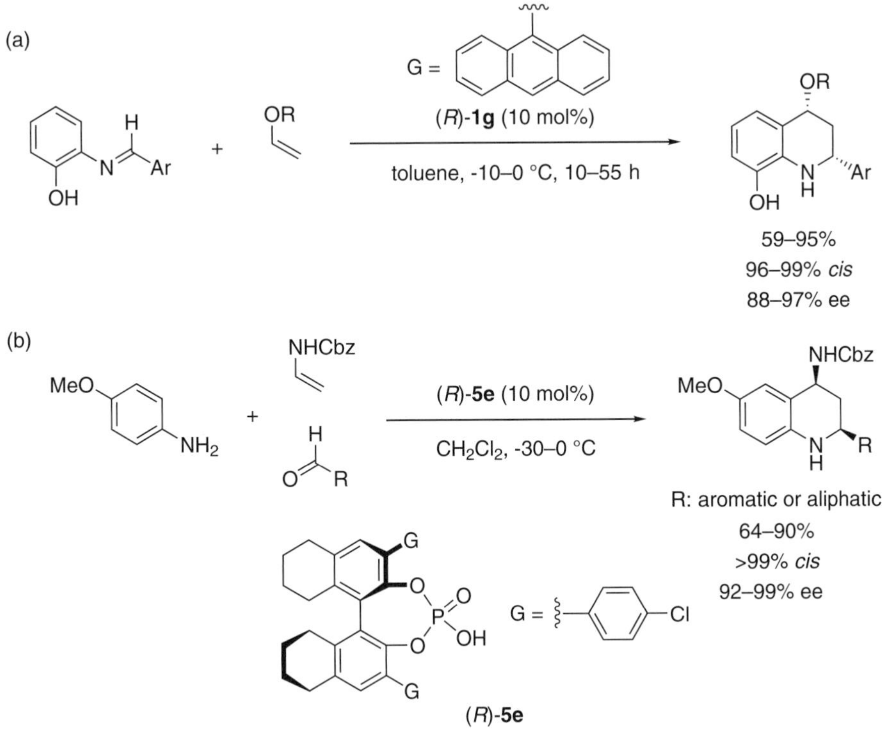

Scheme 3.26 Inverse electron-demand aza-D–A reaction (Povarov reaction).

3.2.4.4 1,3-Dipolar Cycloaddition Reaction

The 1,3-dipolar cycloaddition reaction of nitrones with vinyl ethers provides an efficient route to 1,3-amino alcohols, which can be transformed from the corresponding isoxazolidine products through reductive cleavage of the N–O bond. The development of catalytic enantioselective 1,3-dipolar cycloaddition is hence a vital step toward the efficient synthesis of these amino alcohols in optically active forms. Yamamoto and coworkers successfully developed the enantioselective 1,3-dipolar cycloaddition reaction of diaryl nitrones with ethyl vinyl ethers (Scheme 3.27) [63]. They applied acidic NHTf-substituted chiral phosphoramide catalysts (**6**), originally developed by his research group, to the present dipolar cycloaddition reaction. Chiral phosphoramide **6b** effectively catalyzed the cycloaddition reaction to give the corresponding isoxazolidine derivatives in high yields. The introduction of an electron-withdrawing group to the aromatic ring, Ar^2, is necessary to ensure high enantioselectivity. In addition, *endo*-isomers were furnished as the major diastereomers in the present reaction. The predominance of *endo*-products is in contrast to the formation of *exo*-isomers as major products under Lewis acid catalysis.

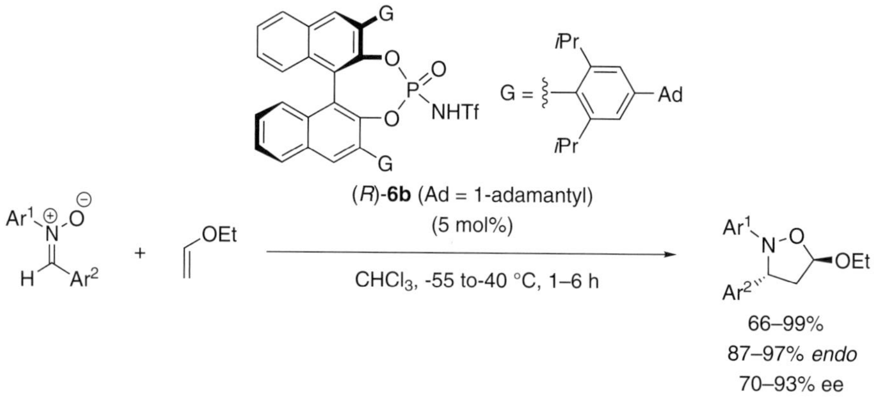

Scheme 3.27 1,3-Dipolar cycloaddition reaction of nitrones with ethyl vinyl ether.

The asymmetric 1,3-dipolar cycloaddition reaction of azomethine ylides to electron-deficient olefins provides chiral pyrrolidine derivatives, an important class of nitrogen heterocycles, as the precursors of biologically active compounds. Gong and coworkers reported the enantioselective catalysis by chiral Brønsted acids in the three-component 1,3-dipolar cycloaddition reaction among aldehydes, α-amino-1,3-dicarbonyl compounds, and maleate as an electron-deficient olefin (Scheme 3.28) [64]. As none of the chiral monophosphoric acids (**1**) provided sufficient enantioselectivities, they newly developed bisphosphoric acid (**7**) derived from linked BINOL, which displayed excellent performance in the present cycloaddition reaction in terms of diastereo- and enantioselectivities and catalytic efficiency. A broad range of aldehydes, including aromatic and α,β-unsaturated ones, could be used in the reaction, giving the corresponding pyrrolidine derivatives with excellent *endo*- and enantioselectivities.

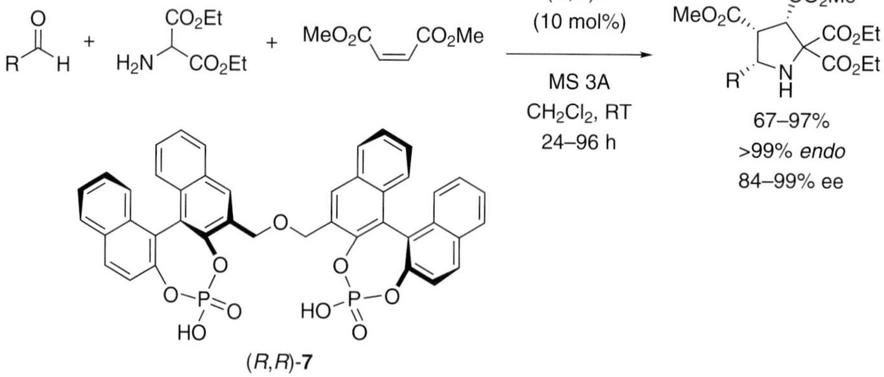

Scheme 3.28 1,3-Dipolar cycloaddition reaction of azomethine ylides with maleate.

3.2.5
Aza–Ene-Type Reactions

Kobayashi and coworkers pioneered the use of enamides or enecarbamates as nucleophiles in enantioselective reactions with either glyoxylates or glyoxylate-derived imines catalyzed by chiral copper complexes [65]. The reaction using enamides or enecarbamates as nucleophilic components, namely, the aza–ene reaction, with imines provides β-amino imines that can be readily transformed into 1,3-diamine derivatives via nucleophilic addition to the imine moiety of the corresponding products.

3.2.5.1 Aza–Ene-Type Reaction of Aldimines with Enecarbamates

Organocatalysis has been proven to be beneficial in many respects. However, one critical drawback inherent to the methodologies reported to date is the inadequate catalytic efficiency. Most organocatalytic reactions are performed at catalyst loads of 10 mol% or more to achieve sufficient chemical yields and to avoid loss of enantioselectivity. To ensure high efficiency, one of the greatest challenges in practical organocatalysis is to decrease the catalyst load [66]. Terada *et al.* demonstrated a highly efficient organocatalytic reaction that uses phosphoric acid with a significantly low catalyst load in the aza–ene-type reaction (Scheme 3.29) [67]. The reaction of aromatic imines with enecarbamates can be accomplished in the presence of 0.1 mol% phosphoric acid catalyst **1g**. *para*- and *meta*-Substitution to aromatic imines, or substitution to fused aromatic and α,β-unsaturated ones, resulted in excellent chemical yields and enantioselectivities, irrespective of the electronic properties of the substituents. Although *ortho*-substitution reduced the catalytic efficiency, giving products in moderate chemical yields, the yields were improved in these cases by increasing the catalyst load to 0.5 mol%. It is noteworthy that the reaction can be performed without considerable loss of enantioselectivity even with a decrease in the catalyst load to as low as 0.05 mol%. The synthetic applicability of the present highly efficient organocatalysis was demonstrated by the reduction of the imine moiety by Red-Al, giving *anti*-1,3-diamine derivatives predominantly.

3.2.5.2 Cascade Transformations Based on Tandem Aza–Ene-Type Reaction

The development of efficient methods to access complex molecules with multiple stereogenic centers continues to be a formidable challenge in both academe and industry. One approach is the use of catalytic enantioselective cascade reactions [68] that have emerged as powerful tools to rapidly increase molecular complexity from simple and readily available starting materials, thus producing enantioenriched compounds in a single operation.

Terada *et al.* successfully applied the aza–ene-type reaction to the cascade transformation by taking advantage of the formation of imine products (Scheme 3.30) [69]. They employed monosubstituted enecarbamates [70] instead of the disubstituted versions and as a result, piperidine derivatives with multiple stereogenic centers were obtained in high stereoselectivities. The acid-catalyzed aza–ene-type reaction of the initial aldimines with monosubstituted enecarbamates afforded aza–ene-type

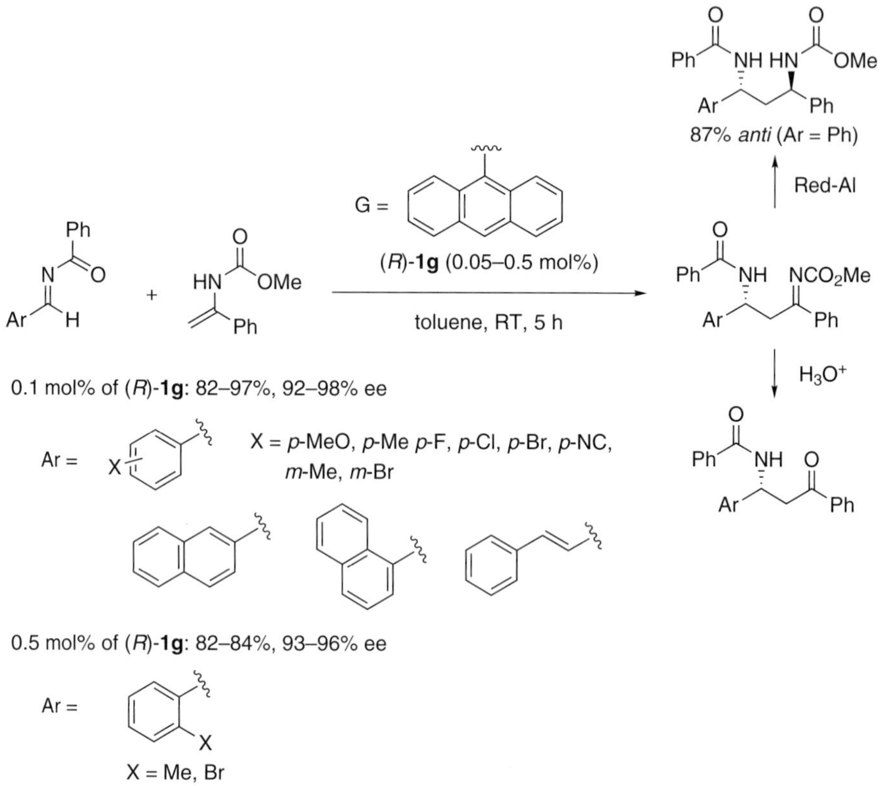

Scheme 3.29 Enantioselective aza–ene-type reaction under low catalyst load.

products of N-acyl aldimines (**E**) as reactive intermediates and **E** underwent further aza–ene-type reaction leading to the subsequent generation of aldimines (**F**). The intramolecular cyclization of intermediate **F** was conducted to terminate the tandem aza–ene-type reaction sequence. It is noteworthy that one stereoisomer was formed exclusively from among the eight possible stereoisomers consisting of four pairs of enantiomers under the influence of phosphoric acid catalyst **1c**. Not only aromatic but also aliphatic aldimines could be used in the present cascade reaction. Moreover, the glyoxylate-derived aldimine could be transformed into the highly functionalized piperidine derivative with excellent enantioselectivity. This cascade methodology allows rapid access to piperidine derivatives with multiple stereogenic centers, as key structural elements of numerous natural products.

3.2.5.3 Two-Carbon Homologation of Hemiaminal Ethers

The homologation of a carbon unit is an important and fundamental methodology in the construction of carbon frameworks in synthetic organic chemistry. Much attention has been devoted to the development of two-carbon homologation using

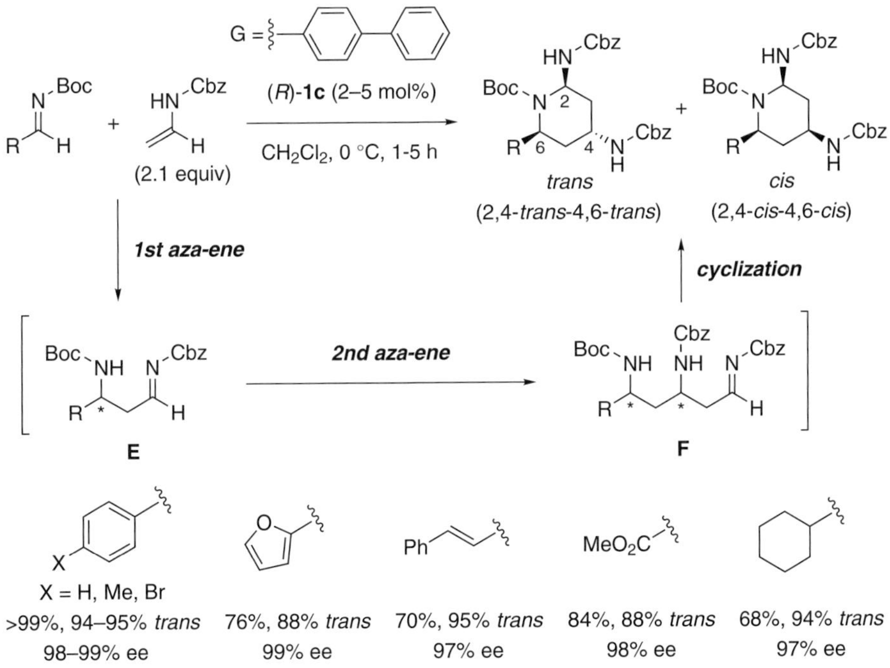

Scheme 3.30 One-pot entry to piperidine derivatives via tandem aza–ene-type reaction/cyclization cascade.

acetaldehyde anion equivalents, as these can be directly utilized in further transformations. From a synthetic viewpoint, monosubstituted enecarbamates are attractive as acetaldehyde anion equivalents for the two-carbon homologation reaction because they are readily available and can provide aldimine products [71]. These aldimines can be directly transformed into 1,3-diamine derivatives, when two-carbon homologation reactions were conducted with imines as the substrate.

Although the high reactivity of aldimine products (**H**) hindered the development of the two-carbon homologation reaction due to overreaction, Terada et al. neatly demonstrated two-carbon homologation using enecarbamates and hemiaminal ethers, instead of imines (**G**), as the substrate (Scheme 3.31) [72]. In the acid-catalyzed reaction of hemiaminal methyl ethers, intermediary and reactive aldimine (**H**) was entrapped by methanol that was generated during the course of imine (**G**) formation. A series of aromatic hemiaminal ethers were employable for the present homologation reaction in the presence of phosphoric acid catalyst **1p**, giving hemiaminal products with high enantioselectivities, although the reactivity of the hemiaminal ethers was considerably dependent on the electronic properties of the substituents on the aromatic ring. Aliphatic hemiaminal ethers were also applicable to the present homologation, but in this case, protection of the nitrogen atom by the more electron-withdrawing trichloroethoxycarbonyl (Troc) group was required to suppress the formation of by-products.

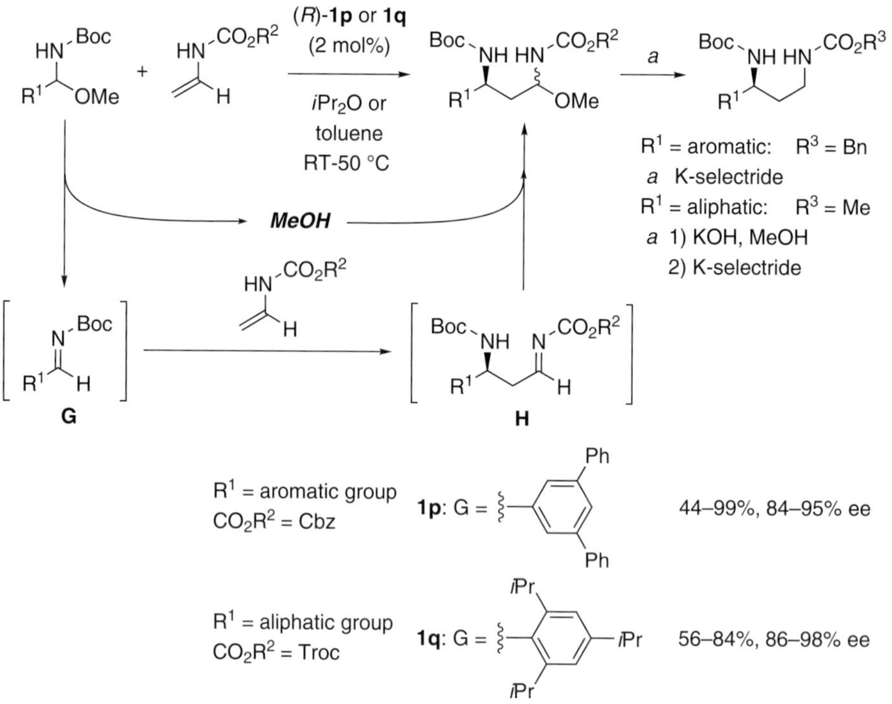

Scheme 3.31 Two-carbon homologation of hemiaminal ethers by enecarbamate.

The present homologation reaction can be applied to a substituted enecarbamate (Scheme 3.32a). Either the *anti*- or *syn*-product could be obtained in a highly diastereoselective manner from the respective geometric isomers of the enecarbamates, and each of the major diastereomers exhibited good to high enantioselectivity. The synthetic utility of the present homologation reaction is highlighted by the sequential transformation of homologation/F–C reaction in one pot (Scheme 3.32b). Catalyst **1q** also accelerated the F–C reaction of hemiaminal ether with indole to afford the desired 1,3-diamine derivatives in good yields and nearly optically pure forms, albeit moderate *syn*-diastereoselectivities. The method enables facile access to highly enantioenriched 1,3-diamine derivatives as pharmaceutically and biologically intriguing molecules.

3.2.5.4 Homocoupling Reaction of Enecarbamates

Tsogoeva and coworkers reported the enantioselective formation of a quaternary stereogenic center bearing the nitrogen atom through the homocoupling reaction of enamides (Scheme 3.33) [73]. The enamides were isomerized *in situ* to ketimines under the influence of phosphoric acid catalyst **1e**, as previously depicted in Figure 3.4. Electrophilic ketimines were generated under equilibrium conditions, of which the homocoupling reaction with nucleophilic enamides proceeded via the aza–ene-type mechanism to give the corresponding imine products [67]. The

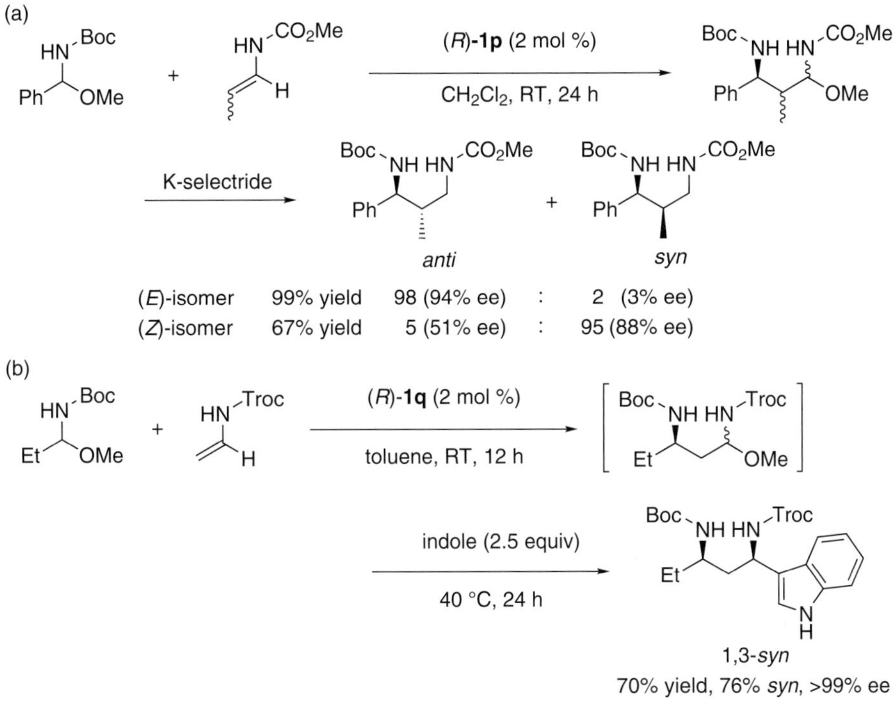

3.2 Carbon–Carbon Bond Forming Reactions | 103

Scheme 3.32 Application of two-carbon homologation reaction.

resultant imines were readily isomerized to an enamine form under acidic conditions, giving enamide products with high enantioselectivities. Hydrolysis of the enamide moiety yielded optically active β-amino ketones having a quaternary carbon atom, which are potentially useful as synthetic building blocks.

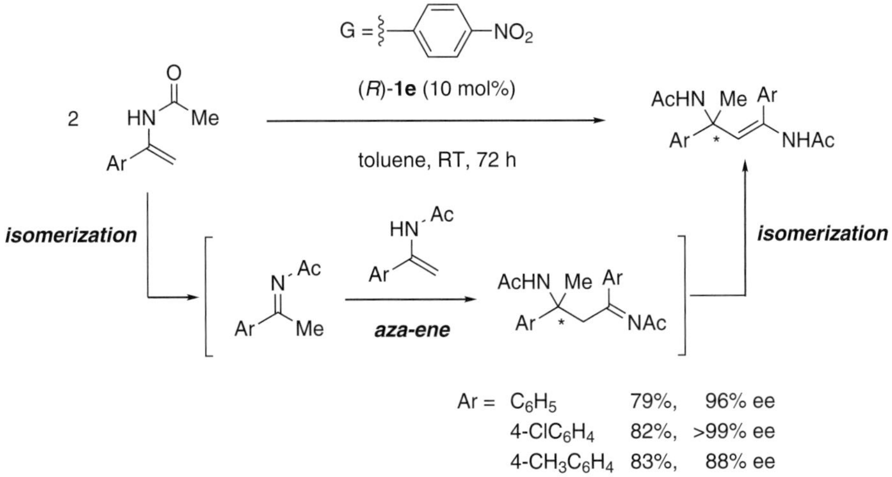

Scheme 3.33 Homocoupling reaction of enamides.

3.2.6
Miscellaneous Reactions

3.2.6.1 Aza-Cope Rearrangement

Sigmatropic rearrangements are one of the most fundamental methods for the construction of the carbon framework and have widespread application in the synthesis of biologically relevant molecules. A variety of enantioselective sigmatropic rearrangements have been established using chiral metal catalysts or organocatalysts [74]. Enantioselective versions of the aza-Cope rearrangement provide efficient routes to optically active homoallylic amines, which are useful precursors for the synthesis of natural products. However, this valuable transformation has never been successfully developed. Recently, and for the first time, Rueping and Antonchick reported the highly enantioselective aza-Cope rearrangement using chiral phosphoric acid catalysts (Scheme 3.34) [75]. The reaction of aromatic aldehydes with achiral homoallylic amine proceeded smoothly in the presence of catalyst 5f and the rearranged products, homoallylic amines, were obtained in good yields with high enantioselectivities. The method furnishes a practical route to the catalytic enantioselective aminoallylation of aromatic aldehydes on the basis of a condensation–rearrangement sequence. Free primary homoallylic amines were obtained by treating the rearranged products with hydroxylamine hydrochloride.

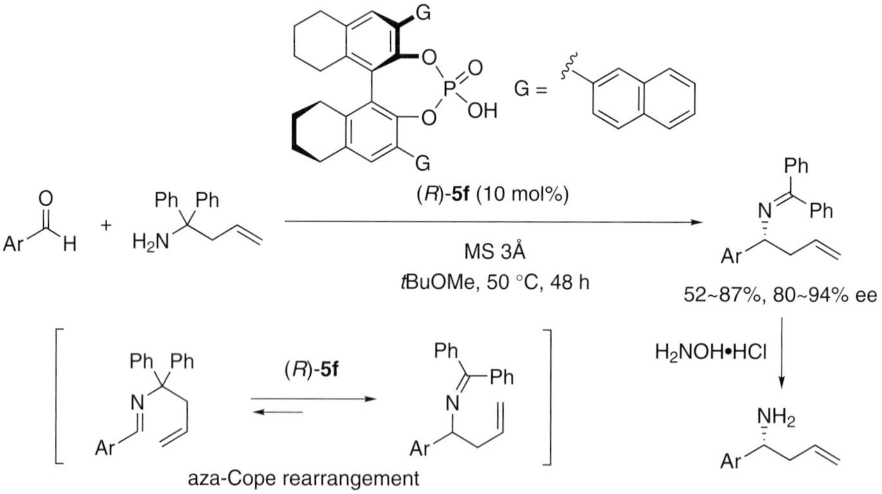

Scheme 3.34 Claisen rearrangement of homoallylamines.

3.2.6.2 Aldol-Type Reaction of Azlactones with Vinyl Ethers

The activation of vinyl ethers by a Brønsted acid catalyst is an extensively utilized and fundamental method in synthetic organic chemistry and is employed in the protection of alcohols and the formation of carbon–carbon bonds, among other processes. In this activation mode, the use of chiral phosphoric acids gives rise to ion pairs of a chiral conjugate base and an oxocarbenium ion via protonation of the vinyl ether. Terada et al. developed chiral conjugate base-controlled enantioselective transformations [76] involving the oxocarbenium ion as the reactive intermediate [77]. They applied the intermediary oxocarbenium to a direct aldol-type reaction of azlactones [78] via their oxazole tautomer and obtained the corresponding products with excellent enantio- and diastereoselectivities (Scheme 3.35) [79]. The method enables efficient access to biologically and pharmaceutically intriguing β-hydroxy-α-amino acid derivatives having a quaternary stereogenic center at the α-carbon atom.

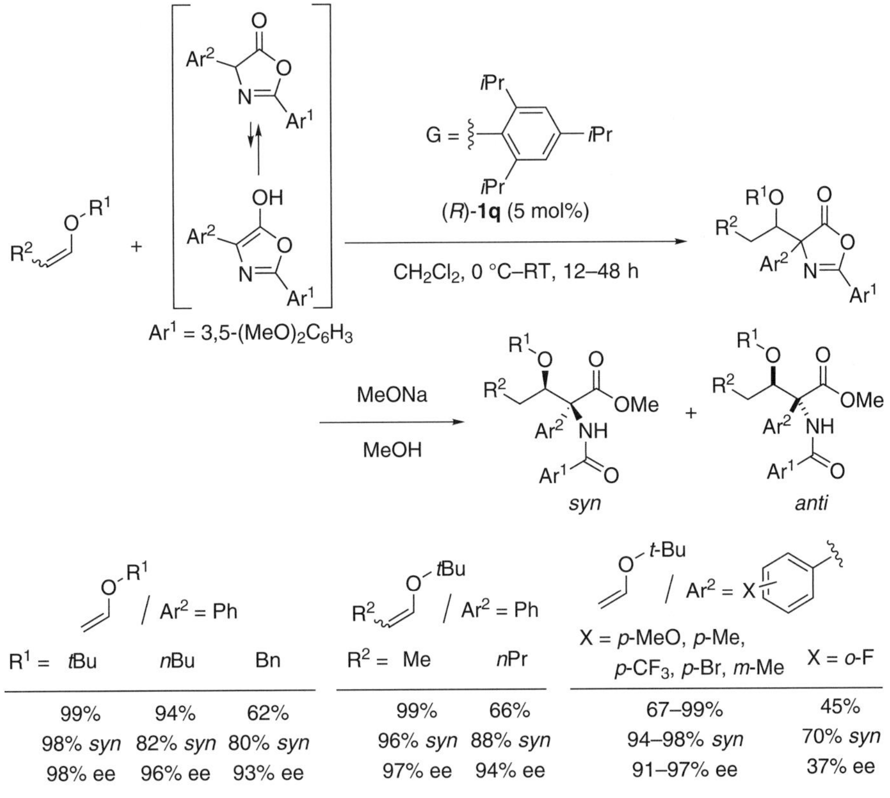

Scheme 3.35 Aldol-type reaction of azlactones via protonation of vinyl ethers.

The electronic manipulation of Ar^1 substituents introduced at the C2 position of azlactone had a significant impact not only on the reactivity of azlactones but also on the stereochemical outcome in terms of both enantio- and diastereoselectivities. When electron-donating methoxy substituents were introduced to the 3,5-positions of the phenyl ring, the corresponding products were obtained in excellent enantio- and diastereoselectivities. The substituent effect of vinyl ethers showed that the sterically demanding tert-butyl ether is important to achieve the high diastereoselectivity. Vinyl ethers with an alkyl group substitution at the terminal position were also applicable, affording the desired products with high enantioselectivities. Azlactones having a series of aromatic groups (Ar^2) revealed uniformly high enantio- and diastereoselectivities for *para-* and *meta-*substituted aromatic rings (Ar^2), irrespective of their electronic properties. However, *ortho-*substitution led to a marked reduction of selectivity and chemical yield. They proposed that the interaction, namely, C—H···O hydrogen bond formation, between the chiral conjugate base and the oxocarbenium ion would allow the reaction to proceed under a chiral environment regulated by the chiral conjugate base, and hence high stereoselectivities were achieved in the present transformation.

3.2.6.3 Cooperative Catalysis by Metal Complexes and Chiral Phosphoric Acids

In the past decade, tremendous progress has been made in the catalysis using small organic molecules, namely, organocatalysis [1]. Meanwhile, catalysis by transition metal complexes has continuously been applied to a broad range of organic transformations and occupies a privileged position in synthetic organic chemistry. Much of the research on catalysis has centered on the use of metal complexes to activate a variety of chemical bonds. In recent years, armed with the idea of taking advantage of both of these catalytic processes, researchers have combined metal complexes and organic molecules in cooperative catalysis, and this has attracted much attention as it could potentially realize unprecedented transformations. Recently, several excellent approaches were established with the chiral phosphoric acid/metal complex binary catalytic system and applied to chiral amine syntheses.

Rueping *et al.* developed a new cooperative process that comprised an enantioselective Brønsted acid and silver complex-catalyzed alkynylation (Scheme 3.36) [80]. The enantioselective alkynylation under binary catalytic conditions was accomplished in the reaction of α-imino ester with a series of aryl-substituted terminal alkynes, in which the α-imino ester and the alkyne were activated by phosphoric acid **1l** and the silver complex, respectively. α-Amino acid products having an alkynyl substituent were obtained in good yields with high enantioselectivities. Although, from a mechanistic viewpoint, an exchange of the metal counteranion from acetate to chiral phosphate, leading to the formation of a chiral silver complex, cannot be ruled out, both catalysts are required for the reaction to proceed and more important, the reaction that proceeded under the influence of a chiral silver complex and an achiral phosphoric acid catalyst resulted in a racemic product. Simultaneous activation of both the α-imino ester and the alkyne by cooperative organic and metallic catalyses is crucial to furnish the α-amino ester products in optically active forms.

Hu *et al.* developed a system that involved cooperative catalysis by a chiral phosphoric acid and an achiral rhodium complex (Scheme 3.37) [81]. They applied the binary catalytic system to a three-component coupling reaction among α-diazoesters, primary alcohols, and aldimines. The steric bulkiness of the primary alcohol had a significant effect on both diastereo- and enantioselectivities. The sterically demanding 9-anthracenemethanol was the best component to give β-amino-α-alkoxy esters with excellent stereoselectivities under the combined and cooperative catalysis by phosphoric acid **1l** and the rhodium complex.

As shown in Schemes 3.36 and 3.37, each reactant was activated by one type of catalyst simultaneously; thus, the metal complex activates nucleophiles while the

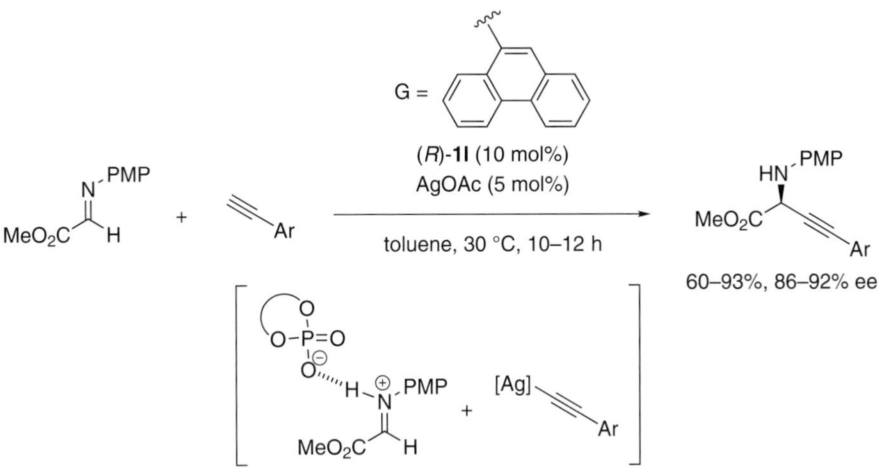

Scheme 3.36 Chiral phosphoric acid and silver complex binary catalytic system for enantioselective alkynylation of α-imino ester.

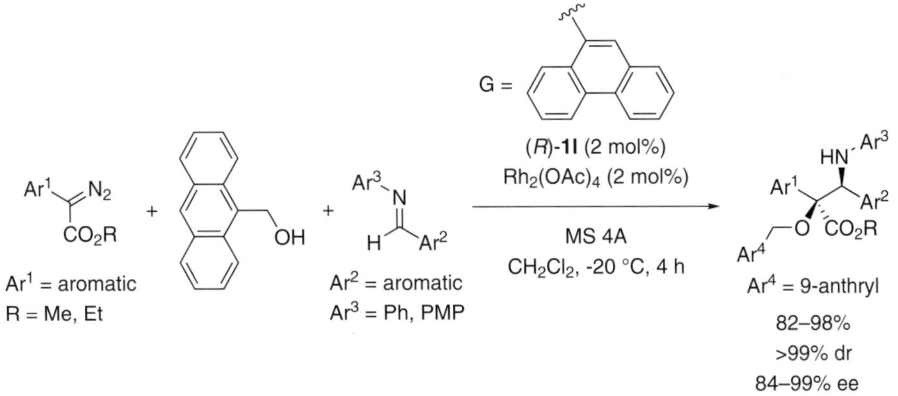

Scheme 3.37 Chiral phosphoric acid and rhodium complex binary catalytic system for three-component coupling reaction.

phosphoric acid activates imines as the electrophilic component. Sorimachi and Terada reported an unprecedented consecutive transformation using a binary catalytic system, that is, relay catalysis for tandem isomerization/carbon–carbon bond formation sequence promoted by a binary catalyst consisting of a ruthenium hydride complex and a Brønsted acid (**8**) (Scheme 3.38) [82]. The reaction of *N*-Boc-protected allylamine with 2-methoxyfuran proceeded smoothly to give a product having a nitrogen functionality at the stereogenic center in good yield. Control experiments revealed that both catalysts, the rhuthenium complex and Brønsted acid **8**, are indispensable to the sequential processes. Although the racemic acid catalyst was employed in this case, an enantioselective version would be applicable to the present relay catalysis, considering recent progress in enantioselective F–C reactions using chiral phosphoric acid catalysts (see Section 3.2.3).

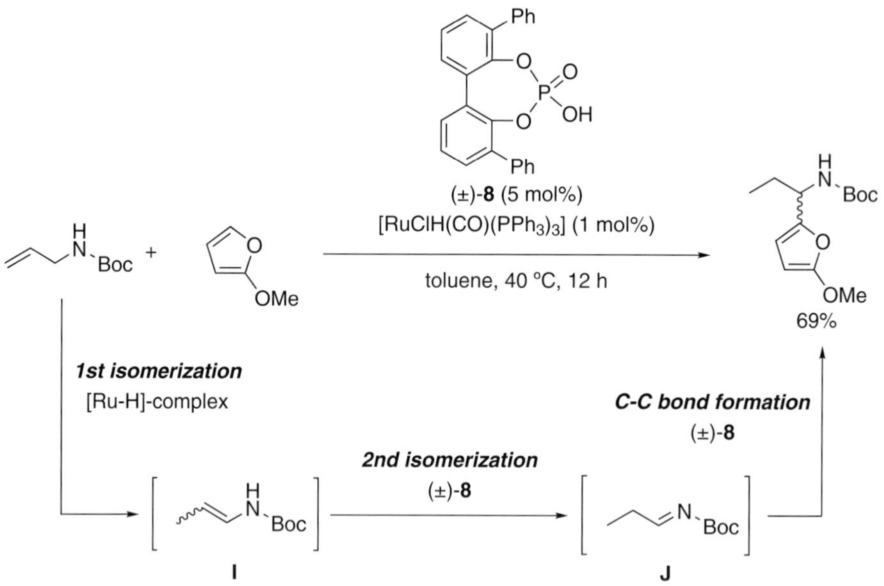

Scheme 3.38 Relay catalysis by ruthenium complex/Brønsted acid binary system.

The present sequential transformation involves a three-step relay catalysis where (i) isomerization of allylcarbamate to enecarbamate (**I**) is catalyzed by the ruthenium hydride complex; (ii) subsequent isomerization of **I** to intermediary imine (**J**) is relayed by acid catalyst **8**; and (iii) the catalytic sequence is terminated by a carbon–carbon bond forming reaction of **J** with the electron-rich aromatic compounds as a nucleophilic component under the influence of **8**. The advantage of the present relay catalysis is that the method enables the generation of reactive imines **J** from readily available allylcarbamates in a one-pot reaction via tandem isomerization.

The enantioselective version of the relay transformation by organic and metallic catalyses was successfully demonstrated by Gong and coworkers (Scheme 3.39) [83]. They accomplished the direct transformation of o-propargylaniline derivatives into tetrahydroquinolines in a highly enantioselective manner through the hydroamination of alkynes/isomerization/enantioselective transfer hydrogenation (see Section 3.3 for details) sequence under the relay catalysis of an achiral Au complex/chiral phosphoric acid binary system.

3.3
Carbon–Hydrogen Bond Forming Reactions

The enantioselective reduction of ketimines is a straightforward approach to synthesize optically active amines. Although catalytic enantioselective hydrogenations and transfer hydrogenations of ketones and olefins have been intensively investigated using chiral metal complexes, the corresponding enantioselective reductions of

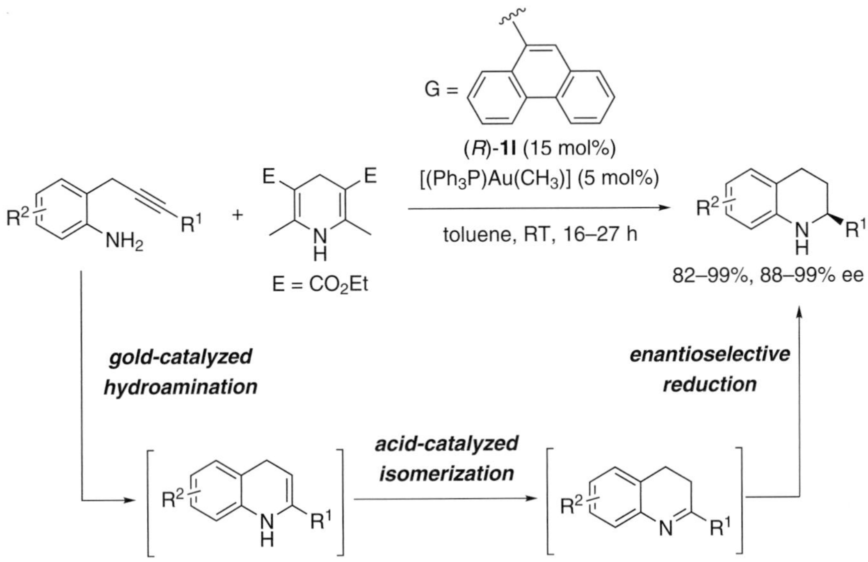

Scheme 3.39 Enantioselective relay catalysis by gold complex/chiral phosphoric acid binary system.

imines are less advanced and hence the development of efficient methods for these transformations is still one of the greatest challenges in the practical synthesis of chiral amines. Most catalytic enantioselective hydrogenations of ketimines have been conducted using common hydrogen sources, such as high-pressure hydrogen gases, di- or trialkylsilylanes, and ammonium formate, under the influence of chiral transition metal complexes [84]. Recently, List and coworkers and MacMillan and coworkers independently reported the organocatalytic transfer hydrogenations of α,β-unsaturated aldehydes using Hantzsch esters, dihydropyridine derivatives, as a biomimetic hydrogen source [85]. Following these excellent reports, Hantzsch esters gained popularity as an efficient hydrogen source for the enantioselective reduction of organic compounds, including imines [86].

3.3.1
Transfer Hydrogenation of Acyclic and Cyclic Imines

In 2005, Rueping et al. reported that chiral phosphoric acids function as an efficient catalyst for the enantioselective reduction of ketimines (Scheme 3.40a-1) [87]. A variety of aryl methyl ketimines were reduced to the corresponding amines in optically active forms using Hantzsch ester as the hydrogenation transfer reagent (HEH) [88]. Subsequently, List and coworkers improved the catalytic efficiency and enantioselectivity by thorough optimization of the substituents (G) that were introduced to the phosphoric acid catalyst (Scheme 3.40a-2) [89]. Almost simultaneously, MacMillan and coworkers successfully developed the enantioselective

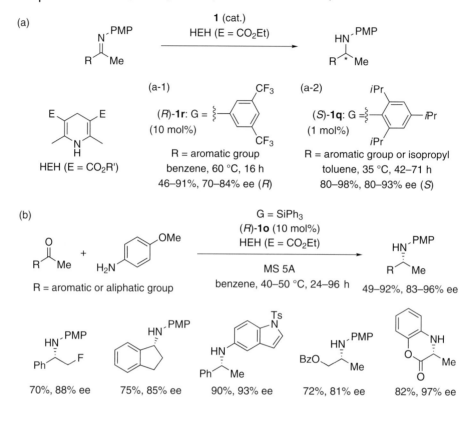

Scheme 3.40 Transfer hydrogenation of acyclic ketimines using Hantzsch ester as a biomimetic hydrogen source.

reductive amination reaction using Hantzsch ester via simple fragment coupling between ketones and aniline derivatives (Scheme 3.40b) [90]. Acid catalyst **1o** having sterically demanding triphenylsilyl groups was optimal for the enantioselectivity. Although the reaction required a methyl ketone subunit, this operationally simple method is applicable not only to aryl but also to alkyl methyl ketones and provides a highly efficient route to structurally diverse chiral amine synthesis. In their conscientious studies on the reductive aminations, MacMillan et al. also reported a couple of examples using cyclic imines as the substrate. Shortly thereafter, a detailed investigation of the transfer hydrogenation of cyclic imines was conducted by Rueping et al. (Scheme 3.41a) [91]. Benzoxazines underwent the transfer hydrogenation in good yields and extremely high enantioselectivities even with a remarkably low catalyst load of **1l** (0.1 mol%). Benzothiazines (Scheme 3.41a) and benzoxazinones (Scheme 3.41b) were also hydrogenated by Hantzsch ester in the presence of the same catalyst **1l** (1 mol%) to give the corresponding products with excellent enantioselectivities.

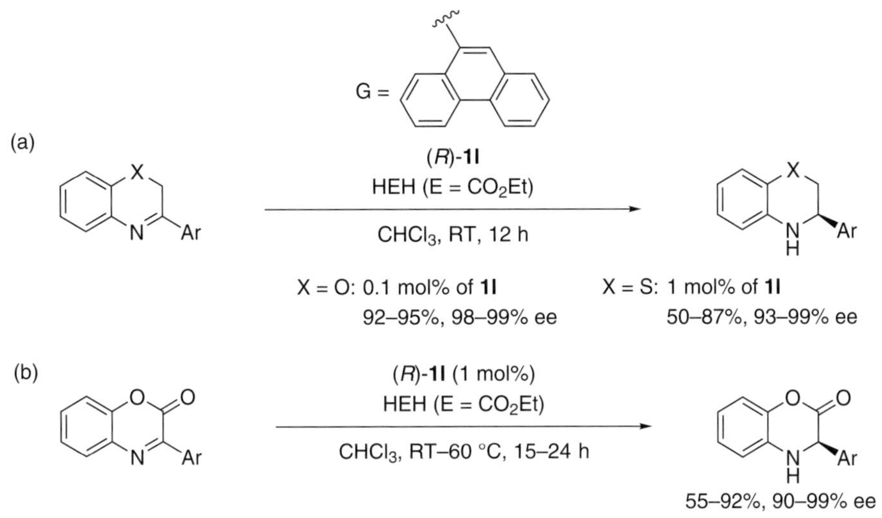

3.3 Carbon–Hydrogen Bond Forming Reactions

(a)

(R)-1I
HEH (E = CO$_2$Et)

CHCl$_3$, RT, 12 h

X = O: 0.1 mol% of 1I
92–95%, 98–99% ee

X = S: 1 mol% of 1I
50–87%, 93–99% ee

(b)

(R)-1I (1 mol%)
HEH (E = CO$_2$Et)

CHCl$_3$, RT–60 °C, 15–24 h

55–92%, 90–99% ee

Scheme 3.41 Transfer hydrogenation of cyclic imines.

Soon after these initial reports, the groups of Antilla [92] and You [93] independently applied the chiral phosphoric acid catalysis to the enantioselective hydrogenation of α-imino esters. The method provides an alternative route to the enantioselective synthesis of α-amino esters. Antilla and coworkers employed a new type of axially chiral phosphoric acid (**9**) derived from VAPOL originally developed by his research group (Scheme 3.42), whereas **1g** was used in You's case. In both cases, excellent enantioselectivities were achieved. You and coworkers further applied the method to the enantioselective reduction of α-imino esters having an alkynyl substituent at the α-position (Scheme 3.43) [94]. Both alkyne and imine moieties were reduced under transfer hydrogenation conditions with an excess amount of

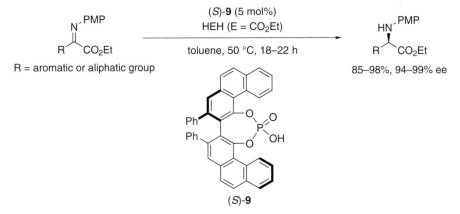

R = aromatic or aliphatic group

(S)-**9** (5 mol%)
HEH (E = CO$_2$Et)

toluene, 50 °C, 18–22 h

85–98%, 94–99% ee

Scheme 3.42 Transfer hydrogenation of α-imino esters.

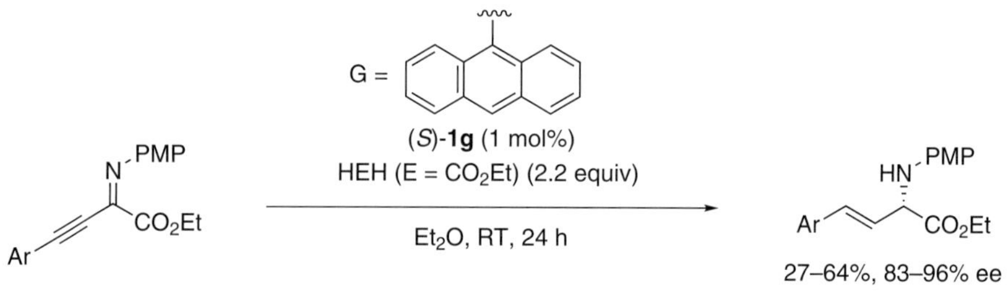

Scheme 3.43 Transfer hydrogenation of α-imino esters having an alkynyl substituent at the α-position.

Hantzsch ester (2.2 equiv) to give β,γ-unsaturated α-amino esters. Although the chemical yields were moderate, high enantio- and *trans*-selectivities were achieved using the same chiral phosphoric acid catalyst **1g**.

As shown in Schemes 3.40, 3.42, and 3.43, the reductive amination of ketones and the hydrogenation of ketimines catalyzed by chiral phosphoric acids were accomplished in high enantioselectivities. However, in these reductions, aryl derivatives, such as PMP, were primarily employed as the nitrogen protective group. The removal of these protective groups usually required somewhat harsh reaction conditions. Li and Antilla circumvented this drawback by using enamides as the substrate (Scheme 3.44) [95]. The reduction of enamides proceeded via tautomerization to intermediary imines to give amine products having an N-acetyl protective group that can be readily removed under standard conditions. They employed acetic acid as the cocatalyst, which allowed the reduction of the catalyst load of **1g** to 1 mol% while maintaining high enantioselectivity. Acetic acid accelerated the tautomerization of the enamide to the imine, while it was inactive in the following reduction and only chiral phosphoric acid **1g** functioned as the active catalyst for the successive transfer hydrogenation of the imine.

List and coworkers developed an excellent approach to synthesize β-branched amines in optically active forms by combining the enamine and reductive amination processes (Scheme 3.45) [96]. The reductive amination of unsymmetrically α,α-disubstituted aldehydes and aniline derivatives proceeded through a tautomerization between imine and enamine forms. Dynamic kinetic resolution occurred under the

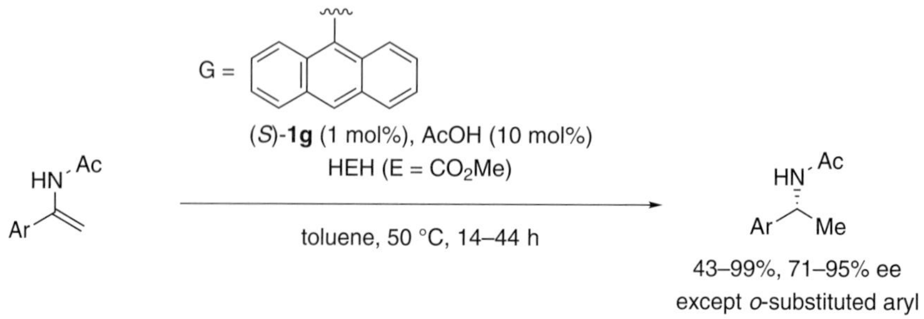

Scheme 3.44 Transfer hydrogenation of enamides.

influence of catalyst (R)-**1q**, where the hydrogenation of (R)-imines was faster than that of the (S)-isomers to give β-branched amines with high enantioselectivities.

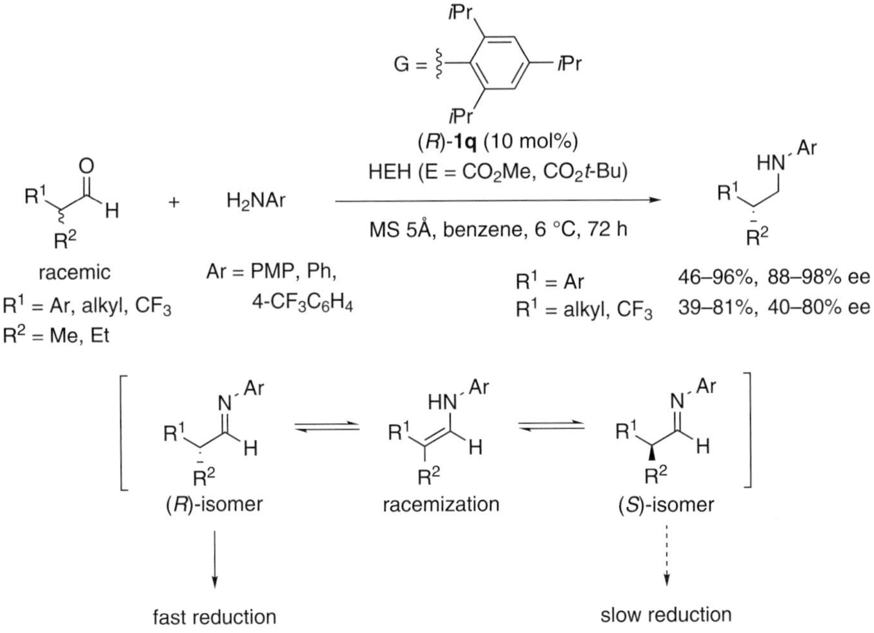

Scheme 3.45 Enantioselective synthesis of β-branched amines through dynamic kinetic resolution.

3.3.2
Cascade Transfer Hydrogenation of Quinoline and Pyridine Derivatives

The asymmetric hydrogenation of nitrogen-substituted heteroaromatic compounds provides a straightforward and attractive route to enantioenriched nitrogen heterocycles. In contrast, the enantioselective hydrogenation of these heteroaromatics using chiral transition metal catalysts has not been established and successful examples of these transformations are limited [97]. The use of Hantzsch dihydropyridine ester as hydrogen source was proven to be beneficial for the asymmetric hydrogenation of these heteroaromatics by Rueping et al. The enantioselective reduction of quinolines to tetrahydroquinolines using Hantzsch ester proceeded smoothly in the presence of chiral phosphoric acid catalyst **1l** (Scheme 3.46) [98]. The method is applicable to a variety of 2-aryl- and 2-alkyl-substituted quinolines and provides direct access to tetrahydroquinoline alkaloids in a highly enantioselective manner. They proposed that the reaction proceeds via a tandem hydrogenation cascade. The first step of the cascade generates enamines (**K**) via 1,4-hydride addition to the quinolines. Under acidic conditions, the subsequent isomerization of **K** provides imines (**L**), which are delivered to the second step of the cascade. The

1,2-hydride addition to imines **L** with generation of the stereogenic center provides enantioenriched tetrahydroquinoline products.

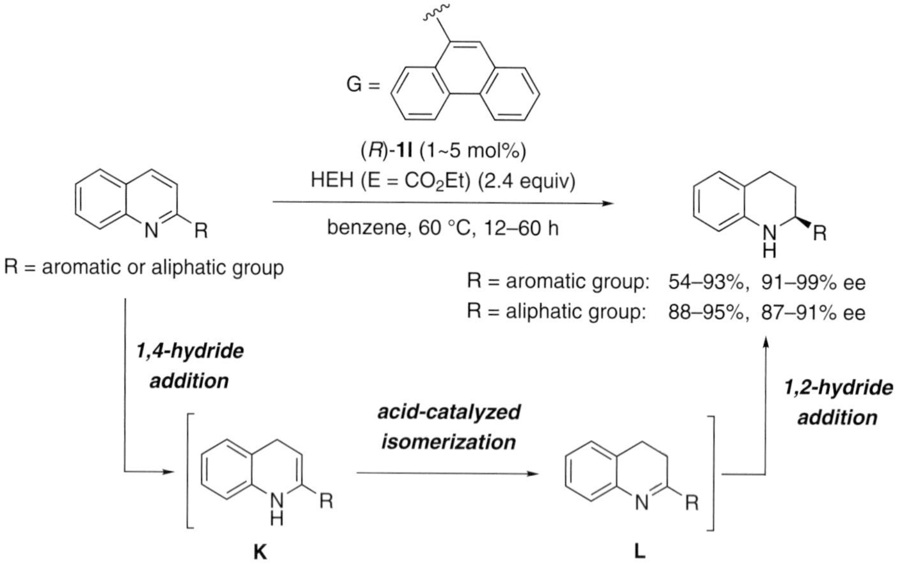

Scheme 3.46 Transfer hydrogenation of 2-substituted quinolines to tetrahydroquinolines.

Du and coworkers reported the enantioselective hydrogenation of quinolines using a new type of axially chiral phosphoric acid catalyst having a bis-BINOL scaffold (Scheme 3.47) [99]. Newly developed catalyst (**10**) showed higher efficiency than the parent mono-BINOL-derived phosphoric acid catalysts. The enantioselectivities of alkyl and aryl 2-substituted quinolines are uniformly high and the corresponding tetrahydroquinolines were obtained quantitatively even when the catalyst load was reduced to as low as 0.2 mol%. 2,3-Disubstituted quinolines were also hydrogenated in high diastereo- and enantioselectivities.

Rueping and Antonchick successively applied the cascade transfer hydrogenation strategy to the enantioselective reduction of pyridine derivatives (Scheme 3.48a) [100]. **1g** functioned as an efficient catalyst for the hydrogenation of pyridine derivatives to furnish the corresponding products, hexahydroquinolinones, with high enantioselectivities. The method allows access to enantioenriched nitrogen heterocycles as useful precursors of various natural products. Metallinos et al. reported the enantioselective reduction of 2-substituted and 2,9-disubstituted 1,10-phenanthrolines (Scheme 3.48b) [101]. Although considerable amounts of *meso*-isomers were formed in the hydrogenation of 2,9-disubstituted 1,10-phenanthrolines, DL-octahydrophenanthrolines were obtained in excellent enantioselectivities.

Rueping et al. further extended the method to the hydrogenation of 3-substituted quinolines (Scheme 3.49) [102]. In the hydrogenation of 3-substituted quinolines, the key step is the enantioselective protonation of intermediary enamine **K'**. The mechanisms of the stereo-determining step are entirely different from those of the

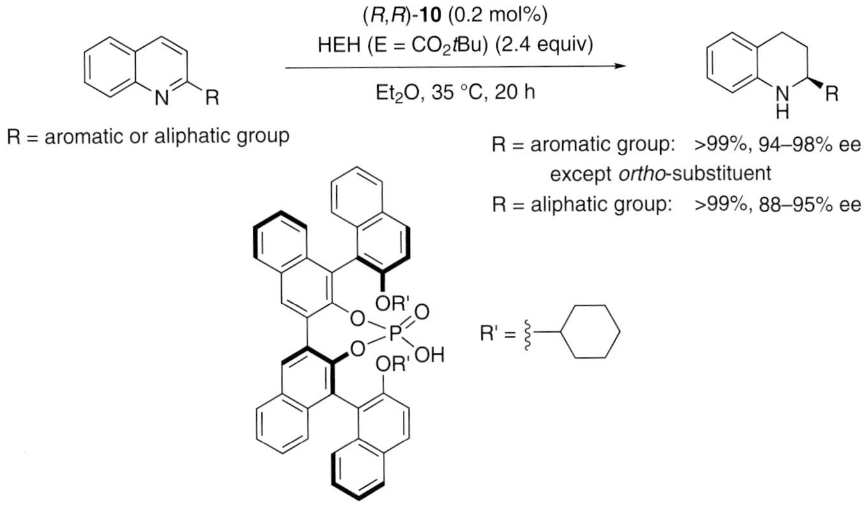

3.3 Carbon–Hydrogen Bond Forming Reactions

Scheme 3.47 Transfer hydrogenation of quinolines using newly developed chiral phosphoric acid catalyst.

asymmetric hydrogenation of 2-substituted quinolines, where the 1,2-hydride addition step is the key to determining the stereochemical outcome (see Scheme 3.46). After thorough optimization of phosphoric acid catalysts, **5b** was found to be the best catalyst, giving 3-substituted tetrahydroquinolines with high enantioselectivities.

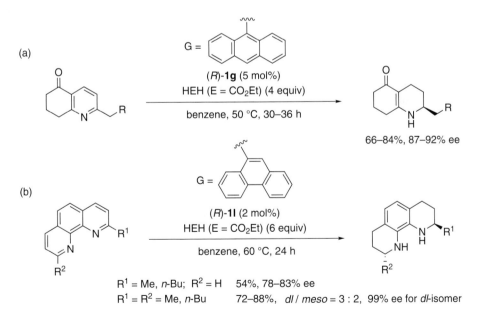

Scheme 3.48 Transfer hydrogenation of pyridine and 1,10-phenanthroline derivatives.

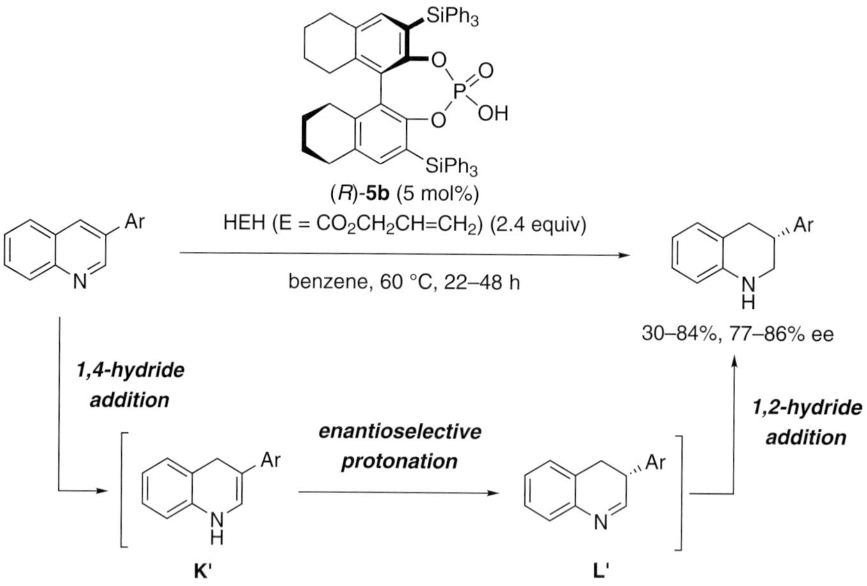

Scheme 3.49 Transfer hydrogenation of 3-substituted quinolines to tetrahydroquinolines.

3.3.3
Application of Transfer Hydrogenation to Cascade Reaction

Cascade transformations involving different organocatalytic processes have emerged as a powerful and efficient method for the rapid construction of complex molecules starting from the simple materials in one pot [68]. In particular, organocatalyses by amines or their salts were intensively applied to cascade transformations, including enamine or iminium ion, as the reactive intermediate [103]. Zhou and List successfully combined chiral phosphoric acid catalysis with these amine catalyses based on enamine and iminium ion mechanisms (Scheme 3.50) [104]. They developed a highly enantioselective synthesis of 3-substituted (hetero)cyclohexylamines from 2,6-diketones via three-step organocatalysis [105], namely, (i) intramolecular aldol condensation via enamine formation; (ii) 1,4-hydride addition under the iminium ion mechanism; and (iii) 1,2-hydride addition cascade catalyzed by chiral phosphoric acid and accelerated by the amine substrate (H_2NAr). Acid catalyst **1q** exhibited excellent performance for the cascade transformation, giving the corresponding 3-substituted (hetero)cyclohexylamines with high enantio- and *cis*-selectivities. The reductive amination of 3-substituted cyclohexanones, the intermediate of the present cascade, generally provides the corresponding *trans*-isomer under standard reductive amination conditions. In contrast, in the present cascade transformation, *cis*-isomers were obtained as the major product.

Rueping and Antonchick developed an organocatalytic multiple cascade sequence using enamines and α,β-unsaturated ketones as substrates to obtain enantioenriched nitrogen heterocycles (Scheme 3.51) [106]. The multiple cascade sequence comprises

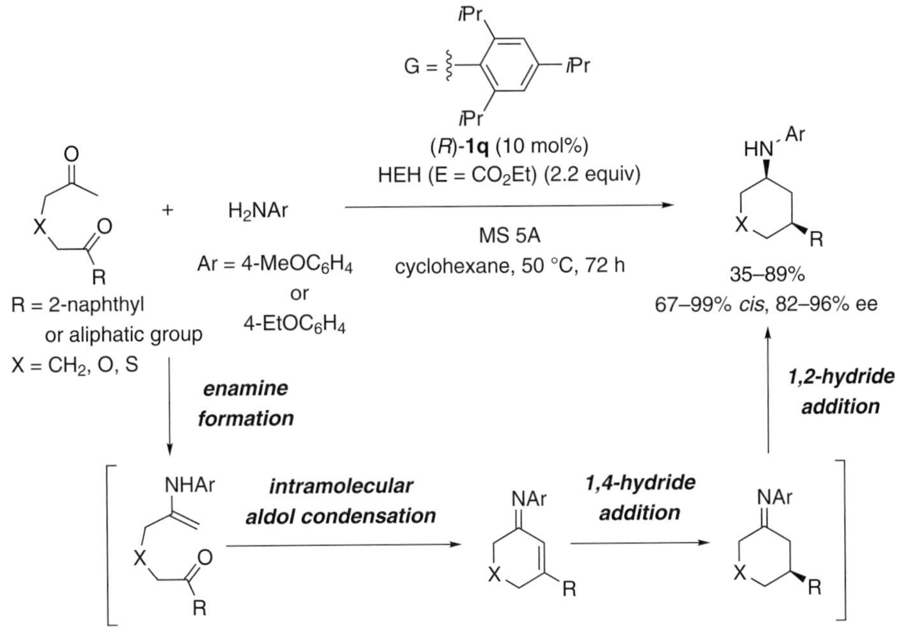

Scheme 3.50 Intramolecular aldol condensation/tandem transfer hydrogenation cascade.

Michael addition, geometrical isomerization, cyclization, dehydration, isomerization, and 1,2-hydride addition, in which each single step is efficiently catalyzed by chiral phosphoric acid catalyst **1g**. The method enables direct, rapid, and efficient access to a diverse set of tetrahydropyridine and azadecalinone derivatives with excellent enantioselectivities.

3.4
Carbon–Heteroatom Bond Forming Reactions

3.4.1
Hydrophosphonylation (Kabachnik–Fields Reaction)

The Kabachnik–Fields reaction, which involves the hydrophosphonylation of phosphites with imines generated *in situ* from carbonyl compounds and amines, is an attractive method for the preparation of α-amino phosphonates. Optically active α-amino phosphonic acids and their phosphonate esters are an attractive class of compounds due to their potent biological activities as nonproteinogenic analogues of α-amino acids. Therefore, considerable attention has been given to their enantioselective synthesis by hydrophosphonylation of preformed imines, using either metal-based catalysts or organocatalysts [107].

Akiyama *et al.* reported that chiral phosphoric acid **1r** functioned as an efficient catalyst for the addition reaction of diisopropyl phosphite with *N*-PMP-protected

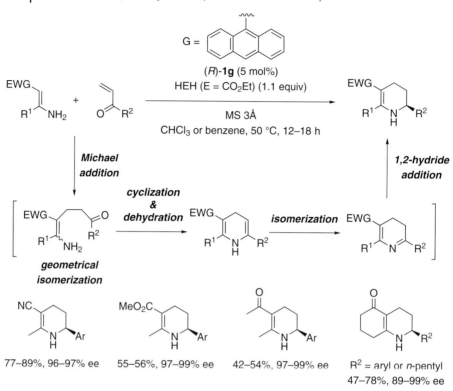

Scheme 3.51 Multiple cascade sequence leading to tetrahydropyridine and azadecalinone derivatives.

aldimines (Scheme 3.52) [108]. Aromatic and α,β-unsaturated imines could be used in the reaction, giving α-amino phosphonate esters in high yields. The use of sterically demanding dialkyl phosphites and aldimines derived from cinnamaldehyde derivatives is essential to achieve high enantioselectivities. The same research group and Yamanaka and coworkers conducted DFT computational analysis of hydrophosphonylation [109]. The chiral phosphoric acid plays a significant role in the activation of both imine and phosphite based on its dual functional catalysis.

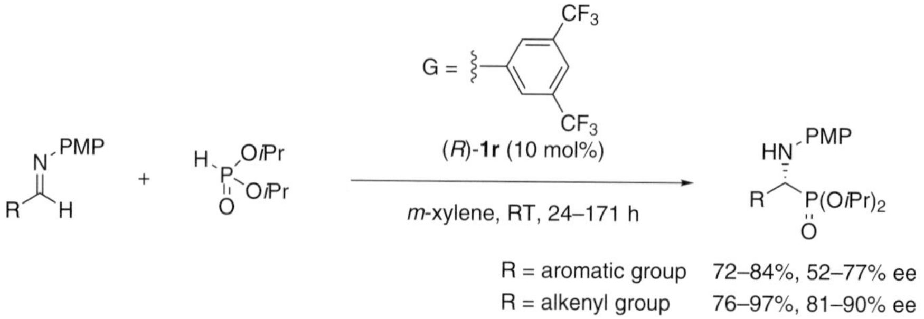

Scheme 3.52 Hydrophosphonylation of imines.

List and coworkers reported an excellent approach to the enantioselective synthesis of β-branched α-amino phosphonates, which involved the extension of the dynamic kinetic resolution strategy (Scheme 3.53) [110] that was previously applied to the enantioselective reductive amination of α-branched aldehydes by his research group (see Scheme 3.45). The method combines dynamic kinetic resolution with the parallel creation of an additional stereogenic center. They successfully accomplished the direct three-component Kabachnik–Fields reaction of 1 equiv each of the racemic aldehyde, p-anisidine, and di(3-pentyl)phosphite in the presence of newly developed chiral phosphoric acid **1t**. The corresponding β-branched α-amino phosphonates were obtained in high diastereo- and enantioselectivities, especially for the aldehydes bearing a secondary alkyl group at the α-position.

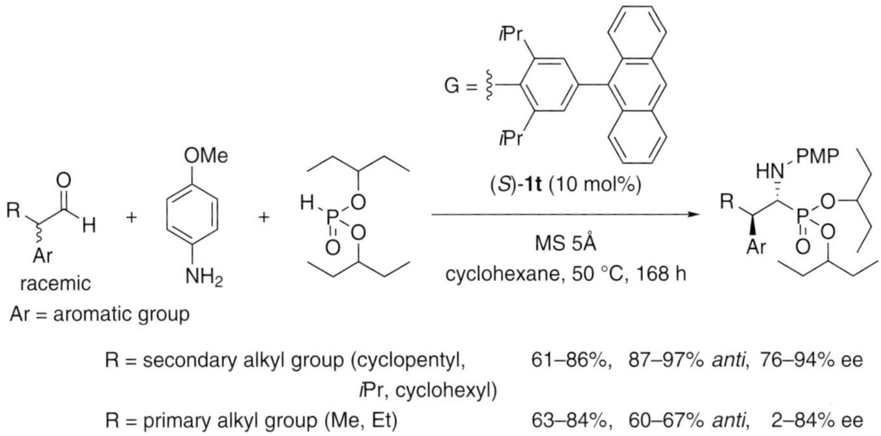

Scheme 3.53 Enantioselective synthesis of β-branched α-amino phosphonates by Kabachnik–Fields reaction.

3.4.2
Formation of (Hemi)Aminals

Aminals, compounds having two amino groups bound to the same carbon atom, are represented in many medicinal agents having versatile therapeutic action, such as proteinase inhibitors and neurotensins. Antilla and coworkers developed an enantioselective synthesis of protected aminals from the amidation reaction of N-Boc imines with a series of sulfonamides catalyzed by chiral phosphoric acids (Scheme 3.54a) [111]. In this novel enantioselective transformation, phosphoric acid **9** exhibited excellent catalytic activity and enantioselectivity in addition to N-Boc aromatic imines. The enantioenriched aminal products were stable upon storage; neither decomposition nor racemization was observed in solution over several days. The same research group reported the enantioselective amidation reaction of N-Boc aromatic imines with phthalimide or its derivatives (Scheme 3.54b) [112].

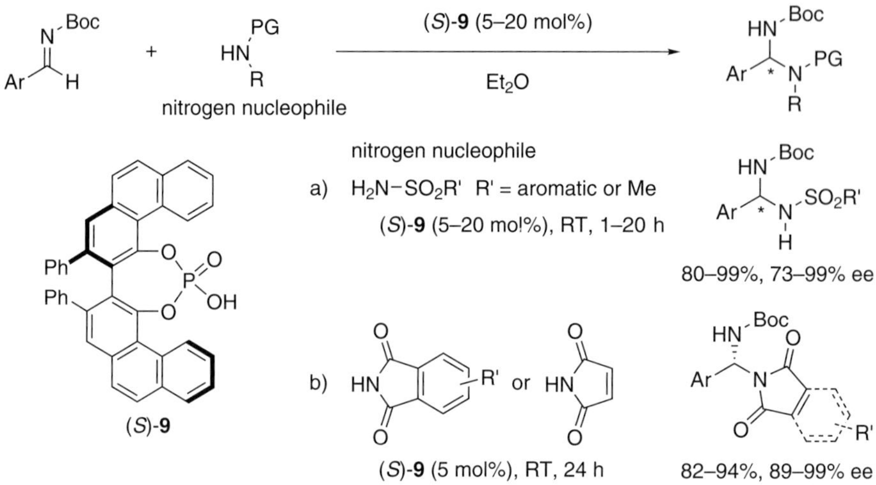

Scheme 3.54 Preparation of aminals via addition of nitrogen nucleophiles to imines.

Cyclic aminals are relatively common structural elements of diverse pharmaceutical compounds. However, no efficient method is available to prepare these compounds. The research groups of List [113] and Rueping [114] independently reported a highly enantioselective direct synthesis of cyclic aminals from aldehydes and o-aminobenzamide derivatives using chiral phosphoric acids (Scheme 3.55). The catalytic reaction involves imine formation/intramolecular amine addition sequence using aldehydes and o-aminobenzamide derivatives. In List's approach, aliphatic aldehydes were mainly employed as the substrate and corresponding dihydroquinazolinones were furnished in excellent enantioselectivities when primary alkyl-substituted aldehydes and phosphoric acid catalyst **1t** were used (Scheme 3.55a). They also applied the enantioselective method to sulfonamide analogues, instead of benzamides. The method provides a practical and efficient route to pharmaceutically relevant compounds, the benzo(thia)diazine class of cyclic aminals. In fact, they synthesized several benzo(thia)diazines, including Aquamox, Thiabutazide, and Penflutizide, with high enantioselectivities using their methodology. Meanwhile, Rueping *et al.* focused their attention on preparing aromatic-substituted cyclic aminals using aromatic aldehydes as the substrate (Scheme 3.55b). Phosphoric acid **1g** was the best catalyst in their case.

Antilla and coworkers further extended their methodology to the addition of an oxygen nucleophile, instead of a nitrogen nucleophile (see Scheme 3.54), to imines (Scheme 3.56) [115]. They successfully developed the catalytic enantioselective addition reaction of alcohols as oxygen nucleophiles with *N*-benzoyl imines using chiral phosphoric acid **1g**. The method provides straightforward access to chiral hemiaminal ethers with high enantioselectivities.

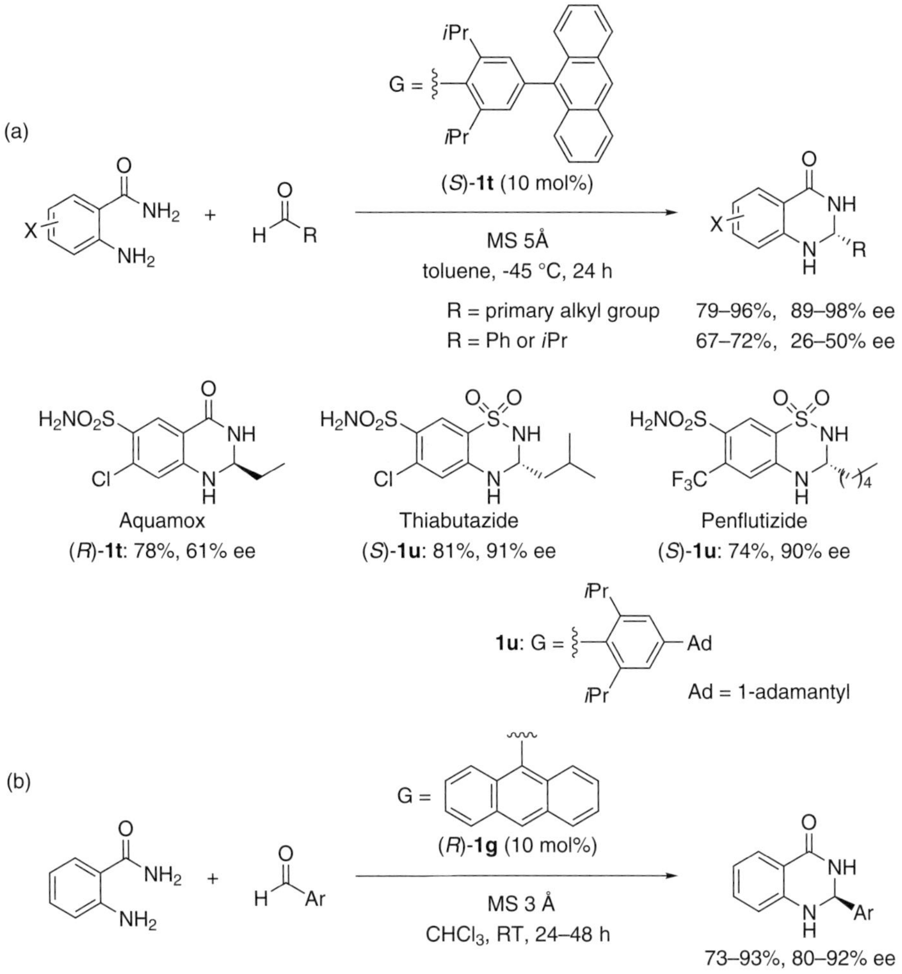

Scheme 3.55 Direct transformation of o-aminobenzamides with aldehydes to cyclic aminals.

3.4.3
Nucleophilic Ring Opening of Aziridines and Related Reactions

The ring opening of aziridines with nitrogen nucleophiles provides an efficient route to a vicinal diamine structural moiety [116] that exists in numerous biologically active compounds and medicinal agents as well as chiral ligands of metal complexes. In this context, considerable effort has been devoted to their enantioselective synthesis. One of the most direct methods for preparing enantioenriched vicinal diamines is the enantioselective desymmetrization of *meso*-aziridines with nitrogen nucleophiles. Antilla and coworkers developed a highly enantioselective synthesis of vicinal diamines based on the desymmetrization strategy using chiral phosphoric acid **9**

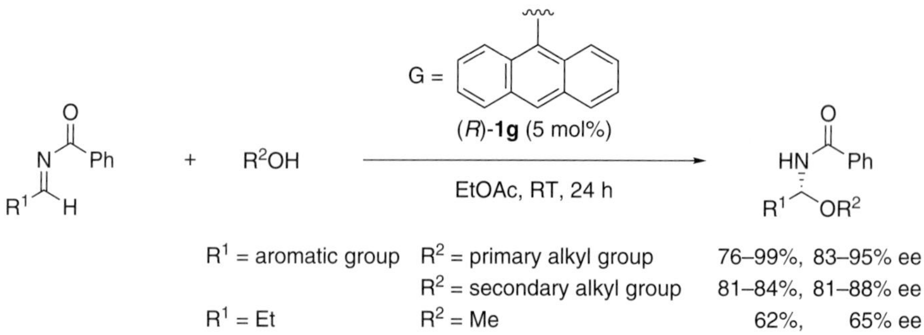

Scheme 3.56 Preparation of hemiaminal ethers via addition of alcohols to imines.

and azidotrimethylsilylane as the nitrogen nucleophile (Scheme 3.57) [117]. A series of *meso*-aziridines derived from (hetero)cyclic and acyclic alkanes having aliphatic or aromatic substituents were applicable to the enantioselective aziridine ring-opening reaction, giving the corresponding diamines with high enantioselectivities. The bis (3,5-trifluoromethyl)benzoyl substituent introduced to the nitrogen atom of aziridines optimized both yield and enantioselectivity. They also proposed that silylated phosphate generated *in situ* by the reaction of chiral phosphoric acid with azidotrimethylsilylane would be a catalytically active species.

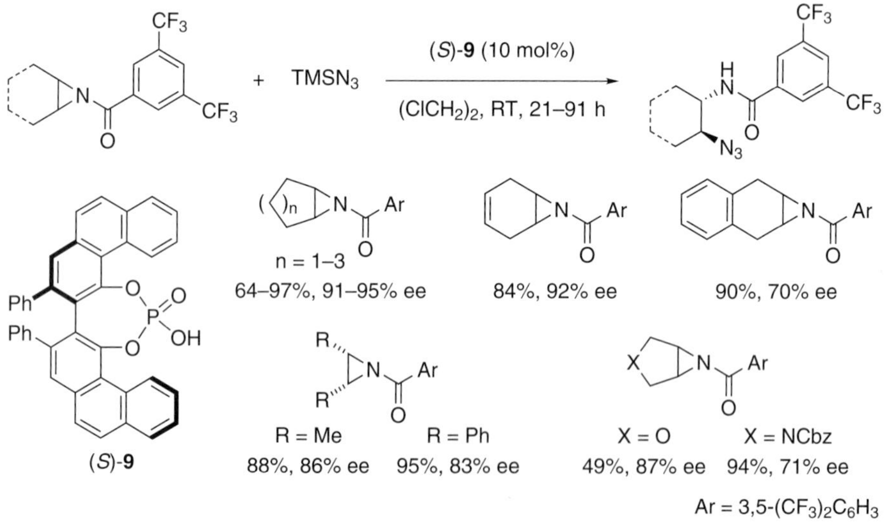

Scheme 3.57 Enantioselective desymmetrization of *meso*-aziridines with azide.

Toste and coworkers successfully demonstrated an enantioselective synthesis of β-alkoxy amines from racemic β-chloro tertiary amines and alcohols as an oxygen nucleophile under the influence of an equimolar amount of silver(I) salt and a catalytic amount of chiral phosphoric acid **1q** (Scheme 3.58) [118]. The methodology is based on chiral counteranion-mediated enantioselective transformation [76]. The reaction proceeded via the asymmetric ring opening of intermediary *meso*-aziridinium ions (**M**) that were generated from the ring closure of β-chloro tertiary amines by silver(I) salts. Subsequent nucleophilic ring opening of the *meso*-aziridinium ion by alcohol provided the corresponding β-alkoxy amines. Secondary, tertiary, and relatively hindered primary alcohols could be used in the reaction, giving β-alkoxy amines with high enantioselectivities. Efficient desymmetrization of *meso*-aziridinium ion is achieved under the influence of the chiral counteranion derived from phosphoric acid **1q**. The resulting catalytic process can be regarded as chiral anion phase-transfer catalysis in analogy to the established chiral cation alternative [119].

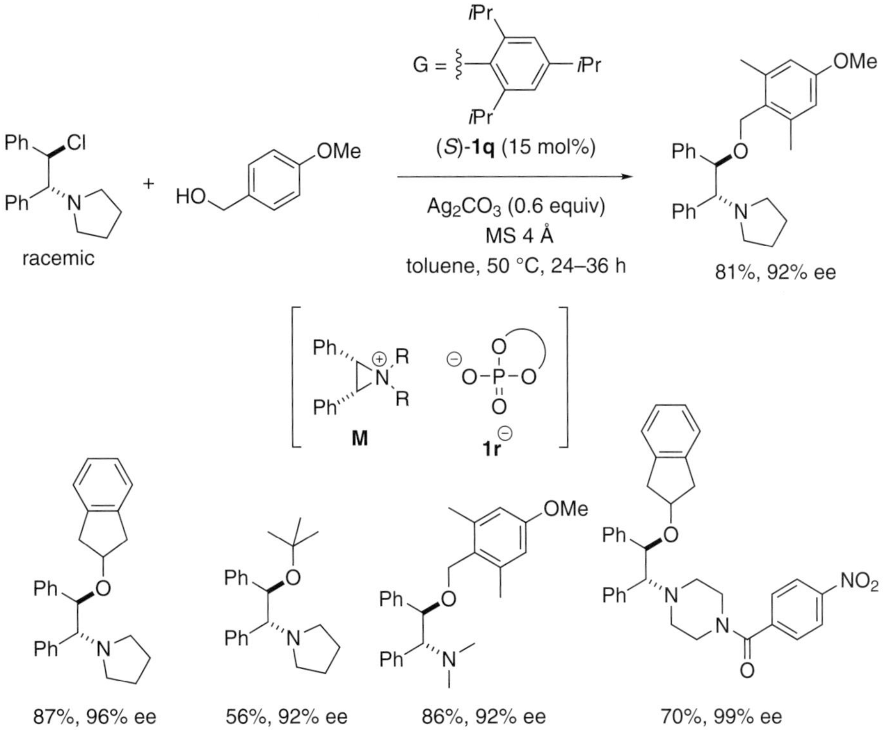

Scheme 3.58 Enantioselective desymmetrization of *meso*-aziridinium ions with alcohols.

3.5 Conclusion

From the contents of this chapter, it was convincingly demonstrated that chiral phosphoric acid catalysts of general type **1** have significantly broadened the scope of the enantioselective synthesis of chiral amines specifically through the nucleophilic addition to imines. Numerous bond forming reactions, including carbon–carbon, carbon–hydrogen, and carbon–heteroatom bond forming reactions, have been successfully established in a highly enantioselective manner by using these acid catalysts. The enantioselective versions of many reactions remain to be developed using chiral phosphoric acids or other types of chiral Brønsted acids [120]. Further elaboration of novel chiral Brønsted acids derived not only from other types of chiral backbones, such as diols, diamines, and amino alcohols, but also from stronger acid functionalities could allow the discovery of more selective and efficient methods for the enantioselective synthesis of nitrogen-containing compounds.

Questions

3.1. When N-aryl imine is treated with Hantzsch ester in the presence of an acid catalyst, amine is formed. What is the driving force of this hydrogen transfer reaction?

3.2. An inverse electron-demand aza-D–A reaction of electron-rich alkenes with N-aryl imines as 2-azadiene (Povarov reaction) provides tetrahydroquinolines. Reactions catalyzed by chiral phosphoric acids yielded different absolute stereochemical outcomes when ethyl vinyl ether and enecarbamate are employed as electron-rich alkenes, although chiral phosphoric acids have the same axial chirality in both cases (see Scheme 3.26).
Propose a plausible explanation for this reversal of enantiofacial selectivity.

3.3. In general, a Lewis acid-catalyzed 1,3-dipolar cycloaddition reaction of nitrones with vinyl ethers provides *exo*-isomers predominantly. In fact, the MeAl-BINOL-catalyzed cycloaddition reaction of ethyl vinyl ether with diaryl nitrone furnished the corresponding product with high *exo*-selectivity [121]. However, phosphoramide **6b** as the chiral Brønsted acid catalyst exhibited high *endo*-selectivity in the same cycloaddition reaction (see Scheme 3.27).

The predominance of *endo*-isomers is in contrast to the formation of *exo*-isomers as the major products under Lewis acid catalysis. Propose a plausible explanation for this difference.

MeAl-(*R*)-binol (10 mol%)   >95% *exo*, 94% ee

(*R*)-**6b** (5 mol%)   97% *endo*, 90% ee

(Ad = 1-adamantyl)

References

1. For recent reviews, see (a) Berkessel, A. and Gröger, H. (2005) *Asymmetric Organocatalysis: From Biomimetic Concepts to Powerful Methods for Asymmetric Synthesis*, Wiley-VCH Verlag GmbH, Weinheim; (b) Dalko, P.I. (ed.) (2007) *Enantioselective Organocatalysis: Reactions and Experimental Procedures*, John Wiley & Sons, New York.

2. For recent reviews, see (a) Akiyama, T., Itoh, J., and Fuchibe, K. (2006) *Adv. Synth. Catal.*, **348**, 999–1010; (b) Doyle, A.G. and Jacobsen, E.N. (2007) *Chem. Rev.*, **107**, 5713–5743; (c) Akiyama, T. (2007) *Chem. Rev.*, **107**, 5744–5758; (d) Yu, X. and Wang, W. (2008) *Chem. Asian J.*, **3**, 516–532.

3. (a) Connon, S.J. (2006) *Angew. Chem., Int. Ed.*, **45**, 3909–3912; (b) Terada, M. (2008) *Chem. Commun.*, 4097–4112.

4. Quin, L.D. (2000) *A Guide to Organophosphorus Chemistry*, John Wiley & Sons, New York, pp. 133–168.

5. Uraguchi, D. and Terada, M. (2004) *J. Am. Chem. Soc.*, **126**, 5356–5357.

6 Akiyama, T., Itoh, J., Yokota, K., and Fuchibe, K. (2004) *Angew. Chem., Int. Ed.*, **43**, 1566–1568.

7 Friestad, G.K. and Mathies, A.K. (2007) *Tetrahedron*, **63**, 2541–2569.

8 For general reviews, see (a) Arend, M., Westermann, B., and Risch, N. (1998) *Angew. Chem., Int. Ed.*, **37**, 1044–1070; (b) Bur, S.K. and Martin, S.F. (2001) *Tetrahedron*, **57**, 3221–3242; (c) Córdova, A. (2004) *Acc. Chem. Res.*, **37**, 102–112.

9 For recent reviews, see (a) Ting, A. and Schaus, S.E. (2007) *Eur. J. Org. Chem.*, 5797–5815; (b) Verkade, J.M.M., van Hemert, L.J.C., Quaedflieg, P.J.L.M., and Rutjes, F.P.J.T. (2008) *Chem. Soc. Rev.*, **37**, 29–41.

10 Yamanaka, M., Itoh, J., Fuchibe, K., and Akiyama, T. (2007) *J. Am. Chem. Soc.*, **129**, 6756–6764.

11 Akiyama, T., Saitoh, Y., Morita, H., and Fuchibe, K. (2005) *Adv. Synth. Catal.*, **347**, 1523–1526.

12 Terada, M., Sorimachi, K., and Uraguchi, D. (2006) *Synlett*, 133–136.

13 Guo, Q.-X., Liu, H., Guo, C., Luo, S.-W., Gu, Y., and Gong, L.-Z. (2007) *J. Am. Chem. Soc.*, **129**, 3790–3791.

14 Rueping, M., Sugiono, E., and Schoepke, F.R. (2007) *Synlett*, 1441–1445.

15 Chen, X.-H., Xu, X.-Y., Liu, H., Cun, L.-F., and Gong, L.-Z. (2006) *J. Am. Chem. Soc.*, **128**, 14802–14803.

16 Gong, L.-Z., Chen, X.-H., and Xu, X.-Y. (2007) *Chem. Eur. J.*, **13**, 8920–8926.

17 (a) Antilla, J.C. and Wulff, W.D. (2000) *Angew. Chem., Int. Ed.*, **39**, 4518–4521; (b) Redlich, M. and Hossain, M.M. (2004) *Tetrahedron Lett.*, **45**, 8987–8990, and references cited therein.

18 Williamsa, A.L. and Johnston, J.N. (2004) *J. Am. Chem. Soc.*, **126**, 1612–1613, and references cited therein.

19 Uraguchi, D., Sorimachi, K., and Terada, M. (2005) *J. Am. Chem. Soc.*, **127**, 9360–9361.

20 Akiyama, T., Suzuki, T., and Mori, K. (2009) *Org. Lett.*, **11**, 2445–2447.

21 (a) Carswell, E.L., Snapper, M.L., and Hoveyda, A.H. (2006) *Angew. Chem., Int. Ed.*, **45**, 7230–7233; (b) Mandai, H., Mandai, K., Snapper, M.L., and Hoveyda, A.H. (2008) *J. Am. Chem. Soc.*, **130**, 17961–17969; (c) Yamaguchi, A., Matsunaga, S., and Shibasaki, M. (2008) *Org. Lett.*, **10**, 2319–2322.

22 Denmark, S.E., Heemetra, J.R. Jr, and Beutner, G.L. (2005) *Angew. Chem., Int. Ed.*, **44**, 4682–4698.

23 Akiyama, T., Honma, Y., Itoh, J., and Fuchibe, K. (2008) *Adv. Synth. Catal.*, **350**, 399–402.

24 Sickert, M. and Schneider, C. (2008) *Angew. Chem., Int. Ed.*, **47**, 3631–3634.

25 Giera, D.S., Sickert, M., and Schneider, C. (2008) *Org. Lett.*, **10**, 4259–4262.

26 (a) Yang, J.W., Stadler, M., and List, B. (2007) *Angew. Chem., Int. Ed.*, **46**, 609–611; (b) Zhang, H., Mitsumori, S., Utsumi, N., Imai, M., Garcia-Delgado, N., Mifsud, M., Albertshofer, K., Cheong, P.H.-Y., Houk, K.N., Tanaka, F., and Barbas, C.F. III (2008) *J. Am. Chem. Soc.*, **130**, 875–886; (c) Kano, T., Yamaguchi, Y., and Maruoka, K. (2009) *Angew. Chem., Int. Ed.*, **48**, 1838–1840, and references cited therein.

27 Terada, M. and Toda, Y. (2009) *J. Am. Chem. Soc.*, **131**, 6354–6355.

28 Frauenrath, H., Arenz, T., Raabe, G., and Zorn, M. (1993) *Angew. Chem., Int. Ed. Engl.*, **32**, 83–85; (b) Arenz, T., Frauenrath, H., Raabe, G., and Zorn, M. (1994) *Liebigs Ann. Chem.*, 931–942; (c) Tayama, E., Otoyama, S., and Isaka, W. (2008) *Chem. Commun.*, 4216–4218.

29 For reviews on enantioselective Strecker reaction, see (a) Gröger, H. (2003) *Chem. Rev.*, **103**, 2795–2827; (b) Shibasaki, M., Kanai, M., and Mita, T. (2008) *Org. React.*, **70**, 1–119; (c) Merino, P., Marques-Lopez, E., Tejero, T., and Herrera, R.P. (2009) *Tetrahedron*, **65**, 1219–1234.

30 Rueping, M., Sugiono, E., and Azap, C. (2006) *Angew. Chem., Int. Ed.*, **45**, 2617–2619.

31 Rueping, M., Sugiono, E., and Moreth, S.A. (2007) *Adv. Synth. Catal.*, **349**, 759–764.

32 For theoretical studies on enantioselective Strecker reaction catalyzed by chiral phosphoric acids, see Simon, L. and Goodman, J.M. (2009) *J. Am. Chem. Soc.*, **131**, 4070–4077.

33 For a review, see Westermann, B. (2003) *Angew. Chem., Int. Ed.*, **42**, 151–153.

34 Rueping, M. and Antonchick, A.P. (2008) *Org. Lett.*, **10**, 1731–1734.

35 (a) Lassaletta, J.-M. and Fernández, R. (1992) *Tetrahedron Lett.*, **33**, 3691–3694; (b) Lassaletta, J.-M., Fernández, R., Martín-Zamora, E., and Díez, E. (1996) *J. Am. Chem. Soc.*, **118**, 7002–7003.

36 Rueping, M., Sugiono, E., Theissman, T., Kuenkel, A., Köckritz, A., Pews-Davtyan, A., Nemati, N., and Beller, M. (2007) *Org. Lett.*, **9**, 1065–1068.

37 For reviews, see (a) Jørgensen, K.A. (2003) *Synthesis*, 1117–1125; (b) Bandini, M., Melloni, A., and Umani-Ronchi, A. (2004) *Angew. Chem., Int. Ed.*, **43**, 550–556; (c) Poulsen, T.B. and Jørgensen, K.A. (2008) *Chem. Rev.*, **108**, 2903–2915.

38 Uraguchi, D., Sorimachi, K., and Terada, M. (2004) *J. Am. Chem. Soc.*, **126**, 11804–11805.

39 For reviews see (a) Joule, J.A. (2001) *Science of Synthesis*, **10**, 361–652; (b) Aygun, A. and Pindur, U. (2003) *Curr. Med. Chem.*, **10**, 1113–1127; (c) Abele, E., Abele, R., Dzenitis, O., and Lukevics, E. (2003) *Chem. Heterocycl. Compd.*, **39**, 825–865; (d) Gribble, G.W. (2003) *Pure Appl. Chem.*, **75**, 1417–1432; (e) Sings, H. and Singh, S. (2003) *Alkaloids*, **60**, 51; (f) Somei, M. and Yamada, F. (2004) *Nat. Prod. Rep.*, **21**, 278–311; (g) Somei, M. and Yamada, F. (2005) *Nat. Prod. Rep.*, **22**, 73–103.

40 Kang, Q., Zhao, Z.-A., and You, S.-L. (2007) *J. Am. Chem. Soc.*, **129**, 1484–1485.

41 Rowland, G.B., Rowland, E.B., Liang, Y., Perman, J.A., and Antilla, J.C. (2007) *Org. Lett.*, **9**, 2609–2611.

42 Terada, M., Yokoyama, S., Sorimachi, K., and Uraguchi, D. (2007) *Adv. Synth. Catal.*, **349**, 1863–1867.

43 Kang, Q., Zhao, Z.-A., and You, S.-L. (2009) *Tetrahedron*, **65**, 1603–1607.

44 Wanner, M.J., Hauwert, P., Schoemaker, H.E., de Gelder, R., van Maarseveen, J.H., and Hiemstra, H. (2008) *Eur. J. Org. Chem.*, 180–185.

45 Li, G., Rowland, G.B., Rowland, E.B., and Antilla, J.C. (2007) *Org. Lett.*, **9**, 4065–4068.

46 Nakamura, S., Sakurai, Y., Kakashima, H., Shibata, N., and Toru, T. (2009) *Synlett*, 1639–1642.

47 Kang, Q., Zheng, X.-J., and You, S.-L. (2008) *Chem. Eur. J.*, **14**, 3539–3542.

48 Enders, D., Narine, A.A., Toulgoat, F., and Bisschops, T. (2008) *Angew. Chem., Int. Ed.*, **47**, 5661–5665.

49 Nakashima, D. and Yamamoto, H. (2006) *J. Am. Chem. Soc.*, **128**, 9626–9627.

50 Terada, M. and Sorimachi, K. (2007) *J. Am. Chem. Soc.*, **129**, 292–293.

51 Terada, M., Tanaka, H., and Sorimachi, K. (2008) *Synlett*, 1661–1664.

52 Jia, Y.-X., Zhong, J., Zhu, S.-F., Zhang, C.-M., and Zhou, Q.-L. (2007) *Angew. Chem., Int. Ed.*, **46**, 5565–5567.

53 For reviews, see (a) Cox, E.D. and Cook, J.M. (1995) *Chem. Rev.*, **95**, 1797–1842; (b) Chrzanowska, M. and Rozwadowska, M.D. (2004) *Chem. Rev.*, **104**, 3341–3370.

54 Seayad, J., Seayad, A.M., and List, B. (2006) *J. Am. Chem. Soc.*, **128**, 1086–1087.

55 Wanner, M.J., van der Haas, R.N.S., de Cuba, K., van Maarseveen, J.H., and Hiemstra, H. (2007) *Angew. Chem., Int. Ed.*, **46**, 7485–7487; also see Sewgobind, N.V., Wanner, M.J., Ingemann, S., de Gelder, R., van Maarseveen, J.H., and Hiemstra, H., (2008) *J. Org. Chem.*, **73**, 6405–6408.

56 Wanner, M.J., Boots, R.N.A., Eradus, B., de Gelder, R., van Maarseveen, J.H., and Hiemstra, H. (2009) *Org. Lett.*, **11**, 2579–2581.

57 Akiyama, T., Tamura, Y., Itoh, J., Morita, H., and Fuchibe, K. (2006) *Synlett*, 141–143.

58 Itoh, J., Fuchibe, K., and Akiyama, T. (2006) *Angew. Chem., Int. Ed.*, **45**, 4796–4798.

59 Liu, H., Cun, L.-F., Mi, A.-Q., Jiang, Y.-Z., and Gong, L.-Z. (2006) *Org. Lett.*, **8**, 6023–6026.

60 Rueping, M. and Azap, C. (2006) *Angew. Chem., Int. Ed.*, **45**, 7832–7835.

61 Akiyama, T., Morita, H., and Fuchibe, K. (2006) *J. Am. Chem. Soc.*, **128**, 13070–13071.

62 Liu, H., Dagousset, G., Masson, G., Retailleau, P., and Zhu, J. (2009) *J. Am. Chem. Soc.*, **131**, 4598–4599.

63 Jiao, P., Nakashima, D., and Yamamoto, H. (2008) *Angew. Chem., Int. Ed.*, **47**, 2411–2413.

64 Chen, X.-H., Zhang, W.-Q., and Gong, L.-Z. (2008) *J. Am. Chem. Soc.*, **130**, 5652–5653.

65 (a) Matsubara, R., Nakamura, Y., and Kobayashi, S. (2004) *Angew. Chem., Int. Ed.*, **43**, 1679–1681; (b) Kiyohara, H., Matsubara, R., and Kobayashi, S. (2006) *Org. Lett.*, **8**, 5333–5335.

66 For some excellent studies of low loading of organocatalyst (less than 0.5 Su, J.T., Vachal, P., and Jacobsen, E.N. (2001) *Adv. Synth. Catal.*, **343**, 197–200; (b) Vachal, P. and Jacobsen, E.N. (2002) *J. Am. Chem. Soc.*, **124**, 10012–10014; (c) Saaby, S., Bella, M., and Jørgensen, A.K. (2004) *J. Am. Chem. Soc.*, **126**, 8120–8121; (d) Shirakawa, S., Yamamoto, K., Kitamura, M., Ooi, T., and Maruoka, K. (2005) *Angew. Chem., Int. Ed.*, **44**, 625–628; (e) Kitamura, M., Shirakawa, S., and Maruoka, K. (2005) *Angew. Chem., Int. Ed.*, **44**, 1549–1551; (f) Terada, M., Ube, H., and Yaguchi, Y. (2006) *J. Am. Chem. Soc.*, **128**, 1454–1455; (g) Terada, M., Nakano, M., and Ube, H. (2006) *J. Am. Chem. Soc.*, **128**, 16044–16045.

67 Terada, M., Machioka, K., and Sorimachi, K. (2006) *Angew. Chem., Int. Ed.*, **45**, 2254–2257.

68 For a recent review, see Enders, D., Grondal, C., and Hüttl, M.R.M. (2007) *Angew. Chem., Int. Ed.*, **46**, 1570–1581.

69 Terada, M., Machioka, K., and Sorimachi, K. (2007) *J. Am. Chem. Soc.*, **129**, 10336–10337.

70 Matsubara, R., Kawai, N., and Kobayashi, S. (2006) *Angew. Chem., Int. Ed.*, **45**, 3814–3816.

71 (a) Gaulon, C., Gizecki, P., Dhal, R., and Dujardin, G. (2002) *Synlett*, 952–956; (b) Prashad, M., Lu, Y., and Repič, O. (2004) *J. Org. Chem.*, **69**, 584–586; (c) Matsubara, R., Kawai, N., and Kobayashi, S. (2006) *Angew. Chem., Int. Ed.*, **45**, 3814–3816.

72 Terada, M., Machioka, K., and Sorimachi, K. (2009) *Angew. Chem., Int. Ed.*, **48**, 2553–2556.

73 Baudequin, C., Zamfir, A., and Tsogoeva, S.B. (2008) *Chem. Commun.*, 4637–4639.

74 For reviews, see (a) Hiersemann, M. and Nubbemeyer, U. (eds) (2007) *The Claisen Rearrangement*, Wiley-VCH Verlag GmbH, Weinheim; (b) Martin Castro, A.M. (2004) *Chem. Rev.*, **104**, 2939–3002.

75 Rueping, M. and Antonchick, A.P. (2008) *Angew. Chem., Int. Ed.*, **47**, 10090–10093.

76 (a) Mayer, S. and List, B. (2006) *Angew. Chem., Int. Ed.*, **45**, 4193–4195, and references cited therein; also see (b) Wang, X. and List, B., (2008) *Angew. Chem., Int. Ed.*, **47**, 1119–1122; (c) Hamilton, G.L., Kanai, T., and Toste, F.D. (2008) *J. Am. Chem. Soc.*, **130**, 14984–14986.

77 Reisman, S.E., Doyle, A.G., and Jacobsen, E.N. (2008) *J. Am. Chem. Soc.*, **130**, 7198–7199.

78 For a review on synthetic utility of azlactones, see Fisk, J.S., Mosey, R.A., and Tepe, J.J. (2007) *Chem. Soc. Rev.*, **36**, 1432–1440.

79 Terada, M., Tanaka, H., and Sorimachi, K. (2009) *J. Am. Chem. Soc.*, **131**, 3430–3431.

80 Rueping, M., Antonchick, A.P., and Brinkmann, C. (2007) *Angew. Chem., Int. Ed.*, **46**, 6903–6906.

81 Hu, W., Xu, X., Zhou, J., Liu, W.-J., Huang, H., Hu, J., Yang, L., and Gong, L.-Z. (2008) *J. Am. Chem. Soc.*, **130**, 7782–7783.

82 Sorimachi, K. and Terada, M. (2008) *J. Am. Chem. Soc.*, **130**, 14452–14453.

83 Han, Z.-Y., Xiao, H., Chen, X.-H., and Gong, L.-Z. *J. Am. Chem. Soc. ASAP*. doi: 10.1021/ja903547q

84 For reviews, see (a) Blaser, H.-U., Malan, C., Pugin, B., Spindler, F., Steiner, H., and Studer, M. (2003) *Adv. Synth. Catal.*, **345**, 103–151; (b) Tang, W. and Zhang, X. (2003) *Chem. Rev.*, **103**, 3029–3070; (c) Riant, O., Mostefaï, N., and Courmarcel, J. (2004) *Synthesis*, 2943–2958.

85 (a) Yang, J.W., Hechavarria Foneca, M.T., Vignola, N., and List, B. (2005) *Angew. Chem., Int. Ed.*, **44**, 110–112; (b) Ouellet, S.G., Tullte, J.B., and MacMillan, D.W.C. (2005) *J. Am. Chem. Soc.*, **127**, 32–33; also see (c) Yang, J.W., Hechavarria Foneca, M.T., and List, B. (2004) *Angew. Chem., Int. Ed.*, **43**, 6660–6662.

86 For reviews, see (a) Ouellet, S.G., Walji, A.M., and Macmillan, D.W.C. (2007) *Acc. Chem. Res.*, **40**, 1327–1339; (b) You, S.-L. (2007) *Chem. Asian J.*, **2**, 820–827.

87 Rueping, M., Sugiono, E., Azap, C., Theissmann, T., and Bolte, M. (2005) *Org. Lett.*, **7**, 3781–3783.

88. For theoretical studies, see Simón, L. and Goodman, J.M. (2008) *J. Am. Chem. Soc.*, **130**, 8741–8747.
89. Hoffmann, S., Seayad, A.M., and List, B. (2005) *Angew. Chem., Int. Ed.*, **44**, 7424–7427.
90. Storer, R.I., Carrera, D.E., Ni, Y., and MacMillan, D.W.C. (2006) *J. Am. Chem. Soc.*, **128**, 84–86.
91. Rueping, M., Antonchick, A.P., and Theissmann, T. (2006) *Angew. Chem., Int. Ed.*, **45**, 6751–6755.
92. Li, G., Liang, Y., and Antilla, J.C. (2007) *J. Am. Chem. Soc.*, **129**, 5830–5831.
93. Kang, Q., Zhao, Z.-A., and You, S.-L. (2007) *Adv. Synth. Catal.*, **349**, 1657–1660.
94. Kang, Q., Zhao, Z.-A., and You, S.-L. (2008) *Org. Lett.*, **10**, 2031–2034.
95. Li, G. and Antilla, J.C. (2009) *Org. Lett.*, **11**, 1075–1078.
96. Hoffmann, S., Nicoletti, M., and List, B. (2006) *J. Am. Chem. Soc.*, **128**, 13074–13075.
97. Glorius, F. (2005) *Org. Biomol. Chem.*, **3**, 4171–4175.
98. Rueping, M., Antonchick, A.P., and Theissmann, T. (2006) *Angew. Chem., Int. Ed.*, **45**, 3683–3686.
99. Guo, Q.-S., Du, D.-M., and Xu, J. (2008) *Angew. Chem., Int. Ed.*, **47**, 759–762.
100. Rueping, M. and Antonchick, A.P. (2007) *Angew. Chem., Int. Ed.*, **46**, 4562–4565.
101. Metallinos, C., Barrett, F.B., and Xu, S. (2008) *Synlett*, 720–724.
102. Rueping, M., Theissmann, T., Raja, S., and Bats, J.W. (2008) *Adv. Synth. Catal.*, **350**, 1001–1006.
103. For reviews, see (a) List, B. (2006) *Chem. Commun.*, 819–824; (b) Palomo, C. and Mielgo, A. (2006) *Angew. Chem., Int. Ed.*, **45**, 7876–7880; (c) Walji, A.M. and MacMillan, D.W.C. (2007) *Synlett*, 1477–1489.
104. Zhou, J. and List, B. (2007) *J. Am. Chem. Soc.*, **129**, 7498–7499.
105. Schrader, W., Handayani, P.P., Zhou, J., and List, B. (2009) *Angew. Chem., Int. Ed.*, **48**, 1463–1466.
106. Rueping, M. and Antonchick, A.P. (2008) *Angew. Chem., Int. Ed.*, **47**, 5836–5838.
107. For a review of enantioselective synthesis of α-amino phosphonate derivatives, see Ma, J.-A. (2006) *Chem. Soc. Rev.*, **35**, 630–636.
108. Akiyama, T., Morita, H., Itoh, J., and Fuchibe, K. (2005) *Org. Lett.*, **7**, 2583–2585.
109. For computational analysis of transition states, see (a) Akiyama, T., Morita, H., Bachu, P., Mori, K., Yamanaka, M., and Hirata, T. (2009) *Tetrahedron*, **65**, 4950–4956; (b) Yamanaka, M. and Hirata, T. (2009) *J. Org. Chem.*, **74**, 3266–3271.
110. Cheng, X., Goddard, R., Buth, G., and List, B. (2008) *Angew. Chem., Int. Ed.*, **47**, 5079–5081.
111. Rowland, G.B., Zhang, H., Rowland, E.B., Chennamadhavuni, S., Wang, Y., and Antilla, J.C. (2005) *J. Am. Chem. Soc.*, **127**, 15696–15697.
112. Liang, Y., Rowland, E.B., Rowland, G.B., Perman, J.A., and Antilla, J.C. (2007) *Chem. Commun.*, 4477–4479.
113. Cheng, X., Vellalath, S., Goddard, R., and List, B. (2008) *J. Am. Chem. Soc.*, **130**, 15786–15787.
114. Rueping, M., Antonchick, A.P., Sugiono, E., and Grenader, K. (2009) *Angew. Chem., Int. Ed.*, **48**, 908–910.
115. Li, G., Fronczek, F.R., and Antilla, J.C. (2008) *J. Am. Chem. Soc.*, **130**, 12216–12217.
116. For a review, see Yudin, A.K. (2006) *Aziridines and Epoxides in Organic Synthesis*, Wiley-VCH Verlag GmbH, Weinheim.
117. Rowland, E.B., Rowland, G.B., Rivera-Otero, E., and Antilla, J.C. (2007) *J. Am. Chem. Soc.*, **129**, 12084–12085.
118. Hamilton, G.L., Kanai, T., and Toste, F.D. (2008) *J. Am. Chem. Soc.*, **130**, 14984–14986.
119. For a review, see Ooi, T. and Maruoka, K. (2007) *Angew. Chem., Int. Ed.*, **46**, 4222–4266.
120. For recent examples of chiral Brønsted acid catalysts, see (a) Hashimoto, T. and Maruoka, K. (2007) *J. Am. Chem. Soc.*, **129**, 10054–10055; (b) Hatano, M., Maki, T., Moriyama, K., Arinobe, M., and Ishihara, K. (2008) *J. Am. Chem. Soc.*, **130**, 16858–16860; (c) Gracía-Gracía, P., Lay, F., Gracía-Gracía, P., Rabalakos, C., and List, B. (2009) *Angew. Chem., Int. Ed.*, **48**, 4363–4366.
121. Simonsen, K.B., Bayón, P., Hazell, R.G., Gothelf, K.V., and Jørgensen, K.A. (1999) *J. Am. Chem. Soc.*, **121**, 3845–3853.

4
Reduction of Imines with Trichlorosilane Catalyzed by Chiral Lewis Bases

Pavel Kočovský and Sigitas Stončius

4.1
Introduction

Obtaining chiral amines from prochiral ketones via reduction of imine intermediates (Scheme 4.1) represents an attractive strategy that opens a straightforward route to valuable building blocks for the pharmaceutical and other fine chemicals industries. The methods that are at present most successful are based on transition metal-catalyzed high-pressure hydrogenation [1, 2], hydrosilylation [1, 3], transfer hydrogenation [4], and organocatalytic reduction with Hantzsch ester [5] or trichlorosilane [6] (see below). Owing to its simplicity and apparent environment friendliness, asymmetric hydrogenation [1, 2] is often regarded as the method of choice, especially in view of the fact that H_2 is the only stoichiometric reagent here. However, even this robust methodology is not entirely free of problems. Thus, to proceed efficiently, high pressure is required by most metal catalysts [1, 2], which is associated with certain risks, especially on an industrial scale, and with increased demands on the equipment for parallel syntheses of libraries of amines on a laboratory scale. Furthermore, metal leaching to the product must be carefully checked and reduced to a ppm level for applications in drug industry and material science. Metal recovery constitutes another common problem as its cost adds to the overall bookkeeping. Finally, transition metals are generally perceived by technologists as capricious entities, since their reactivity may suddenly be reduced by unforeseen factors, such as impurities in the starting materials or solvents, which represents another obstacle hindering their widespread use in manufacturing beyond hydrogenation.

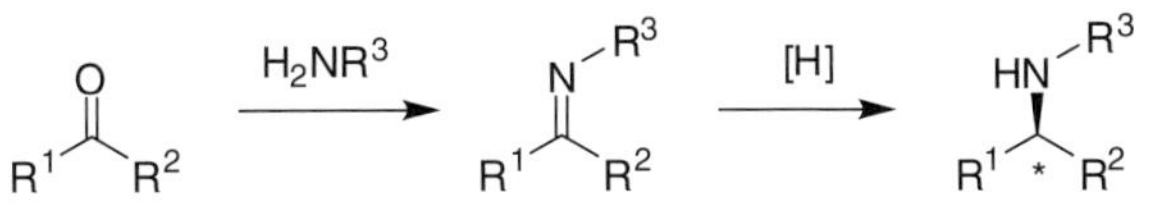

Scheme 4.1 Preparation of amines by asymmetric reduction of imines generated from prochiral ketones.

Chiral Amine Synthesis: Methods, Developments and Applications. Edited by Thomas C. Nugent
Copyright © 2010 WILEY-VCH Verlag GmbH & Co. KGaA, Weinheim
ISBN: 978-3-527-32509-2

By contrast, metal-free organocatalysts [6] can be viewed as an attractive alternative to those processes in which the metal is not vital for the key bond-forming event. The advantages are obvious: no metal leaching to the product, reduced toxicity, and lower cost of the catalysts and their regeneration. Therefore, the prime goal here is to develop new, robust processes that could compete with their established, metal-catalyzed counterparts.

Silanes are widely recognized as efficient reagents for reduction of carbonyl and heterocarbonyl functionality. In the case of alkyl- and arylsilanes, the reaction requires catalysis by Lewis acids or transition metal complexes [1, 3]; however, with more Lewis acidic trichloro- or trialkoxysilanes, an alternative metal-free activation can be accomplished. Thus, it has been demonstrated that extracoordinate silicon hydrides, formed by the coordination of silanes to Lewis bases, such as tertiary amines [7a], DMF [7b] or MeCN, and so on [7], can serve as mild reagents for the reduction of imines to amines [8]. In the case of trichlorosilane, an inexpensive and relatively easy-to-handle reducing reagent, and DMF as a Lewis basic promoter, the intermediacy of hexacoordinate species has been confirmed by ^{29}Si NMR spectroscopy [7b].

4.2
Formamides as Lewis-Basic Organocatalysts in Hydrosilylation of Imines

Since Cl$_3$SiH is known to be activated by DMF and other Lewis bases to effect hydrosilylation of imines (Scheme 4.2) [8], it is hardly surprising that chiral formamides, derived from natural amino acids, emerged as prime candidates for the development of an asymmetric variant of this reaction [8]. It was assumed that, if successful, this approach could become an attractive alternative to the existing enzymatic methods for amine production [9] and to complement another organocatalytic protocol, based on the biomimetic reduction with Hantzsch ester, which is being developed in parallel [5].

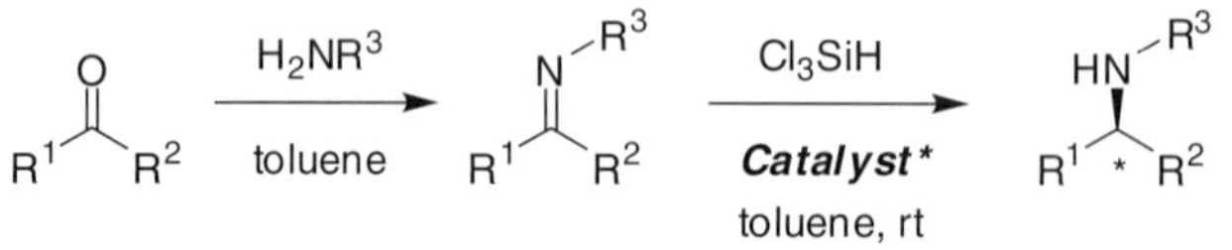

Scheme 4.2 Preparation of amines by asymmetric reduction of imines generated from prochiral ketones. For R^1, R^2, and R^3, see Tables 4.1–4.11.

A selection of the most efficient formamide catalysts based on amino acids is shown in Figures 4.1 and 4.2; representative examples of enantioselective hydrosilylation are listed in Tables 4.1–4.7. The proline-derived anilide **16** and its naphthyl analogue **17**, introduced by Matsumura as the first chiral catalysts [10], exhibited moderate enantioselectivity in the reduction of aromatic ketimines with trichlorosilane at 10 mol% catalyst loading (Table 4.1, entries 1 and 2). The formamide

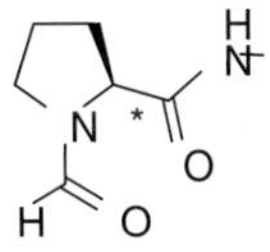

4.2 Formamides as Lewis-Basic Organocatalysts in Hydrosilylation of Imines

(S)-16, R = Ph
(S)-17, R = 1-Naphthyl

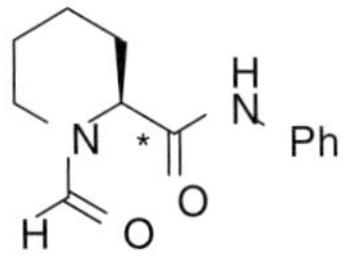

(S)-18

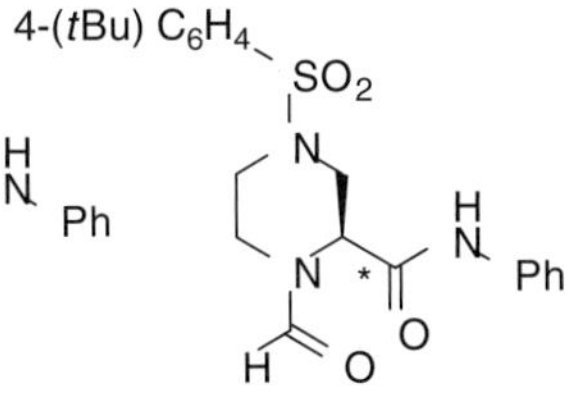

(S)-19

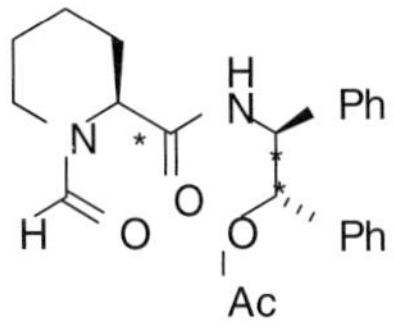

20

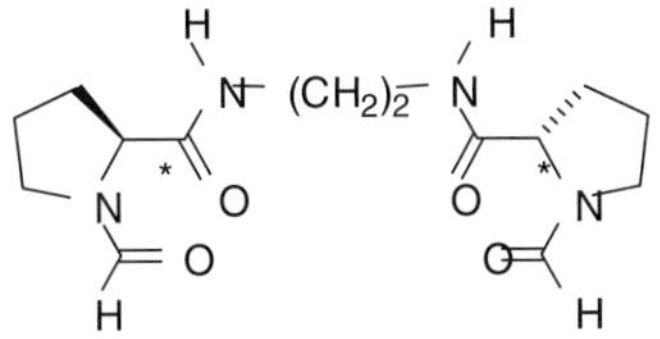

21

Figure 4.1 Formamides derived from cyclic amino acids as catalysts for the asymmetric reduction of imines.

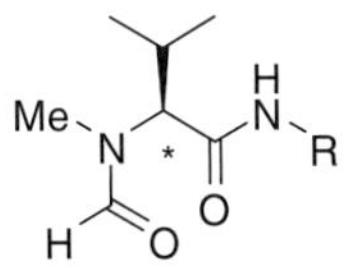

(S)-22, R = C_6H_5
(S)-23, R = 3,5-$Me_2C_6H_3$
(S)-24, R = 3,5-$(iPr)_2C_6H_3$
(S)-25, R = 3,5-$Ph_2C_6H_3$
(S)-26, R = 3,5-$(2\text{-tolyl})_2C_6H_3$
(S)-27, R = 4-$MeOC_6H_4$
(S)-28, R = 3,5-$(MeO)_2C_6H_3$
(S)-29, R = 3,5-$(CF_3)_2C_6H_3$

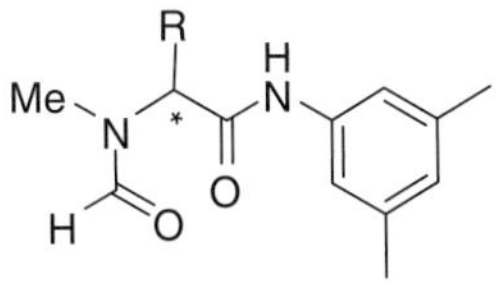

(R)-30, R = chexyl
(S)-31, R = tBu
(R)-32, R = Ph
(S)-33, R = $PhCH_2$
(S)-34, R = Me

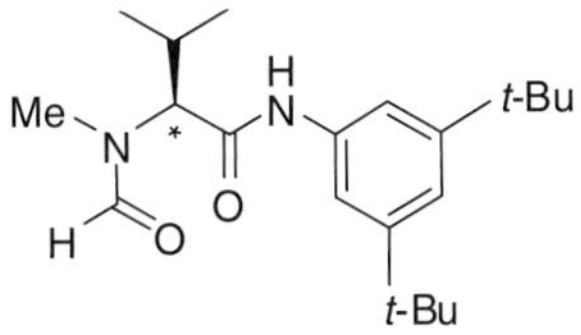

(S)-35 (**Sigamide**)

Figure 4.2 Formamides derived from noncyclic amino acids as catalysts for the asymmetric reduction of imines.

Table 4.1 Reduction of ketimines **6a** and **6b** with trichlorosilane (Scheme 4.2), catalyzed by formamides **16–21** (Figure 4.1).

Entry	Imine	R^1	R^2	R^3	Catalyst (mol%)	Solvent	Temperature (°C)	Yield (%)	ee (%) (config.)
1	6a	Ph	Me	Ph	16 (10)	CH_2Cl_2	Ambient	91	55 (R) [10a]
2	6a	Ph	Me	Ph	17 (10)	CH_2Cl_2	Ambient	52	66 (R) [10a]
3	6a	Ph	Me	Ph	18 (10)	CH_2Cl_2	0	94	73 (R) [11a]
4	6a	Ph	Me	Ph	19 (10)	CH_2Cl_2	−20	95	89 [11b]
5	6b	Ph	Me	PMP[a]	19 (10)	CH_2Cl_2	−20	66	55 [11b]
6	6a	Ph	Me	Ph	20 (5)	CH_2Cl_2	0	87	94 [11a]
7	6b	Ph	Me	PMP	20 (10)	CH_2Cl_2	0	98	92 [11a]
8	6a	Ph	Me	Ph	21 (10)	CH_2Cl_2	0	93	77 (R) [11d]
9	6b	Ph	Me	PMP	21 (10)	CH_2Cl_2	0	77	73 [11d]

a) PMP = p-methoxyphenyl.

functionality proved to be crucial for the activation of Cl_3SiH, as the corresponding acetamides failed to initiate the reaction.

As a follow-up, Sun [11] showed that an expansion of the five-membered ring of the proline moiety to a six-membered ring, as in the pipecolinic derivative **18** (Table 4.1, entry 3) or in its piperazidine analogue **19** (entries 4 and 5), and further elaboration of the carboxylic terminus of the parent amino acid, as in **20** (entries 6 and 7), had a

Table 4.2 Reduction of ketimines **6a**, **6b**, and **6i** with trichlorosilane (Scheme 4.2), catalyzed by formamides **22–35** (Figure 4.2) at room temperature.

Entry	Imine	R^1	R^2	R^3	Catalyst (mol%)	Solvent	Yield (%)	ee (%) (config.)
1	6a	Ph	Me	Ph	22 (10)	CH_2Cl_2	68	79 (S) [12b]
2	6a	Ph	Me	Ph	22 (10)	$CHCl_3$	79	86 (S) [12b]
3	6a	Ph	Me	Ph	22 (10)	MeCN	65	30 (S) [12b]
4	6a	Ph	Me	Ph	23 (10)	$CHCl_3$	70	89 (S) [12b]
5	6a	Ph	Me	Ph	23 (10)	Toluene	81	92 (S) [12b]
6	6b	Ph	Me	PMP[a]	23 (10)	Toluene	85	91 (S) [12b]
7	6b	Ph	Me	PMP	24 (10)	Toluene	99	94 (S) [12c]
8	6b	Ph	Me	PMP	25 (5)	Toluene	92	73 (S) [12k]
9	6b	Ph	Me	PMP	26 (5)	Toluene	98	90 (S) [12k]
10	6a	Ph	Me	Ph	27 (10)	$CHCl_3$	62	85 (S) [12b]
11	6a	Ph	Me	Ph	28 (10)	$CHCl_3$	81	82 (S) [12b]
12	6a	Ph	Me	Ph	29 (10)	$CHCl_3$	88	53 (S) [12b]
13	6b	Ph	Me	PMP	30 (10)	Toluene	95	82 (R) [12b]
14	6b	Ph	Me	PMP	31 (10)	Toluene	95	83 (S) [12b]
15	6a	Ph	Me	Ph	32 (10)	Toluene	84	0 [12b]
16	6b	Ph	Me	PMP	33 (10)	Toluene	84	49 (S) [12b]
17	6i	4-$CF_3C_6H_4$	Me	Ph	34 (10)	Toluene	92	38 (R) [12b]
18	6b	Ph	Me	PMP	35 (5)	Toluene	95	94 (S) [12c]

a) PMP = p-methoxyphenyl.

Table 4.3 Reduction of ketimines **6b–6h** and **7a–7e** with trichlorosilane (Scheme 4.2), catalyzed by Sigamide **35** (Figure 4.2).[a]

Entry	Imine	R^1	R^2	R^3	Yield (%)[b]	ee (%)[c]
1	6b	Ph	Me	PMP[d]	95	94
2	6b	Ph	Me	PMP	92	94[e]
3	6b	Ph	Me	PMP	92	93[f]
4	6c	Ph	Me	3,5-Me$_2$C$_6$H$_3$	89	92
5	6d	4-CF$_3$C$_6$H$_4$	Me	PMP	92	92
6	6e	4-MeOC$_6$H$_4$	Me	PMP	91	91
7	6f	3-(tBu)Me$_2$SiOC$_6$H$_4$	Me	PMP	72	93
8	6g	2-MeC$_6$H$_4$	Me	Ph	90	92[g]
9	6h	2-Naphth	Me	PMP	93	92
10	7a	(E)-PhCH=CH	Me	PMP	94	81
11	7b	(E)-PhCH=C(Me)	Me	PMP	62	84
12	7c	cC$_6$H$_{11}$	Me	PMP	86	85
13	7d	iPr	Me	PMP	83	62
14	7e	tBu	Me	PMP	63 [92[h]]	39 [38[h]]

a) The reaction was carried out at 0.2 mmol scale with 2.0 equiv of Cl$_3$SiH at 18 °C for 16 h in toluene unless stated otherwise; see Ref. [12h].
b) Isolated yield.
c) Amines obtained from **6b**, **6d**, **6e**, and **6h** were (S)-configured; the configuration of the remaining amines is assumed to be (S) in analogy with the rest of the series.
d) PMP = p-methoxyphenyl.
e) With 2.5 mol% of the catalyst.
f) With 1 mol% of the catalyst.
g) Catalyst **23** (Kenamide) was used instead of **35**.
h) In the presence of AcOH (1 equiv).

beneficial effect on the enantioselectivity, which could now be increased up to 94% ee (at 0 °C with 10 mol% catalyst loading) [11a,b,c,h]. On the other hand, the proline-derived dimer **21** (entries 8 and 9), expected to offer an extensive chelation of the reagent, did not provide any particular advantage (≤77% ee) [11d].

In parallel, Malkov and Kočovský have introduced a series of formamides derived from noncyclic N-methyl amino acids (Figure 4.2) [12]. In their studies, the N-methyl substituent remained a constant structural feature (in reminiscence of the DMF pattern), whereas the amide moiety of the original carboxylic terminus of the amino acid was systematically varied. Of the various amino acids, valine derivatives (**22–29** and **35**) proved to be most promising. A detailed investigation of the structure–activity relationship of the catalysts by using a set of model imines **6** allowed the following conclusions to be made [12b]: (1) The valine isopropyl group, as in **22**, represents an optimum in terms of enantioselectivity (Table 4.2, entries 1 and 2). The cyclohexyl analogue **30**, derived from cyclohexyl glycine, exhibited almost identical selectivities (Table 4.2, entry 13), whereas other catalysts with tBu, Ph, CH$_2$Ph, and Me (**31–34**) in place of the original iPr proved to be less enantioselective (entries 14–17) [12b]. (2) The N-methyl group in the formamide moiety is crucial as the corresponding NH derivative exhibited low enantioselectivity (≤35% ee) [12b]. (3)

Table 4.4 Reduction of ketimines **8a–8o** with trichlorosilane (Scheme 4.2), catalyzed by Sigamide 35 (Figure 4.2).[a]

Entry	Imine	R¹	R²	R³	Yield (%)[b]	ee (%)[c]
1	8a	2-Pyridyl	Me	PMP[d]	72	7
2	8b	4-Pyridyl	Me	PMP	85	21
3	8c	2,6-(iPr)₂-4-pyridyl	Me	PMP	~68	78
4	8d	2-Thiazolyl	Me	PMP	83	13
5	8e	5-(iPr)-2-thiazolyl	Me	PMP	71	34
6	8f	1-Me-2-indolyl	Me	PMP	77	91
7	8g	Thiophen-2-yl	Me	PMP	77	89
8	8h	2-Furyl	Me	PMP	62 [57[e]]	56 [62[e]]
9	8i	5-Me-2-furyl	Me	PMP	86	45
10	8j	5-(Me₃Si)-2-furyl	Me	PMP	62	63
11	8k	5-(EtO₂C)-2-furyl	Me	PMP	62	70
12	8l	2-Benzofuryl	Me	PMP	79	70
13	8m	3-Furyl	Me	PMP	60	77
14	8n	3-Benzofuryl	Me	PMP	84	65
15	8o	2,5-Me₂-3-furyl	Me	PMP	90	91

a) The reaction was carried out at 0.2 mmol scale with 2.0 equiv of Cl₃SiH at 18 °C for 16 h in toluene unless stated otherwise; see Ref. [12h].
b) Isolated yield.
c) The resulting amines are assumed to be (S)-configured in analogy with the rest of the series.
d) PMP = p-methoxyphenyl.
e) The reaction was carried out at −20 °C for 24 h.

The formamide function at the N-terminus of valine is another key factor, as evidenced by the behavior of the corresponding acetamide, trifluoroacetamide, carbamate, and urea derivatives, which all reacted sluggishly and gave racemic (or nearly racemic) products [12b]. (4) The carboxyl terminus of the parent valine needs to be converted into an amide with a primary aromatic amine [12]. Amides derived from secondary aromatic amines (e.g., MeNHPh), as well as from nonaromatic congeners (e.g., BuNH₂), proved inferior [12b]. (5) Anilide **22**, the first catalyst of this series, exhibited good enantioselectivities (≤86% ee; Table 4.2, entries 1 and 2); introduction of alkyl groups into the anilide moiety, as in **23** (Kenamide) and **24**, had a positive effect (≤94% ee; entries 4–7) [12b,c], whereas variable enantioselectivities were attained with 3,5-diaryl derivatives **25** and **26** (entries 8 and 9) [12k]. Sigamide (**35**), with two tbutyl groups, was identified as an optimum (≤94% ee at room temperature with 1–5 mol% loading; entry 18) [12c,h]. Other substituents, such as 4-methoxy (**27**), 3,5-dimethoxy (**28**), and 3,5-di(trifluoromethyl) (**29**), had a rather negative effect on the enantioselectivity (entries 10–12) [12b]. (6) Toluene was identified as the solvent of choice, in which highest enantioselectivities were attained, superior to CH₂Cl₂, CHCl₃, and in particular, to MeCN (compare entries 1–3 with 5–7) [12b]. (7) The reduction can be carried out at room temperature; only a marginal increase in enantioselectivity was observed at 0 or −20 °C (at the expense of the reaction rate) [12].

Table 4.5 Reduction of ketimines **9a–9n** with trichlorosilane (Scheme 4.2), catalyzed by Sigamide **35** (Figure 4.2).[a]

Entry	Imine	R¹	R²	R³	Yield (%)[b]	ee (%)[c]
1	9a	Ph	Et	PMP[d]	81	92
2	9b	Ph	$(CH_2)_3CH=CH_2$	PMP	64	84
3	9c	3-ClC$_6$H$_4$	$(CH_2)_3CH=CH_2$	PMP	62	82
4	9d	3-MeOC$_6$H$_4$	$(CH_2)_3CH=CH_2$	PMP	98	90
5	9e	3-MeOC$_6$H$_4$	$(CH_2)_2CH=CH_2$	PMP	99	95
6	9f	Ph	$(CH_2)_2Ph$	PMP	95	95
7	9g	3-(tBu)Me$_2$SiOC$_6$H$_4$	$(CH_2)_2(3,4,5\text{-MeOC}_6H_2)$	PMP	69	96
8	9h	Ph	iPr	PMP	98	97
9	9i	Ph	cPr	PMP	87	95
10	9j	Ph	cBu	PMP	75	94
11	9k	Ph	cHexyl	PMP	83	76
12	9l	Ph	tBu	PMP	18 [46[e]]	10 [10[e]]
13	9m	2-Furyl	iPr	PMP	~75	85
14	9n	4-MeOC$_6$H$_4$	4-CF$_3$C$_6$H$_4$	PMP	79	6

a) The reaction was carried out at 0.2 mmol scale with 2.0 equiv of Cl$_3$SiH at 18 °C for 16 h in toluene unless stated otherwise; see Ref. [12h].
b) Isolated yield.
c) The resulting amines are assumed to be (S)-configured in analogy with the rest of the series.
d) PMP = p-methoxyphenyl.
e) In the presence of AcOH (1 equiv).

The scope of the imine reduction was most systematically investigated with Sigamide (**35**), whereas other catalysts (Figures 4.1 and 4.2) enjoyed a less thorough treatment. Therefore, the scope will be discussed mainly for Sigamide.

The reduction of ketimines (Scheme 4.2 and Tables 4.3–4.7) was carried out under standard conditions [12], that is, with 2 equiv of Cl$_3$SiH (this excess could be considerably lowered for larger scale operations) in toluene (an optimized solvent) at room temperature (15–20 °C) overnight (a compromise to attain both good reaction rates and enantioselectivities) with Sigamide (**35**) as catalyst at 5 mol% loading (although 1 mol% was also shown to be equally effective [12c]) under an argon atmosphere. In some cases, acetic acid (≤1 equiv) was added to the reaction mixture (see below).

Imines **6b–6h**, derived from acetophenone and its congeners, afforded the corresponding amines in high yields and >90% ee (Table 4.3, entries 1–9). Electronic properties of the imine had no effect on the reduction, as documented by the extreme examples of the electron-poor and electron-rich imines **6d** and **6e** (entries 5 and 6). Similarly, an increased steric hindrance, exercised by an *ortho*-substituent (**6g**), proved to be of no consequence (entry 8) [12h].

Distance separation between the aromatic nucleus and the imine moiety, as in the conjugated cinnamyl derivatives **7a** and **7b**, had a minor negative effect on the enantioselectivity (81/84% ee; entries 10 and 11); a noticeable deceleration of the reaction, reflected in the lower isolated yield, was observed for the more sterically hindered methyl homologue **7b** [12h].

Table 4.6 Reduction of ketimines **10a–10i** with trichlorosilane (Scheme 4.2), catalyzed by Sigamide **35** (Figure 4.2).[a]

Entry	Imine	R^1	R^2	R^3	Yield (%)[b]	ee (%)[c]
1	10a	Ph	CH$_2$Cl	PMP[d]	98	96
2	10b	4-ClC$_6$H$_4$	CH$_2$Cl	PMP	87	91
3	10c	4-FC$_6$H$_4$	CH$_2$Cl	PMP	92	94
4	10d[e]	2-ClC$_6$H$_4$	CH$_2$Cl	PMP	54	96
5	10e[e]	2-ClC$_6$H$_4$	CH$_2$Cl	4-ClC$_6$H$_4$	87	95
6	10f[e]	2-ClC$_6$H$_4$	CH$_2$Cl	4-FC$_6$H$_4$	86	95
7	10g[e]	4-MeO-C$_6$H$_4$	CH$_2$Cl	PMP	86	91
8	10h[e]	3-MeOC$_6$H$_4$	CH$_2$Cl	PMP	84	92
9	10i[e]	2-Naphthyl	CH$_2$Cl	PMP	92	91
10	10j[f]	Ph	CH$_2$CN	Ph	97[g]	87
11	10k[f]	Ph	CH$_2$CN	PMP	75[g]	87
12	10l[f]	Ph	CH$_2$CO$_2$Et	PMP	98[g]	89
13	10m[f]	4-FC$_6$H$_4$	CH$_2$CO$_2$Et	PMP	95[g]	90
14	10n[f]	4-MeOC$_6$H$_4$	CH$_2$CO$_2$Et	PMP	80[g]	88
15	10o	Ph	CO$_2$Me	Ph	69[h),i]	59
16	10p	Ph	(CH$_2$)$_2$CO$_2$Me	PMP	38[j]	88

a) The reaction was carried out at 0.2 mmol scale with 2.0 equiv of Cl$_3$SiH at 18 °C for 24 h in toluene unless stated otherwise; see Ref. [12d,g,h].
b) Isolated yield.
c) Amines obtained from **10k** and **10l** were (S)-configured; the configuration of the amines obtained from **10j, 10m, 10n**, and **10p** is assumed to be (S) in analogy with the rest of the series and previous experiments. Note the change in the preference of the substituents in the Cahn–Ingold–Prelog system in the amines obtained from **10a–10i**, for which (R)-configuration is assumed.
d) PMP = p-methoxyphenyl.
e) The imine was generated in situ and reduced without isolation.
f) The imine exists as a minor component in the equilibrium with the corresponding enamine.
g) The reaction was carried out for 48 h in the presence of acetic acid (1 equiv).
h) The reaction was carried out at −20 °C for 24 h, using catalyst **23** (10 mol%).
i) Ref. [12a,b].
j) 2-Methoxy-1-(4′-methoxyphenyl)-5-phenyl-1H-pyrrole was isolated as a by-product (27%), arising by a ring closure of the resulting amino ester.

A complete removal of the conjugated π-system, as in the cyclohexyl analogue **7c**, also resulted in a minor decrease in enantioselectivity (to 85% ee, entry 12) as compared to the phenyl analogue **6b** (94% ee, entry 1). However, it is pertinent to note that this decrease was found to be more substantial when the less bulky catalyst **22** was employed (37% ee at rt and 59% ee at −20 °C) [12b], demonstrating the superiority of Sigamide (**35**). With the less sterically hindered isopropyl derivative **7d**, the enantioselectivity dropped to 62% ee (entry 13). On the other hand, the sterically much more congested tert-butyl derivative **7e** reacted sluggishly and the enantioselectivity was found to further decrease (entry 14) [12h].

The effect of a heteroatom within the substrate was investigated with the aid of the heteroaromatic imines **8a–8o** (Table 4.4). The pyridyl derivatives **8a** and **8b** were reduced cleanly but the products were almost racemic (Table 4.4, entries 1 and 2) [12h], which was attributed to the competing coordination of the pyridine nitrogen to

Table 4.7 Synthesis of esters of α-substituted β-amino acids **40** by reduction of enamines/imines **38/39** with trichlorosilane, catalyzed by Sigamide **35** (Scheme 4.4).[a]

Entry	Imine 39	R	Yield (%)	syn/anti	40, % ee syn; anti
1	39a	Ph	84	>99:1	76
2	39b	4-MeOC$_6$H$_4$	80	>99:1	77
3	39c	4-FC$_6$H$_4$	77	>99:1	73
4	39d	Bn	87	27:73	86; 76
5	39e	Me	85	98:2	77
6	39f	Et	86	95:5	82
7	39g	nBu	84	95:5	76

a) The enamine reduction was carried out on a 0.2 mmol scale with Cl$_3$SiH (2.0 equiv) in the presence acetic acid (1.0 equiv) at rt (18 °C) for 48 h, using catalyst **35** (10 mol%); see Ref. [12g].

the silicon of Cl$_3$SiH [13, 14]. The same effect can account for the very low enantioselectivity observed for the thiazole derivative **8d** (entry 4). On the other hand, the sterically hindered 2,6-diisopropylpyridine derivative **8c**, where the latter coordination is precluded, was reduced to the corresponding amine with 78% ee (entry 3). Introduction of just one isopropyl group, as in **8e**, was less effective (34% ee, entry 5). In contrast to the pyridine-derived imines, the N-methyl-2-indolyl derivative **8f** with a noncoordinating nitrogen atom was reduced with high enantioselectivity (91% ee; entry 6) [12h].

Sulfur as a heteroatom is apparently free of negative effects, as shown by the thiophen-2-yl derivative **8g**, where the enantioselectivity (89% ee, entry 7) was close to that observed for the acetophenone-derived imines (Table 4.3). Furan- and benzofuran-derived imines **8h–8n** were reduced with the efficiency laying between that of the sulfur and nitrogen heterocycles (45–77% ee, entries 8–14); the 3-furyl derivative **8m** exhibited a significantly higher level of asymmetric induction (77% ee, entry 13) than its 2-furyl isomer **8h** (56% and 62% ee at rt and at −20 °C, respectively, entry 8). Again, increasing the steric bulk had a positive effect, as documented by the behavior of the dimethyl derivative **8o** (91% ee, entry 15) [12h].

Extension of the alkyl chain, as in **9a–9e**, had a marginal effect on the enantioselectivity (84–95% ee; Table 4.5, entries 1–5). Derivatives **9f** and **9g** with an aromatic system at the terminus of the alkyl chain exhibited one of the highest enantioselectivities (95–96% ee, entries 6 and 7). Branching in the R^2 group, as in the isopropyl, cyclopropyl, and cyclobutyl derivatives **9h–9j** had a positive effect (94–97% ee, entries 8–10). For the bulkier cyclohexyl derivative **9k**, a minor decrease in enantioselectivity was observed (76% ee, entry 11). On the other hand, reduction of the highly congested *tert*-butyl derivative **9l** proved to be very sluggish and gave an almost racemic product (entry 12). Good enantioselectivity was restored for the furyl isopropyl analogue **9m** (85% ee, entry 13). However, imine **9n**, with two essentially isosteric but electronically opposite aromatic systems, gave an almost racemic product (entry 14), showing that electronic effects [15] play a negligible role compared to the steric effects [12h].

The chloromethyl imines **10a–10i** were employed as representatives of derivatives with a potentially reactive group in the side chain. Their reduction afforded the corresponding amines in >90% ee (Table 4.6, entries 1–9). The resulting amino chlorides were then cyclized to the corresponding aziridines on treatment with *t*-BuOK with retention of the stereochemical integrity [12d]. It is pertinent to note that the preparation of the sensitive chloro imines **10a–10i** from the α-chloro ketones was not entirely free of problems: while the electron-neutral and electron-poor imines **10a–10c** were synthesized and isolated as individual substances, their electron-rich counterparts **10g** and **10h** could not be obtained as pure compounds, since the reaction did not proceed to completion. Therefore, in the latter instances, the imines were generated *in situ* [12d].

The analogous imines with a CH_2CN group (**10j** and **10k**) could not be prepared from β-keto nitriles and anisidine since the products of this reaction exist predominantly in the conjugated enamine form **36j** and **36k** (Scheme 4.3) [12g]. Nevertheless, in the presence of a Brønsted acid, the enamines could be equilibrated with the imine (although the enamine species still largely prevailed in the mixture). Acetic acid (1 equiv) was identified as an optimal additive: the reduction, carried out in its presence, afforded the expected β-amino nitriles in high yields and enantioselectivities (87% ee, entries 10 and 11). A single crystallization of the β-amino nitrile obtained in entry 10 furnished an enantiopure product (99.9% ee). Reduction of the analogous enamines generated from β-keto esters (**36l–36n**), afforded the expected β-amino esters via the corresponding imines **10l–10n** (generated in the reaction mixture by the AcOH-catalyzed equilibration) also in high yields and enantioselectivities (88–90% ee, entries 12–14). Again, a single crystallization of the β-amino ester obtained in entry 12 increased its enantiopurity from 89 to 99.8% ee. Finally, with the lower homologue **10o**, lacking the problem of enamine–imine dichotomy, the enantioselectivity was found to be rather low (59% ee, entry 15) [12b]. On the other hand, the higher homologue **10p** was again reduced with high enantioselectivity (88% ee, entry 16) [12h].

Scheme 4.3 X = CN, CO_2Et (see Table 4.6).

The reduction of imines/enamines **10j–10n/36j–36n** generated from β-keto esters can be regarded as an interesting method for the preparation of β-amino acids. This approach has been extended to the reduction of the imines/enamines derived from α-substituted β-keto esters **37** (Scheme 4.4). In this case, the fast enamine–imine equilibration **38** ⇌ **39** is the key point since imines **39** are chiral but racemic, so that dynamic kinetic resolution is required to operate here [16] (along with the enantioselective reduction) to provide good diastereo- and enantiocontrol. Reduction of α-substituted esters **39a–39g** and the corresponding nitriles was investigated as a probe for the scope of this approach. The main focus here was on α-aryl derivatives, since no suitable methodology is available for this class of β-amino acids (Table 4.7). The α-aryl β-amino esters **40a–40c** were obtained as single diastereoisomers in good yields and respectable enantioselectivities (Table 4.7, entries 1–3); single crystallization significantly improved the enantiopurity of **40a** (to 96% ee, entry 1). Reduction of α-alkyl derivatives **39e–39g** followed the same pattern (entries 5–7); only the α-benzyl analogue **39d** was less diastereoselective (*syn/anti* 1 : 2.7, entry 4). High diastereoselectivity and good enantioselectivity were also observed for the reduction of the corresponding nitriles, though at lower reaction rates [12g].

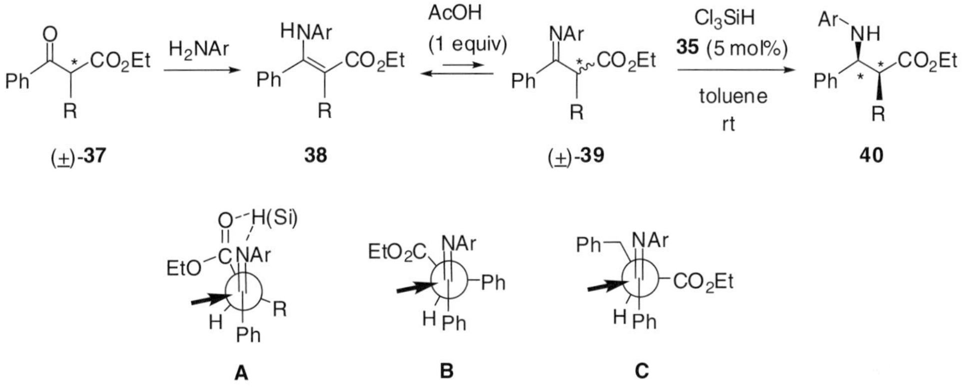

Scheme 4.4 Preparation of α-substituted β-amino acids by asymmetric reduction of imines involving dynamic kinetic resolution. For R, see Table 4.7; Ar = p-MeOC$_6$H$_4$.

The dynamic kinetic resolution operating in the latter reduction is consistent with the model, where the ester carbonyl and the imine group are held together (in a *synperiplanar* conformation) either by hydrogen bonding (in the protonated form) or by chelation to silicon (Scheme 4.4, **A**). In the case of α-aryl derivatives **39a–39c**, featuring high diastereoselectivity, both the chelation (**A**) and the Felkin–Ahn model, in which Ph assumes a perpendicular orientation to the C=N bond (**B**), predict the formation of the same *syn*-diastereoisomer. Predominant formation of the *anti*-isomer in the reduction of **39d** (in 2.7 : 1 ratio) is consistent with conformation **C**, suggesting a small relative difference in the conformational energies of the transition states in this instance. The α-alkyl derivatives **39e–39g** followed the same pattern as their α-aromatic congeners, that is, the **A/B** model [12g].

4.3
Other Amides as Organocatalysts in Hydrosilylation of Imines

Replacement of the formyl group by other amide or urethane units in the case of the valine derivatives resulted in a dramatic decrease in reactivity and enantioselectivity (see above) [12b]. On the other hand, replacement of the formamide moiety in the original proline design **16** by the α-picolinic amide (Figure 4.3), in conjunction with the introduction of a hindered tertiary hydroxyl in place of the amide function of

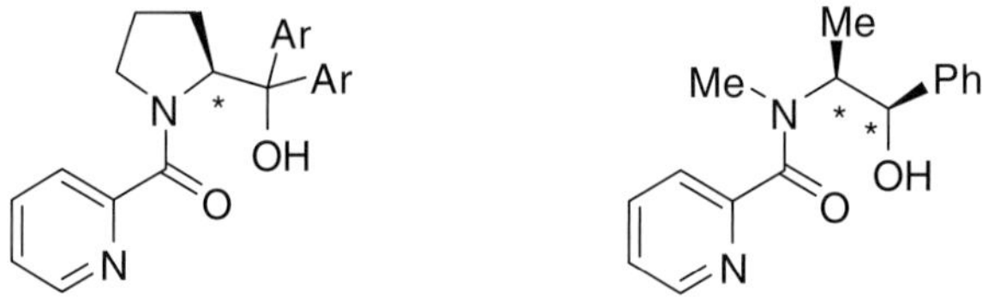

(S)-41, Ar = Ph
(S)-42, Ar = 3,5-Me$_2$C$_6$H$_3$

43

Figure 4.3 α-Picolinic amides as catalysts for the asymmetric reduction of imines.

proline, as in **41** and **42**, had a positive effect on enantioselectivity (≤75% ee; Table 4.8, entries 1–3 and 6) [10b,17b]. Furthermore, the α-picolinic amide **43**, in which ephedrine was employed as the chiral scaffold [17a], exhibited high enantioselectivities (≤93% at −10 °C with 20 mol% catalyst loading; entries 1–8). These amides also appear to hold the promise regarding imine substrates that are ineffective using the earlier discussed catalysts. Thus, reduction of imines **6n–6p**, derived from benzylamine rather than from an aromatic amine, proceeded with an

Table 4.8 Reduction of ketimines **6**, **7**, and **9** with trichlorosilane (Scheme 4.2), catalyzed by the α-picolinic amides **41** and **43** (Figure 4.3).

Entry	Imine	R^1	R^2	R^3	41 (10 mol%)[a]		43 (20 mol%)[b]	
					Yield (%)	ee (%)	Yield (%)	ee (%)
1	6a	Ph	Me	Ph	86	73	88	92
2	6b	Ph	Me	PMP[c]	90	75	90	93
3	6j	4-NO$_2$C$_6$H$_4$	Me	Ph	84	73	85	76
4	6k	4-BrC$_6$H$_4$	Me	Ph			92	87
5	6e	4-MeOC$_6$H$_4$	Me	PMP			82	92
6	6l	4-MeOC$_6$H$_4$	Me	Ph	90	71	91	89
7	6h	2-Naphth	Me	PMP			86	90
8	6m	2-Naphth	Me	Ph			93	90
9	6n	Ph	Me	Bn[d]	67	80	85	80
10	6o	4-MeOC$_6$H$_4$	Me	Bn			85	82
11	6p	2-Naphth	Me	Bn			80	82
12	7c	cC$_6$H$_{11}$	Me	PMP			71	61
13	7f	cC$_6$H$_{11}$	Me	Ph			70	67
14	9a	Ph	Et	PMP			88	84
15	9o	Ph	Et	Ph			91	76

a) The reaction was carried out at 0.3 mmol scale with 1.5 equiv of Cl$_3$SiH at rt for 4 h in dichloromethane unless stated otherwise; see Ref. [10b].
b) The reaction was carried out at 0.2 mmol scale with 2.0 equiv of Cl$_3$SiH at −10 °C for 16 h in chloroform unless stated otherwise; see Ref. [17a].
c) PMP = p-methoxyphenyl.
d) Bn = benzyl.

unprecedentedly high enantioselectivity (≤82% ee; entries 9–11). For comparison, imine **6n** was reduced in the presence of other catalysts, such as **16**, **19**, and **45** (all at 10 mol% loading), with 55, 20, and 45% ee, respectively [11]. On the other hand, with **7c** and the closely related **7f** (i.e., imines derived from a nonaromatic ketone), the enantioselectivity was lower (61 and 67% ee, respectively, entries 12 and 13) than that observed for Sigamide **35** (85% ee; Table 4.3, entry 12) [11].

4.4
Sulfinamides as Organocatalysts in Hydrosilylation of Imines

An interesting catalyst with a fresh design, featuring a Lewis-basic sulfinamide group (Figure 4.4) as the chiral element (**44**), was introduced by Sun and shown to be a success (≤93% ee at −20 °C at 20 mol% catalyst loading; Table 4.9) [11e]. The analogous bissulfinamide **45**, believed to chelate trichlorosilane in a bidentate fashion, exhibited increased enantioselectivities (≤96% ee at −20 °C at 10 mol% catalyst loading; Table 4.9) [11g].

Sulfinamide **46**, developed by the same group, represents the most recent addition to this family. Here, the original formyl group of the proline-derived catalyst **16** was replaced by the t-BuSO moiety, which resulted in high enantioselectivities (≤97% ee at 0 °C with 10 mol% catalyst loading; Table 4.10) [11f]. The striking feature of this catalyst is its efficiency with imines derived from nonaromatic amines, namely, benzyl, allyl, propyl, isobutyl, and p-methoxybenzyl amine (Table 4.10), which renders **46** superior to **43** and dwarfs other catalysts (e.g., **16**, **19**, and **45**). Needless to say, this new feature broadens the synthetic arsenal beyond the N-aryl substrates and opens new synthetic avenues. In comparison, the valine-derived catalyst **47** proved to be less efficient with imines derived from aromatic

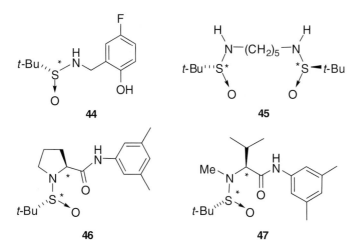

Figure 4.4 Sulfinimide catalysts for the asymmetric reduction of imines.

Table 4.9 Reduction of ketimines **6**, **7**, and **9** with trichlorosilane (Scheme 4.2), catalyzed by the sulfinamides **44** and **45** (Figure 4.4).

Entry	Imine	R^1	R^2	R^3	44 (20 mol%)[a]		45 (10 mol%)[b]	
					Yield (%)	ee (%)	Yield (%)	ee (%)
1	6a	Ph	Me	Ph	92	92	91	96
2	6b	Ph	Me	PMP[c]	97	91	92	92
3	6d	4-CF$_3$C$_6$H$_4$	Me	PMP	80	90	90	83
4	6i	4-CF$_3$C$_6$H$_4$	Me	Ph	93	92	95	95
5	6j	4-NO$_2$C$_6$H$_4$	Me	Ph	94	90	90	93
6	6k	4-BrC$_6$H$_4$	Me	Ph	92	92	92	95
7	6l	4-MeOC$_6$H$_4$	Me	Ph	98	93	83	95
8	6h	2-Naphth	Me	PMP	86	88	86	91
9	6m	2-Naphth	Me	Ph	96	90	77	94
10	7f	cC$_6$H$_{11}$	Me	Ph	78	74	84	75
11	7g	iPr	Me	Ph	78	79	87	82
12	9a	Ph	Et	PMP	84	92	84	92
13	9o	Ph	Et	Ph	92	93	86	94
14	9p	Ph	cPr	Ph			70	93
15	9q	Ph	iBu	Ph	82	86	82	93

a) The reaction was carried out at 0.2 mmol scale with 2.0 equiv of Cl$_3$SiH at $-20\,°C$ for 24–48 h in dichloromethane unless stated otherwise; see Ref. [11e].
b) The reaction was carried out at 0.2 mmol scale with 2.0 equiv of Cl$_3$SiH and 0.3 equiv of 2,6-lutidine at $-20\,°C$ for 16 h in dichloromethane unless stated otherwise; see Ref. [11g].
c) PMP = p-methoxyphenyl.

amines (≤50% ee) (A.V. Malkov, J.L.P. Phillips, C.D. Davies, and P. Kočovský, unpublished results).

4.5
Supported Organocatalysts in Hydrosilylation of Imines

In the development of an efficient catalytic process, facile separation of the product from the catalyst is one of the key technological elements. In organocatalysis, where both the product and the catalyst are small organic molecules, isolation of the desired product is not a trivial task on a large scale and may also become a nuisance in high-throughput parallel chemistry. Since chromatography often represents the only option, immobilization of the catalyst would provide an elegant and practical solution to the problem [18].

The formamide-type catalysts **22–35**, developed by Malkov and Kočovský (Figure 4.2) [12a–d,g,h], proved to be very efficient and, as a result, the loading was reduced to 1–5 mol% (Tables 4.1–4.7). Nevertheless, even this relatively small amount still appears as a contaminant in the product and has to be separated.

Table 4.10 Reduction of ketimines **6** with trichlorosilane (Scheme 4.2), catalyzed by sulfinamide **46** (Figure 4.4).[a]

Entry	Imine	R^1	R^2	R^3	Yield (%)[b]	ee (%)[b]
1	6n	Ph	Me	Bn[c]	98 (90)	96 (97)
2	6o	4-MeO-C$_6$H$_4$	Me	Bn	54 (90)	78 (92)
3	6p	2-Naphth	Me	Bn	96 (83)	96 (98)
4	6q	4-FC$_6$H$_4$	Me	Bn	80 (82)	96 (97)
5	6r	4-CF$_3$C$_6$H$_4$	Me	Bn	94 (93)	98 (96)
6	6s	4-NO$_2$C$_6$H$_4$	Me	Bn	80 (98)	99.6 (94)
7	6t	3-ClC$_6$H$_4$	Me	Bn	98	97
8	6u	Ph	Me	Allyl	82 (80)	92 (93)
9	6v	4-FC$_6$H$_4$	Me	Allyl	88 (83)	90 (96)
10	6w	4-CF$_3$C$_6$H$_4$	Me	Allyl	65	97
11	6x	4-NO$_2$C$_6$H$_4$	Me	Allyl	97	96
12	6y	2-ClC$_6$H$_4$	Me	Allyl	75	97
13	6z	Ph	Me	nPr	67 (60)	66 (90)
14	6z'	Ph	Me	CH$_2$CHMe$_2$	56 (80)	70 (87)
15	6z''	Ph	Me	PMB[d]	85 (83)	93 (95)

a) The reaction was carried out at 0.1 mmol scale with 2.0 equiv of Cl$_3$SiH at 0 °C for 24 h in toluene with 10 mol% catalyst loading unless stated otherwise; see Ref. [11f].
b) Figures in parentheses show the results obtained for a CCl$_4$ solution.
c) Bn = benzyl.
d) PMB = p-methoxybenzyl.

To address these problems, a fluorous tag was appended to the catalyst (**48**) [12c], which reduced the separation to an ordinary filtration through a pad of fluorous silica gel that retained the catalyst, whereas the product was eluted. Subsequent change of the solvent resulted in elution of the catalyst that could be reused. The classical chromatography of the crude mixture after the workup was thus avoided. The presence of the fluorous tag had practically no effect on the catalytic activity (Table 4.11; compare entries 1 with 3–5) [12c]. However, the polyfluorinated starting material required for the synthesis of **48** and the fluorous silica gel are rather expensive, which may become prohibitive for large-scale applications; therefore, other options were also explored.

Further simplification was attained by immobilization of the catalyst on a solid polymer (**49**), so that its separation from the product was reduced to mere mechanical filtration; the resin-bound catalyst could then be used for another run. Of the number of polymeric supports investigated, Merrifield (**49**) and Wang resins were identified as most suitable, exhibiting the same behavior and efficiency [18]. However, owing to the heterogeneous nature of the system, the enantioselectivity of the reaction decreased by 10–15% ee (Table 4.11, entries 6–12). In view of the swelling properties, toluene was found to be less suitable than chloroform (compare entries 6 and 7), which makes the method less environment friendly. Furthermore, a "conditioning effect" was observed for these systems: the second run was always found to be more enantioselective than the first one by ~10% ee and this level was maintained in the subsequent runs (compare entries 7 with 8–12). The latter effect stems from the

Table 4.11 Reduction of ketimines **6** with trichlorosilane, catalyzed by the valine-derived N-methyl formamides (S)-**23**, (S)-**48–52**.[a]

Entry	Catalyst (mol%)	Imine	R^1	R^2	R^3	Solvent	Run	Yield (%)[b]	% ee[c]
1	23 (10)	6a	Ph	Me	Ph	Toluene	1	81	92
2	48 (10)	6b	Ph	Me	PMP[d]	Toluene	1	90	91
3	48 (5)	6d	4-CF$_3$C$_6$H$_4$	Me	PMP	Toluene	1	72	92[e]
4	48 (5)	6e	4-MeOC$_6$H$_4$	Me	PMP	Toluene	1	84	84
5	48 (5)	6h	2-Naphth	Me	PMP	Toluene	1	68	92[e]
6	49 (25)	6b	Ph	Me	PMP	Toluene	1	84	63
7	49 (25)	6b	Ph	Me	PMP	CHCl$_3$	1	80	76
8	49 (25)	6b	Ph	Me	PMP	CHCl$_3$	2	81	82
9	49 (25)	6b	Ph	Me	PMP	CHCl$_3$	3	82	81
10	49 (25)	6b	Ph	Me	PMP	CHCl$_3$	4	80	82
11	49 (25)	6b	Ph	Me	PMP	CHCl$_3$	5	81	82
12	49 (25)	6b	Ph	Me	PMP	CHCl$_3$	6	78	81
13	50 (7)	6b	Ph	Me	PMP	Toluene	1	90	86
14	50 (3)	6b	Ph	Me	PMP	Toluene	1	90	86
15	50 (1)	6b	Ph	Me	PMP	Toluene	1	90	88
16	50 (7)	6d	4-CF$_3$C$_6$H$_4$	Me	PMP	Toluene	1	75	86
17	50 (7)	6e	4-MeOC$_6$H$_4$	Me	PMP	Toluene	1	87	82
18	50 (7)	6h	2-Naphth	Me	PMP	Toluene	1	88	86
19	50 (7)	8g	Thiophen-2-yl			Toluene	1	78	65
20	50 (7)	9i	Ph	cPr	PMP	Toluene	1	71	73
21	50 (7)	9j	Ph	cBu	PMP	Toluene	1	81	75
22	50 (7)	9k	Ph	cHex	PMP	Toluene	1	85	55
23	50 (7)	10a	Ph	CH$_2$Cl	PMP	Toluene	1	70	91
24	50 (7)	10l	Ph	CH$_2$CO$_2$Et	PMP	Toluene	1	77	81
25	51 (10)	6b	Ph	Me	PMP	Toluene	1	89	89
26	51 (5)	6d	4-CF$_3$C$_6$H$_4$	Me	PMP	Toluene	1	81	91
27	51 (5)	6e	4-MeOC$_6$H$_4$	Me	PMP	Toluene	1	78	79
28	51 (5)	6h	2-Naphth	Me	PMP	Toluene	1	68	85
29	51 (5)	10a	Ph	CH$_2$Cl	PMP	Toluene	1	69	94
30	51 (5)	10l	Ph	CH$_2$CO$_2$Et	PMP	Toluene	1	89	82
31	52	6b	Ph	Me	PMP	Toluene	1	90	84
32	52	6b	Ph	Me	PMP	Toluene	2	92	82
33	52	6b	Ph	Me	PMP	Toluene	3	92	80
34	52	6b	Ph	Me	PMP	Toluene	4	89	68

a) The reaction was carried out at 0.2 mmol scale (entries 1–12 and 26–34) or 0.4 mmol scale (entries 13–25) with 2.0 equiv of Cl$_3$SiH at rt for 16 h unless stated otherwise; see Refs [12c,e,f,i,j].
b) Isolated yield.
c) Amines obtained from **6a**, **6b**, **6d**, **6e**, and **6h** were (S)-configured; the configuration of the remaining amines is assumed to be (S) in analogy with the rest of the series. This assignment corresponds to the (R)-configuration in entry 29 due to the change in substituent preference in the Cahn–Ingold–Prelog system.
d) PMP = p-methoxyphenyl.
e) The reaction was carried out at 10 °C.

partial inclusion of the residues of the decomposition products of the excess trichlorosilane, as evidenced, for example, by detecting the Si—O bonds in the regenerated supported catalyst by IR spectroscopy [12e].

To circumvent the problems associated with the resin-supported catalysts, anchoring to a soluble polymeric support was explored as the next step. Recovery and recycling of the soluble catalysts was anticipated to rely on the switch of the solubility induced by changing the solvent polarity [18b,e,g]. Traditionally, polyethylene glycol (PEG) polymers are precipitated by nonpolar solvents, which in this case may also lead to coprecipitation of the polar amine products, thereby affecting the overall efficiency of the process. Therefore, a novel polymeric platform for catalyst immobilization has been developed, featuring an inverted solubility pattern: the catalyst is soluble in a nonpolar and insoluble in a polar medium. The new catalyst **50**, with the benzyl methacrylate polymeric backbone, was prepared via copolymerization of the respective units by means of atom transfer radical polymerization (ATRP) [12i,19]. The catalytic properties of this new catalyst were found to be considerably improved (Table 4.11, entries 13–24) compared to the resin-bound catalyst **49** (entries 6–12), both in terms of enantioselectivity and reproducibility. Toluene was identified as the solvent of choice again. When the reaction was complete, the catalyst was precipitated with methanol and used in subsequent runs with unchanged efficiency. The enantioselectivity of **50** with various substrates was comparable with that of Kenamide **23** and Sigamide **35** (compare Table 4.11, entries 15–24, with Table 4.2, entries 4–6, Table 4.3, entries 1–3, Table 4.4, entry 7, Table 4.5, entries 9–11, and Table 4.6, entries 1 and 12) [12i].

Other types of immobilization, namely, on a dendron (**51**; Table 4.11, entries 25–30) [12j] and on gold nanoparticles (**52**; entries 31–34) [12f], were also successful but did not provide any advantage over **50**. Furthermore, unlike the other supported catalysts **48–51**, deterioration of the performance of the recovered **52** was observed for the subsequent runs (compare entry 31 with 34) [12f] (Figure 4.5). The soluble polymeric catalyst **50** thus remains the supported catalyst of choice.

4.6
Mechanistic Considerations

The experiments reported by Malkov and Kočovský [12h] and by Sun [11a,b,e,f], indicate that the Lewis base-catalyzed imine reduction with trichlorosilane is not affected by isomeric nonhomogeneity of the starting imines. Thus, for example, imines **9h–9j**, which exist as $5:2$ to $5:3$ (E/Z) mixtures, were reduced to the corresponding amines with 94–97% ee (Table 4.5, entries 8–10). Apparently, traces of HCl, naturally present in the moisture sensitive Cl_3SiH, trigger an (E/Z) equilibration of imines **6–10**, which is faster than the reduction (Scheme 4.5).

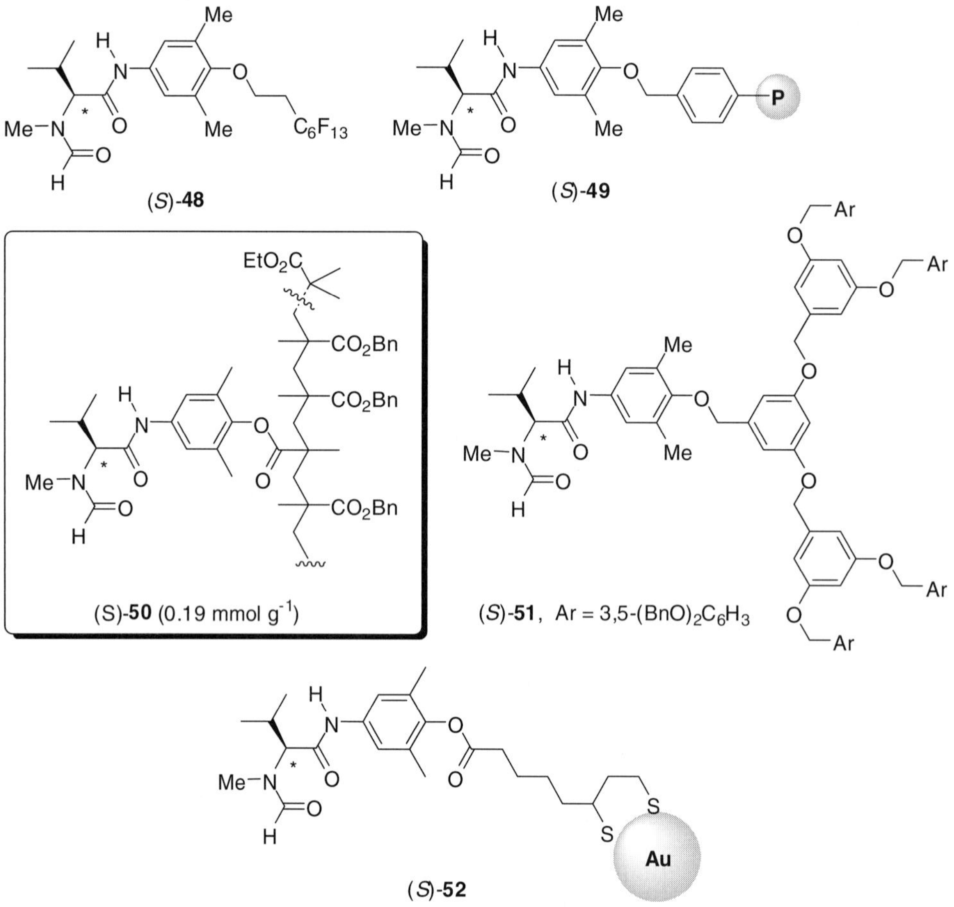

Figure 4.5 Supported organocatalysts for the asymmetric reduction of imines.

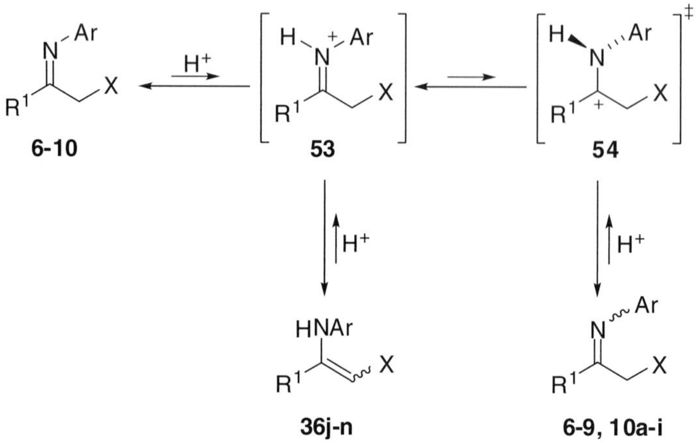

Scheme 4.5 Imine (E/Z) and imine/enamine equilibration.

The generally accepted mechanism for the Brønsted acid-catalyzed isomerization of imines involves the initial protonation of the nitrogen to generate the iminium ion **53** at low concentration, followed either by rotation about the C−N bond (**54**), or via a nucleophilic attack by the acid counterion to produce the corresponding tetrahedral intermediate, which undergoes rotation/proton exchange on nitrogen and subsequent elimination to give the other isomer [20, 21]. Analogous (Z/E) isomerization can be envisaged in the case of β-enamino nitriles and esters **36j–36n**, that is, protonation at the α-position to give the iminium ion **53**, rotation about the C−C bond, and proton elimination. The β-enamino esters **36l–36n** were obtained as the (Z)-isomers, which are apparently more stable due to an intramolecular hydrogen bonding between the N−H and the ester carbonyl group. In the case of **36l**, no signs of isomerization in the presence of acetic acid (2 equiv) could be detected by NMR spectroscopy. By contrast, the β-enamino nitriles **36j** and **36k** readily equilibrate to a mixture of (Z/E) isomers, for example, **36j** isomerizes to a ∼2.6 : 1 mixture in $CDCl_3$ solutions overnight even without an external source of protons [12g,h]. Apparently, in the case of imines **6–8** and **9a–9m**, it is the more stable (E) isomer that is reduced from the re-face to give the (S) enantiomer. Accordingly, for chloro imines **10a–10i**, the (Z) isomer is reduced from the enantiotopic si-face to give the (R) enantiomer (note the change of Cahn–Ingold–Prelog priorities here!). On the other hand, the complete absence of steric preference for (E) or (Z) isomer may account for the loss of enantioselectivity and formation of a racemic product in the case of imine **9n**.

Protonation does not only catalyze the isomerization, but also contributes to the nonselective background reaction by enhancing the electrophilicity of the imines (via **53**). Therefore, the H^+ concentration must be kept relatively low and acetic acid (up to 1 equiv) was identified as an acceptable compromise. Stronger acids promote the nonenantioselective background reaction, whereas addition of bases resulted in a dramatic deceleration [12h].

4.7
Synthetic Applications

The ready formation of esters of β-amino acids by reduction of the corresponding imines/enamines (Table 4.6, entries 12–14), which in turn can be prepared from the readily available β-keto esters, allowed an expedient synthesis of SCH48461 (**56**), a potent, orally active inhibitor of cholesterol absorption [22]. Enamine **36n** (Scheme 4.6) was reduced (via imine **10n**) with Cl_3SiH in the presence of Sigamide (**35**) to afford the β-amino ester **15n** (80% isolated yield, 88% ee), whose treatment with methylmagnesium bromide (acting as a base) produced β-lactam **55** (92%). Enolization of the latter derivative with LDA followed by alkylation with cinnamyl bromide and catalytic hydrogenation afforded **56** in 77% overall yield for the last two steps (S. Stončius, A.V. Malkov, and P. Kočovský, unpublished results).

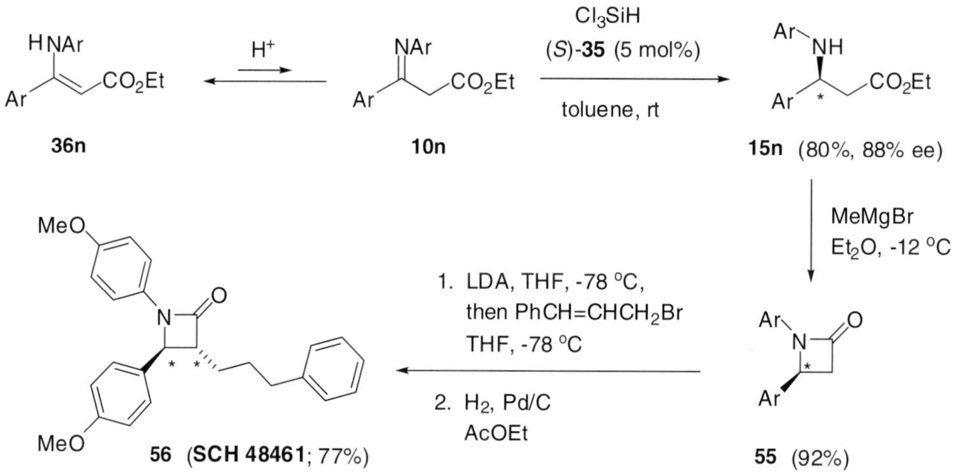

Scheme 4.6 Synthesis of **SCH48461** from β-amino acid **15n**; Ar = p-methoxyphenyl.

With the exception of the sulfinamide **46** and its counterparts (Figure 4.4), the remaining catalysts essentially failed to efficiently promote an asymmetric reduction of the imines derived from nonaromatic amines, which can be regarded as a serious limitation of this methodology. Although in **56** the N-aryl group is actually a prerequisite for its biological activity, enhancing the scope of this approach would require an efficient method for the removal of the aryl group from the nitrogen. With the electron-rich anisidine-derived imines, such as **6b**, this goal can be attained by several oxidative methods. Thus, for instance, $(NH_4)_2Ce(NO_3)_6$ and $PhI(OAc)_2$ have enjoyed a reputation of being efficient and acceptably mild reagents for the removal of the aromatic substituent [23]. For the more sensitive derivatives, such as the amines obtained from chloroimines **10a–10i** (Table 4.6), trichloroisocyanuric acid (TCCA) [24] proved to be the reagent of choice [12d]. Nevertheless, the success here very much depends on the substrate structure, so that the merits of the individual methods may vary.

The asymmetric imine reduction was successfully applied as the key step in the synthesis of N-acetyl colchinol (**60**), whose water soluble phosphate is the potent tubulin-inhibiting, antitumor agent ZD6126 [25] (Scheme 4.7). Reduction of the functionalized imine **9g** with Cl_3SiH, catalyzed by Sigamide (**35**), afforded amine (S)-**14g** (69% yield, 96% ee). In this instance, the N-deprotection turned out to be a particularly difficult task in view of the presence of the trimethoxyphenyl group which, being also electron-rich, competed with the amine deprotection. After numerous experiments, periodic acid was identified as an acceptable reagent, furnishing the primary amine **57** in rather variable yields (45–65%); obviously, further optimization is required. Conversion of **57** into the acetamide **59** proved uneventful, whereas the final oxidative arene–arene coupling required extensive optimization again. Finally, $PhI(O_2CCF_3)_2$ in the presence of BF_3 and CF_3CO_2H and its anhydride [26] afforded N-acetyl colchinol **60** (45%) as

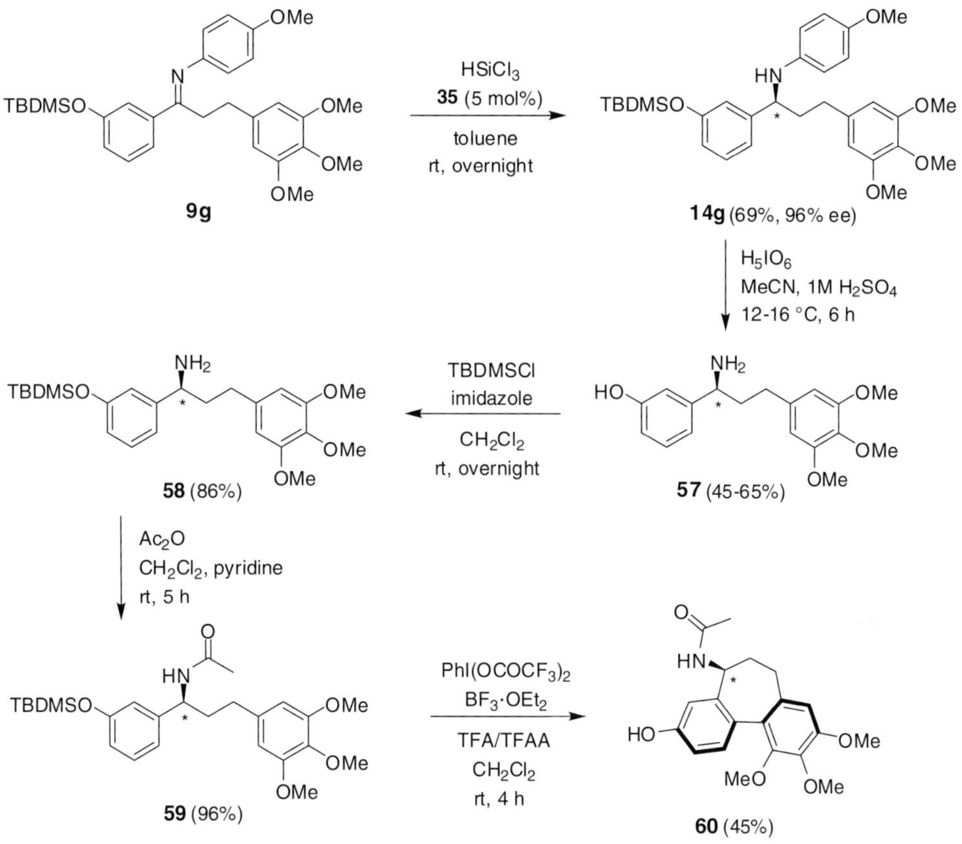

Scheme 4.7 Synthesis of colchinol; TBDMS = (tBu)Me₂Si.

a single diastereoisomer (K. Vranková, A.V. Malkov, and P. Kočovský, unpublished results).

4.8
Conclusions

From the rather limited data reported by Matsumura [10], Sun [11], and Zhang [17], the pipecolinic acid-derived catalyst **20** and the sulfinamide **46** appear to exhibit the highest enantioselectivities, similar to those attained with Sigamide **35**[12]. However, the substrate portfolio [12h] explored with the latter catalyst is much broader (Tables 4.3–4.7) so that a rigorous comparison cannot be made at present. Furthermore, Sigamide has performed consistently well at the standard 5 mol% loading and in selected examples was shown to operate with the same efficiency even when as little as 1 mol% had been used. Most of the catalysts listed in Tables 4.1, 4.2

and 4.8–4.10 were used at 10–20 mol% loading, although some cases with 5 mol% were also reported. Finally, reductions catalyzed by Sigamide **35** were carried out at room temperature, whereas most experiments listed in Tables 4.1, 4.2 and 4.8–4.10 were run at 0 °C or below.

A broad scope of the reduction of ketimines with trichlorosilane catalyzed by Sigamide **35** (Scheme 4.2 and Tables 4.3-4.7) has been demonstrated. High enantioselectivity (typically ≥90% ee) was observed across the spectrum of aromatic, heteroaromatic, and aliphatic substrates, which may contain additional functional groups. The reaction proceeds in toluene at room temperature overnight with 1–5 mol% of the catalyst. Hence, this protocol compares favorably with its alternatives, such as catalytic hydrogenation (which requires a high-pressure equipment) [2] or reduction with Hantzsch dihydropyridine catalyzed by chiral Lewis acids [5]. Furthermore, this catalytic process is complementary to the Cu-catalyzed hydrosilylation developed by Lipshutz [3f–h], which favors a 1,4-addition to α,β-unsaturated imines, whereas Sigamide is 1,2-selective (Table 4.3, entries 10 and 11). Current limitations are relatively few: (1) the reaction exhibits very low enantioselectivity with imines derived from pyridine (Table 4.4, entries 1 and 2) but a remedy to this flaw was found (Table 4.4, entry 3); (2) reduction of ketimines derived from diaryl ketones gives practically racemic products even if the two aryl groups differ in their electronics (Table 4.5, entry 14); (3) most of the current systems only work efficiently with imines derived from aromatic amines (e.g., aniline and anisidine), although the most recent reports suggest that sulfinamides (e.g., **46**) may overcome this limitation (Table 4.10) [11f]. Nevertheless, the anisidine-derived amines can be oxidatively deprotected to produce primary amines [12d, 23, 24]; and (4) imines derived from cyclic ketones exhibit low enantioselectivity, which presumably stems from the rigidity of the cycles that does not allow the substrate to assume the required conformation.

Trichlorosilane, a nonexpensive stoichiometric reducing agent, is relatively easy to handle under anhydrous (but not necessarily anaerobic) conditions. The aqueous workup produces NaCl and SiO_2, two benign inorganics, which in conjunction with the use of toluene as the optimal solvent render this method environmentally acceptable.

4.9
Typical Procedures for the Catalytic Hydrosilylation of Imines

4.9.1
Catalytic Hydrosilylation of Simple Imines [12c,h]

4.9 Typical Procedures for the Catalytic Hydrosilylation of Imines

Trichlorosilane (40 μl, 0.4 mmol, 2 equiv) was added dropwise to a cooled solution (0 °C) of imine **6b** (45 mg, 0.2 mmol, 1 equiv) and catalyst **35** (3.5 mg, 0.01 mmol, 0.05 equiv) in anhydrous toluene (2 ml) under an argon atmosphere and the reaction mixture was allowed to stir at room temperature overnight. The mixture was then diluted with ethyl acetate (5 ml), quenched with a saturated NaHCO$_3$ solution (5 ml), and the layers were separated. The aqueous layer was extracted with AcOEt (2 × 10 ml) and the combined organic extracts were washed with brine (5 ml), dried over anhydrous MgSO$_4$, and evaporated. The residue was purified by flash chromatography on a silica gel column with a hexane–ethyl acetate mixture (10 : 1) to afford (S)-(−)-amine **11b** (43 mg, 95%, 94% ee) as an oil: [α]$_D$ −4.0 (c 1.0, CHCl$_3$).

4.9.2
Catalytic Hydrosilylation of Enamines [12g]

Trichlorosilane (40 μl, 0.4 mmol, 2 equiv) was added dropwise to a cooled solution (0 °C) of enamine **36l** (59.5 mg, 0.2 mmol, 1 equiv), catalyst **35** (3.5 mg, 0.01 mmol, 0.05 equiv), and glacial acetic acid (11.5 μl, 0.2 mmol, 1 equiv) in anhydrous toluene (2 ml) under an argon atmosphere and the reaction mixture was allowed to stir at room temperature for 48 h. The mixture was then diluted with ethyl acetate (5 ml), quenched with a saturated NaHCO$_3$ solution (5 ml), and the layers were separated. The aqueous layer was extracted with AcOEt (2 × 10 ml) and the combined organic extracts were washed with brine (5 ml), dried over anhydrous MgSO$_4$, and evaporated. The residue was purified by flash chromatography on a silica gel column with a petroleum ether–ethyl acetate mixture (9 : 1) to give (S)-(−)-β-amino ester **15l** (59 mg, 98%, 89% ee): [α]$_D$ −5.6 (c 1.0, CHCl$_3$).

Questions

4.1. Discuss the main advantages of the organocatalytic approach to amine synthesis, the environmental issues, and the current limitations of the hydrosilylation method.

4.2. The catalytic hydrosilylation requires the use of Cl$_3$SiH as reagent. Other silanes, such as Me$_3$SiH, are unsuccessful. Discuss the difference in reactivity and give a rationale.

4.3. Discuss the structure of the organocatalysts for hydrosilylation of imines with Cl_3SiH and identify the key structural features. Suggest other functional groups that can be predicted to act in a similar way.

References

1 For a general overview of the reduction of imines, see: the following: (a) Morrison, J.D. (1983) *Asymmetric Synthesis*, vol. 2, Academic, New York; (b) Noyori, R. (1994) *Asymmetric Catalysis in Organic Synthesis*, John Wiley & Sons, New York; (c) Ojima, I. (2000) *Catalytic Asymmetric Synthesis*, 2nd edn, John Wiley & Sons, New York; (d) James, B.R. (1997) *Catal. Today*, **37**, 209; (e) Kobayashi, S. and Ishitani, H. (1999) *Chem. Rev.*, **99**, 1069; (f) Cho, B.T. (2006) *Tetrahedron*, **62**, 7621; (g) Jacobsen, E.N., Pfaltz, A., and Yamamoto, H. (1999) *Comprehensive Asymmetric Catalysis*, vol. I–III, Springer.

2 For recent reports on catalytic hydrogenation (with Ti, Ir, Rh, and Ru), see: Refs [1b–d] and the following: (a) Xiao, D. and Zhang, X. (2001) *Angew. Chem., Int. Ed.*, **40**, 3425; (b) Jiang, X.B., Minnaard, A.J., Hessen, B., Feringa, B.L., Duchateau, A.L.L., Andrien, J.G.O., Boogers, J.A.F., and de Vries, J.G. (2003) *Org. Lett.*, **5**, 1503; (c) Cobley, C.J. and Henschke, J.P. (2003) *Adv. Synth. Catal.*, **345**, 195; (d) Okuda, J., Verch, S., Stürmer, R., and Spaniol, T.S. (2000) *J. Organomet. Chem.*, **605**, 55; (e) Guiu, E., Muñoz, B., Castillón, S., and Claver, C. (2003) *Adv. Synth. Catal.*, **345**, 169; (f) Cobbley, C.J., Foucher, E., Lecouve, J.-P., Lennon, I.C., Ramsden, J.A., and Thominot, G. (2003) *Tetrahedron: Asymmetry*, **14**, 3431; (g) Chi, Y., Zhou, Y.G., and Zhang, X. (2003) *J. Org. Chem.*, **68**, 4120; (h) Bozeio, A.A., Pytkowicz, J., Côté, A., and Charette, A.B. (2003) *J. Am. Chem. Soc.*, **125**, 14260; (i) Trifonova, A., Diesen, J.S., Chapman, C.J., and Andersson, P.G. (2004) *Org. Lett.*, **6**, 3825; (j) Zhu, S.-F., Xie, J.-B., Zhang, Y.-Z., Li, S., and Zhou, Q.-L. (2006) *J. Am. Chem. Soc.*, **128**, 12886; (k) Trifonova, A., Diesen, J.S., and Andersson, P.G. (2006) *Chem. Eur. J.*, **12**, 2318; (l) Cheruku, P., Church, T.L., and Andersson, P.G. (2008) *Chem. Asian J.*, **3**, 1390; for Rh-catalyzed hydrogenation of enamides, see: the following: (m) Hu, X.-P. and Zheng, Z. (2004) *Org. Lett.*, **6**, 3585; (n) Yiang, Q., Gao, W., Deng, J., and Zhang, X. (2006) *Angew. Chem., Int. Ed.*, **45**, 2832; (o) Li, C., Wang, C., Villa-Macros, B., and Xiao, J. (2008) *J. Am. Chem. Soc.*, **130**, 14450; (p) Moessne, C. and Bolm, C. (2005) *Angew. Chem., Int. Ed.*, **44**, 7564; (q) Schinder, P., Kock, G., Pretot, R., Wang, G., Bohnen, F.M., Kruger, C., and Pfaltz, A. (1997) *Chem. Eur. J.*, **3**, 887.

3 For transition metal-catalyzed hydrosilylation, see: (a) Reding, M.T. and Buchwald, S.L. (1998) *J. Org. Chem.*, **63**, 6344; (b) Verdaguer, X., Lange, U.E.W., and Buchwald, S.L. (1998) *Angew. Chem., Int. Ed.*, **37**, 1103; (c) Hansen, M.C. and Buchwald, S.L. (2000) *Org. Lett.*, **2**, 713; (d) Vedejs, E., Trapencieris, P., and Suna, E. (1999) *J. Org. Chem.*, **64**, 6724; (e) Nishikori, H., Yoshihara, R., and Hosomi, A. (2003) *Synlett*, 561; (f) Lipshutz, B.H., Noson, K., and Chrisman, W. (2001) *J. Am. Chem. Soc.*, **123**, 12917; (g) Lipshutz, B.H. and Shimizu, H. (2004) *Angew. Chem., Int. Ed.*, **43**, 2228; (h) Lipshutz, B.H., Frieman, B.A., and Tomaso, A.E. (2006) *Angew. Chem., Int. Ed.*, **45**, 1259.

4 For an overview of transfer hydrogenation of imines, see: Samec, J.S.M., Bäckvall, J.-E., Andersson, P.G., and Brandt, P. (2006) *Chem. Soc. Rev.*, **35**, 237.

5 (a) Singh, S. and Batra, U.K. (1989) *Indian J. Chem., Sect. B*, **28**, 1; (b) Rueping, M., Sugiono, E., Azap, C., Theissmann, T., and Bolte, M. (2005) *Org. Lett.*, **7**, 3781; (c) Rueping, M., Antonchick, A.P., and Theissmann, T. (2006) *Angew. Chem., Int. Ed.*, **45**, 3683; (d) Rueping, M., Antonchik, A.P., and Theissmann, T. (2006) *Angew. Chem., Int. Ed.*, **45**, 6751; (e) Hoffman, S., Seayad, A.M., and List, B. (2005) *Angew.*

Chem., Int. Ed., **44**, 7424; (f) Hoffmann, S., Nicoletti, M., and List, B. (2006) *J. Am. Chem. Soc.*, **128**, 13074; (g) Zhou, J. and List, B. (2007) *J. Am. Chem. Soc.*, **129**, 7498; (h) Storer, R.I., Carrera, D.E., Ni, Y., and MacMillan, D.W.C. (2006) *J. Am. Chem. Soc.*, **128**, 84; (i) Ouellet, S.G., Walji, A.M., and MacMillan, D.W.C. (2007) *Acc. Chem. Res.*, **40**, 1327.

6 (a) Berkessel, A. and Gröger, H. (2005) *Asymmetric Organocatalysis*, Wiley-VCH Verlag GmbH, Weinheim;(b) Dalko, P.I. (ed.) (2007) *Enantioselective Organocatalysis*, Wiley-VCH Verlag GmbH, Weinheim.

7 Activation of Cl_3SiH (a) Benkeser, R.A. and Snyder, D., (1982) *J. Organomet. Chem.*, **225**, 107; (b) Kobayashi, S., Yasuda, M., and Hachiya, I. (1996) *Chem. Lett.*, 407; (c) Iwasaki, F., Onomura, O., Mishima, K., Maki, T., and Matsumura, Y. (1999) *Tetrahedron Lett.*, **40**, 7507;(d) Brook, M.A. (2000) *Silicon in Organic, Organometallic and Polymeric Chemistry*, John Wiley & Sons, New York, pp. 133–136.

8 (a) Kočovský, P. and Malkov, A.V. (2007) Chiral Lewis bases as catalysts, *Enantioselective Organocatalysis* (ed. P.I. Dalko), Wiley-VCH Verlag GmbH, Weinheim, p. 255; (b) Kagan, H. (2007) Organocatalytic enantioselective reduction of olefins, ketones, and imines, in *Enantioselective Organocatalysis* (ed. P.I. Dalko), Wiley-VCH Verlag GmbH, Weinheim, p. 391.

9 For enzymatic approaches, including reduction, see: (a) Alexeeva, M., Enright, A., Dawson, M.J., Mahmoidian, M., and Turner, N.J. (2002) *Angew. Chem., Int. Ed.*, **41**, 3177; (b) Carr, R., Alexeeva, M., Dawson, M.J., Gotor-Fernández, V., Huphrey, C.E., and Turner, N.J. (2005) *ChemBioChem*, **6**, 637; (c) Dunsmore, C.J., Carr, R., Fleming, T., and Turner, N.J. (2006) *J. Am. Chem. Soc.*, **128**, 2224; for a brief overview, see: (d) Alexeeva, M., Carr, R., and Turner, N.J. (2003) *Org. Biomol. Chem.*, **1**, 4133.

10 (a) Iwasaki, F., Onomura, O., Mishima, K., Kanematsu, T., Maki, T., and Matsumura, Y. (2001) *Tetrahedron Lett.*, **42**, 2525; (b) Onomura, O., Kouchi, Y., Iwasaki, F., and Matsumura, Y. (2006) *Tetrahedron Lett.*, **47**, 3751.

11 (a) Wang, Z., Ye, X., Wei, S., Wu, P., Zhang, A., and Sun, J. (2006) *Org. Lett.*, **8**, 999; (b) Wang, Z., Cheng, M., Wu, P., Wei, S., and Sun, J. (2006) *Org. Lett.*, **8**, 3045; (c) Wu, P., Wang, Z., Cheng, M., Zhou, L., and Sun, J. (2008) *Tetrahedron*, **64**, 11304; (d) Wang, Z., Wei, S., Wang, C., and Sun, J. (2007) *Tetrahedron: Asymmetry*, **18**, 705; (e) Pei, D., Wang, Z., Wei, S., Zhang, Y., and Sun, J. (2006) *Org. Lett.*, **8**, 5913; (f) Wang, C., Wu, X., Zhou, L., and Sun, J. (2008) *Chem. Eur. J.*, **14**, 8789; (g) Pei, D., Zhang, Y., Wei, S., Wang, M., and Sun, J. (2008) *Adv. Synth. Catal.*, **350**, 619; (h) Diastereoisomers of **20** have also been explored by Sun but exhibited lower enantioselectivities then **20**.

12 (a) Malkov, A.V., Mariani, A., MacDougall, K.N., and Kočovský, P. (2004) *Org. Lett.*, **6**, 2253; (b) Malkov, A.V., Stončius, S., MacDougall, K.N., Mariani, A., McGeoch, G.D., and Kočovský, P. (2006) *Tetrahedron*, **62**, 264; (c) Malkov, A.V., Figlus, M., Stončius, S., and Kočovský, P. (2007) *J. Org. Chem.*, **72**, 1315; (d) Malkov, A.V., Stončius, S., and Kočovský, P. (2007) *Angew. Chem., Int. Ed.*, **46**, 3722; (e) Malkov, A.V., Figlus, M., and Kočovský, P. (2008) *J. Org. Chem.*, **73**, 3985; (f) Malkov, A.V., Figlus, M., Cooke, G., Caldwell, S.T., Rabani, G., Prestly, M.R., and Kočovský, P. (2009) *Org. Biomol. Chem.*, **7**, 1878; (g) Malkov, A.V., Stončius, S., Vranková, K., Arndt, M., and Kočovský, P. (2008) *Chem. Eur. J.*, **14**, 8082; (h) Malkov, A.V., Vranková, K., Stončius, S., and Kočovský, P. (2009) *J. Org. Chem.*, **74**, 5839; (i) Malkov, A.V., Figlus, M., Prestly, M.R., Rabani, G., Cooke, G., and Kočovský, P. (2009) *Chem. Eur. J.*, **15**, 9651; (j) Cooke, G., Figlus, M., Caldwell, S.T., Walas, D., Sanyal, A., Yesilbag, G., Malkov, A.V., and Kočovský, P. (2009) *Org. Biomol. Chem.*, in press (DOI:10.1039/b916601g); ASAP on October 27; (k) Malkov, A.V., Vranková, K., Sigerson, R., Stončius, S., and Kočovský, P. (2009) *Tetrahedron*, **65**, 9481.

13 Malkov, A.V., Stewart Liddon, A.J.P., Ramírez-López, P., Bendová, L., Haigh, D.,

and Kočovský, P. (2006) *Angew. Chem., Int. Ed.*, **45**, 1432.
14 Fester, G.W., Wagler, J., Brendler, E., Böhme, U., Gerlach, D., and Kroke, E. (2009) *J. Am. Chem. Soc.*, **131**, 6855.
15 For a review on reduction of isosteric, electronically biased ketones, see: Corey, E.J. and Helal, C.J. (1998) *Angew. Chem., Int. Ed.*, **37**, 1987.
16 For leading reviews on DKR, see: (a) Noyori, R., Tokunaga, M., and Kitamura, M. (1995) *Bull. Chem. Soc. Jpn.*, **68**, 36; (b) Ward, R.S. (1995) *Tetrahedron: Asymmetry*, **6**, 1475; (c) Huerta, F.F., Minidis, A.B.E., and Bäckvall, J.-E. (2001) *Chem. Soc. Rev.*, **30**, 321; (d) Pellissier, H. (2003) *Tetrahedron*, **59**, 8291; (e) Pàmies, O. and Bäckvall, J.-E. (2003) *Chem. Rev.*, **103**, 3247; for a recent example of asymmetric reductive amination by DKR, see: Ref. [5f].
17 (a) Zheng, H., Deng, J., Lin, W., and Zhang, X. (2007) *Tetrahedron Lett.*, **48**, 7934; (b) Zheng, H.-J., Chen, W.B., Wu, Z.-J., Deng, J.-G., Lin, W.-Q., Yuan, W.-C., and Zhang, X.-M. (2008) *Chem. Eur. J.*, **14**, 9864; see: also (c) Guizzetti, S., Benaglia, M., Cozzi, F., Rossi, S., and Celentano, G. (2009) *Chirality*, **21**, 233; (d) Gautier, F.M., Jones, S. Martin, S.J. (2009) *Org. Biomol. Chem.*, **7**, 229.
18 For polymer-supported catalysts, see: (a) McNamara, C.A., Dixon, M.J., and Bradley, M. (2002) *Chem. Rev.*, **102**, 3275; (b) Bergbreiter, D.E. (2002) *Chem. Rev.*, **102**, 3345; (c) Heitbaum, M., Glorius, F., and Escher, I. (2006) *Angew. Chem., Int. Ed.*, **45**, 4732; (d) Baker, R.T., Kobayashi, S., and Leitner, W. (2006) *Adv. Synth. Catal.*, **348**, 1337; (e) Cozzi, F. (2006) *Adv. Synth. Catal.*, **348**, 1367; (f) Bergbreiter, D.E. and Sung, S.D. (2006) *Adv. Synth. Catal.*, **348**, 1352; (g) Bergbreiter, D.E. and Li, J. (2004) *Chem. Commun.*, 42.
19 For a critical review on this method, see: Matyjaszewski, K. and Xia, J. (2001) *Chem. Rev.*, **101**, 2921.
20 (a) Jennings, W.B., Al-Showiman, S., Tolley, M.S., and Boyd, D.R. (1975) *J. Chem. Soc., Perkin Trans. 2*, 1535; (b) Johnson, J.E., Morales, N.M., Gorczyca, A.M., Dolliver, D.D., and McAllister, M.A. (2001) *J. Org. Chem.*, **66**, 7979.
21 For a similar discussion of the role of Brønsted acids on the reduction of imines with Hantzsch dihydropyridines, see: Marcelli, T., Hammar, P. and Himo, F. (2008) *Chem. Eur. J.*, **14**, 8562.
22 For a leading reference, see: Rosenblum, S.B., Huynh, T., Afonso, A., Davis, H.R., Yumibe, N., Clader, J.W., and Burnett, D.A. (1998) *J. Med. Chem.*, **41**, 973.
23 (a) Kobayashi, S., Ishitani, H., and Ueno, M. (1998) *J. Am. Chem. Soc.*, **120**, 431; (b) Chi, Y., Zhou, Y.-G., and Zhang, X. (2003) *J. Org. Chem.*, **68**, 4120; (c) Hata, S., Iguchi, M., Iwasawa, T., Yamada, K., and Tomioka, K. (2004) *Org. Lett.*, **6**, 1721; (d) Janey, J.M., Hsiao, Y., and Armstrong, J.D. (2006) *J. Org. Chem.*, **71**, 390; (e) Córdova, A., Notz, W., Zhong, G., Betancort, J.M., and Barbas, C.F. (2002) *J. Am. Chem. Soc.*, **124**, 1842; (f) Ibrahem, I., Casas, J., and Córdova, A. (2004) *Angew. Chem., Int. Ed.*, **43**, 6528; (g) Porter, J.R., Traverse, J.F., Hoveyda, A.H., and Snapper, M.L. (2001) *J. Am. Chem. Soc.*, **123**, 10409; also, for CAN deprotection, see: Refs [5e–f,17b].
24 Verkade, J.M.M., van Hemert, L.J.C., Quaedflieg, P.J.L.M., Alsters, P.L., and van Delft, F.L. and Rutjes, F.P.J.T. (2006) *Tetrahedron Lett.*, **47**, 810.
25 (a) Davis, P.D., Dougherty, G.J., Blakey, D.C., Galbraith, S.M., Tozer, G.M., Holder, A.L., Naylor, M.A., Nolan, J., Stratford, M.R.L., Chaplin, D.J., and Hill, S.A. (2002) *Cancer Res.*, **62**, 7247; (b) Micheletti, G., Poli, M., Borsotti, P., Martinelli, M., Imberti, B., Taraboletti, G., and Giavazzi, R. (2003) *Cancer Res.*, **63**, 1534; for a recent synthesis, see: (c) Besong, G., Billen, D., Dager, I., Kocienski, P., Sliwinski, E., Tai, L.R., and Boyle, F.T. (2008) *Tetrahedron*, **64**, 4700.
26 (a) Besong, G., Jarowicki, K., Kocienski, P., Sliwinski, E., and Boyle, F.T. (2006) *Org. Biomol. Chem.*, **4**, 2193; (b) Besong, G., Billen, D., Dager, I., Kocienski, P., Sliwinski, E., Tai, L.R., and Boyle, F.T. (2008) *Tetrahedron*, **64**, 4700.

5
Catalytic, Enantioselective, Vinylogous Mannich Reactions
Christoph Schneider and Marcel Sickert

5.1
Introduction

The asymmetric Mannich reaction of an enolate and an imine furnishing valuable β-amino carbonyls is a fundamental C–C-bond forming process in organic chemistry that has broad utility in organic synthesis particularly for β-amino acid synthesis [1]. Extending the enolate component into a dienolate offers the opportunity for a bond forming event with an electrophile both at the α- and the γ-positions of this ambident nucleophile (Scheme 5.1).

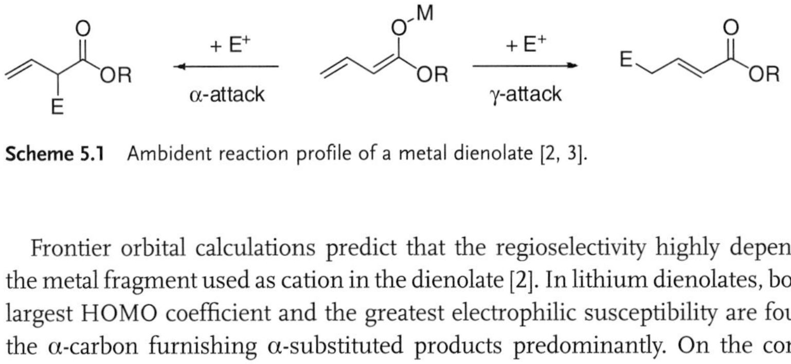

Scheme 5.1 Ambident reaction profile of a metal dienolate [2, 3].

Frontier orbital calculations predict that the regioselectivity highly depends on the metal fragment used as cation in the dienolate [2]. In lithium dienolates, both the largest HOMO coefficient and the greatest electrophilic susceptibility are found at the α-carbon furnishing α-substituted products predominantly. On the contrary, silicon dienolates display larger HOMO coefficients and electrophilic susceptibilities at the γ-position leading in Mukaiyama-type reactions to γ-addition products preferentially (Scheme 5.2) [3]. In addition to electronic effects, steric effects also contribute to the observed regioselectivity with large α-substituents being able to shift the regioselectivity from the α- to the γ-position.

Vinylogous Mannich products have proven to be extremely valuable synthetic intermediates because the additional conjugate double bond may be further elaborated in subsequent chemical transformations.

Chiral Amine Synthesis: Methods, Developments and Applications. Edited by Thomas C. Nugent
Copyright © 2010 WILEY-VCH Verlag GmbH & Co. KGaA, Weinheim
ISBN: 978-3-527-32509-2

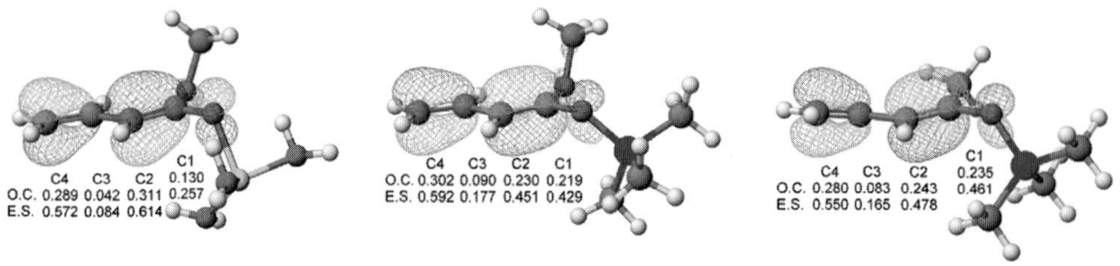

Scheme 5.2 Electron distribution and orbital coefficients of lithium dienolate (left), vinyl silylketene acetal (middle), and silyl dienol ether (right) (reprinted with permission from Wiley-VCH) [3].

In particular, Martin and Bur have exploited the vinylogous Mannich reaction extensively as key step in the total synthesis of alkaloids and other nitrogen heterocycles such as (−)tetrahydroalstonine (**6**), geissoschizine (**7**), and akuammicine (**8**) (Scheme 5.3) [4].

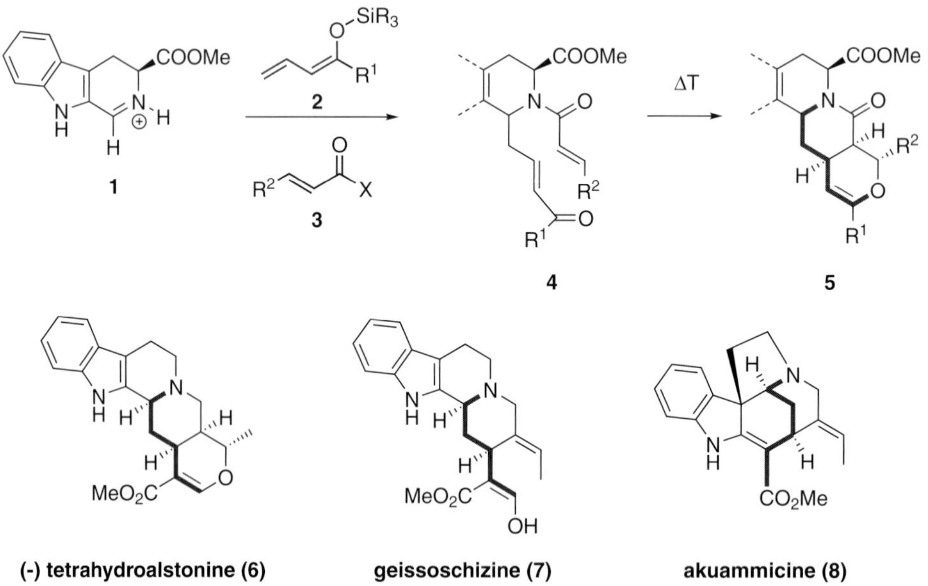

Scheme 5.3 Use of the vinylogous Mannich reaction as a key step in the total synthesis of alkaloids by the Martin group [4].

In this review we will attempt to highlight the most important contributions toward the realization of a catalytic, enantioselective, vinylogous Mannich reaction and show the current state of the art. This chapter is organized in such a way that vinylogous Mannich reactions of preformed silyl dienolates in Mukaiyama-type reactions will be discussed first followed by direct vinylogous Mannich reactions of unmodified substrates.

5.2
Vinylogous Mukaiyama–Mannich Reactions of Silyl Dienolates

The first systematic investigation toward Lewis acid-catalyzed vinylogous Mukaiyama–Mannich reactions was reported by the group of Ojima in 1987 who showed that acyclic vinylketene silyl O,O-acetals **10** reacted with imines activated by stoichiometric amounts of TiCl$_4$ to furnish either 5-amino-2-alkenoates **11** or 5,6-dihydropyridones **12** selectively in excellent yields depending upon the substitution of the silyl dienolate employed (Scheme 5.4) [5]. Although 2-methyl-substituted vinylketene acetal **10a** gave rise to acyclic 5-amino-2-alkenoates **11** exclusively, 3-methyl-substituted vinylketene acetal **10b** furnished 5,6-dihydropyridones **12** as the sole products. No products arising from α-addition of the vinylketene acetal to the imine were obtained.

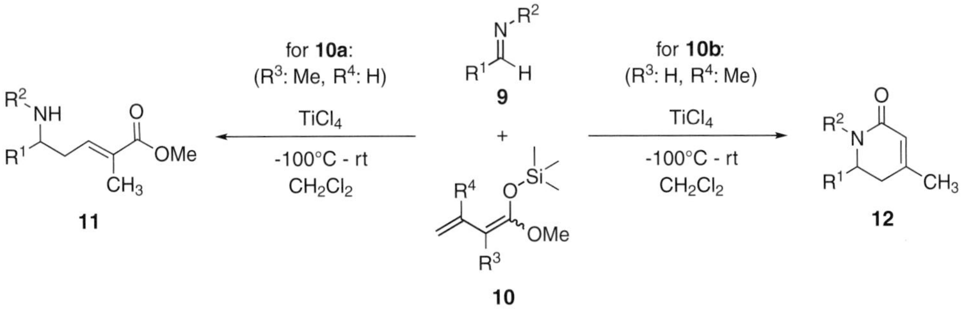

Scheme 5.4 TiCl$_4$-catalyzed vinylogous Mannich reaction of vinylketene O,O-acetals **10** according to Ojima and Brandstadter [5].

Martin and Lopez reported the first example of a catalytic, enantioselective vinylogous Mannich reaction of salicyl imines **13** and 2-silyloxy furans **14** as nucleophiles that are known to have a strong tendency toward γ-selective electrophilic attack [6]. A chiral metal complex formed *in situ* from Ti(O*i*Pr)$_4$ and (S)-BINOL (1 : 2) in ether was employed as catalyst and delivered γ-aminoalkyl-substituted γ-butenolides **15** in good yields, good diastereoselectivity, and typically moderate enantioselectivity (Scheme 5.5). The presence of the chelating hydroxy group in the *ortho*-position within the N-aryl substituent was mandatory for the enantioselectivity of the reaction as other imine substituents delivered the products in racemic form. This finding proved to be quite general as we will see in other vinylogous Mannich reactions and pointed to a highly organized chiral titanium–imine complex as prerequisite for an enantioselective reaction.

Building on this precedence, Hoveyda, Snapper and coworkers established the first highly diastereo- and enantioselective vinylogous Mannich reaction of 2-silyloxy furans **17** with aromatic N-(2-methoxyphenyl)imines **16** [7]. As chiral catalyst they employed a silver phosphine complex (1–5 mol%) prepared *in situ* from commercially available AgOAc and an amino acid-derived, readily prepared Schiff

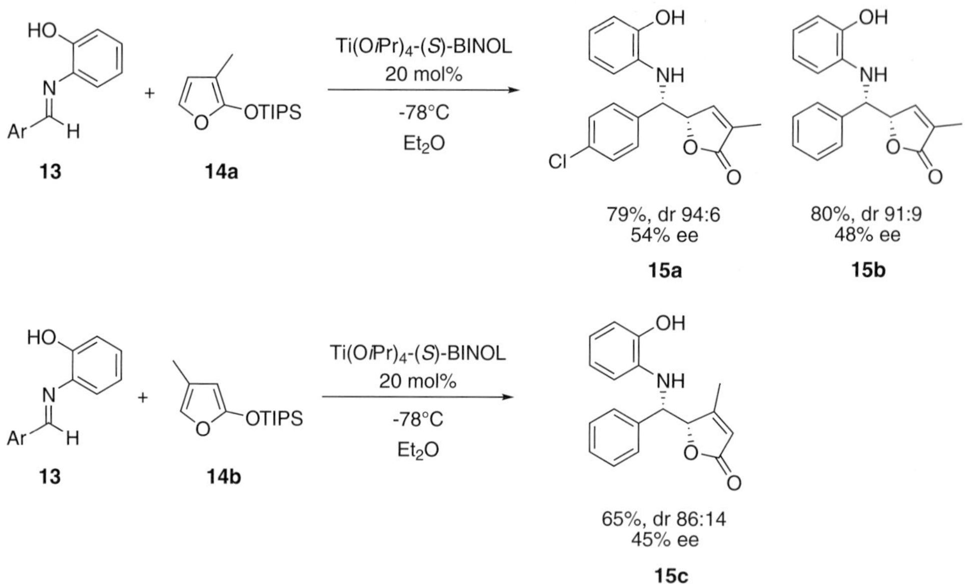

Scheme 5.5 Ti(IV)-BINOLate-catalyzed vinylogous Mannich reaction according to Martin and Lopez [6].

base-phosphine ligand **18**. γ-Aminoalkyl-substituted γ-butenolides **19** were obtained in good yields, with >98% de and typically well above 90% ee and in select cases above 95% ee. The process proved to be highly practical as the reactions could be run in undistilled 2-propanol as solvent and in the air (Table 5.1).

The reaction was extended to more highly substituted silyloxy furans **17b** and **17c** furnishing the products with only slightly diminished enantioselectivities (Scheme 5.6).

Apparently, the chiral Schiff base acts both as a bidentate P,N-ligand to the silver Lewis acid and with the appendent amide linkage serves as a trap for the cationic silicon species generated during the reaction, thereby facilitating the catalytic turnover. Quite similar as in the Martin system, the chelating 2-methoxy substituent in the N-aryl group presumably helps to form a highly organized coordination sphere around the metal center in the transition state of the reaction (Scheme 5.7). At the same time it facilitates the oxidative removal of the N-aryl group with PhI(OAc)$_2$ giving rise to the free amino γ-butenolides in good yields.

The described process could, however, only be successfully applied to aromatic aldimines as the corresponding reactions with aliphatic aldimines led to low yields of the Mannich products. Subsequent investigations by the same group led to the identification of a more robust imine class containing the highly electron-rich *ortho*-thiomethyl-*para*-methoxyphenyl group as N-substituent that is less prone to decomposition [8]. When this type of substituted aniline was employed in a three-component vinylogous Mannich reaction with aliphatic aldehydes and silyloxy furans under otherwise identical reaction conditions, 5-alkyl-5-amino-substituted

5.2 Vinylogous Mukaiyama–Mannich Reactions of Silyl Dienolates

Table 5.1 Silver-catalyzed vinylogous Mannich reaction of 2-silyloxy furan **17a** with aromatic imines according to Hoveyda, Snapper and coworkers [7].

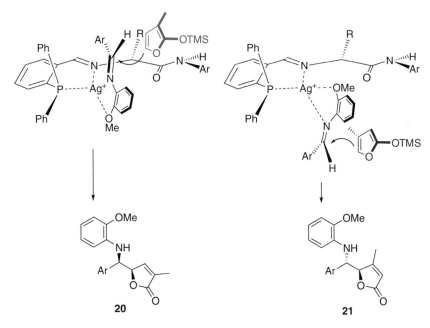

Entry	Ar	Catalysts	Product	Yield (%)	de (%)	ee (%)
1	Ph	18a	19a	82	>98	96
2	p-MeO-C$_6$H$_4$	18a	19b	85	>98	97
3	p-Cl-C$_6$H$_4$	18b	19c	89	>98	93
4	m-NO$_2$-C$_6$H$_4$	18b	19d	75	>98	93
5	o-Me-C$_6$H$_4$	18a	19e	65	>98	94
6	2-Naphtyl	18a	19f	94	>98	>98
7	2-Furyl	18b	19g	78	>98	90

Scheme 5.6 Proposed transition states for the vinylogous Mannich reaction as origin for the enantio- and diastereoselectivity [7].

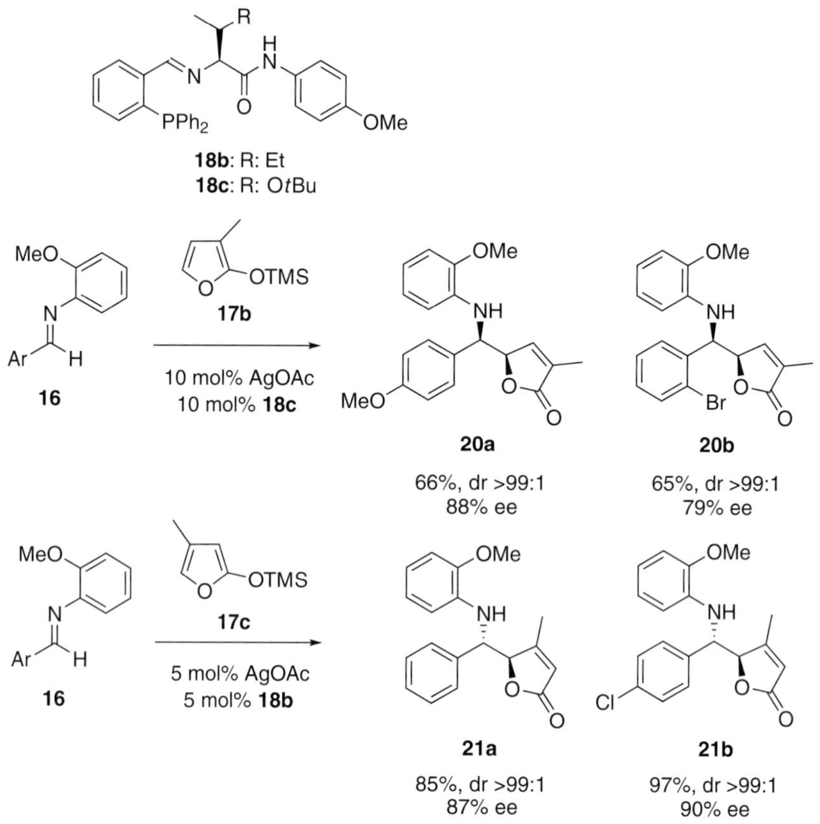

Scheme 5.7 Reaction of methyl-substituted silyloxyfurans **17b** and **17c** [7].

γ-butenolides **22** were obtained in good yields and as essentially single stereoisomers (Table 5.2). In these studies, the *tert*-leucine-derived Schiff base-ligand **18a** gave rise to the best diastereo- and enantioselectivities, and even with the less expensive valine-derived Schiff base ligand, the products were formed with 97% ee. Removal of the *N*-aryl group was readily accomplished with cerium ammonium nitrate followed by acidic hydrolysis of the intermediate aza-quinone.

One of the most remarkable aspects of this reaction is the functional group tolerance exhibited by the catalytic system. Thus, a number of functional aldehydes were submitted to this reaction and furnished vinylogous Mannich products **22e–22i** in moderate yields and excellent stereoselectivity (entries 5–9). The exceptional enantiodifferentiating ability of the chiral silver complex was further demonstrated in doubly stereodifferentiating reactions with chiral aldimines. Irrespective of the configuration of the starting aldehydes, the products were obtained with the identical absolute configuration at the newly generated stereogenic centers with excellent stereoselectivity indicating that the catalyst selectivity completely overrides the substrate selectivity (Scheme 5.8).

Table 5.2 Silver-catalyzed vinylogous Mannich reaction of 2-silyloxy furan **17a** with aliphatic imines according to Hoveyda, Snapper and coworkers [8].

Entry	Aldehyde	Product	Yield (%)	de (%)	ee (%)
1	CyCHO	22a	90	>96	>98
2	iPr-CHO	22b	89	>96	>98
3	tBu-CHO	22c	50	>96	>98
4	nHex-CHO	22d	75	>96	>98
5	MeO$_2$C-(CH$_2$)$_2$-CHO	22e	56	>96	>98
6	BnO-CH$_2$-CHO	22f	51	>96	>98
7	BocNH-CH$_2$-CHO	22g	44	>96	>98
8	(E)-Ph-CH=CH-CHO	22h	32	>96	95
9	Ph-C≡C-CHO	22i	89	>98	>98

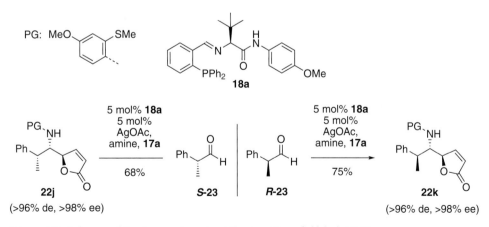

Scheme 5.8 Influence of the stereogenic center at the α-position of aldehyde **23** [8].

Table 5.3 Silver-catalyzed vinylogous Mannich reaction of 2-silyloxy furan **17a** with α-ketoimine esters **24** according to Hoveyda, Snapper and coworkers [9].

Entry	Ar	Product	Yield (%)	anti/syn	ee (%)
1	Ph	25a	88	95:5	92
2	m-MeO-C$_6$H$_4$	25b	95	95:5	93
3	m-Cl-C$_6$H$_4$	25c	72	92:8	87
4	p-Br-C$_6$H$_4$	25d	80	95:5	92
5	p-CF$_3$-C$_6$H$_4$	25e	87	95:5	94
6	o-Br-C$_6$H$_4$	25f	87	<2:98	32
7	2-Naphtyl	25g	81	>98:2	91

This protocol could be directly applied in the addition of 2-silyloxy furans to ketimines and the generation of a quaternary chiral center. Using electronically highly activated α-ketoimine esters **24** as substrates, the same chiral silver phosphine catalyst (10 mol%) as previously used in reactions with aldimines delivered the desired Mannich products in good yields, good diastereoselectivity, and up to 94% ee (Table 5.3) [9].

The vinylogous Mannich reaction of 2-silyloxy furans and imines may also be catalyzed through chiral Brønsted acids, as shown by Akiyama et al. [10]. Previously, Akiyama [11] and Terada [12] had independently discovered that 3,3′-substituted BINOL-based phosphoric acids were excellent Brønsted acids for a broad range of mainly imine addition reactions via protonation of the imines and in situ formation of chiral iminium contact ion pairs. Using the slightly modified phosphoric acid **28** as catalyst carrying additional iodine substituents in the 6,6′-positions, the γ-amino-substituted butenolides **27** were obtained in excellent enantioselectivity and variable diastereoselectivity (Table 5.4).

Aryl-, heteroaryl-, and alkyl-substituted aldimines were effectively converted into the products in typically good yields. Again, ortho-aminophenol was employed as the amine component and the chelating hydroxy group within the imine was proposed to act as an additional hydrogen donor that formed a nine-membered chelate with two hydrogen bonds in the transition state of the reaction (Figure 5.1). Theoretical calculations performed by the Akiyama group supported this scenario [13].

5.2 Vinylogous Mukaiyama–Mannich Reactions of Silyl Dienolates

Table 5.4 Brønsted acid-catalyzed vinylogous Mannich reaction according to Akiyama et al. [10].

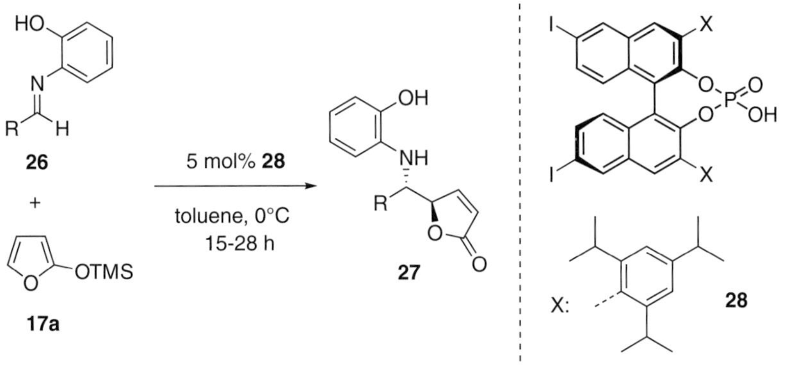

Entry	R	Product	Yield (%)	anti/syn	ee (%)
1	Ph	27a	89	91:9	82
2	p-F-C$_6$H$_4$	27b	100	95:5	87
3	m-NO$_2$-C$_6$H$_4$	27c	86	68:32	96
4	4-Pyridyl	27d	30	94:6	98
5	2-Furyl	27e	77	68:32	89
6	Cyclohexyl	27f	77	88:12	90
7	iPr	27g	84	88:12	92

Thus far, all reported protocols were solely applicable to reactions of 2-silyloxy furans as nucleophiles and concerned the highly stereoselective formation of chiral γ-butenolides. On the basis of the prominent role that this motif plays in natural products and medicinally relevant compounds, this focus is certainly understandable. However, the enantioselective synthesis of acyclic δ-amino α,β-unsaturated carbonyl compounds starting from an acyclic silyl dienolate would be even more desirable as those substrates might actually bear enormous synthetic potential for a broad range of nitrogen-containing compounds in general and nitrogen heterocycles in particular. In acyclic silyl dienolates, however, the α, γ-regioselectivity within the ambident nucleophile is not as easily controlled as in 2-silyloxy furans that further complicates an efficient and selective reaction and might have hampered the development of such a process.

In 2008 our group reported the first example of a catalytic, enantioselective, vinylogous Mannich reaction of acyclic silyl dienolate **30** with imines (Table 5.5) [14].

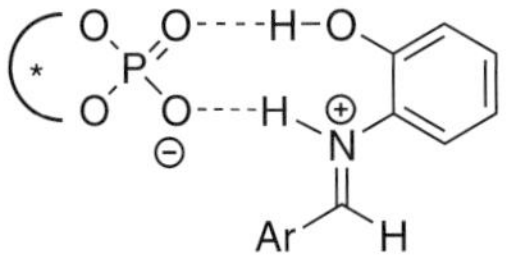

Figure 5.1 Theoretically calculated imine-bound phosphoric acid with two hydrogen bonds according to Akiyama and coworkers [13].

Table 5.5 Brønsted acid-catalyzed vinylogous Mannich reaction of vinylketene O,O-acetal **30** according to Schneider and Sickert [14a].

Entry	R	Product	Time (h)	Yield (%)	ee (%)
1	Ph	31a	8	87	88
2	4-Me-C$_6$H$_4$	31b	4	89	90
3	4-Et-C$_6$H$_4$	31c	3	88	92
4	4-Pent-C$_6$H$_4$	31d	6	92	90
5	4-MeO-C$_6$H$_4$	31e	72	66	82
6	4-F-C$_6$H$_4$	31f	5	93	82
7	4-CN-C$_6$H$_4$	31g	1	94	81
8	3-Cl-C$_6$H$_4$	31h	2	94	82
9	3-Me-C$_6$H$_4$	31i	6	90	79
10	2-Me-C$_6$H$_4$	31j	9	87	80
11	2-Naphthyl	31k	8	90	83
12	3-Thiophenyl	31l	6	92	84
13	3-Furyl	31m	7	88	90
14	tBu	31o	48	83	83

The TBS-group was chosen as silyl fragment within the dienolate to prevent α-attack of the imines on the nucleophile. As chiral catalyst we employed a BINOL-based phosphoric acid of the same type that Akiyama and Terada had established in asymmetric catalysis and found 3,3′-mesityl groups optimal for the enantioselectivity of the reaction. The reactions were run at −30 °C in a solvent mixture of tBuOH, 2-methyl-2-butanol, and THF in equal amounts containing an additional 1 equiv of water.

Aromatic and heteroaromatic aldimines were effectively converted into vinylogous Mannich products **31** with complete γ-regioselectivity and with typically 80–90% ee. The reaction could easily be run in a three-component fashion, starting directly from an aldehyde, para-anisidine, and silyl dienolate **30** obviating the need to prepare the imine in a separate reaction. In contrast to most other protocols that required a salicyl imine moiety in the substrate for selectivity issues, here the amine component within the imine could just be a phenyl group or any para-substituted phenyl group.

Mechanistically we assume that the reaction proceeds via initial imine protonation furnishing the chiral iminium phosphate **33** (Scheme 5.9). This in turn is attacked by silyl dienolate **30** in the central C−C-bond forming event that is controlled by the

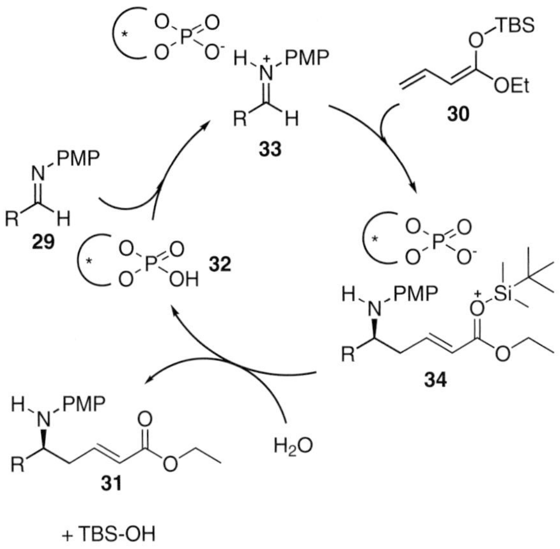

Scheme 5.9 Catalytic cycle for the vinylogous Mannich reaction of vinylketene silyl O,O-acetal **30** developed by Schneider and Sickert [14a].

chiral counterion in an enantioselective fashion. The contact ion pair **34** thus generated is hydrolyzed through the water present in the reaction mixture to furnish TBS-OH, the vinylogous Mannich product **31**, and the chiral Brønsted acid catalyst, thereby closing the catalytic cycle. In ESI-MS/MS-measurements, contact ion pair **34** was clearly detected and fully characterized and the *in situ* formation of TBS-OH was observed by online NMR studies thus strongly supporting this mechanistic scenario.

Subsequent investigations led to the identification of a superior, second-generation catalyst **35** that improved the enantioselectivity of the reaction for most substrates considerably. With a simple change of the *para*-methyl group within the 3,3′-mesityl groups of the BINOL-backbone for a *para*-*tert*-butyl group, most aromatic and heteroaromatic aldimines were now converted into the products with around 90% ee and in select cases with >95% ee (Scheme 5.10) (M. Sickert and C. Schneider, unpublished results).

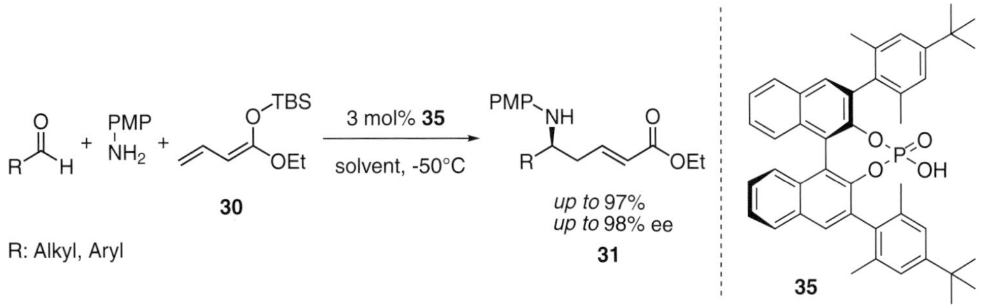

Scheme 5.10 Enantioselective, vinylogous Mannich reaction of silyl dienolate **30** catalyzed by second-generation catalyst **35** (M. Sickert and C. Schneider, unpublished results).

Vinylketene silyl N,O-acetal **36** derived from the β,γ-unsaturated N-acyl piperidine was also shown to participate in vinylogous Mannich reactions to furnish δ-amino α,β-unsaturated amides **37** in good yields and up to 91% ee with just 1 mol% of phosphoric acid **38** (Table 5.6) [14b]. Interestingly, the absolute configuration of the products was opposite to the configuration obtained with phosphoric acid **32** although the catalyst carried the identical R-BINOL-backbone. In order to demonstrate their synthetic potential, the amides were subsequently reduced to either allylic alcohols with the Schwartz reagent Cp$_2$Zr(H)Cl or into the saturated alcohols with LiBHEt$_3$ in good yields.

Subsequent to our report, Carretero and coworkers developed a chiral copper complex that catalyzed highly enantioselective vinylogous Mannich reactions both with 2-silyloxy furans and acyclic silyl dienolates as nucleophiles [15]. The author's group had previously developed the chiral ferrocene-based P,S-ligand Fesulphos in combination with copper(I) salts for Mannich and aza-Diels–Alder reactions. On that basis they were able to show that the reaction of 2-thienylsulfonyl imines **39** and a broad range of acyclic silyl dienolates **40** was efficiently catalyzed by the Cu-Fesulphos catalyst prepared *in situ* from CuBr and the ligand **42**. The products were obtained with complete γ-regioselectivity for most substrates investigated and 89–95% ee for aromatic aldimines and 65–82% ee for aliphatic aldimines (Table 5.7).

Table 5.6 Brønsted acid-catalyzed vinylogous Mannich reaction of vinylketene N,O-acetal **36** according to Schneider and coworkers [14b].

Entry	R	Product	Time (d)	Yield (%)	ee (%)
1	Ph	37a	7	99	90
2	p-Et-C$_6$H$_4$	37b	8	84	80
3	p-F-C$_6$H$_4$	37c	9	85	88
4	m-Me-C$_6$H$_4$	37d	3	82	92
5	m-Cl-C$_6$H$_4$	37e	9	63	82
6	o-Me-C$_6$H$_4$	37f	9	66	80
7	o-Cl-C$_6$H$_4$	37g	9	73	72
8	3-Furyl	37h	9	72	84
9	3-Thiophenyl	37i	9	72	86

Table 5.7 Copper-Fesulphos-catalyzed vinylogous Mannich reaction according to Carretero and coworkers [15].

Entry	R^1	R^2	R^3	Product	Yield (%)	ee (%)
1	Ph	H	H	41a	85	94
2	o-Me-C$_6$H$_4$	H	H	41b	68	93
3	p-MeO-C$_6$H$_4$	H	H	41c	62	94
4	iPr	H	H	41d	72	66
5	Ph	OMe	H	41e	72	83
6	1-Naphtyl	OMe	H	41f	70	91
7	Cyclohexyl	OMe	H	41g	85	75
8	Ph	OMe	Me	41h	75	80
9	2-Naphtyl	OMe	Me	41i	85	77
10	Cyclohexyl	OMe	Me	41j	88	82

This protocol was successfully extended to reactions with 2-silyloxy furan **17a** as the nucleophile and γ-aminoalkyl-γ-butenolides **43** were obtained in high yields, good diastereocontrol, and 88–94% ee (Table 5.8). The 2-thienylsulfonyl group as N-substituent proved to be important both for the reactivity and the selectivity of the reaction presumably by way of chelating the copper catalyst and thereby rigidifying the substrate–catalyst complex in the transition state of the reaction.

Table 5.8 Copper-Fesulphos-catalyzed vinylogous Mannich reaction with silyloxy furans [15].

Entry	R	Product	dr	Yield (%)	ee (%)
1	Ph	43a	93:7	85	88
2	p-Cl-C$_6$H$_4$	43b	90:10	87	90
3	p-MeO-C$_6$H$_4$	43c	81:19	79	94
4	PhCH=CH	43d	91:9	87	95

5.3
Direct Vinylogous Mannich Reactions of Unmodified Substrates

In the context of atom economy, it has been a general goal in modern organic chemistry to use unmodified substrates for important C–C-bond forming reactions instead of latent silyl enolates that have to be prepared in a separate step and require the use of stoichiometric amounts of a silicon fragment. In the area of catalytic enantioselective vinylogous Mannich reactions, there has been substantial progress toward the realization of such highly desirable reactions.

In 2004 Terada and his group revealed that 2-methoxyfuran (45) that can be considered as a vinylketene O,O-acetal was sufficiently nucleophilic to engage Boc-imines 44 in Brønsted acid-catalyzed Friedel–Crafts-type reaction to furnish γ-aminoalkyl-substituted furans 46 in excellent yields and up to 97% ee (Table 5.9) [16]. As already explained, phosphoric acid 47 protonates the imine, thus generating a chiral contact ion pair the anionic counterion of which controls the enantioselectivity of the C–C bond forming event. Although typically 2 mol% of the Brønsted acid were

Table 5.9 Organocatalytic Aza-Friedel–Crafts reaction of furan 45 according to Terada and coworkers [16].

Entry	R	Product	Yield (%)	ee (%)
1	Ph	46a	95	97
2	p-Cl-C$_6$H$_4$	46b	88	97
3	p-MeO-C$_6$H$_4$	46c	95	96
4	m-Me-C$_6$H$_4$	46d	80	94
5	m-Br-C$_6$H$_4$	46e	89	96
6	o-Me-C$_6$H$_4$	46f	84	94
7	o-Br-C$_6$H$_4$	46g	85	91
8	2-Furyl	46h	94	86
9	2-Naphtyl	46i	93	96

employed, only 0.5 mol% turned out to be sufficient for a gram-scale reaction delivering the product in identical yield and ee.

The first direct vinylogous Mannich reaction employing α,α-dicyanoalkenes as the CH-acidic substrates was reported by the group of Chen in 2007 [17]. As chiral catalyst they employed a bifunctional thiourea **51** with Brønsted acidic and basic properties that activated the imine via hydrogen bonding from the thiourea component and deprotonated the CH-acidic substrate with the amine moiety at the same time. A broad range of different aromatic and heteroaromatic Boc-aldimines and α,α-dicyanoalkenes **48** were effectively converted into the products **50** with exceptional regio-, diastereo-, and enantioselectivities delivering basically single isomers in excellent yields (Scheme 5.11). This reaction is certainly a prominent example of the power of asymmetric two-center catalysis [18] in a bifunctional molecule where both the nucleophilic and the electrophilic components are activated at the same time through the same catalyst leading to a highly organized transition state within the reaction and hence excellent stereoselectivity.

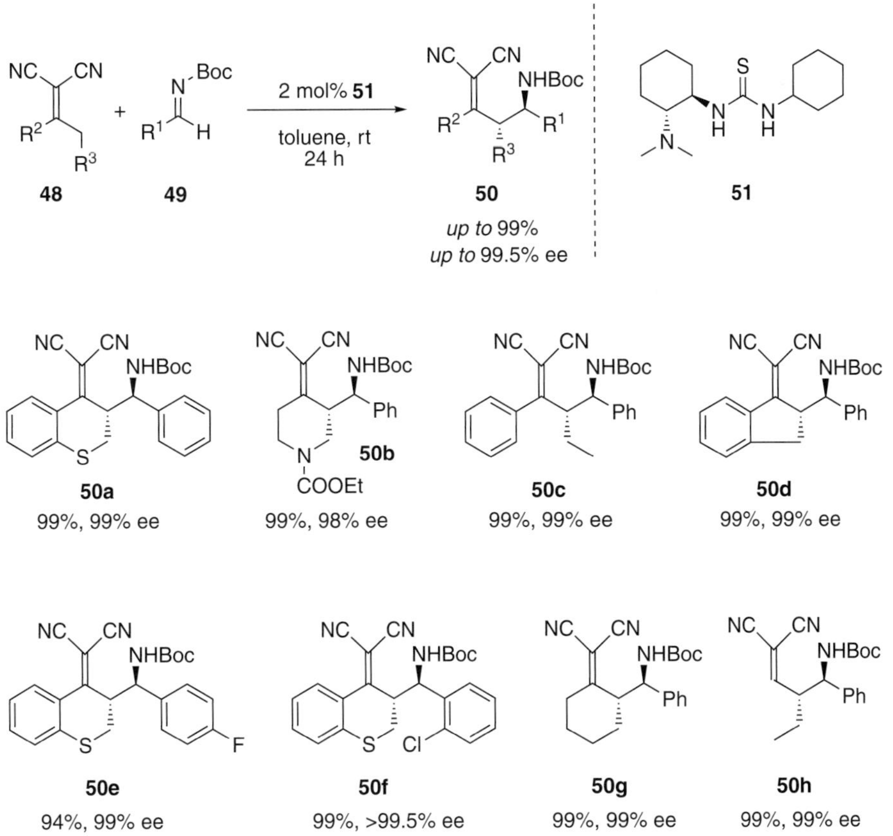

Scheme 5.11 Organocatalytic, direct, vinylogous Mannich reaction of α,α-dicyanoalkenes **48** according to Chen and coworkers [17].

Jørgensen and Niess took advantage of the same CH-acidic substrate class and reacted α,α-dicyanoalkenes **48** with α-amido sulfones **52** as imine precursors under phase-transfer conditions to furnish vinylogous Mannich products **53** in good yields and enantioselectivity [19]. As phase-transfer catalyst they employed the proline-derived, quaternary spiro-ammonium salt **54** that in combination with the stoichiometric base K_3PO_4 in toluene solution gave optimal results. A broad range of different α,α-dicyanoalkenes (aromatic and aliphatic) could be employed in reactions with aromatic and heteroaromatic α-amidosulfones and delivered the products with typically complete *anti*-diastereoselectivity and 74–94% ee (Scheme 5.12).

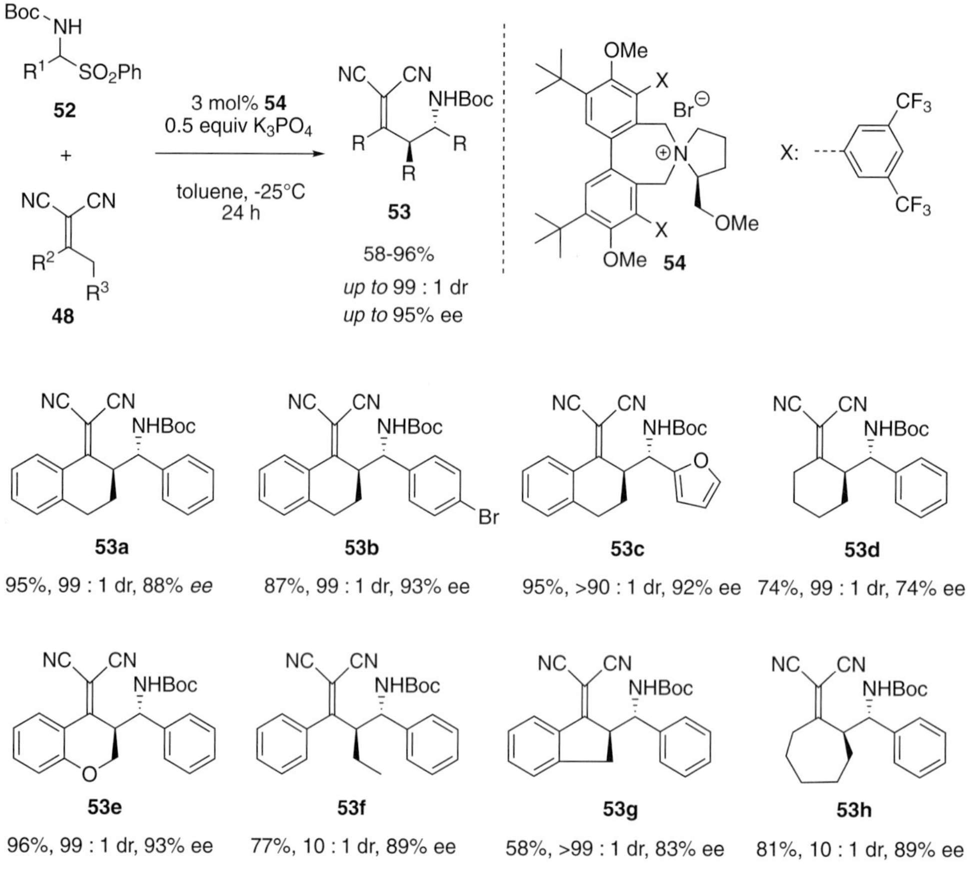

Scheme 5.12 Phase transfer-catalyzed, direct, vinylogous Mannich reaction of α,α-dicyanoalkenes **48** according to Jørgensen and Niess [19].

As the only other substrate class that reacted in direct, enantioselective, vinylogous Mannich reactions so far, Shibasaki and coworkers have identified γ-butenolides **56** as CH-acidic compounds that were submitted to reactions with *N*-diphenylphosphinoyl imines **55** (Table 5.10) [20].

Table 5.10 Lanthanum pybox-catalyzed, direct, vinylogous Mannich reaction of γ-butenolides **56** according to Shibasaki and coworkers [20].

Entry	R^1	R^2	Product	Time (h)	Yield (%)	dr	ee (%)
1	Ph	H	57a	72	83	96:4	83
2	p-Me-C$_6$H$_4$	H	57b	72	89	96:4	84
3	p-Cl-C$_6$H$_4$	H	57c	45	99	95:5	83
4	o-Me-C$_6$H$_4$	H	57d	69	89	91:9	79
5	iBu	H	57e	31	84	86:14	78
6	nBu	H	57f	43	86	82:18	72
7	Ph	Me	57g	62	82	>97:3	68

Upon treatment with a chiral lanthanum(III)-pybox catalyst prepared *in situ* from La(OTf)$_3$ and the L-alanine-derived (*S*,*S*)-Me-pybox ligand **58** in combination with tetramethylethylenediamine (TMEDA) as organic base and triflic acid, vinylogous enolization of the γ-butenolide skeleton was achieved and a highly γ-regioselective vinylogous Mannich reaction occurred delivering γ-aminoalkyl-substituted γ-butenolides **57** with 68–84% ee and up to 97:3 diastereoselectivity. Aromatic and heteroaromatic aldimines performed more enantioselectively than aliphatic aldimines that, however, could also be employed with 72–78% ee. A chiral lanthanum dienolate was postulated to be involved in the enantioselective C–C-bond forming event (Scheme 5.13).

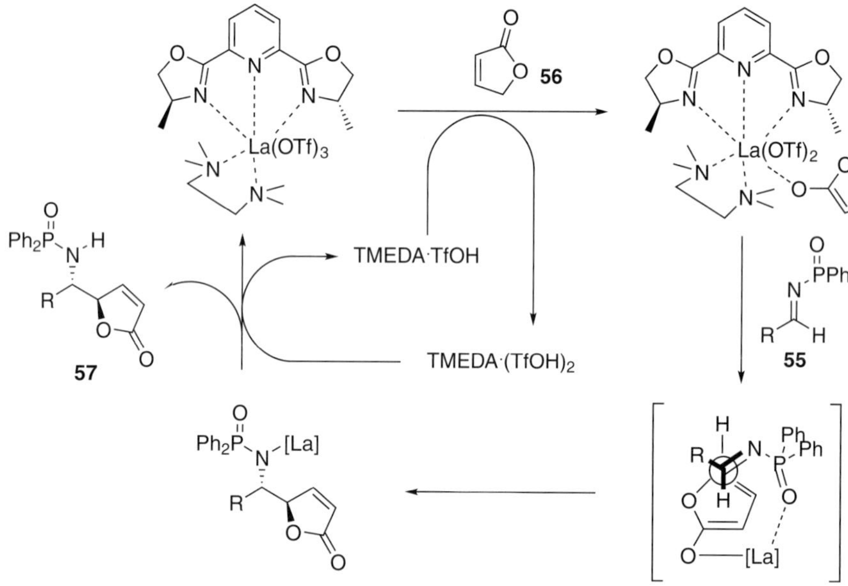

Scheme 5.13 Postulated reaction pathway of the lanthanum-pybox-catalyzed, direct, vinylogous Mannich reaction of γ-butenolide **56** according to Shibasaki and coworkers [20].

5.4
Miscellaneous

A conceptually different entry into the preparation of dienolates and their use in vinylogous Mannich reactions was developed by Lautens and coworkers. Vinyloxiranes can effectively be ring opened with catalytic amounts of a Lewis acid to furnish β,γ-unsaturated aldehydes that are in equilibrium with their dienol tautomer. As such this transformation involved an "umpolung" of the formerly electrophilic vinyl epoxide to the nucleophilic dienol that can be treated with electrophiles.

Lautens *et al.* treated vinyloxirane **60** with 10 mol% Sc(OTf)$_3$ and *N*-benzhydryl-α-iminoesters **59** and obtained δ-amino-α,β-unsaturated aldehydes **61** via the rearrangement-Mannich sequence in generally good yields that may further be converted into pipecolic esters upon hydrogenation (Scheme 5.14) [21]. Aromatic imines may be employed in place of the glyoxyl imines with somewhat diminished yields. No catalytic, enantioselective protocol has yet been developed according to this scheme. However, an auxiliary-based asymmetric version was realized that took advantage of the well-known stereodirecting ability of 8-phenylmenthol. Reaction of vinyloxirane **60** and 8-phenylmenthyl glyoxylate **59c** furnished amino aldehyde **61c** in good yield and excellent diastereoselectivity of 11 : 1.

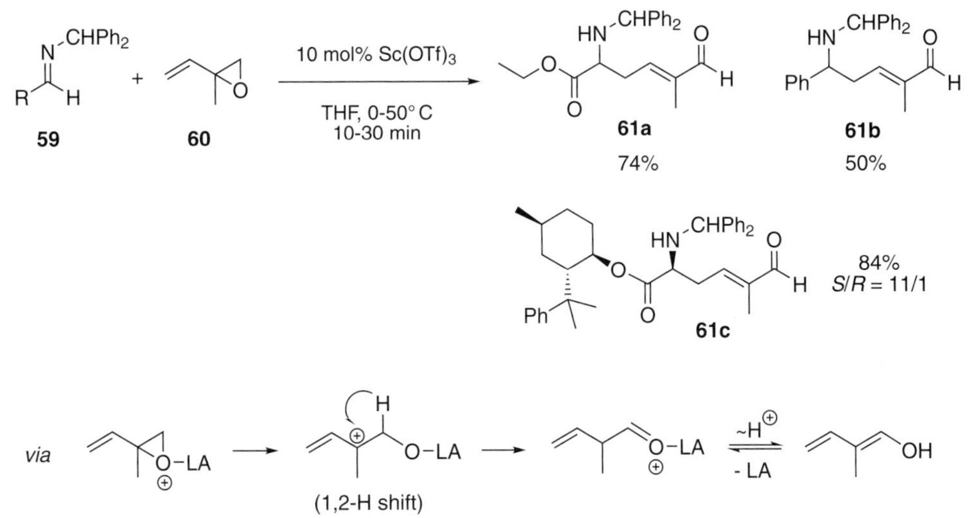

Scheme 5.14 Sc(OTf)$_3$-catalyzed vinylogous Mannich-type reaction of vinyloxirane **60** and imines **59** according to Lautens et al. [21].

5.5
Conclusion

The catalytic, enantioselective, vinylogous Mannich reaction has recently emerged as a very powerful tool in organic synthesis for the assembly of highly functionalized and optically enriched δ-amino carbonyl compounds. Two distinctly different strategies have been developed. The first approach calls for the reaction of preformed silyl dienolates as latent metal dienolates that react in a chiral Lewis acid- or Brønsted acid-catalyzed Mukaiyama-type reaction with imines. Alternatively, unmodified CH-acidic substrates such as α,α-dicyanoalkenes or γ-butenolides were used in vinylogous Mannich reactions that upon deprotonation with a basic residue in the catalytic system generate chiral dienolates *in situ*.

With both strategies, remarkable progress has been made toward the realization of highly enantioselective, vinylogous Mannich reactions furnishing products with almost perfect enantioselectivity in select cases. Nevertheless, there is still much room for improvement as the substrate scope for many of the reactions is quite limited and for the most part only silyloxy furans and α,α-dicyanoalkenes can be employed successfully as nucleophiles. In addition, the enantioselectivity observed in the reactions is not generally high for a broader substrate range. Future work in this area will certainly focus mainly on these two aspects.

Acknowledgments

We gratefully acknowledge the valuable intellectual and practical contributions of our coworkers David Giera, Michael Boomhoff, Susann Krake, and Falko Abels in this research area and the generous financial support of the Deutsche Forschungsgemeinschaft through the priority research program "Organocatalysis" (SPP 1179).

Questions

5.1. Propose an explanation for the better performance of *ortho*-thiomethyl *para*-methoxy aniline-derived imines in the silver phosphine-catalyzed vinylogous Mannich reactions of silyloxy furans with aliphatic imines according to Snapper and Hoveyda in contrast to the standard *ortho*-methoxy aniline-derived imines (Table 5.2).

5.2. Phosphoric acids **32** and **35** have been successfully employed as chiral Brønsted acid catalysts for vinylogous Mannich reactions of acyclic silyl dienolates with imines (Table 5.5 and Scheme 5.10). Why does the exchange of a *para*-methyl group for a *tert*-butyl group on the 3,3'-aryl groups within the BINOL-backbone have such a pronounced effect on the enantioselectivity of the reaction although this position appears to be quite remote from the reaction center?

5.3. Propose a transition state for the direct vinylogous Mannich reaction with the bifunctional thiourea catalyst **51** reported by Chen and his group that accounts for the observed relative and absolute configurations of the products (Scheme 5.11).

References

1 Reviews: (a) Arend, M., Westermann, B., and Risch, N., (1998) *Angew. Chem.*, **110**, 1096–1122; (1998) *Angew. Chem., Int. Ed.*, **37**, 1044–1070; (b) Denmark, S.E. and Nicaise, O.J.-C. (1999) *Comprehensive Asymmetric Catalysis* (eds E.N. Jacobsen, A. Pfaltz, and H. Yamamoto), Springer-Verlag, p. 923; (c) Arend, M. (1999) *Angew. Chem.*, **111**, 3047–3049; (1999) *Angew. Chem., Int. Ed.*, **38**, 2873–2874; (d) Kobayashi, S. and Ishitani, H. (1999) *Chem. Rev.*, **99**, 1069–1094; (e) Córdova, A. (2004) *Acc. Chem. Res.*, **37**, 102–112; (f) Ting, A. and Schaus, S.E. (2007) *Eur. J. Org. Chem.*, 5797–5815.

2 Fukui, K., Yonezawa, T., Nagata, C., and Shingu, H. (1954) *J. Chem. Phys.*, **22**, 1433–1441.

3 Denmark, S.E., Heemstra, J.R., and Beutner, G.L. (2005) *Angew. Chem.*, **117**, 4760–4777; (2005) *Angew. Chem.*, **44**, 4682–4698.

4 Reviews: (a) Martin, S.F., (2002) *Acc. Chem. Res.*, **35**, 895–904; (b) Bur, S.K. and Martin, S.F. (2001) *Tetrahedron*, **57**, 3221–3242.

5 Brandstadter, S.H. and Ojima, I. (1987) *Tetrahedron Lett.*, **28**, 613–616; for a single prior example, see (a) Oida, T., Tanimoto, S., Ikehira, H., and Okano, M. (1983) *Bull. Chem. Soc. Jpn.*, **56**, 645–646; (b) Danishefsky, S., Prisbylla, M., and Lipisko, B. (1980) *Tetrahedron Lett.*, **21**, 805–808.

6 Martin, S.F. and Lopez, O.D. (1999) *Tetrahedron Lett.*, **40**, 8949–8953.

7 Carswell, E.L., Snapper, M.L., and Hoveyda, A.H. (2006) *Angew. Chem.*, **118**, 7388–7391; (2006) *Angew. Chem.*, **45**, 7230–7233.

8 Mandai, H., Mandai, K., Snapper, M.L., and Hoveyda, A.H. (2008) *J. Am. Chem. Soc.*, **130**, 17961–17969.

9 Wieland, L.C., Vieira, E.M., Snapper, M.L., and Hoveyda, A.H. (2009) *J. Am. Chem. Soc.*, **131**, 570–576.

10 Akiyama, T., Honma, Y., Itoh, J., and Fuchibe, K. (2008) *Adv. Synth. Cat.*, **350**, 399–402.

11 Review: Akiyama, T. (2007) *Chem. Rev.*, **107**, 5744–5758, and reference cited therein.

12 Review: Terada, M. (2008) *Chem. Commun.*, 4097–4112, and reference cited therein.

13 Yamanaka, M., Itoh, J., Fuchibe, K., and Akiyama, T. (2007) *J. Am. Chem. Soc.*, **129**, 6756–6764.

14 (a) Sickert, M. and Schneider, C. (2008) *Angew. Chem.*, **120**, 3687–3690; (2008) *Angew. Chem.*, **47**, 3631–3634; (b) Giera, D.S., Sickert, M., and Schneider, C. (2008) *Org. Lett.*, **10**, 4259–4262.

15 González, A.S., Arrayás, R.G., Rivero, M.R., and Carretero, J.C. (2008) *Org. Lett.*, **10**, 4335–4337.

16 Uraguchi, D., Sorimachi, K., and Terada, M. (2004) *J. Am. Chem. Soc.*, **126**, 11804–11905.

17 Liu, T.-Y., Cui, H.-L., Long, J., Li, B.-J., Wu, Y., Ding, L.-S., and Chen, Y.-C. (2007) *J. Am. Chem. Soc.*, **129**, 1878–1879.

18 Ma, J.-A. and Cahard, D. (2004) *Angew. Chem.*, **116**, 4666–4683; (2004) *Angew. Chem.*, **43**, 4566–4583.

19 Niess, B. and Jørgensen, K.A. (2007) *Chem. Commun.*, 1620–1622.

20 Yamaguchi, A., Matsunaga, S., and Shibasaki, M. (2008) *Org. Lett.*, **10**, 2319–2322.

21 Lautens, M., Tayama, E., and Nguyen, D. (2004) *Org. Lett.*, **6**, 345–347.

6
Chiral Amines from Transition-Metal-Mediated Hydrogenation and Transfer Hydrogenation

Tamara L. Church and Pher G. Andersson

6.1
Scope and Related Publications

This chapter focuses specifically on the synthesis of amines of the form HNR^*R^1, $NR^*R^1R^2$, and $HN(X)R^*$ (R^* = chiral alkyl group; R^1, R^2 = alkyl or aryl groups; X = heteroatom such as N, S, or P) via the asymmetric hydrogenation of imines, enamines, and iminiums. Thus, we will discuss the formation of amines that have (i) chirality α to the nitrogen, and (ii) no carbonyl group α to nitrogen. Many reported examples of this reaction use substrates with C=N moieties in a ring, so cyclic amines will receive considerable attention. We will discuss in detail the formation of chiral amines using the asymmetric reduction of aza-aromatic substrates, which has been a popular topic in recent years, and will cover the use of N-carbonyl groups to facilitate this reaction. However, the hydrogenations of nonaromatic α,β-unsaturated amides will not be discussed. Though chiral amides ($HN(R^*)C(O)R^1$) can be produced highly enantioselectively using these reactions, the sheer number of publications on the topic places it outside the scope of this chapter. These hydrogenations are, however, well covered in a number of existing books and review articles [1]. In addition, we will not cover the related process, reductive amination, which is discussed elsewhere in this book.

6.2
Chiral Amines with a Disubstituted Nitrogen Atom, HNR^*R^1

6.2.1
Direct Asymmetric Hydrogenation of Alkyl- and Aryl-Substituted Imines

Though less developed than the analogous reductions of olefins and ketones, the iridium-catalyzed asymmetric hydrogenation of imines (especially N-benzyl and N-aryl amines) has received considerable study. It has therefore been the subject of

Chiral Amine Synthesis: Methods, Developments and Applications. Edited by Thomas C. Nugent
Copyright © 2010 WILEY-VCH Verlag GmbH & Co. KGaA, Weinheim
ISBN: 978-3-527-32509-2

dedicated books [2, 3] and discussed in books on imine hydrogenation [4], as well as on asymmetric hydrogenation [5–7] and its industrial application [8]. The asymmetric hydrogenation of imines has also been discussed in reviews covering asymmetric imine reduction [9], chiral amine synthesis [10–12], additives and cocatalysts in asymmetric catalysis [13, 14], asymmetric hydrogenation in the production of commercial and fine chemicals [15–17] and of pharmaceutical ingredients [18, 19], as well as reviews focusing on the applications of various ligand classes [20–27]. The industrial synthesis of the commercial biocide (S)-metolachlor, which features an iridium-catalyzed imine hydrogenation as a key step, has also been the subject of multiple reviews [28–30]. The wide availability of reviews and books on the field of asymmetric imine hydrogenation obviate a detailed discussion of its development here. Instead, we will use four common substrates to direct a brief overview of the catalysts available for the enantioselective hydrogenation of N-alkyl and N-aryl imines, and then focus on a few recent topics of particular relevance to amine synthesis. For more thorough accounts of the development of this field, the reader is referred to Refs [2, 3].

6.2.1.1 Development

Alkyl- and aryl-substituted imines have received the most attention as substrates for asymmetric hydrogenation, and the development of the field can therefore be outlined by examining their reductions. These are usually catalyzed by chiral complexes of titanium, ruthenium, rhodium, or iridium, though gold catalysts have also recently proven useful for this purpose [31]. New catalysts are generally tested for the reductions of substrates **A–D** (Scheme 6.1).

The earliest rhodium-based asymmetric-imine-hydrogenation catalysts [32–34] demonstrated the viability of the reaction, but suffered from irreproducible yields and/or low enantioselectivities. Thirty years of subsequent research have produced a few rhodium-based systems that offer >90% ee in the hydrogenations of **B** [35, 36] and/or **D** [37]. In particular, mixtures of [Rh(COD)Cl]$_2$ (COD = 1,5-cyclooctadiene) and enantiomerically pure, sulfonated bis(diphenylphosphino)pentane ligands (*vide infra*) reduce **B** very selectively, though the number of sulfonate groups on the ligand strongly affects a system's enantioselectivity [35, 36, 38]. Additives such as amines [39, 40], iodide salts [40–43], silver salts with water [37], and even micelle-forming amphiphilic compounds [42] have been shown to affect the selectivity of rhodium-catalyzed asymmetric imine hydrogenation, though they can sometimes be detrimental [40, 44]. Achiral anions can even reverse the sense of selectivity in some cases [42].

Despite being widely applied in asymmetric olefin and ketone hydrogenations, ruthenium-based catalysts have only recently begun to show high enantioselectivity in the direct hydrogenation of imines. They are, however, widely used in asymmetric transfer hydrogenation (see Section 2.4). James and coworkers demonstrated modest enantioselectivites for the direct reduction of **B** with a variety of ruthenium-diphosphine compounds, achieving a maximum ee (27%) with Ru$_2$Cl$_5$(**1**)$_2$ (Figure 6.1) [45]. Newer ruthenium catalysts for asymmetric hydrogenation are often based on Noyori's [*trans*-Cl$_2$Ru(diphosphine)(diamine)] motif and are activated with base

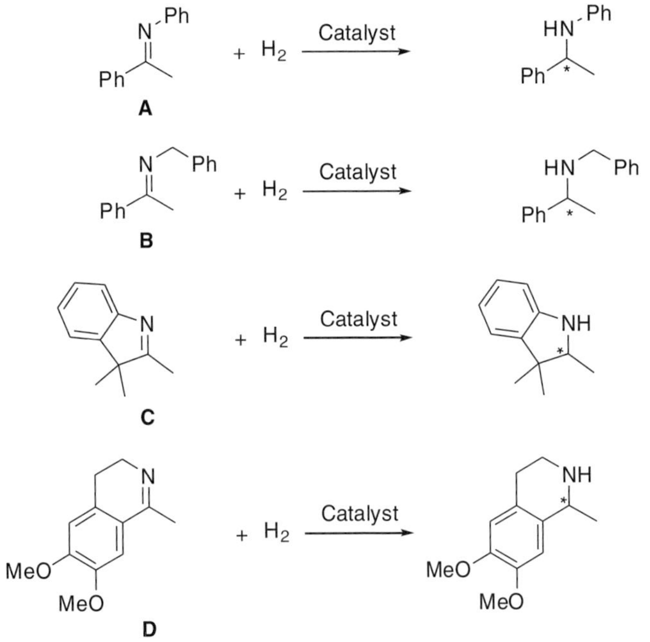

6.2 Chiral Amines with a Disubstituted Nitrogen Atom, HNR*R¹

Scheme 6.1 Common test substrates for the metal-catalyzed hydrogenations of imines.

(generally KOt-Bu) to form an active hydrogenation catalyst *in situ* [46]. Cobley and Henschke used a library of these complexes to identify catalysts that yielded good or very good selectivity for several test substrates (ee = 92, 62, 88, and 79% for the reductions of **A**, **B**, **C**, and **D**, respectively); each substrate had a different optimal catalyst [47]. Another [*trans*-Cl$_2$Ru(diphosphine)(diamine)] catalyst recently showed better enantioselectivity for **B** and **D** (82 and 89% ee, respectively), but not for **A** (71%) [48]. Enantioselective imine reductions have also been demonstrated with complexes of the form [RuHCl(diphosphine)(diamine)]/KOi-Pr (70% ee for **A**, 60% ee for **B**) [49], [Cp*RuCl]$_4$/diamine/KOt-Bu (Cp* = pentamethylcyclopentadienyl; 51% ee for **A**) [50], and [*trans*-Cl$_2$Ru(*mer*-diamine-phosphine)(DMSO)]/KOt-Bu (37% ee for **C**) [51].

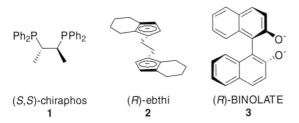

(S,S)-chiraphos
1

(R)-ebthi
2

(R)-BINOLATE
3

Figure 6.1 Selected ligands for use in ruthenium- (**1**) and titanium-catalyzed (**2**, **3**) asymmetric imine hydrogenation.

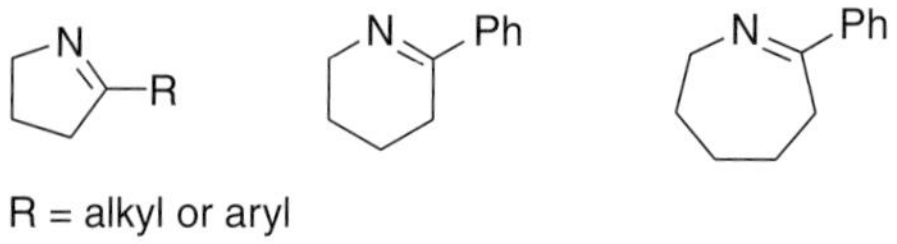

R = alkyl or aryl

Figure 6.2 Cyclic imines hydrogenated in ≥95% ee by Willoughby and Buchwald [55].

There are very few titanium-based catalysts for imine hydrogenation. Willoughby and Buchwald developed (R,R,R)-[(2)Ti(3)] (Figure 6.1), which can be activated with nBuLi and PhSiH$_3$ to produce a catalyst for asymmetric imine hydrogenation [52]. They also examined the scope [53] and mechanism [54] of this catalyst in detail. (R,R,R)-[(2)Ti(3)] produced good enantioselectivities for acyclic imines, but was particularly impressive in the reduction of cyclic imines, yielding 95–99% ee in the reductions of 2-substituted imines in five-, six-, and seven-membered rings (Figure 6.2) [55]. The enantioselectivity of imine hydrogenation by these catalysts was correlated to the E/Z ratio of the starting imine. Thus the acyclic imines, which are difficult to obtain in stereopure form, were reduced less selectively than their endocyclic analogues, which are necessarily stereopure.

Brintzinger and coworkers produced a titanium complex with a binaphthyl-linked *ansa*-metallocene ligand [56]. Upon activation with nBuLi, this complex catalyzed the reductions of **B** and 2-phenyl-1-pyrroline in 76 and 98% ee, respectively. Titanium catalysts with chiral amido-cyclopentadienyl ligands have also been reported, but were considerably less stereoselective than the metallocene-based ones [57].

Iridium is by far the most popular metal used in catalysts for direct, asymmetric imine hydrogenation. Among the advantages of iridium-based catalysts for this reaction is the apparent insensitivity of their stereoselectivity to the E/Z ratio of the starting material [58]. Chiral iridium-diphosphine catalysts for imine reduction began to appear in the literature around 1990 [59]. Osborn [60, 61] obtained moderate enantioselectivities by screening various diphosphine ligands in the iridium-catalyzed hydrogenation of acyclic and cyclic substrates. Improvements in the stereoselectivity were achieved using iodide salts [62–64], protic amines [65], 1,3-indandione [64], or imides [64, 66] as additives. In the past decade, several chiral iridium-diphosphine complexes have been applied to imine hydrogenation [67–73], and a few give >90% ee for analogues of **A** with substituents on the N-aryl moiety [69, 71, 73]. Other ligands with two phosphorus donors, such as diphosphites [74], diphosphinites [74, 75], and mixed phosphine-phosphinite ligands [76, 77], have also been used in moderately selective, iridium-based catalysts for imine hydrogenation. Chiral monodentate phosphorus donors have also been used [78–81]; these generally work best with added (achiral) N or P donors [78, 79, 81]. Pfaltz and coworkers introduced chiral phosphino-oxazoline ligands to iridium-catalyzed imine hydrogenation in 1997 [58], and many more have since been developed [82–89]. A handful of these catalyze the reduction of **A** (or very similar substrates) in ≥90% ee [82–84, 86, 89]. Chiral ligands that contain N and P donors but no oxazoline unit have also been used in asymmetric iridium-catalyzed imine hydrogenation [90, 91], as have oxazoline-thiols [92], N-sulfonated diamines [37, 93], and phosphine-olefins [94]. Unfortunately, the effects

Table 6.1 Metals used in catalysts for direct asymmetric imine hydrogenation.

Metal	Advantages	Disadvantages
Ti	• Extremely selective for cyclic imines.	• Catalysts are extremely air- and moisture-sensitive. • Enantioselectivity depends on the E/Z purity of the starting imine.
Ru	• Availability of a wide range of chiral diamines and diphosphines simplifies combinatorial-style searches for the most suitable catalyst for a given substrate.	• In general, less enantioselective than iridium-based catalysts.
Rh	• High enantioselectivity available.	• Many combinations of ligand and additive may have to be tested before high selectivity is obtained.
Ir	• High enantioselectivity available and may be improved with additives. • Selectivity generally insensitive to E/Z isomer ratios in substrate. • Many reported ligands.	• Many combinations of ligand and additive may have to be tested before high selectivity is obtained.
Au	• High enantioselectivity available.	• Only one report to date.

of additives on the rate and stereoselectivity of iridium-catalyzed imine hydrogenation are not very predictable; often, a range of additives must be tested to optimize the outcome in a given system.

A brief comparison of the advantages and disadvantages of titanium, ruthenium, rhodium, iridium, and gold catalysts for the asymmetric hydrogenation of alkyl- and aryl-substituted imines is given in Table 6.1.

6.2.1.1.1 **A Representative Synthesis**   It would be inappropriate, and near impossible, to review asymmetric imine hydrogenation without discussing (*S*)-metolachlor. A great deal of progress in the iridium-catalyzed asymmetric hydrogenation of imines has been inspired by the industrial synthesis of (*S*)-metolachlor and by the extremely well-documented development of this synthesis [28–30]. The key step in the commercial synthesis of (*S*)-metolachlor is the hydrogenation of "MEA imine," **E** (Scheme 6.2) [67].

6.2.1.2 **Pressure in the Asymmetric Hydrogenation of Alkyl- and Aryl-Substituted Imines**
The utility of any synthetic method depends upon its ease of use. The applicability of reactions involving gases can be limited, especially in smaller laboratories, by the need for high pressures. Thus, low- or ambient-pressure asymmetric imine hydrogenations are attractive. Rhodium catalysts usually reduce standard imines at room temperature under 50–70 bar H_2. An exception is Scorrano's $[Rh(nbd)(5)]^+[ClO_4]^-$ (nbd = norbornadiene) system (Figure 6.3), which hydrogenated **B** under 1 bar H_2, though not in useful stereoselectivities [32]. Ruthenium catalysts are most often used under at least 10 bar H_2, though good enantioselectivity (60–70% ee for **A** and **B**) has

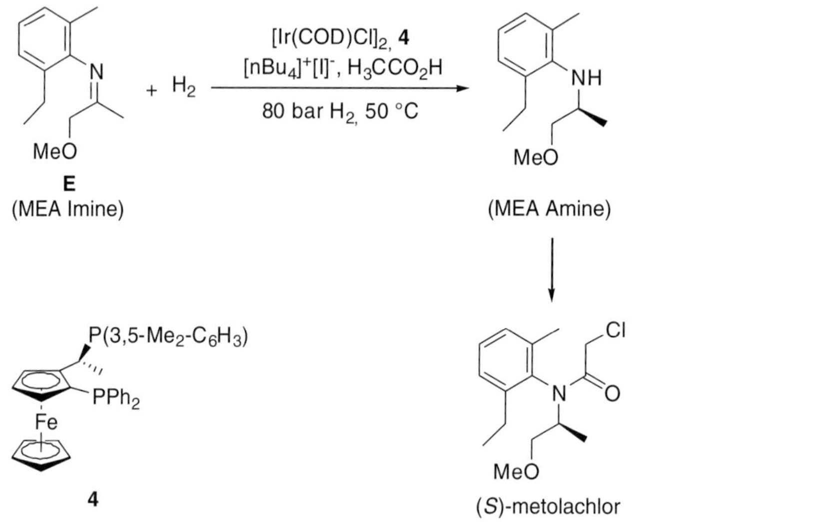

6 Chiral Amines from Transition-Metal-Mediated Hydrogenation and Transfer Hydrogenation

Scheme 6.2 The iridium-catalyzed asymmetric hydrogenation of MEA imine, **E**, is part of the industrial synthesis of (S)-metolachlor.

been reported with as low as 3 bar H_2 [49]. The compromise in that case was reaction time; full conversions were obtained after 36 h at room temperature. Buchwald's titanium catalysts hydrogenated cyclic imines under 5.5 bar H_2 at 65 °C with excellent selectivity (95–99% ee), and good isolated yields could be obtained for some substrates after as little as 8 h, though most required 24 h [52–55]. The gold catalysts reported recently by Corma and coworkers were both active and selective at 4 bar and 45 °C [31]. Iridium has received the most study in asymmetric imine hydrogenation and is present in several catalysts that hydrogenate imines selectively at ambient pressure. As early as 1997, Pfaltz and coworkers reported that $[(6b)Ir(COD)]^+[PF_6]^-$ (Figure 6.3) reduced **A** and **B** with "very similar" enantioselectivities at 1 and 100 bar H_2; no comment was made regarding the effect of H_2 pressure on reaction times or conversions [58]. James and coworkers found that H_2 pressure (in the range 1–55 bar) significantly affected the rate, but not the enantioselectivity, of the hydrogenation of **B** by $[(7)Ir(COD)]^+[PF_6]^-$ [85]. Dervisi et al. produced a catalyst, $[Ir(COD)(10)]^+[PF_6]^-$,

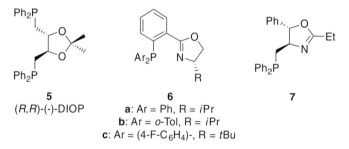

Figure 6.3 Ligands used in the low-pressure hydrogenation of imines.

Table 6.2 Asymmetric hydrogenations of 1-(N-arylimino)-1-(aryl)-ethanes under atmospheric pressure of H$_2$.

Entry	R^1	R^2	Conversion (ee)$^{a)}$ [% (%)]		
			L* = 8$^{[b]}$	L* = 9$^{[c]}$	L* = 10$^{[d]}$
			X = BAr$_F^-$	X = BAr$_F^-$	X = PF$_6^-$
			tBuOMe, 10 °C	CH$_2$Cl$_2$, rt	CH$_2$Cl$_2$, rt
			+ 4 Å MS		
1	H	H	>99.5 (93)	91 (86)	99 (84)
2	MeO	H	—$^{e)}$	93 (86)	100 (94)
3	Cl	H	>99.5 (97)	99 (83)	—$^{e)}$
4	H	MeO	>99.5 (94)	98 (69)	100 (81)
5	H	Cl	>99.5 (90)	—$^{e)}$	99 (80)

a) Conversion to and ee of the chiral amine product.
b) See Ref. [89]. Ar = 3,5-Me$_2$-C$_6$H$_3$.
c) See Ref. [71].
d) See Ref. [69].
e) Not determined.

that was actually inhibited by high H$_2$ pressures; it was nearly inactive under 10 bar H$_2$ [69]. This catalyst hydrogenated **A** and substituted analogues in 80–94% ee (Table 6.2), but reduced **B** to near-racemic amine and was unreactive toward **C**. Two other systems published in 2006 hydrogenated imines under 1 bar H$_2$. Imamoto et al. reported that [(9)Ir(COD)]$^+$[BAr$_F$]$^-$ hydrogenated **A** and its analogues in 69–99% ee (Table 6.2) [71]; and Q.-L. Zhou and coworkers hydrogenated the same type of compounds in 91–97% ee, albeit at 10 °C, with [(8)Ir(COD)]$^+$[BAr$_F$]$^-$ (Table 6.2) [89]. Neither Imamoto nor Zhou tested their catalysts on substrates without N-aryl substituents.

6.2.1.3 Reducing the Environmental Impact of the Reaction

Direct hydrogenation reactions have one intrinsic environmentally friendly attribute – all of the atoms in both the substrate and the H$_2$ gas appear in the product amine, so these reactions are completely atom economic. In addition, a highly stereoselective route to a chiral molecule can avoid the material-intensive and wasteful separation of enantiomers. Nevertheless, asymmetric imine hydrogenation has several features whose environmental impact could be improved. Most importantly, it is generally catalyzed by heavy metal complexes (which are usually discarded after a single use), and the popular iridium-catalyzed reaction is almost always performed in CH_2Cl_2 solution. In fact, the asymmetric hydrogenations of olefins, ketones, and imines have been identified by the pharmaceutical industry as being reactions for which "green" improvements are very desirable [95]. Although there are very few selective catalysts for asymmetric imine hydrogenation that are not based on heavy metals (titanium-based systems, see above), efforts have been made to perform both the rhodium- and iridium-catalyzed reactions in greener solvents, and even to immobilize and reuse the catalysts.

In 1989, Sinou and coworkers designed sulfonated versions of the bis(diphenylphosphino)pentane ligand (bdpp, **11b**, Scheme 6.3). When coordinated to rhodium, these produced water-soluble catalysts for asymmetric imine hydrogenation [38]. The authors hydrogenated unsaturated compounds in aqueous/organic mixtures and isolated the products by simple extraction of the organic phase. Though they focused mostly on the reduction of enamides, the authors did report the hydrogenation of **B** in 34–58% ee. Bakos et al. [35] and Lensink and de Vries [38, 96] further examined sulfonated bdpp ligands in rhodium-catalyzed asymmetric imine hydrogenations in H_2O/EtOAc (Scheme 6.3). Though the former group used ligand mixtures with varying degrees of sulfonation, and the latter used epimeric mixtures of phosphines, both groups obtained very high ee values in the asymmetric hydrogenations of **B** (up to 94 and 96%, respectively) and related imines. Lensink and de Vries proposed that a single catalyst with both superior activity and enantioselectivity was dominating catalysis in these mixtures. Indeed, the rate and enantioselectivity of imine hydrogenation by rhodium catalysts with sulfonated bdpp ligands depended strongly on the number of sulfonate groups on the ligands; the monosulfonated ligand (**11a**,

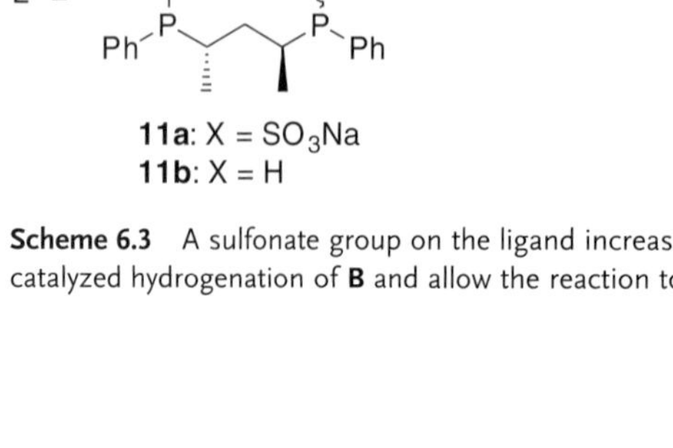

Scheme 6.3 A sulfonate group on the ligand increases the enantioselectivity in the rhodium-catalyzed hydrogenation of **B** and allow the reaction to be performed in H_2O/EtOAc.

Scheme 6.3) gave the fastest and most selective catalysis, whereas the disulfonated ligand produced a catalyst that was nearly unselective. Thus **11a** forms the fast, selective catalyst. de Vries and coworkers investigated this dramatic stereoselectivity in the hydrogenation of **B** [96]. They found that the high ee values obtained with the [Rh(COD)Cl]$_2$/**11a** system could not be reproduced using [Rh(COD)Cl]$_2$ or [Rh(COD)$_2$]$^+$[BF$_4$]$^-$ with unsubstituted **11b** and camphor sulfonate in MeOH. Thus they concluded that the improved enantioselectivity observed with ligand **11a** could not be attributed to an anion effect. However, Buriak and Osborn studied the effect of added sulfonate anions (including camphor sulfonate) on the [(**11b**)Rh(COD)]$^+$[ClO$_4$]$^-$-catalyzed asymmetric hydrogenation of imines in benzene and did observe significant increases in enantioselectivity [42]. The opposing results of de Vries and Osborn were obtained in very different solvents (polar, protic methanol and nonpolar, aprotic benzene, respectively), suggesting that ionic interactions could be responsible for the effect of sulfonate salts, but that these effects are, quite reasonably, different in benzene and methanol. Based on a series of NMR structural studies and on the comparative effects of various anions, Buriak and Osborn proposed that both the binding ability and the lability of the sulfonate anion combined to influence the mechanism of imine hydrogenation. Thus, in modifying ligand **11b** to obtain catalysts soluble in EtOAc/H$_2$O, Sinou and coworkers fortuitously added an anion whose properties were well balanced to steer the rhodium-catalyzed asymmetric hydrogenation of imines. Unfortunately, neither Bakos and coworkers nor Lensink and de Vries attempted to reuse their aqueous catalyst solutions for subsequent hydrogenations, so it is not clear whether these could be repeatedly separated and used in asymmetric imine hydrogenation. Sinou and coworkers were able to reuse their catalyst solutions following the Rh/**11a**-catalyzed hydrogenations of enamides [38], but did not describe attempts to do so after the hydrogenation of imine **B**. Unfortunately, extracting the Rh/**11a** catalyst in the aqueous phase after imine hydrogenation might not be possible; Lensink and de Vries noted that it dissolved in the EtOAc phase rather than the aqueous phase [36]. However, the wide variety of water-soluble phosphine ligands available is auspicious for further progress in this area [97].

More recently, attempts have been made to use greener solvents for iridium-catalyzed asymmetric imine hydrogenation and to immobilize iridium catalysts for reuse. Leitner and coworkers examined supercritical CO$_2$ as a solvent for the Pfaltz group's [(**6a**)Ir(COD)]$^+$[PF$_6$]$^-$ imine-hydrogenation catalyst (see Figure 6.3), which was originally used in CH$_2$Cl$_2$ (Table 6.3, entry 1) [88]. To improve the catalyst solubility in scCO$_2$, the authors examined ligands with distal perfluorohexane groups as well as with various counterions. Modifying the phosphino-oxazoline ligands produced active and enantioselective catalysts, but was less convenient than changing the catalyst anion to tetrakis(3,5-bis(trifluoromethyl)phenyl)borate, BAr$_F^-$. Pfaltz and coworkers had already examined the resulting catalyst, [(**6a**)Ir(COD)]$^+$[BAr$_F$]$^-$, for asymmetric olefin reduction [98]. This catalyst hydrogenated **A** to full conversion and in 87% ee after 5 h at 40 °C in scCO$_2$. Contrary to the case in CH$_2$Cl$_2$ solution, the enantioselectivity of imine reduction by [(**6a**)Ir(COD)]$^+$ depended strongly on the catalyst anion; the [BAr$_F$]$^-$ salt was a much more selective catalyst than the [BPh$_4$]$^-$ or

Table 6.3 Asymmetric hydrogenations of **A** using [(**6a**)Ir(COD)]$^+$[X]$^-$ catalysts in CH$_2$Cl$_2$ and in alternative solvents.

Ph–N=C(Ph)(CH$_3$) + H$_2$ $\xrightarrow{[(\mathbf{6a})\text{Ir(COD)}]^+[X]^-,\ \text{solvent}}$ Ph–NH–C*H(Ph)(CH$_3$)

Entry	X	Solvent	Use #						
			1	2	3	4	5	6	7
			Conversion (ee)$^{a)}$ [% (%)]						
1$^{b)}$	PF$_6^-$	CH$_2$Cl$_2$	100 (70)	—$^{c)}$	—$^{c)}$	—$^{c)}$	—$^{c)}$	—$^{c)}$	—$^{c)}$
2$^{d)}$	PF$_6^-$	scCO$_2$/[BMIM]$^+$[PF$_6$]$^-$	>99 (62)	>99 (62)	99 (62)	95 (64)	>99 (62)	81 (52)	>99 (61)
			Approximate time to completion (ee)$^{e)}$ [h (%)]						
3$^{f)}$	BAr$_F^-$	scCO$_2$	3 (79)	3 (81)	3 (77)	4 (74)	22 (71)	42 (76)	65 (70)

a) Conversion to and ee of phenyl 2-phenylethylamine.
b) See Ref. [58]. Isolated yield was 97%.
c) Not determined.
d) See Ref. [87].
e) Time to complete conversion of phenyl 2-phenylethylamine and ee of the product.
f) See Ref. [88].

[PF$_6$]$^-$ salts. The chiral amine produced from imine hydrogenation by [(**6a**)Ir(COD)]$^+$[BAr$_F$]$^-$ in scCO$_2$ could be extracted from the reaction and collected on a cold trap by purging with compressed CO$_2$. The product collected in this manner contained <5 ppm Ir, and the still-active catalyst remained in the scCO$_2$ phase. The catalyst could then be reused three times with very little loss of activity or enantioselectivity; subsequent runs were still enantioselective but required more time (see Table 6.3, entry 3).

Pfaltz and Leitner subsequently addressed the problem of diminishing rates in successive hydrogenations in scCO$_2$ by combining that solvent with ionic liquids [87]. The complex [(**6a**)Ir(COD)]$^+$[PF$_6$]$^-$ hydrogenated **A** in a mixture of an ionic liquid ([BMIM]$^+$[PF$_6$]$^-$, BMIM (1-butyl-3-methylimidazolium)) and scCO$_2$ without losing activity or selectivity after seven reaction cycles (Table 6.3, entry 2). Other ionic liquids could be used as solvents, but often at the expense of enantioselectivity. Better selectivity (78% ee) in a single hydrogenation was obtained in scCO$_2$/[EMIM]$^+$[BAr$_F$]$^-$ ([EMIM]$^+$ = 1-ethyl-3-methylimidazolium). The scCO$_2$ component was essential to the rate of hydrogenation in all of the solvent systems examined; NMR spectroscopic experiments suggested that it solubilized H$_2$ in the reaction mixture. The ionic liquid component stabilized the catalyst not only to repeated reaction cycles, but also to prolonged exposure to air. Thus, using the asymmetric hydrogenation catalyst [(**6a**)Ir(COD)]$^+$[PF$_6$]$^-$ in a [BMIM]$^+$[PF$_6$]$^-$/scCO$_2$ solvent system (rather than CH$_2$Cl$_2$) made it both more stable and recyclable, and still allowed it to reduce **A** with moderate stereoselectivity. Higher enantioselectivities and an expanded substrate scope will improve the utility of this system.

Given the growing popularity of immobilized, enantioselective catalysts based on known homogeneous catalysts and solid supports [99–103], it is not surprising that iridium-based hydrogenation catalysts have been immobilized on solid supports. As part of efforts to optimize the isolation of MEA amine after hydrogenation, Pugin *et al.* immobilized the xyliphos ligand **4** (Scheme 6.2) on silica and on polystyrene [104]. These solid-supported ligands were combined with [Ir(COD)Cl]$_2$, acetic acid, and Bu$_4$NI to form active catalysts for the asymmetric hydrogenation of **E**. The silica-bound ligand in particular formed a very active catalyst that hydrogenated **E** at up to 20 000 mmol E/(mmol Ir h) and turned over as many as 195 000 times. Although the catalyst was active, selective (ee = 77 ± 2%), and separable by filtration, it could not be reused; no reaction was observed when additional substrate was added directly to the reaction mixture. The homogeneous catalyst, on the other hand, hydrogenated an additional portion of substrate but at a significantly lower rate. Based on a series of control experiments using the homogeneous catalyst and its derivatives, the authors concluded that the high concentration of catalyst sites on the support favored catalyst deactivation by dimerization, limiting the catalysts' activity and longevity. Though these investigations were ultimately halted in favor of the homogeneous catalyst, they demonstrated that silica- and polystyrene-bound ligands could support active and selective catalysts for imine hydrogenation. Pugin and coworkers also produced a water-extractable version of the xyliphos ligand, but did not discuss its reuse.

Building on their work with achiral iridium- and rhodium-based imine hydrogenations on clay supports [105], Salagre and coworkers reported the immobilization of [((S,S)-**11b**)Ir(COD)]$^+$[PF$_6$]$^-$, a catalyst that hydrogenates **B** to racemic amine, onto montmorillonite K-10 clay [68]. The resulting complex actually hydrogenated **B** somewhat stereoselectively. The enantioselectivity was low (18% ee) in the first reaction, but actually improved upon reuse, giving 23, 48, and 59% ee in the three subsequent runs. Unfortunately, this increasing enatioselectivity came at the price of activity; in the fourth hydrogenation, only 6% conversion was achieved after 24 h (cf. 99% in the first use). Metal leaching from the supported catalysts was observed. The authors discovered that exposure to O$_2$ was at least partially responsible for the increasing enantioselectivity of the reaction and that treating the catalyst with O$_2$ for 10 min before the first use allowed it to hydrogenate **B** in 40% ee without discernible loss in activity. Even the oxygen-treated catalyst became more enantioselective on repeated use, giving 54% ee in the second run (albeit with reduced activity). A control experiment showed that pretreatment with O$_2$ did not render the homogeneous [((S,S)-**11b**)Ir(COD)]$^+$[PF$_6$]$^-$ catalyst stereoselective. Though the enantioselectivities achieved with this system were low to moderate, it stands out as a supported, reusable iridium catalyst for asymmetric imine hydrogenation.

Active and stable catalysts for the hydrogenations of **B** and 2-phenyl-1-pyrroline were also obtained by supporting (salen)Pd and (salen)Au (salen = ligand of the form N,N'-bis(3,5-di-*tert*-butylsalicylidene)-chiral diamine) catalysts on mesoporous silica and on zeolites [106, 107]. Though both active and recyclable without loss of activity [107], these catalysts were only slightly enantioselective; the ee values were reported to be 5–15%.

6.2.2
Direct Asymmetric Hydrogenation of Heteroaromatics

The omnipresence of substituted azacycles in Nature can hardly be exaggerated. Thus, their chiral synthesis is highly desirable, and the catalytic asymmetric hydrogenation of heteroaromatics could be a useful tool for this purpose. Very stereoselective, high-yielding hydrogenation catalysts for these substrates have appeared only recently. Though the resonance stability of heteroaromatics can be a hindrance for this reaction [108], some heteroaromatics can be hydrogenated in excellent yields and ee values. The asymmetric hydrogenation of heterocycles has been the subject of a few recent reviews [109–111] and a review section [112, 113].

6.2.2.1 Quinolines and Isoquinolines

Consistent with the idea that aromaticity can be a serious obstacle to the hydrogenation of heteroaromatics, the first of these substrates to be reduced asymmetrically were benzo-fused compounds, in which the heteroaromatic rings are less stabilized than their monocyclic analogues [114]. In particular, the asymmetric hydrogenations of quinolines to give useful 1,2,3,4-tetrahydroquinolines [115] have seen considerable recent progress, mostly with iridium-based catalysts. In 2003, Y.-G. Zhou and coworkers published the first highly enantioselective hydrogenation of substituted quinolines [116]. They found that 2-methylquinoline could be hydrogenated in 94% ee using a catalyst generated in situ from [Ir(COD)Cl]$_2$ and (R)-MeO-Biphep (**12**, Figure 6.4), and iodine (I$_2$/Ir = 10) in toluene (Table 6.4, entry 1). Inspired by this work, several other groups produced active and selective iridium catalysts for the asymmetric hydrogenation of substituted quinolines [117–128]. Table 6.4 shows the reduction of 2-methylquinoline by these catalysts; the relevant ligands are given in Figure 6.4. Most systems combine chiral diphosphines with [Ir(COD)Cl]$_2$ and I$_2$ to form the active catalyst in situ. Y.-G. Zhou and coworkers combined [Ir(COD)Cl]$_2$ and I$_2$ with the chiral, ferrocene-based N∩P ligand **15** to produce a quinoline-hydrogenation catalyst that gave comparable yields, but slightly (5–10%) lower ee values than their earlier system did for most substrates (Table 6.4, entry 2); substrates with bulky groups vicinal to the N (e.g., 2-phenyl- and 2-(2-phenylethyl)-quinoline) were hydrogenated in particularly low enantioselectivities [117]. Lemaire and coworkers used a BINAP derivative (**23**) as a ligand (entry 15) [129] and Reetz and Li found high enantioselectivities using BINOL-derived phosphinites (**20**) as ligands (entry 10) [122]. They also found that achiral phosphine and phosphate additives could improve enantioselectivity in the asymmetric hydrogenation of more challenging quinoline substrates. Minnaard and coworkers reported that achiral phosphine additives also improved the enantioselectivity of their quinoline-hydrogenation system, a mixture of [Ir(COD)Cl]$_2$/**22**/piperidine hydrochloride (entry 14) [125].

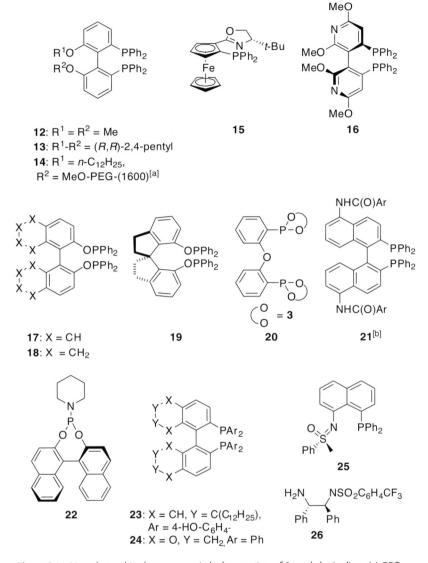

Figure 6.4 Ligands used in the asymmetric hydrogenation of 2-methylquinoline. (a) PEG (1600) = a poly(ethylene glycol) chain having M_n = 1600. (b) = 3,5-bis(3,5-dibenzyloxyphenylmethoxy)phenyl.

Chan and coworkers applied catalysts based on (R)-P-Phos (**16**, Table 6.4, entries 3 and 4) [118], BINAPO (**17** and **18**, entries 5 and 6) [119], and atropisomeric (**13**, entry 7) [120] ligands to the reaction. Ligand **16** was particularly interesting because its polar nature allowed the catalyst to be immobilized and recycled. After using [Ir(COD)Cl]$_2$/**16**/I$_2$ to catalyze the hydrogenation of 2-methylquinoline in biphasic mixtures of hexane and dimethyl poly(ethylene glycol) (dmpeg), the authors decanted off the

6 Chiral Amines from Transition-Metal-Mediated Hydrogenation and Transfer Hydrogenation

Table 6.4 Asymmetric hydrogenations of 2-methylquinoline by various catalyst systems.

Ir/L*, additive, 40–60 bar H_2, rt

Entry	Ir Source	L*	Additive	S/A/Ir[a]	Conv. (ee)[b] [% (%)]
1[c),d)]	[Ir(COD)Cl]$_2$	12	I$_2$	100:10:1	94 (94)
2[e)]	[Ir(COD)Cl]$_2$	15	I$_2$	100:5:1	>95 (90)
3[d),f),g)]	[Ir(COD)Cl]$_2$	16	I$_2$	100:5:1	98 (89)
4[f),g),h)]	[Ir(COD)Cl]$_2$	16	I$_2$	100:5:1	99 (88)
5[i)]	[Ir(COD)Cl]$_2$	17	I$_2$	100:10:1	>99 (81)
6[i),j)]	[Ir(COD)Cl]$_2$	18	I$_2$	100:10:1	>99 (95)
7[k)]	[Ir(COD)Cl]$_2$	13	I$_2$	100:10:1	>99 (89)
8[g),l)]	[Ir(COD)Cl]$_2$	19	I$_2$	1000:10:1	>99 (92)
9[g),l),m)]	[Ir(COD)Cl]$_2$	19	I$_2$	1000:10:1	40 (86)
10[n)]	[Ir(COD)Cl]$_2$	20	I$_2$	100:1:1	>96 (92)
11[o)]	[Ir(COD)Cl]$_2$	21	I$_2$	100:10:1	>95 (90)
12[p)]	[Ir(COD)Cl]$_2$	14	I$_2$	100:5:1	>95 (91)
13[m),p)]	[Ir(COD)Cl]$_2$	14	I$_2$	100:5:1	95 (84)
14[q)]	[Ir(COD)Cl]$_2$	22	P(o-tol)$_3$ + pipHCl	50:1:5:1	>99 (89)
15[r)]	[Ir(COD)Cl]$_2$	23	I$_2$	100:5:1	89 (70)
16[s)]	[L*Ir(COD)]$^+$[BAr$_F$]$^-$	25	—	100:0:1	>95 (87)
17[d),t)]	[Cp*Ir(L*)(OTf)]	26	TFA	500:50:1	99 (98)
18[u)]	[Ir(I)(H)(O$_2$CMe)(L*)] + HBr	24	I$_2$	100:10:1	>99 (91)

a) Mole ratio of substrate to additive to catalyst.
b) Conversion to and ee of 2-methyl-1,2,3,4-tetramethylquinoline product.
c) See Ref. [116].
d) Isolated yield is given in place of conversion.
e) See Ref. [117].
f) See Ref. [118].
g) Solvent was a mixture of dmpeg and hexane.
h) On the eighth use.
i) See Ref. [119].
j) Isolated yield was 98%.
k) See Ref. [120].
l) See Ref. [121].
m) On the fifth use.
n) See Ref. [122].
o) See Ref. [123].
p) See Ref. [124].
q) See Ref. [125]. T = 60 °C, pipHCl = piperidine hydrochloride.
r) See Ref. [129].
s) See Ref. [126]. T = 60 °C.
t) See Ref. [127]. TFA = trifluoroacetic acid.
u) See Ref. [128].

product-containing hexane layer, then washed the dmpeg layer twice with hexane. The chiral 2-methyl-1,2,3,4-tetrahydroquinoline was isolated from the hexane layer, and the catalyst-containing dmpeg layer was recharged with hexane and substrate for reuse. After eight uses, no attrition in efficiency or enantioselectivity was observed (cf. entries 3 and 4). They also used a phosphine ligand with a spirobiindane backbone (**19**, entries 8 and 9) to achieve efficient asymmetric hydrogenation at very low catalyst loading; this catalyst could also be reused, but was considerably less active and slightly less enantioselective after multiple reuses [121]. The less polar ligand **19** may have made the catalyst slightly soluble in hexanes, causing it to leach away over time. It is also possible that the catalyst formed from [Ir(COD)Cl]$_2$/**19**/I$_2$ was less stable than that from [Ir(COD)Cl]$_2$/**16**/I$_2$ or that the different catalyst loadings used in the two systems affected their rates of decomposition. Fan and coworkers developed another catalyst-recycling strategy based on hexane extraction [123]. They employed catalysts based on the dendritic diphosphine ligand **21** (and higher and lower order dendrimers) for the asymmetric hydrogenation of quinolines, and found these catalysts to be active even at low catalyst loadings (entry 10). After performing the hydrogenations in THF, they added hexane to precipitate the catalyst and recover it by filtration. They recycled the catalyst with minimal loss of enantioselectivity, though its activity suffered after more than four uses. More recently, Wang and Y.-G. Zhou reported the ligand **14**, a modification of the MeO-Biphep ligand **12** [124]. The long poly(ethylene glycol) chain of **14** prevents it from dissolving in saturated hydrocarbons, meaning that it can also be extracted from hydrogenation reaction mixtures by filtration after the addition of hexane. The catalyst was not completely soluble in toluene, so the recycling experiments were carried out in CH$_2$Cl$_2$/toluene. Contrary to the case of the dendrimer-ligated catalysts, the [Ir(COD)Cl]$_2$/**14**/I$_2$ catalyst lost some enantioselectivity, but almost no activity, after five uses.

Discrete, ligated iridium complexes have also been used for the asymmetric hydrogenation of quinolines. Lu and Bolm designed the chiral-at-sulfur ligand **25**, whose corresponding iridium complex [Ir(**25**)(COD)]$^+$[BAr$_F$]$^-$ hydrogenated quinolines in good enantioselectivities (Table 6.4, entry 16) [126]. This catalyst did not require an additive; rather, it produced near-racemic product when hydrogenating 2-methylquinoline in the presence of I$_2$. Fan and coworkers synthesized the phosphine-free complex [Cp*Ir(OTf)(**26**)], which was also very stereoselective in the hydrogenation of quinolines, especially slightly below room temperature (entry 17) [127]. The catalyst was most active in the presence of trifluoroacetic acid, which did not affect its enantioselectivity. The most notable advantage of this catalyst was its relative insensitivity to O$_2$; the hydrogenation reaction could be set up in air without using degassed solvents. Finally, Genet and coworkers activated the Ir^{III} precursor [Ir(I)(H)(O$_2$CMe)(**24**)] with HBr to produce a selective catalyst for quinoline hydrogenation (entry 18) [128, 130].

The catalysts shown in Table 6.4 have also been used to hydrogenate a variety of quinolines with substituents at the 2 and 6 positions; some of the highest enantioselectivities for common substrates are shown in Table 6.5. Attempts to hydrogenate 3- and 4-substituted quinolines enantioselectively have been much less successful; in general, they are reduced to racemic 3-substituted 1,2,3,4-tetrahydroquinolines [116].

Table 6.5 Highly enantioselective hydrogenations of other quinolines.

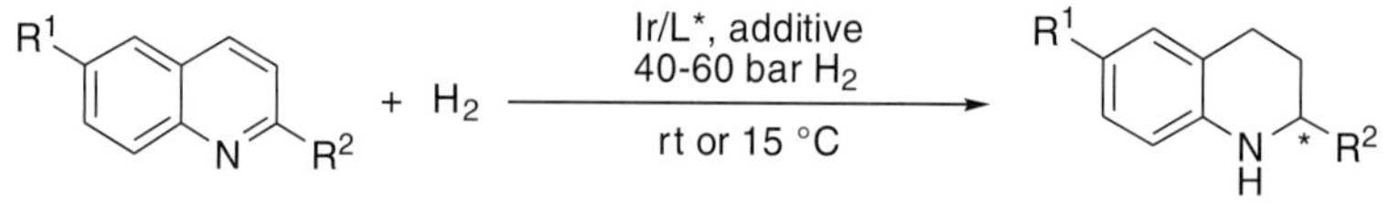

Entry	R¹	R²	Catalyst system	T [°C]	Yield (ee)[a] [% (%)]
1[b],[c]	H	Me	[Cp*Ir(26)(OTf)]/TFA	rt	70 (97)
2[b]	H	nPentyl	[Cp*Ir(26)(OTf)]/TFA	15	98 (96)
3[d]	H	CH_2CPh_2OH	[Ir(COD)Cl]$_2$/(14)/I$_2$	rt	97 (91)
4[b]	H	CH_2(1-HO-Cy)	[Cp*Ir(26)(OTf)]/TFA	15	97 (99)
5[e]	H	Ph	[Ir(COD)Cl]$_2$/(18)/I$_2$	rt	96 (84)
6[f]	F	Me	[Ir(COD)Cl]$_2$/(12)/I$_2$	rt	88 (96)
7[b]	Me	Me	[Cp*Ir(26)(OTf)]/TFA	15	>99 (97)
8[b]	MeO	Me	[Cp*Ir(26)(OTf)]/TFA	15	97 (97)

a) Isolated yield and ee of the desired product.
b) See Ref. [127].
c) Conversion is given in place of isolated yield.
d) See Ref. [124].
e) See Ref. [119].
f) See Ref. [116].

Though most catalysts for asymmetric quinoline hydrogenation are iridium-based, an extremely selective ruthenium-based system has been reported. Chan and coworkers used [(p-cymene)Ru(26)(OTf)], an analogue of the iridium catalyst [Cp*Ir(OTf)(26)] (see Table 6.4, entry 17), to hydrogenate various 2-substituted quinolines in [BMIM]$^+$[PF$_6$]$^-$ solution [131]. Remarkably, the ionic liquid improved both the stability and enantioselectivity of the catalyst (compared to MeOH solution); it could be stored in [BMIM]$^+$[PF$_6$]$^-$ solution under air for a month without losing activity or enantioselectivity. The catalyst was recycled using hexane extraction and could be reused repeatedly with almost no loss of enantioselectivity or activity over five uses. Even after eight uses, the catalyst hydrogenated 2-methylquinoline in 80% yield and 97% ee (Scheme 6.4).

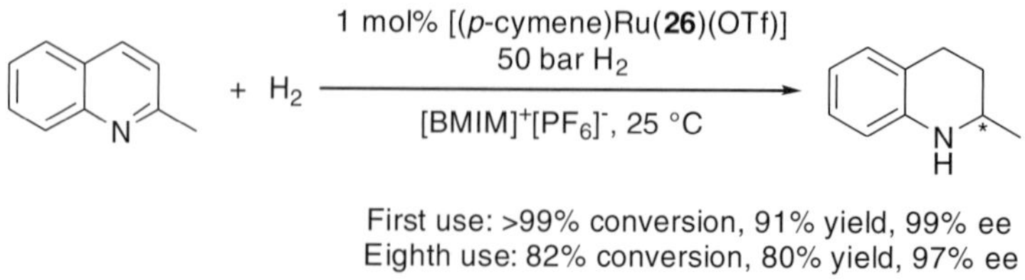

First use: >99% conversion, 91% yield, 99% ee
Eighth use: 82% conversion, 80% yield, 97% ee

Scheme 6.4 Enantioselective, ruthenium-catalyzed hydrogenation of 2-methylquinoline in ionic liquid.

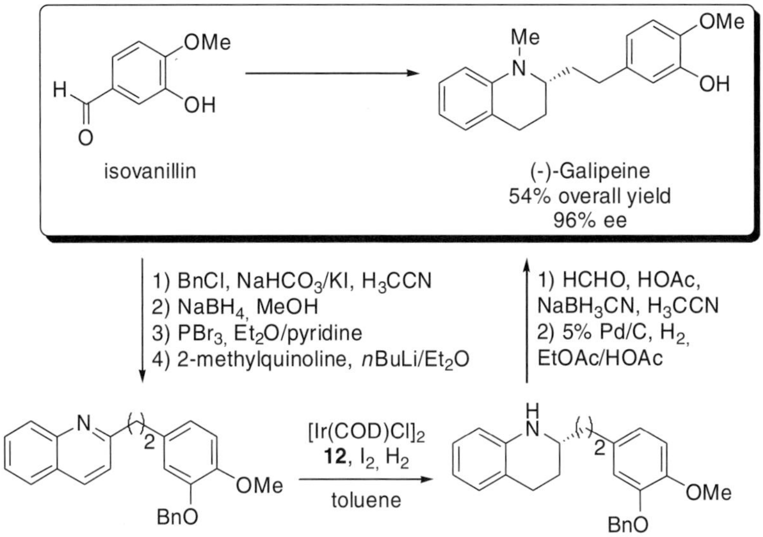

Scheme 6.5 Total synthesis of (−)-galipeine, employing the iridium-catalyzed asymmetric hydrogenation of a quinoline as a key step.

6.2.2.1.1 Quinolines – A Representative Synthesis Y.-G. Zhou and Yang applied their catalyst system to the enantioselective total synthesis of (−)-galipeine [132], a biologically active tetrahydroquinone alkaloid isolated from the bark of the *Galipea officinalis* tree in 1999 [133]. They began with two commercially available, achiral starting materials, isovanillin and 2-methylquinoline (Scheme 6.5). After coupling and protecting these in a total of four steps, they produced achiral 2-[2-[4-methoxy-3-(phenylmethoxy)phenyl]ethyl]quinoline. The enantioselective hydrogenation of this compound with [Ir(COD)Cl]$_2$/**12**/I$_2$ gave the corresponding chiral 2-substituted 1,2,3,4-tetrahydroquinoline, which was N-methylated and finally deprotected to give (−)-galipeine in 54% overall yield and 96% ee.

In an extension of their iridium-catalyzed asymmetric hydrogenations of quinolines, Y.-G. Zhou and coworkers published the asymmetric hydrogenation of quinolines that were activated by *in situ* reaction with benzyl chloroformate [134]. The activated quinolines were hydrogenated in 88–90% ee using [Ir(COD)Cl]$_2$/**27** (Table 6.6). The carbobenzyloxy group was attached to the nitrogen atom in the hydrogenation product, but could be removed by hydrogenation over Pd/C. A noncoordinating base was required to scavenge HCl formed from the chloroformate. In the presence of 4 Å molecular sieves, low conversion and a reversal of the product configuration were observed. Though it gave very good enantioselectivity in the hydrogenation of quinolines, this catalyst system is most notable for its ability to enantioselectively hydrogenate 1-substituted isoquinolines (Table 6.6). Only the imine bonds of the isoquinolines were reduced in the hydrogenation, so 1,2-dihydroisoquinolines were formed. The reaction was more enantioselective in the presence of lithium salts. Although the yield and ee values for these isoquinoline reductions were lower than for quinoline reductions by the same catalyst system, they

Table 6.6 Asymmetric hydrogenations of 1-substituted isoquinolines in the presence of chloroformates and lithium salts [134].

Reaction: 1-substituted isoquinoline + H_2 → 1,2-dihydroisoquinoline-N-carboxylate

Conditions: [Ir(COD)Cl]$_2$, **27**, ClCO$_2$R^3, Li$_2$CO$_3$, LiBF$_4$, THF, rt, 12-15 h, 41 bar H_2

Entry	R^1	R^2	R^3	Yield (ee)$^{a)}$ [% (%)]
1	H	Me	Me	85 (90)
2	H	Me	Bn	87 (83)
3	H	nBu	Me	87 (60)$^{b)}$
4	H	Bn	Me	83 (10)$^{b)}$
5	H	Ph	Me	57 (82)
6	MeO	Me	Me	57 (63)

Ligand **27**: substituted biaryl bis(PPh$_2$) ligand

a) Isolated yield and ee of the 1,2-dihydroisoquinoline product.
b) Product converted to the corresponding tetrahydroisoquinoline for ee determination.

were good in many cases. This is, to our knowledge, the only asymmetric hydrogenation of these heteroaromatics reported to date. The impetus to asymmetrically reduce isoquinolines is somewhat attenuated by the fact that their four-electron-reduction products, 1,2,3,4-tetrahydroisoquinolines, are accessible from the transfer hydrogenation of 3,4-dihydroisoquinolines (see below). In addition, Noyori and coworkers have reported the hydrogenation of isoquinoline-type rings with exocyclic olefin moieties, N-acyl-1-alkylidenetetrahydroisoquinolines, to yield chiral N-acylated 1,2,3,4-tetrahydroisoquinolines [135].

6.2.2.1.2 Isoquinolines – A Representative Synthesis To demonstrate the utility of their isoquinoline-hydrogenation procedure, Y.-G. Zhou and coworkers synthesized the naturally occurring alkaloid 1,2-dimethyl-1,2,3,4-tetrahydroisoquinoline (Scheme 6.6) [134]. They activated 2-methylisoquinoline with methyl chloroformate and hydrogenated it to methyl 1-methyl-1,2-dihydroisoquinoline-2-carboxylate in 78% ee. The remaining double bond in the heterocyclic ring was hydrogenated using Pd/C, and the carboxylate moiety was reduced with LiAlH$_4$ to yield the product in 50% overall yield.

Thus although 2-substituted quinolines have been reduced to chiral 1,2,3,4-tetrahydroquinolines with good enantioselectivity, there remains considerable progress to be made in this area. Some 2-substituted quinolines can be hydrogenated in >95% ee, but others, such as 2-phenylquinoline, give considerably lower values. More important, none of the catalyst systems that hydrogenate quinolines highly

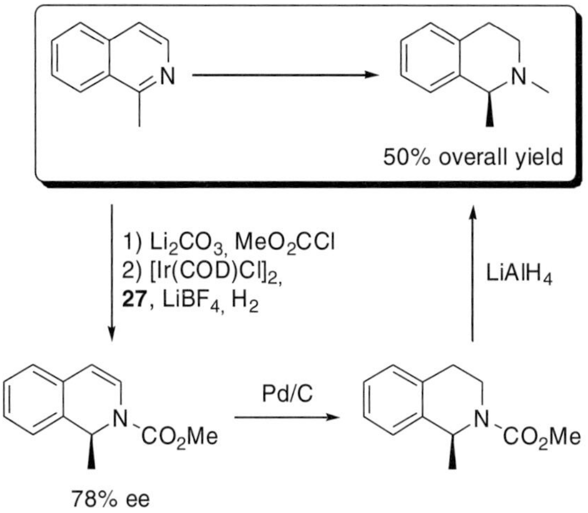

Scheme 6.6 Synthesis of 1,2-dimethyl-1,2,3,4-tetrahydroisoquinoline via iridium-catalyzed asymmetric hydrogenation.

stereoselectively have been demonstrated for substrates with more than one substituent on the heterocyclic ring. The asymmetric hydrogenation of the related isoquinolines is largely unexplored.

6.2.2.2 Quinoxalines

Compared to the asymmetric hydrogenation of 2-substituted quinolines, the enantioselective reduction of prochiral quinoxalines has received very little study. In fact, only the 2-methyl variant has been subjected to asymmetric hydrogenation. The first asymmetric hydrogenation of 2-methylquinoxaline (giving the product in 3% ee) was reported in 1987 [136], but no highly selective catalysts for this reaction were reported until 1998 [137]. Bianchini and coworkers used the orthometalated iridium dihydride [Ir(*fac-exo-*(R)-**28**)H$_2$] to hydrogenate 2-methylquinoxaline, forming 2-methyl-1,2,3,4-tetrahydroquinoxaline in up to 90% ee (Table 6.7, entry 1). This high selectivity was achieved with a fairly mild H$_2$ pressure (5 bar) but at a high temperature (100 °C). Unfortunately, the high ee value was only available at moderate conversion (54%); when the authors increased the pressure in order to improve the reaction yield, the product had lower ee (entry 2). The few other reports of the asymmetric hydrogenation of 2-methylquinoxaline involve a diverse set of catalysts. Cobley and coworkers applied a library of [*trans*-Cl$_2$ Ru(diamine)(diphosphine)]/KO*t*-Bu catalysts to the reaction; the best reduced 1000 equiv substrate to 99% conversion and 73% ee (entry 3) [47, 138]. This catalyst was useful at low loading and hydrogenated neat 2-methylquinoxaline, eliminating the need for solvent altogether. Bianchini and coworkers used an iridium catalyst with the neutral diphosphine ligand **31** to catalyze the reduction in MeOH (entry 4) [139], but this catalyst was both

Table 6.7 Asymmetric hydrogenation of 2-methylquinoxaline.

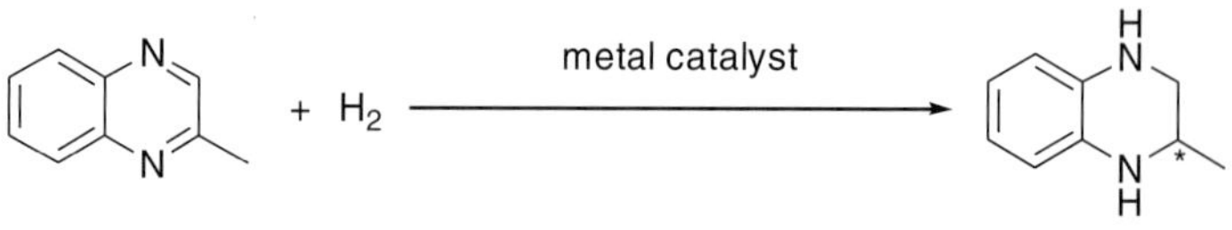

Entry	Catalyst system	S/M[a]	Solv.[b]	T [°C]	P(H$_2$) [bar]	Conv. (ee)[c] [% (%)]
1[d],[e]	[Ir(fac-exo-(R)-28)H$_2$]	100:1	MeOH	100	5	54 (90)
2[d],[e]	[Ir(fac-exo-(R)-28)H$_2$]	100:1	MeOH	100	13	85 (75)
3[f]	[Ru(29)(30)Cl$_2$]/KOt-Bu	1000:1	None	50	30	>99 (69)
4[g]	[Ir(31)(COD)]$^+$[OTf]$^-$	100:1	MeOH	40	20	41 (23)
5[e],[h]	[Rh(COD)Cl]$_2$/5	100:1	MeOH	100	5	60(8)
6[i]	[Ir(COD)Cl]$_2$/13/I$_2$	100:1	THF	rt	48	99 (80)

a) Substrate-to-metal ratio.
b) Solvent.
c) Conversion to and ee of 2-methyl-1,2,3,4-tetrahydroquinoline.
d) See Ref. [137].
e) Isolated yield is given instead of conversion.
f) See Ref. [47].
g) See Ref. [139].
h) See Ref. [140].
i) I$_2$/Ir = 5: 1; see Ref. [120].

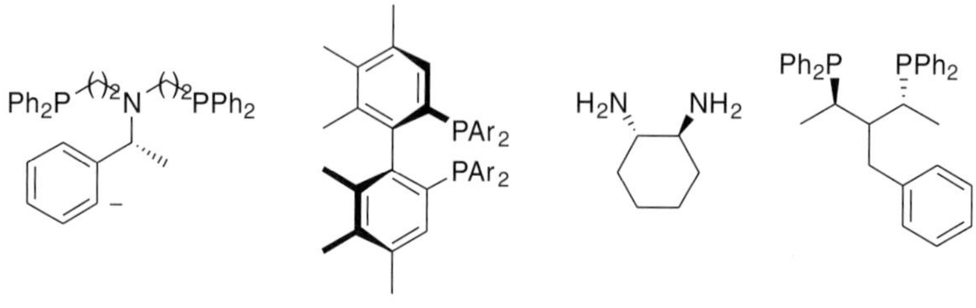

(R)-28

29
Ar = 3,5-Me$_2$-C$_6$H$_3$-

30

31

less active and less selective than their previous system. Brunner and Rosenboem attempted the reaction using [Rh(COD)Cl]$_2$/5, but obtained almost no enantioselectivity [140]. This catalyst was, however, useful for the diastereoselective reduction of folic acid, which has a 2-substituted 1,4-diazabenzene ring reminiscent of the quinoxalines. Most recently, Chan and coworkers applied their [Ir(COD)Cl]$_2$/13/I$_2$ catalyst (which produces good enantioselectivities for the hydrogenation of 2-substituted quinolines) to 2-methylquinoxaline, which it reduced in 80% ee (entry 6).

6.2.2.3 Pyridines

The asymmetric hydrogenations of substituted pyridines would produce chiral piperidine moieties, which are found in many biologically active compounds [141],

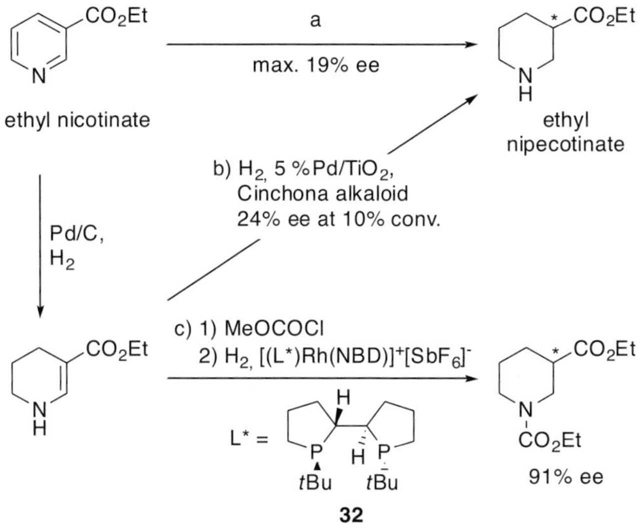

6.2 Chiral Amines with a Disubstituted Nitrogen Atom, HNR*R¹

Scheme 6.7 Net enantioselective hydrogenation of ethyl nicotinate by (a) direct asymmetric reduction, (b) achiral partial hydrogenation followed by asymmetric hydrogenation, and (c) achiral partial hydrogenation followed by functionalization and then asymmetric hydrogenation.

and the reaction is made even more attractive by the fact that numerous pyridine derivatives are commercially available. Unfortunately, the asymmetric hydrogenation of pyridines is no easy task, and the direct reaction has yet to be accomplished. Nevertheless, inventive methods for hydrogenating pyridine derivatives to chiral products have been created.

Several groups have accomplished the asymmetric reduction of ethyl nicotinate to ethyl nipecotinate. The direct hydrogenation of ethyl nicotinate (Scheme 6.7a) has been studied with both homogeneous [142] and heterogeneous [143] catalysts, yielding low ee values. Two- and three-step strategies have also been developed, and both begin with the Pd/C-catalyzed hydrogenation of ethyl nicotinate to ethyl 1,4,5,6-tetrahydro-3-pyridinecarboxylate over Pd/C. This could then be hydrogenated directly to ethyl nipecotinate by a cinchona-modified Pd/TiO_2 catalyst (Scheme 6.7b) [144], or protected at the N position to form a carbamate that could be hydrogenated enantioselectively by a rhodium catalyst, $[(32)Rh(nbd)]^+[SbF_6]^-$ (Scheme 6.7c) [145]. The three-step method, though not technically an enantioselective hydrogenation of an aza-aromatic, is a highly selective reduction of ethyl nicotinate to an ethyl nipecotinate derivative.

Glorius and coworkers used chiral 1,3-oxazolidin-2-ones as auxiliaries to direct the highly enantioselective reduction of pyridines by an achiral, heterogeneous catalyst (Scheme 6.8) [146]. Though auxiliaries are usually not ideal from the standpoint of atom economy, the 1,3-oxazolidin-2-one was removed during the hydrogenation, avoiding the need for a separate removal step, and could be recovered and recycled. Using this system, Glorius and coworkers could install up to three stereocenters simultaneously; only the substituent in the 3-position (R^4 in Scheme 6.8) was racemic

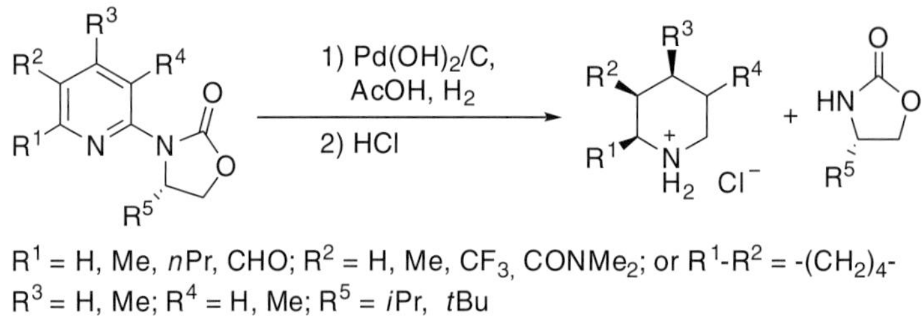
1) Pd(OH)2/C, AcOH, H2
2) HCl
R1 = H, Me, nPr, CHO; R2 = H, Me, CF3, CONMe2; or R1-R2 = -(CH2)4-
R3 = H, Me; R4 = H, Me; R5 = iPr, tBu

Scheme 6.8 The auxiliary-directed asymmetric hydrogenation of pyridines.

in the final product. An acid was required to protonate and activate the pyridine moiety and protect the catalyst from the piperidine product.

The complement to Glorius' system, that is, one in which prochiral pyridines are activated by an achiral auxiliary and hydrogenated by a chiral catalyst, was published by Legault and Charette in 2005 [147]. They converted a series of pyridines to *N*-iminopyridinium ylides and hydrogenated these in very good yields and ee values using the Pfaltz-type catalyst $[(\mathbf{6c})Ir(COD)]^+[BAr_F]^-$ in the presence of catalytic amounts of I_2 (Table 6.8). Different iridium catalysts gave widely varying outcomes for the reaction. No reaction occurred in the absence of I_2, which the authors speculated was serving to oxidize the Ir^I precatalyst to an active Ir^{III} catalyst. The hydrogenation products were crystalline solids whose enantiopurity could easily be improved by a single recrystallization prior to deprotection of the piperidine NH moiety using Raney nickel or lithium/ammonia. Therefore, this reaction provides a route to chiral piperidines from pyridines, using a chiral catalyst to impart stereo-

Table 6.8 Asymmetric hydrogenation of activated pyridines with a chiral iridium catalyst.

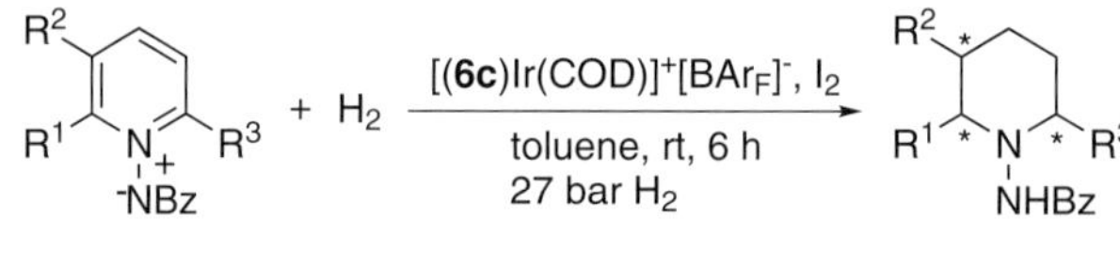
+ H2
[(6c)Ir(COD)]+[BArF]-, I2
toluene, rt, 6 h
27 bar H2

Entry	R¹	R²	R³	Yield (ee)[a] [% (%)]
1	Me	H	H	98 (90)
2	Et	H	H	96 (83)
3	*n*Pr	H	H	98 (84)
4	Bn	H	H	97 (58)
5	CH_2OBn	H	H	85 (76)
6	$(CH_2)_3OBn$	H	H	88 (88)
7	Me	Me	H	91 (54)[b]
8	Me	H	Me	92 (85)[c]

a) Isolated yield and ee of the piperidine product.
b) Diastereomeric ratio: >95 : 5.
c) Diastereomeric ratio: 57 : 43; ee values for the two products were 86 and 84%, respectively.

Table 6.9 Asymmetric hydrogenation of 7,8-dihydroquinolin-5(6H)-ones to 2,3,4,6,7,8-hexahydro-5(1H)-quinolines.

Entry	R	n	Yield (ee)[b] [% (%)]
1	Me	1	80 (86)
2	nPentyl	1	98 (97)
3	iPr	1	87 (84)
4	Ph	1	57 (92)
5	Ph(CH$_2$)$_2$	1	76 (92)
6	Bn	1	95 (85)
7	Ph	0	>95(21)[c]
8	nBu	0	>95 (45)[c]

a) Molar ratio of substrate to I$_2$ to Ir.
b) Isolated yield and ee of the desired product.
c) Conversion is given in place of isolated yield.

selectivity and an achiral auxiliary to activate the substrate. The main limitations of Legault and Charette's system came in the hydrogenation of disubstituted substrates (entries 7 and 8), and in particular substrates with substituents in the 3-position (entry 7).

Pyridines with electron-withdrawing substituents, and in particular the 7,8-dihydroquinolin-5(6H)-ones, have recently been hydrogenated without the aid of an auxiliary [148]. Y.-G. Zhou and coworkers reported the hydrogenation of these heterocycles to 2,3,4,6,7,8-hexahydro-5(1H)-quinolines (Table 6.9) using [Ir(COD)Cl]$_2$, I$_2$, and chiral diphosphines; they found the best results with ligand **12** in benzene. The system produced very good enantioselectivities and yield for substrates with a series of hydrocarbon substituents at the 2-position, although the 2-phenyl-substituted compound was hydrogenated to lower conversion than most. Two 5,6-dihydro-7H-cyclopenta[b]pyridin-7-one substrates were reduced to full conversion, though in much lower enantioselectivities (Table 6.9, entries 7 and 8). Finally, three pyridines without fused rings were also tested; both ethyl 2,6-dimethylnicotinate and 1-methyl-6-pentyl-3-phenylsulfonylpyridine were inert under the reaction conditions, whereas 3-cyano-2,6-lutidine was reduced in 21% conversion and 85% ee. It is not yet known whether this catalyst system can selectively install multiple stereocenters simultaneously.

Remarkable entries into asymmetric pyridine hydrogenation have been made, but there remain some unresolved issues; the most serious is that the stereochemistry at the 3-position cannot yet be controlled.

6.2.3
Direct Asymmetric Hydrogenation of "Activated" Imines

N-Alkyl and N-aryl imines have received the most attention in the literature, but significant research has also been performed in the area of "activated" imines, that is, imines with electron-withdrawing substituents at N. In addition to having different reactivity from N-alkyl and N-aryl imines, these compounds are intrinsically open to further functionalization after hydrogenation. The first of these compounds to be reduced enantioselectively were the N-tosyl amines. In contrast to the related reaction with N-alkyl and N-aryl amines, the asymmetric hydrogenations of N-tosyl amines are most often catalyzed by complexes of palladium.

In 1986, Okamoto and coworkers reported the catalytic hydrogenation of methyl N-(p-toluenesulfonyl)-1-imino-1-phenylacetate using a series of cobalt complexes [149]. The HCl or formic acid salt of quinine was added to direct the stereochemistry and produced optical yields up to 20%. The sense of enantioselectivity (R or S) depended on the solvent. Four years later, Oppolzer and coworkers reported two syntheses of the crystalline sultams (R)- and (S)-**35**, which they designed for use as chiral auxiliaries in asymmetric syntheses [150]. In one synthesis, the auxiliary was produced in enantiomerically pure form by methylating saccharine to the benzisothiazole **34** and hydrogenating it (Scheme 6.9). The hydrogenation reaction was readily catalyzed by $Ru_2Cl_4(33)_2NEt_3$. Interestingly, this catalyst system proved less active and selective when applied to the asymmetric hydrogenation of a similar, but acyclic substrate. Charette and Giroux attempted the hydrogenation of E-(N-tosyl) acetophenone imine with $Ru_2Cl_4(33)_2NEt_3$, obtaining low yields and a maximum ee of 20% [151]. They obtained much better yields and enantioselectivities using [Ru (OAc)$_2$(**33**)]; N-tosyl acetophenone imine was reduced in 62% ee, and 84% ee was obtained using the homologous 4-methyl-N-(1-phenylpropylidene) benzenesulfonamide as a substrate. Lennon and coworkers used [trans-Cl$_2$Ru(R)-**33**)(R,R)-**36**)] to

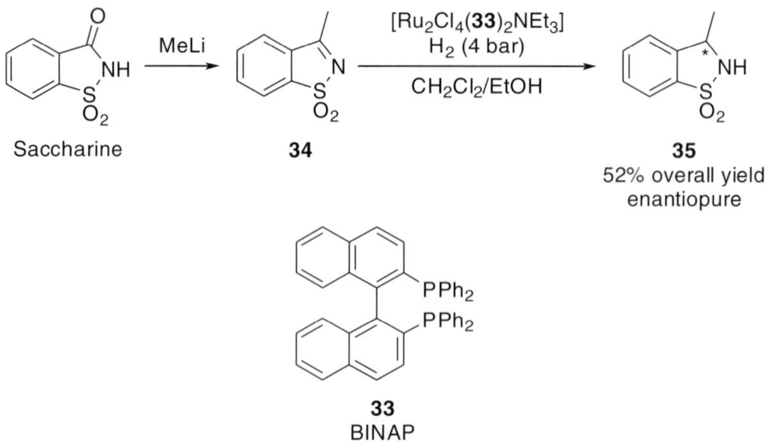

Scheme 6.9 Asymmetric imine hydrogenation to produce the sultams **35**.

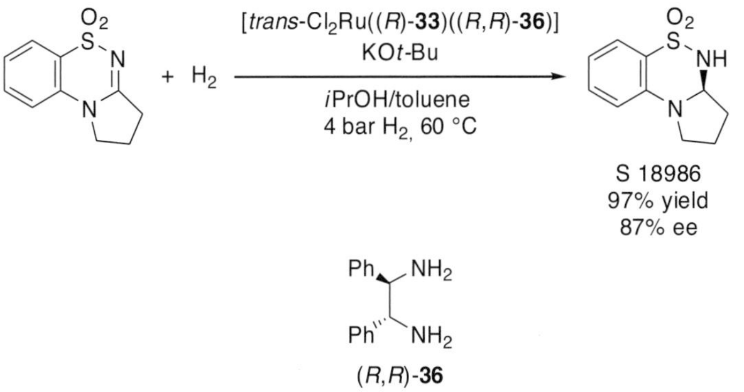

Scheme 6.10 Asymmetric hydrogenation of an activated imine to yield the AMPA receptor modulator S 18986.

hydrogenate a cyclic N-sulfonyl imine to S 18986 in 87% ee (Scheme 6.10) [152a]. The S enantiomer of this compound increases the sensitivity of AMPA-selective glutamate receptors, and thus may be useful in treating memory and cognitive disorders; the R enantiomer is inactive toward the same receptors [152b].

Zhang and coworkers significantly improved the enantioselectivity and substrate scope of catalytic asymmetric N-tosylimine hydrogenation by introducing Pd and Rh to the reaction [153]. Burk and Feaster had previously applied Rh catalysts to the asymmetric hydrogenation of N-aroylhydrazones [154], and Amii and coworkers had used palladium catalysts to asymmetrically reduce highly electron-deficient α-carbonyl imines [155–157]. Thus, Zhang and coworkers screened a variety of rhodium/diphosphine and palladium/diphosphine catalysts in the reduction of E-(N-tosyl) acetophenone imine. Excellent conversions and enantioselectivities were obtained with Pd(O$_2$CCF$_3$)$_2$/32, which was then applied to a series of N-tosyl imines (Scheme 6.11), an isothiazole dioxide and two exocyclic N-tosyl imines. Zhang's success in the asymmetric hydrogenation of N-tosyl imines prompted other researchers to apply palladium/diphosphine catalysts to similar substrates. Y.-G. Zhou and coworkers applied Pd(O$_2$CCF$_3$)$_2$ with various diphosphines to the asymmetric reductions of diphenylphosphinyl-protected imines, dihydroisothiazole dioxides, oxathiazole dioxides, benzoxathiazine dioxides, and benzothiadiazine dioxides

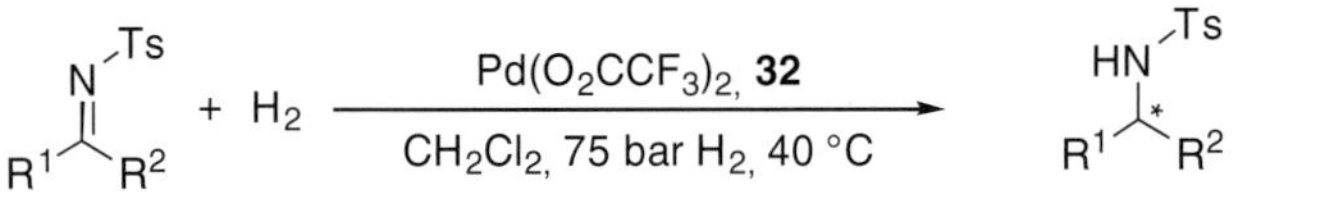

R^1 = tBu, X-C$_6$H$_4$ (X = H, 4-Me, 4-F, 4-MeO, 3-MeO, 4-Cl, 3-Cl), 1-Naphthyl, 2-Naphthyl R^2 = Me, Et

>99% conversion
75–>99% ee

Scheme 6.11 Asymmetric hydrogenation of N-tosylated imines by a palladium/diphosphine catalyst.

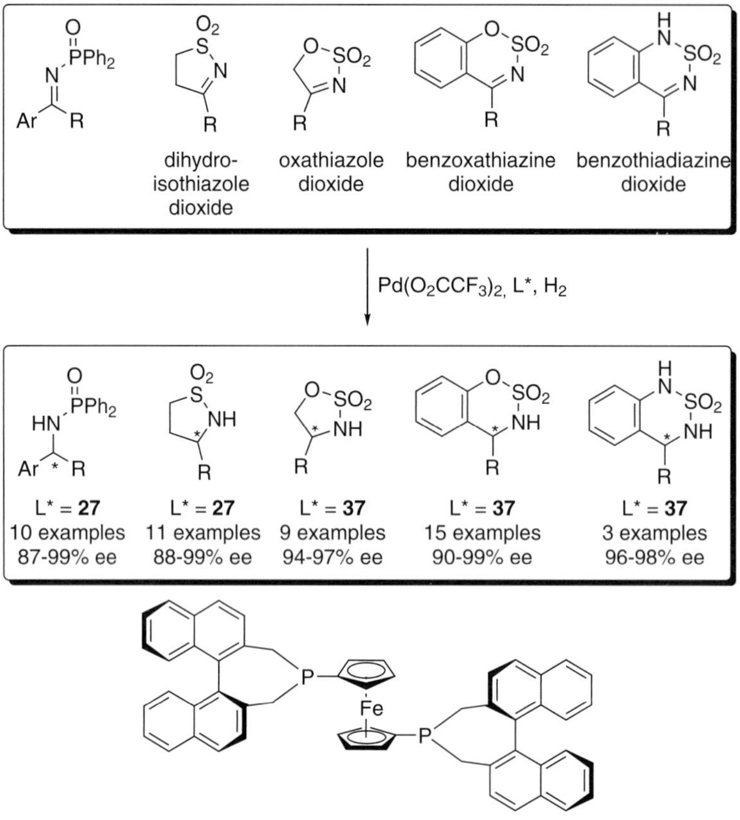

Scheme 6.12 Asymmetric hydrogenation of electron-poor imines.

(Scheme 6.12) [158–160]. They also demonstrated that benzoxathiazine dioxides could be used to synthesize chiral amino alcohols by LiAlH$_4$ reduction of the sulfamidate formed from hydrogenation.

6.2.4
Asymmetric Transfer Hydrogenation of Imines

Transfer hydrogenation is the catalyst-mediated reduction of an unsaturated moiety using an equivalent of H$_2$ that has been abstracted from a donor organic molecule. Molecular H$_2$ is not involved, so transfer hydrogenation is a convenient process for small-scale reductions in laboratories that lack hydrogenation equipment. In some cases, it is also much more chemoselective than direct hydrogenation (see below), making it a valuable tool for hydrogenating imines in the presence of other unsaturated groups. Of course, these advantages come at the cost of atom efficiency; most of the sacrificial H$_2$ donor does not become a part of the product. The asymmetric transfer hydrogenation of ketones has seen intense study, but the analogous reduction of imines is less common. It has been discussed in books dealing with broader topics such as asymmetric transfer hydrogenation [161], asymmetric imine hydrogenation [2], transfer hydrogenation catalyzed by ruthenium complexes [7], and the transfer hydrogenation of imines [162]. It has also been covered in a dedicated review [163], as well as reviews on the topics of asymmetric transfer hydrogenation [164–168], asymmetric imine reduction [9], and asymmetric amine synthesis [10].

Noyori and coworkers brought imines into the arena of asymmetric transfer hydrogenation in 1996 [169]. They applied the [(arene)Ru(amine-amido)Cl]/HCO_2H/NEt_3 system, which they had developed for the asymmetric transfer hydrogenation of ketones [170], to prochiral imines, and obtained remarkable results. Imine **D** was used for catalyst optimization studies, and these revealed that the reaction was extremely sensitive to the particular amine-amido ligand used; only those with one RNH_2 and one $ArSO_2NR^-$ terminus produced active, selective catalysts (Table 6.10) [171]. The reaction worked well in polar, aprotic solvents, but

Table 6.10 Asymmetric transfer hydrogenation of imines using HCO_2H/NEt_3 as the hydrogen source.

Entry	Substrate	Catalyst		Solv.[a]	Yield (ee)[b] [% (%)]
		Ar	Arene		
1	B	Mes[c]	Benzene	CH_2Cl_2	72 (77)
2		1-Naphthyl	Benzene	CH_2Cl_2	90 (89)
3	D	4-MeC$_6$H$_4$	p-Cymene[d]	H_3CCN	>99 (95)
4[e]		4-MeC$_6$H$_4$	p-Cymene[d]	DMF[f]	86 (97)

a) Solvent.
b) Isolated yield and ee of amine product.
c) Mes = 2,4,6-Me$_3$C$_6$H$_2$.
d) p-Cymene = 4-iso-propyltoluene.
e) Only the imine group was reduced.
f) DMF = dimethylformamide.

was not useful in ethers or alcohols. The catalysis was very slow in neat HCO_2H/NEt_3. Endocyclic amines (entries 3 and 4) were generally reduced in high yield and >90% ee; slightly lower ee values were obtained with exocyclic (entry 2) or acyclic imines (entry 1). Although this catalyst system is known to hydrogenate ketones enantioselectively, it is much more reactive toward imines. In fact, the quantitative transfer hydrogenation of **D** could be performed in acetone *solution*, and only 3% of the acetone was reduced to *i*-PrOH. This chemoselectivity is extremely attractive for the reduction of substrates containing both ketone and imine groups.

Mao and Baker designed two enantiomeric rhodium analogues of Noyori's catalyst by replacing the arene ligand on ruthenium with a pentamethylcyclopentadienyl (Cp*) ligand on rhodium [172]. The resulting [Cp*RhCl(**26**)] compounds were highly active catalysts for the asymmetric transfer hydrogenation of imines using HCO_2H/NEt_3 as the hydrogen source; only 10 min was required to reduce **D** in 96% conversion (Table 6.11). The catalysts gave high ee values for the reduction of 1-alkyl-3,4-dihydroisoquinolines (entry 3) and moderate ee values for that of benzoisothiazole dioxides (entry 2), but struggled to induce any enantioselectivity when reducing 1-aryl-3,4-dihydroisoquinolines (entry 4) or acyclic imines (entry 1). Blackmond and coworkers performed a kinetic analysis of this catalyst system and showed that its activity could be further improved by the slow addition of HCO_2H [173].

Table 6.11 Rhodium-catalyzed asymmetric transfer hydrogenation of imines.

Entry	Substrate	t [min]	Solvent	Conv. (ee)[a] [% (%)]
1	B	10	CH_2Cl_2	88 (8)
2	(benzisothiazole dioxide, N-Bn)	40	CH_2Cl_2	93 (68)
3	D	10	H_3CCN	96 (89)
4	(1-phenyl-3,4-dihydroisoquinoline)	180	H_3CCN	90 (4)

a) Conversion to and ee of amine product.

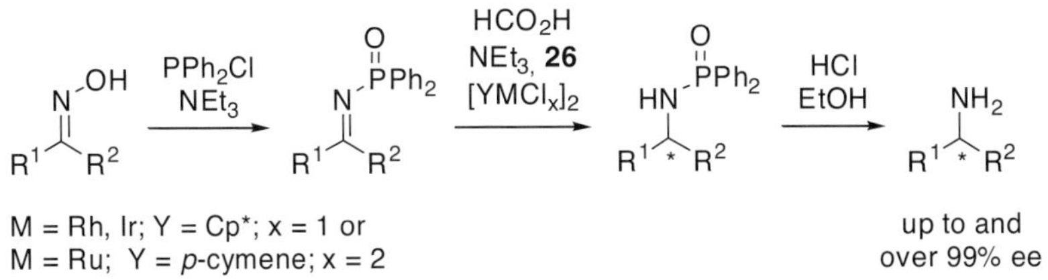

6.2 Chiral Amines with a Disubstituted Nitrogen Atom, HNR*R¹

Scheme 6.13 Improved selectivity in the asymmetric transfer hydrogenation of acyclic olefins using −P(O)Ph$_2$ protecting groups.

To deal with the problem of E/Z isomerism and its effect on asymmetric transfer hydrogenation, researchers at Avecia synthesized acyclic imines that were protected with −P(O)Ph$_2$ groups [174]. These bulky groups caused the protected imines to exist predominantly or exclusively as a single isomer, making them better substrates for metal-catalyzed asymmetric transfer hydrogenation. The resulting chiral, protected amines could be obtained in high to excellent ee (84 - > 99% in the examples given) and could then be deprotected in acidic ethanol without loss of stereochemistry (Scheme 6.13). The Avecia group produced their protected imines from the corresponding oximes, though a synthesis beginning with the corresponding ketones is available [175, 176].

6.2.4.1 Reducing the Environmental Impact of the Reaction

Much like the direct hydrogenation of imines, the transfer hydrogenation of imines has seen a number of recent attempts to use less toxic solvents and to recycle catalysts, and many of the same methods employed to improve the former reaction have also been applied to the latter.

Zhu, and coworkers modified the Noyori system for asymmetric transfer hydrogenation in order to perform the reaction in water [177]. They designed **38** [178], a sulfonated version of Noyori's ligand **26** for use in aqueous solution (Figure 6.5). They replaced the donor system of HCO$_2$H/NEt$_3$ with HCO$_2$Na, which can produce formic acid *in situ*. They combined this ligand and donor with several metal precursors and surfactant additives in water, and tested the resulting solutions as catalysts for the transfer hydrogenation of **D**. They found that [RuCl$_2$(p-cymene)]$_2$, **38**, HCO$_2$Na, and cetyltrimethylammonium bromide (CTAB) best combined activity and selectivity (all but one ee value was ≥90%) for that reaction. Iridium and rhodium precursors were both quite enantioselective, but less active than the ruthenium one. The choice and amount of surfactant had noticeable effects on the stereoselectivity of

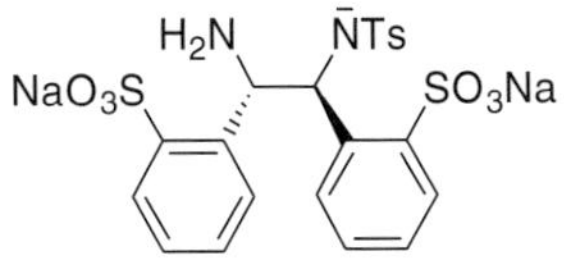

Figure 6.5 A sulfonated version of ligand **26**.

transfer hydrogenation. Though it was active and selective for cyclic amines, including those with N-SO$_2$ moieties, the [RuCl$_2$(p-cymene)]$_2$/**38**/CTAB/HCO$_2$Na system did not react with N-(tert-butylsulfonyl) acetophenone imine. Imine **B** decomposed completely upon attempted reduction. In most reactions, the organic product was obtained by repeated extraction with CH$_2$Cl$_2$, which mitigates the environmentally friendly nature of the process, but the authors were also able to extract with Et$_2$O/hexane and reuse the catalyst after adding more HCO$_2$Na and substrate. After four uses, the catalyst system showed no significant loss of enantioselectivity, although a small decline in yield was noted. Two other reports on the asymmetric transfer hydrogenation of imines in water appeared in 2007. Zhu and coworkers reported an amine-functionalized relative of **26** that could be used for the rhodium-catalyzed reaction [179], and Canivet and Süss-Fink synthesized a variety of [(amine-amido)(arene)RuOH$_2$]$^+$[BF$_4$]$^-$ salts as catalysts for the reaction [180]. Both used HCO$_2$Na as the hydrogen source. The former catalyst system was more selective for the hydrogenation of **D**, but the latter was able to catalyze the reduction of acyclic **B** in up to 91% ee. In addition, Zhu and Deng again used CH$_2$Cl$_2$ to isolate the amine products after hydrogenation, whereas Canivet and Süss-Fink used Et$_2$O. Neither group discussed the reuse of their catalysts, though Canivet and Süss-Fink recovered their ruthenium aquo complexes unchanged after the reaction.

There have been multiple efforts toward supported catalysts for asymmetric transfer hydrogenation, and the 4-position on the aryl sulfonate group of **26** has proven a convenient site for functionalization. Thus far, this ligand has been supported on dendrimers [181, 182], polystyrenes [183], silica gel [184], mesoporous siliceous foam [185], and mesoporous siliceous foam modified with magnetic particles [186]. The resulting modified ligands have been used in combination with ruthenium, rhodium, and iridium to catalyze the asymmetric transfer of imines and, more commonly, ketones.

The first supported catalysts for the asymmetric transfer hydrogenation of imines were the dendrimeric ones. Chen et al. built oligoamide dendrimers, and later oligoether dendrimers, with **26**-like groups at the termini and used these as ligands for ruthenium-based transfer-hydrogenation catalysts [181, 182]. Though they focused mostly on the asymmetric reductions of ketones, the authors found that the dendrimer-based catalysts hydrogenated benzoisothiazole **34** to conversions and enantioselectivities similar to those obtained with the original ligand **26**; the oligoamide dendrimer was more selective than the oligoether one. The catalyst formed from the oligoamide dendrimer-bound ligand could be reused in the asymmetric hydrogenation of ketones [181]; recycling experiments were not attempted with the imine substrate.

Haraguchi et al. recently supported **26** on three different types of cross-linked polystyrenes – standard cross-linked polystyrene, cross-linked polystyrene with –SO$_3$Na on some aryl groups, and cross-linked polystyrene with –SO$_3^-$[BnNBu$_3$]$^+$ on some aryl groups [183]. Each of these three ligands combined with [RuCl$_2$(p-cymene)]$_2$ to catalyze the asymmetric transfer hydrogenation of different imines under different conditions. The complex supported on cross-linked polystyrene was useful for the reduction of **B** (up to 86% ee at rt) or **D** (92% ee at rt) by HCO$_2$H/NEt$_3$ in

CH$_2$Cl$_2$ solution and was reused for several reductions of **B**. After three reuses, the catalyst maintained its activity, but the enantiopurity of the products decreased by ~10%. The complex did not catalyze the reductions of **A**, **C**, or 2-methylquinoxaline, and it was inactive in aqueous solution. On the other hand, the catalysts supported on cross-linked polystyrenes with hydrophilic −SO$_3$Na or amphiphilic −SO$_3^-$[BnNBu$_3$]$^+$ groups were active for transfer hydrogenation in water using HCO$_2$Na as the hydrogen source, though only cyclic substrates could be used with these systems. Substrate **D** was hydrogenated in 89% ee by these catalysts, and the reaction was faster in the presence of the hydrophilic catalyst. The selectivity of this catalyst system could be improved upon the addition of cetyltrimethylammonium chloride (**D** reduced in 94% ee). Adding benzyltributylammonium chloride slightly reduced the enantioselectivity. These catalysts were not reused.

Finally, ligand **26** has been grafted onto three types of silica materials. Tu and coworkers developed a synthesis of silica-bound **26** (Scheme 6.14) and applied it first to the asymmetric transfer hydrogenation of ketones [187] and later to imine **34** [184]. The catalyst could be recovered by washing with CH$_2$Cl$_2$, and though it lost some activity after repeated uses, little change in stereoselectivity was observed after seven uses. Huang and Ying used two different methods to attach **26** to silica gel and also grafted it onto siliceous mesocellular foam [185], a form of SiO$_2$ with uniformly sized, interconnected pores. Working with **D** as a substrate, they found that the catalyst supported on mesoporous silica foam was not only more enantioselective than the one supported on silica gel, but was also more stable. After six uses, the catalyst still hydrogenated **D** in 95–100% yield and 90–91% ee. They also performed a control experiment in which they used a physical mixture of siliceous mesoporous foam, [RuCl$_2$(p-cymene)]$_2$, and **26** as the catalyst; this catalyst was active and selective in the first transfer hydrogenation run, but could not be reused. Li and coworkers modified

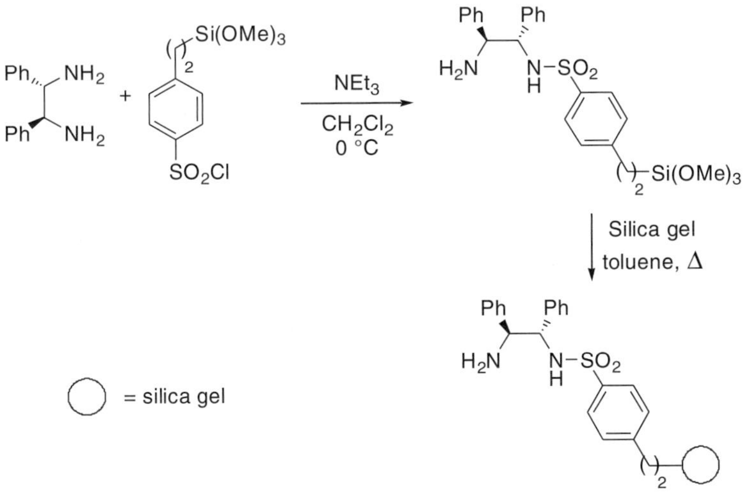

Scheme 6.14 Synthesis of a silica-bound version of **26**.

Huang and Ying's approach by grafting magnetic γ-Fe$_2$O$_3$ onto spherical siliceous mesoporous foam particles before attaching the ligand [186]. This design feature was intended to make the catalyst magnetic and therefore easier to recover. Indeed, the new catalyst was recoverable, though it was washed with CH$_2$Cl$_2$ between runs, as were the catalysts bound to silica gel and unmodified siliceous mesoporous foam. ICP analysis showed that the magnetically modified supported catalyst lost ~11% of the bound ruthenium after nine uses; this compares well to the nonmagnetic supported catalyst that lost ~11% of the bound ruthenium after six uses. The activities and selectivities of silica-bound catalysts for ruthenium-catalyzed asymmetric transfer hydrogenation are summarized in Table 6.12.

Table 6.12 Reusable, silica-bound catalysts for the asymmetric transfer hydrogenation of imines.

Entry		Substrate	First use Conv. (ee)[a] [% (%)]		nth use	
				n	Conv. (ee)[a] [% (%)]	
1[b]	Silica gel		>99 (93)	7	96 (90)[c]	
2[d]	Siliceous mesoporous foam		95–100 (90–91)	6	95–100 (90–91)	
3[e]	Siliceous mesoporous foam w/magnetic particles		99 (94)	9	99 (90)	

a) Conversion to and ee of the desired amine product.
b) See Ref. [184].
c) Reaction time increased from 1 to 7 h.
d) See Ref. [185]. Reaction time was 12 h.
e) See Ref. [186]. Reaction time was 1.5–7 h.

Though very active and stereoselective solid-bound or water-soluble catalysts for asymmetric transfer hydrogenation exist, there remains considerable potential for improving the environmental impact of this reaction. The solid-bound catalysts, though recyclable, are extracted with a chlorinated solvent before reuse, as are the water-soluble catalysts in many cases. However, at least one group has reported success in extracting reaction products using Et_2O [180]. The imine substrate scope of these catalysts also remains largely unexplored, though many have been tested more thoroughly for ketone substrates.

6.2.4.2 Syntheses Using the Asymmetric Transfer Hydrogenation of Imines as a Key Step

Noyori and coworkers' success in the asymmetric transfer hydrogenation of **D** inspired several groups to apply his system to the syntheses of 1,2,3,4-tetrahydroisoquinolines and 1,2,3,4-tetrahydro-β-carbolines. In some early examples, Vedejs *et al.* used a Noyori catalyst to synthesize new asymmetric proton donors [188] and Sheldon and coworkers produced a key intermediate in the synthesis of morphine [189]. A group of researchers at Glaxo Wellcome published the syntheses of the neuromuscular blocker GW0430 and related compounds using transfer hydrogenation as a key step [190, 191]. Tietze *et al.* used a strategy of oxidizing racemic 1,2,3,4-tetrahydroisoquinolines to the imines, then reducing them with Noyori's catalyst to form chiral versions, in the synthesis of several naturally occurring and novel alkaloids [192, 193]. Czarnocki and coworkers have produced several 1,2,3,4-tetrahydro-β-carbolines from this reaction [194], including some tetracyclic examples [195] and the antiworm agent (*R*)-praziquantel [196].

Ahn *et al.* applied the asymmetric transfer hydrogenation of benzoisothiazole dioxides to the formation of chiral sultams [197] that are useful as chiral auxiliaries [198].

6.3
Chiral Amines with Trisubstituted Nitrogen, $NR^*R^1R^2$

Imine hydrogenation produces an N−H bond, so the chiral amines produced by this method will necessarily have mono- or disubstituted N atoms. Chiral $NR^*R^1R^2$ amines can, however, be produced by asymmetric hydrogenation, using *N,N*-disubstituted iminium ions or enamines as substrates. Neither of these substrate classes has been studied to the same extent that imines have, but both can be hydrogenated enantioselectively.

6.3.1
Hydrogenation and Transfer Hydrogenation of *N,N*-Disubstituted Iminiums

Magee and Norton reported the direct, ruthenium-catalyzed asymmetric hydrogenation of several *in situ* generated iminium salts to ammoniums with modest ee values;

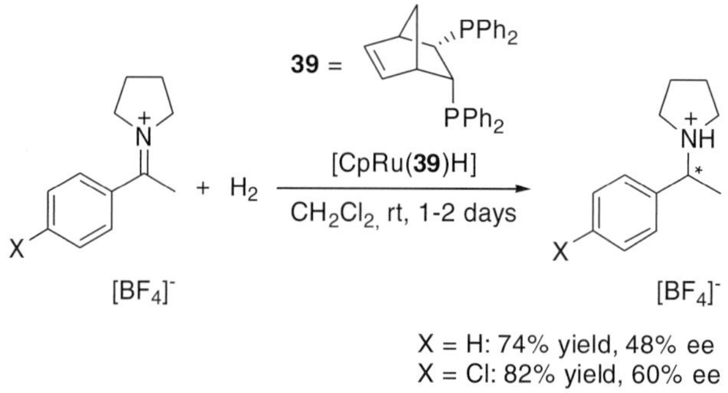

Scheme 6.15 Ruthenium-catalyzed asymmetric hydrogenation of pyrrolidinium salts.

X = H: 74% yield, 48% ee
X = Cl: 82% yield, 60% ee

the product amines were isolated after basic workup. Two of their substrates were pyrrolidinium salts and therefore produced chiral N-trisubstituted amines (Scheme 6.15) [199]. Norton's group subsequently pursued the mechanism of H_2 transfer in the ruthenium-catalyzed hydrogenation of iminiums [200, 201].

Zhu and coworkers have applied asymmetric transfer hydrogenation to N-alkyl 3,4-dihydroisoquinolinium salts, producing two tertiary amines highly stereoselectively [177]. Working in aqueous solution, they used Noyori's chiral [((R,R)-26)RuCl(p-cymene)] complex as a catalyst, HCO_2Na as a hydrogen source, and CTAB as a surfactant. N-Benzyl-1-phenyl-6,7-dimethoxyisoquinolinium bromide was reduced in 98% ee, and the related 1-methyl-substituted substrate was reduced in 90% ee (Scheme 6.16). The same reactions could be carried out using a sulfonated version of ligand **26**. After basic workup, these hydrogenations yielded chiral 1-substituted N-benzyl-6,7-dimethoxy-1,2,3,4-tetrahydroisoquinolines, meaning that this strategy may be an alternative to the asymmetric hydrogenation of isoquinolines; more asymmetric examples are required to demonstrate its utility.

Czarnocki and coworkers have applied the asymmetric hydrogenation of iminium ions to the synthesis of some naturally occurring alkaloids [202, 203]. In a representative example, they hydrogenated compound **40**, synthesized in two high-yielding steps, with [((S,S)-26)RuCl(C₆H₆)] as the catalyst and HCO_2H/NEt_3 as the hydrogen source (Scheme 6.17). The result was (+)-crispine A, an alkaloid that was isolated from the welted thistle plant in 2002 [204]. Czarnocki and coworkers' synthesis gave (+)-crispine A in 92% ee [202, 203].

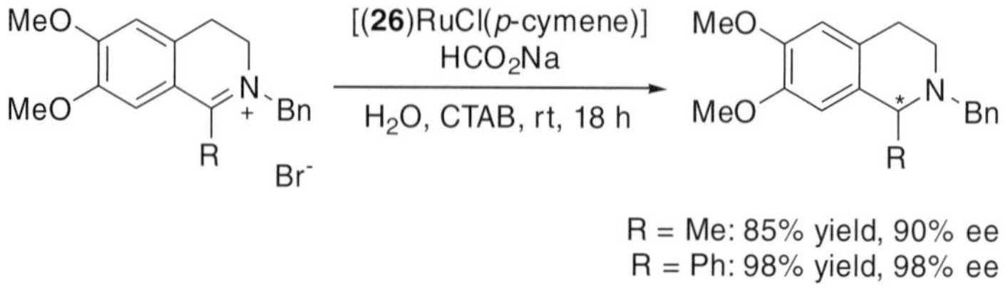

R = Me: 85% yield, 90% ee
R = Ph: 98% yield, 98% ee

Scheme 6.16 The ruthenium-catalyzed transfer hydrogenation of N-benzyl 3,4-dihydroisoquinolinium ions to chiral N-benzyl-1,2,3,4-tetrahydroisoquinolines.

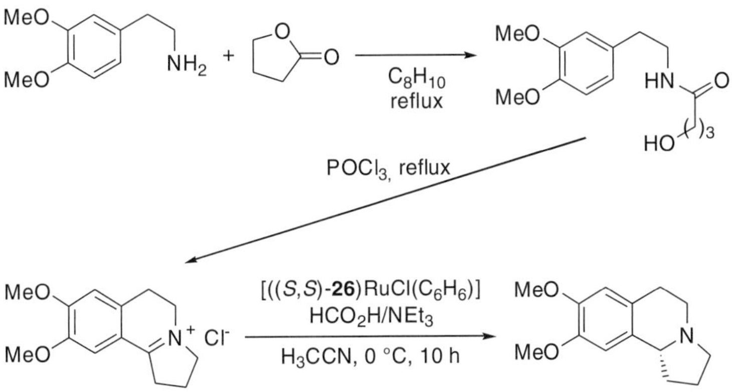

Scheme 6.17 Czarnocki and coworkers used asymmetric transfer hydrogenation to convert an iminium salt to the naturally occurring alkaloid (+)-crispine A.

6.3.2
Hydrogenation and Transfer Hydrogenation of Enamines

Lee and Buchwald published the first asymmetric hydrogenation of enamines [205], using the Buchwald group's (S,S,S)-[(2)Ti(3)] catalyst. Their chiral titanocene catalyst hydrogenated a range of aromatic enamines, and one aliphatic one, to chiral tertiary amines, giving high ee values in all cases (Table 6.13). The catalyst did not tolerate an aromatic bromide. As with titanium-catalyzed amine hydrogenation, this reaction was highly air sensitive and required somewhat high catalyst loading (5 mol%), meaning that it is unlikely to find routine use in amine synthesis. Nevertheless, it occurred under mild conditions (atmospheric pressure or 5 bar H_2) and gave the chiral products in good isolated yields.

The rhodium-catalyzed asymmetric hydrogenations of β-amino-α,β-unsaturated esters and amides (i.e., β-carbonyl enamides) have also been reported and can be very stereoselective (≥95% ee) [206–209]. However, this method has only been applied to the synthesis of mono- and disubstituted amines. In one case, evidence suggests that the reaction is actually a hydrogenation of the substrate's imine tautomer (which would preclude its use for the synthesis of trisubstituted amines) [208], whereas another β-enamine amide hydrogenation under very similar conditions appears to operate predominantly by enamine hydrogenation [209]. Thus, this reaction may hold promise for the synthesis of chiral NR*R¹R² amines, at least in some cases. Rhodium catalysts have also been applied to the asymmetric hydrogenation of enamines that have no additional coordinating groups. Börner and coworkers published the rhodium/diphosphine-catalyzed hydrogenation of otherwise-unfunctionalized enamines to tertiary amines in up to 72% ee [210]. Q.-L. Zhou and

Table 6.13 Asymmetric, titanium-catalyzed hydrogenation of imines

$$\underset{R^1}{\overset{R^2\diagdown N\diagup R^3}{\bigg|\bigg|}} + H_2 \quad \xrightarrow{(S,S,S)\text{-}[(2)\text{Ti}(3)]}{\text{THF, 24 h}} \quad \underset{R^1}{\overset{R^2\diagdown N\diagup R^3}{\bigg|*}}$$

Entry	R^1	R^2	R^3	Cond.[a]	Yield (ee)[b] [% (%)]
1	Ph	-(CH$_2$)$_4$-		I	75 (92)
2	4-MeO-C$_6$H$_4$	-(CH$_2$)$_4$-		I	72 (92)
3	4-Cl-C$_6$H$_4$	-(CH$_2$)$_4$-		I	89 (89)
4	4-Br-C$_6$H$_4$	-(CH$_2$)$_4$-		I	0 (—^{c})
5	2-Napthyl	-(CH$_2$)$_4$-		I	77 (96)
6	2-Me-C$_6$H$_4$	-(CH$_2$)$_4$-		II	87 (98)
7	1-Cyclohexenyl	-(CH$_2$)$_4$-		II	72 (95)
8	4-MeO-C$_6$H$_4$	(CH$_2$)$_2$O(CH$_2$)$_2$-		II	88 (91)
9	Ph	Me	Bn	II	83 (96)
10	Ph	Et	Et	II	78 (94)

a) Reaction conditions. I = 1 bar, rt; II = 5 bar, 65 °C.
b) Isolated yield and ee of amine product.

coworkers obtained better enantioselectivities using a rhodium catalyst in combination with iodine (2 mol% with respect to substrate) and acetic acid (20 mol%) [211]. Kalck et al. recently described the asymmetric hydrogenation of enamines by rhodium and iridium catalysts in ionic liquid solution; a wide range of ee values were reported [212].

In the past years, three groups have independently reported iridium-catalyzed asymmetric hydrogenations of enamines [213–215]. The groups of Andersson and Pfaltz both used discrete cationic complexes of the form [L*Ir(COD)]$^+$[BAr$_F$]$^-$ (L* = chiral phosphino-oxazoline) to catalyze the reaction, and both encountered substantial difficulties in producing consistently high conversions and ee values. Andersson and coworkers used a catalyst based on the bicycle-supported ligand **41** to hydrogenate 1-(aryl)vinyl enamines (Table 6.14) [213] and achieved excellent conversions and moderate to good enantioselectivities in the hydrogenation of substrates with acyclic amine functionalities. Enamines with cyclic amine groups, however, were reduced in moderate yields and low enantioselectivities. Baeza and Pfaltz used catalysts based on a few different ligands [215]. They found that [(**6a**)Ir(COD)]$^+$[BAr$_F$]$^-$ gave high ee values in the hydrogenation of terminal enamines with N-aryl groups at −20 °C, but was less enantioselective when N-benzyl substrates were used, whereas the converse was observed for [(**42**)Ir(COD)]$^+$[BAr$_F$]$^-$. The substitution patterns on the substrate strongly influenced the conversions obtained in the reaction. Baeza and Pfaltz also hydrogenated 1,2-disubstituted enamines in moderate enantioselectivities, though these reactions required catalysts with different ligands in order to maximize conversion and selectivity. Thus, the Andersson and

6.3 Chiral Amines with Trisubstituted Nitrogen, NR*R¹R²

Table 6.14 Asymmetric enamine hydrogenation using [L*Ir(COD)]⁺[BArF]⁻ catalysts.

$$\underset{Ph}{R^1_{\diagdown N} R^2} + H_2 \xrightarrow[CH_2Cl_2, 5-6\ h]{[L^*Ir(COD)]^+[BArF]^-} \underset{Ph}{R^1_{\diagdown N} R^2}$$

Entry	L*	T [°C]	P(H₂) [bar]	R¹	R²	Conv. (ee)ᵃ⁾ [% (%)]
1ᵇ⁾	41	rt	50	Et	Et	>99 (84)
2ᶜ⁾	42	−20	10	Et	Et	>99 (54)
3ᵇ⁾	41	rt	50	Ph	Me	>99 (91)
4ᶜ⁾	6a	−20	10	Ph	Me	98 (93)
5ᶜ⁾	42	−20	10	Bn	Me	>99 (93)
6ᵇ⁾	41	rt	50	-(CH₂)₄-		66 (33)
7ᶜ⁾	6a	−20	10	-(CH₂)₄-		>99 (44)

Structures **41** (N-P(o-tol) bicyclic with oxazoline–iPr) and **42** (Cy₂P, Bn Bn Bn oxazoline-Ph).

a) Conversion to and ee of product.
b) See Ref. [213].
c) See Ref. [215].

Pfaltz systems represent promising beginnings for the asymmetric hydrogenation of enamines by discrete, cationic iridium catalysts, but both require more development before they can find wide use in amine synthesis. In particular, these systems are still highly substrate-dependent, meaning that a new catalyst screening is necessary for each new substrate.

Q.-L. Zhou and coworkers used an [Ir(COD)Cl]₂/**43**/I₂ system to hydrogenate enamines stereoselectively. In contrast to the Andersson and Pfaltz systems, which struggled to selectively hydrogenate enamines with cyclic amines, Zhou's catalyst was only demonstrated for enamines in which both the N and double bond were part of a ring, and gave high enantioselectivity only when the amine was part of a five-membered ring (Scheme 6.18). The system was quite selective for the hydrogenation of 2-substituted N-aryl-4,5-dihydropyrroles, giving ee values of 89–97%. An N-butyl-substituted analogue was reduced in 72% ee. The authors also applied their catalyst system to the synthesis of (+)-crispine A by deprotonating compound **40** (see Scheme 6.19a) to the enamine **44**, then reducing it to crispine A in 90% ee. Transfer hydrogenation of **44** has also been applied to the enantioselective synthesis of (+)-crispine A (Scheme 6.19b) [203].

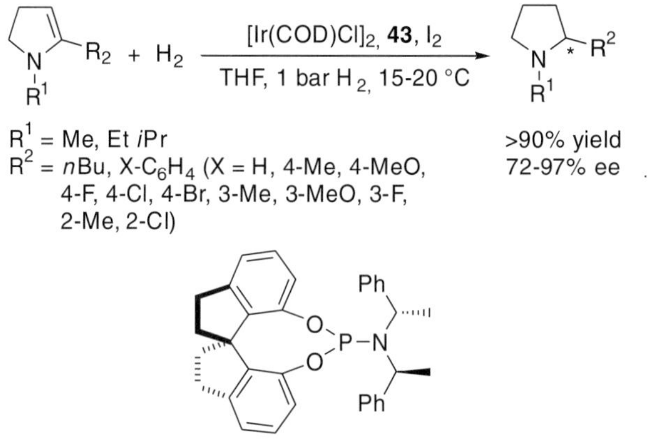

Scheme 6.18 Asymmetric hydrogenation of 2-substituted N-alkyl-4,5-dihydropyrroles by an iridium catalyst.

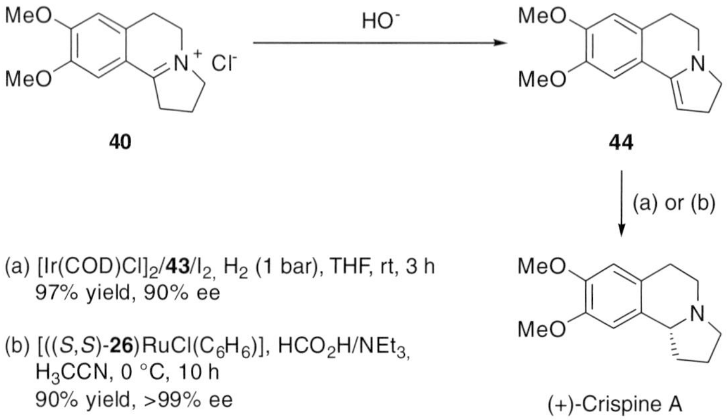

R^1 = Me, Et, iPr
R^2 = nBu, X-C$_6$H$_4$ (X = H, 4-Me, 4-MeO, 4-F, 4-Cl, 4-Br, 3-Me, 3-MeO, 3-F, 2-Me, 2-Cl)

>90% yield
72–97% ee

Conditions: [Ir(COD)Cl]$_2$, **43**, I$_2$, THF, 1 bar H$_2$, 15–20 °C

(a) [Ir(COD)Cl]$_2$/**43**/I$_2$, H$_2$ (1 bar), THF, rt, 3 h
97% yield, 90% ee

(b) [((S,S)-**26**)RuCl(C$_6$H$_6$)], HCO$_2$H/NEt$_3$, H$_3$CCN, 0 °C, 10 h
90% yield, >99% ee

(+)-Crispine A

Scheme 6.19 Hydrogenations of enamine **44** to form (+)-crispine A.

6.4
Conclusion

The past 35 years have seen both the asymmetric hydrogenation and asymmetric transfer hydrogenation of imines develop into useful methods for the synthesis of chiral amines. Particularly, focused research over the past 15 years has led to highly enantioselective examples of both reaction types and has added aza-aromatics, activated imines, and iminium cations to their purview. In addition, the asymmetric hydrogenation and asymmetric transfer hydrogenation of imines have both been applied to total syntheses. Because they are necessarily isomerically pure, cyclic imines have been reduced in the highest enantioselectivities. The asymmetric reductions of acyclic imines have been improved using iridium-based catalysts for direct hydrogenation or activating substituents on N in transfer hydrogenation.

Though efforts to reduce the environmental impacts of asymmetric imine hydrogenation and transfer hydrogenation have been made, more progress is desirable in this field. The development of reusable catalysts that perform well in less toxic solvents would be particularly useful. Success in this area has been achieved in the hydrogenation of 2-substituted quinolines.

Questions

6.1. Reductive amination is often used for the alkylation of amines:

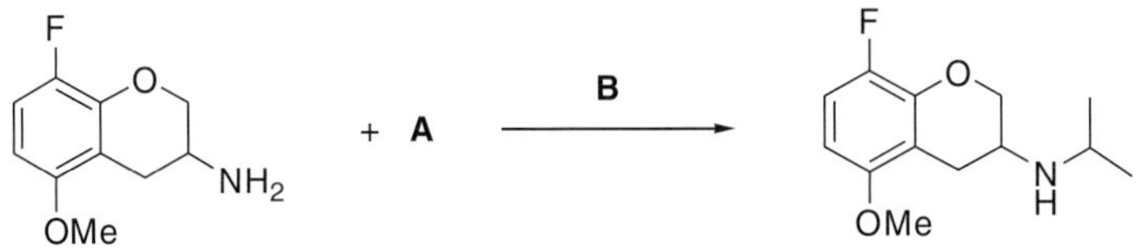

(a) Identify the reactant **A** required in the reaction above and suggest a suitable reducing agent **B**.

(b) Draw the structure of the intermediate that is formed during the reaction.

6.2. Compound **D** is an intermediate in the production of the anti-HIV drug indinavir and is synthesized as shown below. In addition to **D**, compound **E** is also formed in small amounts (ratio **D:E** = 97 : 3).

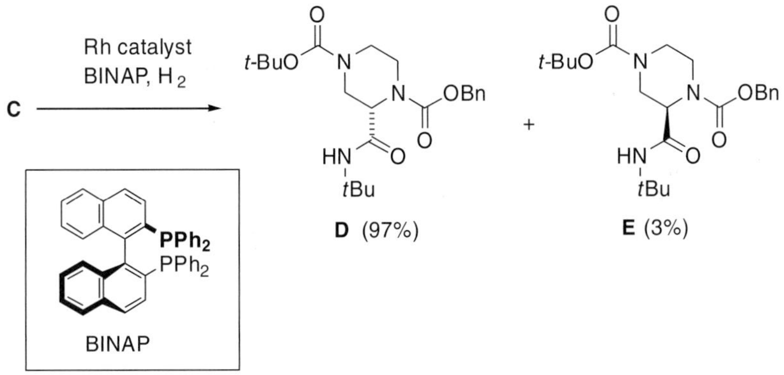

(a) Give the structure of the starting material **C**.

(b) What is the ee (enantiomeric excess) of the product in the reaction?

References

1. (a) Ager, D.J. (2007) *Handbook of Homogeneous Hydrogenation I* (eds J. de Vries and C.J. Elsevier), Wiley-VCH Verlag GmbH, Weinheim; (b) Knowles, W.S. (2002) *Angew. Chem., Int. Ed.*, **41**, 1998; (c) Noyori, R. (2002) *Angew. Chem., Int. Ed.*, **41**, 2008; (d) Brown, J.M. (1999) *Comprehensive Asymmetric Catalysis I* (eds E.N. Jacobsen, A. Pfaltz, and H. Yamamoto), Springer, Berlin; (e) Genet, J.P. (2008) *Modern Reduction Methods* (eds P.G. Andersson and I.J. Munslow), Wiley-VCH Verlag GmbH, Weinheim.
2. Blaser, H.-U. and Spindler, F. (1999) *Comprehensive Asymmetric Catalysis I* (eds E.N. Jacobsen, A. Pfaltz, and H. Yamamoto), Springer, Berlin.
3. Spindler, F. and Blaser, H.-U. (2007) *Handbook of Homogeneous Hydrogenation III* (eds J. de Vries and C.J. Elsevier), Wiley-VCH Verlag GmbH, Weinheim.
4. Claver, C. and Fernandez, E. (2008) *Modern Reduction Methods* (eds P.G. Andersson and I.J. Munslow), Wiley-VCH Verlag GmbH, Weinheim.
5. Ohkuma, T., Kitamura, M., and Noyori, R. (2000) *Catalytic Asymmetric Synthesis*, 2nd edn (ed. I. Ojima), Wiley-VCH Verlag GmbH, Weinheim.
6. Chi, Y., Tang, W., and Zhang, X. (2005) *Modern Rhodium-Catalyzed Organic Reactions* (ed. P.A. Evans), Wiley-VCH Verlag GmbH, Weinheim.
7. Kitamura, M. and Noyori, R. (2004) *Ruthenium in Organic Synthesis* (ed. S-.I. Murahashi), Wiley-VCH Verlag GmbH, Weinheim.
8. Blaser, H.-U., Spindler, F., and Thommen, M. (2007) *Handbook of Homogeneous Hydrogenation I* (eds J. de Vries and C.J. Elsevier), Wiley-VCH Verlag GmbH, Weinheim.
9. Brunel, J.M. (2003) *Rec. Res. Dev. Org. Chem.*, **7**, 155.
10. James, B.R. (1997) *Catal. Today*, **37**, 209.
11. Johansson, A. (1995) *Contemp. Org. Synth.*, **2**, 393.
12. Tararov, V.I. and Börner, A. (2005) *Synlett*, 203.
13. Vogl, E.M., Gröger, H., and Shibasaki, M. (1999) *Angew. Chem., Int. Ed.*, **38**, 1570.
14. Fagnou, K. and Lautens, M. (2002) *Angew. Chem., Int. Ed.*, **41**, 26.
15. Blaser, H.-U., Malan, C., Pugin, B., Spindler, F., Steiner, H., and Studer, M. (2003) *Adv. Synth. Catal.*, **345**, 103.
16. Lennon, I.C. and Moran, P.H. (2004) *Chim. Oggi, Supplement: Chiral Catalysis*, 34.
17. Blaser, H.-U., Pugin, B., and Spindler, F. (2005) *J. Mol. Catal. A Chem.*, **231**, 1.
18. Farina, V., Reeves, J.T., Senanayake, C.H., and Song, J.J. (2006) *Chem. Rev.*, **106**, 2734.
19. Lennon, I.C. and Moran, P.H. (2003) *Curr. Opin. Drug Discov. Dev.*, **6**, 855.
20. Bolm, C. (2008) *NATO Sci. Ser., II*, **246**, 85.
21. Wu, J. and Chan, A.S.C. (2006) *Acc. Chem. Res.*, **39**, 711.
22. Gómez Arrayás, R., Adrio, J., and Carretero, J.C. (2006) *Angew. Chem., Int. Ed.*, **45**, 7674.
23. Chelucci, G., Orru, G., and Pinna, G.A. (2003) *Tetrahedron*, **59**, 9471.
24. Diéguez, M., Pàmies, O., Ruiz, A., Díaz, Y., Castillón, S., and Claver, C. (2004) *Coord. Chem. Rev.*, **248**, 2165.
25. Togni, A. (1996) *Angew. Chem., Int. Ed. Engl.*, **35**, 1475.
26. Diéguez, M., Pàmies, O., and Claver, C. (2004) *Chem. Rev.*, **104**, 3189.
27. Xie, J.-H. and Zhou, Q.-L. (2008) *Acc. Chem. Res.*, **41**, 581.
28. Blaser, H.-U., Buser, H.-P., Coers, K., Hanreich, R., Jalett, H.-P., Jelsch, E., Pugin, B., Schneider, H.-D., Spindler, F., and Wegmann, A. (1999) *Chimia*, **53**, 275.
29. Blaser, H.U. and Spindler, F. (1997) *Chimia*, **51**, 297.
30. Blaser, H.-U. (2002) *Adv. Synth. Catal.*, **344**, 17.
31. González-Arellano, C., Corma, A., Iglesias, M., and Sánchez, F. (2005) *Chem. Commun.*, 3451.
32. Levi, A., Modena, G., and Scorrano, G. (1975) *J. Chem. Soc., Chem. Commun.*, 6.
33. Vastag, S., Heil, B., Törös, S., and Markó, L. (1977) *Transition Met. Chem.*, **2**, 58.
34. Bakos, J., Tóth, I., Heil, B., and Markó, L. (1985) *J. Organomet. Chem.*, **279**, 23.
35. Bakos, J., Orosz, A., Heil, B., Laghmari, M., Lhoste, P., and Sinou, D. (1991) *J. Chem. Soc., Chem. Commun.*, 1684.

36 Lensink, C. and De Vries, J.G. (1992) *Tetrahedron: Asymmetry*, **3**, 235.
37 Li, C., Wang, C., Villa-Marcos, B., and Xiao, J. (2008) *J. Am. Chem. Soc.*, **130**, 14450.
38 Amrani, Y., Lecomte, L., Sinou, D., Bakos, J., Toth, I., and Heil, B. (1989) *Organometallics*, **8**, 542.
39 Bakos, J., Tóth, I., Heil, B., Szalontai, G., Párkányi, L., and Fülöp, V. (1989) *J. Organomet. Chem.*, **370**, 263.
40 Tararov, V.I., Kadyrov, R., Riermeier, T.H., Holz, J., and Börner, A. (1999) *Tetrahedron: Asymmetry*, **10**, 4009.
41 Vastag, S., Bakos, J., Törös, S., Takach, N.E., King, R.B. Heil, B., and Markó, L. (1984) *J. Mol. Catal.*, **22**, 283.
42 Buriak, J.M. and Osborn, J.A. (1996) *Organometallics*, **15**, 3161.
43 Kang, G.J., Cullen, W.R., Fryzuk, M.D., James, B.R., and Kutney, J.P. (1988) *J. Chem. Soc., Chem. Commun.*, 1466.
44 Cullen, W.R., Fryzuk, M.D., James, B.R., Kutney, J.P., Kang, G.J., Herb, G., Thorburn, I.S., and Spogliarich, R. (1990) *J. Mol. Catal.*, **62**, 243.
45 Fogg, D.E., James, B.R., and Kilner, M. (1994) *Inorg. Chim. Acta*, **222**, 85.
46 Ohkuma, T., Ooka, H., Hashiguchi, S., Ikariya, T., and Noyori, R. (1995) *J. Am. Chem. Soc.*, **117**, 2675.
47 Cobley, C.J. and Henschke, J.P. (2003) *Adv. Synth. Catal.*, **345**, 195.
48 Jackson, M. and Lennon, I.C. (2007) *Tetrahedron Lett.*, **48**, 1831.
49 Abdur-Rashid, K., Lough, A.J., and Morris, R.H. (2001) *Organometallics*, **20**, 1047.
50 Cheruku, P., Church, T.L., and Andersson, P.G. (2008) *Chem.- Asian J.*, **3**, 1390.
51 Clarke, M.L., Diáz-Valenzuela, M.B., and Slawin, A.M.Z. (2007) *Organometallics*, **26**, 16.
52 Willoughby, C.A. and Buchwald, S.L. (1992) *J. Am. Chem. Soc.*, **114**, 7562.
53 Willoughby, C.A. and Buchwald, S.L. (1994) *J. Am. Chem. Soc.*, **116**, 8952.
54 Willoughby, C.A. and Buchwald, S.L. (1994) *J. Am. Chem. Soc.*, **116**, 11703.
55 Willoughby, C.A. and Buchwald, S.L. (1993) *J. Org. Chem.*, **58**, 7627.
56 Ringwald, M., Stürmer, R., and Brintzinger, H.H. (1999) *J. Am. Chem. Soc.*, **121**, 1524.
57 Okuda, J., Verch, S., Sturmer, R., and Spaniol, T.P. (2000) *J. Organomet. Chem.*, **605**, 55.
58 Schnider, P., Koch, G., Pretôt, R., Wang, G., Bohnen, F.M., Kruger, C., and Pfaltz, A. (1997) *Chem.-Eur. J.*, **3**, 887.
59 (a) Ng Cheong Chan, Y. and Osborn, J.A. (1990) *J. Am. Chem. Soc.*, **112**, 9400; (b) Spindler, F., Pugin, B., and Blaser, H.-U. (1990) *Angew. Chem., Int. Ed. Engl.*, **29**, 558.
60 Sablong, R. and Osborn, J.A. (1996) *Tetrahedron: Asymmetry*, **7**, 3059.
61 Sablong, R. and Osborn, J.A. (1996) *Tetrahedron Lett.*, **37**, 4937.
62 Morimoto, T., Nakajima, N., and Achiwa, K. (1994) *Chem. Pharm. Bull.*, **42**, 1951.
63 Morimoto, T., Nakajima, N., and Achiwa, K. (1995) *Synlett*, 748.
64 Zhu, G. and Zhang, X. (1998) *Tetrahedron: Asymmetry*, **9**, 2415.
65 Tani, K., Onouchi, J.-i., Yamagata, T., and Kataoka, Y. (1995) *Chem. Lett.*, 955.
66 Morimoto, T. and Achiwa, K. (1995) *Tetrahedron: Asymmetry*, **6**, 2661.
67 Blaser, H.U., Buser, H.P., Häusel, R., Jalett, H.P., and Spindler, F. (2001) *J. Organomet. Chem.*, **621**, 34.
68 Margalef-Català, R., Claver, C., Salagre, P., and Fernández, E. (2000) *Tetrahedron: Asymmetry*, **11**, 1469.
69 Dervisi, A., Carcedo, C., and Ooi, L.-l. (2006) *Adv. Synth. Catal.*, **348**, 175.
70 Yamagata, T., Tadaoka, H., Nagata, M., Hirao, T., Kataoka, Y., Ratovelomanana-Vidal, V., Genet, J.P., and Mashima, K. (2006) *Organometallics*, **25**, 2505.
71 Imamoto, T., Iwadate, N., and Yoshida, K. (2006) *Org. Lett.*, **8**, 2289.
72 Reetz, M.T., Beuttenmuller, E.W., Goddard, R., and Pasto, M. (1999) *Tetrahedron Lett.*, **40**, 4977.
73 Xiao, D. and Zhang, X. (2001) *Angew. Chem., Int. Ed.*, **40**, 3425.
74 Guiu, E., Muñoz, B., Castillón, S., and Claver, C. (2003) *Adv. Synth. Catal.*, **345**, 169.
75 Guiu, E., Aghmiz, M., Díaz, Y., Claver, C., Meseguer, B., Militzer, C., and Castillón, S. (2006) *Eur. J. Org. Chem.*, 627.

76 Vargas, S., Rubio, M., Suárez, A., and Pizzano, A. (2005) *Tetrahedron Lett.*, **46**, 2049.
77 Vargas, S., Rubio, M., Suárez, A., del Rio, D., Álvarez, E., and Pizzano, A. (2006) *Organometallics*, **25**, 961.
78 Reetz, M.T. and Bondarev, O. (2007) *Angew. Chem., Int. Ed.*, **46**, 4523.
79 Faller, J.W., Milheiro, S.C., and Parr, J. (2006) *J. Organomet. Chem.*, **691**, 4945.
80 Murai, T., Inaji, S., Morishita, K., Shibahara, F., Tokunaga, M., Obora, Y., and Tsuji, Y. (2006) *Chem. Lett.*, **35**, 1424.
81 Jiang, X.-b., Minnaard, A.J., Hessen, B., Feringa, B.L., Duchateau, A.L.L., Andrien, J.G.O., Boogers, J.A.F., and de Vries, J.G. (2003) *Org. Lett.*, **5**, 1503.
82 Blanc, C., Agbossou-Niedercorn, F., and Nowogrocki, G. (2004) *Tetrahedron: Asymmetry*, **15**, 2159.
83 Trifonova, A., Diesen, J.S., Chapman, C.J., and Andersson, P.G. (2004) *Org. Lett.*, **6**, 3825.
84 Trifonova, A., Diesen, J.S., and Andersson, P.G. (2006) *Chem.-Eur. J.*, **12**, 2318.
85 Ezhova, M.B., Patrick, B.O., James, B.R., Waller, F.J., and Ford, M.E. (2004) *J. Mol. Catal. A Chem.*, **224**, 71.
86 Cheemala, M.N. and Knochel, P. (2007) *Org. Lett.*, **9**, 3089.
87 Solinas, M., Pfaltz, A., Cozzi, P.G., and Leitner, W. (2004) *J. Am. Chem. Soc.*, **126**, 16142.
88 Kainz, S., Brinkmann, A., Leitner, W., and Pfaltz, A. (1999) *J. Am. Chem. Soc.*, **121**, 6421.
89 Zhu, S.-F., Xie, J.-B., Zhang, Y.-Z., Li, S., and Zhou, Q.-L. (2006) *J. Am. Chem. Soc.*, **128**, 12886.
90 Moessner, C. and Bolm, C. (2005) *Angew. Chem., Int. Ed.*, **44**, 7564.
91 Guiu, E., Claver, C., Benet-Buchholz, J., and Castillón, S. (2004) *Tetrahedron: Asymmetr.y*, **15**, 3365.
92 Rethore, C., Riobe, F., Fourmigue, M., Avarvari, N., Suisse, I., and Agbossou-Niedercorn, F. (2007) *Tetrahedron: Asymmetry*, **18**, 1877.
93 Shirai, S.-y., Nara, H., Kayaki, Y., and Ikariya, T. (2009) *Organometallics*, **28**, 802.
94 Maire, P., Deblon, S., Breher, F., Geier, J., Böhler, C., Rüegger, H., Schönberg, H., and Grützmacher, H. (2004) *Chem.-Eur. J.*, **10**, 4198.
95 Constable, D.J.C., Dunn, P.J., Hayler, J.D., Humphrey, G.R., Leazer, J.L., Jr., Linderman, R.J., Lorenz, K., Manley, J., Pearlman, B.A., Wells, A., Zaks, A., and Zhang, T.Y. (2007) *Green Chem.*, **9**, 411.
96 Lensink, C., Rijnberg, E., and de Vries, J.G. (1997) *J. Mol. Catal. A Chem.*, **116**, 199.
97 Pinault, N. and Bruce, D.W. (2003) *Coord. Chem. Rev.*, **241**, 1.
98 Lightfoot, A., Schnider, P., and Pfaltz, A. (1998) *Angew. Chem., Int. Ed.*, **37**, 2897.
99 Trindade, A.F., Gois, P.M.P., and Afonso, C.A.M. (2009) *Chem. Rev.*, **109**, 418.
100 Fraile, J.M., García, J.I., Herrerías, C.I., Mayoral, J.A., and Pires, E. (2009) *Chem. Soc. Rev.*, **38**, 695.
101 Barbaro, P. (2006) *Chem.-Eur. J.*, **12**, 5666.
102 Li, C. (2004) *Catal. Rev.*, **46**, 419.
103 Fan, Q.-H., Li, Y.-M., and Chan, A.S.C. (2002) *Chem. Rev.*, **102**, 3385.
104 Pugin, B., Landert, H., Spindler, F., and Blaser, H.-U. (2002) *Adv. Synth. Catal.*, **344**, 974.
105 Claver, C., Fernández, E., Margalef-Català, R., Medina, F., Salagre, P., and Sueiras, J.E. (2001) *J. Catal.*, **201**, 70.
106 González-Arellano, C., Corma, A., Iglesias, M., and Sánchez, F. (2008) *Eur. J. Inorg. Chem.*, 1107.
107 Ayala, V., Corma, A., Iglesias, M., Rincón, J.A., and Sánchez, F. (2004) *J. Catal.*, **224**, 170.
108 For a comparative index of aromaticity in heterocycles, see Bird, C.W. (1992) *Tetrahedron*, **48**, 335. A recent review of aromaticity in heterocycles can be found in: Balaban, A.T., Oniciu, D.C., and Katritzky, A.R., (2004) *Chem. Rev.*, **104**, 2777.
109 Zhou, Y.-G. (2007) *Acc. Chem. Res.*, **40**, 1357.
110 Kuwano, R. (2008) *Heterocycles*, **76**, 909.
111 Glorius, F. (2005) *Org. Biomol. Chem.*, **3**, 4171.
112 Church, T.L. and Andersson, P.G. (2008) *Coord. Chem. Rev.*, **252**, 513.
113 The asymmetric hydrogenations of N-acetyl, N-carbamate, and N-sufonyl indoles are asymmetric hydrogenations of enamides, not imines, and will therefore not be covered here. For reports of this reaction, see: (a) Kuwano, R. and Kashiwabara, M., (2006) *Org. Lett.*, **8**,

2653; (b) Kuwano, R., Kashiwabara, M., Sato, K., Ito, T., Kaneda, K., and Ito, Y. (2006) *Tetrahedron: Asymmetry*, **17**, 521.

114 Compare, for example, the aromaticity index for pyridine versus for the pyridine ring of quinoline. See Matito, E., Duran, M., and Solà, M. (2005) *J. Chem. Phys.*, **122**, 014109.

115 Katritzky, A.R., Rachwal, S., and Rachwal, B. (1996) *Tetrahedron*, **52**, 15031.

116 Wang, W.-B., Lu, S.-M., Yang, P.-Y., Han, X.-W., and Zhou, Y.-G. (2003) *J. Am. Chem. Soc.*, **125**, 10536.

117 Lu, S.-M., Han, X.-W., and Zhou, Y.-G. (2004) *Adv. Synth. Catal.*, **346**, 909.

118 Xu, L., Lam, K.H., Ji, J., Wu, J., Fan, Q.-H., Lo, W.-H., and Chan, A.S.C. (2005) *Chem. Commun.*, 1390.

119 Lam, K.H., Xu, L., Feng, L., Fan, Q.-H., Lam, F.L., Lo, W.-h., and Chan, A.S.C. (2005) *Adv. Synth. Catal.*, **347**, 1755.

120 Qiu, L., Kwong, F.Y., Wu, J., Lam, W.H., Chan, S., Yu, W.-Y., Li, Y.-M., Guo, R., Zhou, Z., and Chan, A.S.C. (2006) *J. Am. Chem. Soc.*, **128**, 5955.

121 Tang, W.-J., Zhu, S.-F., Xu, L.-J., Zhou, Q.-L., Fan, Q.-H., Zhou, H.-F., Lam, K., and Chan, A.S.C. (2007) *Chem. Commun.*, 613.

122 Reetz, M.T. and Li, X. (2006) *Chem. Commun.*, 2159.

123 Wang, Z.-J., Deng, G.-J., Li, Y., He, Y.-M., Tang, W.-J., and Fan, Q.-H. (2007) *Org. Lett.*, **9**, 1243.

124 Wang, X.-B. and Zhou, Y.-G. (2008) *J. Org. Chem.*, **73**, 5640.

125 Mršić, N., Lefort, L., Boogers, J.A.F., Minnaard, A.J., Feringa, B.L., and de Vries, J.G. (2008) *Adv. Synth. Catal.*, **350**, 1081.

126 Lu, S.-M. and Bolm, C. (2008) *Adv. Synth. Catal.*, **350**, 1101.

127 Li, Z.-W., Wang, T.-L., He, Y.-M., Wang, Z.-J., Fan, Q.-H., Pan, J., and Xu, L.-J. (2008) *Org. Lett.*, **10**, 5265.

128 Deport, C., Buchotte, M., Abecassis, K., Tadaoka, H., Ayad, T., Ohshima, T., Genet, J.-P., Mashima, K., and Ratovelomanana-Vidal, V. (2007) *Synlett*, 2743.

129 Jahjah, M., Alame, M., Pellet-Rostaing, S., and Lemaire, M. (2007) *Tetrahedron: Asymmetry*, **18**, 2305.

130 Genet and coworkers hydrogenated 2-phenylquinoline to (S)-2-phenyl-1,2,3,4-tetrahydroquinoline using phosphine-supported, iodo-bridged diiridium complexes. These catalysts gave moderate ee values for this reaction, but were more useful for the enantioselective hydrogenation of nonaromatic cyclic imines. See Yamagata, T. Tadaoka, H. Nagata, M. Hirao, T. Kataoka, Y. Ratovelomanana-Vidal, V. Genet, J.P., and Mashima, K. (2006) *Organometallics*, **25**, 2505.

131 Zhou, H., Li, Z., Wang, Z., Wang, T., Xu, L., He, Y., Fan, Q.-H., Pan, J., Gu, L., and Chan, A.S.C. (2008) *Angew. Chem., Int. Ed.*, **47**, 8464.

132 Yang, P.-Y. and Zhou, Y.-G. (2004) *Tetrahedron: Asymmetry*, **15**, 1145.

133 Jacquemond-Collet, I., Hannedouche, S., Fabre, N., Fourasté, I., and Moulis, C. (1999) *Phytochemistry*, **51**, 1167.

134 Lu, S.-M., Wang, Y.-Q., Han, X.-W., and Zhou, Y.-G. (2006) *Angew. Chem., Int. Ed.*, **45**, 2260.

135 Noyori, R., Ohta, M., Hsiao, Y., Kitamura, M., Ohta, T., and Takaya, H. (1986) *J. Am. Chem. Soc.*, **108**, 7117.

136 Murata, S., Sugimoto, T., and Matsuura, S. (1987) *Heterocycles*, **26**, 763.

137 Bianchini, C., Barbaro, P., Scapacci, G., Farnetti, E., and Graziani, M. (1998) *Organometallics*, **17**, 3308.

138 Henschke, J.P., Burk, M.J., Malan, C.G., Herzberg, D., Peterson, J.A., Wildsmith, A.J., Cobley, C.J., and Casy, G. (2003) *Adv. Synth. Catal.*, **345**, 300.

139 Bianchini, C., Barbaro, P., and Scapacci, G. (2001) *J. Organomet. Chem.*, **621**, 26.

140 Brunner, H. and Rosenboem, S. (2000) *Monatsh. Chem.*, **131**, 1371.

141 Buffat, M.G.P. (2004) *Tetrahedron*, **60**, 1701.

142 Studer, M., Wedemeyer-Exl, C., Spindler, F., and Blaser, H.-U. (2000) *Monatsh. Chem.*, **131**, 1335.

143 Raynor, S.A., Thomas, J.M., Raja, R., Johnson, B.F.G., Bell, R.G., and Mantle, M.D. (2000) *Chem. Commun.*, 1925.

144 Blaser, H.U., Hönig, H., Studer, M., and Wedemeyer-Exl, C. (1999) *J. Mol. Catal. A Chem.*, **139**, 253.

145 Lei, A., Chen, M., He, M., and Zhang, X. (2006) *Eur. J. Org. Chem.*, 4343.

146 Glorius, F., Spielkamp, N., Holle, S., Goddard, R., and Lehmann, C.W. (2004) *Angew. Chem., Int. Ed.*, **43**, 2850.

147 Legault, C.Y. and Charette, A.B. (2005) *J. Am. Chem. Soc.*, **127**, 8966.
148 Wang, X.-B., Zeng, W., and Zhou, Y.-G. (2008) *Tetrahedron Lett.*, **49**, 4922.
149 Kobayashi, K., Okamoto, T., Oida, T., and Tanimoto, S. (1986) *Chem. Lett.*, 2031.
150 Oppolzer, W., Wills, M., Starkemann, C., and Bernardinelli, G. (1990) *Tetrahedron Lett.*, **31**, 4117.
151 Charette, A.B. and Giroux, A. (1996) *Tetrahedron Lett.*, **37**, 6669.
152 (a) Cobley, C.J., Foucher, E., Lecouve, J.-P., Lennon, I.C., Ramsden, J.A., and Thominot, G. (2003) *Tetrahedron: Asymmetry*, **14**, 3431; (b) Desos, P., Serkiz, B., Morain, P., Lepagnol, J., and Cordi, A. (1996) *Bioorg. Med. Chem. Lett.*, **6**, 3003.
153 Yang, Q., Shang, G., Gao, W., Deng, J., and Zhang, X. (2006) *Angew. Chem., Int. Ed.*, **45**, 3832.
154 Burk, M.J. and Feaster, J.E. (1992) *J. Am. Chem. Soc.*, **114**, 6266.
155 Abe, H., Amii, H., and Uneyama, K. (2001) *Org. Lett.*, **3**, 313.
156 Suzuki, A., Mae, M., Amii, H., and Uneyama, K. (2004) *J. Org. Chem.*, **69**, 5132.
157 Zhang and coworkers later examined this substrate class with rhodium catalysts. See Shang, G., Yang, Q., and Zhang, X. (2006) *Angew. Chem., Int. Ed.*, **45**, 6360.
158 Wang, Y.-Q., Yu, C.-B., Wang, D.-W., Wang, X.-B., and Zhou, Y.-G. (2008) *Org. Lett.*, **10**, 2071.
159 Wang, Y.-Q., Lu, S.-M., and Zhou, Y.-G. (2007) *J. Org. Chem.*, **72**, 3729.
160 Wang, Y.-Q. and Zhou, Y.-G. (2006) *Synlett*, 1189.
161 Blacker, A.J. (ed.) (2007) *Handbook of Homogeneous Hydrogenation III* (eds J. de Vries and C.J. Elsevier), Wiley-VCH Verlag GmbH, Weinheim.
162 Wills, M. (2008) *Modern Reduction Methods* (eds P.G. Andersson and I.J. Munslow), Wiley-VCH Verlag GmbH, Weinheim.
163 Roszkowski, P. and Czarnocki, Z. (2007) *Mini Rev. Org. Chem.*, **4**, 190.
164 Palmer, M.J. and Wills, M. (1999) *Tetrahedron: Asymmetry*, **10**, 2045.
165 Noyori, R. and Hashiguchi, S. (1997) *Acc. Chem. Res.*, **30**, 97.
166 Gladiali, S. and Alberico, E. (2006) *Chem. Soc. Rev.*, **35**, 226.
167 Noyori, R., Yamakawa, M., and Hashiguchi, S. (2001) *J. Org. Chem.*, **66**, 7931.
168 Noyori, R. (2002) *Angew. Chem., Int. Ed.*, **41**, 2008.
169 Uematsu, N., Fujii, A., Hashiguchi, S., Ikariya, T., and Noyori, R. (1996) *J. Am. Chem. Soc.*, **118**, 4916.
170 Fujii, A., Hashiguchi, S., Uematsu, N., Ikariya, T., and Noyori, R. (1996) *J. Am. Chem. Soc.*, **118**, 2521.
171 Ligands with one R_2NH- terminus and one $ArSO_2NR-$ terminus have only recently been used in the asymmetric transfer hydrogenation of imines. See Martins, J.E.D., Clarkson, G.J., Wills, M. (2009) *Org. Lett*, **11**, 847.
172 Mao, J. and Baker, D.C. (1999) *Org. Lett.*, **1**, 841.
173 Blackmond, D.G., Ropic, M., and Stefinovic, M. (2006) *Org. Process Res. Dev.*, **10**, 457.
174 Martin, J. and Campbell, L.A. (2001) WO 0112574 (A1), 44 pp., Avecia Limited, UK.
175 Wipf, P. and Stephenson, C.R.J. (2003) *Org. Lett.*, **5**, 2449.
176 Jennings, W.B. and Lovely, C.J. (1991) *Tetrahedron*, **47**, 5561.
177 Wu, J., Wang, F., Ma, Y., Cui, X., Cun, L., Zhu, J., Deng, J., and Yu, B. (2006) *Chem. Commun.*, 1766.
178 Ma, Y., Liu, H., Chen, L., Cui, X., Zhu, J., and Deng, J. (2003) *Org. Lett.*, **5**, 2103.
179 Li, L., Wu, J., Wang, F., Liao, J., Zhang, H., Lian, C., Zhu, J., and Deng, J. (2007) *Green Chem.*, **9**, 23.
180 Canivet, J. and Süss-Fink, G. (2007) *Green Chem.*, **9**, 391.
181 Chen, Y.-C., Wu, T.-F., Deng, J.-G., Liu, H., Cui, X., Zhu, J., Jiang, Y.-Z., Choi, M.C.K., and Chan, A.S.C. (2002) *J. Org. Chem.*, **67**, 5301.
182 Chen, Y.-C., Wu, T.-F., Jiang, L., Deng, J.-G., Liu, H., Zhu, J., and Jiang, Y.-Z. (2005) *J. Org. Chem.*, **70**, 1006.
183 Haraguchi, N., Tsuru, K., Arakawa, Y., and Itsuno, S. (2009) *Org. Biomol. Chem.*, **7**, 69.
184 Liu, P.-N., Gu, P.-M., Deng, J.-G., Tu, Y.-Q., and Ma, Y.-P. (2005) *Eur. J. Org. Chem.*, 3221.

185 Huang, X. and Ying, J.Y. (2007) *Chem. Commun.*, 1825.
186 Li, J., Zhang, Y., Han, D., Gao, Q., and Li, C. (2009) *J. Mol. Catal. A Chem.*, **298**, 31.
187 Liu, P.N., Gu, P.M., Wang, F., and Tu, Y.Q. (2004) *Org. Lett.*, **6**, 169.
188 Vedejs, E., Trapencieris, P., and Suna, E. (1999) *J. Org. Chem.*, **64**, 6724.
189 Meuzelaar, G.J., Van Vliet, M.C.A., Maat, L., and Sheldon, R.A. (1999) *Eur. J. Org. Chem.*, 2315.
190 Kaldor, I., Feldman, P.L., Mook, R.A., Jr., Ray, J.A., Samano, V., Sefler, A.M., Thompson, J.B., Travis, B.R., and Boros, E.E. (2001) *J. Org. Chem.*, **66**, 3495.
191 Samano, V., Ray, J.A., Thompson, J.B., Mook, R.A., Jr., Jung, D.K., Koble, C.S., Martin, M.T., Bigham, E.C., Regitz, C.S., Feldman, P.L., and Boros, E.E. (1999) *Org. Lett.*, **1**, 1993.
192 Tietze, L.F., Zhou, Y., and Topken, E. (2000) *Eur. J. Org. Chem.*, 2247.
193 Tietze, L.F., Rackelmann, N., and Müller, I. (2004) *Chem. Eur. J.*, **10**, 2722.
194 Roszkowski, P., Wojtasiewicz, K., Leniewski, A., Maurin, J.K., Lis, T., and Czarnocki, Z. (2005) *J. Mol. Catal. A Chem.*, **232**, 143.
195 Szawkaøo, J. and Czarnocki, Z. (2005) *Monatsh. Chem.*, **136**, 1619.
196 Roszkowski, P., Maurin, J.K., and Czarnocki, Z. (2006) *Tetrahedron: Asymmetry*, **17**, 1415.
197 Ahn, K.H., Ham, C., Kim, S.-K., and Cho, C.-W. (1997) *J. Org. Chem.*, **62**, 7047.
198 Ahn, K.H., Kim, S.-K., and Ham, C. (1998) *Tetrahedron Lett.*, **39**, 6321.
199 Magee, M.P. and Norton, J.R. (2001) *J. Am. Chem. Soc.*, **123**, 1778.
200 Guan, H., Iimura, M., Magee, M.P., Norton, J.R., and Janak, K.E. (2003) *Organometallics*, **22**, 4084.
201 Guan, H., Iimura, M., Magee, M.P., Norton, J.R., and Zhu, G. (2005) *J. Am. Chem. Soc.*, **127**, 7805.
202 Szawkalo, J., Czarnocki, S.J., Zawadzka, A., Wojtasiewicz, K., Leniewski, A., Maurin, J.K., Czarnocki, Z., and Drabowicz, J. (2007) *Tetrahedron: Asymmetry*, **18**, 406.
203 Szawkalo, J., Zawadzka, A., Wojtasiewicz, K., Leniewski, A., Drabowicz, J., and Czarnocki, Z. (2005) *Tetrahedron: Asymmetry*, **16**, 3619.
204 Zhang, Q., Tu, G., Zhao, Y., and Cheng, T. (2002) *Tetrahedron*, **58**, 6795.
205 Lee, N.E. and Buchwald, S.L. (1994) *J. Am. Chem. Soc.*, **116**, 5985.
206 Kubryk, M. and Hansen, K.B. (2006) *Tetrahedron: Asymmetry*, **17**, 205.
207 Dai, Q., Yang, W., and Zhang, X. (2005) *Org. Lett.*, **7**, 5343.
208 Hsiao, Y., Rivera, N.R., Rosner, T., Krska, S.W., Njolito, E., Wang, F., Sun, Y., Armstrong, J.D. III, Grabowski, E.J.J., Tillyer, R.D., Spindler, F., and Malan, C. (2004) *J. Am. Chem. Soc.*, **126**, 9918.
209 Zhong, Y.-L., Krska, S.W., Zhou, H., Reamer, R.A., Lee, J., Sun, Y., and Askin, D. (2009) *Org. Lett.*, **11**, 369.
210 Tararov, V.I., Kadyrov, R., Riermeier, T.H., Holz, J., and Börner, A. (2000) *Tetrahedron Lett.*, **41**, 2351.
211 Hou, G.-H., Xie, J.-H., Wang, L.-X., and Zhou, Q.-L. (2006) *J. Am. Chem. Soc.*, **128**, 11774.
212 Kalck, P., Urrutigoiety, M., Bachelier, A., Preti, M., and Riviere, P. (2008) WO2008096075 (A2), 29 pp., Holis Technologies Fr.
213 Cheruku, P., Church, T.L., Trifonova, A., Wartmann, T., and Andersson, P.G. (2008) *Tetrahedron Lett.*, **49**, 7290.
214 Hou, G.-H., Xie, J.-H., Yan, P.-C., and Zhou, Q.-L. (2009) *J. Am. Chem. Soc.*, **131**, 1366.
215 Baeza, A. and Pfaltz, A. (2009) *Chem. Eur. J.*, **15**, 2266.

7
Asymmetric Reductive Amination
Thomas C. Nugent

7.1
Introduction

Despite its long history and common usage of the term, reductive amination has remained underdeveloped. Many studies are available regarding racemic reductive amination, but only Börner and Tararov have provided summaries of the asymmetric version [1]. In this chapter we provide an overview of the available methods and achievements since the first demonstrated enantioselective reductive amination by Blaser in 1999 [2].

Compared to the large volume of reports detailing asymmetric imine reduction, only a small number of publications are available regarding asymmetric reductive amination. Reductive amination remains less developed than the field of imine reduction due in great part to the incompatibility of transition metal hydride catalysts in the presence of ketones (alcohol by-product formation) and/or catalyst inhibition arising from amine starting material or amine product complexation with a catalytically active transition metal species [3]. Future method development will surely require striking the right balance between these competing nonproductive reaction pathways.

Before discussing the relevant literature, a few qualifiers are mentioned and discussed. Reductive amination is sometimes incorrectly associated with the reduction of preformed imines and derivatives thereof. Reductive amination, by definition, is only the one-pot conversion of a ketone to an amine in which a reductant coexists in the presence of a ketone starting material [4]. The term *indirect reductive amination* is occasionally used, but can be more tersely and accurately described as *imine reduction* [5]. The reader will note that depending on the perspective one takes for the transformation, the term *reductive amination* is applied when considering the reaction course of the ketone starting material, and the term *reductive alkylation* is applied when describing the starting amine's conversion to the product amine.

Included in this section are procedures that are not classified as reductive amination, but are also not imine reductions in the sense that an imine has been isolated, here isolation means a work-up or distillation has been performed before the

Chiral Amine Synthesis: Methods, Developments and Applications. Edited by Thomas C. Nugent
Copyright © 2010 WILEY-VCH Verlag GmbH & Co. KGaA, Weinheim
ISBN: 978-3-527-32509-2

reduction step. As a consequence, we include several one-pot procedures in which imine formation is performed *without the presence of the reductant*; once the ketone is fully consumed, no workup is performed, but instead the reductant is added to complete the reaction sequence. These two-stage, no workup procedures are by definition not considered reductive aminations [4], but importantly they do represent a significant improvement in reaction step efficiency versus imine formation, isolation, and reduction thereof, and are therefore included in this section.

The enantioselective reductive amination of ketoacid substrates has been demonstrated and provides amino acids that are beyond the scope of this review [6]. Enzymatic-based reductive amination is now possible and allows nonamino acid chiral amine synthesis, however, this field of study is also beyond the scope of this material [7]. Finally, much of the material discussed here also appeared in a recent review of ours on the general subject of chiral amine synthesis.

7.2
Transition Metal-Mediated Homogeneous Reductive Amination

The first reported example of enantioselective reductive amination was that of Blaser et al. at Solvias (Scheme 7.7) [2]. At the time, 1999, they were still tweaking the industrial process for metolachlor, the active ingredient of the herbicide Dual®, and examined its synthesis via the reductive amination of methoxyacetone with 2-methyl-5-ethyl-aniline (MEA, limiting reagent). Working at the 100 mmol scale, they showed that a very low loading of an Ir–xyliphos complex, under 80 bar (1160 psi) H_2, neat, 50 °C, and 14 h, were optimal. By doing so, a 76% ee with full conversion was achieved.

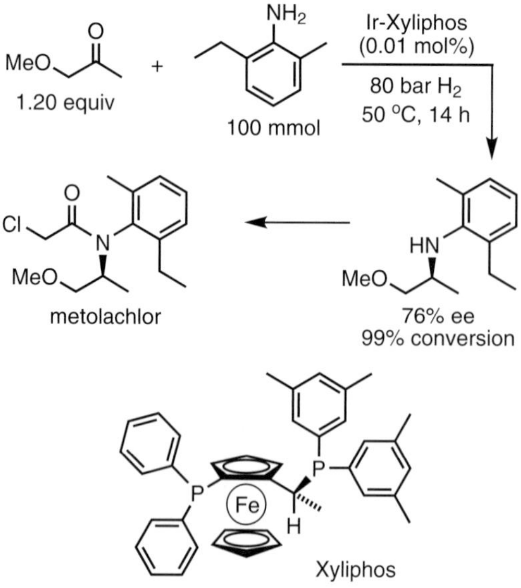

Scheme 7.1 The first enantioselective reductive amination: synthesis of metolachlor.

The reaction was not as accommodating as just described and additionally required a cosolvent (10 ml of cyclohexane per 100 mmol of MEA) and an acid, which screening identified as methanesulfonic acid (0.2 ml per 100 mmol of MEA). In addition, tetrabutylammonium iodide (20 mg) was required. The combination of these additives enhanced the rate of reaction presumably by aiding solubility in this two phase reaction.

Although this was an outstanding first example of enantioselective reductive amination that achieved the project's ee target, the final industrial process actually employed the corresponding preformed and isolated imine because of the advantage of reducing by one-hundredth the catalyst loading compared to the reductive amination process outlined here [8].

In 2003, Zhang investigated the reductive amination of aryl–alkyl ketones (limiting reagent) with p-anisidine (1.2 equiv) [9]. The optimal conditions were noted when using 1.0 mol% of an Ir–f–Binaphane complex, 69 bar (1000 psi) H_2, CH_2Cl_2, 25 °C, and 10 h (Scheme 7.2). Unique and advantageous to this reductive amination system was a >99% yield for every substrate studied, but the ee's varied. For example, acetophenone, phenyl–ethyl ketone, and phenyl–^{n}butyl ketone lead to progressively lower ee: 94%, 85%, and 79%, respectively. A trend was also established for methyl-substituted acetophenones: p-CH_3-Ph (96% ee), m-CH_3-Ph (89% ee), and o-CH_3-Ph (44% ee). In general, para-substitution on the aromatic ring of acetophenone lead to high ee: p-CH_3O-Ph (95% ee), p-F-Ph (93% ee), p-Cl-Ph (92% ee), and p-Br-Ph (94% ee).

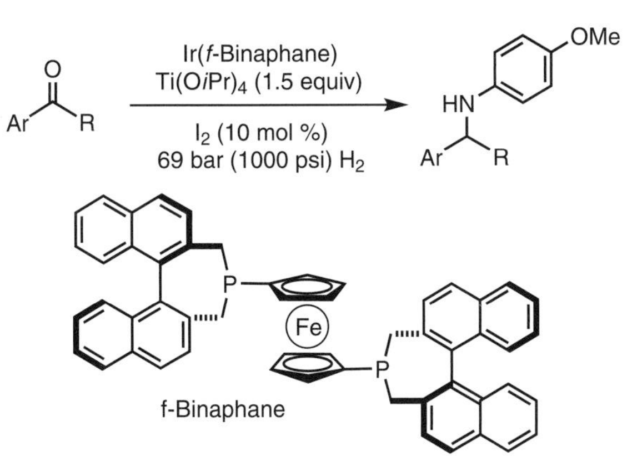

Scheme 7.2 First Ir-based reductive amination.

Importantly, Zhang noted that the reaction did not proceed without the addition of iodine [10]. Furthermore use of Ti(OiPr)$_4$ was not trivial, and replacement by MgSO$_4$, 4 Å MS, or TsOH was not helpful. They additionally noted that replacing the chiral f-Binaphane ligand with BINAP or BIPHEP resulted in poor ee. The source of nitrogen is important because of the desire to ultimately provide a chiral primary amine; screening studies with benzylamine, aniline, o-anisidine, m-anisidine, p-anisidine, and 2,6-dimethylaniline showed p-anisidine to be optimal.

The independent and combined efforts of the industrial group of Kadyrov (Degussa AG, now Evoniks) and the academic group of Börner have provided important advancements in the field of α-chiral amine synthesis [11]. Reporting on the enantioselective reductive amination of 1- and 2-acetylnaphthalene, phenyl ethyl ketone, and substituted acetophenone derivatives using Leuckart–Wallach transfer hydrogenation conditions, they showed that the corresponding primary amine could be isolated in high yield and ee (Scheme 7.3). For example, acetophenone provided a 92% yield with 95% ee, and aromatic-substituted derivatives provide similar results: m-CH_3-Ph (74% yield, 89% ee), p-CH_3-Ph (93% yield, 93% ee), p-CH_3O-Ph (83% yield, 95% ee), p-Cl-Ph (93% yield, 92% ee), p-Br-Ph (56% yield, 91% ee), and p-NO_2-Ph (92% yield, 95% ee), additional substrates are noted in Scheme 7.3. These results were possible when employing 0.5–1.0 mol% of a Ru–BINAP (and ligand derivatives of BINAP) complex, excess NH_3/HCO_2NH_4, MeOH, 85 °C, and ~20 h in a pressure vessel.

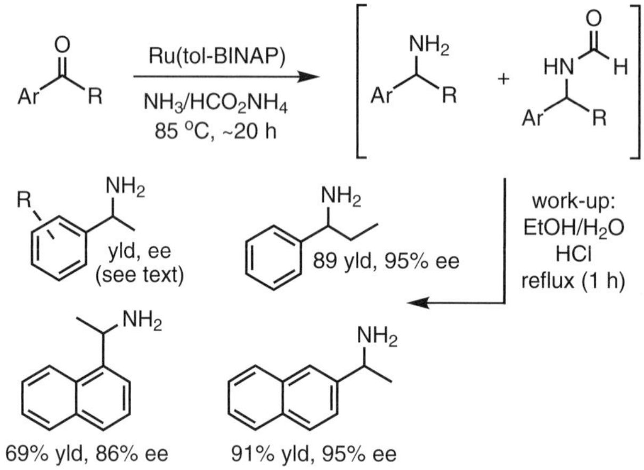

Scheme 7.3 Ru-based reductive amination providing primary amines.

The reaction actually produces a mixture of the primary amine and the corresponding formyl derivative, but the amine product is exclusively isolated after the crude product is treated with HCl (Scheme 7.3). As of this writing, the method provides low yields and ee for cyclic aromatic substrates, for example, 1-indanone (6% yield, no ee reported, chiral Ru catalyst) [11b]. Regarding aliphatic ketones, for example, 2-octanone (44% yield, 24% ee, chiral Ru catalyst [11a] or 37% yield with an achiral Rh catalyst [11b]). The same authors have recently made inroads concerning the use of aromatic ketones and molecular hydrogen [12], although the transfer hydrogenation method presented here appears to be superior as of now.

The development of these methods for alkanone-based substrates would represent another precedent-setting breakthrough by Kadyrov and Börner. The overall importance of these findings is the exceptional reaction step efficiency (ketone to primary amine in one pot); unfortunately, this feature of amine synthesis is not openly

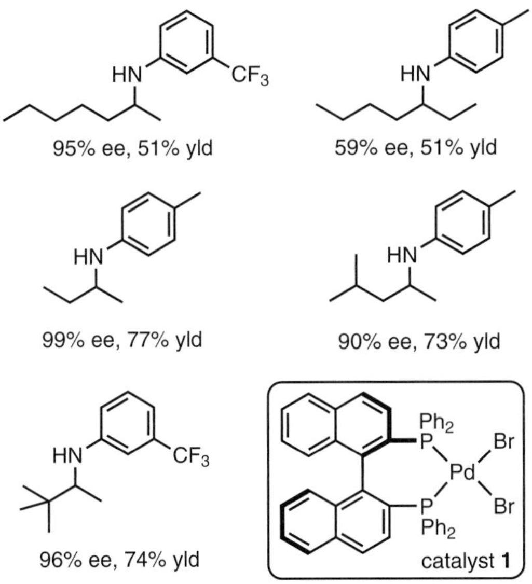

Figure 7.1 First Pd-based reductive amination.

(Structures shown: 95% ee, 51% yld; 59% ee, 51% yld; 99% ee, 77% yld; 90% ee, 73% yld; 96% ee, 74% yld; catalyst **1**)

discussed in the literature, but of critical importance regarding the advancement of chiral amine synthesis in general.

An elegant and refreshing approach was demonstrated in 2009 by Rubio-Pérez. Here, 13 different alkyl–alkyl and aryl–alkyl ketones (limiting reagent) were reductively aminated in the presence of 1.5 equiv of an aniline derivative, most notably *para*-methylaniline or *meta*-trifluoromethylaniline, employing 2.5 mol% of Pd catalyst **1**, 55 bar (800 psi) H_2, $CHCl_3$, 70 °C, and 24 h (Figure 7.1) [13]. It was additionally crucial to add 5 Å molecular sieves (150 mg per 1.0 mmol of ketone substrate). The system was very effective for aliphatic ketones, major highlights are displayed in Figure 7.1, which is important because of the severe lack of methods applicable to this class of ketone substrates. Aryl–alkyl ketones, acetophenone derivatives, and phenyl–ethyl ketone performed less convincingly with ee's ranging from 34–43% and yields ranging from 53–67%. In addition, 2,3-butanedione was selectively mono-aminated in very good yield (85%), but with poor enantioselectivity (20% ee).

In 2009, Xiao employed a half-sandwich Cp*Ir(III) complex with different chiral-monosulfonated DPEN ligands (Figure 7.2), earlier developed for imine reduction by the same group, in combination with a BINOL-based chiral phosphate counteranion (TRIP anion), to reductively aminate 24 acetophenone derivatives (1.2 equiv) with *p*-anisidine (limiting reagent) under the conditions of 1.0 mol% of the indicated Ir catalyst (Table 7.1), 5 bar (73 psi) H_2, toluene, 35 °C, and 15–24 h [14].

Additionally required were 4 Å MS (200 mg for 0.5 mmol of *p*-anisidine) and 5.0 mol% of the chiral phosphoric acid derivative (Figure 7.1, TRIP or XH), which were crucial for allowing complete conversion, suppression of alcohol by-product formation, and high ee. The enantioselectivity for 18 aryl–alkyl ketones are shown in

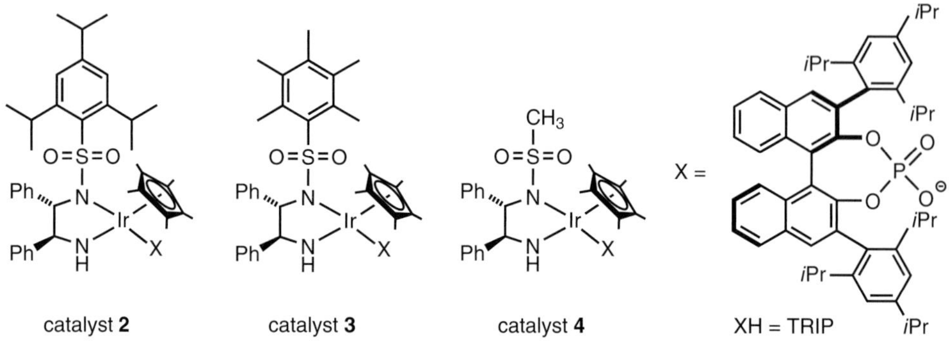

7 Asymmetric Reductive Amination

Figure 7.2 Ir catalysts with TRIP ligands for reductive amination.

Table 7.1 and in general it can be stated that excellent isolated yields were obtained (>90%), using Xiao's method. Regarding the "ortho"-substituted acetophenone examples (Table 7.1, last row), 1-acetylnaphthalene (not shown) was also an excellent substrate (90% yield, 87% ee). In addition to these successes with acetophenone derivatives, two aryl–ethyl ketones were examined and provided similar high ee values.

These results are outstanding when considering that a reductive amination strategy has been employed and are even more impressive because aliphatic ketones, which are historically avoided because of poor yield and ee, are also acceptable

Table 7.1 Ee values for Ir-based reductive amination of acetophenone derivatives with p-anisidine.

R substituent on Substrate		CH_3	iBu	CH_3O	Cl	Br	F	CF_3	NO_2	CN
Substrate	Catalyst									
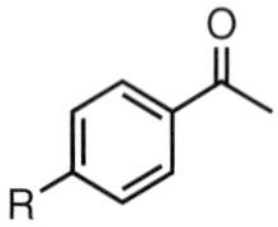 para	2[a]	97	95	95	95	94	95	91	88	86
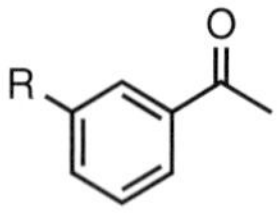 meta	2[a]	94	—	94	—	94	—	93	81	
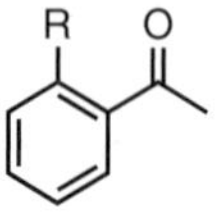 ortho	4[a]	91	—	86	83		96[b]			

a) See Figure 7.2.
b) See Figure 7.2, catalyst **3** was used.

substrates. For the eight alkyl–methyl ketones examined (not shown), the Figure 7.2 catalysts were sufficient to allow reductive amination, thus the addition of TRIP was not required (aryl–alkyl ketones require the addition of TRIP). In summary, the Xiao substrate examples are impressive, ranging from aryl-alkyl ketones that can even contain non-conjugated alkene moieties; ees are high (generally >90%) with good to excellent yield (79–91%). Researchers have examined similar reaction conditions before and failed, the key innovations of Xiao would appear to be finding a catalyst that is moisture and acid resilient, preferentially reduces imines or iminium ions over the starting ketone, and importantly is not "trapped" by the amine product.

7.3
Enantioselective Organocatalytic Reductive Amination

Examining a variety of chiral BINOL-based phosphoric acid catalysts in 2005, List found TRIP (Scheme 7.4) to be optimal (1.0 mol%) for the catalytic protonation of preformed imines of p-anisidine [15].

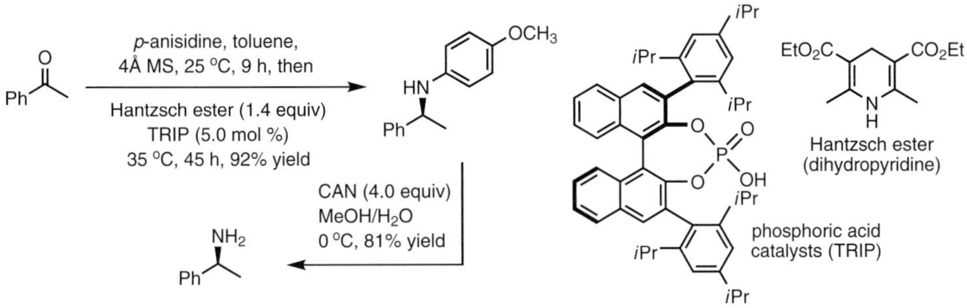

Scheme 7.4 TRIP-based two-stage, no workup ketone to amine synthesis.

The resulting achiral iminium cations, with chiral phosphate counteranion, were then enantioselectively reduced using an achiral Hantzsch ester (dihydropyridine) providing enantioenriched amines. During this imine reduction study, one example was shown in which acetophenone and p-anisidine [16] were prestirred in the presence of toluene and 4 Å molecular sieves [17] for 9 h (imine formation), after which the temperature was raised to 35 °C, and the Hantzsch ester (1.4 equiv) and phosphoric acid (TRIP, 5 mol%) were added to give the amine product in 88% ee over an additional 45 h. This is an exciting observation and while not a reductive amination, it is an operational improvement over simple imine reduction which requires imine isolation.

Follow-up studies regarding the further development of the just noted one-pot, two-stage amine synthesis from ketones (Scheme 7.4), or of attempts to examine reductive amination conditions themselves have, to the best of our knowledge, yet to be elaborated on. Regardless, List did make an additional study in which reductive amination was incorporated into a triple organocatalyzed cascade reaction (Scheme 7.5) allowing the formation of cis-3-substituted cyclohexylamines from 2,6-diketones. The reaction conditions called for 1.5 equiv of p-anisidine (or derivatives thereof) with the chiral phosphoric acid TRIP (10 mol%) in cyclohexane with 5 Å MS (250 mg per 0.25 mmol of diketone) at 50 °C for 72 h. A large number of R groups were acceptable, ranging from aliphatic (linear, α-branched, and β-branched) to aromatic and provided good yield, good diastereoselectivity, and good to excellent enantioselectivity. While very specialized, it is also a wonderful example of ingenuity and provides products that are of pharmaceutical interest as chiral building blocks.

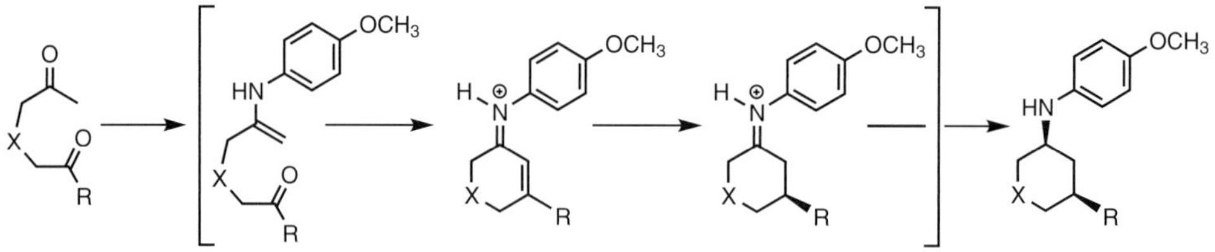

Scheme 7.5 Organocatalytic cascade featuring reductive amination with TRIP.

β-chiral amines are not included in this review, but we remind the reader that List has extended his TRIP/p-anisidine system to an elegant dynamic kinetic reductive amination protocol for α-branched aldehydes, which provides β-chiral amines [18]. A computational investigation of the stereochemical pathway for β-chiral amine formation has been reported on and is noteworthy [19].

During an overlapping time frame as the work of List, MacMillan devised the first, and to date only, intermolecular organocatalytic enantioselective reductive amination (reductant and ketone coexisting) with a Hantzsch ester [20]. Where List took advantage of TRIP, MacMillan found that an analogous phosphoric acid, albeit with triphenylsilyl groups, was critical for allowing reductive amination (Figure 7.3). Using p-anisidine as the limiting reagent, in the presence of Hantzsch ester (1.2 equiv), the acid catalyst (10 mol%), ketone (3.0 equiv), and 5 Å molecular sieves (1.0 g per 1.0 mmol of p-anisidine) at 40–50 °C in benzene (0.1 M), was optimal. When acetophenone was examined, only 24 h was required for the complete reaction, but all other substrates required 72–96 h when using p-anisidine. Although we only show the p-anisidine reductive amination products here (Figure 7.3), MacMillan additionally showed that a diverse set of aryl and heteroaromatic amines served equally well for the enantioselective reductive amination of aryl–alkyl and alkyl–alkyl ketones.

In 2007 Kočovský and Malkov reported an alternative organocatalytic approach, taking advantage of trichlorosilane in combination with a chiral formamide ligand (now referred to as Sigamide) for chloroamine formation (Scheme 7.6) and subsequent conversion to highly enantioenriched aziridines (not shown) [21].

Using their earlier developed organocatalytic methods for imine reduction, they applied similar conditions for the enantioselective reduction of α-chloroimines. Some of the α-chloroimines were unstable or sensitive to purification, prompting

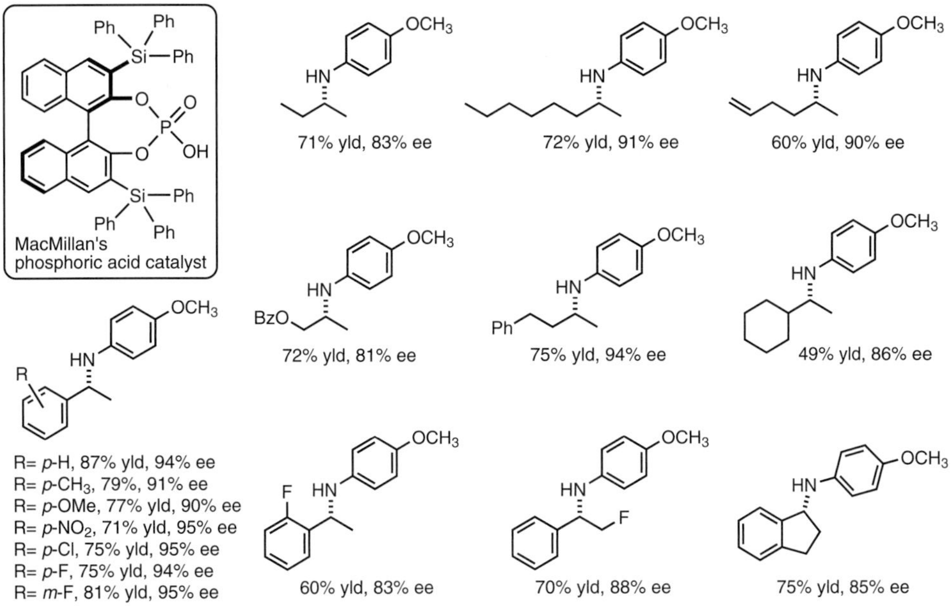

7.3 Enantioselective Organocatalytic Reductive Amination

Figure 7.3 Organocatalytic reductive amination – product examples.

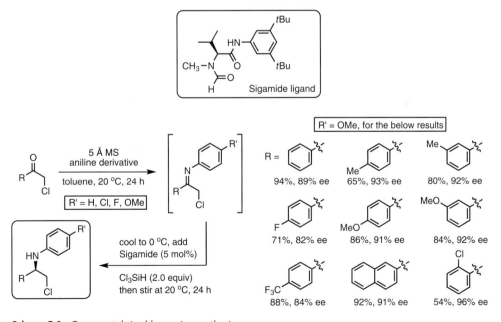

Scheme 7.6 Organocatalytic chloroamine synthesis.

them to develop a one-pot, two-stage, no workup, ketone to amine synthesis. By combining an aniline derivative (limiting reagent), an α-chloroketone, and 5 Å molecular sieves (200 mg per 0.2 mmol of aniline derivative) in toluene (0.1 M) for 24 h, they were able to form the imine *in situ*. After confirming full consumption of the ketone, the reducing agent, Cl_3SiH, and Sigamide ligand were added (Scheme 7.6).

7.4
Diastereoselective Reductive Amination

Auxiliary methods can hold current value when the auxiliary is inexpensive, available in both enantiomeric forms, and, for amine synthesis, can be incorporated with concomitant generation of the new stereogenic α-chiral amine center. Reductive amination with chiral ammonia equivalents not only holds this potential but is now a proven and established method that allows chiral primary amine synthesis in two reaction steps (reductive amination and hydrogenolysis) from prochiral ketones. The approach is of interest because of its overall reaction step efficiency.

Regarding the use of chiral amine auxiliaries for reductive amination, only (R)- and (S)-phenylethylamine (also referred to as α-methylbenzylamine) and (R)- and (S)-*tert*-butylsulfinylamide have gained widespread acceptance as useful chiral ammonia equivalents. The strategy is therefore to use these chiral amines to induce a new chiral amine stereogenic center on a ketone substrate. An alternative approach is to use a chiral ketone with an achiral source of nitrogen, for example, benzylamine. This latter approach is less often employed, but nonetheless equally important and discussed first.

7.4.1
Stereoselective Reductive Amination with Chiral Ketones

Substance P receptor antagonists have been extensively sought after for alleviating a range of ailments, for example, gastrointestinal and psychotic disorders and inflammation-related diseases, and many of the manufactured drug antagonists share a common advanced core: (2S,3S)-*cis*-2-benzhydryl-3-aminoquinuclidine or (2S,3S)-**1** (Scheme 7.7). In 2004 Nugent reported the first successful stereoselective reductive amination of an α-chiral stereochemically labile ketone (Scheme 7.7), allowing a greatly improved synthesis of (2S,3S)-**1** to come forth [22, 23].

The new reductive amination conditions came from recognizing earlier developed achiral reductive amination conditions employing the mild Lewis acid $Ti(OiPr)_4$ [24], this removed the α-chiral ketone racemization problem associated with classical Brønsted acid reductive amination procedures, and by replacing the previously used hydride reagents ($NaBH_3CN$ and $NaBH_4$) [24, 25] with the more environment friendly $Pt-C/H_2$ or $Pd-C/H_2$ reducing systems. For the shown chiral ketone substrate (Scheme 7.7), reaction under the conditions of $Ti(OiPr)_4$ (1.2 equiv), $BnNH_2$ (1.1 equiv), Pt-C (0.4 mol% Pt), 4.1 bar (60 psi) H_2, THF, 25 °C, over 12 h provided the best overall result in terms of the combined outcome of preservation of ketone stereo-

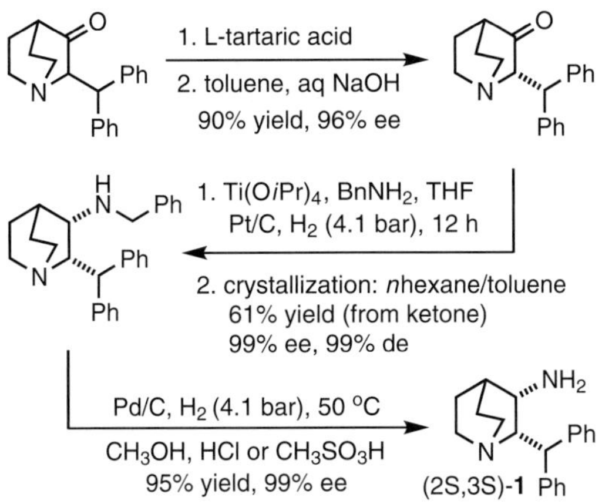

7.4 Diastereoselective Reductive Amination

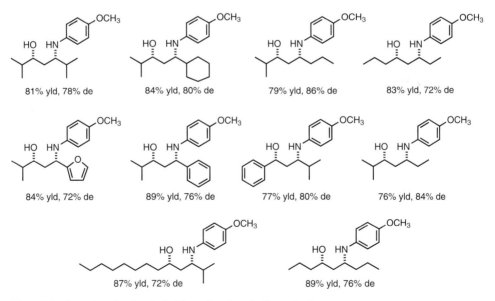

Scheme 7.7 Key quinuclidine intermediate synthesis via stereoselective reductive amination.

chemistry, diastereoselectivity, yield, and rate of reaction, culminating in the most stepwise and yield-efficient synthesis of this chiral quinuclidine to date.

Menche recently established an effective two-step strategy for synthesizing chiral 1,3-amino alcohols [26]. The first step, an enantioselective aldol reaction, provides chiral β-hydroxy ketones, which are then reductively aminated under the conditions of p-anisidine (2.0 equiv), Ti(OiPr)$_4$ (1.4 equiv), CH$_3$CN (−20 °C), in the presence of polymethylhydrosilane (2.2 equiv) over 48 h. A variety of examples are shown in Figure 7.4.

Figure 7.4 Examples of aminoalcohol formation via reductive amination.

7 Asymmetric Reductive Amination

Lopinavir (Scheme 7.8) and Ritonavir (not shown) are HIV-protease inhibitors that are often used in combination to treat HIV, and Menche adroitly applied his new methodology to the core 1,4 chiral diamine building block, which is shared by both of these marketed drugs. The synthesis is outlined in Scheme 7.8 and represents the shortest synthesis of this key advanced intermediate to date.

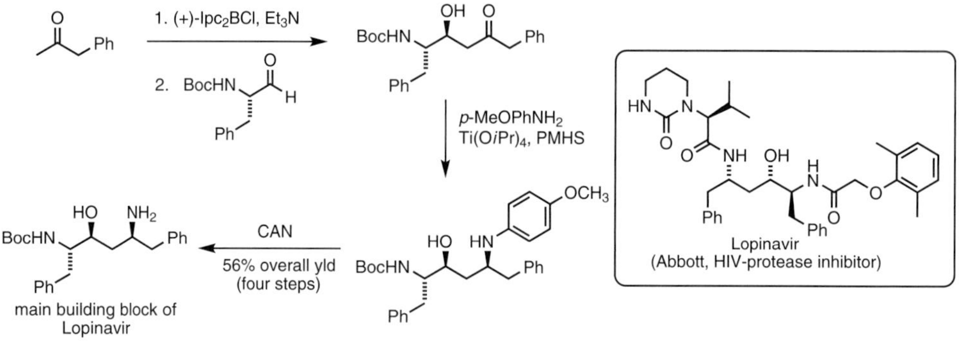

Scheme 7.8 Application of aldol/reduction amination strategy – formal synthesis of lopinavir.

Extending the work of Kim [27], Letourneux and Brunel derivatized 3β-acetoxy-7-keto-5α-cholestane (Scheme 7.9), with ammonia, methylamine, 10 acyclic diamines, and spermine (a tetraamine) forming the corresponding β-aminosteroids [28]. The amines were tested for their gram-positive bacteria activity versus the known and structurally related squalamine (not shown).

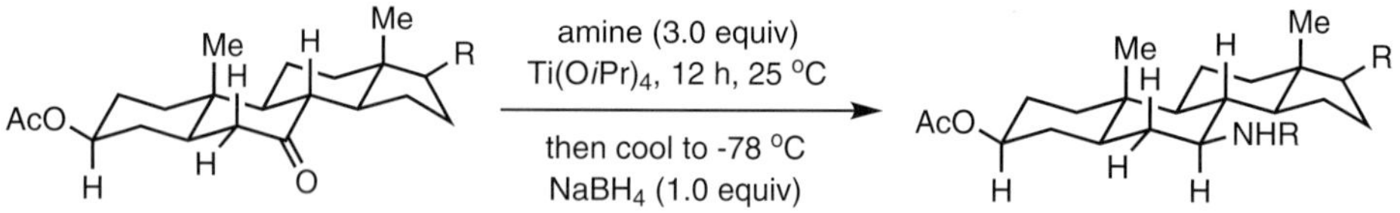

Scheme 7.9 Two-stage, no workup ketone to amine synthesis.

Optimal conditions required the α-chiral ketone (steroid) to be stirred with Ti(O*i*Pr)$_4$ (1.3 equiv) and the amine (3.0 equiv) in a dilute solution of MeOH (0.08 M) over 12 h. The reaction was then cooled to −78 °C, NaBH$_4$ (1.0 equiv) added, and stirred for 2 h [29]. Yields were generally poor (6–45%), except when methylamine (77% yield) and 1,2-diethylamine (61% yield) were used. For all 13 examples, the diastereoselectivity is reported to be excellent (>95%). No mention was made of the lability of the α-chiral stereogenic center adjacent to the ketone, presumably it is resilient to epimerization due to the conformational rigidity afforded by the extended steroid skeleton.

Although the above examples represent a limited breadth of demonstrated application, the general method would appear to hold promise due to its acceptability of labile α-chiral ketones.

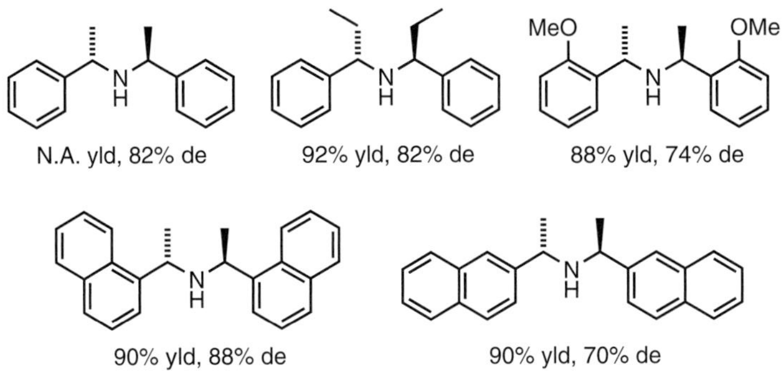

Figure 7.5 C$_2$-symmetrical secondary amines via reductive amination.

7.4.2
The Phenylethylamine Auxiliary and Stereoselective Reductive Amination

In 2004 Alexakis independently reported on similar reaction conditions (Ti(O*i*Pr)$_4$/Pd-C/H$_2$) as Nugent [22b, 23], albeit when using chiral amines, for example, phenylethylamine (PEA), phenylpropylamine, and so on, to reductively aminate skeleton matching achiral ketones [30]. By doing so, he synthesized a set of five C$_2$-symmetrical secondary amines from aryl–alkyl ketones (Figure 7.5). The optimal conditions called for neat reaction conditions, equal molar quantities of the achiral ketone and chiral amine, Ti(O*i*Pr)$_4$ (3.0 equiv), Pd-C (0.5 mol%), and 1.0 bar (14.5 psi) H$_2$. No reaction times were reported.

Nugent has published several manuscripts examining a two-step strategy allowing prochiral ketones to be converted into enantioenriched chiral primary amines (Scheme 7.10). The method, initially reported on in 2005 [31], relies on reductive amination with (*R*)- or (*S*)-PEA followed by hydrogenolysis, and has been optimized for a large variety of ketone substrate classes. The key innovation was identification of the most efficient Lewis acid/heterogeneous hydrogenation catalyst combination, which in turn allowed the reductive amination of previously unreactive, sterically hindered ketones in good to excellent yield [31, 32] and led to significantly improved stereoselectivity for alkyl–methyl ketone substrates [33].

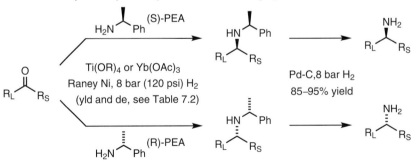

Scheme 7.10 Two-step chiral primary amine synthesis – reductive amination followed by hydrogenolysis.

Table 7.2 Raney-Ni substrate classes and optimal acid catalysts for stereoselective reductive amination with phenylethylamine (Scheme 7.10 – Step 1).

Ketone classes	Specific examples	Yield (%)	de (%)	Acid cat.[a]	Comment
R_L–CO–CH_3	$R_L = iPr$	78	98	Ti(OiPr)$_4$	Viable alternative AcOH[b]
	$R_L = $ chexyl	90	98	Ti(OiPr)$_4$	Viable alternative AcOH[b]
	$R_L = $ Ph	85	95	Ti(OiPr)$_4$	Other catalysts – lower yield[c]
R_L–CO–R_S	$R_L = $ Ph; $R_S = nPr$	92	94	Ti(OiPr)$_4$[d]	Other catalysts <30% product yield
	$R_L = iPr$; $R_S = nPr$	75	87	Ti(OiPr)$_4$[e]	Other cat <20% product yield
	$R_L = iPr$; $R_S = nBu$	80	88	Ti(OiPr)$_4$[e]	Other cat <20% product yield
R_M–CO–CH_3	$R_M = iBu$	79	93	Ti(OiPr)$_4$	Viable alternative AcOH[f]
	$R_M = $ –CH_2CH_2Ph	87	89	Yb(OAc)$_3$	Other catalysts - low de
R_S–CO–CH_3	$R_S = nhexyl$	86	87	Yb(OAc)$_3$	Other catalysts - low de
	$R_S = nbutyl$	82	85	Yb(OAc)$_3$	Other catalysts - low de

a) Unless otherwise noted, reaction conditions are as follows: Ti(OiPr)$_4$ (1.25 equiv) or Yb(OAc)$_3$ (0.80–1.1 equiv), Raney-Ni (100 wt% of limiting reagent – ketone), (S)- or (R)-PEA (1.1 equiv), 22 °C, 8 bar (120 psi) H$_2$, 12 h.
b) The use of 20 mol% AcOH, in MeOH, allows very similar results, but only at 50 °C and 20 bar (290 psi) H$_2$.
c) Optimal conditions with AcOH (20 mol%), 50 °C, 30 bar (435 psi), and MeOH provide 55% yield, 93% de.
d) Reaction conditions: 35 °C, 8 bar (120 psi) H$_2$, and 15 h.
e) Reaction conditions: 3.0 equiv Ti(OiPr), 60 °C, 8 bar (120 psi) H$_2$, 36 h.
f) The use of 20 mol% AcOH in MeOH allows very similar results, but only at elevated temperature (50 °C).

One way of organizing the findings of Nugent is regarding the optimal heterogeneous hydrogenation catalyst. For example, Raney-Ni, the most broadly applicable catalyst, allows acyclic alkyl–alkyl' and aryl–alkyl ketones lacking an sp^3-hybridized quaternary carbon, for example, *tert*-butyl moiety, attached to the carbonyl carbon, to be reductively aminated with good yield and high diastereoselectivity (Table 7.2). Pt-C is the optimal hydrogenation catalyst for those alkyl–alkyl' ketone substrates containing sp^3-hybridized quaternary carbons adjacent to the carbonyl moiety (Figure 7.6) [32], while Pd-C is optimal for cyclic aryl–alkyl ketones, for example, tetralone and benzsuberone (Figure 7.6) [32]. Hydrogenolysis of the reductive amination product provides the primary amine in very high yielding with uncompromised ee.

When comparing the stereoselectivity of the reductive amination products of (R)- or (S)-PEA, when using Ti(OiPr)$_4$ (1.25 equiv) or Yb(OAc)$_3$ (10 mol%) or Y(OAc)$_3$ (15 mol%) or Ce(OAc)$_3$ (15 mol%) or Brønsted acids (catalytic or stoichiometric, e.g., AcOH), the de of the amine product is the same. Furthermore, taking the ketones used for these reductive amination studies and intentionally preforming and isolating the (R)- or (S)-PEA ketimines (Dean–Stark trap synthesis) and reducing them in the same manner, albeit without the presence of the Lewis acid or Brønsted acid, the de is the same as that found for the reductive amination [34].

In stark contrast to these stereochemical trends, 2-alkanones without branching at the α- or β-carbons, for example, 2-octanone or benzylacetone (Table 7.2), can be

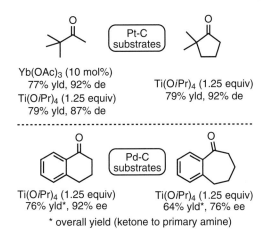

Figure 7.6 Substrates requiring Pd-C or Pt-C with phenylethylamine.

reductively aminated with significantly higher diastereoselectivity when using stoichiometric quantities (80–110 mol%) of Yb(OAc)$_3$ [33].

Finally, some general observations were noted by Nugent regarding reductive amination with phenylethylamine that are likely to be general in nature. Failure to have the optimal Lewis acid or Brønsted acid, or no acid at all, results in gross alcohol by-product formation. For example, regarding 2-alkanones, alcohol by-product formation is suppressed below 3% when employing catalytic quantities of Lewis acids [Yb(OAc)$_3$ (10 mol%) or Y(OAc)$_3$ (15 mol%) or Ce(OAc)$_3$ (15 mol%)] or catalytic or stoichiometric quantities of a weak Brønsted acid, for example, AcOH [33]. Application of the above-mentioned Lewis acids or Brønsted acids in catalytic or stoichiometric quantities for the reductive amination of aryl–alkyl ketones or very sterically hindered alkanones, for example, i-propyl n-propyl ketone, results in gross alcohol formation, for these substrates Ti(OiPr)$_4$ is required (Table 7.2) [32].

GlaxoSmithKline chemists Gudmundsson and Xie recently reported on a new drug candidate for the treatment of human papillomavirus (HPV) associated with infection of the genital mucosa, in high-risk scenarios cervical cancer results [35]. The initial approach focused on racemic primary amine formation followed by preparative chiral supercritical fluid chromatography (SFC) to separate the enantiomers. To enable racemic primary amine formation (not shown), the ketone (structure shown in Scheme 7.11) was reductively aminated with NH$_4$OAc/NaBH$_3$CN over 15 h at 60 °C, providing the amine in 52% yield. This initial resolution approach failed because solubility problems hampered the use of SFC separation, and attempts at classical chemical resolution with an array of chiral acids failed to provide high ee with high yield recovery.

Moving to asymmetric methods, they found Noyori's enantioselective transfer hydrogenation employing HCO$_2$NH$_4$, Ru(II), and DPEN-based ligands reductively aminated (not shown) the same ketone providing the enantioenriched primary amine in 80% ee and 60% yield. They additionally synthesized the corresponding

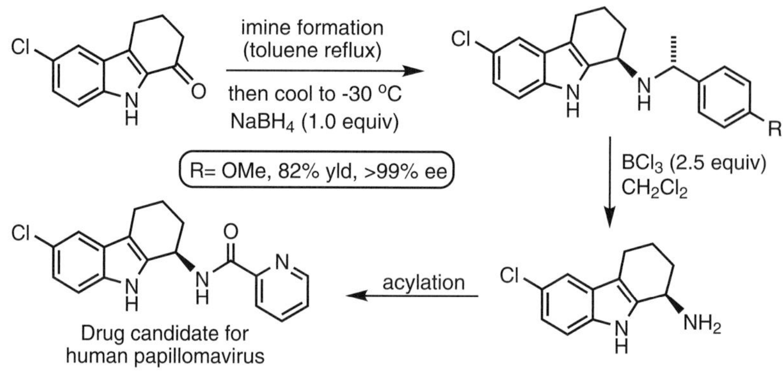

Scheme 7.11 Reductive amination as a key step in drug development.

imine, from ammonia, and then applied the earlier discussed Kadyrov conditions (Section 7.2). With this imine substrate, the modified Kadyrov conditions provided an 82% ee, but with an unacceptable impurity profile.

Gudmundsson and Xie then turned their attention to the often used industrial chiral ammonia equivalents: (R)- and (S)-1-(phenyl)ethylamine (PEA), which are also commonly referred to as (R)- or (S)-α-methylbenzylamine (α-MBA). Reductive amination with (R)-PEA or the (R)-1-(4-methoxyphenyl)ethylamine derivative resulted in the desired major diastereomer, 90% de (Scheme 7.11, R = H) and 92% de (Scheme 7.11, R = OMe) respectively. On the kilogram scale, they found it more reliable to use a two-stage, no workup procedure, first forming the imine (toluene/Dean–Stark trap, 10 h) and then without workup for (R)-1-(4-methoxyphenyl)ethylamine, added EtOH, cooled to −30 °C, and stirred for 12 h with NaBH$_4$ (1.0 equiv). It is typical that process development chemists design and screen workup conditions allowing the product to be obtained in higher de or ee than the reaction provides. This case was no different and attention to the workup allowed the desired diastereomer to be isolated in multi-kilogram quantities in >99% ee as the HCl salt for both (R)-1-(phenyl)ethylamine (86% yield) and (R)-1-(4-methoxyphenyl)ethylamine (82% yield). The corresponding primary amine was their target, but attempts at N-debenzylation via the common hydrogenolysis route (Pd, Ru, or Rh/H$_2$) failed due to the sensitivity of the tetrahydrocarbazole moiety. This problem was overcome, at the 13.5 kg scale, by treatment with BCl$_3$ (2.5 equiv) in CH$_2$Cl$_2$, resulting in an 81% yield of the primary amine in >99% ee.

7.4.3
The *tert*-Butylsulfinamide Auxiliary and Stereoselective Reductive Amination

In 1999, the group of Ellman demonstrated a two-stage, no workup ketone to amine synthesis [36]. This was achieved by adding the (S)- or (R)-*tert*-butylsulfinamide auxiliary, [t-BuS(O)NH$_2$], limiting reagent, to the ketone (1.2 equiv) and Ti(OEt)$_4$ (2.0 equiv) in THF. After heating (60–75 °C, usually 10–19 h), the resulting sulfinyl imine

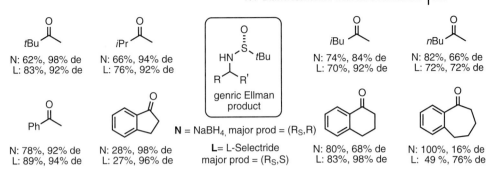

Figure 7.7 Stereodivergent amine synthesis: reductant-directed diastereomeric product formation with the (R)-tert-butylsulfinamide auxiliary.

(TLC examination) is cooled to −48 °C and then transferred via cannula to a −48 °C THF suspension of NaBH$_4$ (4 equiv). In 2007, he updated the substrate breadth and reported how the method could be simplified to a stereodivergent method requiring only one enantiomer of the chiral amine auxiliary, tert-butylsulfinylamide, by using either NaBH$_4$ (4.0 equiv) or L-Selectride (3.0 equiv) to provide the epimeric sulfinyl amine product (Figure 7.7) [37]. Regardless of the reductant used, −48 °C was the optimal temperature for reduction. In addition to the shown ketone substrates, p-NO$_2$ and p-MeO tetralone substrates (not shown) also provided the sulfinyl amine product in good yield and high de. The corresponding primary amine is easily obtained by removal of the sulfinyl auxiliary under acid conditions, commonly HCl in dioxane at room temperature.

It is well established that control of peptide conformation allows the tertiary structure vital for enzyme activity. Unnatural α- and β-amino acids are particularly useful in this regard for designing peptides with controlled conformation and thus targeted function, but amino acid mimics have also been put to good effect in this regard. This was the goal of Pannecoucke when he successfully synthesized enantiopure monofluorinated allylic amines as site-specific amino acid mimics in peptides (Scheme 7.12) [38].

Employing the stereodivergent strategy of Ellman to α-fluoroenones, a one-pot ketone to amine synthesis (two-stage procedure, no workup) allowed the production of either diastereomer from only one chiral amine source, (S)-tert-butylsulfinamide (Scheme 7.12) Pannecoucke reported that no benefit came from first isolating the chiral sulfinyl imine, thus the sulfinyl imine was formed (2 h of THF reflux), subsequently cooled (−78 °C), and then reduced using a source of hydride. The reduction step was complete within 1–2 h, regardless of the chosen reductant. This approach provided moderate to good yield (46–86% yield) with excellent diastereoselectivity (94–98%). In most cases, the diastereomeric excess could be increased via silica gel chromatography. The reported "R" groups were aromatic (p-MeOPh-) and alkyl (PhCH$_2$CH$_2$- and TBDPSiOCH$_2$CH$_2$- and TBDPSiOCH$_2$CH(CH$_3$)-).

Coordinating reductants, for example, NaBH$_4$, BH$_3$, 9-BBN, and DIBAL-H, provided the same stereochemistry as that of the stereogenic sulfur atom of the

242 | 7 Asymmetric Reductive Amination

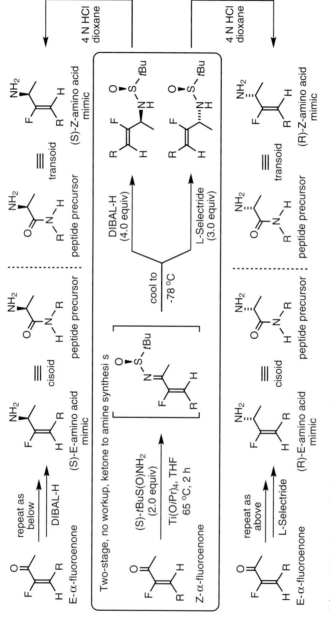

Scheme 7.12 Functionalized chiral amine synthesis: amino acid mimics for peptides.

tert-butylsulfinamide auxiliary, while "noncoordinating" reductants, for example, LiBHEt$_3$, K-Selectride, and L-Selectride provided the epimeric product. DIBAL-H and L-Selectride were the reductants of choice, allowing, after HCl in dioxane removal of the sulfinyl moiety, either enantiomer of the primary amine to be formed.

7.5 Conclusions

Two important, but rarely discussed, challenges regarding chiral amine synthesis are evaluation of reaction step efficiency from commodity chemical (starting material) to final chiral primary amine product; and, second, the ability to install a nitrogen atom while leaving a broad array of coexisting functional groups unchanged. The former point is addressed by using a reductive amination strategy, but the latter still needs to be explored. Future demonstration of broad functional group compatibility, taken with the standard concerns of good yield and high enantiomeric excess, will allow asymmetric reductive amination to be considered as a reliable and preparative method capable of delivering functional group rich chiral amine building blocks that are "drug" like.

Questions

7.1. (a) Draw the mechanism for the reductive amination of 2-butane with (R)-phenylethylamine in the presence of acetic acid, Raney-Ni/H$_2$.
(b) What role is Ti(iOPr)$_4$ fulfilling when it replaces a Brønsted acid in reductive amination?

7.2. Provide two features of reductive amination that have stalled progress when using transition metals.

7.3. (a) Three main sources of nitrogen are used by the methodologies outlined in this chapter: aniline and its substituted derivatives, phenylethylamine, and tert-butylsulfinamide. Which is considered genotoxic?
(b) The introduction of chiral amines into advanced building blocks must be compatible with coexisting functional groups on the ketone starting material. Additionally, a chiral primary amine is often required for further elaboration to a chiral amine-based pharmaceutical drugs or natural product. The reductive amination of ketones with aniline derivatives (e.g., p-anisidine), phenylethylamine, and tert-butylsulfinamide provides facile access to chiral amine products, but these products do not become valuable until they are deprotected to the corresponding chiral primary amines. Provide the three different deprotection conditions required for the mentioned sources of nitrogen.

References

1. (a) Tararov, V.I. and Börner, A. (2005) *Synlett*, 203–211; (b) Tararov, V.I., Kadyrov, R., Riermeier, T.H., Fischer, C., and Börner, A. (2004) *Adv. Synth. Catal.*, **346**, 561–565.
2. Blaser, H.-U., Buser, H.-P., Jalett, H.-P., Pugin, B., and Spindler, F. (1999) *Synlett*, 867–868.
3. (a) Pouy, M.J., Leitner, A., Weix, D.J., Ueno, S., and Hartwig, J.F. (2007) *Org. Lett.*, **9**, 3949–3952; (b) Hou, G.-H., Xie, J.-H., Wang, L.-X., and Zhou, Q.-L. (2006) *J. Am. Chem. Soc.*, **128**, 11774–11775; (c) Blaser, H.-U., Buser, H.-P., Coers, K., Hanreich, R., Jalett, H.-P., Jelsch, E., Pugin, B., Schneider, H.-D., Spindler, F., and Wegmann, A. (1999) *Chimia*, **53**, 275–280; (d) Markó, L. and Bakos, J. (1974) *J. Organomet. Chem.*, **81**, 411–414.
4. For literature pertaining to the origins and definition of reductive amination, see (a) Emerson, W.S. (1948) *Org. React.*, **4**, 174; (b) Moore, M.L. (1949) *Org. React.*, **5**, 301; (c) Smith, M.B. and March, J. (eds.) (2001) *March's Advanced Organic Chemistry*, 5th edn, John Wiley & Sons, Inc., New York, pp. 1187–1189.
5. Similarly, direct asymmetric reductive amination is discouraged and should be replaced by the shorter and more accurate phrase asymmetric reductive amination or enantioselective reductive amination.
6. (a) Kadyrov, R., Riermeier, T.H., Dingerdissen, U., Tararov, V.I., and Börner, A., (2003) *J. Org. Chem.*, **68**, 4067–4070; (b) Kitamura, M., Lee, D., Hayashi, S., Tanaka, S., and Yoshimura, M. (2002) *J. Org. Chem.*, **67**, 8685–8687.
7. (a) Koszelewski, D., Lavandera, I., Clay, D., Guebitz, G.M., Rozzell, D., and Kroutil, W. (2008) *Angew. Chem., Int. Ed.*, **47**, 9337–9340; (b) Höhne, M., Kühl, S., Robins, K., and Bornscheuer, U.T. (2008) *ChemBioChem*, **9**, 363–365.
8. To fully appreciate a great story of perseverance and know-how at Solvias regarding the dynamics of developing a catalytic enantioselective process for industrial scale production, see Blaser, H.-U. (2002) *Adv. Synth. Catal.*, **344**, 17–31.
9. Chi, Y., Zhou, Y.-G., and Zhang, X. (2003) *J. Org. Chem.*, **68**, 4120–4122.
10. The reader is referred to the following citations to better appreciate the role of iodine: (a) Xiao, D. and Zhang, X. (2001) *Angew. Chem., Int. Ed.*, **40**, 3425–3428; (b) Spindler, F. and Blaser, H.-U. (1999) *Enantiomer*, **4**, 557–568; (c) Togni, A. (1996) *Angew. Chem., Int. Ed. Engl.*, **35**, 1475–1477.
11. (a) Kadyrov, R. and Riermeier, T.H. (2003) *Angew. Chem., Int. Ed.*, **42**, 5472–5474; (b) Börner, A., Dingerdissen, U., Kadyrov, R., Riermeier, T.H., and Tararov, V. (2004) Method for the production of amines by reductive amination of carbonyl compounds under transferhydrogenation conditions, U.S. Patent No. 2004267051, Degussa AG, Germany.
12. Riermeier, T., Haack, K.-J., Dingerdissen, U., Boerner, A., Tararov, V., and Kadyrov, R. (2005) Method for producing amines by homogeneously catalyzed reductive amination of carbonyl compounds, U.S. Patent No. 6884887, Degussa AG, Germany.
13. Rubio-Pérez, L., Pérez-Flores, F.J., Sharma, P., Velasco, L., and Cabrera, A. (2009) *Org. Lett.*, **11**, 265–268.
14. Li, C., Villa-Marcos, B., and Xiao, J. (2009) *J. Am. Chem. Soc.*, **131**, 6967–6969.
15. Hoffmann, S., Seayad, A.M., and List, B. (2005) *Angew. Chem., Int. Ed.*, **44**, 7424–7427.
16. We were unable to locate the number of equiv used in the text or Supporting Information.
17. We were unable to locate the wt% of 4 Å molecular sieves used in the text or the Supporting Information.
18. Hoffmann, S., Nicoletti, M., and List, B. (2006) *J. Am. Chem. Soc.*, **128**, 13074–13075.
19. Marcelli, T., Hammar, P., and Himo, F. (2009) *Adv. Synth. Catal.*, **351**, 525–529.
20. Storer, R.I., Carrera, D.E., Ni, Y., and MacMillan, D.W.C. (2006) *J. Am. Chem. Soc*, **128**, 84–86.
21. Malkov, A.V., Stončius, S., and Kočovský, P. (2007) *Angew. Chem., Int. Ed.*, **46**, 3722–3724.

22 (a) Nugent, T.C. and Seemayer, R. (2006) *Org. Process. Res. Dev.*, **10**, 142–148; (b) Nugent, T.C. and Seemayer, R. (2004) Process for the preparation of (S,S)-cis-2-benzhydryl-3-benzylaminoquinuclidine, Patent No. WO2004035575, Pfizer Products, Inc. and DSM Pharmaceuticals, Inc.

23 The method was used as a trade secret within Catalytica/Pfizer since 1998, see citation [19] of reference [22a].

24 Mattson, R.J., Pham, K.M., Leuck, D.J., and Cowen, K.A. (1990) *J. Org. Chem.*, **55**, 2552–2554.

25 For recent publications by Bhattacharyya, see (a) Bhattacharyya, S. and Kumpaty, H. (2005) *J. Synthesis*, 2205–2209; (b) Miriyala, B., Bhattacharyya, S., and Williamson, J.S. (2004) *Tetrahedron*, **60**, 1463–1471; (c) Neidigh, K.A., Avery, M.A., Williamson, J.S., and Bhattacharyya, S. (1998) *J. Chem. Soc., Perkin Trans.1*, 2527–2532.

26 Menche, D., Arikan, F., Li, J., and Rudolph, S. (2007) *Org. Lett.*, **9**, 267–270.

27 Kim, H.S., Cho, N.J., and Khan, S.N. (2008) 7-alpha-aminosteroid derivatives or pharmaceutically acceptable salts thereof, preparation method thereof and composition for anticancer or antibiotics containing the same as an active ingredient, Patent No. WO038965 A1 (South Korea).

28 Loncle, C., Salmi, C., Letourneux, Y., and Brunel, J.M. (2007) *Tetrahedron*, **63**, 12968–12974.

29 For an example of another chiral ketone reductively aminated with $Ti(OiPr)_4$ or $Al(OiPr)_3$ with hydride reagents, see Dhainaut, J., Leon, P., Lhermitte, F., and Oddon, G.(May 2003) A process which is useful for converting the carbonyl function in position 4″ of the cladinose unit of an aza-macrolide into an amine derivative, U.S. Patent No. 6,562,953 (Merial, France).

30 Alexakis, A., Gille, S., Prian, F., Rosset, S., and Ditrich, K. (2004) *Tetrahedron Lett.*, **45**, 1449–1451.

31 Nugent, T.C., Wakchaure, V.N., Ghosh, A.K., and Mohanty, R.R. (2005) *Org. Lett.*, **7**, 4967–4970.

32 Nugent, T.C., Ghosh, A.K., Wakchaure, V.N., and Mohanty, R.R. (2006) *Adv. Synth. & Catal.*, **348**, 1289–1299.

33 Nugent, T.C., El-Shazly, M., and Wakchaure, V.N. (2008) *J. Org. Chem.*, **73**, 1297–1305.

34 The outlined Nugent reductive amination protocol with (R)- or (S)-PEA and prochiral alkyl–alkyl′ and aryl–alkyl ketones (acyclic or cyclic) allows higher yields and shorter reaction times than the previously practiced two-step strategy via isolated (R)- or (S)-PEA, see: Moss, N., Gauthier, J., and Ferland, J.-M., (1995) *Synlett*, 142–144; (b) Cimarelli, C. and Palmieri, G. (2000) *Tetrahedron: Asymmetry*, **11**, 2555–2563.

35 Boggs, S.D., Cobb, J.D., Gudmundsson, K.S., Jones, L.A., Matsuoka, R.T., Millar, A., Patterson, D.E., Samano, V., Trone, M.D., Xie, S., and Zhou, X.-M. (2007) *Org. Process Res. Dev.*, **11**, 539–545.

36 Borg, G., Cogan, D.A., and Ellman, J.A. (1999) *Tetrahedron Lett.*, **40**, 6709–6712.

37 Tanuwidjaja, J., Peltier, H.M., and Ellman, J.A. (2007) *J. Org. Chem.*, **72**, 626–629.

38 Dutheuil, G., Couve-Bonnaire, S., and Pannecoucke, X. (2007) *Angew. Chem., Int. Ed.*, **46**, 1290–1292.

8
Enantioselective Hydrogenation of Enamines with Monodentate Phosphorus Ligands

Qin-Lin Zhou and Jian-Hua Xie

8.1
Introduction

Chiral amines are widely used as chiral ligands, auxiliaries, resolving agents, and important intermediates in synthetic chemistry. Asymmetric hydrogenation of enamines such as enamides is an atom economic and straightforward protocol for the preparation of chiral amines. In 1972, Kagan reported the first example of asymmetric hydrogenation of enamides with chiral Rh-DIOP complex as a catalyst [1]. This opened a new window for the asymmetric synthesis of chiral N-acyl amines although the enantioselectivity of the reaction was not more than 78% ee (Scheme 8.1). When Noyori et al. introduced chiral Ru-BINAP, it significantly increased the level of enantioselectivity to 99.5% ee in the asymmetric hydrogenation of N-acyl-1-alkylidenetetrahydroisoquinolines [2]. However, this ruthenium-catalyzed asymmetric hydrogenation is very sensitive to the stereochemistry of the enamide, with only the (Z)-isomer of the enamide being hydrogenated while the (E)-isomer was recovered intact (Scheme 8.2). Another breakthrough in this field came in 1996 when Burk et al. showed that chiral rhodium complexes bearing a diphosphine ligand Me-DuPHOS or Me-BPE could hydrogenate α-arylenamides to yield a wide variety of valuable α-1-arylethylamine derivatives with high enantioselectivities (up to 98.5% ee) (Scheme 8.3) [3]. The merit of this hydrogenation is that both (E)- and (Z)-isomers of β-substituted enamides can be hydrogenated, thus enlarging the substrate scope of the reaction. Subsequently, many efficient diphosphines, such as BICP (bis(diphenylphosphino)dicyclopentane) [4], PennPhos P,

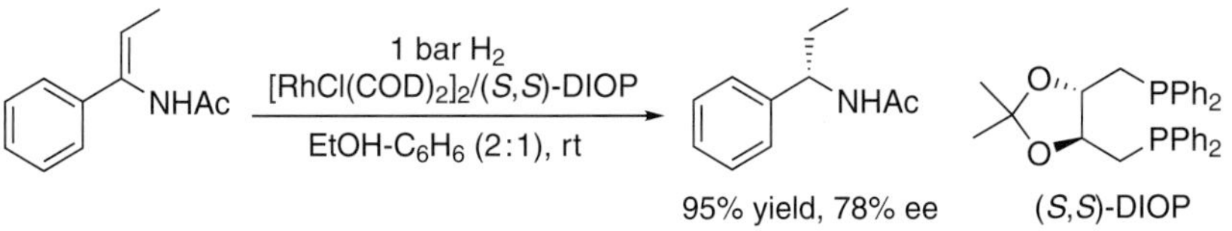

Scheme 8.1 Rh-DIOP catalyzed asymmetric hydrogenation of enamides.

Chiral Amine Synthesis: Methods, Developments and Applications. Edited by Thomas C. Nugent
Copyright © 2010 WILEY-VCH Verlag GmbH & Co. KGaA, Weinheim
ISBN: 978-3-527-32509-2

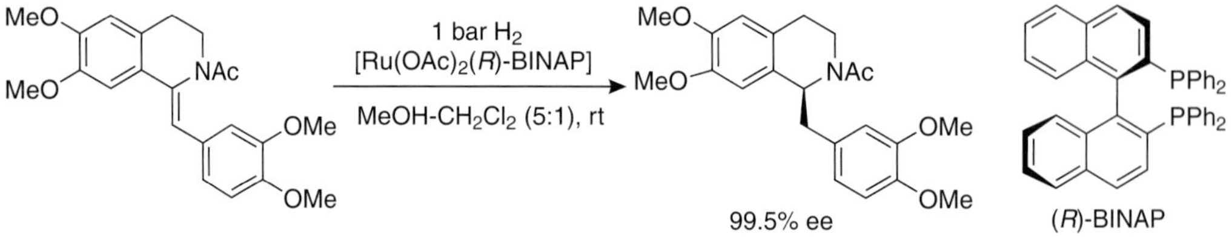

Scheme 8.2 Ru-BINAP catalyzed asymmetric hydrogenation of enamides.

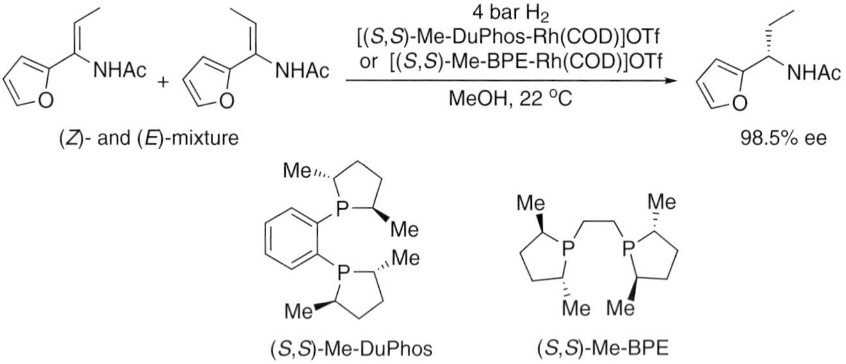

Scheme 8.3 Rh-DuPHOS/BPE-catalyzed asymmetric hydrogenation of enamides.

P,P'-1,2-phenylenebis(endo-2,5-dialkyl-7-phosphabicyclo[2,2,1]heptane) [5], BDPAB (2,2'-bis(diphenylphosphinoamino)-1,1'-binaphthyl) [6], SpirOP ((1R,5R,6R)-1,6-bis(diphenylphosphinoxy)spiro[4.4]nonane) [7], TangPhos ((1S,1'S,2R,2'R)-1,1'-di-tert-butyl-(2,2')-diphospholane) [8], were reported to be efficient ligands in the rhodium-catalyzed asymmetric hydrogenation of enamides.

During the very early stages of enamide asymmetric hydrogenation research, monodentate chiral phosphines were employed. For example, Knowles et al. achieved an ee value of 90% in the hydrogenation of α-dehydroamino acids with a rhodium complex of ligand CAMP in 1972 [9]. However, in the following 30 years, all ligands that gave a high degree of enantiocontrol in the rhodium-catalyzed asymmetric hydrogenation of enamides are bidentate diphosphines. The major reason for this situation is the influence of the traditional belief that excellent enantioselectivity usually comes from a chelating catalyst bearing a bidentate ligand since the strong coordination of the bidentate ligand could prevent the rotation of the donor atom–metal bond in the catalyst [10].

At the beginning of 2000s, three groups (Pringle [11], Reetz [12], and de Vries and Feringa [13]) independently reported the use of monodentate chiral phosphorus ligands for the rhodium-catalyzed asymmetric hydrogenation of functionalized olefins. The monodentate chiral ligands such as phosphonites, phosphites, and phosphoramidites (Figure 8.1) induced remarkably high enantioselectivities, which are comparable to those obtained using the best bidentate diphosphine ligands in the rhodium-catalyzed asymmetric hydrogenation of α-dehydroamino acid derivatives and itaconic acid derivatives. These monodentate chiral phosphorus ligands are all

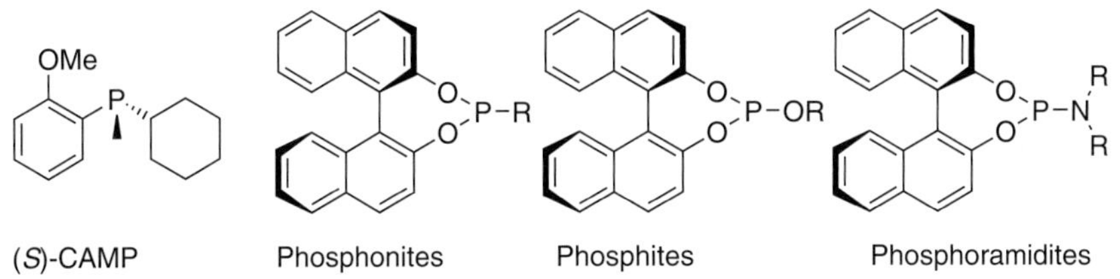

(S)-CAMP   Phosphonites   Phosphites   Phosphoramidites

Figure 8.1 Monodentate phosphorus ligands (R = alkyl or aryl).

based on a binaphthyl skeleton. Since then, more and more efficient chiral monodentate phosphorus ligands were developed and some of them have shown high chiral inducements in the rhodium-catalyzed asymmetric hydrogenation of enamides.

In this chapter, we will focus on the synthesis of chiral amines by catalytic asymmetric hydrogenation of the corresponding enamides and N,N-dialkyl enamines with monodentate chiral phosphorus ligands. Thus, only the monodentated phosphorus ligands, which gave good enantioselectivity in the asymmetric hydrogenation of enamides and N,N-dialkyl enamines, are discussed here. Monodentate phosphorus ligands have been successfully applied to the asymmetric hydrogenations of N-acetyl α-/β-dehydroamino acid derivatives to form chiral amino acids [14], but are not discussed here because the focus of this book is on nonamino acid-based chiral amines.

8.2
Asymmetric Hydrogenation of Enamides

The first example of enantioselective hydrogenation of enamides with monodentate phosphorus ligands for the synthesis of chiral amines was reported by Zhou and coworkers in 2002 [15]. By using the rhodium complex bearing a chiral spirobiindane phosphoramidite ligand SIPHOS, a variety of enamides can be hydrogenated to produce the corresponding amines with excellent enantioselectivities. Chan [16] and de Vries and Feringa [17] independently demonstrated that the monophosphoramidite ligands with a binaphthyl backbone were also efficient ligands for the rhodium-catalyzed asymmetric hydrogenation of enamides. The monophosphite ligands with a binaphthyl structure were introduced into this asymmetric hydrogenation reaction by Reetz et al. [18]. Initiated by these pioneering works, many monodentate phosphoramidites, phosphites, phosphonites, and phosphines have been developed for the synthesis of chiral amines by rhodium-catalyzed asymmetric hydrogenation of enamides.

8.2.1
Chiral Monodentate Phosphoramidite Ligands

Chiral monophosphoramidites were the first type of highly efficient monodentate ligands reported in the rhodium-catalyzed asymmetric hydrogenation of enamides. Monophosphoramidites can be prepared in different ways. The most common route is the treatment of enantiomerically pure diol with phosphorus trichloride (PCl_3), followed by addition of secondary amine or its lithiated equivalent [19].

In 2002, Zhou and coworkers designed and synthesized a new type of spiro phosphoamidites **1** based on spirobiindane backbone [20]. These spiro monophosphoramidite ligands **1**, especially SIPHOS (**1a**), showed high enantioselectivities (up to 99.7% ee) in the rhodium-catalyzed asymmetric hydrogenation of arylenamides (Scheme 8.4) [15]. The reaction was performed under typical conditions with rhodium catalyst, usually produced *in situ* from $[Rh(COD)_2]BF_4$ with 2 equiv of

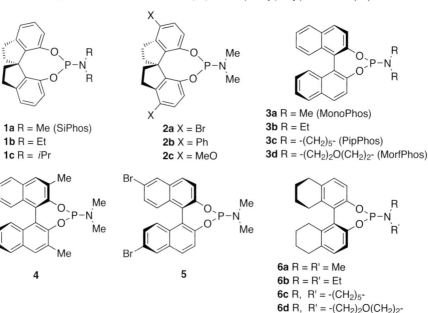

Ligand	Conditions	ee [%]	Ligand	Conditions	ee [%]
1a	50 bar H_2, toluene, 5 °C	98.7 (96)[a]	3c	25 bar H_2, THF, rt	99[b]
1b	10 bar H_2, toluene, 5 °C	57	3d	25 bar H_2, THF, rt	>99[b]
1c	10 bar H_2, toluene, 5 °C	38	4	15 bar H_2, THF, rt	44[b,c]
2a	50 bar H_2, toluene, 5 °C	97.1	5	15 bar H_2, THF, rt	89[b,c]
2b	50 bar H_2, toluene, 5 °C	97.7	6a	20 bar H_2, THF, 0 °C	96.2
2c	50 bar H_2, toluene, 5 °C	95.0	6b	20 bar H_2, THF, 0 °C	91
3a	20 bar H_2, CH_2Cl_2, -20 °C	95	6c	25 bar H_2, THF, rt	98[b]
3b	20 bar H_2, CH_2Cl_2, -5 °C	99	6d	25 bar H_2, THF, rt	97[b]

[a]10 bar H_2. [b]S/C = 50. [c]ee value of *N*-(1-(4-chlorophenyl)ethyl)acetamide (**8e**).

1a R = Me (SiPhos)
1b R = Et
1c R = iPr

2a X = Br
2b X = Ph
2c X = MeO

3a R = Me (MonoPhos)
3b R = Et
3c R = -(CH$_2$)$_5$- (PipPhos)
3d R = -(CH$_2$)$_2$O(CH$_2$)$_2$- (MorfPhos)

4

5

6a R = R' = Me
6b R = R' = Et
6c R, R' = -(CH$_2$)$_5$-
6d R, R' = -(CH$_2$)$_2$O(CH$_2$)$_2$-

SIPHOS. The degree of asymmetric induction of the reaction was found to be solvent dependent, with toluene providing the highest selectivities. Furthermore, a small dialkylamino group on the phosphorus atom of the monophosphoramidite is necessary for obtaining high enantioselectivity. When the dialkylamino group was changed from dimethylamino (**1a**) to diethylamino (**1b**) and diisopropylamino (**1c**), the enantioselectivity of the catalyst dramatically decreased from 96% ee to 57% ee to 38% ee, respectively, during the asymmetric hydrogenation of the model substrate N-(1-phenylvinyl)acetamide (**7a**) (Scheme 8.4).

Subsequent study on the substituent effect of SIPHOS ligand on the enantioselectivity showed that the introduction of either electron-donating or electron-withdrawing groups at the 4- and 4′-position of SIPHOS did not increase the enantioselectivity of ligand. For example, by using ligands **2a** (X = Br), **2b** (X = Ph), and **2c** (X = MeO), the enantioselectivities of the hydrogenation of **7a** were 97.1, 97.7, and 95.0% ee, respectively [21]. A variety of arylenamides can be hydrogenated with rhodium complex of SIPHOS ligand to produce the corresponding N-acetyl amines **8** (Scheme 8.4) with high enantioselectivities (91.1–99.3% ee). The electronic nature of the phenyl ring of the enamide had a little influence on the enantioselectivity of the reaction (Table 8.1). Chiral 1-aminoindanes are key intermediates for chiral drugs such as rasagiline for Parkinson's disease [22]. The rhodium complex containing a SIPHOS ligand **1a** is also an efficient catalyst for the asymmetric hydrogenation of cyclic enamides (Scheme 8.5). Under the conditions of 100 bar H_2 at 0 °C in toluene, N-(1,2-dehydro-1-indanyl)acetamides **11a–11c** were hydrogenated to the corresponding cyclic amines **12a–12c** with 94, 88, and 95% ee, respectively [23].

The monophosphoramidite ligands with a binaphthyl structure also showed excellent enantioselectivity in the rhodium-catalyzed asymmetric hydrogenation of enamides. Chan [16] and de Vries and Feringa [17] independently reported that the rhodium complex bearing a MonoPhos (**3a**, Scheme 8.4) was an efficient catalyst for the asymmetric hydrogenation of enamides. Under the conditions of 1 mol% catalyst generated *in situ* from [Rh(COD)$_2$]BF$_4$ with 2 equiv of MonoPhos in dichloromethane, under 300 psi H_2 at low reaction temperature (-20 °C), the substrate **7a** was reduced to N-acetyl amine **8a** with 95% ee (Scheme 8.4). Compared with this result, the monophosphoramidite ligands with a rigid spirobiindane backbone showed an advantage regarding chiral induction. The reason for this might be that the spiro monophosphoramidite SIPHOS ligand provides a more efficient chiral environment around the rhodium atom of the catalyst [15]. Introducing two methyl groups at the 3- and 3′-positions of MonoPhos, ligand **4** (Scheme 8.4), led to a dramatic drop in enantioselectivity. For the hydrogenation of N-(1-(4-chlorophenyl)vinyl)acetamide (**7e**, Table 8.1) with ligand **4**, the product amine **8e** was produced in only 44% ee with the opposite configuration compared to the product obtained with MonoPhos ligand. In contrast, the introduction of substituents such as Br at the 6- and 6′-positions of MonoPhos, ligand **5**, has little effect on the enantioselectivity [24].

Chan and coworkers found that the enantioselectivity of the hydrogenation of **7a** was significantly increased to 99% ee by replacing the dimethylamino group in the MonoPhos ligand with a diethylamino group (ligand **3b**) (Scheme 8.4) [25]. Systematic study of the effect of different dialkylamino groups, of the MonoPhos

Table 8.1 Asymmetric hydrogenation of terminal arylenamides with chiral monophosphoramidites.

7a X = H 7f X = 4-CF$_3$ 9a X = O 8a X = H 8f X = 4-CF$_3$ 10a X = O
7b X = 4-CH$_3$ 7g X = 3-CH$_3$ 9b X = S 8b X = 4-CH$_3$ 8g X = 3-CH$_3$ 10b X = S
7c X = 4-CH$_3$O 7h X = 3-CH$_3$O 8c X = 4-CH$_3$O 8h X = 3-CH$_3$O
7d X = 4-Br 7i X = 3,4-(CH)$_4$ 8d X = 4-Br 8i X = 3,4-(CH)$_4$
7e X = 4-Cl 8e X = 4-Cl

Product	1a[a]	3a[b]	3b[c]	3c[d]	3d[d]	6a[e]	17[f]	18[g]	21[h]	22[i]
8a	98.7	95	99	99	>99	96	99		97.6	98.1
8b	99.7	92	98			95	>99	97	99.3	98.4
8c		90	98	99		92	98	97	97.4	95.7
8d	99.5	96	99			98	99		99.7	95.3
8e	99.3	92		99	99		99	94	99.8	95
8f	98.9	96	99.6				99			92.2
8g	91.6	93	98			95				
8h		96	98			98				
8i		70					99.5		98.4	96.8
10a	98.7	92								
10b	95.8	94								
References	[15]	[16]	[25]	[26]	[26]	[27]	[28]	[29]	[30]	[31]

a) Reactions carried out with S/C = 100, in toluene, pH$_2$ = 50 bar, 5 °C, 12 h.
b) Reactions carried out with S/C = 100, in CH$_2$Cl$_2$, pH$_2$ = 20 bar, −20 °C, 8 h.
c) Reactions carried out with S/C = 100, in THF, pH$_2$ = 20 bar, 5 °C, 4–16 h.
d) Reactions carried out with S/C = 100, in CH$_2$Cl$_2$, pH$_2$ = 25 bar, rt, 8 h.
e) Reactions carried out with S/C = 100, in THF, pH$_2$ = 20 bar, −10 °C, 6–18 h.
f) Reactions carried out with S/C = 100, in CH$_2$Cl$_2$, pH$_2$ = 10 bar, rt, 20 h.
g) Reactions carried out with S/C = 100, in CH$_2$Cl$_2$, pH$_2$ = 25 bar, rt, 16 h.
h) Reactions carried out with S/C = 100, in CH$_2$Cl$_2$, pH$_2$ = 40 bar, rt, 2 h.
i) Reactions carried out with S/C = 100, in CH$_2$Cl$_2$, pH$_2$ = 10 bar, rt, 2 h.

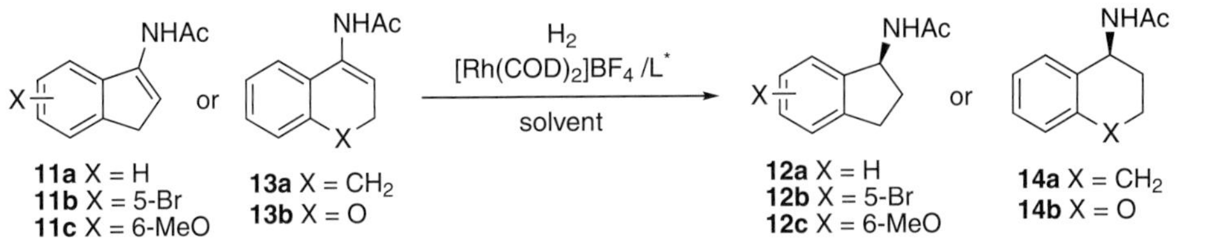

8.2 Asymmetric Hydrogenation of Enamides

11a X = H
11b X = 5-Br
11c X = 6-MeO

13a X = CH$_2$
13b X = O

12a X = H
12b X = 5-Br
12c X = 6-MeO

14a X = CH$_2$
14b X = O

Ligand	Conditions	Product
(S)-SIPHOS (**1a**):	S/C = 100, toluene, 100 bar H$_2$, -5 °C	**12a** (94% ee), **12b** (88% ee), **12c** (95% ee)
(S)-PipPhos (**3c**):	S/C = 50, CH$_2$Cl$_2$, 55 bar H$_2$, -20 °C	**12a** (98% ee), **14a** (98% ee), **14b** (>99% ee)
(S)-MorfPhos (**3d**):	S/C = 50, CH$_2$Cl$_2$, 55 bar H$_2$, -20 °C	**12a** (97% ee), **14a** (97% ee), **14b** (>99% ee)
(S)-**6c**:	S/C = 50, CH$_2$Cl$_2$, 55 bar H$_2$, -20 °C	**12a** (82% ee), **14a** (76% ee), **14b** (99% ee)
(S)-**6d**:	S/C = 50, CH$_2$Cl$_2$, 25 bar H$_2$, rt	**12a** (89% ee), **14a** (88% ee), **14b** (>99% ee)

Scheme 8.5 Asymmetric hydrogenation of cyclic enamides with monophosphoramidite ligands.

ligand, by de Vries and Feringa showed that a piperidyl (PipPhos, **3c**) or morpholine (MorfPhos, **3d**) moiety was the choice of dialkylamino group for the binaphthyl-type monophosphoramidite ligands **3**, improving the enantioselectivity of the reaction to 99% ee at room temperature [26]. With PipPhos or MorfPhos ligands, cyclic enamides **11a**, **13a**, and **13b** (Scheme 8.5) can also be easily hydrogenated to chiral cyclic amines **12a**, **14a**, and **14b** with 97–99% ee under 55 bar H$_2$ in the presence of 2 mol% rhodium catalysts. Furthermore, the rhodium complexes of PipPhos and/or MorfPhos are efficient catalysts for the asymmetric hydrogenation of the (Z)-isomer of β-substituted enamides **15b**, giving the amine product **16b**, with 96 and 98% ee, respectively (Scheme 8.6). However, these catalysts are inefficient for the hydrogenation of the (E)-isomer of **15b**, providing very low enantioselectivity (<30% ee). In the asymmetric hydrogenation of α-alkylenamide **19**, PipPhos was showed to be efficient, providing the product **20** with 82% ee under 55 bar H$_2$ at −20 °C (Scheme 8.7).

The rigid backbone of the monophosphoramidite ligand has improved the enantioselectivity of the reaction. This was further verified by the ligands with an octahydrobinaphthyl backbone. In the asymmetric hydrogenation of N-(1-phenylvinyl)acetamide (**7a**), 96.2% ee of enantioselectivity was obtained by Chan with ligand **6a** (Scheme 8.4), which was better than that obtained with Monophos (**3a**) [27]. De Vries and Feringa also applied monophosphoramidite ligands based on an octahydrobinaphthyl backbone in the rhodium-catalyzed asymmetric synthesis of chiral N-acetyl amines [26]. They found that ligands **6c** (with a P-piperidyl) and **6d** (with a P-morpholine) were highly enantioselective for the hydrogenation of enamide **7a** (Scheme 8.4), cyclic enamides (Scheme 8.5), and β-substituted acyclic enamides (Scheme 8.6), providing the corresponding chiral amines with up to 99% ee.

Zheng and coworkers introduced a series of chiral ferrocenyl amines into the binaphthyl-type monophosphoramidite ligands to improve the enantioselectivity of hydrogenation reaction. They demonstrated that the monophosphoramidite ligand

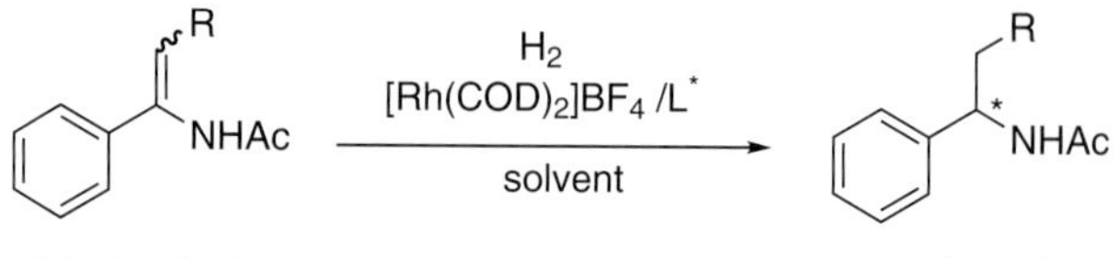

8 Enantioselective Hydrogenation of Enamines with Monodentate Phosphorus Ligands

(Z) or (E)-**15a** R = Me
(Z) or (E)-**15b** R = Et

16a R = Me
16b R = Et

Ligand	Substrate	Conditions	Product
(S)-PipPhos (**3c**):	(Z)-**15b**	S/C = 50, CH_2Cl_2, 25 bar H_2, rt	**16b** (96%ee)
(S)-MorfPhos (**3d**):	(Z)-**15b**	S/C = 50, CH_2Cl_2, 25 bar H_2, rt	**16b** (98%ee)
(S)-**6c**:	(Z)-**15b**	S/C = 50, CH_2Cl_2, 25 bar H_2, rt	**16b** (97%ee)
(S)-**6d**:	(Z)-**15b**	S/C = 50, CH_2Cl_2, 5 bar H_2, rt	**16b** (99%ee)
(R_c,S_a)-**17**	(Z)/(E)-**15a**	S/C = 100, CH_2Cl_2, 10 bar H_2, rt	**16a** (94%ee)
(R_c,S_a)-**17**	(Z)/(E)-**15b**	S/C = 100, CH_2Cl_2, 10 bar H_2, rt	**16b** (96%ee)
(S,S)-**18**:	(Z)-**15b**	S/C = 100, CH_2Cl_2, 5 bar H_2, rt	**16b** (99%ee)
(S,S)-**18**:	(Z)-**15b**	S/C = 100, CH_2Cl_2, 5 bar H_2, rt	**16b** (90%ee)
(S,S)-**18**:	(Z)/(E)-**15b**	S/C = 100, CH_2Cl_2, 5 bar H_2, rt	**16b** (92%ee)

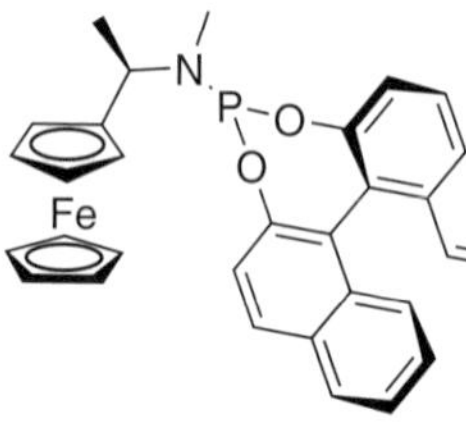

(R_c,S_a)-**17**

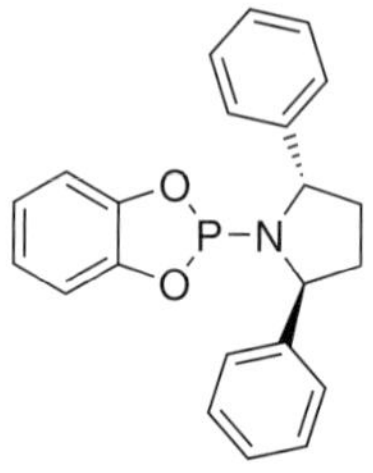

18

Scheme 8.6 Asymmetric hydrogenation of β-substituted enamides with monophosphoramidite ligands.

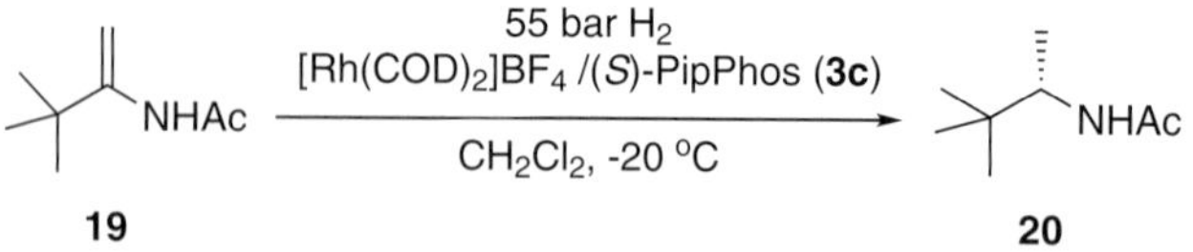

19 → **20**
55 bar H_2
$[Rh(COD)_2]BF_4$ /(S)-PipPhos (**3c**)
CH_2Cl_2, -20 °C
82% ee, 97% conv.

Scheme 8.7 Asymmetric hydrogenation of alkylenamide **19** with Rh/(S)-PipPhos (**3c**).

(R_c,S_a)-**17** (Scheme 8.6) with a (R)-N-methyl α-ferrocenylethylamine moiety has an excellent enantioselectivity in the rhodium-catalyzed asymmetric hydrogenation of enamides (Scheme 8.6) [28]. There are two chiral centers in the ligands **17**. The chiralities in the ligand (R_c,R_a)-**17** containing a (R)-N-methyl α-ferrocenylethylamine and a (R)-binaphthyl moiety were mismatched. For example, in the rhodium-catalyzed hydrogenation of enamide **7a**, the ligand (R_c,S_a)-**17** gave the product (R)-**8a** in 99% ee, whereas the ligand (R_c,R_a)-**17** gave the product (S)-**8a** in 86% ee. With ligand (R_c,S_a)-**17**, a series of α-arylenamides have been hydrogenated to the corresponding amines with up to 99.5% ee in dichloromethane under 10 bar

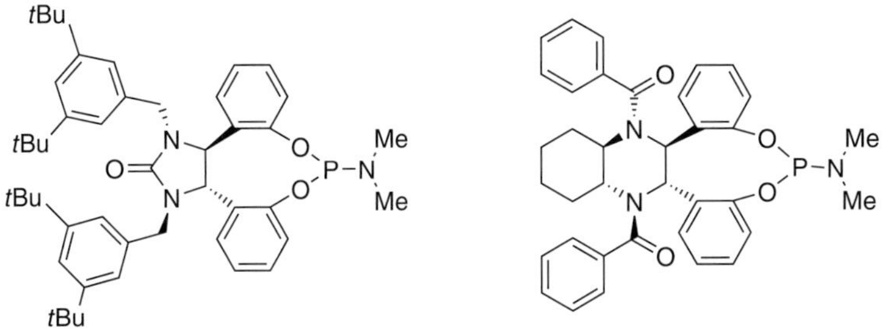

Figure 8.2 Monodentate phosphoramidites with N-heterocycle backbones.

21 (DpenPhos)   22 (CydamPhos)

hydrogen pressure at room temperature (Table 8.1). The rhodium complex of ligand (R_c,S_a)-**17** was also efficient catalyst for the hydrogenation of the Z/E mixture of β-substituted enamides **15a** and **15b**, giving the N-acetyl amines **16a** and **16b** with 94 and 96% ee, respectively (Scheme 8.6).

The monophosphoramidite ligand **18** (Scheme 8.6) was synthesized from an achiral catechol and a chiral amine [29]. The rhodium catalyst with ligand **18** afforded a comparable enantioselectivity with other monodentated phosphoramidites (96.5–97% ee) in the asymmetric hydrogenation of arylenamides **7** (Table 8.1). A high enantioselectivity (99% ee) was observed with ligand **18** in the hydrogenation of (Z)-β-substituted enamide (Z)-**15b** (EtOAc (Ac = acetyl), 25 bar, rt). In the hydrogenation of (E)-**15b**, the ligand **18** also gave the product **16b** in 90% ee (CH_2Cl_2, 5 bar, rt) with the same product configuration. This result showed that the rhodium complex of ligand **18** could be an efficient catalyst for the asymmetric hydrogenation of Z/E mixture of β-substituted enamides. The hydrogenation of Z/E mixture of **15b** (2 : 3) yielded the N-acetyl amine **16b** with 93% ee (CH_2Cl_2, 25 bar, rt), which was the calculated average of the enantioselectivities for both isomers.

Recently, Ding and Liu reported a new class of phosphoramidites, DpenPhos [30]. These monophosphoramidites were synthesized by the reaction of a chiral diol, derived from enantiomerically pure 1,2-di(2-dimethoxylphenyl)-1,2-ethylenediamine, with hexamethylphosphorus triamide (HMPT) or hexaethylphosphorus triamide in good yields. The ligand **21** (Figure 8.2) having 3,5-di-*tert*-butylbenzyl groups on the amide nitrogens and a dimethylamino group on the phosphorus atom has a high enantioselectivity in rhodium-catalyzed asymmetric hydrogenation of arylenamides **7**. Under the optimized conditions (S/C = 100, CH_2Cl_2, 40 atm H_2, rt, 2 h), a wide range of arylenamides **7** can be hydrogenated to the corresponding chiral amines with 96.1–99.8% ee by using ligand **21** (Table 8.1). However, the synthesis process of DpenPhos ligand was rather tedious; to overcome this drawback and keep the advantages of excellent enantioselectivity and fine-turning capability of DpenPhos ligand, Ding and coworkers developed another type of monophosphoramidite ligand, CydamPhos (**22**) (Figure 8.2) [31]. Ligand CydamPhos was synthesized from the readily accessible enantiomerically pure *trans*-1,2-diaminocyclohexane and salicylaldehyde derivatives in three steps with good yield. The ligand **22** was demonstrated to be efficient in rhodium-catalyzed asymmetric hydrogenation of arylenamides **7** with enantioselectivities of 91.8–98.4% ee (Table 8.1). This result is inferior to that obtained with DpenPhos ligand **21** and showed that the five-membered heterocyclic ring of the backbone in DpenPhos ligand is better than six-membered heterocyclic ring for increasing chiral inducement of the ligands.

The excellent performance in rhodium-catalyzed asymmetric hydrogenations as well as the easy preparation of monodentate phosphoramidite ligands tempted chemists to immobilize or self-assemble the chiral catalysts to be recyclable. Of course, once the monomers of the phosphoramidite ligands are polymerized or linked via self-assembly, they are no longer monodentate. However, the nature of the coordination of immobilized ligands to the central metal of the catalysts more closely resemble to the monodentate phosphoramidites. Using this concept, Ding and Wang used linkers such as a single methylene unit to make a bis-MonoPhos ligand (Scheme 8.8) [32], capable of self-assembly to polymer acting catalysts **24** by reacting

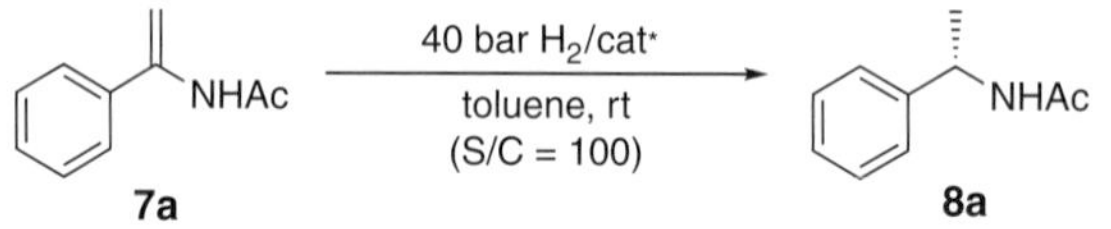

Cat*	24a	24b	24c	26
ee [%]	97.3	96.8	95.9	91.0

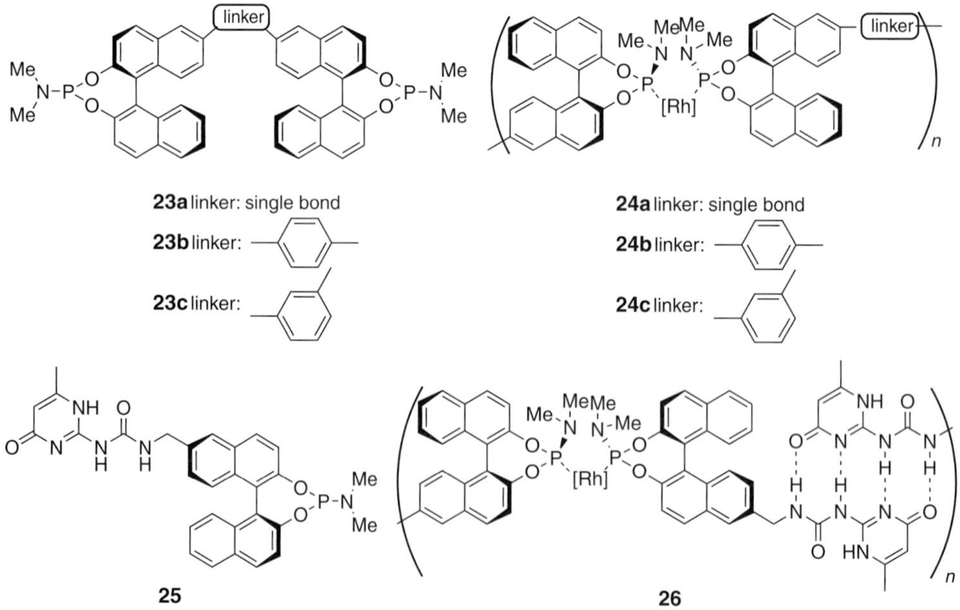

Scheme 8.8 Immobilized phosphoramidites and their rhodium catalysts.

with the catalyst precursor [Rh(COD)$_2$]BF$_4$. The self-supporting catalysts **24** gave high enantioselectivities (**24a**, 97.3% ee; **24b**, 96.8% ee; **24c**, 95.9% ee) in the asymmetric hydrogenation of arylenamide **7a** under 40 bar H$_2$ at 25 °C in toluene, which are comparable with those obtained by using the analogous monomeric ligands. A self-supporting catalyst **26** (Scheme 8.8), obtained by reaction of the rhodium precursor [Rh(COD)$_2$]BF$_4$ with monophosphoramidite ligand **25** through a noncovalent self-assembly method (hydrogen bonding and metal coordination), was also developed [33]. When the catalyst **26** was applied in catalytic asymmetric hydrogenation of arylenamide **7a** under 40 bar H$_2$ pressure in toluene at room temperature, N-acetyl amine **8a** was obtained with 91% ee. However, polymer-based catalysts are normally reused, but no data regarding the ability to use this approach in a reiterative catalyst manner was commented on for catalyst **26**.

Dendrimer-supported monophosphoramidites are also a choice for making monodentate ligands into recyclable catalysts. Fan and coworkers designed a series of dendrimer-supported monophosphoramidite ligands **27a–27d** based on binaphthyl backbone, which have high enantioselectivities in the rhodium-catalyzed asymmetric hydrogenation of arylenamides **7** (Scheme 8.9) [34]. A unique positive dentritic effect on enantioselectivity was observed in the reaction. When the dendrimer generation of the ligand **27** was increased from the first generation (**27a**, G$_0$) to the third generation (**27d**, G$_3$), the enantioselectivity of the product **8** was remarkably increased from 47–61% ee to 90–94% ee in the rhodium-catalyzed asymmetric hydrogenation of enamides **7**.

8.2.2
Chiral Monodentate Phosphite Ligands

Monodentate phosphites are another type of prominent monodentate phosphorus ligands applied in asymmetric hydrogenation of enamides for the synthesis of chiral amines. Chiral monodentate phosphites can be easily prepared from a chiral diol and an alcohol. Generally, the chiral diol was first reacted with a phosphorus trichloride to form a phosphorochloridite, followed by the reaction with an appropriate alcohol to yield a chiral monodentate phosphite [35]. The reaction of an alcohol with phosphorus trichloride to yield a phosphorodichloridite, which was then treated with a chiral diol, is also a good procedure for the synthesis of chiral monodentate phosphites [36].

The application of monophosphites in rhodium-catalyzed asymmetric hydrogenation of enamides was first reported by Reetz *et al.* in 2002 [18]. Under the conditions of catalysts generated *in situ* from 0.2 mol% [Rh(COD)$_2$]BF$_4$ and 0.4 mol% ligands **28** and 60 bar H$_2$ at 30 °C in dichloromethane, the model substrate **7a** was hydrogenated with up to 95.3% ee (Scheme 8.10). The alkoxy group on the phosphorus atom of the ligand was demonstrated to be important for obtaining high enantioselectivity. The ligands derived from alcohols cyclopentol (**28c**, 93.7% ee), benzyl alcohol (**28d**, 94.2% ee), neopentyl alcohol (**28g**, 95.3% ee), as well as chiral alcohols 2-butanol (**28e**, 93.1% ee) and 1-phenylethanol (**28i**, 94.1% ee; **28j**, 94.9% ee) gave good results, whereas the ligand derived from methanol (**28a**) afforded only moderate enantioselectivity

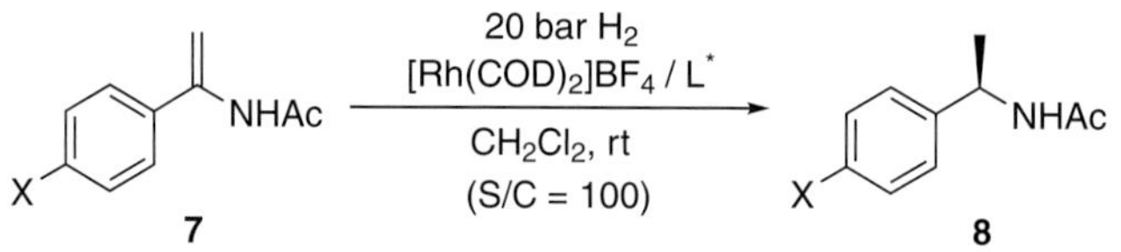

Product	X	27a	27b	27c	27d
8a	H	60% ee	78% ee	86% ee	90% ee
8b	Me	60% ee	71% ee	78% ee	94% ee
8d	Br	61% ee	73% ee	78% ee	94% ee
8e	Cl	47% ee	60% ee	78% ee	92% ee

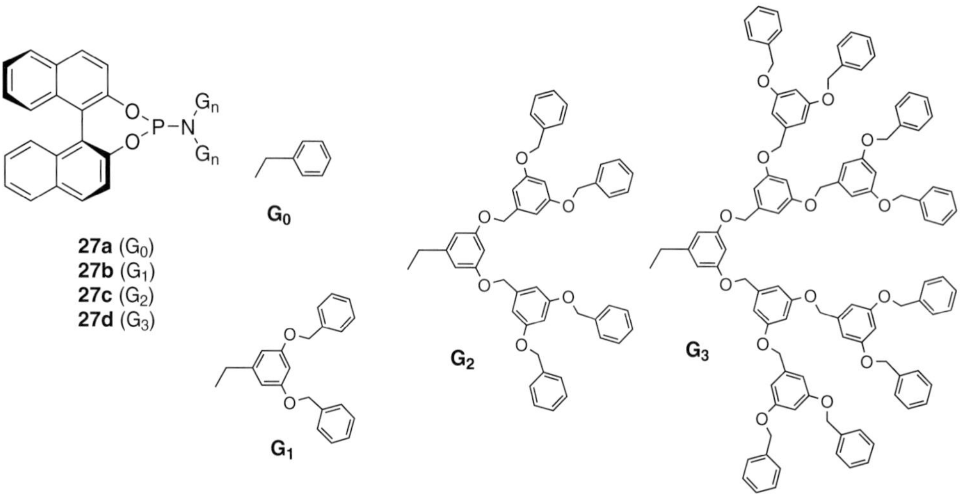

27a (G$_0$)
27b (G$_1$)
27c (G$_2$)
27d (G$_3$)

Scheme 8.9 Asymmetric hydrogenation of arylenamides with dendritic phosphoramidite ligands.

(76% ee). The chirality on the alcohol has a small role in the enantioselectivity, for instance, the ligands (R_a,S)-**28h** and (S_a,S)-**28i** gave the product with 92 and 94.1% ee, respectively. With mononphosphite ligand **28j**, different arylenamides were hydrogenated to the corresponding amines with 94.9–97.0% ee (Table 8.2). Rhodium complex of monophosphite ligand **28j** was also an efficient catalyst for the asymmetric hydrogenation of β-substituted arylenamide **15a**, 97% ee for (Z)-isomer and 76.2% ee for the (E)-isomer (Scheme 8.11).

Compared with the simple monophosphites, the monodentate phosphites derived from the easily available carbohydrates afforded considerably higher enantioselectivities in the asymmetric hydrogenation of enamides. Zheng and coworkers developed a series of binaphthyl-carbohydrate monodentate phosphites from D-fructose and D-glucose. The phosphite **29** (Scheme 8.11) exhibited a better enantioselectivity in the rhodium-catalyzed asymmetric hydrogenation of arylenamides [37]. Fructose and glucose-derived phosphites have several chiral centers and

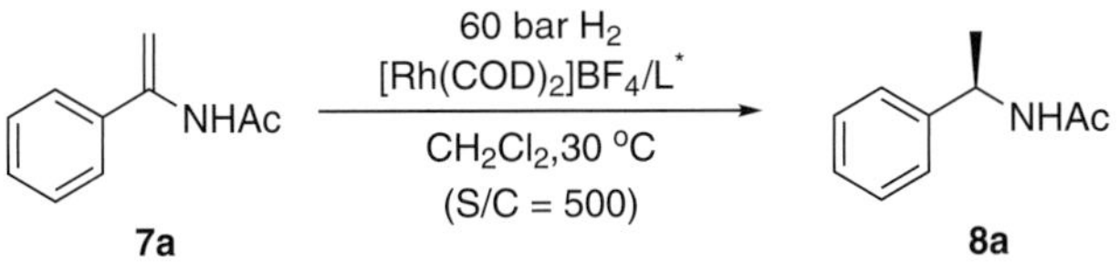

8.2 Asymmetric Hydrogenation of Enamides

Scheme 8.10 Asymmetric hydrogenation of arylenamide **7a** with monophosphites **28** Bn benzyl.

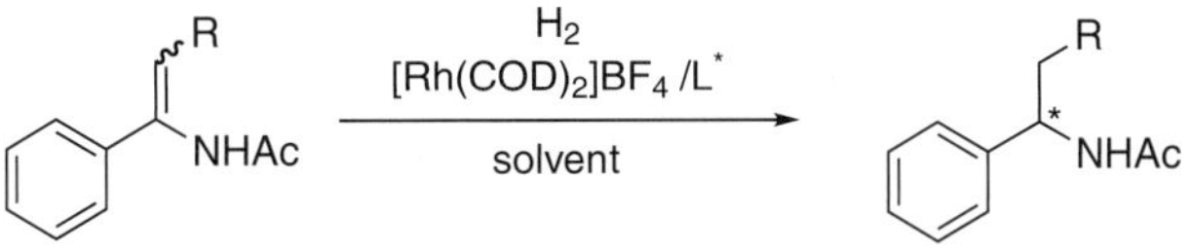

Reaction: **7a** → **8a** under 60 bar H_2, [Rh(COD)$_2$]BF$_4$/L*, CH$_2$Cl$_2$, 30 °C (S/C = 500)

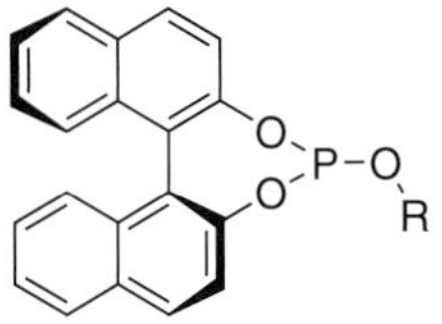

Ligand	ee [%]	Ligand	ee [%]
28a	76.0 (R)	28g	95.3 (R)
28b	89.4 (R)	28h	92.0 (R)
28c	93.7 (R)	28i	94.1 (S)
28d	94.2 (R)	28j	94.9 (R)
28e	93.1 (S)	28l	86.0 (R)
28f	93.7 (S)	28m	92.7 (S)

28a R = Me
28b R = CH(CH$_3$)$_2$
28c R = c-C$_5$H$_9$
28d R = Bn
28e R = CH$_2$CH(CH$_3$)$_2$
28f R = CH$_2$CH(C$_2$H$_5$)$_2$
28g R = CH$_2$C(CH$_3$)$_3$
28h R = (S)-CH(CH$_3$)(C$_2$H$_5$)
28i R = (S)-CH(CH$_3$)(C$_2$H$_5$) (R-BINOL)
28j R = (R)-CH(Ph)(CH$_3$)
28l R = CH$_2$CH$_2$OCH$_3$
28m R = CH$_2$CH$_2$Cl

Reaction: (Z) or (E)-**15a** R = Me / (Z) or (E)-**15b** R = Et → **16a** R = Me / **16b** R = Et, H$_2$, [Rh(COD)$_2$]BF$_4$/L*, solvent

Ligand	Substrate	Conditions	Product
(S$_a$,R)-28j	(Z)-15a (95%)	S/C = 500, CH$_2$Cl$_2$, 60 bar H$_2$, rt	16a (97.0% ee)
(S$_a$,R)-28j	(E)-15a (84%)	S/C = 500, CH$_2$Cl$_2$, 60 bar H$_2$, rt	16a (76.2% ee)
29	(Z)/(E)-15a	S/C = 100, CH$_2$Cl$_2$, 10 bar H$_2$, rt	16a (96.7% ee)
30	(Z)/(E)-15a	S/C = 100, CH$_2$Cl$_2$, 10 bar H$_2$, 20 °C	16a (99.2% ee)
31	(Z)/(E)-15a	S/C = 100, CH$_2$Cl$_2$, 10 bar H$_2$, 20 °C	16a (97% ee)
31	(Z)/(E)-15b	S/C = 100, CH$_2$Cl$_2$, 10 bar H$_2$, 20 °C	16b (99.0% ee)

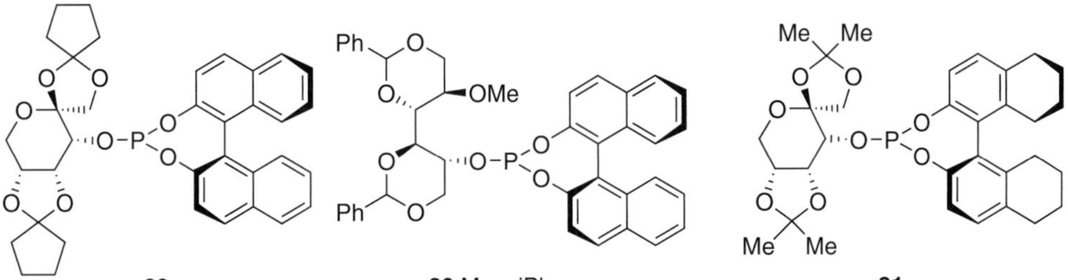

29   30 ManniPhos   31

Scheme 8.11 Asymmetric hydrogenation of β-substituted enamides with monophosphite ligands.

Table 8.2 Asymmetric hydrogenation of terminal α-arylenamides **7** with chiral monophosphite ligands.

7a X = H	7e X = 4-Cl		8a X = H	8e X = 4-Cl
7b X = 4-CH$_3$	7f X = 4-CF$_3$		8b X = 4-CH$_3$	8f X = 4-CF$_3$
7c X = 4-CH$_3$O	7i X = 3,4-(CH)$_4$		8c X = 4-CH$_3$O	8i X = 3,4-(CH)$_4$
7d X = 4-Br	7j X = 4-F		8d X = 4-Br	8j X = 4-F

Product	28j[a]	29[b]	30[c]	31[b]	32a[d]
8a	94.9	95.0	99.8	99.0	99.0
8b					98
8c		95.9	99.5	98.9	99
8d		98.5	99.9	99.6	98
8e	95.8	98.5	99.7	99.5	98
8f		98.5	99.9	99.3	
8i		96.9	99.5	99.5	
8j		96.5	99.7	99.3	97
References	[18]	[37]	[38]	[40]	[41]

a) Reactions carried out with S/C = 500, in CH$_2$Cl$_2$, pH$_2$ = 60 bar, 30 °C, 20 h.
b) Reactions carried out with S/C = 100, in CH$_2$Cl$_2$, pH$_2$ = 10 bar, room temperature, 12 h.
c) Reactions carried out with S/C = 100, in CH$_2$Cl$_2$, pH$_2$ = 10 bar, 20 °C, 12 h.
d) Reactions carried out with S/C = 100, in CH$_2$Cl$_2$, pH$_2$ = 10 bar, rt, 20 h.

the phosphite **29** with a (R)-binaphthyl and a "D-fructose" moiety was demonstrated to be a configuration-matched ligand. Under the optimized conditions (S/C = 100, CH$_2$Cl$_2$, 10 bar H$_2$, rt), a series of chiral N-acetyl arylamines **8** could be obtained with high enantioselectivities (95.0–98.5% ee) by rhodium-catalyzed asymmetric hydrogenation of arylenamides **7** with phosphite ligand **29** (Table 8.2). Generally, the arylenamides with electron-withdrawing substituents on the phenyl ring gave higher enantioselectivity than those with electron-donating substituents. The rhodium catalyst containing a phosphite **29** was also efficient for the asymmetric hydrogenation of Z/E mixture of β-substituted arylenamide **15a**, providing N-acetyl arylamine **16a** with 96.7% ee (Scheme 8.11). The D-mannitol-based monophosphite **30** (ManniPhos) was found to be excellent ligand in the asymmetric hydrogenation of α-arylenamides [38].

The rhodium-catalyzed asymmetric hydrogenation of terminal arylenamides **7** by using monophosphite ligand **30** provided chiral N-acetyl arylamines **8** with 79.1–99.9% ee (Table 8.2). The electronic nature of the substitute on the phenyl ring has little effect on the enantioselectivity of reaction, but the substitution on the *ortho*-position of phenyl ring, for instance, N-(1-(2-chlorophenyl)vinyl)acetamide (79.1% ee), resulted in a lower enantioselectivity. Ligand **30** provided extremely high enantioselectivity (99.2% ee) for the hydrogenation of the Z/E mixture of β-substituted arylenamide **15a** (Scheme 8.11). This is the best result obtained to date for the asymmetric hydrogenation of β-substituted arylenamides. Furthermore, the monophosphite **30** (ManniPhos) was also an efficient ligand for the asymmetric hydrogenation of cyclic enamides. For example, the asymmetric hydrogenation of cyclic enamide **11a** catalyzed by Rh/**30** yielded chiral N-(2,3-dihydro-1H-inden-1-yl)-acetamide (**12a**) with 96.0% ee (Scheme 8.12). It should be noted that the rhodium catalyst bearing ligand **30** has a high activity, it can produce the chiral N-acetyl arylamine **8a** with 99.5% ee at 0.1% mol catalyst loading (S/C = 1000, 100% conversion) and 95.9% ee at 0.02 mol% catalyst loading (S/C = 5000, 88% conversion). The chiral binaphthyl moiety in the ligand **30** is indispensable for obtaining high enantioselectivity. The replacement of the chiral binaphthyl structure in the monophosphite **30** with an achiral biphenyl structure led to an extremely low enantioselectivity (49.7% ee) in rhodium-catalyzed asymmetric hydrogenation of arylenamides. In another example, the chiral monophosphite with a chiral biphenyl backbone developed by Rampf and coworker also gave a low enantioselectivity in asymmetric hydrogenation of enamides [39].

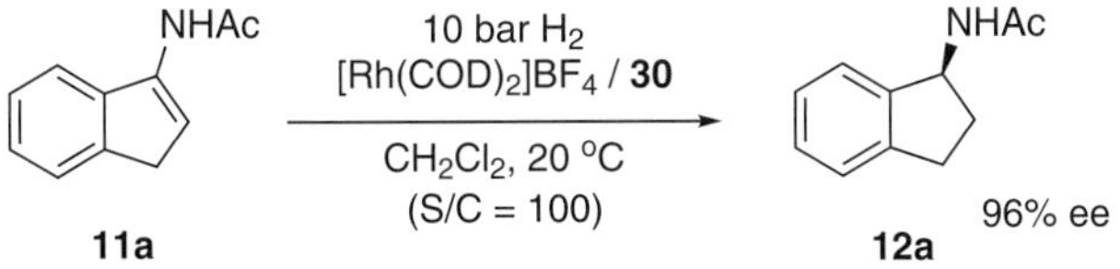

Scheme 8.12 Asymmetric hydrogenation of N-(1H-inden-1-yl)acetamide (**11a**) with monophosphite **30**.

A variety of octahydrobinaphthyl-based carbohydrate monodentate phosphites have been synthesized [40]. Among them, the monophosphite **31** (Scheme 8.11) derived from D-fructose showed the highest level of asymmetric induction in rhodium-catalyzed hydrogenation of α-arylenamides **7** (97.3–99.6% ee). In the asymmetric hydrogenation of β-substituted Z/E mixtures of arylenamides **15a** and **15b**, the ligand **31** gave the N-acetyl arylamines **16a** and **16b** with 97.3 and 99.0% ee, respectively (Scheme 8.11). The ee values obtained by octahydrobinaphthyl-based monophosphite **31** are little lower than those obtained with the binaphthyl-based ManniPhos (monophosphites **30**, Table 8.2).

Recently, Zheng and coworkers reported a type of recoverable and soluble PEGpoly(ethyleneglycol) monomethyl ether-derived polymer monophosphites (MeOPEG-monophosphites) **32a–32c** (Scheme 8.13), which were highly efficient ligands for the rhodium-catalyzed asymmetric hydrogenation of α-arylenamides [41]. For example, the MeOPEG-monophosphite **32a** gave 96–99% ee in the hydrogenations of α-arylenamides **7** (Table 8.2) and 97% ee in the hydrogenation of Z/E mixture of β-substituted enamide **15b**. The rhodium catalyst bearing MeOPEG-monophosphite ligand **32a** can be recycled four times without seriously diminishing enantioselectivity in the asymmetric hydrogenation of model substrate **7a** (Scheme 8.13). Polymer-supported monophosphite ligands **33** and **34**, developed by Chen, also showed very high enantioselectivities (95.7–96.4% ee) in the hydrogenation of α-arylenamide **7a** [42].

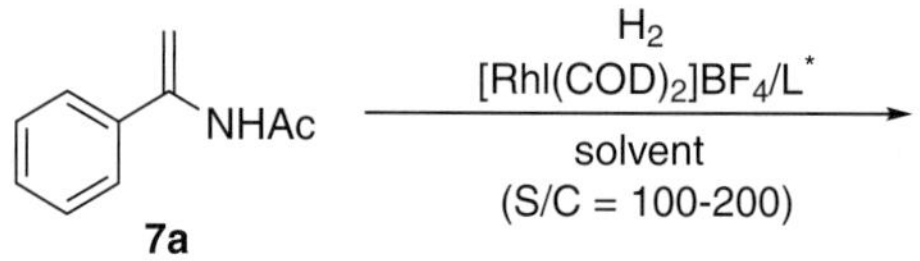

8 Enantioselective Hydrogenation of Enamines with Monodentate Phosphorus Ligands

7a →[H$_2$, [RhI(COD)$_2$]BF$_4$/L*, solvent, (S/C = 100-200)]→ **8a**

Ligand	Conditions	ee [%]
32a	S/C = 100, CH$_2$Cl$_2$, 10 bar H$_2$, rt	98
32b	S/C = 100, CH$_2$Cl$_2$, 10 bar H$_2$, rt	97
32c	S/C = 100, CH$_2$Cl$_2$, 10 bar H$_2$, rt	93
33a	S/C = 200, CH$_2$Cl$_2$, 20 bar H$_2$, rt	96.4
33b	S/C = 200, CH$_2$Cl$_2$, 20 bar H$_2$, rt	95.6
34a	S/C = 200, CH$_2$Cl$_2$, 20 bar H$_2$, rt	95.7
34b	S/C = 200, CH$_2$Cl$_2$, 20 bar H$_2$, rt	95.0

Run	Conversion [%]	ee [%]
1	100	97
2	100	97
3	100	96
4	100	91

Note: with Rh/**32b** catalyst, S/C = 100, CH$_2$Cl$_2$, 10 bar H$_2$, rt.

(S)-MeOPEG-monophasphites

32a M.W.$_{PEG}$ = 1100
32b M.W.$_{PEG}$ = 2000
32c M.W.$_{PEG}$ = 5000

33a R = Ph
33b R = H

34a PEG$_{2000}$
34b PEG$_{5000}$

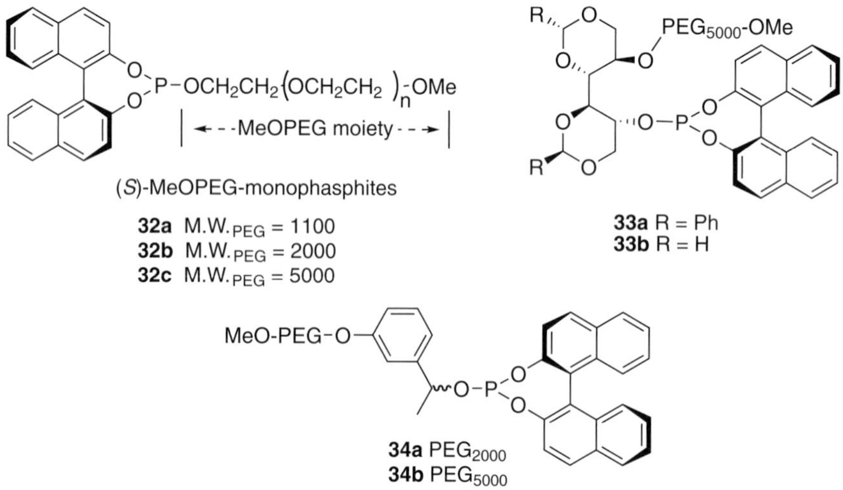

Scheme 8.13 Asymmetric hydrogenation of arylenamides with PEG-supported monophosphite ligands.

8.2.3
Other Chiral Monodentate Phosphorus Ligands

In addition to monophosphoramidites and monophosphites, monophosphonites and monophosphines have also been applied as chiral ligands for rhodium-catalyzed asymmetric hydrogenation of enamides. Reetz et al. applied monophosphonites ligands **35a** (Scheme 8.14) in the rhodium-catalyzed asymmetric hydrogenation of arylenamides [43]. Under the conditions of 0.2 mol% catalyst generated in situ from [Rh(COD)$_2$]BF$_4$, 1.3 bar of hydrogen at 30 °C in DCM for 20 h, arylenamide **7a** was hydrogenated with monophosphonite ligand **35a** to produce amine **8a** in 75.6% ee [43, 44]. Beller and coworkers achieved a higher enantioselectivity (93% ee) with monophosphine ligand **36a** in the same reaction under 2.5 bar pressure of hydrogen

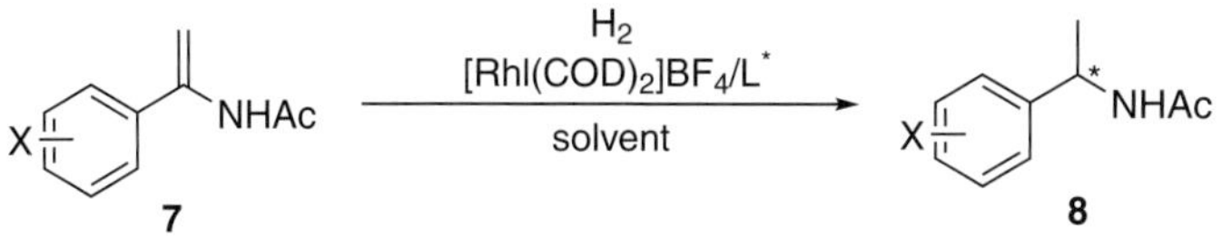

8.2 Asymmetric Hydrogenation of Enamides

7
- 7a X = H
- 7c X = 4-MeO
- 7f X = 4-CF$_3$
- 7h X = 3-MeO
- 7j X = 4-F

8
- 8a X = H
- 8c X = 4-MeO
- 8f X = 4-CF$_3$
- 8h X = 3-MeO
- 8j X = 4-F

Reagents: H$_2$, [RhI(COD)$_2$]BF$_4$/L*, solvent

Ligand	Substrate	Conditions	Product
35a	7a	S/C = 500, CH$_2$Cl$_2$, 1.5 bar H$_2$, 20 °C	8a (75.6% ee)
36a	7a	S/C = 500, CH$_2$Cl$_2$, 1.5 bar H$_2$, 20 °C	8a (14.0% ee)
36a	7a	S/C =100, toluene, 2.5 bar H$_2$, 30 °C	8a (93% ee)
36a	7c	S/C =100, toluene, 2.5 bar H$_2$, 30 °C	8c (91% ee)
36a	7f	S/C =100, toluene, 2.5 bar H$_2$, 30 °C	8f (78% ee)
36a	7h	S/C =100, toluene, 2.5 bar H$_2$, 30 °C	8h (95% ee)
36a	7j	S/C =100, toluene, 2.5 bar H$_2$, 30 °C	8j (86% ee)

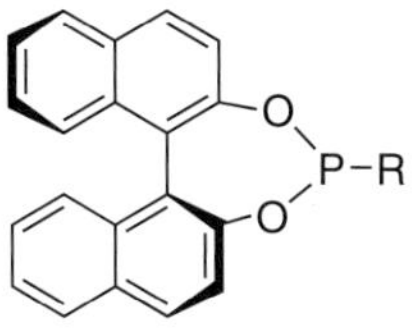

35a R = Me
35b R = tBu

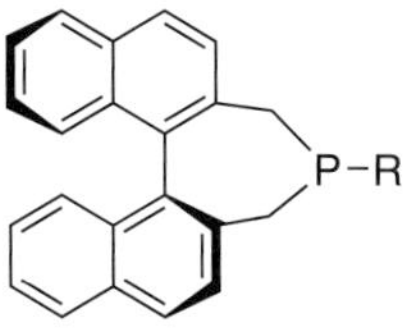

36a R = Ph
36b R = tBu

Scheme 8.14 Asymmetric hydrogenation of enamides with monodentate phosphonite and phosphine ligands.

in toluene at 30 °C with 1 mol% catalyst [45]. In Beller's report, the enantioselectivity depends on the nature of the substituents at the phosphorus atom of the ligand and the nature of the enamide substrates. The aryl-substituted phosphines are better chiral ligands than the alkyl-substituted ones, with the monophosphine **36a** being the most efficient ligand.

Furthermore, secondary monophosphine ligand, (2S,5S)-2,5-diphenylphospholane, was also used in the rhodium-catalyzed asymmetric hydrogenation of β-substituted enamide for the synthesis of chiral N-acetyl amines, albeit with low enantioselectivity (<28% ee) [46].

8.2.4
Mixed Chiral Monodentate Phosphorus Ligands

A mechanistic study of rhodium-catalyzed olefin hydrogenation using monodentate ligands (phosphoramidites, phosphites, or phosphonites) has shown that two monophosphorus ligands are bound to the metal in the transition state of the reaction [47]. Thus, the use of a mixture of two chiral monodentate ligands L^a and L^b in rhodium-catalyzed hydrogenation would lead to the formations of two homo-combination catalysts [RhL^{a}L^a and RhL^{b}L^b] and one heterocombination catalyst [RhL^{a}L^b]. These three species can be formed in various ratios and are in equilibrium with one another (Scheme 8.15). If the heterocombination catalyst [RhL^{a}L^b] dominates and/or shows higher activity and enantioselectivity, a superior catalytic profile can be expected from the undefined mixture just mentioned.

$$RhL^aL^b \rightleftharpoons RhL^aL^a + RhL^bL^b$$

heterocombination .... homocombination

Scheme 8.15 Catalytic complexes using mixed monophosphorus ligands.

The first attempt to use a mixture of chiral monophosphorus ligands to control enantioselectivity in asymmetric hydrogenation was reported by Xiao and Chen [48]. They used a mixture of the monodentate phosphite ligands derived from bisphenol and chiral alcohol in the asymmetric hydrogenation of dimethyl itaconate, however no enhancement on the enantioselectivity was observed by using mixed ligands. Reetz [43] and Feringa [49] independently introduced mixtures of chiral monodentate phosphorus ligands such as phosphites, phosphonites, and phosphoramidites with a binaphthyl backbone into the rhodium-catalyzed hydrogenation of olefins such as α-dehydroamino esters, dimethyl itaconate, and enamines. Higher enantioselectivities have been achieved by using mixed chiral phosphorus ligands in a number of hydrogenation reactions. For example, in the rhodium-catalyzed asymmetric hydrogenation of arylenamide **7a**, 96.1% ee was obtained by a combination of **35a** (R = Me) and **35b** (R = *t*Bu), while the homocatalysts Rh/**35a** and Rh/**35b** yielded the corresponding hydrogenation product **8a** with 75.6 and 13.2% ee, respectively (Scheme 8.16). [43, 44]

The application of a mixture of a chiral and an achiral monophosphorus ligand for the rhodium-catalyzed asymmetric hydrogenation of enamides was tested by Beller and coworkers [45]. By using a mixture of a chiral monophosphine **36a** and an achiral ligand tris(4-methoxyphenyl)phosphine [P(4-MeOC$_6$H$_4$)$_3$] (1:1), the N-(1-phenylvinyl)acetamide (**7a**) was hydrogenated to amine **8a** with 88% ee, but this enantioselectivity is inferior to that obtained with single monophosphine **36a** (93% ee).

8.3
Asymmetric Hydrogenation of N,N-Dialkyl Enamines

In contrast to the hydrogenation of N-acetyl enamides, there are very few examples of successful asymmetric hydrogenation of N,N-dialkyl enamines, which provides a direct approach to the synthesis of chiral tertiary amines. The reason is that an N-acetyl group in the enamides is considered indispensable for the substrates to form a chelate complex with the metal of catalyst in transition state, giving good reactivity and enantioselectivity, while there is no N-acetyl group in N,N-dialkyl enamines.

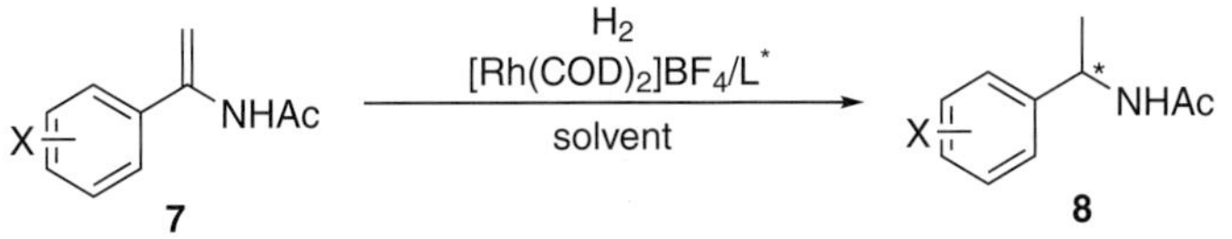

8.3 Asymmetric Hydrogenation of N,N-Dialkyl Enamines

Scheme 8.16 Asymmetric hydrogenation of enamides with mixed chiral monodentate phosphorus ligands.

7a X = H
7e X = 4-Cl
7i X = 3,4-(CH)$_4$

8a X = H
8e X = 4-Cl
8i X = 3,4-(CH)$_4$

substrate	L^a	L^b	product	ee [%] RhL^{a}L^a	RhL^{a}L^b	RhL^{b}L^b
7a	28a	35b	8a	76.0	95.0	13.2
7a	28a	36b	8a	76.0	89	24.2
7a	28d	35b	8a	91.4	97.4	13.2
7a	35a	35b	8a	75.6	96.1	13.2
7a	35a	35b	8a	75.6	87.2	24.2
7e	35a	36b	8e		95.0	
7i	35a	35b	8i	78.2	97.0	<3
7i	28a	35b	8i	76.0	97.2	<3

conditions: CH$_2$Cl$_2$, 1.3 or 1.5 bar H$_2$, 30 °C, L^a : L^b = 1:1

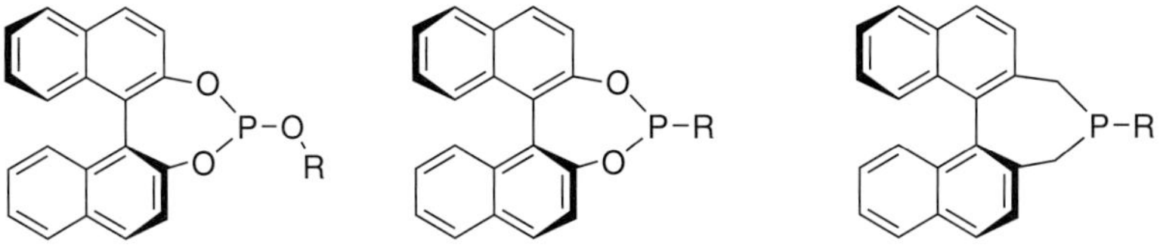

28a R = Me
28b R = tBu
28d R = Bn

35a R = Me
35b R = tBu

36a R = Ph
36b R = tBu

Because of this, N,N-dialkyl enamines would be considered as belonging to one of, if not the most, the difficult classes of substrates regarding catalytic asymmetric hydrogenation in terms of enantioselectivity.

The first example of catalytic asymmetric hydrogenation of N,N-dialkyl enamines was reported by Buchwald and Lee in 1994. By using 5 mol% chiral *ansa*-titanocene catalyst [(S,S,S)-(EBTHI)TiO-binaphtho] (EBTHI = ethylenebis(tetrahydroindenyl)), they achieved excellent enantioselectivities (up to 98% ee) in the hydrogenation of (1-arylvinyl)amines [50]. In 2000, Böner used chiral rhodium diphosphine complexes for the hydrogenation of 2-N-piperidinylethylbenzene and 2-alkyl-1,3,3-trimethyleneindoline and obtained the tertiary amines in moderate enantiomeric excesses [51].

The successful asymmetric hydrogenation of N,N-dialkyl enamines with chiral monodentate phosphorus ligands was first reported by Zhou and coworkers in 2006. They achieved high enantioselectivities for the rhodium-catalyzed asymmetric hydrogenation of 1-(1,2-diarylvinyl)pyrrolidines (Schemes 8.17 and 8.18) for the synthesis of chiral tertiary amines [52]. In the presence of 2 mol% I_2 and 20 mol% HOAc as additives and the rhodium catalyst generated *in situ* from 1 mol% [Rh (COD)$_2$]BF$_4$ and 2.2 mol% chiral monophosphonite **39c** (Scheme 8.17), the enamine 1-(1,2-diphenylvinyl)pyrrolidine (**37a**) was hydrogenated smoothly in THF under 10 bar of hydrogen pressure at room temperature to provide chiral amine **38a** in 100% conversion with 87% ee within 12 h (Scheme 8.18). Of all the chiral ligands tested, the spiro monophosphonite **39c**, with a sterically bulky *tert*-butyl on the phosphorus atom of the ligand, was the best and the corresponding binaphthyl-based monophosphonite **41** gave a lower ee value (Scheme 8.17). However, the bidentate phosphorus ligands such as BINAP, SDD (7,7'-bis(diphenylphosphino)-1,1'-spirobiindane), and JosiPhos (1-[2-(diphenylphosphino)ferrocenyl]ethyldicyclohexylphosphine) were inefficient for this reaction, providing low conversions (<40%) and poor enantioselectivities (<10% ee). The enantioselectivity of the reaction was also sensitive to the nature of substituents of enamine substrates. Generally, the substrates containing electron-donating substituents on Ar^1 and/or electron-withdrawing substituents on Ar^2 gave higher enantioselectivity (Scheme 8.18). The highest enantioselectivity (99.9% ee) was achieved in the hydrogenation of enamine **37i** having a 4-F on Ar^2.

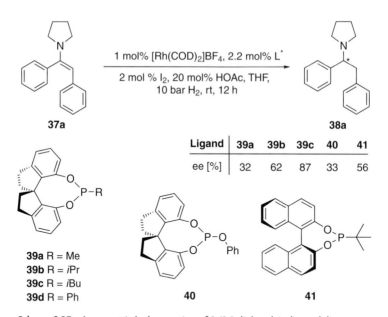

Ligand	39a	39b	39c	40	41
ee [%]	32	62	87	33	56

Scheme 8.17 Asymmetric hydrogenation of 1-(1,2-diphenylvinyl)pyrrolidine.

An iridium catalyst containing a spiro phosphoramidite ligand has recently been developed by Zhou and coworkers, which was highly efficient for the asymmetric hydrogenation of simple cyclic enamines **42** [53]. Comparing with other monodentate phosphorus ligands such as monophosphoramidites **47** with a binaphthyl backbone and the bidentate phosphorus ligands such as BINAP, SDP, SynPhos

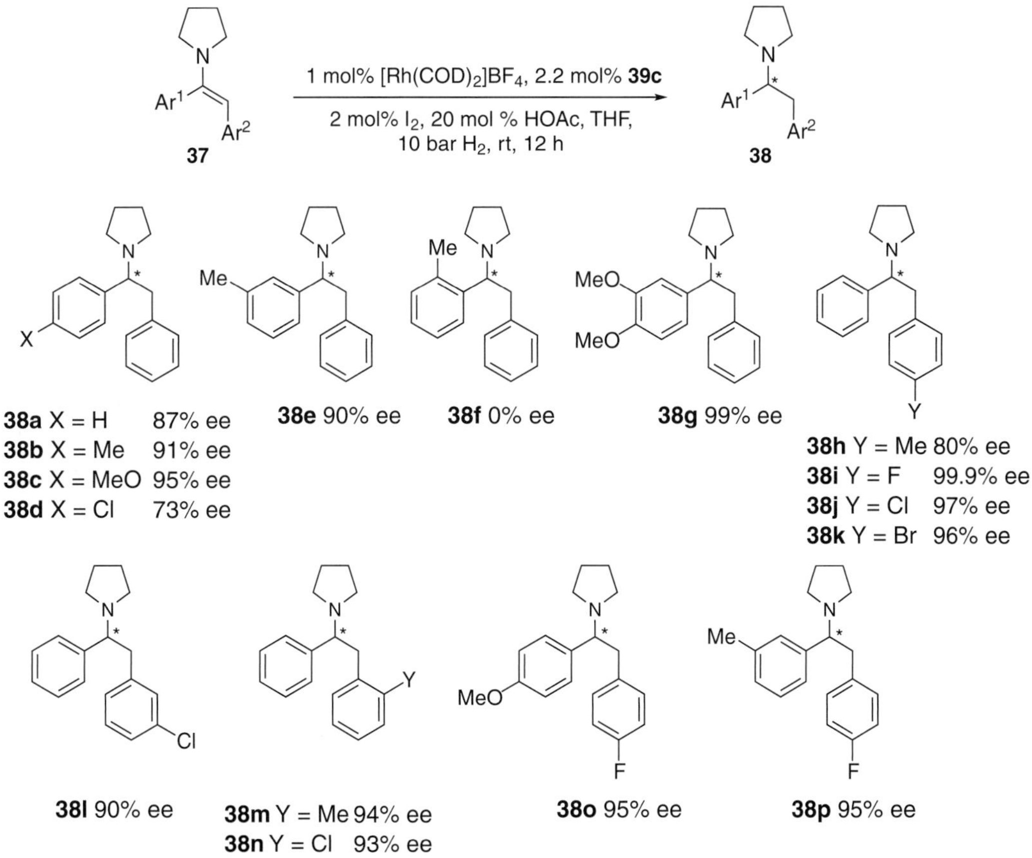

8.3 Asymmetric Hydrogenation of N,N-Dialkyl Enamines

Scheme 8.18 Asymmetric hydrogenation of 1-(1,2-diarylvinyl)pyrrolidines with Rh/(R)-**39c**.

Reaction: **37** → **38** with 1 mol% [Rh(COD)$_2$]BF$_4$, 2.2 mol% **39c**, 2 mol% I$_2$, 20 mol % HOAc, THF, 10 bar H$_2$, rt, 12 h.

38a X = H   87% ee
38b X = Me   91% ee
38c X = MeO 95% ee
38d X = Cl   73% ee

38e 90% ee

38f 0% ee

38g 99% ee

38h Y = Me 80% ee
38i Y = F   99.9% ee
38j Y = Cl 97% ee
38k Y = Br 96% ee

38l 90% ee

38m Y = Me 94% ee
38n Y = Cl 93% ee

38o 95% ee

38p 95% ee

([(5,6)(5′,6′)-bis(etylenedioxy)biphenyl-2,2′-diyl]bis(diphenylphosphine)), and Josi-Phos, the monophosphoramidite ligand (R_a,S,S)-**44** was proven to be the most efficient in the hydrogenation of 1-methyl-5-phenyl-2,3-dihydro-1H-pyrrole (**42a**), producing the cyclic amine **43a** with 94% ee (Scheme 8.19). A variety of enamines **42** with a five-membered ring could be hydrogenated to the cyclic amines **43** with 72–97% ee by the catalyst Ir/(R_a,S,S)-**44** (Scheme 8.20). However, this catalyst is less efficient for the hydrogenation of the enamine with a six-membered ring.

The catalyst Ir/(R_a,S,S)-**44** has been successfully applied to the synthesis of chiral tricyclic amines (Scheme 8.21). The hydrogenation of tricyclic enamine such as 2,3,5,6-tetrahydropyrrolo[2,1-a]isoquinolines **48a** (X = H) and **48b** (X = MeO) catalyzed by Ir/(R_a,S,S)-**44** afforded the corresponding tricyclic amines **48a** and **48b** with 88 and 90% ee, respectively. This reaction provided a convenient approach to the isoquinoline alkaloid crispine A (**48b**) [54], which was isolated from Carduus crispus, Linn. (welted thistle) and has significant cytotoxic activities [55].

42a **43a**

Ligand	ee [%]
(*R*)-SIPHOS **(1a)**	44 (*S*)
(*R*)-**39c**	71 (*S*)
(*R*)-**40**	18 (*S*)
(R_a,*S*,*S*)-**44**	94 (*S*)
(R_a,*R*,*R*)-**45**	57 (*S*)
(S_a,*S*,*S*)-**46**	70 (*S*)
(S_a,*R*,*R*)-**47**	88 (*R*)

(R_a,*S*,*S*)-**44**

(R_a,*R*,*R*)-**45** (S_a,*S*,*S*)-**46** (S_a,*R*,*R*)-**47**

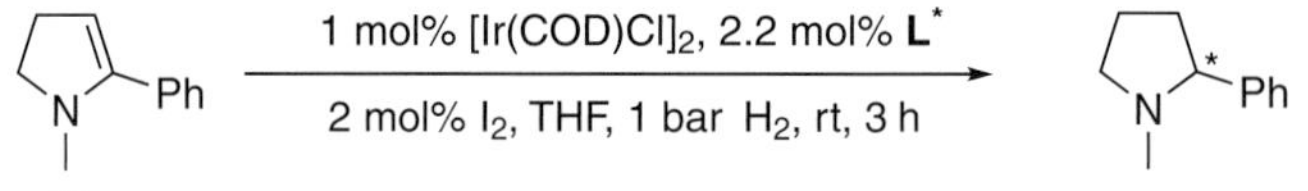

Scheme 8.19 Asymmetric hydrogenation of 1-methyl-5-phenyl-2,3-dihydro-1*H*-pyrrole.

42 **43**

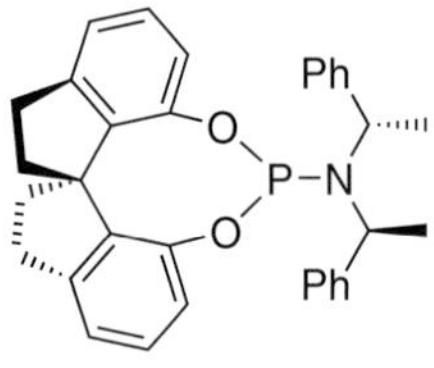

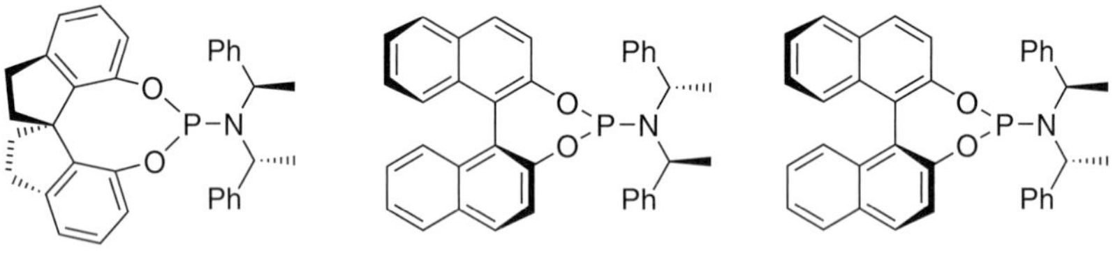

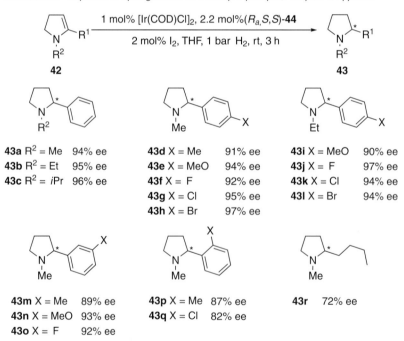

43a R^2 = Me 94% ee
43b R^2 = Et 95% ee
43c R^2 = *i*Pr 96% ee

43d X = Me 91% ee
43e X = MeO 94% ee
43f X = F 92% ee
43g X = Cl 95% ee
43h X = Br 97% ee

43i X = MeO 90% ee
43j X = F 97% ee
43k X = Cl 94% ee
43l X = Br 94% ee

43m X = Me 89% ee
43n X = MeO 93% ee
43o X = F 92% ee

43p X = Me 87% ee
43q X = Cl 82% ee

43r 72% ee

Scheme 8.20 Asymmetric hydrogenation of 1-alkyl-5-aryl-2,3-dihydro-1*H*-pyrroles with Ir/(R_a,*S*,*S*)-**44**.

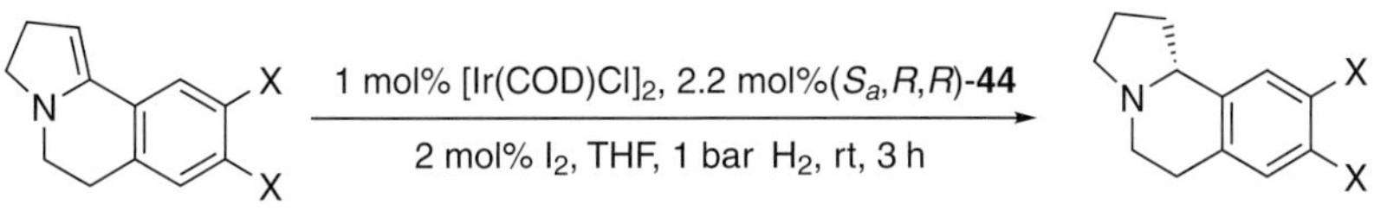

Scheme 8.21 Asymmetric hydrogenation of tricyclic enamines with Ir/(R_a,S,S)-**44**.

48a X = H
48b X = MeO

49a X = H   82% ee
49b X = MeO
(crispine A) 90% ee (97% yield)

Reaction conditions: 1 mol% [Ir(COD)Cl]$_2$, 2.2 mol% (S_a,R,R)-**44**, 2 mol% I$_2$, THF, 1 bar H$_2$, rt, 3 h

8.4 Conclusion and Outlook

Compared with bidentate phosphorus ligands, monodentate phosphorus ligands applied to the catalytic asymmetric hydrogenation of enamines, for the synthesis of chiral amines or amine derivatives, is still a primary stage. However, in recent years, a substantial progress has been made in the asymmetric hydrogenation with chiral monophosphorus ligands and some of them, such as SIPHOS (**1a**), PipPhos (**3c**), MorfPhos (**3d**), and ManniPhos (**30**), have become prominent ligands in the asymmetric hydrogenation of enamides. The catalytic asymmetric hydrogenation of unfunctionalized enamines such as N,N-dialkyl enamines with monodentate phosphorus ligands has also been developed. In such cases, the monodentate spiro phosphonite **39c** and phosphoramidite (R_a,S,S)-**44** showed excellent enantioselectivities.

The diversity requirement of chiral amines in the synthesis of natural products and chiral drugs is everlasting and the most studies about the catalytic asymmetric hydrogenation of enamines have dealt with simple substrates to date. Hence, it is necessary to explore highly efficient enantioselective protocol to provide more complex and also industrially useful chiral amines. We are confident that the easily accessible and changeable monodentate phosphorus ligands will find a wide application in this field.

Questions

8.1. The polymerized or linked monodentate phosphoramidites are no longer monodentate phosphorus ligands, but why were they also discussed in this chapter as monodentate phosphorus ligands in the asymmetric hydrogenation of enamines?

8.2. Why is the asymmetric hydrogenation of N,N-dialkyl enamines more difficult than the asymmetric hydrogenation of N-acetyl enamides?

8.3. In the Rh-catalyzed asymmetric hydrogenation of enamides, why can the mixed two different chiral monodentate phosphorus ligands improve the enantioselectivity of reaction in some cases?

References

1 Kagan, H.B. and Dang, T.-P. (1972) *J. Am. Chem. Soc.*, **94**, 6429–6433.
2 Noyori, R., Ohta, M., Hsiao, Y., Kitamura, M., Ohta, T., and Takaya, H. (1986) *J. Am. Chem. Soc.*, **108**, 7117–7119.
3 Burk, M.J., Wang, Y.M., and Lee, J.R. (1996) *J. Am. Chem. Soc.*, **118**, 5142–5143.
4 Zhu, G. and Zhang, X. (1998) *J. Org. Chem.*, **63**, 9590–9593.
5 Zhang, Z., Zhu, G., Jiang, Q., Xiao, D., and Zhang, X. (1999) *J. Org. Chem.*, **64**, 1774–1775.
6 Zhang, F.Y., Pai, C.-C., and Chan, A.S.C. (1998) *J. Am. Chem. Soc.*, **120**, 5808–5809.
7 Hu, W., Yan, M., Lau, C.-P., Yang, S.M., Chan, A.S.C., Jiang, Y., and Mi, A. (1999) *Tetrahedron Lett.*, **40**, 973–976.
8 Tang, W. and Zhang, X. (2002) *Angew. Chem., Int. Ed.*, **41**, 1612–1614.
9 Knowles, W., Sabacky, M.J., and Vineyyard, B.D. (1972) *J. Chem. Soc. Chem. Commun.*, 10–11.
10 Guo, H., Ding, K., and Dai, L. (2003) *Chin. Sci. Bull.*, **49**, 2003–2016.
11 Claver, C., Fernandez, E., Gillon, A., Heslop, K., Hyett, D.J., Martorell, A., Orpen, A.J., and Pringle, P.G. (2000) *Chem. Commun.*, 961–962.
12 (a) Reetz, M.T. and Mehler, G. (2000) *Angew. Chem., Int. Ed.*, **39**, 3889–3890; (b) Reetz, M.T. and Sell, T. (2000) *Tetrahedron Lett.*, **41**, 6333–6336.
13 van den Berg, M., Minnaard, A.J., Schudde, E.P., van Esch, J., de Vries, A.H.M., de Vries, J.G., and Feringa, B.L. (2000) *J. Am. Chem. Soc.*, **122**, 11539–11540.
14 Jerphagnon, T., Renaud, J.-L., and Bruneau, C. (2004) *Tetrahedron: Asymmetry*, **15**, 2101–2111.
15 Hu, A.-G., Fu, Y., Xie, J.-H., Zhou, H., Wang, L.-X., and Zhou, Q.-L. (2002) *Angew. Chem., Int. Ed.*, **41**, 2348–2350.
16 Jia, X., Guo, R., Li, X., Yao, X., and Chan, A.S.C. (2002) *Tetrahedron Lett.*, **43**, 5541–5544.
17 van den Berg, M., Haak, R.M., Minnaard, A.J., de Vries, A.H.M., de Vries, J.G., and Feringa, B.L. (2002) *Adv. Synth. Catal.*, **344**, 1003–1007.
18 Reetz, M.T., Mehler, G., Meiswinkel, A., and Sell, T. (2002) *Tetrahedron Lett.*, **43**, 7941–7943.
19 de Vries, A.H.M., Meetsma, A., and Feringa, B.L. (1996) *Angew. Chem., Int. Ed. Engl.*, **35**, 2374–2376.
20 Fu, Y., Xie, J.-H., Hu, A.-G., Zhou, H., Wang, L.-X., and Zhou, Q.-L. (2002) *Chem. Commun.*, 480–481.
21 Zhu, S.-F., Fu, Y., Xie, J.-H., Liu, B., Xing, L., and Zhou, Q.-L. (2003) *Tetrahedron: Asymmetry*, **14**, 3219–3224.
22 Oldfield, V., Keating, G.M., and Perry, C.M. (2007) *Drugs*, **67**, 1725–1747.
23 Fu, Y., Guo, X.-X., Zhu, S.-F., Hu, A.-G., Xie, J.-H., and Zhou, Q.-L. (2004) *J. Org. Chem.*, **69**, 4648–4655.
24 van den Berg, M., Minnaard, A.J., Haak, R.M., Leeman, M., Schudde, E.P., Meetsma, A., Feringa, B.L., de Vries, A.H.M., Maljaars, C.E.P., Willans, C.E., Hyett, D., Boogers, J.A.F., Henderickx, H.J.W., and de Vries, J.G. (2003) *Adv. Synth. Catal.*, **345**, 308–323.
25 Jia, X., Li, X., Xu, L., Shi, Q., Yao, X., and Chan, A.S.C. (2003) *J. Org. Chem.*, **68**, 4539–4541.
26 Bernsman, H., van den Berg, M., Hoen, R., Minnaard, A.J., Mehler, G., Reetz, M.T., de Vries, J.G., and Feringa, B.L. (2005) *J. Org. Chem.*, **70**, 943–951.
27 Li, X., Jia, X., Lu, G., Au-Yeung, T.T.-L., Lam, K.-H., Lo, T.W.H., and Chan, A.S.C. (2003) *Tetrahedron: Asymmetry*, **14**, 2687–2691.
28 Zeng, Q.-H., Hu, X.-P., Duan, Z.-C., Liang, X.-M., and Zheng, Z. (2006) *J. Org. Chem.*, **71**, 393–396.
29 Hoen, R., van den Berg, M., Bernsmann, H., Minnaard, A.J., de Vries, J.G., and Feringa, B.L. (2004) *Org. Lett.*, **6**, 1433–1436.
30 Liu, Y. and Ding, K. (2005) *J. Am. Chem. Soc.*, **127**, 10488–10489.
31 Zhao, B., Wang, Z., and Ding, K. (2006) *Adv. Synth. Catal.*, **348**, 1049-1057.
32 Wang, X. and Ding, K. (2004) *J. Am. Chem. Soc.*, **126**, 10524–10525.
33 Shi, L., Wang, X., Sandoval, C.A., Li, M., Qi, Q., Li, Z., and Ding, K. (2006) *Angew. Chem., Int. Ed.*, **45**, 4108–4112.

34 Zhang, F., Li, Y., Li, Z.-W., He, Y.-M., Zhu, S.-F., Fan, Q.-H., and Zhou, Q.-L. (2008) *Chem. Commun.*, 6048–6050.
35 Pastor, S.D., Shum, S.P., Rodebaugh, R.K., Debellis, A.D., and Clarke, F.H. (1993) *Helv. Chim. Acta*, **76**, 900–914.
36 Brunel, J.M. and Buono, G. (1993) *J. Org. Chem.*, **58**, 7313–7314.
37 Huang, H., Zheng, Z., Luo, H., Bai, C., Hu, X., and Chen, H. (2003) *Org. Lett.*, **5**, 4137–4139.
38 Huang, H., Zhuo, Z., Luo, H., Bai, C., Hu, X., and Chen, H. (2004) *J. Org. Chem.*, **69**, 2355–2361.
39 Hannen, P., Militzer, H.-C., Vogl, E.M., and Rampf, F.A. (2003) *Chem. Commun.*, 2210–2211.
40 Huang, H., Liu, X., Chen, H., and Zheng, Z. (2005) *Tetrahedron: Asymmetry*, **16**, 693–697.
41 Hu, X.-P., Huang, J.-D., Zeng, Q.-H., and Zheng, Z. (2006) *Chem. Commun.*, 293–295.
42 Chen, W., Roberts, S.N., and Whittall, J. (2006) *Tetrahedron Lett.*, **47**, 4263–4266.
43 Reetz, M.T., Sell, T., Meisweinkel, A., and Mehler, G. (2003) *Angew. Chem., Int. Ed.*, **42**, 790–793.
44 Reetz, M.T., Mehler, G., and Meiswinkel, A. (2004) *Tetrahedron: Asymmetry*, **15**, 2165–2167.
45 Enthaler, S., Hagemann, B., Junge, K., Erre, G., and Beller, M. (2006) *Eur. J. Org. Chem.*, 2912–2917.
46 Galland, A., Dobrota, C., Toffano, M., and Fiaud, J.-C. (2006) *Tetrahedron: Asymmetry*, **17**, 2354–2357.
47 Reetz, M.T., Meiswinkel, A., Mehler, G., Angermund, K., Graf, M., Thiel, W., Mynott, R., and Blackmond, D. (2005) *J. Am. Chem. Soc.*, **127**, 10305–11013
48 Chen, W. and Xiao, J. (2001) *Tetrahedron Lett.*, **42**, 8737–8742.
49 Peña, D., Minnaard, A.J., Boogers, J.A.F., de Vries, A.H.M., de Vries, J.G., and Feringa, B.L. (2003) *Org. Biomol. Chem.*, **1**, 1087–1089.
50 Lee, N.E. and Buchwald, S.L. (1994) *J. Am. Chem. Soc.*, **116**, 5985–5986.
51 Tararov, V.I., Kadyrov, R., Riermeier, T.H., Holz, J., and Böner, A. (2000) *Tetrahedron Lett.*, **41**, 2351–2355.
52 Hou, G.-H., Xie, J.-H., Wang, L.-X., and Zhou, Q.-L. (2006) *J. Am. Chem. Soc.*, **128**, 11774–11775.
53 Hou, G.-H., Xie, J.-H., Yan, P.-C., and Zhou, Q.-L. (2009) *J. Am. Chem. Soc.*, **131**, 1366–1367.
54 Wu, T.R. and Chong, J.M. (2006) *J. Am. Chem. Soc.*, **128**, 9646–9647.
55 Zhang, Q., Tu, G., Zhao, Y., and Cheng, T. (2002) *Tetrahedron*, **58**, 6795–6798.

9
Bidentate Ligands for Enantioselective Enamide Reduction
Xiang-Ping Hu and Zhuo Zheng

9.1
Introduction

In 1972, Kagan and Dang reported the first Rh-catalyzed asymmetric hydrogenation of an enamide [1] (Scheme 9.1), opening up opportunities for convenient access to chiral amines that are important building blocks for organic synthesis as well as resolving agents, chiral auxiliaries, and ligands for many useful transformations.

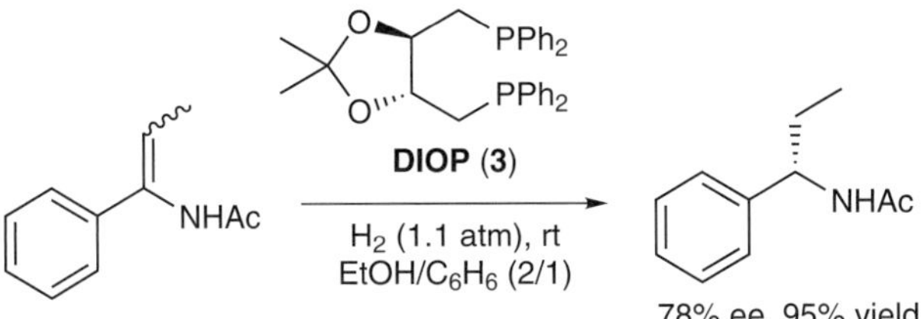

Scheme 9.1 The first catalytic asymmetric hydrogenation of enamide **1a** reported by Kagan.

Following this pioneering work, a number of studies on the Rh- or Ru-catalyzed asymmetric hydrogenation of enamides have been published. Now, catalytic asymmetric hydrogenation of enamides is arguably one of the most powerful methods for the preparation of chiral amines. The breakthrough in this area is largely due to the successful development of efficient chiral bidentate phosphorus ligands such as DuPHOS, BINAP, TangPhos, PPFAPhos, and others. In this chapter, we will mainly focus on the catalytic asymmetric hydrogenation of enamides with chiral chelating phosphorus ligands. Although there are an increasing number of monodentate P-containing ligands that have been found to be effective for ruthenium, rhodium, and iridium-catalyzed asymmetric hydrogenation of enamides in the past few years, they will not be discussed here. For details of these chemistries, the reader should refer to the chapter detailing the use of monodentate phosphorus-based ligands.

Chiral Amine Synthesis: Methods, Developments and Applications. Edited by Thomas C. Nugent
Copyright © 2010 WILEY-VCH Verlag GmbH & Co. KGaA, Weinheim
ISBN: 978-3-527-32509-2

9.2
Catalytic Enantioselective Hydrogenation of Enamides

9.2.1
Synthesis of Enamides

The accessibility of prochiral precursors is a key factor to a successful hydrogenation process from a practical point. There are several common and efficient methods for the formation of various enamides. An example is the preparation of N-acyl-α-arylenamides as illustrated in Scheme 9.2.

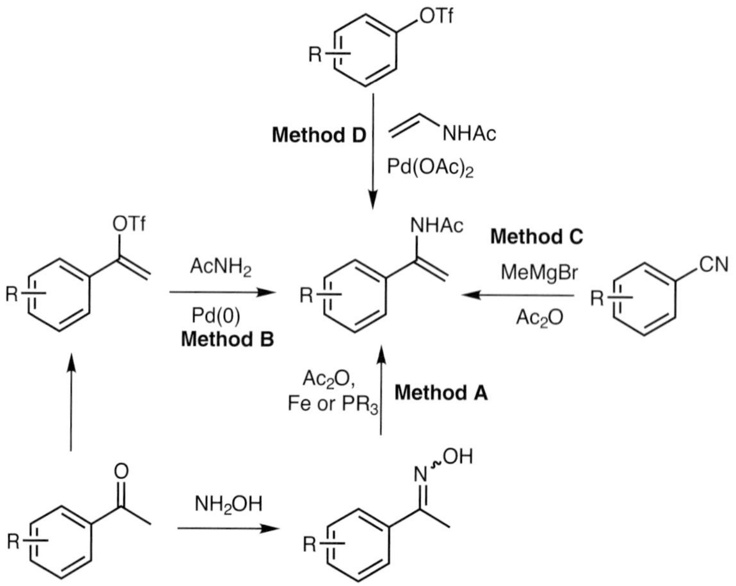

Scheme 9.2 General methods for the synthesis of N-acyl-α-arylenamides.

The present methods for the preparation of N-acyl-α-arylenamides include (i) reductive acylation of ketoximes with iron metal [2] or phosphines [3] in the presence of acyl chlorides or acyl anhydrides (Method A); (ii) metal-catalyzed coupling of vinyl halides [4], triflates [5], or tosylates [6] with amides (Method B); (iii) reaction of imine intermediates derived from nitriles with acyl chlorides or anhydrides (Method C) [7]; and (iv) Pd-catalyzed coupling of aryl tosylates with N-acyl vinylamines (Method D) [8].

Some cyclic enamides or enecarbamates can be easily and directly prepared by the condensation of primary amides or carbamates with cyclic ketones in the presence of *para*-toluenesulfonic acid as catalyst and by using a Dean–Stark apparatus to remove water (Scheme 9.3) [9].

The enamides with an exocyclic double bond have been extensively used in the catalytic asymmetric hydrogenation. These compounds can be prepared by some different methods as illustrated in Scheme 9.4 [10] and Scheme 9.5 [11].

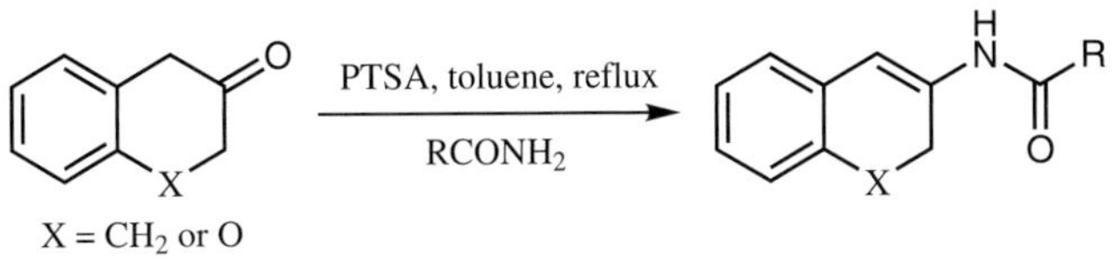

9.2 Catalytic Enantioselective Hydrogenation of Enamides

Scheme 9.3 Synthesis of cyclic enamides.

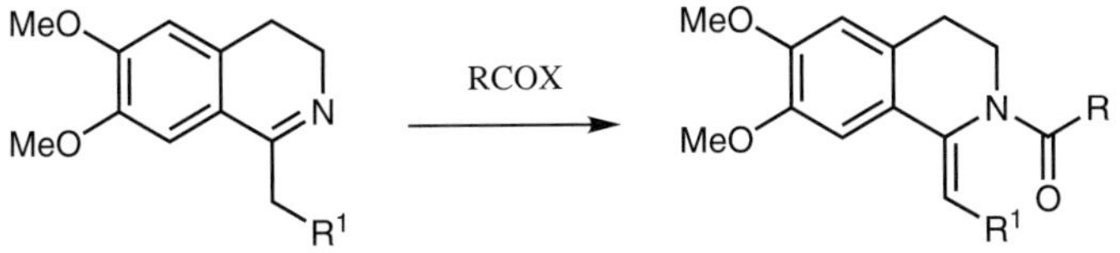

Scheme 9.4 Synthesis of exocyclic enamides.

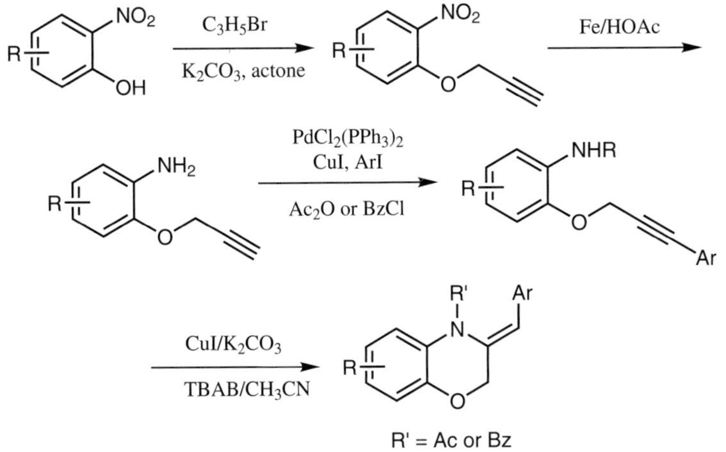

Scheme 9.5 Synthesis of exocyclic enamides.

The synthesis of N-phthaloyl enamides has been reported by a remarkably general method for aerobic oxidative amination of unactivated alkyl olefins as shown in Scheme 9.6 [12]. From a practical synthesis point of view, the phthalimide can not only serve as a directing group for asymmetric hydrogenation but can also be removed under mild conditions.

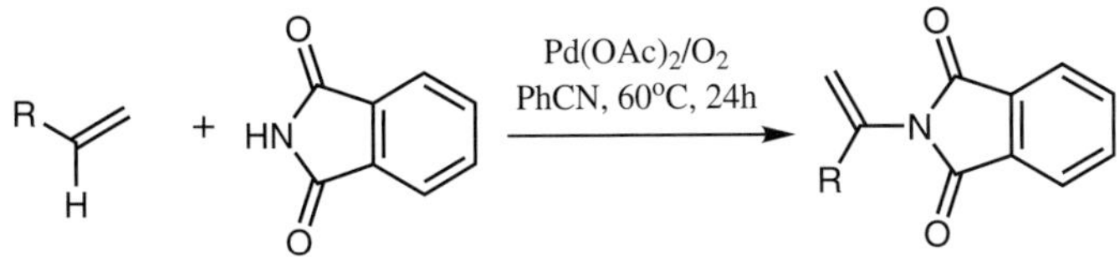

Scheme 9.6 Synthesis of N-phthaloyl enamides.

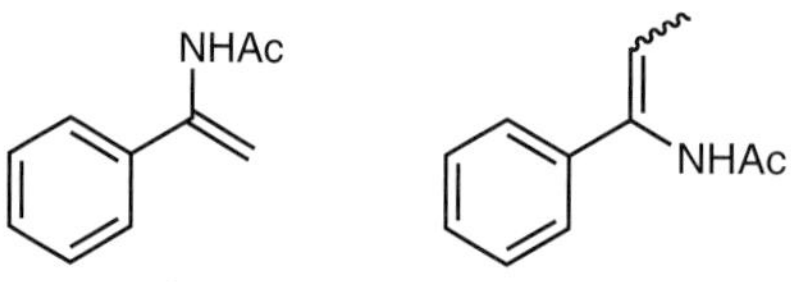

Figure 9.1 The most widely used model substrates **4a** and **1a** for ligand evaluation.

9.2.2
Catalytic Asymmetric Hydrogenation of Acyclic Enamides

The hydrogenation of acyclic enamides has made great progress in past decades. In particular, N-acyl-α-arylenamides have been used as a model substrate class for the evaluation of the efficiency of newly developed ligands in catalytic asymmetric hydrogenation. It should be borne in mind that for most new ligands, only experimentals with selected model test substrates carried out under standard conditions are available, with very few having been applied to industrially relevant problems and to challenging substrate class. Therefore, in this chapter, we only concisely list the optimal results (>95% ee) for these new ligands, and detail the important advances in some recent development of the hydrogenation of practical or challenging substrates. We will provide the progress herein according to the type of ligands used (Figure 9.1).

9.2.2.1 Chiral Phospholane Ligands for Rh-Catalyzed Asymmetric Hydrogenation

In comparison with the hydrogenation of α-dehydroamino acid derivatives, research regarding the hydrogenation of enamides was somewhat sluggish with few satisfactory results being reported until 1996 when phospholane ligands DuPHOS and BPE were employed in this reaction [13]. In this significant report, Burk found that the cationic Rh catalyst based on Me-DuPHOS (**5**) and Me-BPE (**6**) effected the hydrogenation of N-acetyl-α-arylenamides to yield a wide variety of α-arylethylamine derivatives with high enantioselectivities (up to 97.8% ee). More important, these catalysts have the ability to tolerate β-substituents in both (*E*)- and (*Z*)-positions of enamides, thus allowing the production of a diverse array of α-arylalkylamines through hydrogenation of isomeric mixtures of enamide substrates (Scheme 9.7).

Following this important work, a large number of chiral phospholane ligands have since been found to show good to excellent enantioselectivities in the Rh-catalyzed asymmetric hydrogenation of enamides. In this context, Zhang's group has contributed most of the important advances after the successful application of DuPHOS and BPE in this area. Efficient ligands developed within Zhang's group include TangPhos (**7**) [14], DuanPhos (**8**) [15], Binaphane (**11**) [16], Ketalphos (**14**) [17], and so on (Figure 9.2). In particular, with TangPhos, TONs as high as 10,000 can be achieved in the hydrogenation of 1,1-disubstituted enamides, and for the β-branched enamides, excellent selectivities were obtained no matter whether a single isomer or an (*E/Z*) isomeric mixture was used. Figure 9.2 and Table 9.1 detail the most selective phospholane-based ligands (>95% ee) that have been used in the hydrogenation of model substrates, namely, N-(1-phenylethenyl)acetamide (**4a**) and N-(1-phenylpropenyl)acetamide (**1a**) (Figure 9.1).

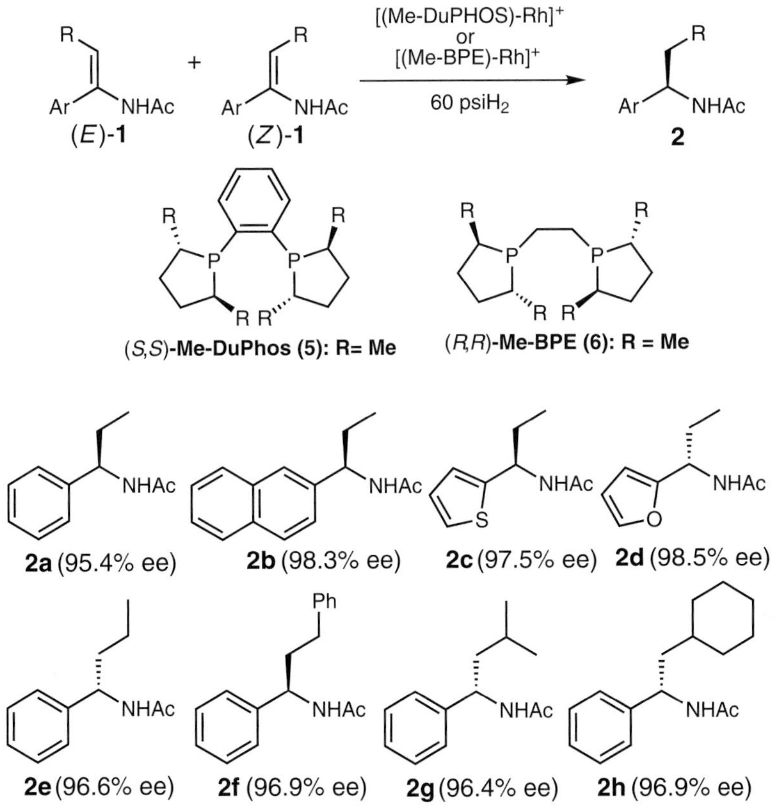

9.2 Catalytic Enantioselective Hydrogenation of Enamides

Scheme 9.7 Rh-catalyzed asymmetric hydrogenation of enamides with Me-DuPHOS (5) and Me-BPE (6).

In addition to α-arylenamides, some α-alkylenamides such as α-*tert*-butylenamide (16) and α-adamantylenamide (17) (Figure 9.3) have also been hydrogenated in high enantioselectivities (>99% ee) and activities (TON 5000, TOF >625 h^{-1}) with Rh-Me-DuPHOS [23]. Interestingly, these bulky alkylenamides are reduced with the opposite sense of induction. A computational modeling study reported by Landis and Feldgus suggested that the hydrogenation of α-alkyl and α-arylenamides involves the different coordination pathways [24].

Although excellent enantioselectivities have been obtained in the hydrogenation of α-aryl-β-alkylenamides (1) with various chiral ligands, the asymmetric hydrogenation of the structurally similar α-alkyl-β-arylenamides surprisingly received less attention. In an early example, using an Rh/DIPAMP complex, only moderate enantioselectivity (50% ee) was achieved [25]. Very recently, Zhang has found that a series of (Z)-α-alkyl-β-arylenamides (18) can be hydrogenated in excellent ee values (up to 99% ee) by using the Rh/Tangphos catalytic system (Scheme 9.8). In sharp contrast, much lower ee values were observed for (E)-isomer in EtOAc, albeit with the same sense of product chirality as that obtained from (Z)-isomer [26].

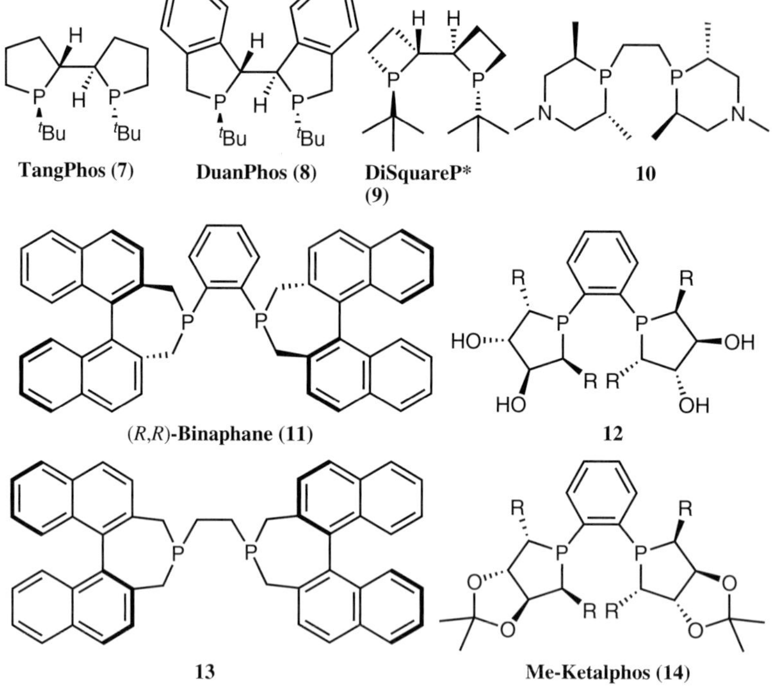

TangPhos (7)   **DuanPhos (8)**   **DiSquareP* (9)**   **10**

(R,R)-Binaphane (11)   **12**

13   **Me-Ketalphos (14)**

Figure 9.2 Pholpholane ligands for asymmetric hydrogenation.

This hydrogenation method is a practical way for the preparation of β-arylisopropylamines, an important class of chiral compounds with valuable pharmaceutical applications (Figure 9.4). For example, deacylation of the chiral product (S)-**19a** leads directly to (S)-amphetamine (**20**), which is a useful stimulant with strong biological and physiological effects. Additional modification of (R)-**19a** will result in selegiline (**22**) for the treatment of Alzheimer's disease. Asymmetric hydrogenation of (Z)-**19c** will also provide practical access to important chiral drugs such as formoterol (**21**) and tamsulosin (**23**) [26].

By using an Rh-catalyst containing a ligand from electronic-rich TangPhos or DuanPhos, Zhang has successfully hydrogenated a range of N-phthaloyl α-arylenamides **24** in excellent enantioselectivities (up to 99% ee) (Scheme 9.9). However, much lower enantioselectivities were observed with α-alkylenamides and α-arylenamides bearing an *ortho*-substituent on the aromatic ring [27].

9.2.2.2 Chiral 1,4-Diphosphine Ligands for Rh-Catalyzed Asymmetric Hydrogenation

In 1972, Kagan and Dang reported the first catalytic asymmetric hydrogenation of enamides with an Rh/DIOP catalyst, however, the enantioselectivity is not good as some other chiral bisphosphines [1]. A possible explanation for this observation is that the seven-membered chelate ring of DIOP bound to a transition metal is too conformationally flexible. This drawback attracted extensive research efforts focused on modifying the DIOP skeleton, leading to a diversity of 1,4-diphosphine ligands

Table 9.1 Rh-catalyzed asymmetric hydrogenation of model substrates with phospholane ligands.

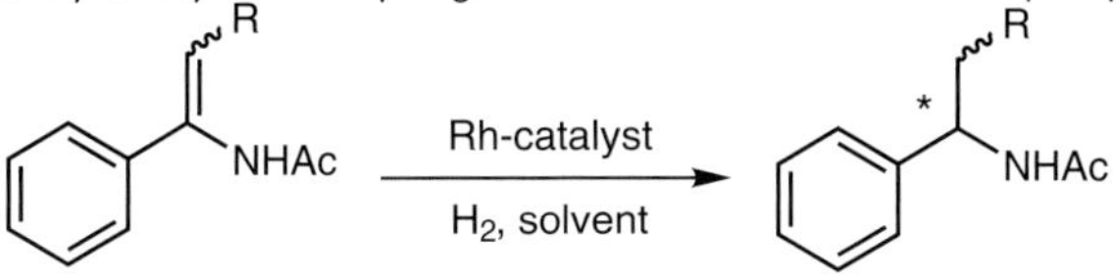

4a: R = H
1a: R = Me

15a: R = H
2a: R = Me

Ligand	Substrate	S/C ratio	Reaction conditions	% ee (config.)	References
(R,R)-Me-BPE	4a	500	MeOH, 22 °C, 60 psi H_2	95.2 (R)	[13]
(R,R)-Me-BPE	(Z/E)-1a	500	MeOH, 22 °C, 60 psi H_2	95.4 (R)	[13]
(R,R)-Ph-BPE	4a	5000	MeOH, 25 °C, 10 bar H_2	99	[18]
TangPhos (7)	4a	10,000	MeOH, rt, 20 psi H_2	99.3 (R)	[14]
TangPhos (7)	(Z/E)-1a	100	MeOH, rt, 20 psi H_2	98 (R)	[14]
DuanPhos (8)	4a	100	MeOH, rt, 20 psi H_2	>99 (R)	[15]
DiSquareP* (9)	4a	100	MeOH, rt, 2 atm H_2	>99 (R)	[19]
DiSquareP* (9)	(Z)-1a	100	MeOH, rt, 2 atm H_2	>99 (R)	[19]
10	4a	100	CH_2Cl_2, rt, 15 psi H_2	96 (R)	[20]
10	(Z/E)-1a	100	CH_2Cl_2, rt, 15 psi H_2	99 (R)	[20]
11	(Z/E)-1a	100	CH_2Cl_2, rt, 20 psi H_2	99.1 (S)	[17]
12	4a	100	MeOH, rt, 10 atm H_2	96 (S)	[21]
13	(Z/E)-1a	100	MeOH, rt, 40 psi H_2	97.5 (R)	[22]
Me-Ketalphos (14)	(Z/E)-1a	100	CH_2Cl_2, rt, 10 atm H_2	98 (S)	[16]

exhibiting excellent enantioselectivities in the hydrogenation of enamides. One such modification is the development of (R,R)-BICP by Zhang [28]. This ligand contains two cyclopentane rings in its backbone that are present to restrict its conformational flexibility. The key feature of the Rh/(R,R)-BICP catalytic system is that the enantioselectivity of the hydrogenation reaction is not sensitive to the geometry of the starting enamides, providing a wide range of arylalkylamines in 90.5–95.2% ee [29]. A further application of the Rh/(R,R)-BICP catalytic system is in the synthesis of chiral β-amino alcohols involving asymmetric hydrogenation of α-arylenamides with a MOM-protected β-hydroxy group since the preparation and isolation of the isomerically pure E- and Z-isomers are difficult. With this methodology, a variety of N-acetyl and O-MOM-protected arylglycinols were obtained in quantitative yields with high enantiomeric excesses (90–99% ee) (Scheme 9.10) [30].

Considering that a metal-DIOP complex is conformationally flexible and the stereogenic centers may be too far from the substrate, Kagan synthesized a modified DIOP ligand **28** in which the stereogenic centers are closer to the phosphorus atom [31].

Figure 9.3 α-Alkylenamides to have been reduced with phospholane ligands.

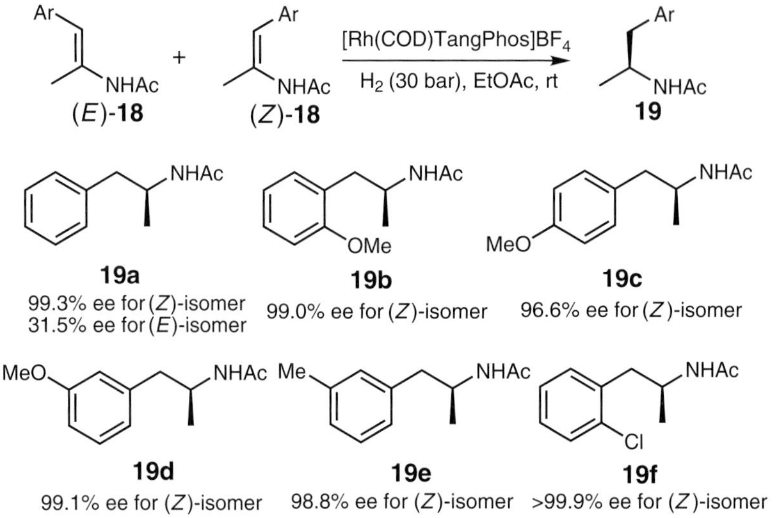

9 Bidentate Ligands for Enantioselective Enamide Reduction

Scheme 9.8 Rh-catalyzed asymmetric hydrogenation of α-alkyl-β-arylenamides **18** with TangPhos.

However, enantioselectivity for the Rh-catalyzed asymmetric hydrogenation of enamides with this ligand is lower than that obtained with the corresponding Rh-DIOP complex. Zhang speculated that the two methyl groups in **28** may locate in axial positions in the chelating ring resulting in an unfavorable conformation for enantioselective asymmetric reactions [32]. As a result, Zhang [32] and Rajanbabu [33] independently reported a new bisphosphine ligand (R,S,S,R)-DIOP* (**29**) in which methyl groups along with all other substituents are oriented in equatorial positions (Figure 9.5). A following investigation on the Rh-catalyzed asymmetric hydrogenation of α-arylenamides with (R,S,S,R)-DIOP* (**29**) disclosed that this new ligand displayed extremely high enantios-electivities (97.3 to >99% ee), far superior to that obtained with DIOP.

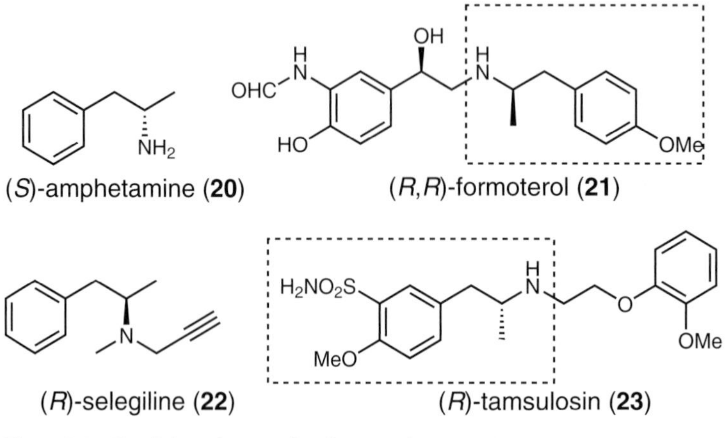

Figure 9.4 Chiral drugs bearing β-arylisopropylamine units.

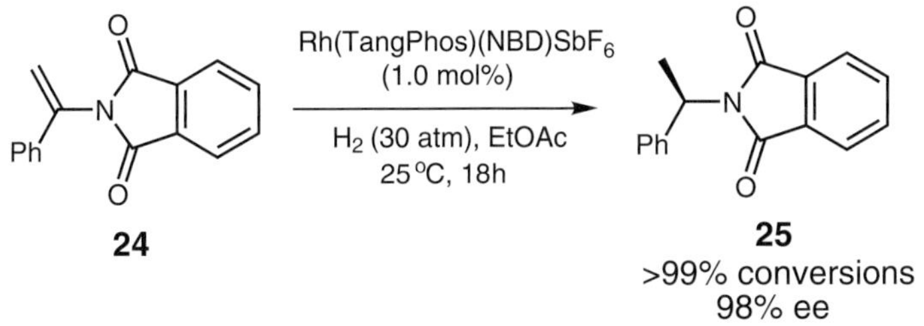

9.2 Catalytic Enantioselective Hydrogenation of Enamides | 281

Scheme 9.9 Rh-catalyzed asymmetric hydrogenation of N-phthaloyl α-arylenamides with TangPhos and DuanPhos.

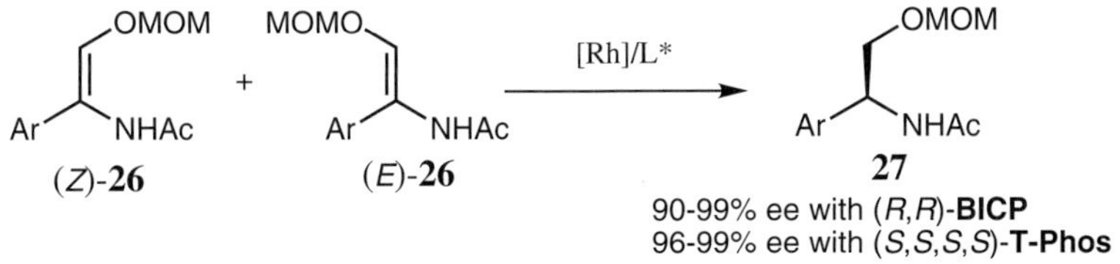

Scheme 9.10 Rh-catalyzed asymmetric hydrogenation of MOM-protected β-hydroxy-α-arylenamides **26** with (R,R)-BICP and (S,S,S,S)-T-Phos.

Using "BDA" [34] and "Dispoke" [35] methodologies developed by Ley for 1,2-diol protections as the key step, Zhang prepared a series of chiral 1,4-diphenylphosphines with the 1,4-dioxane backbone from tartaric acid such as T-Phos (**31**) and SK-Phos (**32**) (Figure 9.6) [36]. These ligands are highly effective for the asymmetric hydrogenation of enamides and MOM-protected β-hydroxyl enamides, providing a wide range of chiral amines and β-amino alcohols in up to 99% ee.

Lee developed a new type of 1,4-diphosphane ligands (BDPMI, **33**) with an imidazolidin-2-one backbone, based on the hypothesis that the consecutive *gauche* steric interactions between the N-substituents and phosphanylmethyl groups may restrict the conformational flexibility of the seven-membered metal-chelating ring (Figure 9.7) [37]. Indeed, the optimal BDPMI ligand showed excellent enantios-

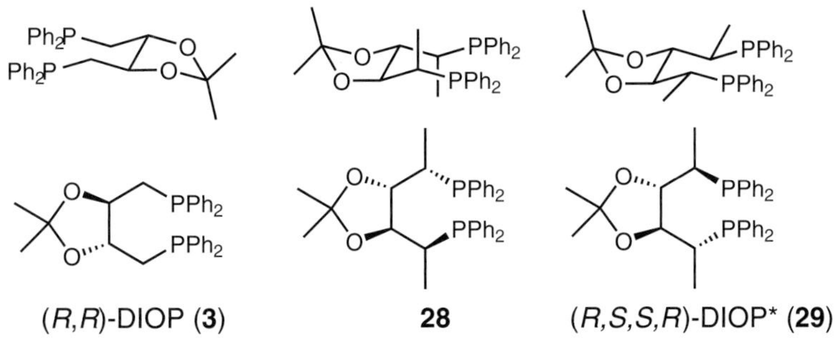

Figure 9.5 Conformational analysis on the ligand design.

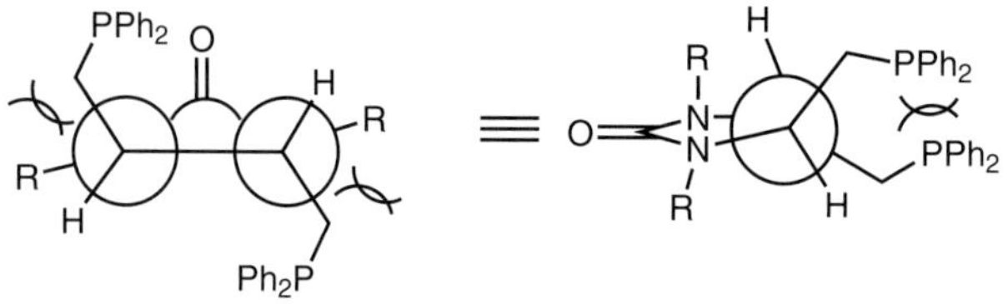

Figure 9.6 Chiral bisphosphane ligands based on DIOP modifications for asymmetric hydrogenation.

electivities (>99% ee) in the Rh-catalyzed asymmetric hydrogenation of α-arylenamides and β-alkyl-α-arylenamides. The introduction of configurationally beneficial α-methyl substituents to the diphenylphosphino groups (BDPMI* 34) in the BDPMI ligand led to an improved enantioselectivity [38]. To facilitate the separation and subsequent reuse of the Rh/(S,S)-BDPMI catalyst, Lee immobilized this catalyst in ionic liquids (ILs) by the introduction of two 1,2-dimethylimidazolium salt tags resembling closely the IL reaction medium into BDPMI skeleton [39]. The resulting immobilized catalyst **35** maintained high enantioselectivity (97% ee) as the parent catalyst, and could be reused several times for the hydrogenation of 1-phenylenamide **4a** without significant loss of catalytic efficiency (Table 9.2).

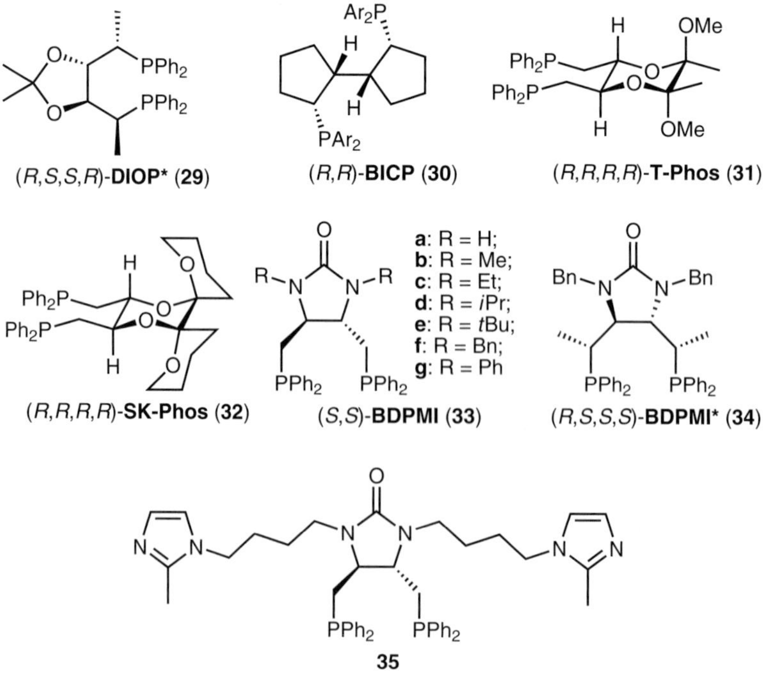

(R,S,S,R)-DIOP* (29)

(R,R)-BICP (30)

(R,R,R,R)-T-Phos (31)

(R,R,R,R)-SK-Phos (32)

(S,S)-BDPMI (33)
a: R = H;
b: R = Me;
c: R = Et;
d: R = iPr;
e: R = tBu;
f: R = Bn;
g: R = Ph

(R,S,S,S)-BDPMI* (34)

35

Figure 9.7 Newman projection of (S,S)-BDPMI **33** showing possible *gauche* interactions.

9.2 Catalytic Enantioselective Hydrogenation of Enamides

Table 9.2 Rh-catalyzed asymmetric hydrogenation of model substrates with 1,4-bisphosphine ligands.

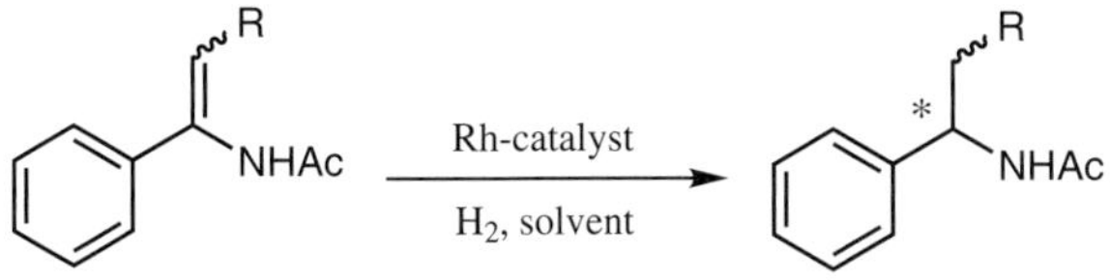

4a: R = H
1a: R = Me

15a: R = H
2a: R = Me

Ligand	Substrate	S/C ratio	Reaction conditions	% ee (config.)	References
(R,S,S,R)-DIOP*	4a	50	MeOH, rt, 1.1 bar H_2	98.8 (R)	[32, 33]
(R,S,S,R)-DIOP*	(Z/E)-1a	50	MeOH, rt, 10 bar H_2	97.3 (R)	[32, 33]
(R,R)-BICP	(Z/E)-1a	100	Toluene, rt, 40 psi H_2	95.0	[29]
(R,R,R,R)-T-Phos	(Z/E)-1a	100	MeOH, rt, 40 psi H_2	98 (S)	[36]
(R,R,R,R)-SK-Phos	(Z/E)-1a	100	MeOH, rt, 40 psi H_2	97 (S)	[36]
(S,S)-BDPMI (b)	4a	100	CH_2Cl_2, rt, 1 atm H_2	98.5 (R)	[37]
(S,S)-BDPMI (c)	4a	100	CH_2Cl_2, rt, 1 atm H_2	97.3 (R)	[37]
(S,S)-BDPMI (d)	4a	100	CH_2Cl_2, rt, 1 atm H_2	95.7 (R)	[37]
(S,S)-BDPMI (e)	4a	100	CH_2Cl_2, rt, 1 atm H_2	97.2 (R)	[37]
(S,S)-BDPMI (f)	4a	100	CH_2Cl_2, rt, 1 atm H_2	96.2 (R)	[37]
(S,S)-BDPMI (b)	(Z/E)-1a	100	CH_2Cl_2, rt, 1 atm H_2	>99.0 (R)	[37]
(R,S,S,S)-BDPMI*	4a	100	CH_2Cl_2, rt, 1 atm H_2	98.6 (R)	[38]
35	4a	100	[bmim][SbF$_6$]/iPrOH	97.5 (R)	[39]

9.2.2.3 Bisaminophosphine Ligands for Rh-Catalyzed Asymmetric Hydrogenation

In 1998, Chan developed a type of chiral Rh-catalysts containing 2,2′-bis(diphenylphosphinoamino)-1,1′-binaphthyl (BDPAB, **36**) and 2,2′-bis(diphenylphosphinoamino)-5,5′,6,6′,7,7′,8,8′-octahydro-1,1′-binaphthyl (H$_8$-BDPAB, **37**) (Figure 9.8) [40]. An important advantage of the use of this class of Rh-catalysts is that the ligands can be easily prepared from the corresponding diamines. It was clearly observed that the enantioselectivities of the hydrogenation of α-arylenamides catalyzed by Rh-(R)-H$_8$-

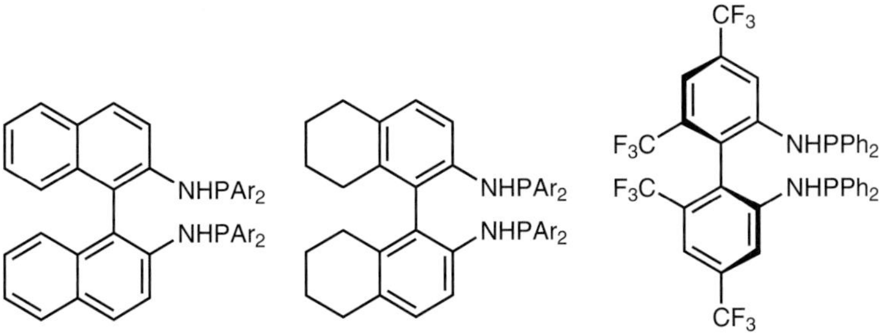

Figure 9.8 Chiral biphosphonamidite ligands for asymmetric hydrogenation.

BDPAB were consistently higher than those from the same reaction with Rh-(R)-BDPAB catalyst. Rh-(R)-H$_8$-BDPAB catalyst was extremely effective in the Rh-catalyzed asymmetric hydrogenation of α-arylenamides, leading to various chiral α-arylethylamine derivatives with excellent enantioselectivities (up to 99.0% ee).

Very recently, Wang has prepared a new class of bis(aminophosphine) ligands 38 based on a new biphenyldiamine: 4,4′,6,6′-tetrakis-trifluoromethyl- biphenyl-2,2′-diamine (TF-BIPHAM) (Figure 9.8) [41]. These bis(aminophosphine) ligands exhibited excellent to almost perfect enantioselectivities in the Rh-catalyzed asymmetric hydrogenation of enamides (97.6–99.9% ee) even at a catalyst loading of 0.1 mol%.

9.2.2.4 Unsymmetrical Hybrid Phosphorus-Containing Ligands for Rh-Catalyzed Asymmetric Hydrogenation

A recent advance in the ligand design for asymmetric hydrogenation is the use of unsymmetrical hybrid chiral bidentate phosphorus-containing ligands. In this area, we and some other groups have developed a diversity of chiral phosphine-aminophosphine ligands, phosphine-phosphoramidite ligands, phosphine-phosphite ligands, and so on. Some representative results in the Rh-catalyzed asymmetric hydrogenation of model enamide substrates 4a and 1a are shown in Table 9.3.

Table 9.3 Rh-catalyzed asymmetric hydrogenation of model substrates with unsymmetrical hybrid phosphorus-containing ligands.

4a: R = H
1a: R = Me

15a: R = H
2a: R = Me

Ligand	Substrate	S/C ratio	Reaction conditions	% ee (config.)	References
39b	4a	100	THF, rt, 300 psi H$_2$	96.5	[43]
39b	4a	1000	THF, rt, 300 psi H$_2$	95.8	[43]
40b	4a	100	CH$_2$Cl$_2$, rt, 10 atm H$_2$	93.4	[44]
(S_c,R_p,S_a)-41	4a	5000	CH$_2$Cl$_2$, rt, 10 bar H$_2$	99.3 (R)	[45]
(S_c,S_p,S_a)-41	4a	100	CH$_2$Cl$_2$, rt, 10 bar H$_2$	99.6 (S)	[45]
44	4a	100	CH$_2$Cl$_2$, rt, 10 bar H$_2$	96.7 (R)	[46]
45b	4a	100	CH$_2$Cl$_2$, rt, 10 bar H$_2$	99.5 (R)	[48]
46	4a	100	CH$_2$Cl$_2$, rt, 10 atm H$_2$	99.7 (S)	[50]
47	4a	100	CH$_2$Cl$_2$, rt, 10 atm H$_2$	96 (R)	[51]
47	1a	100	CH$_2$Cl$_2$, rt, 10 atm H$_2$	96 (R)	[51]
48a	4a	1000	CH$_2$Cl$_2$, rt, 10 bar H$_2$	99 (R)	[52]
49	4a	100	CH$_2$Cl$_2$, rt, 1 bar H$_2$	99 (R)	[53]
49	4a	1000	CH$_2$Cl$_2$, rt, 1 bar H$_2$	98 (R)	[53]
52	4a	200	MeOH, rt, 100 psi H$_2$	96.3 (S)	[54]
53	4a	100	THF, rt, 20 bar H$_2$	98 (R)	[56]
54	4a	100	THF, rt, 35 bar H$_2$	95 (S)	[57]

Figure 9.9 Chiral phosphine-aminophosphine ligands for asymmetric hydrogenation.

(S_c,R_p)-BoPhoz (39)
a: Ar = Ph; b: Ar = 3,5-$(CF_3)_2C_6H_3$

(R_c)-HW-Phos (40)
a: Ar = Ph; b: Ar = 3,5-$F_2C_6H_3$;
c: Ar = 4-$CF_3C_6H_4$; d: Ar = 3,5-$(CF_3)_2C_6H_3$

In 2002, Boza reported the first class of phosphine-aminophosphine ligands, BoPhoz (**39a**), which displayed excellent enantioselectivities in the rhodium-catalyzed asymmetric hydrogenation of α-dehydroamino acid derivatives, itaconic acids, and α-ketoesters [42]. However, this ligand class is inefficient in the hydrogenation of enamides in terms of the enantioselectivity, and gave relatively poor ee values (~70% ee). A subsequent investigation on the optimization of ligand structure, performed by Chan [43], discovered that the introduction of a CF_3 group (**39b**) into the phenyl rings of aminophosphino moiety could significantly improve the enantioselectivity of the rhodium-catalyzed asymmetric hydrogenation of enamides (Figure 9.9). With this catalytic system, a variety of α-arylenamides were quantitatively converted to the corresponding chiral products in ee values ranging 92.1–99.7%. Interestingly, this optimal ligand showed somewhat lower enantioselectivities in the hydrogenation of α-arylenamides with an electron-withdrawing group in the 4-position of the phenyl ring when the hydrogenation was carried out at room temperature; however, equally high enantioselectivities were achieved at 5 °C no matter what kind of substituent is present in the substrates. Very recently, we have also developed a new class of phosphine-aminophosphine ligands (**40**) based on a chiral 1,2,3,4-tetrahydro-1-naphthylamine backbone (Figure 9.9) [44]. A systematic study on the effect of ligand structure suggested that ligand **40b** with two F-atoms in the 3,5-positions of the phenyl ring of the aminophosphino moiety gave the best results, in which 93.4–97.0% ee were obtained in the hydrogenation of various enamides.

In 2004, we introduced a new family of highly unsymmetrical phosphine-phosphoramidite ligands (PPFAPhos **41**, Figure 9.10), which contain a planar-chiral ferrocenyl backbone and an axial-chiral binaphthyl moiety, for the Rh-catalyzed asymmetric hydrogenation of α-arylenamides [45]. When a rhodium catalyst containing (S_c,R_p,S_a)-PPFAPhos with a (S_a)-binaphthyl moiety was used in the hydrogenation of substrate **4a**, a significant increase in the ee value of the product to 99.6% was obtained in comparison with the result using Bophoz, which has the same (S_c,R_p)-ferrocenyl backbone. In sharp contrast, however, ligand (S_c,R_p,R_a)-PPFAPhos with a (R_a)-binaphthyl fragment gave only 10.6% ee and favored a hydrogenation product with a configuration opposite to that obtained with (S_c,R_p,S_a)-PPFAPhos. This result suggested that the introduction of a chiral binaphthyl moiety into the planar ferrocenyl backbone strongly influenced the catalytic activity and enantioselectivity. Using (S_c,S_p,R_a)-PPFAPhos, a hydrogenation product was obtained with 99.6% ee but with a chirality opposite to that obtained with (S_c,R_p,S_a)-PPFAPhos. In contrast,

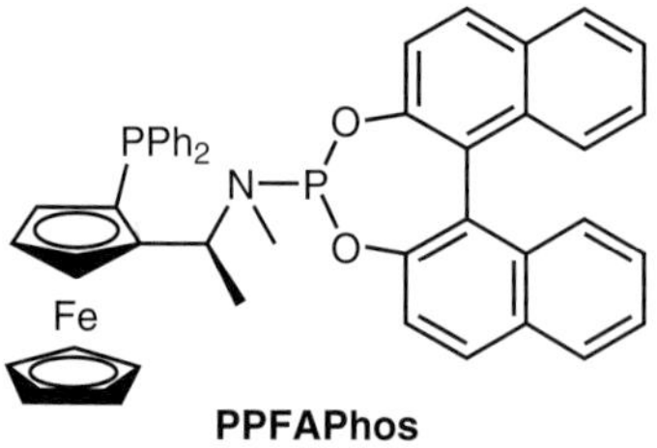

Figure 9.10 Chiral ferrocene-based phosphine-phosphoramidite ligands (PPFAPhos) developed within our group.

Structures shown:
- PPFAPhos (S_c,R_p,S_a)-**41**, (S_c,R_p,R_a)-**41**
- PPFAPhos (S_c,S_p,R_a)-**41**, (S_c,S_p,S_a)-**41** 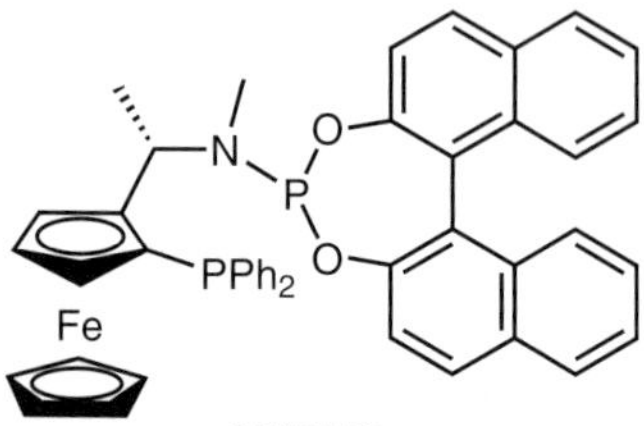
- (S_c,R_p)-**42** 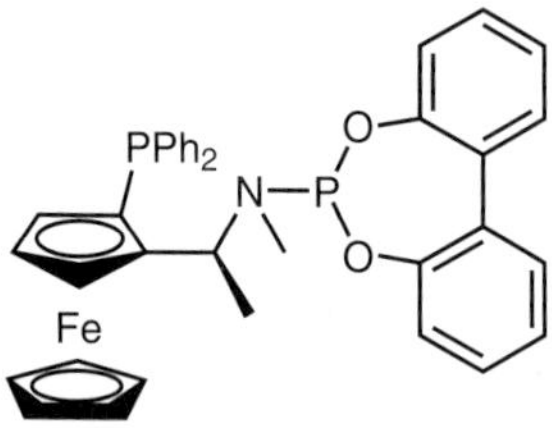
- (S_c,R_p)-**43** 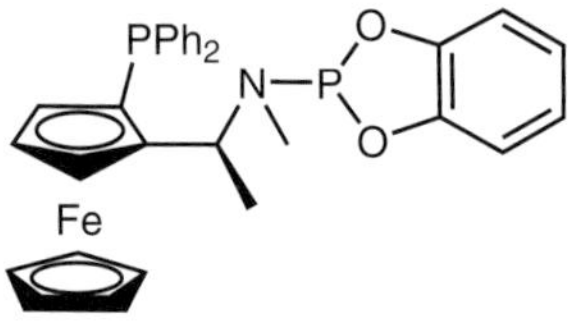
- (S_c,R_p,S_a)-PPFAPhos-H$_8$ (**44**) 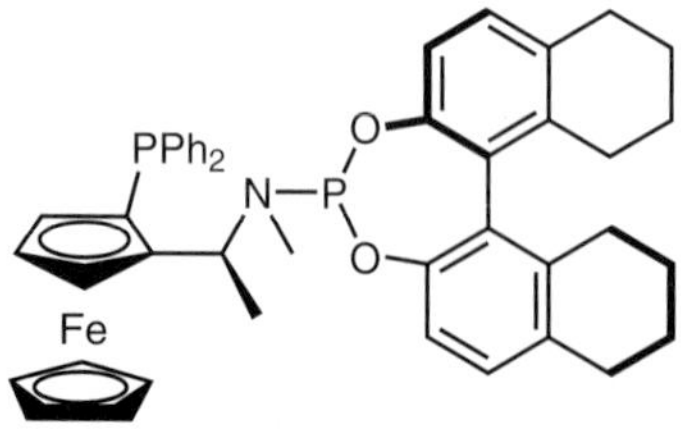

(S_c,S_p,S_a)-PPFAPhos only gave a hydrogenation product with 82.6% ee in the same configuration as that obtained with (S_c,R_p,S_a)-PPFAPhos. When ligands **42** and **43** with an achiral phosphoramidite moiety were used in the model reaction, only moderate enantioselectivity was obtained. These results indicate that the binaphthyl moiety plays a crucial role in the enantioselectivity and controls the chirality of the hydrogenation products. The matched stereogenic elements are S_c,R_p,S_a or S_c,S_p,R_a. The hydrogenation of other α-arylenamides also gave very high ee values. The subsequent investigation proved that the H$_8$-BINOL-based phosphine-phosphoramidite ligand **44** also displayed high enantioselectivities [46]. A similar approach was reported by Chan's group using different diastereomers at nearly the same time [47].

Although high efficiency of PPFAPhos is observed in the Rh-catalyzed asymmetric hydrogenation of various enamides, high cost in the tedious synthesis of this ligand class has seriously prevented its practical application in asymmetric catalysis. To overcome this shortcoming, in 2006, we developed a new class of phosphine-phosphoramidite ligands, PEAPhos (**45**), bearing a similar structure to PPFAPhos (Figure 9.11) [48]. PEAPhos can be easily prepared from commercially available and

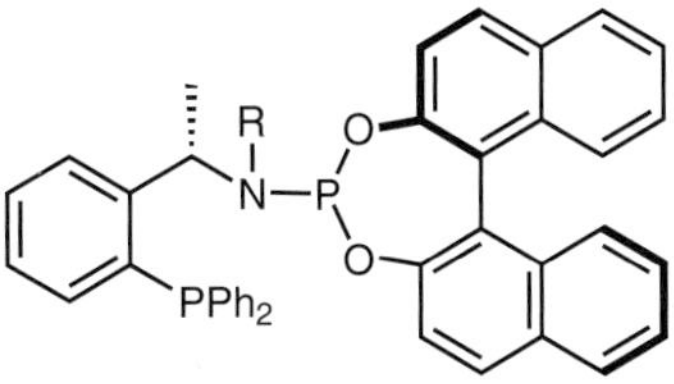

9.2 Catalytic Enantioselective Hydrogenation of Enamides

(S_c,S_a)-**PEAPhos-H**: R = H (**45a**)
(S_c,S_a)-**PEAPhos-Me**: R = Me (**45b**)

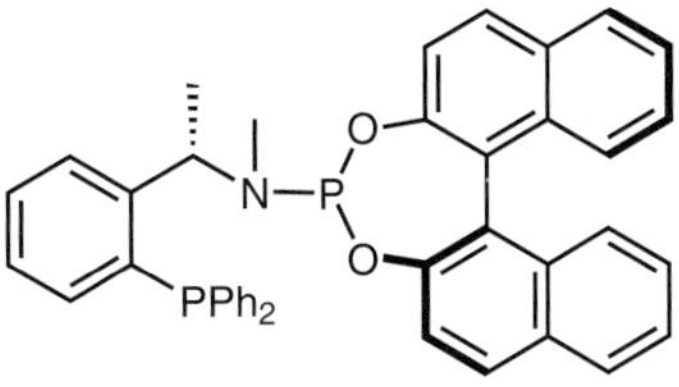

(S_c,R_a)-**PEAPhos-H**: R = H (**45c**)
(S_c,R_a)-**PEAPhos-Me**: R = Me (**45d**)

Figure 9.11 Chiral α-phenylethylamine-derived phosphine-phosphoramidite ligands (PEAPhos) developed within our group.

inexpensive (S)-α-phenylethylamine through a two-step transformation. The research revealed that PEAPhos is also highly efficient for the Rh-catalyzed enantioselective hydrogenation of enamides and up to 99.9% ee was obtained. Most interestingly, the central chirality in the phenylethylamine backbone decides the absolute configuration of the hydrogenation product no matter the (R)- or (S)-configuration of binaphthyl moiety, contrary to the results obtained by phosphoramidite-containing ligands reported so far.

Following this work, we developed another new phosphine-phosphoramidite ligand based on chiral 1,2,3,4-tetrahydro-1-naphthylamine, THNAPhos (**46**) (Figure 9.12) [49]. THNAPhos has a similar but more rigid backbone in comparison with PEAPhos. As expected, THNAPhos exhibited excellent enantioselectivities in the Rh-catalyzed asymmetric hydrogenation of a variety of α-arylenamides [50]. Especially, for the hydrogenation of β-alkyl-α-arylenamides **1**, THNAPhos also gave an ee value of over 99%. Further increasing the rigidity of ligand by replacing

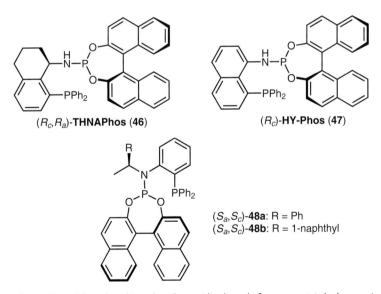

(R_c,R_a)-**THNAPhos** (**46**)

(R_c)-**HY-Phos** (**47**)

(S_a,S_c)-**48a**: R = Ph
(S_a,S_c)-**48b**: R = 1-naphthyl

Figure 9.12 Other phosphine-phosphoramidite ligands for asymmetric hydrogenation.

1,2,3,4-tetrahydro-1-naphthylamino moiety with 1-naphthylamine backbone (HY-Phos **47**, Figure 9.12), however, resulted in the decreased enantioselectivity of 96% ee [51]. Very recently, Franciò and Leitner reported a set of novel phosphine-phosphoramidite ligands **48** possessing two elements of chirality prepared through a modular synthetic approach (Figure 9.12). The ligands bearing (S)-central and (S)-axial chiralities are unique in providing enantioselectivities (>99% ee) in the Rh-catalyzed asymmetric hydrogenation of the model substrate **4a** [52].

Recently, Zhang has reported a new class of triphosphorus bidentate phosphine-phosphoramidite ligands, which displayed excellent enantioselectivities (up to 99% ee) in the hydrogenation of di- and tri-substituted enamides [53]. In particular, this new ligand class showed unprecedented enantioselectivities in the hydrogenation of *ortho*-substituted α-arylenamides and an α-(1-naphthyl)enamide, thereby solving a long-standing problem in the hydrogenation of enamides (Scheme 9.11).

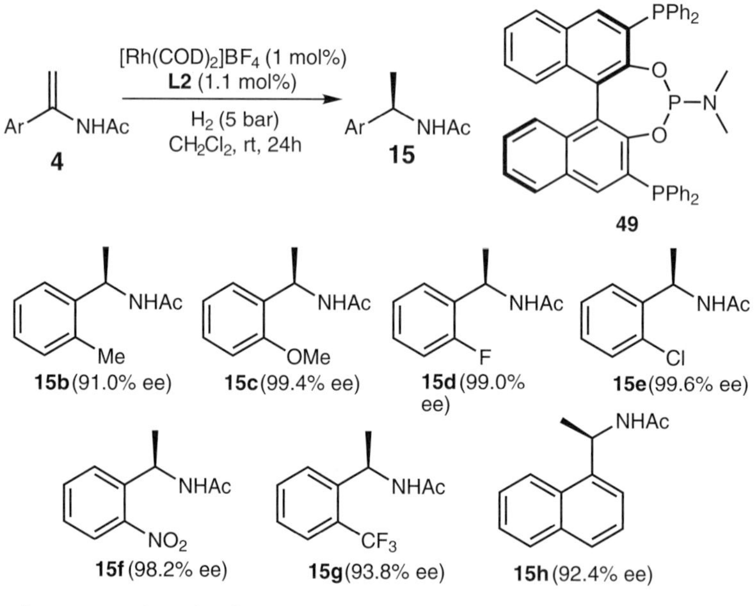

Scheme 9.11 Rh-catalyzed asymmetric hydrogenation of *ortho*-substituted arylenamides with a triphosphorus bidentate phosphine-phosphoramidite ligand **49**.

The potential application of this catalytic system in industry lies in the synthesis of (R)-α-(1-naphthyl)ethylamine (**50**), a key precursor to Cinacalcet hydrochloride (**51**) for the treatment of hyperparathyroidism and hypercalcemia, via the hydrogenation of α-(1-naphthyl)enamide at a decreased catalyst loading (0.1 mol%) in TFE under 80 bar of H_2 within 24 h (Scheme 9.12) [53].

In comparison with the extensive application of phosphine-phosphoramidite ligands, there are few successful examples on the use of phosphine-phosphite ligands in the Rh-catalyzed asymmetric hydrogenation of enamides (Figure 9.13). In 2008, Vidal-Ferran reported that a phosphine-phosphite ligand **53** with the matched chiralities gave good enantioselectivity in this model hydrogenation [54].

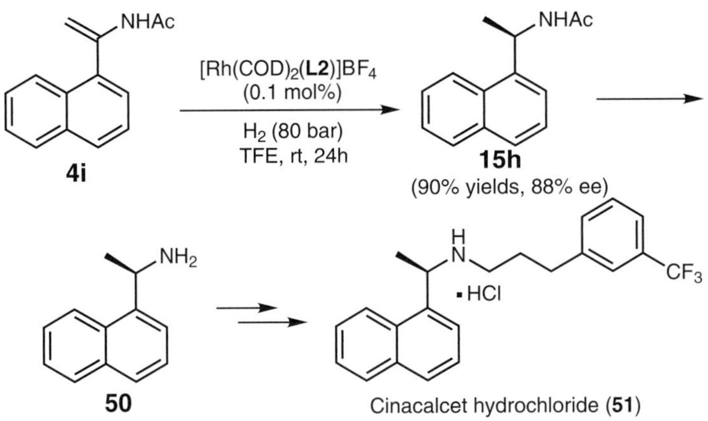

9.2 Catalytic Enantioselective Hydrogenation of Enamides

Scheme 9.12 Asymmetric hydrogenation route to Cinacalcet hydrochloride.

Ferrocene-based 1,5-diphosphine ligands, Taniaphos, gave excellent enantioselectivities for various metal-catalyzed asymmetric reactions [55]. However, these ligands are not very efficient in the Rh-catalyzed asymmetric hydrogenation of enamides. Chen found that the introduction of P-chirality in this ligand skeleton can significantly enhance the enantioselective discrimination. Indeed, P-chiral ligand **52** gave superior results to the corresponding Taniaphos (96.3% ee versus 86.4% ee) [56]. Another type of unsymmetrical hybrid bidentate ligands successfully employed in the catalytic asymmetric hydrogenation of enamides is reported by Evans, in which a phosphonite-thioether ligand **54** was found to show 95% ee in the Rh-catalyzed asymmetric hydrogenation of **4a** [57].

9.2.3
Catalytic Asymmetric Hydrogenation of Cyclic Enamides

In comparison with the hydrogenation of acyclic enamides, there are fewer successful examples on the catalytic asymmetric hydrogenation of cyclic enamides

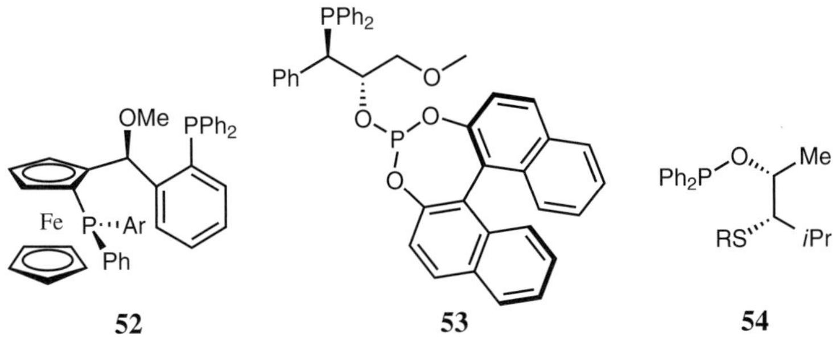

Figure 9.13 Other bidentate ligands for asymmetric hydrogenation.

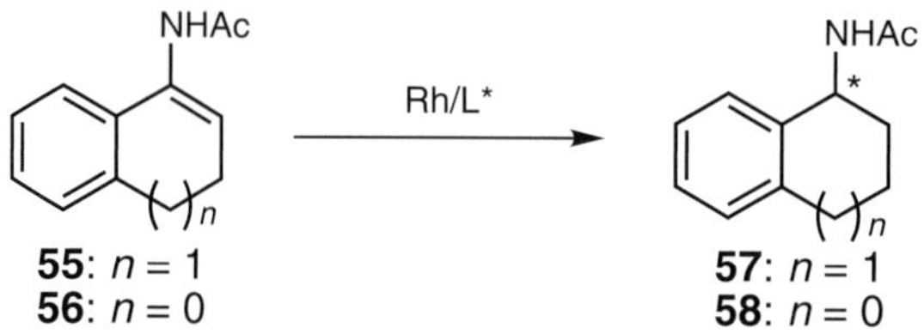

Scheme 9.13 Asymmetric hydrogenation of cyclic enamides with an endocyclic double bond derived from α-tetralone and α-indanone.

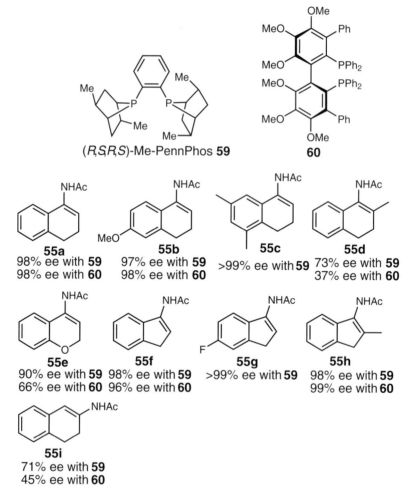

Scheme 9.14 Asymmetric hydrogenation of cyclic enamides using the Rh/Me-PennPhos or Rh/(S)-o-Ph-hexaMeO-BIPHEP catalyst.

occurring in the literatures despite the potential importance of this process for the synthesis of biologically active compounds. The cyclic enamides used in the catalytic hydrogenation can be classified into two categories according to the different positions of double bond in the substrates: one with an endocyclic double bond and the other with an exocyclic double bond.

The typical cyclic enamides with an endocyclic double bond are derived from α-tetralone and α-indanone (Scheme 9.13). Although the structures of these two substrates **55** and **56** are similar to the acyclic substrates with a β-substituent, the hydrogenation of these cyclic enamides proves to be far more difficult than their acyclic analogues. Most of the bidentate phosphorus-containing ligands being highly efficient in the catalytic hydrogenation of acyclic enamides showed low to moderate enantioselectivities. For example, Burk reported that an unexpectedly low enantioselectivity (0% ee) was achieved with an Rh-Me-DuPHOS complex, although excellent results were obtained in the hydrogenation of an indanone-derived enamide [23]. Under the similar conditions, the Rh-Me-BPE catalyst gave 69% ee, however, lowering the reaction temperature to 0 °C improved the enantioselectivity to 92% ee [23]. Using an Rh-BICP catalyst provided only modest enantioselectivities (60–70% ee) [29].

One breakthrough in this area occurred in 1999 when Zhang employed Me-PennPhos in the Rh-catalyzed asymmetric hydrogenation of cyclic enamides [58]. Excellent enantioselectivity and catalytic activity were achieved even when the ratio of N-(3,4-dihydro-1-naphthyl)acetamide/Rh-Me-PennPhos was 2000/1. Hydrogenation of several cyclic enamides derived from α-tetralones and α-indanones with Rh-Me-PennPhos complex gave high enantioselectivities regardless of the substituents on the aromatic ring (Scheme 9.14). More interestingly, a tetra-substituted five-membered cyclic enamide also gave high enantioselectivities. A similar result was reported by the same group with an Rh/(S)-o-Ph-hexaMeO-BIPHEP catalyst [59]. However, both of Me-PennPhos and o-Ph-hexaMeO-BIPHEP are inefficient in the hydrogenation of the enamide derived from β-tetralone in terms

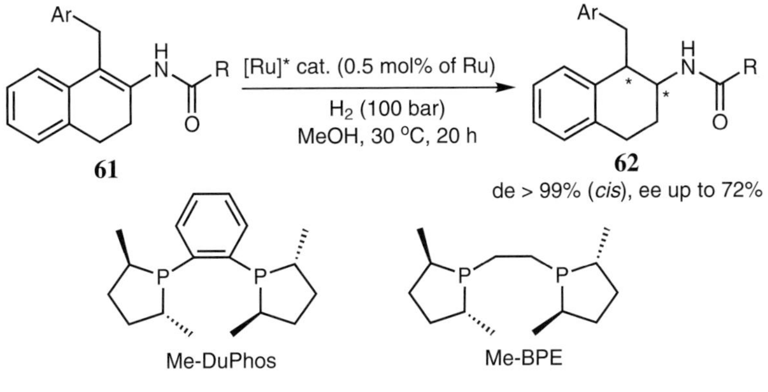

Scheme 9.15 Asymmetric hydrogenation of tetra-substituted cyclic enamides using the Ru-catalyst containing Me-DuPHOS or Me-BPE.

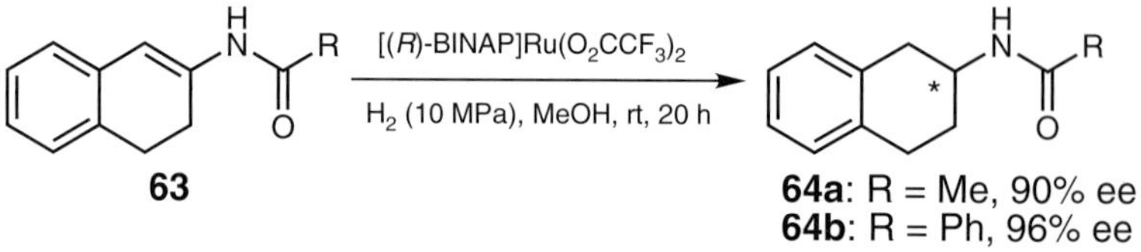

Scheme 9.16 Asymmetric hydrogenation of cyclic enamides derived from β-tetralone using the Ru-catalyst.

of the enantioselectivity (71% ee with Me-PennPhos and 45% ee with o-Ph-hexaMeO-BIPHEP).

In comparison with the hydrogenation of cyclic enamides derived from α-indanone and α-tetralone, there are few studies on the successful hydrogenation of enamides prepared from β-tetralones. Bruneau reported the first ruthenium-catalyzed enantioselective hydrogenation of tetra-substituted enamides derived from racemic α-substituted-β-tetralones into optically active amides possessing two stereogenic centers [60]. The ruthenium catalysts are generated *in situ* by protonation with 2 equiv of tetrafluoroboric acid of a mixture of (1,5-cyclooctadiene)Ru(methallyl)$_2$ and 1 equiv of optically active diphosphine, Me-DuPHOS or Me-BPE. In all cases,

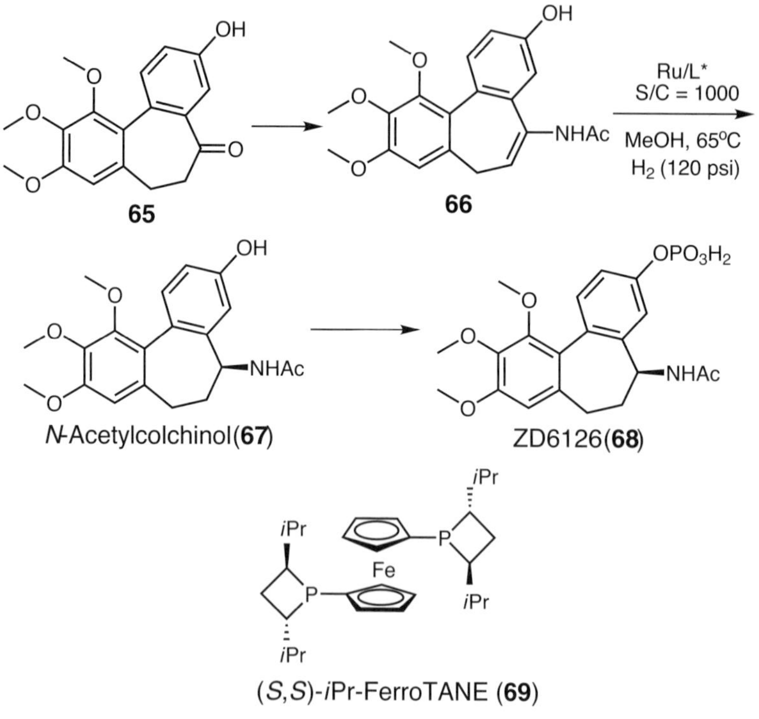

Scheme 9.17 Synthesis of N-acetylcolchinol via asymmetric hydrogenation of cyclic enamide intermediate **66**.

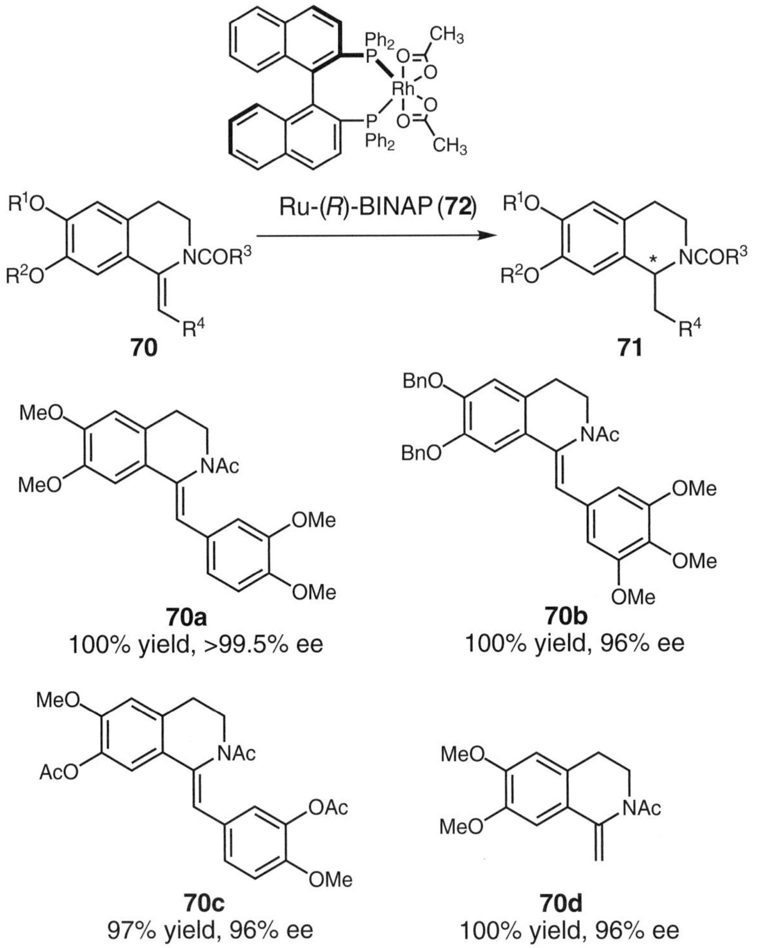

9.2 Catalytic Enantioselective Hydrogenation of Enamides

Scheme 9.18 Asymmetric hydrogenation of (Z)-N-acyl-1-alkylidenetetra- hydroisoquinolines.

the reaction of enamides with 0.5 mol% of ruthenium catalyst went to completion at 30 °C within 20 h and afforded amides as the single *cis*-diastereoisomers in good yields (95–98%) and ee's up to 72% (Scheme 9.15).

More recently, Bruneau found that the tri-substituted cyclic enamides derived from β-tetralone and its analogues can be efficiently hydrogenated in

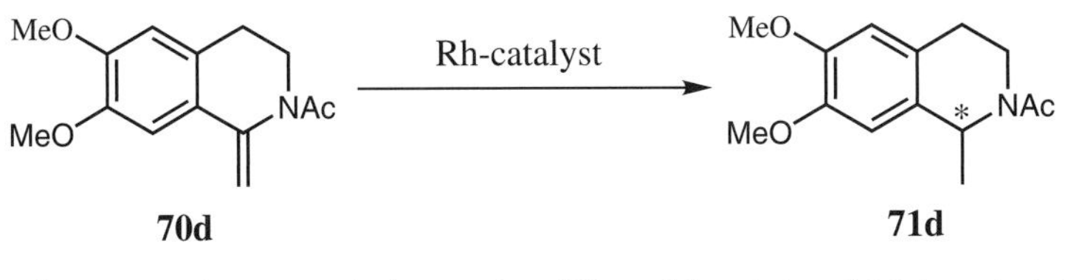

Scheme 9.19 Asymmetric hydrogenation of N-acetyl-1-methylene-1,2,3,4-tetrahydroisoquinoline.

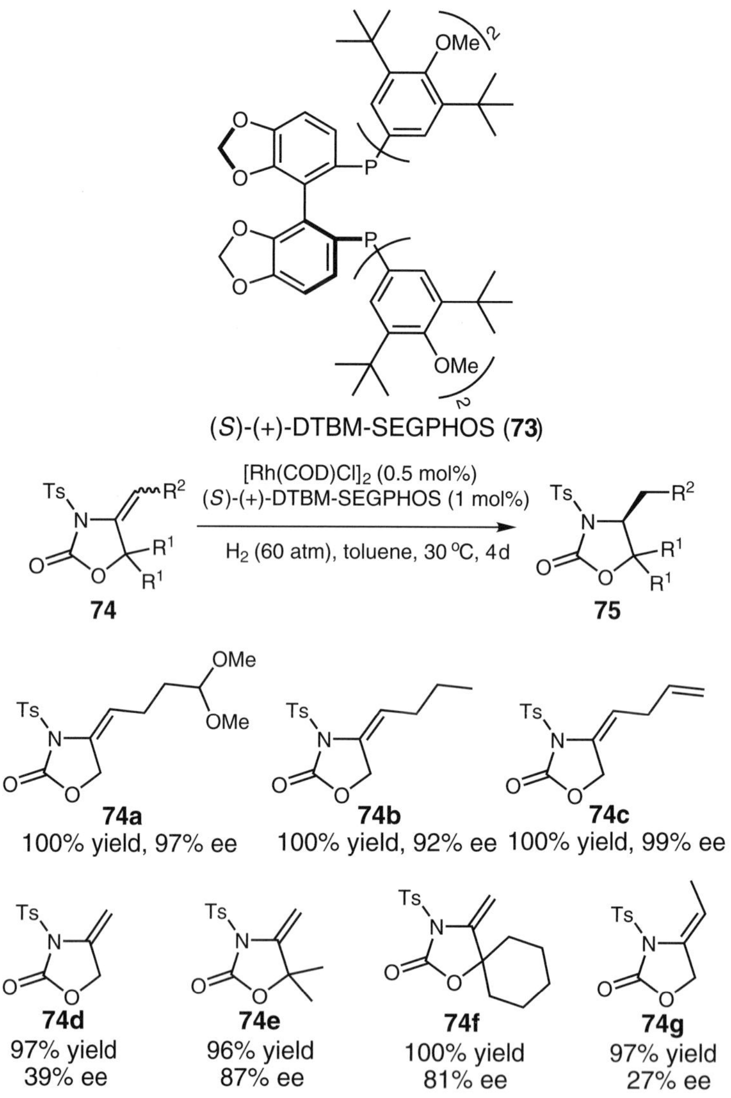

(S)-(+)-DTBM-SEGPHOS (73)

Scheme 9.20 Asymmetric hydrogenation of tri-substituted or di-substituted exocyclic double bond of N-tosyl-4-alkylidene-1,3-oxazolidin-2-ones.

the presence of ruthenium catalysts, in which up to 96% ee was obtained (Scheme 9.16) [61].

A synthesis of N-acetylcolchinol (**67**), a key intermediate in the synthesis of the drug substance ZD6126 (**68**), has been recently developed by a Ru-catalyzed asymmetric hydrogenation of prochiral enamide **66** bearing an endocyclic double bond with phospholane ligands [62]. The research disclosed that (S,S)-iPr-Ferro-

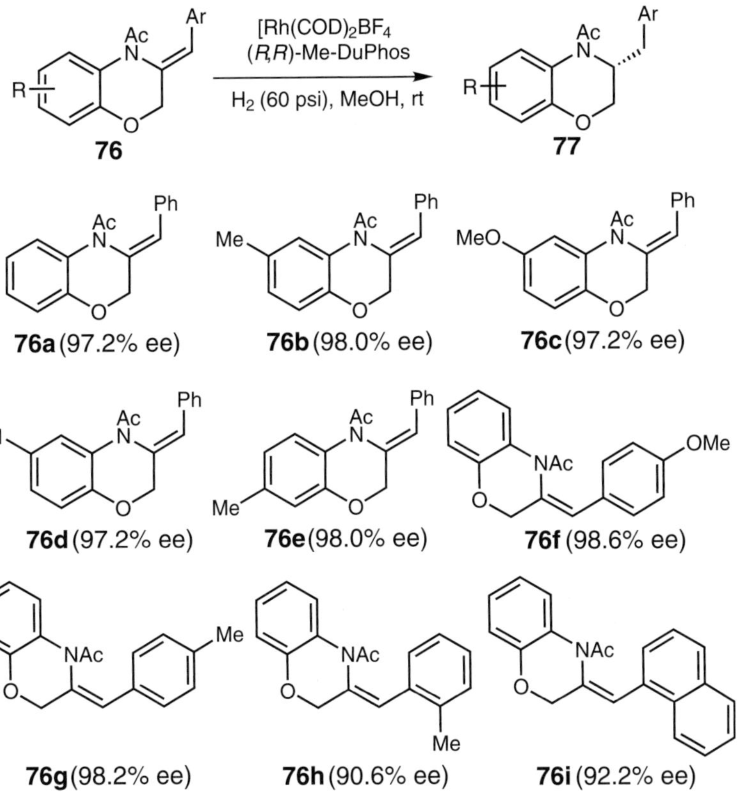

9.2 Catalytic Enantioselective Hydrogenation of Enamides

Scheme 9.21 Asymmetric hydrogenation of exocyclic double bond of (Z)-3-arylidene-4-acyl-3,4-dihydro-2H-benzoxazines.

TANE (**69**) gave the best result of up to 91.6% ee when the hydrogenation was performed at a molar S/C of 1000 under a H_2 pressure of 120 psi at 65 °C (Scheme 9.17).

In 1986, Noyori reported a hexa-coordinate BINAP-Ru(II) complex that can efficiently catalyze the asymmetric hydrogenation of (Z)-N-acyl-1-alkylidenetetra- hydroisoquinolines (**70**) in the presence of 0.5–1 mol% of the catalyst loadings in a 5 : 1 mixture of ethanol and dichloromethane under 1–4 atm of hydrogen at 23 °C (Scheme 9.18) [63]. However, the corresponding E-enamide substrates were inert to this catalytic system under the same hydrogenation conditions.

Zhang examined the Rh-catalyzed asymmetric hydrogenation of N-acetyl-1-methylene-1,2,3,4-tetrahydroisoquinoline **70d** with some diphosphine ligands developed within his group (Scheme 9.19). (R,R)-BICP showed high catalytic activity for the hydrogenation of **70d**, however, the enantioselectivity was moderate [29]. With (R,R,R,R)-T-Phos, the hydrogenation gave the corresponding tetrahydroisoquinoline in 94% ee [36]. TangPhos was also highly effective for this hydrogenation providing the hydrogenation product in quantitative yield and with 97% ee [16].

Lu and coworkers found that the asymmetric hydrogenation of the tri-substituted or di-substituted exocyclic double bond of N-tosyl-4-alkylidene-1,3-oxazolidin-2-ones (**74**), which are enamides in nature, proceeds smoothly under the catalysis of neutral [Rh(COD)Cl]$_2$ and (S)-(+)-DTBM-SEGPHOS (**73**) with nearly quantitative yields and high enantioselectivities (Scheme 9.20) [64]. This method provided a novel way to prepare chiral N-tosyloxazolidinones (**75**) with high enantiomeric excess, which can be converted into amino acids, amino alcohols, and piperidine derivatives.

By using an Rh/(R,R)-Me-DuPHOS catalytic system, Zhou has successfully hydrogenated a new class of exocyclic enamides, (Z)-3-arylidene-4-acyl-3,4-dihydro-2H-benzoxazines (**76**) [11]. Hydrogenation gave high enantioselectivities (up to 98.6% ee) regardless of the substituents on the aromatic ring of 1,4-benzoxazines (Scheme 9.21). In the case of the compounds with an *ortho*-substituent of aryl on exocyclic double bond, a slightly lower enantioselectivity was obtained.

9.3
Conclusions

The catalytic asymmetric hydrogenation of enamides is now arguably one of the most powerful methods for the preparation of chiral amines. Access to a diverse set of chiral phosphorus ligands has allowed enantioselective catalytic hydrogenation of enamides to achieve unprecedented versatility and synthetic utility. Despite these major advances, many challenges with regard to practical applications remain. Further efforts in the quest for new efficient chiral phosphorus ligands, as well as new practical applications of this methodology in both laboratory and industry, are needed.

Questions

9.1. In the catalytic asymmetric hydrogenation of enamides, which are the most used central metals? (For undergraduate student)

9.2. There are still some challenging enamide substrates for the catalytic asymmetric hydrogenation. Can you give some examples and describe their recent advances?

9.3. Although the catalytic asymmetric hydrogenation of enamides has made great progress in the past decades, there are few successful examples in industry. Can you give the possible reasons?

References

1. (a) Kagan, H.B. and Dang, T.-P. (1972) *J. Am. Chem. Soc.*, **94**, 6429; (b) Burk, M.J., Casy, G., and Johnson, N.B. (1998) *J. Org. Chem.*, **63**, 6084.
2. Zhu, G., Casalnuovo, A.L., and Zhang, X. (1998) *J. Org. Chem.*, **63**, 8100.
3. Zhao, H., Vandenbossche, C.P., Koenig, S.G., Singh, S.P., and Bakale, R.P. (2008) *Org. Lett.*, **10**, 505.
4. (a) Coleman, R.S. and Liu, P.-H. (2004) *Org. Lett.*, **6**, 577; (b) Jiang, L., Job, G.E., Klapars, A., and Buchwald, S.L. (2003) *Org. Lett.*, **5**, 3667; (c) Shen, R. and Porco, J.A. (2000) *Org. Lett.*, **2**, 1333; (d) Ogawa, T., Kiji, T., Hayami, K., and Suzuki, H. (1991) *Chem. Lett.*, 1443.
5. (a) Wallace, D.J., Klauber, D.J., Chen, C., and Volante, R.P. (2003) *Org. Lett.*, **5**, 4749; (b) Harrison, P. and Meek, G. (2004) *Tetrahedron Lett.*, **45**, 9277.
6. Klapars, A., Campos, K.R., Chen, C., and Volante, R.P. (2005) *Org. Lett.*, **7**, 1185.
7. (a) van den Berg, M., Haak, R.M., Minnaard, A.J., de Vries, A.H.M., de Vries, J.G., and Feringa, B.L. (2002) *Adv. Synth. Catal.*, **344**, 1003; (b) Savarin, C.G., Boice, G.N., Murray, J.A., Corley, E., DiMichele, L., and Hughes, D. (2006) *Org. Lett.*, **8**, 3903.
8. Harrison, P. and Meek, G. (2004) *Tetrahedron Lett.*, **45**, 9277.
9. (a) Dupau, P., Le Gendre, P., Bruneau, C., and Dixneuf, P.H. (1999) *Synlett*, 1832; (b) Dupau, P., Bruneau, C., and Dixneuf, P.H. (1999) *Tetrahedron: Asymmetry*, **10**, 3467.
10. Kitamura, M., Hsiao, Y., Ohta, M., Tsukamoto, M., Ohta, T., Takaya, H., and Noyori, R. (1994) *J. Org. Chem.*, **59**, 297.
11. Zhou, Y.-G., Yang, P.-Y., and Han, X.-W. (2005) *J. Org. Chem.*, **70**, 1679.
12. (a) Timokhin, V.I., Anastasi, N.R., and Stahl, S.S. (2003) *J. Am. Chem. Soc.*, **125**, 12996; (b) Brice, J.L., Harang, J.E., Timokhin, V.I., Anastasi, N.R., and Stahl, S.S. (2005) *J. Am. Chem. Soc.*, **127**, 2868.
13. Burk, M.J., Wang, Y.M., and Lee, J.R. (1996) *J. Am. Chem. Soc.*, **118**, 5142.
14. Tang, W. and Zhang, X. (2002) *Angew. Chem., Int. Ed.*, **41**, 1612.
15. Liu, D. and Zhang, X. (2005) *Eur. J. Org. Chem.*, 646.
16. Xiao, D., Zhang, Z., and Zhang, X. (1999) *Org. Lett.*, **1**, 1679.
17. Dai, Q., Wang, C.-J., and Zhang, X. (2006) *Tetrahedron*, **62**, 868.
18. Pilkington, C.J. and Zanotti-Gerosa, A. (2003) *Org. Lett.*, **5**, 1273.
19. Imamoto, T., Oohara, N., and Takahashi, H. (2004) *Synthesis*, 1353.
20. Yan, Y. and Zhang, X. (2006) *Tetrahedron Lett.*, **47**, 1567.
21. Li, W., Zhang, Z., Xiao, D., and Zhang, X. (2000) *J. Org. Chem.*, **65**, 3489.
22. Chi, Y. and Zhang, X. (2002) *Tetrahedron Lett.*, **43**, 4849.
23. Burk, M.J., Casy, G., and Johnson, N.B. (1998) *J. Org. Chem.*, **63**, 6804.
24. Feldgus, S. and Landis, C.R. (2001) *Organometallics*, **20**, 2374.
25. Bachman, G.L. and Vineyard, B.D. (1977) Ger. Offen. DE2638072.
26. Chen, J., Zhang, W., Geng, H., Li, W., Hou, G., Lei, A., and Zhang, X. (2009) *Angew. Chem., Int. Ed.*, **48**, 800.
27. Yang, Q., Gao, W., Deng, J., and Zhang, X. (2006) *Tetrahedron Lett.*, **47**, 821.
28. Zhu, G., Cao, P., Jiang, Q., and Zhang, X. (1997) *J. Am. Chem. Soc.*, **119**, 1799.
29. Zhu, G. and Zhang, X. (1998) *J. Org. Chem.*, **63**, 9590.
30. Zhu, G., Casalnuovo, A.L., and Zhang, X. (1998) *J. Org. Chem.*, **63**, 8100.
31. Kagan, H.B., Fiaud, J.C., Hoornaert, C., Meyer, D., and Poulin, J.C. (1979) *Bull. Soc. Chim. Belg.*, **88**, 923.
32. Li, W. and Zhang, X. (2000) *J. Org. Chem.*, **65**, 5871.
33. Yan, Y.-Y. and RajanBabu, T.V. (2000) *Org. Lett.*, **2**, 4137.
34. Ley, S.V., Downham, R., Edwards, P.J., Innes, J.E., and Woods, M. (1995) *Contemp. Org. Synth.*, **2**, 365.
35. Fujita, M., Laine, D., and Ley, S.V. (1999) *J. Chem. Soc., Perkin Trans. 1*, 1647.
36. Li, W., Waldkirch, J.P., and Zhang, X. (2002) *J. Org. Chem.*, **67**, 7618.
37. Lee, S., Zhang, Y.J., Song, C.E., Lee, J.K., and Choi, J.H. (2002) *Angew. Chem., Int. Ed.*, **41**, 847.

38 Zhang, Y.J., Kim, K.Y., Park, J.H., Song, C.E., Lee, K., Lah, M.S., and Lee, S. (2005) *Adv. Synth. Catal.*, **347**, 563.
39 Lee, S., Zhang, Y.J., Piao, J.Y., Yoon, H., Song, C.E., Choi, J.H., and Hong, J. (2003) *Chem. Commun.*, 2624.
40 Zhang, F.-Y., Pai, C.-C., and Chan, A.S.C. (1998) *J. Am. Chem. Soc.*, **120**, 5808.
41 Wang, C.-J., Gao, F., and Liang, G. (2008) *Org. Lett.*, **10**, 4711.
42 (a) Boaz, N.W., Debenham, S.D., Mackenzie, E.B., and Large, S.E. (2002) *Org. Lett.*, **4**, 2421; (b) Boaz, N.W., Mackenzie, E.B., Debenham, S.D., Large, S.E., and Ponasik, J.A. Jr (2005) *J. Org. Chem.*, **70**, 1872.
43 Li, X., Jia, X., Xu, L., Kok, S.H.L., Yip, C.W., and Chan, A.S.C. (2005) *Adv. Synth. Catal.*, **347**, 1904.
44 Qiu, M., Hu, X.-P., Huang, J.-D., Wang, D.-Y., Deng, J., Yu, S.-B., Duan, Z.-C., and Zheng, Z. (2008) *Adv. Synth. Catal.*, **350**, 2683.
45 Hu, X.-P. and Zheng, Z. (2004) *Org. Lett.*, **6**, 3585.
46 Zeng, Q.-H., Hu, X.-P., Duan, Z.-C., Liang, X.-M., and Zheng, Z. (2005) *Tetrahedron: Asymmetry*, **16**, 1233.
47 Jia, X., Li, X., Lam, W.S., Kok, S.H.L., Xu, L., Lu, G., Yeung, C.-H., and Chan, A.S.C. (2004) *Tetrahedron: Asymmetry*, **15**, 2273.
48 Huang, J.-D., Hu, X.-P., Duan, Z.-C., Zeng, Q.-H., Yu, S.-B., Deng, J., Wang, D.-Y., and Zheng, Z. (2006) *Org. Lett.*, **8**, 4367.
49 Wang, D.-Y., Hu, X.-P., Huang, J.-D., Deng, J., Yu, S.-B., Duan, Z.-C., Xu, X.-F., and Zheng, Z. (2007) *Angew. Chem., Int. Ed.*, **46**, 7810.
50 Qiu, M., Hu, X.-P., Wang, D.-Y., Deng, J., Huang, J.-D., Yu, S.-B., Duan, Z.-C., and Zheng, Z. (2008) *Adv. Synth. Catal.*, **350**, 1413.
51 Yu, S.-B., Huang, J.-D., Wang, D.-Y., Hu, X.-P., Deng, J., Duan, Z.-C., and Zheng, Z. (2008) *Tetrahedron: Asymmetry*, **19**, 1862.
52 Eggenstein, M., Thomas, A., Theuerkauf, J., Franciò, G., and Leitner, W. (2009) *Adv. Synth. Catal.*, **351**, 725.
53 Zhang, W. and Zhang, X. (2006) *Angew. Chem., Int. Ed.*, **45**, 5515.
54 Fernández-Pérez, H., Pericàs, M.A., and Vidal-Ferran, A. (2008) *Adv. Synth. Catal.*, **350**, 1984.
55 (a) Ireland, T., Grossheimann, G., Wieser-Jeunesse, C., and Knochel, P. (1999) *Angew. Chem., Int. Ed.*, **38**, 3212; (b) Lotz, M., Polborn, K., and Knochel, P. (2002) *Angew. Chem., Int. Ed.*, **41**, 4708.
56 Chen, W., Roberts, S.M., Whittall, J., and Steiner, A. (2006) *Chem. Commun.*, 2916.
57 Evans, D.A., Michael, F.E., Tedrow, J.S., and Campos, K.R. (2003) *J. Am. Chem. Soc.*, **125**, 3534.
58 Zhang, Z., Zhu, G., Jiang, Q., Xiao, D., and Zhang, X. (1999) *J. Org. Chem.*, **64**, 1774.
59 Tang, W., Chi, Y., and Zhang, X. (2002) *Org. Lett.*, **4**, 1695.
60 Dupau, P., Bruneau, C., and Dixneuf, P.H. (2001) *Adv. Synth. Catal.*, **343**, 331.
61 Renaud, J.L., Dupau, P., Hay, A.-E., Guingouain, M., Dixneuf, P.H., and Bruneau, C. (2003) *Adv. Synth. Catal.*, **345**, 230.
62 (a) Lennon, I.C., Ramsden, J.A., Brear, C.J., Broady, S.D., and Muir, J.C. (2007) *Tetrahedron Lett.*, **48**, 4623; (b) Broady, S.D., Golden, M.D., Leonard, J., Muir, J.C., and Maudet, M. (2007) *Tetrahedron Lett.*, **48**, 4627.
63 Noyori, R., Ohta, M., Hsiao, Y., Kitamura, M., Ohta, T., and Takaya, H. (1986) *J. Am. Chem. Soc.*, **108**, 7117.
64 Shen, Z., Lu, X., and Lei, A. (2006) *Tetrahedron*, **62**, 9237.

10
Enantioselective Reduction of Nitrogen-Based Heteroaromatic Compounds

Da-Wei Wang, Yong-Gui Zhou, Qing-An Chen, and Duo-Sheng Wang

Optically active amines are fundamentally important synthetic intermediates and structural components of biologically active natural products, drugs, and pharmaceuticals. Asymmetric hydrogenation of nitrogen-based heteroaromatic compounds is one of the most attractive and efficient approaches to enantiomerically pure amines, especially for numerous saturated or partially saturated chiral cyclic amines, whose synthesis by direct cyclization is often difficult. In the past decades, some progresses have been achieved in the hydrogenation of nitrogen-containing heteroaromatic compounds [1]. For example, quinolines, quinoxalines, pyridines, indoles, and pyrroles have been hydrogenated successfully with over 90% enantiomeric excess (ee). In this chapter, we focus on the synthesis of chiral amines via enantioselective hydrogenation of nitrogen-based heteroaromatic compounds.

10.1
Asymmetric Hydrogenation of Quinolines

10.1.1
Ir- and Ru-Catalyzed Asymmetric Hydrogenation of Quinolines

Among these studies on asymmetric hydrogenation of nitrogen-based heteroaromatic compounds, the hydrogenation of quinolines was studied extensively and elaborately. In 2003, Zhou and coworkers reported the first highly enantioselective hydrogenation of quinolines with high enantioselectivities [2]. They employed [Ir(COD)Cl]$_2$/MeO-BiPhep as catalyst using iodine as additive, while the hydrogenation reaction could not take place in the absence of iodine. Their studies showed that this reaction was highly solvent dependent, and toluene was the best solvent.

Subsequently, some commercially available chiral bidentate phosphine ligands were tested for the asymmetric hydrogenation of quinolines, and (S)-MeO-BiPhep was the best ligand with 94% ee (Scheme 10.1). The optimal conditions are [Ir(COD)Cl]$_2$/MeO-BiPhep/I$_2$ in toluene at 700 psi H$_2$.

Chiral Amine Synthesis: Methods, Developments and Applications. Edited by Thomas C. Nugent
Copyright © 2010 WILEY-VCH Verlag GmbH & Co. KGaA, Weinheim
ISBN: 978-3-527-32509-2

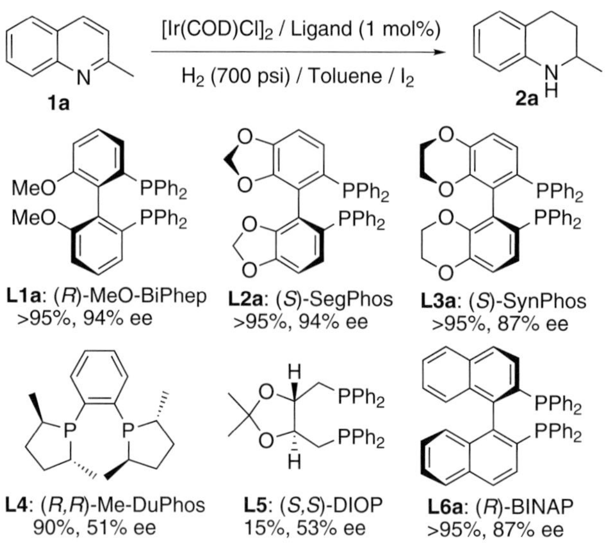

10 Enantioselective Reduction of Nitrogen-Based Heteroaromatic Compounds

1a → **2a**

[Ir(COD)Cl]$_2$ / Ligand (1 mol%)
H$_2$ (700 psi) / Toluene / I$_2$

L1a: (R)-MeO-BiPhep
>95%, 94% ee

L2a: (S)-SegPhos
>95%, 94% ee

L3a: (S)-SynPhos
>95%, 87% ee

L4: (R,R)-Me-DuPhos
90%, 51% ee

L5: (S,S)-DIOP
15%, 53% ee

L6a: (R)-BINAP
>95%, 87% ee

Scheme 10.1 Effect of ligands on conversion and ee.

Having established the optimal condition, the scope of the Ir-catalyzed asymmetric hydrogenation of quinoline derivatives was explored. A variety of 2-substituted and 2,6-disubstituted quinoline derivatives were hydrogenated using Ir/MeO-Biphep/I$_2$ as the catalyst. Several 2-alkyl-substituted quinolines were hydrogenated with high enantioselectivities (>92% ee) regardless of the length of side chain (Table 10.1, entries 1–6). 2-Arenethyl-substituted quinolines also gave excellent asymmetric induction (entries 8–10). Interestingly, for 2-aryl-substituted quinolines, slightly lower enantioselectivity was obtained (entry 14). Their catalytic system can tolerate hydroxyl group (entries 15–19). C=C double bond in side chain of substrate (entry 5) can be hydrogenated under their standard conditions. It is a pity that very low enantioselectivity and reactivity were obtained for 3- and 4-substituted quinoline derivatives (0 and 1% ee for 3- and 4-methylquinoline, respectively) and the reason for this is not clear.

The [Ir(COD)Cl]$_2$/MeO-BiPhep/I$_2$ system was also efficient for enantioselective hydrogenation of 2-benzylquinoline derivatives [3]. Under the above optimized condition, all the 2-benzylquinolines were hydrogenated completely to give the corresponding 1,2,3,4-tetrahydro-benzylquinoline derivatives (Scheme 10.2). Excellent enantioselectivities and high yields were obtained regardless of the electronic properties and steric hindrance of substituent groups. 2-Benzyl-6-fluoroquinoline gave the highest enantioselectivity (96% ee).

Zhou and coworkers also studied the asymmetric hydrogenation of 2-functionalized quinolines using [Ir(COD)Cl]$_2$/MeO-BiPhep/I$_2$ system [3]. The reaction was conducted at ambient temperature under H$_2$ pressure of 800 psi in the presence of 1.0 mol% of catalysts prepared *in situ* from [Ir(COD)Cl]$_2$ and 2.2 equiv of chiral ligand. As illustrated in Table 10.2, a variety of 2-functionalized quinoline derivatives

Table 10.1 Ir-catalyzed asymmetric hydrogenation of quinolines by Zhou.

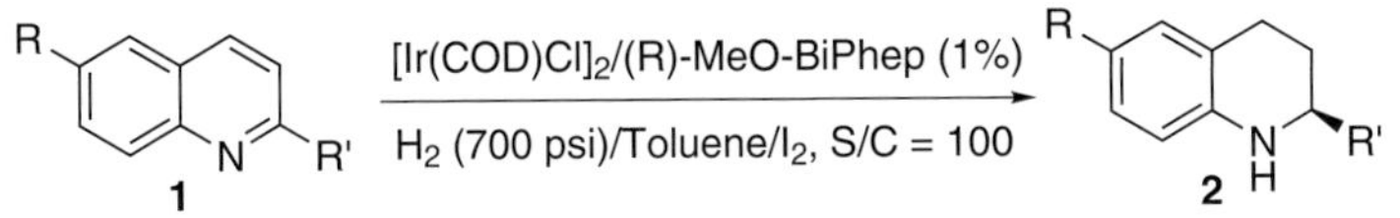

$$\text{1} \xrightarrow[\text{H}_2 \text{ (700 psi)/Toluene/I}_2\text{, S/C = 100}]{\text{[Ir(COD)Cl]}_2\text{/(R)-MeO-BiPhep (1\%)}} \text{2}$$

Entry	R/R'	Yield (%)	ee (%)
1	H/Me	94	94 (R)
2	H/Et	88	96 (R)
3	H/nPr	92	93 (R)
4	H/nBu	86	92 (R)
5	H/3-butenyl[a]	91	92 (R)
6	H/npentyl	92	94 (R)
7	H/iPr	92	94 (S)
8	H/phenethyl	94	93 (R)
9	H/3,4-(OCH$_2$O)C$_6$H$_3$(CH$_2$)$_2$-	88	93 (R)
10	H/3,4-(MeO)$_2$C$_6$H$_3$(CH$_2$)$_2$-	86	96 (R)
11	F/Me	88	96 (R)
12	Me/Me	91	91 (R)
13	MeO/Me	89	84 (R)
14	H/Ph	95	72 (S)
15	H/Me$_2$CH(OH)CH$_2$-	87	94 (S)
16	H/cC$_6$H$_{11}$(OH)CH$_2$-	89	92 (S)
17	H/Ph$_2$CH(OH)CH$_2$-	94	91 (S)
18	H/CH$_2$OH	83	75 (S)
19	H/CH$_2$OCOCH$_3$	90	87 (S)

a) C=C was hydrogenated.

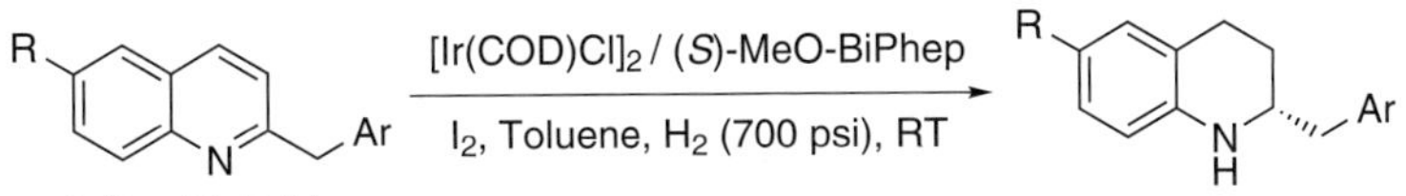

3: R = H, F, Me
Ar = C$_6$H$_5$, 2-CH$_3$C$_6$H$_4$, 1-C$_{10}$H$_7$,
CF$_3$C$_6$H$_4$, 4-FC$_6$H$_4$, 3,4-(MeO)$_2$C$_6$H$_3$

4: 88–96% ee
9 examples

Reagents: [Ir(COD)Cl]$_2$ / (S)-MeO-BiPhep, I$_2$, Toluene, H$_2$ (700 psi), RT

Scheme 10.2 Ir-catalyzed asymmetric hydrogenation of 2-benzylquinolines.

could be successfully hydrogenated to afford their corresponding derivatives. For the quinolines bearing alkyl ketones, the ee's were slightly affected by the electronic properties and steric hindrance of substituents, such as for methyl ketone and propyl ketone, 90 and 84% ee were obtained, respectively (Table 10.2, entries 2–3). With quinolines bearing arylketone groups, the ee's of all the products were excellent regardless of the electronic properties and steric hindrance (Table 10.2, entries 1, 4–10, 12–14). Interestingly, the system could tolerate the esters, amide, benzenesulfonyl groups, and all these substrates were transformed to the corresponding

Table 10.2 Ir-catalyzed hydrogenation of 2-functionalized quinolines.

Substrate **5** (2-R²-substituted quinoline with R¹) → Product **6** (tetrahydroquinoline) using [Ir(COD)Cl]₂ / (S)-MeO-BiPhep, I₂, Benzene, RT, H₂ (800 psi).

Entry	R^1/R^2	Yield (%)	ee (%), config.
1	H/COPh	91 (6a)	96 (R)
2	H/COMe	93 (6b)	90 (R)
3	H/CO(nPr)	91 (6c)	84 (R)
4	H/CO(p-MeOPh)	84 (6d)	83 (R)
5	H/CO(o-MeOPh)	78 (6e)	95 (R)
6	H/CO(p-MePh)	89 (6f)	95 (R)
7	H/CO(o-MePh)	97 (6g)	96 (R)
8	H/CO(p-iPrPh)	97 (6h)	95 (R)
9	H/CO(p-CF₃Ph)	90 (6i)	95 (R)
10	H/CO(1-naphthyl)	89 (6j)	95 (R)
11	H/CO(CH₂)₂Ph	90 (6k)	87 (R)
12	Me/COPh	82 (6l)	94 (R)
13	F/COPh	92 (6m)	96 (R)
14	H/CO(3,4-(MeO)₂Ph)	95 (6n)	94 (R)
15	H/p-MeOPhCH=CH[a]	80 (6o)	95 (S)
16	H/COOMe	88 (6p)	82 (R)
17	H/COOEt	93 (6q)	92 (R)
18	H/CONEt₂	98 (6r)	80 (R)
19	H/SO₂Ph	97 (6s)	90 (R)
20	H/(CH₂)₃OTBS	90 (6t)	94 (S)
21	H/(CH₂)₄OTBS	65 (6u)	89 (S)

a) The double bond was also hydrogenated.

tetrahydroquinoline derivatives with 80–92% ee's (Table 10.2, entries 16–19). It is noted that substrates with hydroxyl by TBS protected could also be hydrogenated smoothly with high enantioselectivities (Table 10.2, entries 20–21).

Following their seminal work, Zhou group showed the usefulness of ferrocene phosphinite-oxazoline ligands as N,P ligands in the Ir-catalyzed asymmetric hydrogenation of quinolines and up to 92% ee was obtained (Scheme 10.3) [4]. The influence of the relative configuration of the chirality in the oxazoline ring and the planar chirality of the ferrocene ring was studied. Central chirality dominates the absolute configuration of the products. The best result was obtained by using **L7** as a chiral ligand. In 2005, they found that S,P ligands were also effective for the hydrogenation of quinolines [5]. The ligands S,P-**9** and S,P-**10** with same central chirality and opposite planar chirality gave product with the same absolute configuration. Interestingly, if a bulky trimethylsilyl (TMS) group (S,P-**11** and S,P-**12**) was introduced to the Cp ring of S,P-**9** and S,P-**10** ligands, hydrogenation products with opposite absolute configuration were obtained in moderate enantioselectivity.

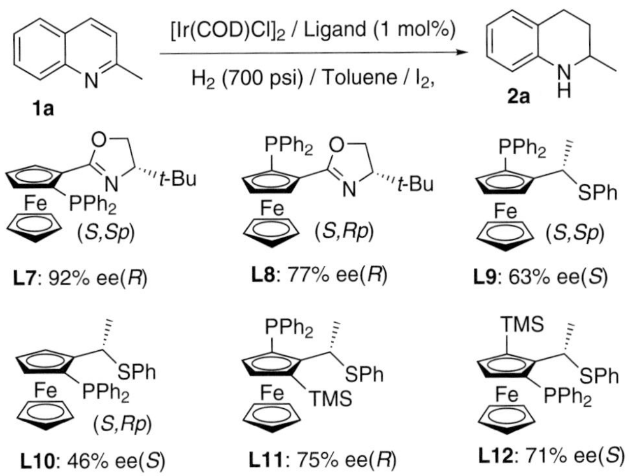

10.1 Asymmetric Hydrogenation of Quinolines

Scheme 10.3 Asymmetric hydrogenation of quinolines using ferrocene-derived N,P and S,P ligands.

After the initial work of Zhou's group on iridium-catalyzed enantioselective hydrogenation of quinoline derivatives, some other groups, such as Fan, Chan, Xu, Reetz, Rueping, Du and Bolm et al. have reported their results on asymmetric hydrogenation of quinoline derivatives.

In 2005, Fan and coworkers developed a highly effective and air-stable catalyst system Ir/P-Phos/I_2 for the asymmetric hydrogenation of quinoline derivatives (Scheme 10.4) [6]. They found that THF was the best solvent that gave the highest enantioselectivity (92% ee). The reactions were carried out at room temperature, and a series of quinoline derivatives were examined with full conversions and excellent enantioselectivity. More important, the catalyst could be effectively immobilized in DMPEG with retained reactivity and enantioselectivity in eight catalytic runs.

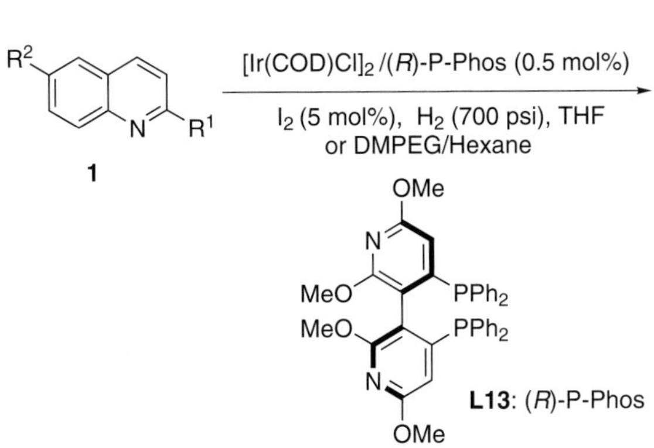

Scheme 10.4 Asymmetric hydrogenation of quinolines by Fan and Chan.

10 Enantioselective Reduction of Nitrogen-Based Heteroaromatic Compounds

In the same year, Xu and coworkers showed the usefulness of Ir-**L14** catalyst in the asymmetric hydrogenation of quinoline derivatives [7]. They found that better enantioselectivities were obtained using DMPEG/hexane as reaction medium than THF. The highest enantioselectivity was 97% ee. However, the recycling of the catalyst proved to be difficult (Scheme 10.5).

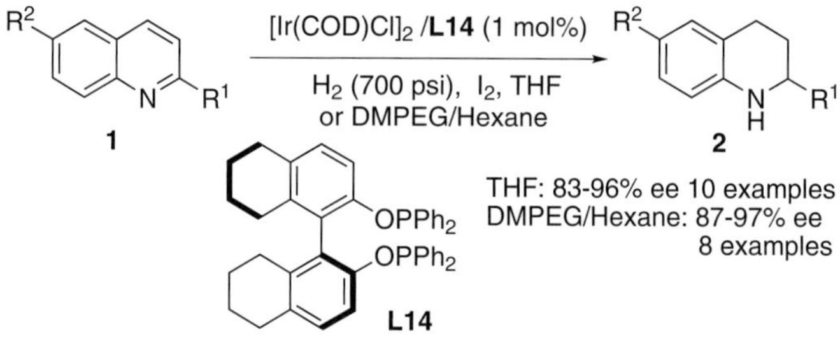

Scheme 10.5 Asymmetric hydrogenation of quinolines by Xu, Fan, and Chan.

In 2006, Chan group prepared a series of chiral diphosphine ligands denoted as PQ-Phos by employing the atropdiastereoselective Ullmann coupling and ring closure reactions as key steps. These PQ-Phos ligands have been successfully applied in the catalytic asymmetric hydrogenation of quinolines with 72–94% ee (Scheme 10.6) [8].

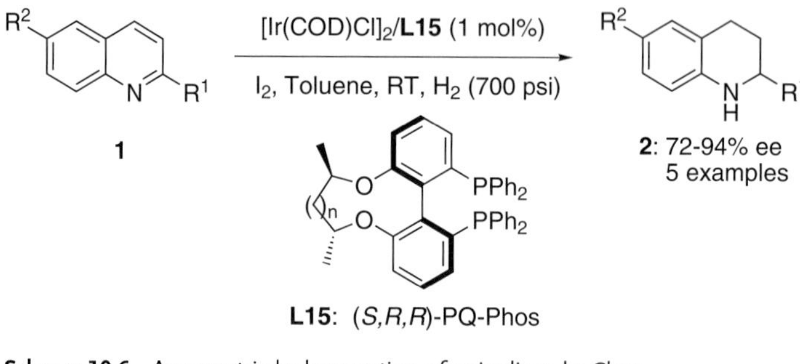

Scheme 10.6 Asymmetric hydrogenation of quinolines by Chan.

Reetz and Li studied asymmetric hydrogenation of quinolines using iridium complex with a chiral BINOL-derived diphosphonite ligand with an achiral diphenyl ether backbone as catalyst, achiral P-ligands serving as possible additives (Scheme 10.7) [9]. Under the optimized conditions, 2-substituted and 2,6-disubstituted quinoline derivatives were hydrogenated with good to excellent enantioselectivities (73–96% ee). Subsequently, they examined the effect of mixture of **L16** and achiral P-ligands such as **L16a–L16e** on the enantioselectivity, and slight improvements were observed in the results: **L16a** (92% ee), **L16b** (90% ee), **L16c** (94% ee), **L16d** (94% ee), **L16e** (94% ee), while 92% ee was obtained with **L16** only.

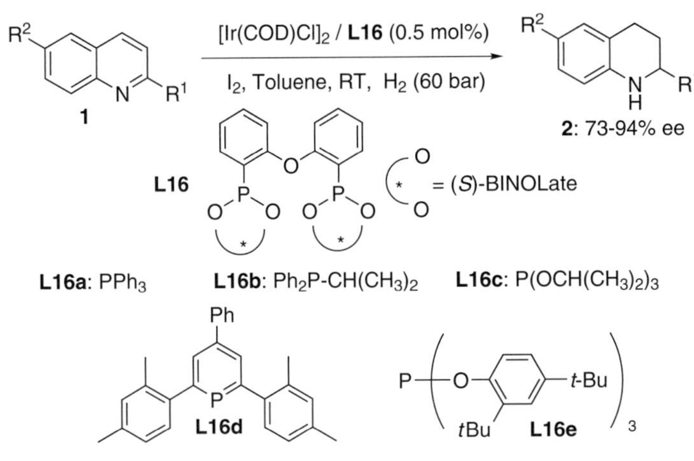

10.1 Asymmetric Hydrogenation of Quinolines

Scheme 10.7 Asymmetric hydrogenation of quinolines by Reetz.

In 2006, Genet and Mashima developed a new and convenient one-pot reaction to prepare mononuclear halide–carboxylate iridium(III) complexes and cationic triply halogen-bridged dinuclear iridium(III) complexes of BINAP, p-TolBINAP, and SynPhos. These iridium(III) complexes were tested as catalyst precursors for asymmetric hydrogenation of 2-phenylquinoline (Scheme 10.8) [10]. Cationic iododinuclear p-TolBINAP (S)-**L6e** and SynPhos (S)-**L3c** complexes had better enantios-

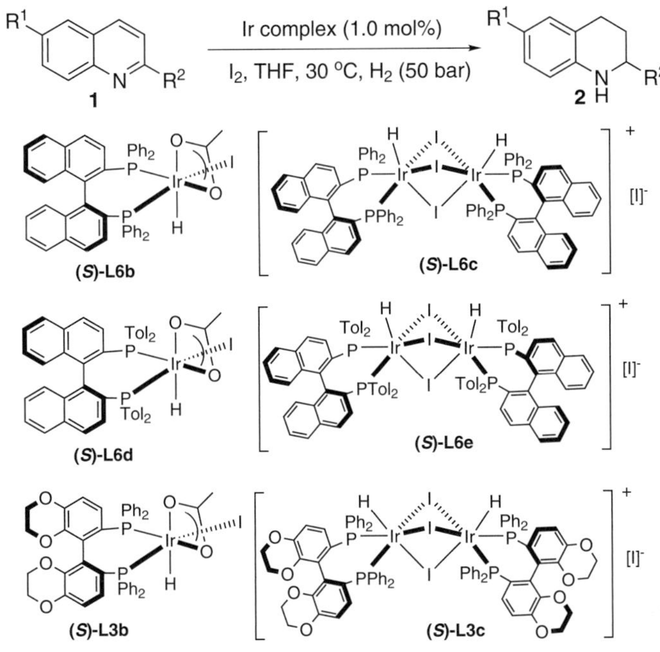

Scheme 10.8 Asymmetric hydrogenation of quinolines by Genet and Mashima.

electivity for asymmetric hydrogenation of 2-phenylquinoline. However, only moderate enantioselectivities were detected. In 2007, they showed that these Ir catalysts based on SynPhos and DifluorPhos were also efficient for asymmetric hydrogenation of 2-alkylquinolines [11]. Excellent conversions (up to 100%) and good enantioselectivities (up to 92%) were obtained in the hydrogenation of 2-substituted and 2,6-disubstituted quinoline derivatives.

The iridium-catalyzed asymmetric hydrogenation of quinolines in the presence of iodine catalyst loading is relatively high, which may be due to the catalyst deactivation during the reaction. It was reported that the Ir complexes could form irreversibly the inactive dimer or trimer by hydride-bridged bonds, which retards the reaction [12].

In 2007, Fan and coworkers applied the encapsulation of iridium complex into a dendrimer framework to reduce dimerization (S)-GnDenBINAP) [13]. Thus, high enantioselectivities with significantly high catalytic activities (TOF up to 3450 h^{-1}) and productivities (TON up to 43 000) have been achieved, and the third-generation catalyst could be recovered by precipitation and filtration and reused at least six times without loss of reactivity and enantioselectivity (Scheme 10.9). The strategy of Fan was a successful example in avoiding the formation of inactive dimmer in iridium-catalyzed asymmetric hydrogenation of quinolines.

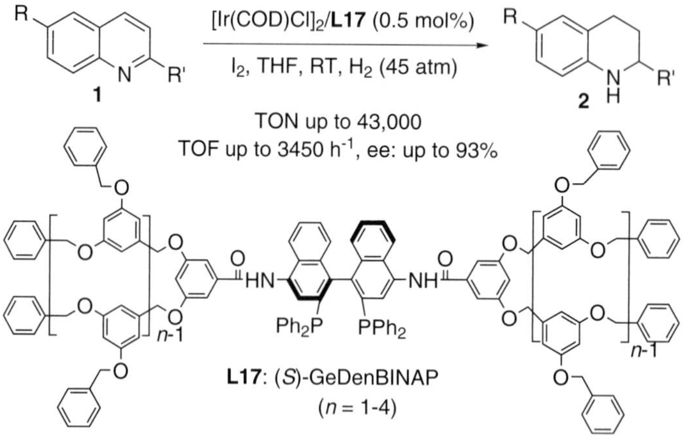

Scheme 10.9 Asymmetric hydrogenation of quinolines by Xu, Fan, and Chan.

In 2007, Xu and coworkers displayed the usefulness of chiral diphosphinite ligands based on 1,1′-spirobiindane backbone with high substrate/catalyst ratio (up to 5000) and up to 94% ee (Scheme 10.10) [14]. Interestingly, this reaction was performed smoothly in a DMPEG–hexane biphasic system, resulting in efficient separation and recycling of the catalyst. The enantioselectivity was retained quite well after four runs. However, the conversion was dropped to approximate 40%, which might be due to the decomposition or a certain degree of leaching of the catalyst in the course of recycling.

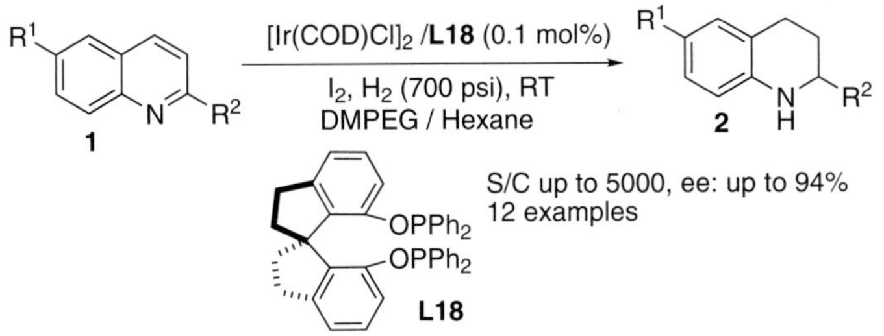

10.1 Asymmetric Hydrogenation of Quinolines

Scheme 10.10 Asymmetric hydrogenation of quinolines by Xu, Fan, and Chan.

Then, Lam and coworkers reported that the iridium complexes with C_2-symmetric ligands such as Xyl-P-Phos, Cl–MeO–BiPhep, SynPhos, and DifluorPhos are effective catalysts for the asymmetric hydrogenation of quinolines [15]. After establishing a general condition, H_2 (700 psi), THF (or THF/CH_2Cl_2), a variety of quinoline derivatives were investigated. In general, all the quinolines were hydrogenated to give the corresponding 1,2,3,4-tetrahydroquinolines in full conversion with up to 95% ee (Scheme 10.11). It was noted that the catalysts could be recovered and reused for at least three cycles in DMPEG/hexane biphasic system.

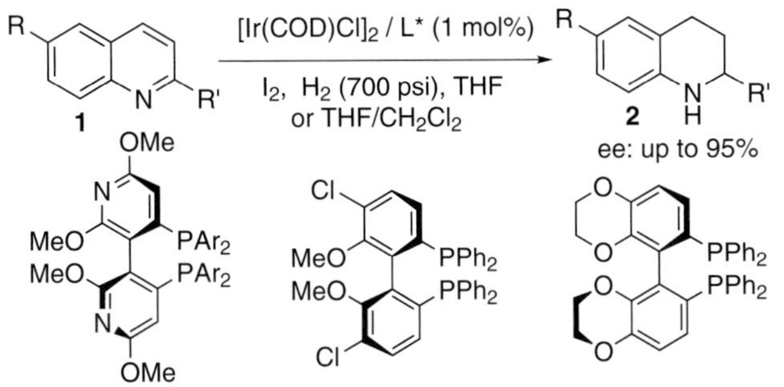

Scheme 10.11 Asymmetric hydrogenation of quinolines by Lam, Fan, and Chan.

Pellet-Rostaing and coworkers have devised an effective approach to synthesize more electron-donating BINAP chiral ligands and examined their catalytic effect on the asymmetric hydrogenation of 2-methylquinoline [16], and 89% conversion with 70% ee was obtained (Scheme 10.12).

In 2008, Vries group reported asymmetric hydrogenation of quinolines catalyzed by iridium complexes based on monodentate BINOL-derived phosphoramidites PipPhos. They used tri-*ortho*-tolylphosphine and/or chloride salts as additives, and enantioselectivities were strongly enhanced to 89% ee (Scheme 10.13) [17]. Toluene and DCM were the best solvents, and the reaction was carried out at 60 °C for 24 h in the pressure of 50 bar H_2, and a series of 2-substituted and 2,6-disubstituted quinolines were examined with excellent

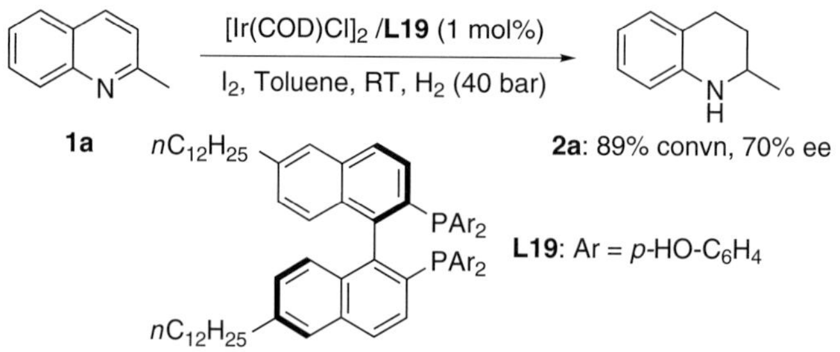

10 Enantioselective Reduction of Nitrogen-Based Heteroaromatic Compounds

Scheme 10.12 Asymmetric hydrogenation of quinolines by Pellet-Rostaing and Lemaire.

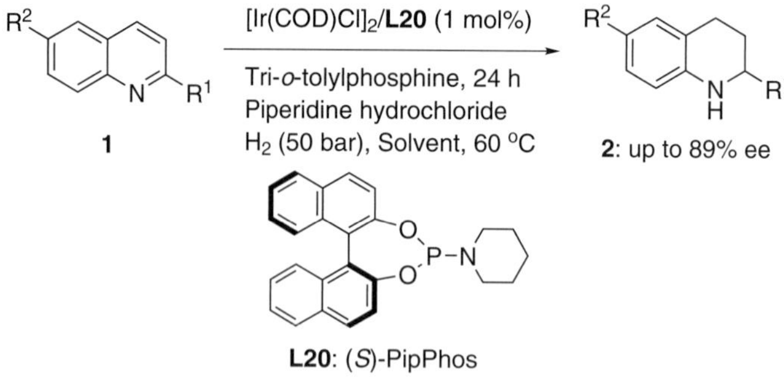

Scheme 10.13 Asymmetric hydrogenation of quinolines by Vries.

conversions and 76–89% ee. It should be noted that iodine is not necessary in this hydrogenation reaction.

Bolm and Lu synthesized a series of naphthalene-bridged P,N-type sulfoximine ligands and their iridium complexes. Asymmetric hydrogenation of quinolines was explored using these iridium complexes as catalyst, and 55–92% ee with moderate conversions were detected [18]. It was noted that iodine was also not necessary in the hydrogenation of quinolines, and the enantioselectivity was reduced when iodine was added (Scheme 10.14).

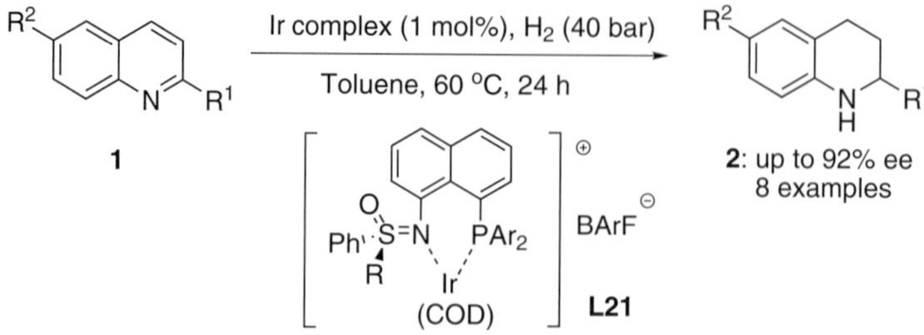

Scheme 10.14 Asymmetric hydrogenation of quinolines by Bolm.

The iridium-catalyzed asymmetric hydrogenation of quinolines provides a convenient and efficient way to enantiomerically pure amines with bisphosphines, N,P-ligands and S,P-ligands. However, the reactivities and enantioselectivities were generally substrate dependent. There is no omnipotent ligand for every substrate. In 2008, Zhou and Wang devised and divergently synthesized a series of tunable axial chiral bisphosphine ligands based on (S)-MeO-Biphep by introduction of different substituents at the 6,6′-positions of the biaryl backbone (Scheme 10.15) [19]. The iridium complex with these ligands has been successfully applied in asymmetric hydrogenation of quinolines, and it was found the ligand ($R_1 = n\text{-}C_{12}H_{25}$, $R_2 = $ MeO-PEG-1600) could gave the best result (92% ee). More important, the catalyst system could be reused in five runs, and a slightly lower enantioselectivity (84% ee) with 95% conversion was obtained.

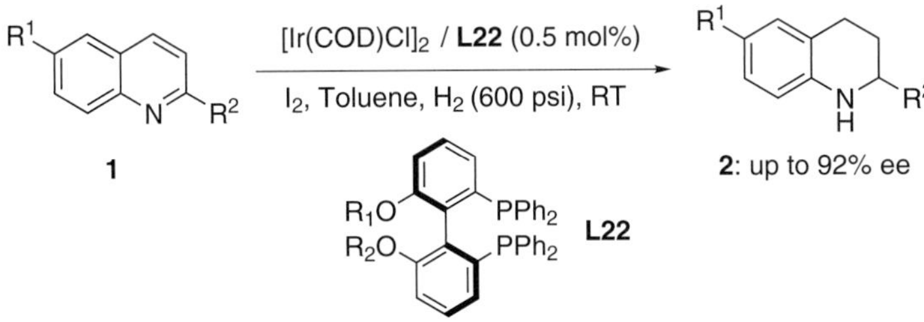

Scheme 10.15 Asymmetric hydrogenation of quinolines by Zhou.

In 2008, Fan and Xu developed an air-stable and phosphine-free Ir catalyst for the asymmetric hydrogenation of quinolines [20]. They used chiral cationic Cp*Ir(OTf)(CF$_3$TsDPEN) complex as catalyst [21]. The reaction proceeded smoothly in undegassed methanol with no need for inert gas protection and afforded the 1,2,3,4-tetrahydroquinoline derivatives in up to 99% ee (Table 10.3). The counterion of iridium catalyst is very important, OTf gave high reactivity and enantioselectivity, and no reactivity was observed for chloride. It is noted that it is one of the best results of asymmetric hydrogenation of quinolines.

Leitner and coworkers disclosed new phosphine phosphoramidite ligands containing two elements of chirality, which can be prepared via a modular approach. These ligands proved to be very efficient for asymmetry of 2-substituted quinolines [22]. The hydrogenation reaction proceeded smoothly with up to 97% ee and full conversion (Scheme 10.16).

The above catalytic systems for the asymmetric hydrogenation of quinolines are mainly iridium catalysts. In 2008, Fan and coworkers developed recyclable phosphine-free chiral cationic ruthenium-catalyzed asymmetric hydrogenation of quinolines [23]. They found that the phosphine-free cationic Ru/TsDPEN catalyst exhibited unprecedented reactivity and high enantioselectivity in the hydrogenation of quinolines in neat ionic liquid. The results were very excellent and enantioselec-

Table 10.3 Asymmetric hydrogenation of quinolines by Lam, Fan, and Chan.

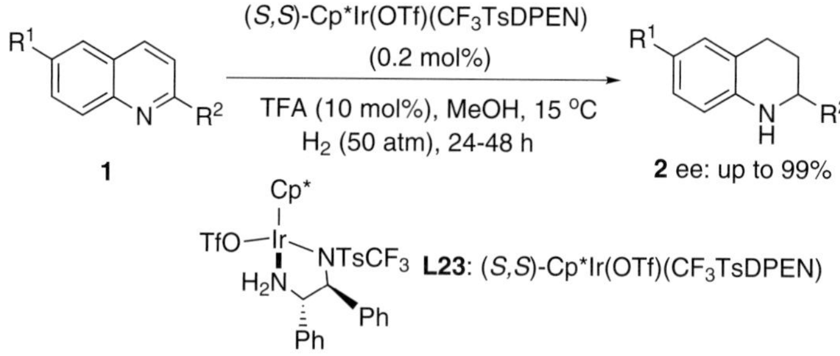

Reaction: Quinoline **1** (R^1, R^2) → tetrahydroquinoline **2** using (S,S)-Cp*Ir(OTf)(CF$_3$TsDPEN) (0.2 mol%), TFA (10 mol%), MeOH, 15 °C, H$_2$ (50 atm), 24–48 h; ee: up to 99%.

L23: (S,S)-Cp*Ir(OTf)(CF$_3$TsDPEN)

Entry	R^1/R^2	Yield (%)	ee (%)
1	H/Me	99	98
2	H/Et	98	98
3	H/nPr	98	96
4	H/nBu	97	97
5	H/npentyl	98	96
6	H/phenethyl	97	97
7	H/3,4-(OCH$_2$O)C$_6$H$_3$(CH$_2$)$_2$-	97	97
8	H/3,4-(MeO)$_2$C$_6$H$_3$(CH$_2$)$_2$-	96	96
9	F/Me	98	94
10	Me/Me	95	97
11	MeO/Me	97	97
12	H/Ph	90	79
13	H/Me$_2$CH(OH)CH$_2$-	97	99
14	H/cC$_6$H$_{11}$(OH)CH$_2$-	97	99

tivities of all the products were 96–99% ee. The use of ionic liquid not only facilitated the recycle, but also enhanced the stability and selectivity of the catalyst (Table 10.4). It should be noted that it is the first example of Ru-catalyzed asymmetric hydrogenation of quinolines.

Scheme: Quinoline **1** (R^1, R^2) → tetrahydroquinoline **2** using [Ir(COD)Cl]$_2$ / **L24** (0.5 mol%), RT, I$_2$ (5 mol%), Toluene, H$_2$ (40 bar); up to 97% ee, 3 examples.

L24a: R = Ph
L24b: R = naphthyl
L24c: R = Ph

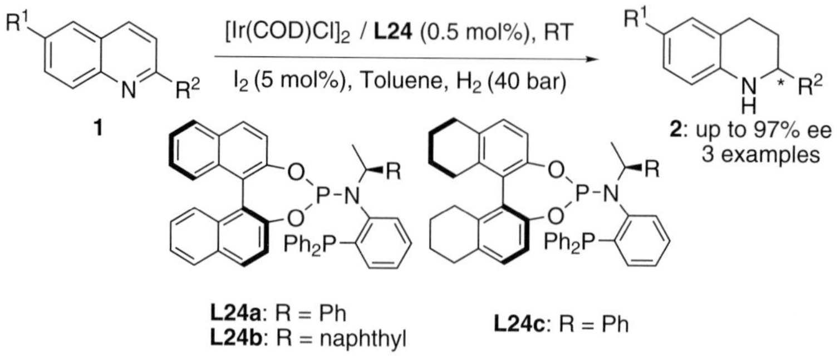

Scheme 10.16 Asymmetric hydrogenation of quinolines by Leitner.

Table 10.4 Asymmetric ruthenium-catalyzed hydrogenation of quinolines by Fan.

Entry	R¹/R²	Yield (%)	ee (%)
1	H/Me	96	99
2	H/Et	95	98
3	H/nPr	95	99
4	H/nBu	95	98
5	H/$n$pentyl	97	98
6	H/phenethyl	96	97
7	H/3,4-(OCH$_2$O)C$_6$H$_3$(CH$_2$)$_2$-	96	97
8	H/3,4-(MeO)$_2$C$_6$H$_3$(CH$_2$)$_2$-	96	98
9	F/Me	92	98
10	Me/Me	94	99
11	MeO/Me	91	99
12	MeCO/Me	87	96
13	H/Me$_2$CH(OH)CH$_2$-	92	99
14	H/cC$_6$H$_{11}$(OH)CH$_2$-	93	99

Iridium-catalyzed asymmetric hydrogenation of quinolines provides a convenient route to synthesize optically active tetrahydroquinoline derivatives from the very cheap starting materials quinolines. In 2003, the group of Zhou found that their methodology could be successfully applied to the asymmetric synthesis of tetrahydroquinoline alkaloids and chiral drugs (Scheme 10.17). For example, the hydrogenation product of 6-fluoro-2-methylquinoline is the key intermediate of antibacterial agent of flumequine [2]. Furthermore, some naturally occurring tetrahydroquinoline alkaloids such as angustrureine, galipinine, and cuspareine were easily synthesized by N-methylation of hydrogenated products with high overall yields [2]. In 2004, Zhou and Yang reported the total synthesis of alkaloid (−)-galipeine [24], which contains a free phenol hydroxyl. (−)-Galipeine has been synthesized from isovanillin by protection of the hydroxyl group in seven steps with 54% overall yield. Absolute configurations of (+)-angustureine, (−)-galipinine, and (−)-galipeine could be assigned through their enantioselective synthesis.

To further understand the hydrogenation of quinolines, Zhou and coworkers explored the mechanistic studies and found that iodine activated the catalyst in the hydrogenation of quinolines, which is accorded with the works of Osborn and

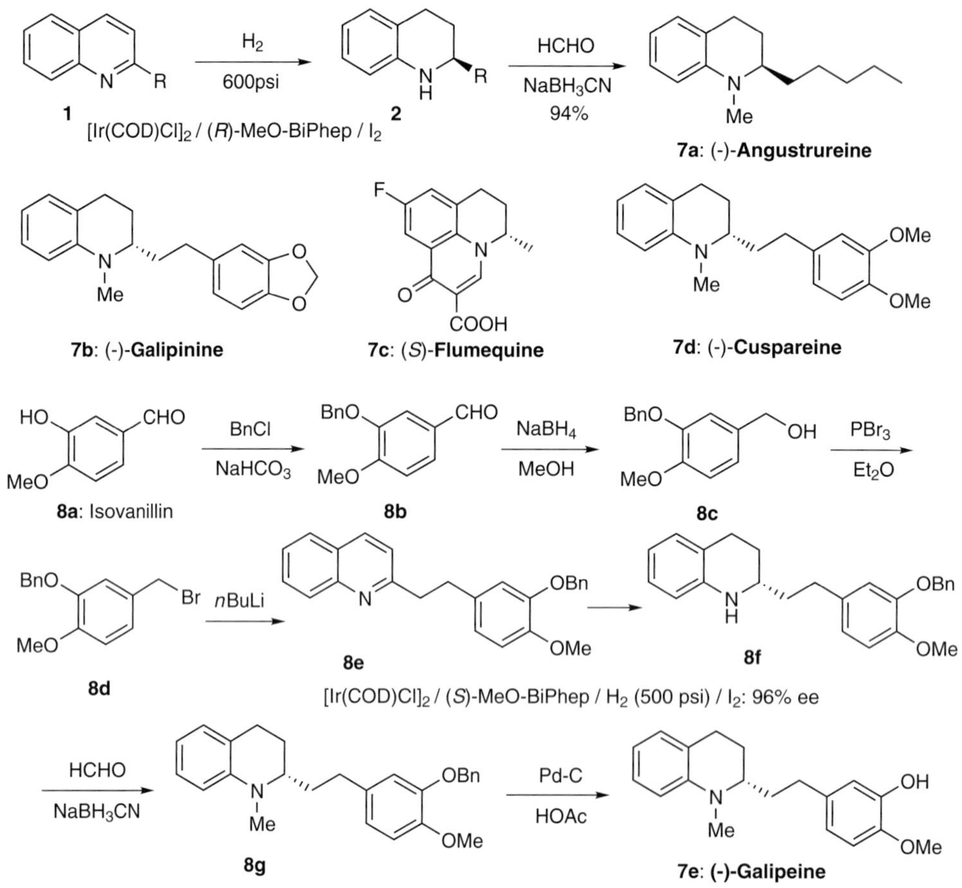

Scheme 10.17 Asymmetric synthesis of tetrahydroquinoline alkaloids and drug.

Dorta [25]. Their mechanistic studies on hydrogenation sequence of C=C and C=N bonds of quinolines revealed that the hydrogenation mechanism of quinolines involves 1,4-hydride addition to quinoline, isomerization, and 1,2-hydride addition, and the catalytic active species may be Ir(III) complex with chloride and iodide [3].

The synthesis of reaction intermediate is also very important and sometimes could provide a direct proof to support the mechanism. To obtain the stable intermediate of this reaction, they selected 2-functionalized quinoline **5a** as the starting material (Scheme 10.18). The results showed that the steady intermediate **9b** was achieved after isomerization from the intermediate **9a** when **5a** was treated using Pd/C with hydrogen in MeOH. After the synthesis of the intermediate **9b**, the hydrogenation reaction was carried out, and the desired product was obtained with 96% ee [3], which was the same enantioselectivity as direct hydrogenation of compound **5a**. They can also detect the existence of intermediate **9b** in the direct hydrogenation of compound **5a** with a low pressure of hydrogen and in a shorter reaction time in

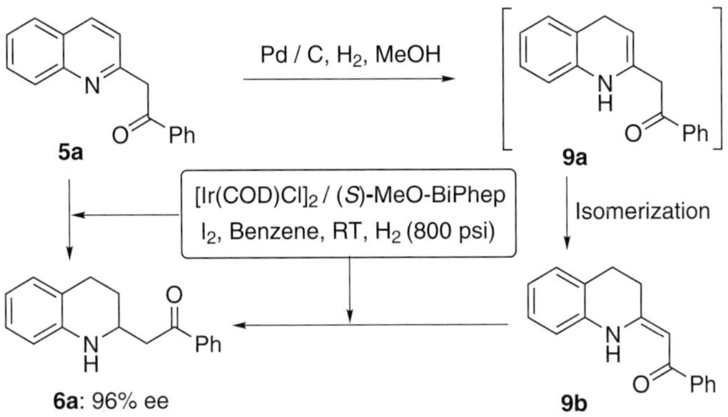

Scheme 10.18 The synthesis and hydrogenation of intermediate enamine **9b**.

their hydrogenation reaction. Subsequent computational result suggested that the first step, 1,4-hydride addition, was more favorable than 1,2-hydride addition.

Subsequently, Zhou envisioned that the catalytic process of asymmetric hydrogenation of quinolines with I_2 as an additive might be a cascade reaction involving a 1,4-hydride addition, isomerization, and 1,2-hydride addition (Scheme 10.19) [3], based on the analysis of the above hydrogenation phenomena of quinolines and the suggestions of the group Zhang [26] and Rueping [27]. They thought the oxidative addition of I_2 to Ir(I) to generate Ir(III) species is also possible in iridium-catalyzed asymmetric hydrogenation of quinolines in the presence of iodine, and the active catalytic species might be the Ir(III) hydrido chloro iodo complex. Therefore, a plausible mechanism was suggested by them as follows (Scheme 10.19). The oxidative addition of I_2 to the Ir(I) species precursor **A** generates the Ir(III) species, and the subsequent heterolytic cleavage of H_2 can occur to form the Ir(III)-H species **B** with hydrogen iodide elimination. The quinoline substrate could coordinate with Ir(III) species **B** (the I and Cl were omitted for clearness), and then 1,4-hydride transfer affords the intermediate **D**. Subsequently, the heterolytic cleavage of H_2 with the intermediate **D** gives an enamine **F** and regenerates the Ir(III)-H species **B**. The enamine **F** isomerizes to yield imine **G** that might be catalyzed by the generated HI acting as a strong Brønsted acid, which was also explained by Rueping. Imine intermediate **G** could coordinate with Ir(III)-H species **B** to form the intermediate **H**, followed by the insertion and sigma-bond metathesis to release the product 1,2,3,4-tetrahydroquinolines **P** to complete the catalytic cycle [3].

Zhou explored the asymmetric hydrogenation of 2,3-disubstituted quinolines. They thought that the hydrogenation mechanism of 2,3-disubstituted quinolines was somewhat different from that of 2-substituted quinolines (Scheme 10.20) [3]. For the hydrogenation of 2-substituted quinoline, the hydrogenation of C=N bond is the enantioselectivity-controlled step (Scheme 10.19, **H–I**), while the enantioselectivity-controlled step of 2,3-disubstituted quinolines is the isomerization of enamine to imine and the hydrogenation of C=N bond, which is in fact a dynamic kinetic

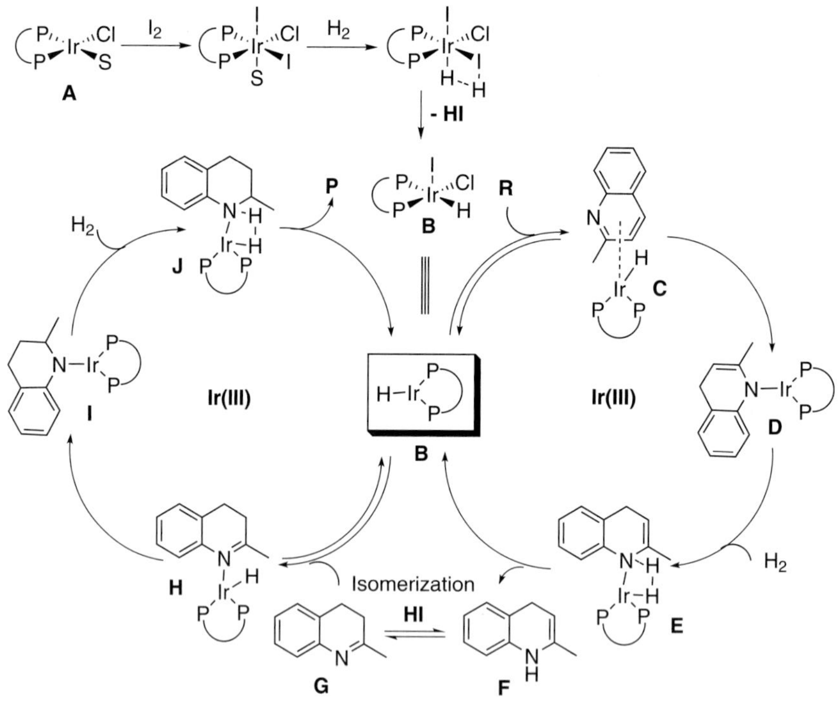

314 | *10 Enantioselective Reduction of Nitrogen-Based Heteroaromatic Compounds*

Scheme 10.19 The proposed mechanism for Ir-catalyzed hydrogenation of quinolines.

resolution process. To obtain high enantioselectivity, it should meet the equation $K_{iso} \gg K_{hy}$. They thought that high temperature could accelerate the rate of isomerization (K_{iso}) of (S)-**G** and (R)-**G**, and low pressure of hydrogen gas can decrease the rate of hydrogenation (K_{hy}). Therefore, the asymmetric hydrogenation reactions should be performed in high reaction temperature and low hydrogen

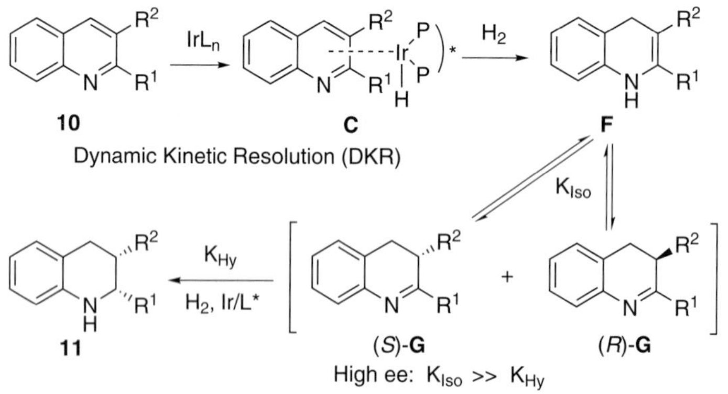

Scheme 10.20 The possible mechanism of Ir-catalyzed asymmetric hydrogenation of 2,3-disubstituted quinolines.

pressure. Their experiments showed that the best combination of reaction temperature and pressure of hydrogenation was 70 °C and 40 psi of hydrogen pressure in THF.

Under the optimized conditions, they investigated a variety of 2,3-disubstituted quinoline derivatives. In general, 2,3-alkyl-disubstituted quinolines were hydrogenated smoothly to give the corresponding 1,2,3,4-tetrahydroquinolines with good conversions and in 73–86% ee (Table 10.5, entries 1–11). For 2,3,6-trisubstituted quinolines, the reactions proceeded well, and excellent reactivities with good enantioselectivities were obtained (Table 10.5, entries 9–11). Interestingly, for the cyclic compounds **11l** and **11n**, excellent reactivities and diastereoselectivities (up to >20: 1, major in *cis* isomer) were obtained (Table 10.5, entries 12,14), which was complementary for the *trans* configuration reported by Du using chiral phosphoric acid as catalyst [28]. The successful hydrogenation of 2,3-disubstituted quinolines provided a new evidence for the hydrogenation of quinolines mechanism suggested by them.

During their studies on the hydrogenation mechanism of quinolines, Zhou and coworkers found that the dehydroaromatization reactions of 1,4-dihydropyridines (Hantzsch esters) could be realized using their catalytic system, and hydrogen gas was produced in this reaction [29]. Subsequently, they reported the

Table 10.5 Iridium-catalyzed asymmetric hydrogenation of 2,3-disubstituted quinolines.

Entry	$R^1/R^2/R^3$	Yield (%)	Syn/anti	ee (%)
1	H/Me/Me	92 (**11a**)	>20: 1	73
2	H/Me/Et	93 (**11b**)	>20: 1	85
3	H/Me/iPr	94 (**11c**)	>20: 1	86
4	H/Me/nBu	94 (**11d**)	>20: 1	83
5	H/Me/npentyl	91 (**11e**)	>20: 1	83
6	H/Me/3-butenyl	90 (**11d**)[a]	>20: 1	83
7	H/Me/phenethyl	97 (**11g**)	>20: 1	80
8	H/Me/benzyl	98 (**11h**)	>20: 1	81
9	Me/Me/Et	91 (**11i**)	>20: 1	84
10	F/Me/Et	89 (**11j**)	>20: 1	83
11	MeO/Me/Et	76 (**11k**)	>20: 1	85
12	H/(CH$_2$)$_4$	96 (**11l**)	>20: 1	39
13	H/Me/Ph	90 (**11m**)	>20: 1	38
14		93 (**11n**)	>20: 1	86

a) The product was **11d**.

iridium-catalyzed asymmetric transfer hydrogenation of quinolines using Hantzsch esters as hydrogen source by the combination of the two reactions (Scheme 10.21). A mild asymmetric transfer hydrogenation of quinolines was realized smoothly using [Ir(COD)Cl]$_2$/(S)-SegPhos/I$_2$ in the presence of Hantzsch esters with up to 88% ee [30], which avoids the use of compressed hydrogen gas.

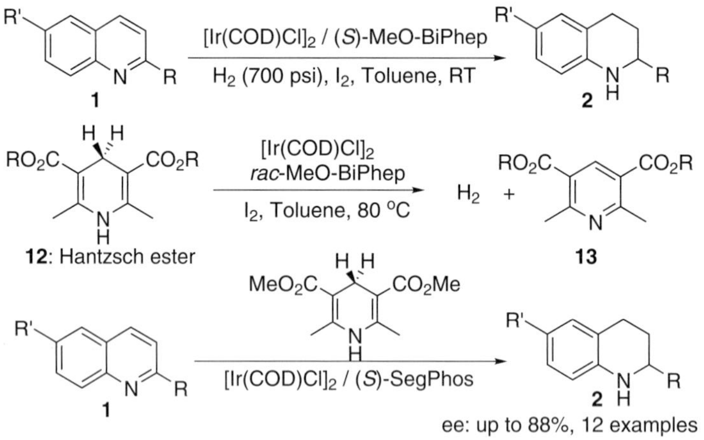

Scheme 10.21 Asymmetric transfer hydrogenation of quinolines.

The asymmetric hydrogenation of quinolines has been achieved with great success in obtaining chiral amines using Ir catalyst, while iodine is necessary that activated the complex of Ir and ligands. If iodine was not added, no hydrogenation was detected. But this strategy is not effective for the assorted isoquinoline and pyridine derivatives. In 2006, Zhou group developed a new strategy for the asymmetric hydrogenation of quinolines activated by chloroformates (Scheme 10.22) [31]. The chloroformates were crucial for hydrogenation of quinolines, and the reasons suggested by them was as follow: (1) aromaticity was destroyed partially by the formation of quinolinium salts, (2) the activating reagent bonded to the N-atom may avoid the catalyst poison, and (3) attached CO$_2$R is probably important for the coordination between substrate and catalyst.

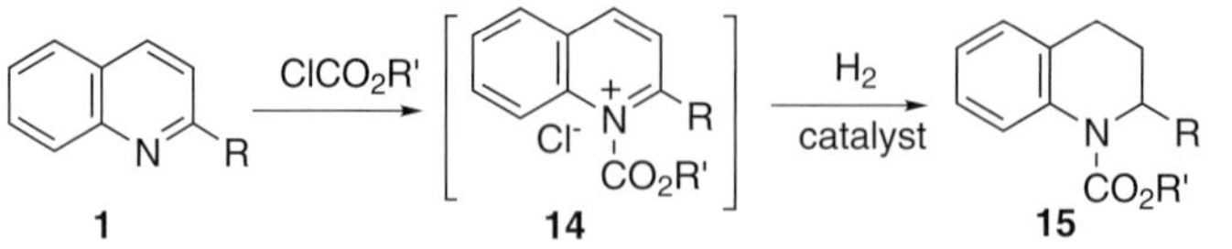

Scheme 10.22 Asymmetric hydrogenation of quinolines activated by chloroformates.

Table 10.6 Hydrogenation of quinolines activated by $ClCO_2Bn$.

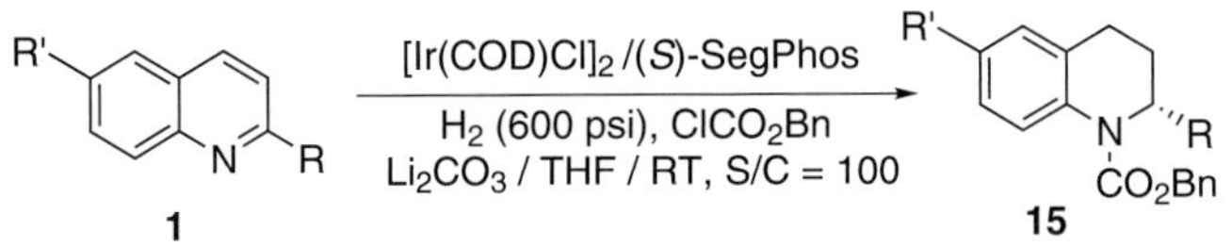
R'
N
R
1
[Ir(COD)Cl]2 /(S)-SegPhos
H2 (600 psi), ClCO2Bn
Li2CO3 / THF / RT, S/C = 100
R'
N
R
CO2Bn
15

Entry	R′/R	Yield (%)	ee (%)
1	H/Me	90	90 (*S*)
2	H/Et	85	90 (*S*)
3	H/*n*Pr	80	90 (*S*)
4	H/*n*Bu	88	89 (*S*)
5	H/*n*pentyl	91	89 (*S*)
6	F/Me	83	89 (*S*)
7	Me/Me	90	89 (*S*)
8	MeO/Me[a)]	92	90 (*S*)
9	H/Ph	41	80 (*R*)
10	H/phenethyl	86	90 (*S*)
11	H/3,4-$(MeO)_2C_6H_3(CH_2)_2$	80	90 (*S*)
12	H/3-MeO-4-$BnOC_6H_3(CH_2)_2$	88	88 (*S*)

a) Reaction at 50 °C.

Therefore, the addition of base to neutralize hydrogen chloride is necessary for a full conversion. A survey of base/solvent profile revealed that the combination of Li_2CO_3 and THF gave complete conversion. The optimal conditions Ir/(*S*)-SegPhos/$ClCO_2Bn/Li_2CO_3$, a variety of 2-substituted quinolines, were hydrogenated with 80–90% ee (Table 10.6) [31].

Asymmetric hydrogenation of quinolines activated by chloroformates also provided a convenient route to synthesize optically active tetrahydroquinoline alkaloids (Scheme 10.23). They also applied this methodology to total synthesis of some alkaloids [31]. For instance, reduction of hydrogenation products with $LiAlH_4$ in Et_2O gives the *N*-methylation products in high yields, which are the naturally occurring tetrahydroquinoline alkaloids.

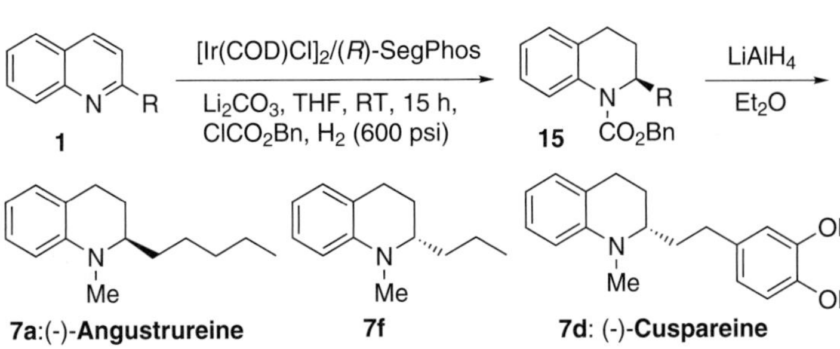
N
R
1
[Ir(COD)Cl]2/(R)-SegPhos
Li2CO3, THF, RT, 15 h,
ClCO2Bn, H2 (600 psi)
N
R
15
CO2Bn
LiAlH4
Et2O
N
Me
7a:(-)-Angustrureine
N
Me
7f
N
Me
OMe
OMe
7d: (-)-Cuspareine

Scheme 10.23 Asymmetric synthesis of tetrahydroquinoline alkaloids.

10.1.2
Organocatalyzed Asymmetric Transfer Hydrogenation of Quinolines

Recently, enantioselective organocatalysis has reached maturity with an impressive and steadily increasing number of publications, which provides a promising picture for their possibilities in organic chemistry. Very recently, biomimetic highly enantioselective organocatalyzed transfer hydrogenation of α,β-unsaturated carbonyl compounds and imines has been reported by List, MacMillan, and Rueping using Hantzsch esters as hydrogen source. In 2006, Rueping group realized the first example of a metal-free enantioselective reduction of heteroaromatic compounds using this strategy [32]. Catalysts are sterically congested chiral phosphoric acids derived from BINOL. The reaction was carried out in benzene at 60 °C with 2 mol% of chiral Brønsted acids as organocatalyst with Hantzsch dihydropyridine (HEH) as hydrogen source. Good to excellent enantioselectivities (87 ≥ 99% ee) with excellent yields were obtained for 2-substituted quinolines (Table 10.7), especially for 2-aryl

Table 10.7 Asymmetric reduction of quinolines by Rueping.

Entry	R	Time (h)	Yield (%)	ee (%)
1	Phenyl	12	92	97
2	2-Fluorophenyl	30	93	98
3	2-Methylphenyl	48	54	91
4	2,4-Dimethylphenyl	60	65	97
5	2-Naphthyl	12	93	>99
6	3-Bromophenyl	18	92	98
7	4-(CF_3)C_6H_3	30	91	>99
8	1,1'-Biphenyl-4-yl	12	91	>99
9	4-Methoxyphenyl	12	90	98
10	2-Furyl	12	93	91
11	Chloromethyl	12	91	88
12	nButyl	12	91	87
13	nPentyl	12	88	90
14	2-Phenthyl	12	90	90
15	H/3,4-(OCH_2O)C_6H_3(CH_2)$_2$-	12	94	91
16	H/3,4-(MeO)$_2C_6H_3$(CH_2)$_2$-	12	95	90

quinolines that provided an alternative method to chiral tetrahydroquinoline derivatives.

Since a general transfer hydrogenation method of quinolines by chiral Brønsted acids as organocatalyst was developed, Rueping and coworkers have successfully applied this methodology to the total synthesis of alkaloids within three steps. The angustrureine, galipinine, and cuspareine have been synthesis successfully in good yields (Scheme 10.24).

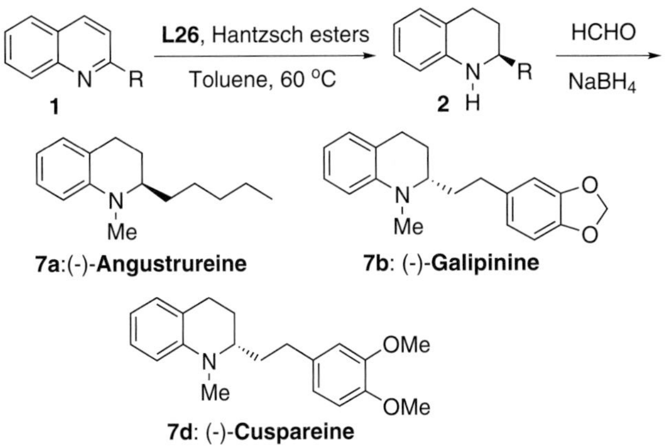

Scheme 10.24 Asymmetric synthesis of tetrahydroquinoline alkaloids by asymmetric transfer hydrogenation methods.

Then, the same group extended this strategy to transfer hydrogenation of 3-substituted quinolines with up to 86% ee (Scheme 10.25) [27]. They thought the transfer hydrogenation of 3-substituted quinolines was also a cascade reaction involving 1,4-hydride addition, enantioselective isomerization, and 1,4-hydride addition. The key step was the enantioselective isomerization of the enamine to imine, which was different from that of the hydrogenation of 2-substituted quinolines, while the key step of 2-substituted quinolines was reduction of C=N bond. After determining the optimized condition, they explored the scope of 3-substituted quinolines. Moderate to good yields (30–84%) and enantioselectivities (77–86% ee) were obtained in benzene at 60 °C.

In 2008, Du and coworkers designed and synthesized novel double axially chiral phosphoric acid catalysts based on BINOL [28]. Subsequently, these catalysts have been successfully applied in asymmetric transfer hydrogenation of 2-substitued (Table 10.8) and 2,3-disubstitued quinolines (Scheme 10.26). They found that ether was the best solvent. For 2-substitued quinolines, up to 98% ee was obtained when the substitutent of catalyst was cyclohexanyl.

More important, 2,3-disubstitued quinolines (Scheme 10.26) were also reduced. Excellent diastereoselectivities (>20: 1) with high enantioselectivities (up to 92% ee)

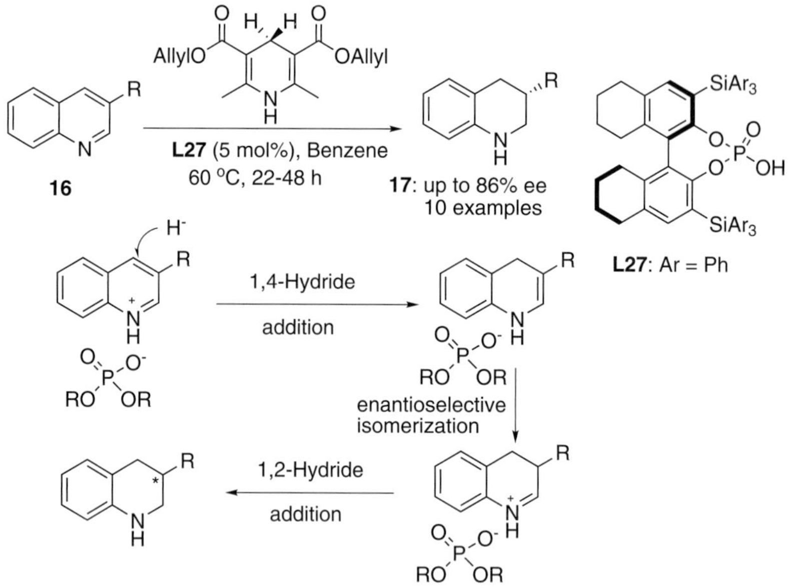

Scheme 10.25 Asymmetric reduction of 3-substituted quinolines by Rueping.

were obtained for the transfer hydrogenation of 2,3-disubstitued quinolines (three examples) [28].

10.2
Asymmetric Hydrogenation of Isoquinolines

In 2006, Zhou group realized the asymmetric hydrogenation of quinolines using chloroformates as the activator [31]. Then, they extended this strategy to asymmetric hydrogenation of isoquinoline derivatives and realized the first example of asymmetric hydrogenation of isoquinolines. The base is necessary for a full conversion suggested by them that could neutralize hydrogen chloride produced by chloroformates and isoquinolines. They found that the ee slightly increased when additive $LiBF_4$ or LiOTf was added. The hydrogenation results of some isoquinolines under optimal conditions are summarized in Table 10.9. In general, all the isoquinolines were hydrogenated to give the corresponding dihydroisoquinolines. Moderate to good enantioselectivities and good yields were obtained regardless of the electronic properties and steric hindrance of substituent groups.

Asymmetric hydrogenation of isoquinolines activated by chloroformates also provides a convenient route to synthesize optically active isoquinoline alkaloids. Zhou used this methodology to synthesize (S)-(−)-carnegine **21b** successfully, which is the naturally occurring tetrahydroisoquinoline alkaloid (Scheme 10.27) [31]. The

Table 10.8 Asymmetric reduction of 2-substitued quinolines by Du.

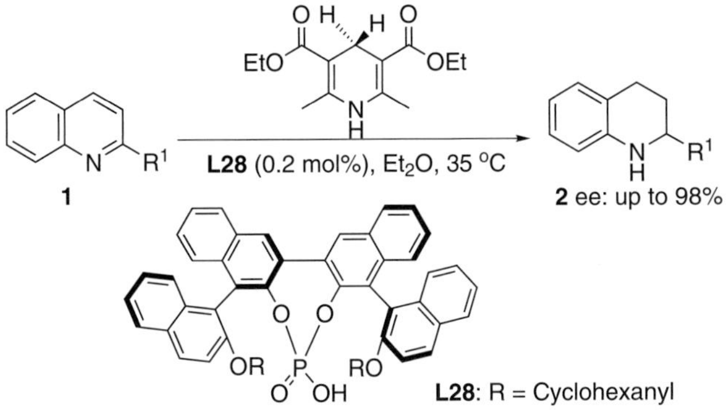

Entry	R¹	Yield (%)	ee (%)
1	C_6H_5	>99	96
2	4-BrC_6H_5	>99	97
3	4-ClC_6H_5	96	98
4	4-FC_6H_5	>99	97
5	4-$CH_3C_6H_5$	>99	94
6	2-$CH_3C_6H_5$	52	86
7	4-$CH_3OC_6H_5$	>99	95
8	2-naphthyl	>99	97
9	4-$CF_3C_6H_5$	>99	98
10	1,1'-Biphenyl-4-yl	>99	97
11	CH_3	>99	88
12	$CH_3CH_2CH_2$	>99	94
13	$CH_3(CH_2)_2CH_2$	>99	94
14	$CH_3(CH_2)_3CH_2$	>99	92
15	Cyclohexyl	>99	90
16	$C_6H_5CH_2CH_2$	>99	93
17	3,4-$(OCH_2O)C_6H_3(CH_2)_2$-	>99	90
18	3,4-$(MeO)_2C_6H_3(CH_2)_2$-	>99	95

hydrogenated products were treated by Pd/C in MeOH with hydrogen gas to afford the corresponding 1,2,3,4-tetrahydro-isoquinoline derivatives, followed by reduction by LiAlH₄ to gave the N-methylation products in good yields.

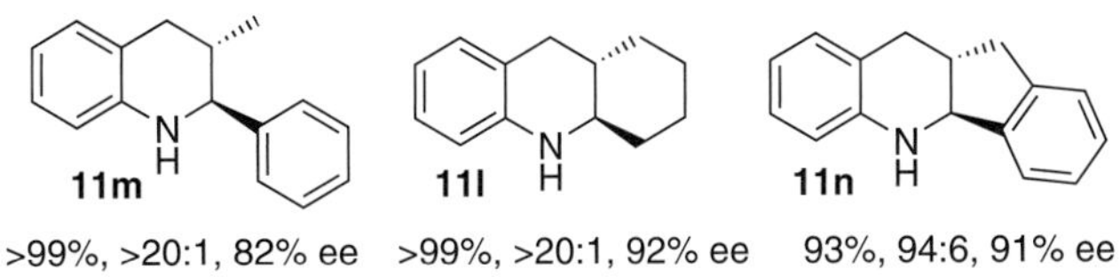

11m >99%, >20:1, 82% ee
11l >99%, >20:1, 92% ee
11n 93%, 94:6, 91% ee

Scheme 10.26 Asymmetric reduction of 2,3-disubstitued quinolines by Du.

Table 10.9 Hydrogenation of isoquinolines activated by chloroformates

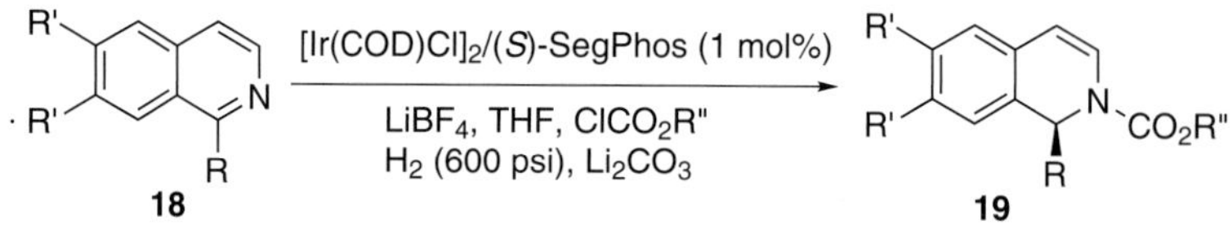

Entry	R'/R	R''	Yield (%)	ee (%)
1	H/Me	Me	85	80 (S)
2	H/Me	Bn	87	83 (S)
3	H/Ph	Me	57	82 (S)
4	H/Ph	Bn	49	83 (S)
5	MeO/Me	Me	44	63 (S)
6[a]	MeO/Me	Bn	46	65 (S)
7	H/Et	Me	85	62 (S)
8	H/n-Bu	Me	87	60 (S)
9	H/Bn	Me	83	10 (S)

a) Without LiBF$_4$.

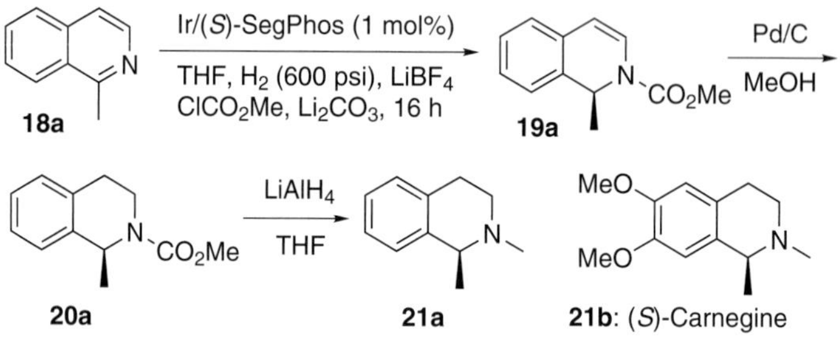

Scheme 10.27 Asymmetric synthesis of isoquinoline alkaloids.

10.3
Asymmetric Hydrogenation of Indoles

The asymmetric hydrogenation of substituted indoles is one of the most straightly access to optically pure chiral amines derived from indolines. Asymmetric hydrogenation of heteroaromatic compounds is a formidable issue and was kept unexplored until very recently. Indoles, for their unique structural features, were the heteroaromatics followed 2-methylquinoxaline to realize the highly enantioselective hydrogenation [33]. First, their reactivity and enantioselectivity can be modulated by introducing different protecting groups at the nitrogen atom of indoles, and those groups may act as directing group for high asymmetric induction. Second, indoles possess only one C=C out of the benzene ring and can be hydrogenated with one hydrogen gas molecule. The lack of plural C=C simplifies the pathways of catalytic hydrogenation and might lead to facilitate the design of chiral catalyst and realize their hydrogenation readily.

In 2000, Kuwano et al. reported the first precedent of asymmetric hydrogenation of 2-substituted indoles [34]. N-Acetyl-2-butylindole (**22a**) was chosen as the model substrate to search for optimal reaction conditions (Table 10.10). As in their preceding work, they found that Rh(acac)(cod) with bisphosphine ligands exhibited good activity for the hydrogenation of the matrix indoles [35]. Therefore, they examined various commercially available bisphosphine ligands with Rh(acac)(cod) as the metal precursor. For most ligands, N-acetyl-2-butylindoline (**23a**) was obtained in racemic form. Fortunately, with *trans*-chelating chiral bisphosphine ligand (S,S)-(R,R)-PhTRAP [36], 85% ee and 75% conversion were attained. Further investigation showed that the rhodium precursor [Rh(nbd)$_2$]SbF$_6$ was superior to Rh(acac)(cod). With the addition of base, high activity and enantioselectivity were achieved (full conversion and up to 94% ee) (Table 10.10). Then, the effect of base and pressure of H$_2$ was examined with [Rh(nbd)$_2$]SbF$_6$/PhTRAP as the catalyst. The reactions were carried out in 2-propanol at 60 °C under 5.0 Mpa of H$_2$ for 2 h. Et$_3$N and Cs$_2$CO$_3$ were effective in view of the reactivity and the enantioselectivity (entries 2 and 3). The reaction can also be carried out at slightly lower pressure (entry 6). When the hydrogen pressure is increased to 10.0 Mpa, increasing the substrate-to-catalyst ratio to 1000/1, the reaction can also proceed smoothly without loss of enantioselectivity (entry 7).

Table 10.10 Asymmetric hydrogenation of N-acetyl-2-butylindole (**22a**).

Entry	Base	H$_2$ (Mpa)	Conversion (%)	ee (%)
1	None	5.0	Trace	7 (S)
2	Et$_3$N	5.0	100	94 (R)
3	Cs$_2$CO$_3$	5.0	100	94 (R)
4	K$_2$CO$_3$	5.0	44	76 (S)
5	Pyridine	5.0	0	—
6	Cs$_2$CO$_3$	1.0	100	92 (R)
7	Cs$_2$CO$_3$	10.0	100	93 (R)

Table 10.11 Asymmetric hydrogenation of N-acetyl indoles (22a–h).

$$\text{22a-h} \xrightarrow[\text{5.0 MPa H}_2,\ i\text{PrOH, 60 °C, 2 h}]{\substack{[\text{Rh(nbd)}_2]\text{SbF}_5\ (1.0\ \text{mol\%}) \\ (S,S)\text{-}(R,R)\text{-PhTRAP}\ (1.05\ \text{mol\%}) \\ \text{Cs}_2\text{CO}_3\ (10\ \text{mol\%})}} \text{23a-h}$$

Entry	R^1/R^2	Yield [%]	Ee [%]
1	H/Bu (22a)	92 (23a)	94 (R)
2	H/iBu (22b)	91 (23b)	91 (R)
3	H/Ph (22c)	91 (23c)	87 (R)
4	H/CO$_2$Me (22d)	95 (23d)	95 (S)
5	5-Me/Bu (22e)	94 (23e)	94 (R)
6	5-CF$_3$/Bu (22f)	84 (23f)	92 (R)
7	6-CF$_3$/Bu (22g)	83 (23g)	92 (R)
8	6-MeO/Bu (22h)	98 (23h)	94 (R)

Subsequently, the generality of substrates of this catalyst system was examined, and a variety of 2-substituted indoles were hydrogenated to afford the corresponding indolines with high enantioselectivity and isolated yields. 2-Isobutylindole (22b) and 2-phenylindole (22c) proceeded with 91 and 87% ee, respectively (Table 10.11, entries 2–3). For indole-2-carboxylate 22d, the reaction conditions were changed in order to attain high enantiomeric excess (entry 4, 95% ee), higher temperature, hydrogen pressure, and Et$_3$N instead of Cs$_2$CO$_3$ as the base was required. The steric and electronic properties of the substituent on the fused aromatic ring of 22 had little effect on the enantioselectivities. Indoles with both electron-withdrawing and electron-donating groups at 5- or 6-position of the benzene ring were hydrogenated with good enantioselectivities (entries 5–8, 92–94% ee). The hydrogenation of 3-substituent indole (N-acetyl-3-methylindole) was also tried, under the above conditions. N-Acetyl-3-methylindoline was obtained with 37% isolated yield and 86% ee, accompanied by a majority of alcoholysis product of 3-methyl indole.

In 2004, by screening the effect of different protecting groups, such as acetyl (Ac), tert-butoxycarbonyl (Boc), tosyl (Ts), methanesulfonyl (Ms), and trifluoromethylsulfonyl (Tf), under the optimal conditions, tosyl-protected 3-substituted indoles showed the best reactivity and enantioselectivity, and the hydrogenation of N-tosyl-3-methylindole (24a) gave 96% yield and 98% ee (Table 10.12, entry 1) [37]. As shown in Table 10.12, various 3-substituted N-tosylindoles (24b–24f) were hydrogenated by the Rh–PhTRAP complex with excellent enantioselectivities (entries 2–6, 95–98% ee). This catalytic hydrogenation system was even compatible with functional groups such as silyl-protected alcohol, tert-butyl ester, and Boc-protected amine (entries 4–6).

Table 10.12 Asymmetric hydrogenation of N-tosylindoles (**24a–24f**).

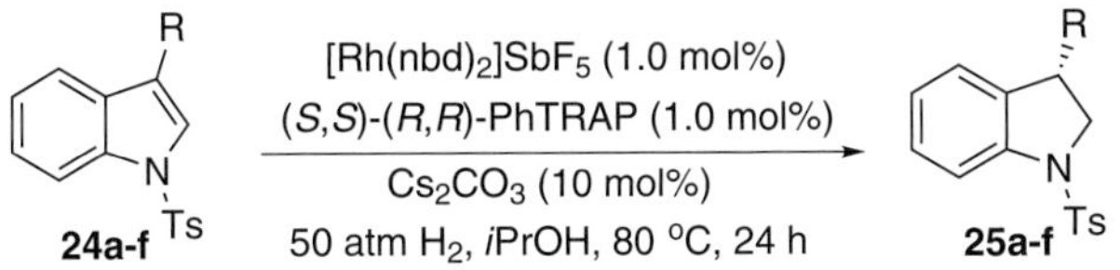

[Rh(nbd)$_2$]SbF$_5$ (1.0 mol%)
(S,S)-(R,R)-PhTRAP (1.0 mol%)
Cs$_2$CO$_3$ (10 mol%)
50 atm H$_2$, iPrOH, 80 °C, 24 h

Entry	R	Yield (%)	ee (%)
1	Me (**24a**)	96 (**25a**)	98 (S)
2	iPr (**24b**)	94 (**25b**)	97 (S)
3	Ph (**24c**)	93 (**25c**)	96 (S)
4	CH$_2$CH$_2$OTBS (**24d**)	94 (**25d**)	98 (S)
5	CH$_2$CH$_2$CO$_2$tBu (**24e**)	93 (**25e**)	97 (S)
6	CH$_2$CH$_2$NHBoc (**24f**)	71 (**25f**)	95 (S)

This method was successfully applied to the total synthesis of chiral indoline alkaloid **25h**, which is Wierenga's synthetic intermediate for the left-hand segment of the antitumor agent (+)-CC-1065. As shown in Scheme 10.28, the asymmetric hydrogenation of N-methanesulfonyl indole (**24g**) to the corresponding indoline (**25g**) with 93% ee was the key step.

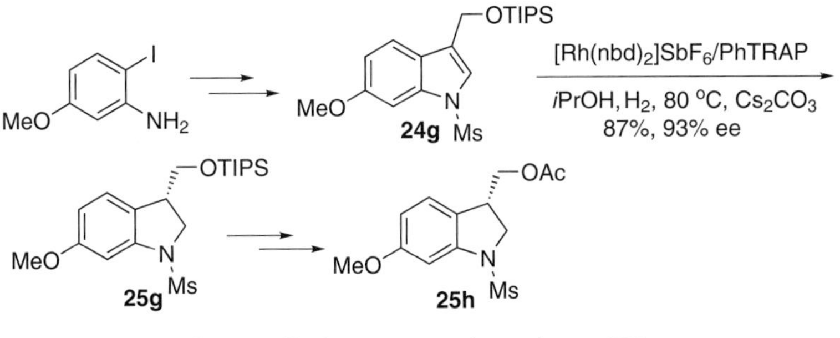

Scheme 10.28 Application of hydrogenation in the synthesis of **25h**.

Though the asymmetric hydrogenation of 2- and 3-substituted indoles with N-acetyl and N-tosyl as protective groups, respectively, has achieved great progress [38], there still exist some challenging task such as changing the protecting group and the reduction of 2,3-disubstituted indoles. Compared to the former protecting groups, tert-butoxycarbonyl is more readily attached to the indoles and removed from the products. This readily removable protecting group will facilitate the synthesis of chiral amines. Kuwano group devoted their continuous effort to this field, and, recently, they reported the asymmetric reduction of N-Boc indoles with Ru–PhTRAP catalyst with great success [39]. As for 2-substituted indoles, regardless of the electronic property of the substituent at the 5-position of **26** (Table 10.13), indoles

Table 10.13 Ru-catalyzed asymmetric hydrogenation of N-Boc-indoles (**26a–26h**).

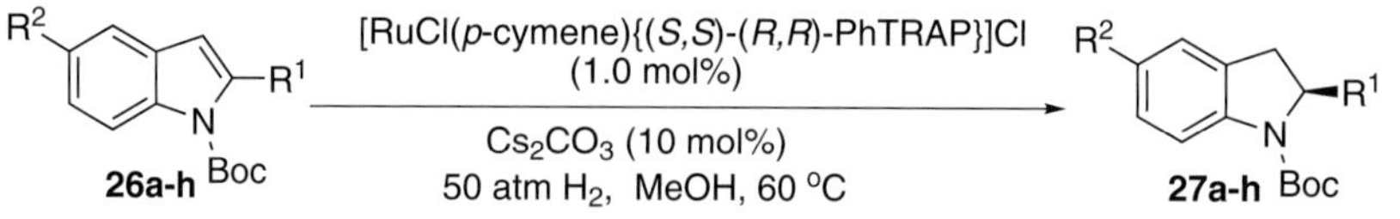

Entry	R¹/R²	Yield (%)	ee (%)
1	Me/H (**26a**)	99 (**27a**)	95 (R)
2	Me/OMe (**26b**)	97 (**27b**)	91 (R)
3	Me/F (**26c**)	96 (**27c**)	90 (R)
4	Bu/H (**26d**)	94 (**27d**)	92 (S)
5	cC$_6$H$_{11}$/H (**26e**)	92 (**27e**)	87 (R)
6	Ph/H (**26f**)	99 (**27f**)	95 (R)
7	p-F-C$_6$H$_4$/H (**26g**)	95 (**27g**)	93 (R)
8	CO$_2$Me/H (**26h**)	91 (**27h**)	90 (S)

bearing alkyl, aryl, and ester groups were all transferred to products with excellent enantioselectivities.

Fortunately, this Ru catalyst was also effective for 3-substituted N-Boc-indoles, and the 3-methyl and 3-phenyl indoles **28a** and **28b** were hydrogenated to the corresponding indolines with 87 and 94% ee, respectively [39]. This catalyst system was also extended to hydrogenation of 2,3-disubstituted indoles, and only the cis-2,3-dimethylindoline (−)-**31** was observed with 59% yield and 72% ee (Scheme 10.29).

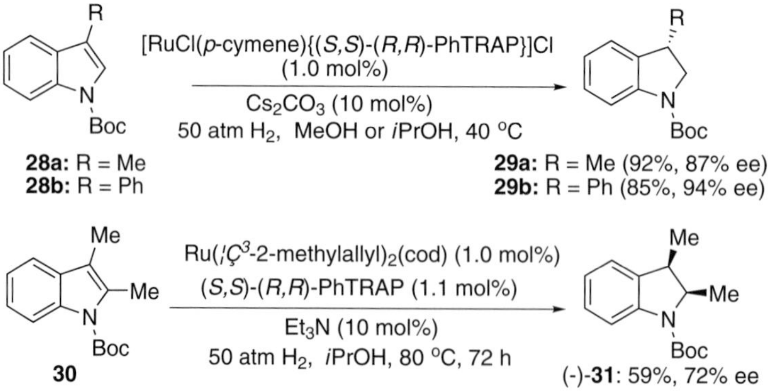

Scheme 10.29 Asymmetric hydrogenation of 3-substituted and 2,3-disubstituted indoles.

Based on the above results, asymmetric hydrogenation of indoles provides a convenient route to the asymmetric synthesis of chiral N-protecting indolines, which are the kind of important chiral amines. Those promising methods may be applied to the synthesis of indoline-based alkaloids.

10.4
Asymmetric Hydrogenation of Pyrroles

The asymmetric hydrogenation of pyrroles is a useful method to optically active pyrrolidines that are building blocks of pyrroline-based alkaloids and other biologically active compounds. It was not until very recently that the catalytic asymmetric hydrogenation of pyrroles has become the truth.

In 2001, Tungler and coworkers described the diastereoselective hydrogenation of N-(1'-methylpyrrole-2'-acetyl)-(S)-proline methyl ester (**32**) using the Rh/C as catalyst [40]. By introducing (S)-proline moiety as chiral auxiliary, high asymmetric induction was obtained. When **32** was subjected to 5% Rh/C catalyst in methanol with 20 bar H_2, the reduced product **33** was obtained with full conversion and 95% de. This substrate induced asymmetric reduction and was effective only to (2-pyrrolyl) acetic acid derivatives (Scheme 10.30).

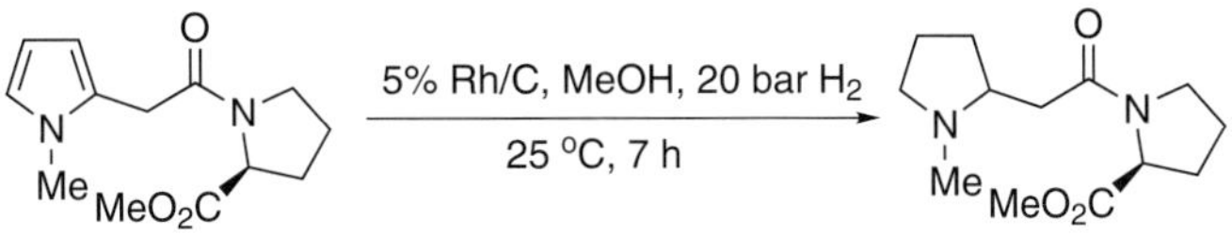

Scheme 10.30 Asymmetric hydrogenation of N-Me-pyrroles by Tungler.

Kuwano group kept their passion in exploring the asymmetric hydrogenation of five-membered heteraromatics, and very recently they reported the first highly enantioselective hydrogenation of pyrroles with catalyst prepared *in situ* from Ru (η^3-methylallyl)$_2$(cod) and (S,S)-(R,R)-PhTRAP. N-Boc-pyrrole-2-carboxylate **34a** was chosen as model substrate, which was hydrogenated to complete saturated product with 92% yield and 79% ee [41]. Subsequently, they explored asymmetric hydrogenation of various 2,3,5-trisubstituted pyrroles, which can be regarded as analogue of 2-substituted indoles (Table 10.14). Substrates **34b–34c** can be hydrogenated to the fully saturated indolines **35b–35c** with all the three substituents R^1, R^2, and R^3 located *cis* to each other (entries 2 and 3). The asymmetric hydrogenation of 4,5,6,7-tetrahydroindole **34d** mainly produced monohydrogenation product **36d** with 70% yield and 95% ee (entry 4). For the 2,3,5-triarylpyrroles (**34e–34i**) bearing three bulky groups, only one less steric hindrance C=C was hydrogenated with up to 99% ee (entries 5–9).

They pointed out that this hydrogenation process proceeds stepwise with successive 1,2-additions of hydrogen to two C=C. The less hindered double bond was first hydrogenated, followed by the reduction of the remaining C=C (Scheme 10.31). As for the asymmetric hydrogenation of **34a**, the first hydrogen molecule was added to produce the monohydrogenation achiral intermediate **36a**. Subsequently, cyclic enamide **36a** was hydrogenated under the same condition to give the chiral pyrroline product **35a** in high enantioselectivity. In contrast, 2,3,5-trisubstituted

Table 10.14 Asymmetric hydrogenation of N-Boc-pyrroles.

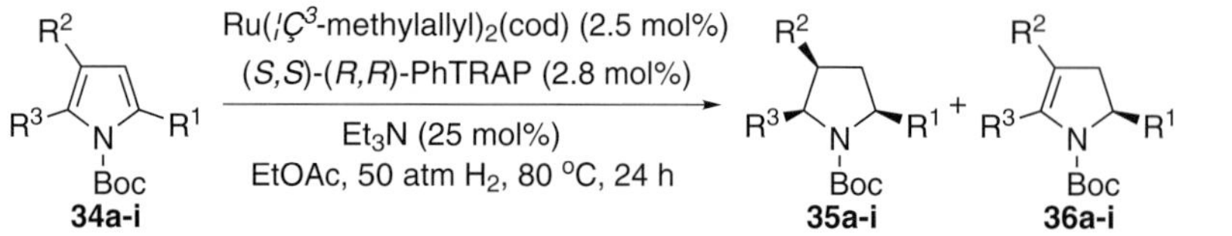

Ru(η^3-methylallyl)$_2$(cod) (2.5 mol%)
(S,S)-(R,R)-PhTRAP (2.8 mol%)
Et$_3$N (25 mol%)
EtOAc, 50 atm H$_2$, 80 °C, 24 h

Entry	R^1/R^2/R^3	35 : 36	Yield (%)	ee (%)
1	CO$_2$Me/H/H (34a)	100 : 0	92 (35a)	79 (35a)
2	CO$_2$Me/Me/Me (34b)	100 : 0	85 (35b)	96 (35b)
3	Ph/C$_3$H$_7$/Me (34c)	100 : 0	96 (35c)	93 (35c)
4	CO$_2$Me/-(CH$_2$)$_4$- (34d)	16 : 84	8 (35d)	91 (35d)
			70 (36d)	95 (36d)
5	Ph/Ph/Ph (34e)	0 : 100	99 (36e)	99 (36e)
6	p-F-C$_6$H$_4$/Ph/Ph (34f)	0 : 100	99 (36f)	99 (36f)
7	p-MeO-C$_6$H$_4$/Ph/Ph (34g)	0 : 100	96 (36g)	98 (36g)
8	Ph/Ph/p-CF$_3$-C$_6$H$_4$ (34h)	0 : 100	99 (36h)	99 (36h)
9	Ph/Ph/p-MeO-C$_6$H$_4$ (34i)	0 : 100	97 (36i)	99 (36i)

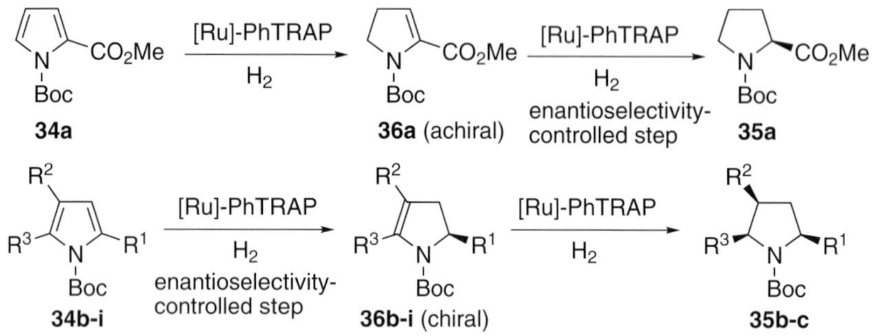

Scheme 10.31 Pathway of asymmetric hydrogenation of indoles (34a–34i) by Kuwano.

pyrroles **34b–34i** was hydrogenated enantioselectively to obtain chiral cyclic enamide **36b–36i** in the initial step. Following this enantioselectivity-controlled step, the chiral center created at 5-position of **36b–36c** would direct the stereoselective reduction of the remaining double bond to give the cis-trisubstituted **35b–35c** as sole products. As for **36e–36i** with three bulky aryl groups, unsaturated products were obtained due to steric hindrance. The different enantiocontrolled step suggested that the Ru–PhTPAP catalytic system is effective for hydrogenation of pyrroles rather than cyclic enamides [41]. The asymmetric hydrogenation of pyrroles provides a direct route to the synthesis of pyrroline-based chiral amines.

10.5
Asymmetric Hydrogenation of Quinoxalines

The asymmetric hydrogenation of quinoxalines is a challenging task and provides a convenient way to chiral tetrahydroquinoxalines, which are of great biological interest. In 1987, the asymmetric hydrogenation of quinoxalines was reported by Murata, who used Rh[(S,S)-DIOP]H as the catalyst in hydrogenation of 2-methylquinoxaline [42]. Although a dismal 3% ee was obtained, this work is the first example of asymmetric hydrogenation of aromatic compounds (Scheme 10.32).

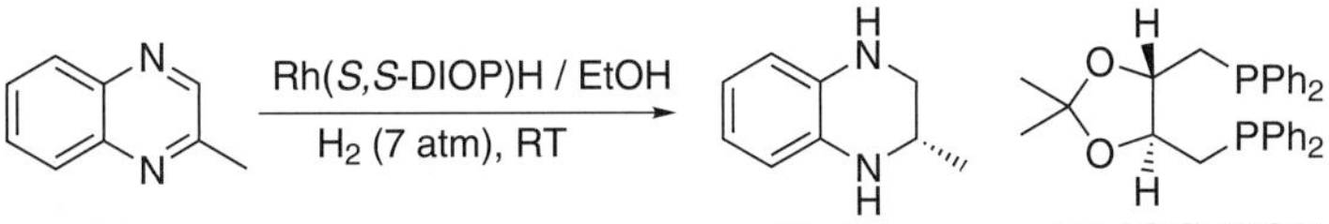

Scheme 10.32 Hydrogenation of 2-methylquinoxaline catalyzed by Rh[(S,S)-DIOP]H.

In 1998, Bianchini group achieved up to 90% ee with 54% yield in hydrogenation of 2-methylquinoxaline using an orthometalated dihydride–iridium complex (1 mol%) as the catalyst precursor in MeOH. The highest yield (97%) was obtained in i-PrOH but with a significant drop in enantioselectivity (73% ee) [43]. The same group employed [(R,R)-BDPBzPIr(COD)]OTf and [(R,R)-BDPBzPRh(NBD)]OTf complexes (1 mol%) for asymmetric hydrogenation of 2-methylquinoxaline with 23 and 11% ee, respectively (Table 10.15) [44].

In 2003, Henschke and coworkers described asymmetric hydrogenation of 2-methylquinoxaline with Noyori's catalyst RuCl$_2$(diphosphine)(diamine). In most case, moderate enantioselectivities and excellent conversions were obtained within 20 h with S/C 1000/1 [45]. The combination of Xyl-HexaPHEMP and (S,S)-DACH afforded the best enantioselectivity (73% ee) with full conversion.

Recently, Chan prepared Ir–PQ-Phos complex and applied it to the catalytic enantioselective hydrogenation of 2-methylquinoxaline [8], and up to 80% ee was obtained in THF with full conversion in the presence of I$_2$.

10.6
Asymmetric Hydrogenation of Pyridine Derivatives

The asymmetric hydrogenation of pyridine derivatives is a practical and direct method to chiral piperidine derivatives that are important synthetic intermediates and the structural unit of many biologically active compounds. Recently, some attempts to develop enantioselective hydrogenation of pyridine derivatives have been made.

In 1999, Studer and coworkers reported a two-step procedure for the enantioselective preparation of chiral ethyl nipecotinate [46] that consisted of an efficient partial

Table 10.15 Asymmetric hydrogenation of 2-methylquinoxaline.

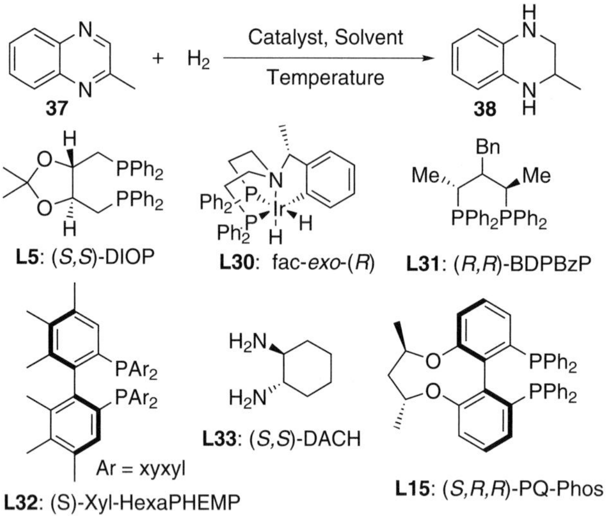

Entry	Catalyst	Solvent	Conversion (%)	ee (%)
1	(+)-(DIOP)RhH	C_2H_5OH	72	3
2	L30	CH_3OH	54	90
3	L30	i-PrOH	97	73
4	[L31Ir(COD)]OTf	CH_3OH	41	23
5	[L31Rh(NBD)]OTf	CH_3OH	93	11
6	$RuCl_2$/L32/(S,S)-DACH	t-BuOH	99	73
7	[Ir(COD)Cl]$_2$/L15/I_2	THF	>99	80

hydrogenation of ethyl nicotinate **39** with Pd/TiO$_2$ and a subsequent asymmetric hydrogenation using a cinchona-modified heterogeneous Pd/TiO$_2$ as catalyst. The highest enantioselectivity (24% ee) was obtained with Pd/TiO$_2$ in a DMF/ H$_2$O/AcOH system (Scheme 10.33).

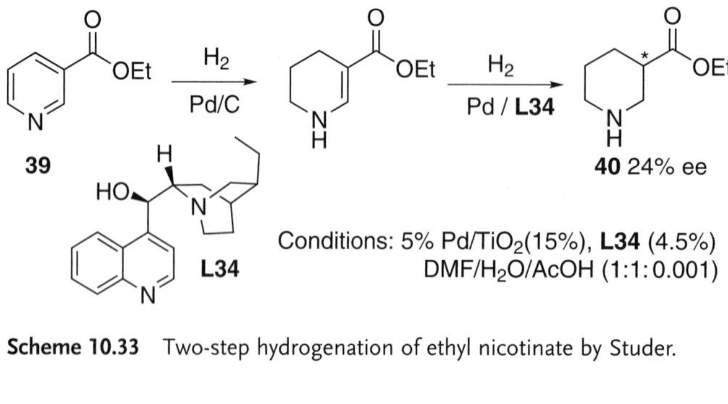

Conditions: 5% Pd/TiO$_2$(15%), **L34** (4.5%)
DMF/H$_2$O/AcOH (1:1:0.001)

Scheme 10.33 Two-step hydrogenation of ethyl nicotinate by Studer.

10.6 Asymmetric Hydrogenation of Pyridine Derivatives

Studer's group also explored homogeneous asymmetric hydrogenation of pyridine derivatives in 2000 [47]. The best enantioselectivity (27% ee, 41% yields) was obtained for the hydrogenation of simple monosubstituted pyridines using the Rh(NBD)$_2$BF$_4$/L5 as catalyst. High temperature, pressure, and high catalyst loading (5 mol%) are necessary for full conversion (Scheme 10.34).

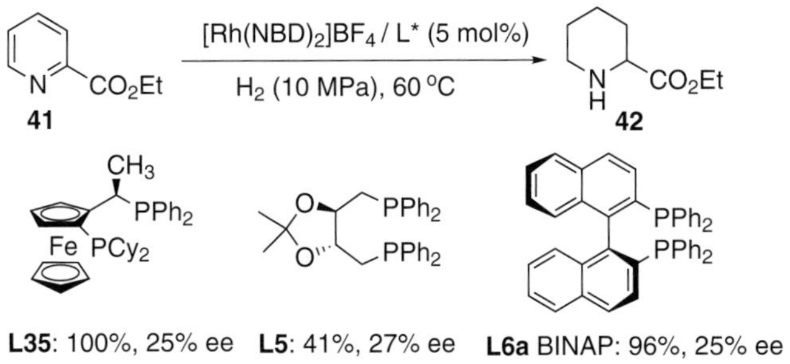

L35: 100%, 25% ee   **L5**: 41%, 27% ee   **L6a** BINAP: 96%, 25% ee

Scheme 10.34 Homogeneous enantioselective hydrogenation of pyridines by Studer.

Thomas and Johnson described a direct asymmetric hydrogenation of ethyl nicotinate using Pd-ferrocenyl catalyst anchored within MCM-41 [48]. The confined catalyst (5%) displayed a good activity (TON = 291), but no increase in enantioselectivity (17% ee) was observed compared to the corresponding homogeneous catalyst (Scheme 10.35).

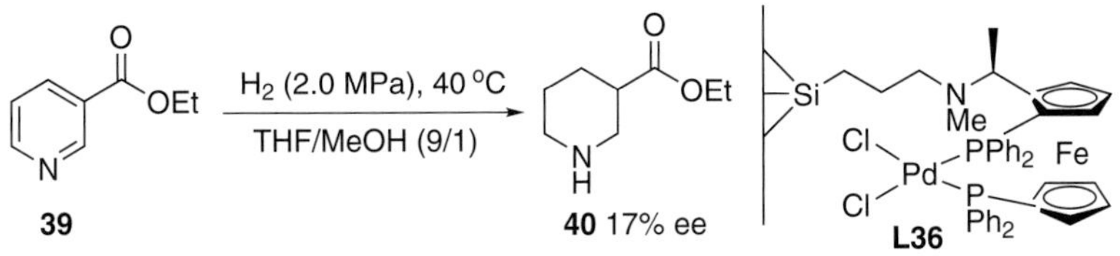

Scheme 10.35 Hydrogenation of ethyl nicotinate using a chiral heterogeneous catalyst.

With the help of chiral auxiliary substitutent, the pyridine derivatives can be diastereoselectively hydrogenated. In 2000, Hegedus group reported a diastereoselective heterogeneous hydrogenation of N-nicotinoyl-(S)-proline methyl ester **43** with up to 94% de and complete conversion [49]. However, this result was not reproducible and the maximum de was corrected to 30% (Scheme 10.36).

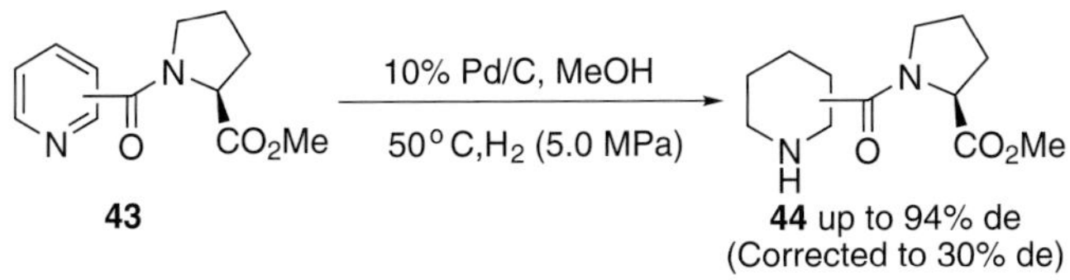

44 up to 94% de
(Corrected to 30% de)

Scheme 10.36 Diastereoselective hydrogenation of pyridine derivatives **43**.

10 Enantioselective Reduction of Nitrogen-Based Heteroaromatic Compounds

Two year later, Pinel group used (S)-proline and (S)-pyroglutamic esters as chiral auxiliaries for the diastereoselective hydrogenation of 2-methylnicotinic acid with up to 35% de [50]. They tried to improve diastereoselectivity of these reactions, but no better result was obtained (Scheme 10.37).

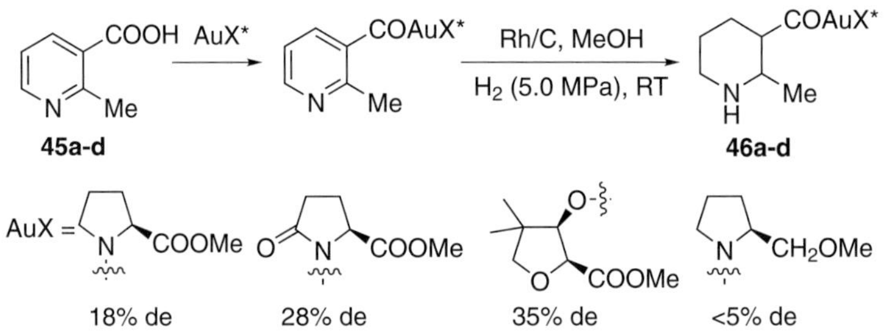

Scheme 10.37 Diastereoselective hydrogenation of nicotinic derivatives by Pinel.

Chiral oxazolidinones are also common chiral auxiliary used in the organic synthesis to provide chiral compounds. In 2004, Glorius et al. described an example of the efficient asymmetric hydrogenation of pyridines through the introduction of a chiral oxazolidinone auxiliary to the 2-position of the pyridine derivatives [52].

Substrates **48** could be easily synthesized from 2-bromo- or chloropyridines **47** and the oxazolidinone by copper catalysis (Scheme 10.38) [53]. The hydrogenation of pyridine derivatives **48** was carried out in acetic acid under a hydrogen atmosphere of 100 bar with commercially available achiral heterogeneous catalysts. Acetic acid not only activates the pyridine substrates **48** but also protonates the piperidine products **50** to bind the piperidine's free electron pair, and thereby suppressing the catalyst poisoning. Pd(OH)$_2$/C emerged as the best catalyst for hydrogenation, and excellent enantioselectivities were obtained after removal of the chiral auxiliary (85–98% ee) (Table 10.16). The high diastereoselectivity was ascribed to strong hydrogen bonding

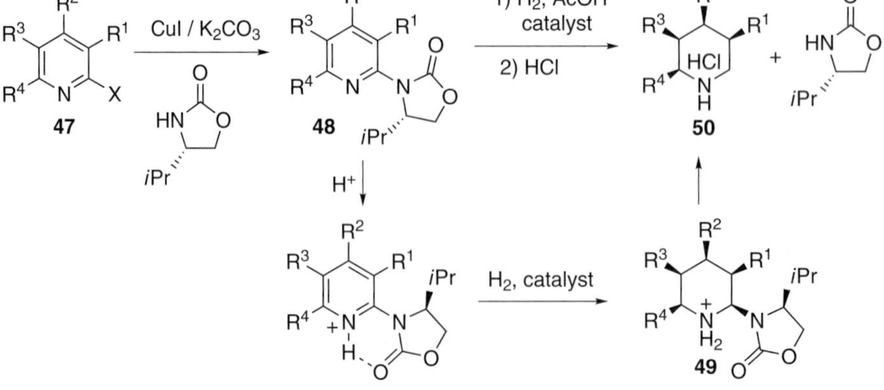

Scheme 10.38 Asymmetric hydrogenation of 2-oxazolidinone-substituted pyridines.

Table 10.16 Asymmetric hydrogenation of 2-oxazolidinone-substituted pyridines.

Entry	R	Yield (%)	ee (%)
1	6-Me	93	91
2	6-nPr	95	95
3	6-CHO	64	97
4	5-Me	90	98
5	5-CF$_3$	93	95
6	5-CONMe$_2$	92	85

between the pyridinium and the oxazolidinone moiety in acetic acid. In addition, the simple purification procedure makes this method more practical for the synthesis of chiral-substituted piperidine derivatives, which are precursors for the synthesis of various natural products. The piperidinium hydrochloride and the auxiliary can be separated readily by extraction, and chiral chiral auxiliary can be recovered and reused.

An achiral auxiliary-based method is another efficient strategy to activate pyridine derivatives for asymmetric hydrogenation of pyridines. In 2005, Charette and Legault reported catalytic asymmetric hydrogenation of N-iminopyridinium ylides to provide enantioenriched-substituted piperidine derivatives [54]. N-Benzoyliminopyridinium ylide activates the substrate and offers a secondary coordinating group. The use of a catalytic amount of iodine was crucial to obtain high yields. Presumably, this was ascribed to iodine's assistance in the oxidation of an Ir(I) to a more catalytically active Ir(III) species [55]. Different catalysts were screened, and cationic iridium complexes of phosphinooxazoline, with tetrakis (3,5-bis(trifluoromethyl)phenyl)borate (BArF) as the counterion, were identified as the optimum catalyst, providing **52** with up to 90% ee (Table 10.17). Moreover, the reduced products are highly crystalline solids that can readily be enriched by single recrystallization from ethyl acetate (enriched to 97% ee). The hydrogenation adducts obtained can be converted to the corresponding piperidine derivatives using Raney nickel or lithium in ammonia to cleave the N–N bond. The ease of introduction and the removal of the auxiliary make the auxiliary-based method more useful for the synthesis of chiral amines.

Zhang and coworkers reported a highly asymmetric hydrogenation of 3-substituted pyridine derivatives using Rh(NBD)(Tang-Phos)SbF$_6$ as the catalyst with high enantioselectivity (48–99% ee) in three steps [56]. This process consists of an efficient partial hydrogenation of nicotinate with Pd/C, acylation of the intermediate, and a subsequent highly enantioselective homogeneous hydrogenation of enamides **53** using Rh(NBD)(TangPhos)SbF$_6$ as the catalyst (Scheme 10.39).

Table 10.17 Charette's enantioselective hydrogenation of N-iminopyridinium ylides.

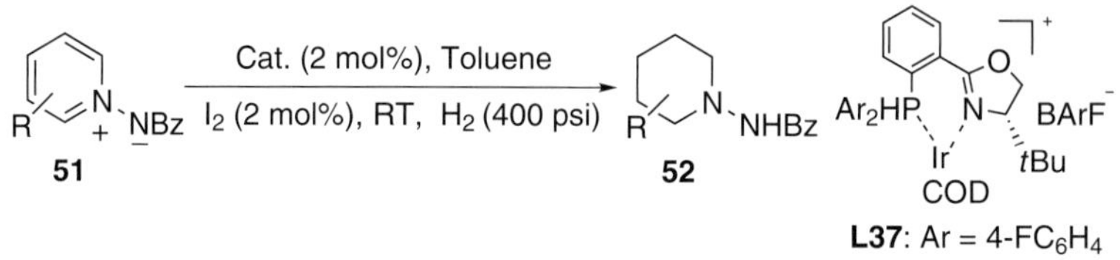

Entry	R	Yield (%)	ee (%)
1	2-Me	98	90
2	2-Et	96	83
3	2-nPr	98	84
4	2-Bn	97	58
5	2-CH$_2$OBn	85	76
6	2-(CH$_2$)$_3$OBn	88	88
7	2,3-Me$_2$	91	54

In 2008, Zhou and coworkers reported Ir-catalyzed asymmetric hydrogenation of pyridine derivatives, 7,8-dihydro-quinolin-5(6H)-ones, with excellent enantioselectivities (up to 97% ee) with the catalyst used in the asymmetric hydrogenation of quinolines [57]. The success of asymmetric hydrogenation of 7,8-dihydro-quinolin-5(6H)-ones **55** was attributed to the fact that the strong electron-withdrawing character of carbonyl reduces the inhibitory effect of the product. Therefore, other trisubstituted pyridine derivatives with other electron-withdrawing groups were also tested (Scheme 10.40). The catalyst system displayed no catalytic activities on substrates bearing ester or sulfonyl groups **57a** and **57b**. Cyano-substituted pyridine derivative **57c** gave low reactivity and moderate 85% enantioselcetivity. For the asymmetric hydrogenation of pyridine with 2-acyl group **57d**, full conversion and low enantioselectivity (45% ee) were obtained.

Organocatalysts are inexpensive, readily available, and nontoxic. The products are free from the contamination of metals, so various highly enantioselective, metal-free reactions have been developed to obtain biologically active compounds based on

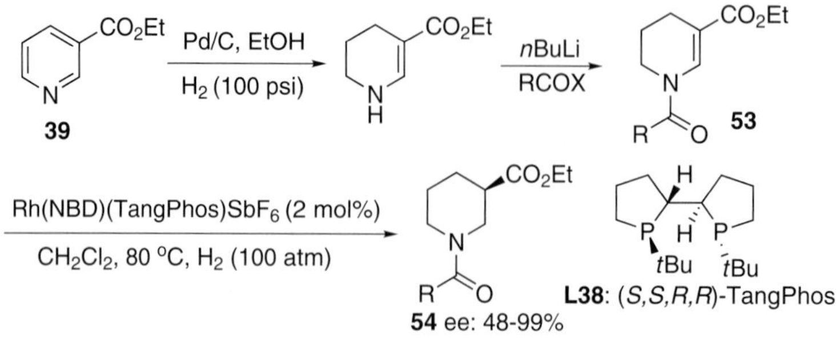

Scheme 10.39 Asymmetric hydrogenation process of 3-substituted pyridine derivatives.

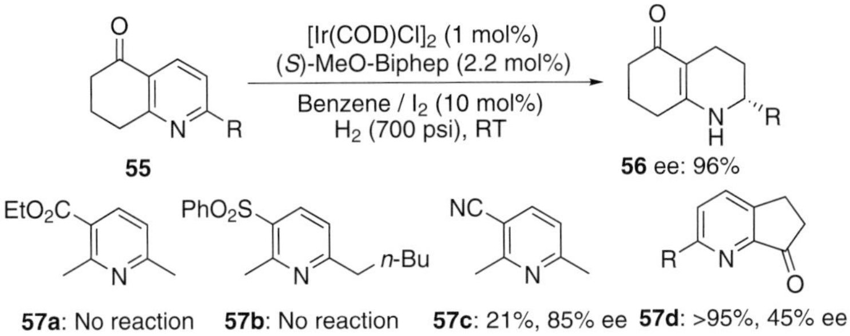

10.6 Asymmetric Hydrogenation of Pyridine Derivatives

Scheme 10.40 Ir-catalyzed hydrogenation of trisubstituted pyridine derivatives.

the application of organocatalysts. Very recently, Rueping and Antonchick have employed chiral Brønsted acids in catalytic enantioselective hydrogenation of tri-substituted pyridine derivatives with excellent enantioselectivities [58], in which Hantzsch dihydropyridine acted as hydrogen sources. The highest selectivitity was obtained with catalyst **L26b**, which provided trisubstituted pyrolidine derivatives in up to 92% ee (Scheme 10.41).

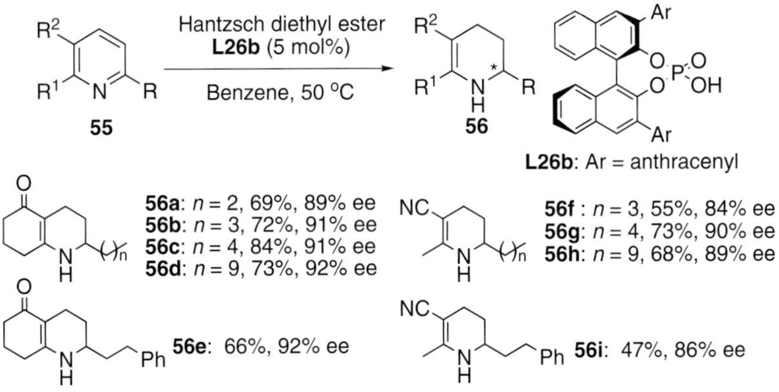

Scheme 10.41 Brønsted acid-catalyzed enantioselective reduction of pyridines.

A plausible mechanism for this stereoselective hydrogenation is depicted in Scheme 10.42 [58]. The first step in the asymmetric cascade hydrogenation is the protonation of the pyridine **55** through the Brønsted acid catalyst **L26b** to give the iminium ion **A**. Subsequent transfer of the first hydride from the Hantzsch ester generates the pyridinium salt **B** that gives rise to the iminium ion **C** by an isomerization. A second hydride transfer results in the desired product **56**, and subsequent proton transfer will then recycle the Brønsted acid **L26b** and generate a second equivalent of the Hantzsch pyridine.

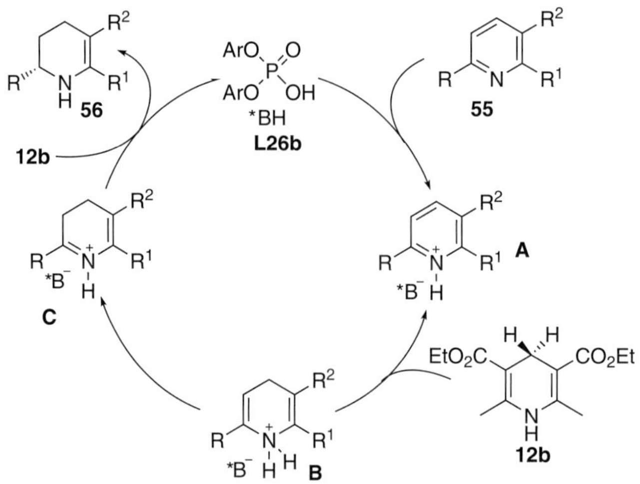

Scheme 10.42 Mechanistic proposal for organocatalytic reduction of pyridines.

10.7
Summary and Outlook

This review focused on recent advances in asymmetric reduction of nitrogen-based heteroaromatic compounds. Two kinds of systems were developed for hydrogenation of quinolines. One is the highly active iridium catalyst Ir/diphosphine/I_2, in which the iodine additive is crucial for activity and enantioselectivity. The other is Ir/diphosphine/Li_2CO_3 in the presence of chloroformates, which activates quinolines by the formation of quinolinium salt. The latter can also be applied to asymmetric hydrogenation of isoquinolines and quinolines. A highly enantioselective organocatalytic asymmetric transfer hydrogenation of quinoline derivatives was developed using a chiral BINOL-derived Brønsted acid as the catalyst. Protected indoles can be hydrogenated efficiently using Rh or Ru complexes with trans chelating bisphosphines ligands. Pyrroles can be hydrogenated using Ru/TRAP complexes with excellent enantioselectivity. Quinoxalines can be hydrogenated using Ru and Ir complexes with moderate enantioselectivity. Pyridine derivatives with a chiral auxiliary and an achiral auxiliary can be efficiently hydrogenated using heterogeneous catalysts and homogeneous Ir/N–P catalysts in the presence of iodine, respectively. Pyridines with electron-withdrawing group at 3-position can be successfully reduced using chiral organophosphoric acid or Ir complexes in the presence of iodine.

Although some promising results have been obtained in asymmetric reduction of nitrogen-based heteroaromatic compounds, it is especially important to point out that these methods are still far from being mature and are still in its infancy. The reported examples in this field are confined to bicyclic heteroaromatic compounds, such as quinoline, isoquinoline, quinoxaline, and indole; the aromatic stabilization of heteroaromatic ring is reduced, thus increasing the reactivity of these substrates toward hydrogenation. For the monocyclic heteroaromatic compounds, only furan derivatives gave the excellent enantioselectivity. Low enantioselectivity was obtained for direct hydrogenation of simple pyridine derivatives. Future attention will focus on the following. First, development of new activation strategy for nitrogen-based heteroaromatic compounds is desirable, and superacids and Lewis acid may be the best choice. Second, explorations of new highly active homogeneous catalyst, new metal precursors, chiral ligands, and additive effect should be screened extensively. Third, Rueping and Du's organcatalytic hydrogenation of quinolines and special pyridines opened a new window to asymmetric reduction of heteroaromatic compounds; therefore, design and development of new organocatalyst type are good direction due to simple operation of organocatalytic process and the compatibility of functional group. Fourth, chiral-modified heterogeneous catalysts should also be explored for asymmetric hydrogenation of nitrogen-based heteroaromatic compounds due to simple purification procedure and recyclability of chiral catalysts. Fifth, an understanding of mechanistic details of these successful reactions might eventually lead to next generation of general asymmetric reduction methods for aromatic compounds containing nitrogen.

Questions

10.1. How many equivalent of hydrogen gas does hydrogenation of pyridine to piperidine need?

10.2. Please write the mechanism of Ir-catalyzed asymmetric hydrogenation of quinoline derivatives in the presence of iodine? (D.-W. Wang, X.-B.Wang, D.-S.Wang, S.-M. Lu, Y.-G. Zhou, *J. Org. Chem.* **2009**, *74*, 2780)

10.3. For the Ir-catalyzed asymmetric hydrogenation of 2-substituted quinolines and 2,3-disubstituted quinolines, please write the mechanism and point out difference in enantioselectivity-controlled step between 2-substituted quinolines and 2,3-disubstituted quinolines, and suggest a strategy for the improvement of enantioselectivity of hydrogenation of 2,3-disubstituted quinolines. (M. Rueping, T. Theissmann, S. Raja, J.W.P. Bats, *Adv. Synth. Catal.* **2008**, *350*, 1001 and D.-W. Wang, X.-B.Wang, D.-S.Wang, S.-M. Lu, Y.-G. Zhou, *J. Org. Chem.* **2009**, *74*, 2780)

References

1 (a) Rylander, P.N. (1979) *Catalytic Hydrogenation in Organic Synthesis*, Academic Press, New York, p. 175; (b) Lu, S.-M., Han, X.-W., Zhou, Y.-G. (2005) *Chin. J. Org. Chem.*, **25**, 634–640; (c) Glorius, F. (2005) *Org. Biomol. Chem.*, **3**, 4171–4175; (d) Dyson, P.J. (2003) *Dalton Trans.*, **15**, 2964–2974; (e) de Vries,

J.G. and Elsevier, C.J. (eds) (2007) *The Handbook of Homogeneous Hydrogenation*, Wiley-VCH, Weiheim; (f) Zhou, Y.-G. (2007) *Acc. Chem. Res.*, **40**, 1357–1366.

2. Wang, W.-B., Lu, S.-M., Yang, P.-Y., Han, X.-W., and Zhou, Y.-G. (2003) *J. Am. Chem. Soc.*, **125**, 10536–10537.

3. Wang, D.-W., Wang, X.-B., Wang, D.-S., Lu, S.-M., and Zhou, Y.-G. (2009) *J. Org. Chem.*, **74**, 2780–2787.

4. Lu, S.-M., Han, X.-W., and Zhou, Y.-G. (2004) *Adv. Synth. Catal.*, **346**, 909–912.

5. Zhao, Y.-J., Wang, Y.-Q., and Zhou, Y.-G. (2005) *Chin. J. Catal.*, **26**, 737–739.

6. (a) Xu, L.J., Lam, K.H., Ji, J.X., Fan, Q.-H., Lo, W.-H., and Chan, A.S.C. (2005) *Chem. Commun.*, 1390–1392; (b) Wu, J. and Chan, A.S.C. (2006) *Acc. Chem. Res.*, **39**, 711–720.

7. Lam, K.H., Xu, L.J., Feng, L.C., Fan, Q.-H., Lam, F.L., Lo, W.-H., and Chan, A.S.C. (2005) *Adv. Synth. Catal.*, **347**, 1755–1758.

8. Qiu, L., Kwong, F.Y., Wu, J., Lam, W.H., Chan, S., Yu, W.-Y., Li, Y.-M., Guo, R., Zhou, Z., and Chan, A.S.C. (2006) *J. Am. Chem. Soc.*, **128**, 5955–5965.

9. Reetz, M.T. and Li, X.-G. (2006) *Chem. Commun.*, 2159–2160.

10. Yamagata, T., Tadaoka, H., Nagata, M., Hirao, T., Kataoka, Y., Ratovelomanana-Vidal, V., Genet, J.P., and Mashima, K. (2006) *Organometallics*, **25**, 2505–2513.

11. Deport, C., Buchotte, M., Abecassis, K., Tadaoka, H., Ayad, T., Ohshima, T., Genet, J.-P., Mashima, K., and Ratovelomanans-Vidal, V. (2007) *Synlett*, 2743–2747.

12. Blaser, H.-U., Pugin, B., Spindler, F., and Togni, A. (2002) *C. R. Chim.*, **5**, 379.

13. Wang, Z.-J., Deng, G.-J., Li, Y., He, Y.-M., Tang, W.-J., and Fan, Q.-H. (2007) *Org. Lett.*, **9**, 1243–1246.

14. Tang, W.-J., Zhu, S.-F., Xu, L.-J., Zhou, Q.-L., Fan, Q.-H., Zhou, H.-F., Lam, K., and Chan, A.S.C. (2007) *Chem. Commun.*, 613–615.

15. Chan, S.-H., Lam, K.-H., Li, Y.-M., Xu, L.-J., Tang, W.-J., Lam, F.-L., Lo, W.-H., Yu, W.-Y., Fan, Q.-H., and Chan, A.S.C. (2007) *Tetrahedron: Asymmetry*, **18**, 2625–2631.

16. Jahjah, M., Alame, M., Pellet-Rostaing, S., and Lemaire, M. (2007) *Tetrahedron: Asymmetry*, **18**, 2305–2312.

17. Mrsic, N., Lefort, L., Boogers, J.A.F., Minnaard, A.J., Feringa, B.L., and Vries, J.G. (2008) *Adv. Synth. Catal.*, **350**, 1081–1089.

18. Lu, S.-M. and Bolm, C. (2008) *Adv. Synth. Catal.*, **350**, 1101–1105.

19. Wang, X.-B. and Zhou, Y.-G. (2008) *J. Org. Chem.*, **73**, 5640–5642.

20. Li, Z.-W., Wang, T.-L., He, Y.-M., Wang, Z.-J., Fan, Q.-H., Pan, J., and Xu, L.-J. (2008) *Org. Lett.*, **10**, 5265–5268.

21. Ohkuma, T., Utsumi, N., Tsutsumi, K., Murata, K., Sandoval, C., and Noyori, R. (2006) *J. Am. Chem. Soc.*, **128**, 8724–8725.

22. Eggenstein, M., Thomas, A., Theuerkauf, J., Francio, G., and Leitner, W. (2009) *Adv. Synth. Catal.*, **351**, 725–732.

23. Zhou, H.-F., Li, Z.-W., Wang, Z.-J., Wang, T.-L., Xu, L.-J., He, Y.-M., Fan, Q.-H., Pan, J., Gu, L.-Q., and Chan, A.S.C. (2008) *Angew. Chem., Int. Ed.*, **47**, 8464–8467.

24. Yang, P.-Y. and Zhou, Y.-G. (2004) *Tetrahedron: Asymmetry*, **15**, 1145–1149.

25. (a) Chan, Y.N.C. and Osborn, J.A. (1990) *J. Am. Chem. Soc.*, **112**, 9400–9401; (b) Dorta, R., Broggini, D., Stoop, R., Ruegger, H., Spindler, F., and Togni, A. (2004) *Chem. Eur. J.*, **10**, 267–278.

26. (a) Xiao, D.-M. and Zhang, X.-M. (2001) *Angew. Chem., Int. Ed.*, **40**, 3425–3428; (b) Chi, Y.-X., Zhou, Y.-G., and Zhang, X.-M. (2003) *J. Org. Chem.*, **68**, 4120–4122.

27. Rueping, M., Theissmann, T., Raja, S., and Bats, J.W.P. (2008) *Adv. Synth. Catal.*, **350**, 1001–1006.

28. Guo, Q.-S., Du, D.-M., and Xu, J. (2008) *Angew. Chem., Int. Ed.*, **47**, 759–762.

29. Lu, S.-M., Wang, Y.-Q., Han, X.-W., and Zhou, Y.-G. (2005) *Chin. J. Catal.*, **26**, 287–290.

30. Wang, D.-W., Zeng, W., and Zhou, Y.-G. (2007) *Tetrahedron: Asymmetry*, **18**, 1103–1107.

31. Lu, S.-M., Wang, Y.-Q., Han, X.-W., and Zhou, Y.-G. (2006) *Angew. Chem., Int. Ed.*, **45**, 2260–2263.

32. Rueping, M., Antonchick, A.P., and Theissmann, T. (2006) *Angew. Chem., Int. Ed.*, **45**, 3683–3686.

33. Kuwano, R. (2008) *Heterocycles*, **76**, 909–922.

34 Kuwano, R., Sato, K., Kurokawa, T., Karube, D., and Ito, Y. (2000) *J. Am. Chem. Soc.*, **122**, 7614–7615.
35 Kuwano, R., Sato, K., and Ito, Y. (2000) *Chem. Lett.*, 428–429.
36 (a) Sawamura, M., Hamashima, H., and Ito, Y. (1991) *Tetrahedron: Asymmetry*, **2**, 593–596; (b) Sawamura, M., Hamashima, H., Sugawara, M., Kuwano, R., and Ito, Y. (1995) *Organometallics*, **14**, 4549–4558; (c) Kuwanoand, R. and Sawamura, M. (2007) in *Catalysts for Fine Chemical Synthesis*, vol. **5**, Regio- and Stereo-Controlled Oxidations and Reductions (eds S.M. Roberts and J. Whittall), John Wiley & Sons, Inc., West Sussex, pp. 73–86.
37 Kuwano, R., Kaneda, K., Ito, T., Sato, K., Kurokawa, T., and Ito, Y. (2004) *Org. Lett.*, **6**, 2213–2215.
38 Kuwano, R., Kashiwabara, M., Sato, K., Ito, T., Kaneda, K., and Ito, Y. (2006) *Tetrahedron: Asymmetry*, **17**, 521–535.
39 Kuwano, R. and Kashiwabara, M. (2006) *Org. Lett.*, **8**, 2653–2655.
40 Háda, V., Tungler, A., and Szepesy, L. (2001) *Appl. Catal. A*, **210**, 165–171.
41 Kuwano, R., Kashiwabara, M., Ohsumi, M., and Kusano, H. (2008) *J. Am. Chem. Soc.*, **130**, 808–809.
42 Murata, S., Sugimoto, T., and Matsuura, S. (1987) *Heterocycles*, **26**, 763–766.
43 Bianchini, C., Barbaro, P., Scapacci, G., Farnetti, E., and Graziani, M. (1998) *Organometallics*, **17**, 3308–3310.
44 Bianchini, C., Barabro, P., and Scapacci, G. (2001) *J. Organomet. Chem.*, **621**, 26–33.
45 (a) Cobley, C.J. and Henschke, J.P. (2003) *Adv. Synth. Catal.*, **345**, 195–201; (b) Henschke, J.P., Burk, M.J., Malan, C.G., Herzberg, D., Peterson, J.A., and Wildsmith, A.J., Cobley, C. J., Casy, G. (2003) *Adv. Synth. Catal.*, **345**, 300–307.
46 Blaser, H.-U., Honig, H., Studer, M., and Wedemeyer-Exl, C. (1999) *J. Mol. Catal. A: Chem.*, **139**, 253–257.
47 Studer, M., Wedemeyer-Exl, C. Spindler, F., and Blaser, H.U. (2000) *Monatsh. Chem.*, **131**, 1335–1343.
48 Raynor, S.A., Thomas, J.M., Raja, R., Johnson, B.F.G., Bell, R.G., and Mantle, M.D. (2000) *Chem. Commun.*, **19**, 1925–1926.
49 Hegedus, L., Hada, V. Tungler A., Mathe, T., and Szepesy, L. (2000) *Appl. Catal. A*, **201**, 107–114.
50 Douja, N., Besson, M. Gallezot, P., and Pinel, C. (2002) *J. Mol. Catal. A: Chem.*, **186**, 145–151.
51 Ouja, N., Malacea, R. Banciu, M., Besson, M., and Pinel, C. (2003) *Tetrahedron Lett.*, **44**, 6991–6993.
52 Glorius, F., Spielkamp, N. Holle, S., Goddard, R., and Lehman, C.W. (2004) *Angew. Chem., Int. Ed.*, **43**, 2850–2852.
53 Klapars, A., Huang, X., and Buchwald, S.L. (2002) *J. Am. Chem. Soc.*, **124**, 7421–7428.
54 Legault, C.Y. and Charette, A.B. (2005) *J. Am. Chem. Soc.*, **127**, 8966–8967.
55 For more detailed mechanistic insights, see Legault, C.Y., Charette, A.B., and Cozzi, P.G. (2008) *Heterocycles*, **76**, 1271–1283.
56 Lei, A., Chen, M. He, M., and Zhang, X. (2006) *Eur. J. Org. Chem.*, 4343–4347.
57 Wang, X.-B., Zeng, W., and Zhou, Y.-G. (2008) *Tetrahedron Lett.*, **49**, 4922–4924.
58 Rueping, M. and Antonchick, A.P. (2007) *Angew. Chem., Int. Ed.*, **46**, 4562–4566.

11
Asymmetric Hydroamination
Alexander L. Reznichenko and Kai C. Hultzsch

11.1
Introduction: Synthesis of Amines via Hydroamination

The addition of an amine N–H bond across an unsaturated carbon–carbon linkage – the so-called hydroamination, allows a facile and highly atom-economical access to industrially relevant nitrogen-containing basic and fine chemicals as well as naturally occurring alkaloid skeletons [1–3]. Significant research efforts have led to the development of efficient catalyst systems for inter- and intramolecular hydroamination reactions (Eqs 11.1 and 11.2). Many catalytic systems based on transition metals (groups 3–5 and 8–10, as well as lanthanides and actinides) and main group metals (alkali and alkaline earth metals), as well as Brønsted acids, have been developed over the past six decades, especially the last decade has seen significant progress. However, the majority of these catalyst systems are confined to a limited set of substrates, for example, requiring activated multiple C–C bonds, such as 1,3-dienes, vinyl arenes, allenes, or ring-strained olefins. Because of its potential relevance for the synthesis of pharmaceuticals and biological active molecules, most of which are chiral, the development of chiral catalysts for stereoselective hydroamination (including kinetic resolution of chiral amines) has drawn increasing attention over the last decade [4–7].

Equations 11.1 and 11.2. Generic inter- and intramolecular hydroamination reactions.

$$\text{R}^1\text{R}^2\text{C=CR}^1\text{R}^2 + \text{R}^1\text{R}^2\text{N-H} \longrightarrow \text{product} \quad (11.1)$$

$$\text{alkenyl amine} \longrightarrow \text{cyclic amine} \quad (11.2)$$

Chiral Amine Synthesis: Methods, Developments and Applications. Edited by Thomas C. Nugent
Copyright © 2010 WILEY-VCH Verlag GmbH & Co. KGaA, Weinheim
ISBN: 978-3-527-32509-2

The direct addition of amines to alkenes is thermodynamically feasible ($\Delta G^0 \approx -14.7$ kJ mol^{-1} for the addition of ammonia to ethylene) [8, 9] with a slightly exothermic to thermoneutral reaction enthalpy. However, this seemingly simple reaction is hampered by a high activation barrier caused by an electrostatic repulsion between the nitrogen lone pair of the approaching amine and the π-bond of the electron-rich olefin. The activation barrier cannot be overcome by increasing the reaction temperature because the negative reaction entropy shifts the equilibrium toward the starting materials. Therefore, amines add commonly only to activated, electron-deficient alkenes (e.g., vinyl ethers or Michael acceptors) in the absence of a catalyst. For the scope of this overview, such aza-Michael reactions will not be considered and the reader is referred to pertinent reviews in the literature [10, 11].

11.2
Hydroamination of Simple, Nonactivated Alkenes

11.2.1
Intermolecular Hydroamination of Simple Alkenes

Although several significant contributions to the field of intermolecular hydroamination of simple alkenes have been made during the last decade, the asymmetric hydroamination of *nonactivated* alkenes with *unprotected* amines has yet to be elaborated and the first asymmetric intermolecular hydroamination of terminal, nonactivated alkenes with cyclic ureas has been reported only recently (Scheme 11.1) [12]. The Markovnikov addition proceeded with up to 78% ee utilizing the axial chiral MeOBIPHEP-ligated bis(gold(I))-catalyst system (*S*)-**1**. Unfortunately, the reaction required a large excess of the alkene substrate and lower alkene loadings led to diminished enantioselectivities.

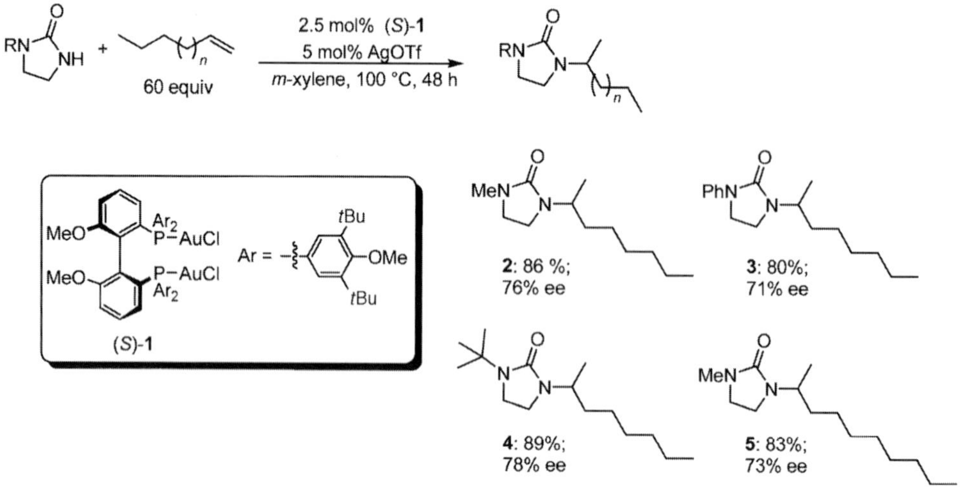

Scheme 11.1 Gold-catalyzed asymmetric hydroamination of alkenes with cyclic ureas [12].

Since more reactive alkenes, such as vinyl arenes or sterically strained polycycles, react more readily in the hydroamination reaction, several asymmetric hydroamination reactions utilizing these substrates have been disclosed. Weakly basic anilines can react with vinyl arenes to give the Markovnikov addition products **6** and **7** with good yields and enantioselectivities in the presence of a chiral phosphine ligand Pd complex as demonstrated by Hartwig (Eq. 11.3) [13] and later by Hii (Eq. 11.4) [14].

Equations 11.3 and 11.4. Palladium-catalyzed asymmetric Markovnikov addition of aniline to vinyl arenes [13, 14].

$$F_3C\text{-vinylarene} + \text{PhNH}_2 \xrightarrow[\text{toluene, 25 °C, 72 h}]{10\text{ mol\%} [\{(R)\text{-BINAP}\}Pd(OTf)_2]} \text{(S)-6: 80\%; 81\% ee} \quad (11.3)$$

$$\text{styrene} + \text{PhNH}_2 \xrightarrow[\text{neat, 80 °C, 72 h}]{2\text{ mol\%} [\{(R)\text{-BINAP}\}Pd(MeCN)(H_2O)](OTf)_2} \text{7: 93\%; 70\% ee} \quad (11.4)$$

The proposed mechanism for this process involves an insertion of the vinyl arene into a palladium hydride species, followed by a nucleophilic attack of the amine on the resulting η^3-benzylic palladium intermediate (Scheme 11.2) [15–17]. The high Markovnikov regioselectivity results from the electronically favored secondary insertion of the vinyl arene in the palladium hydride bond. Similar mechanistic models have been suggested for the late transition metal-catalyzed hydroamination of allenes [18–20], dienes [17, 21–23], and trienes [24]. Generally, the η^3-allyl species were identified as the resting state of the catalyst during the hydroamination reaction, indicating that nucleophilic attack is rate determining [15, 23].

Two mechanistically plausible scenarios for nucleophilic attack on the η^3-benzyl palladium species seem feasible. Formation of the C−N bond could occur either via external attack of the amine through inversion of configuration at the carbon stereocenter, or alternatively the amine could coordinate to palladium followed by an internal attack on the η^3-benzyl ligand. Mechanistic investigations [15] using stoichiometric amounts of the enantio- and diastereomerically pure η^3-benzyl palladium complex [{(R)-Tol-BINAP}{η^3-1-(2-naphthyl)ethyl}Pd](OTf) (**8**) revealed that the reaction with aniline produced predominantly (R)-N-1-(2-naphthyl)ethylaniline (R)-**9**), consistent with external nucleophilic attack (Scheme 11.3) [15]. However, it was noted that the catalytic reaction of [{(R)-Tol-BINAP}Pd(OTf)$_2$] with vinyl arenes and amines produced preferentially the opposite enantiomeric (S)-amine hydroamination

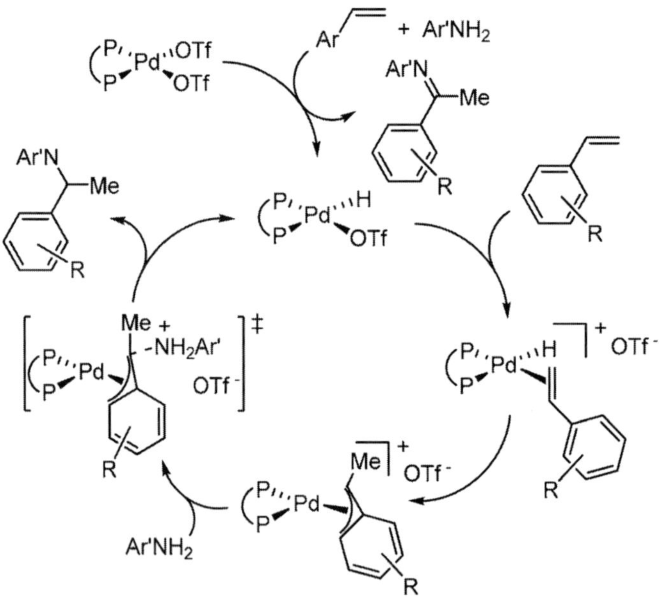

Scheme 11.2 Proposed mechanism for the palladium-catalyzed hydroamination of vinyl arenes.

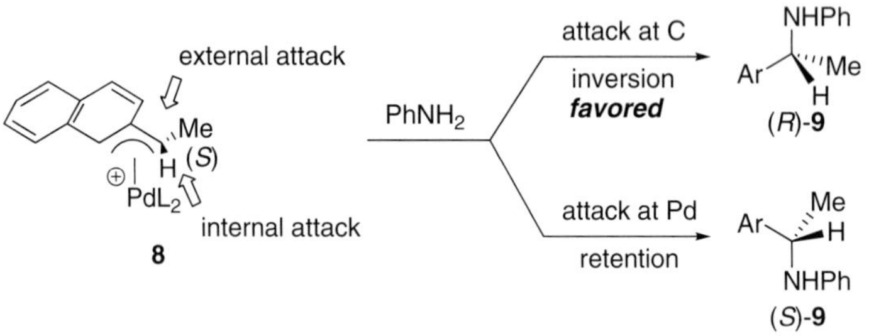

Scheme 11.3 Preferential external attack of the aniline nucleophile leads to inversion of stereochemistry at the benzyl palladium intermediate.

product, indicating, in analogy to asymmetric rhodium-catalyzed hydrogenation [25], that the minor diastereomer of the catalytic active benzyl palladium species was responsible for the majority of product formed in the catalytic process.

The range of amines involved may be expanded to more basic alkylamines (Eq. 11.5) [16]. Compound **10** was obtained in moderate yield and enantioselectivity utilizing the (R,R)-Et-FerroTANE ligand. Note that almost stoichiometric amounts of a strong Brønsted acid are required to afford the Markovnikov hydroamination product of the vinyl arene.

Equation 11.5. Palladium-catalyzed asymmetric Markovnikov addition of an alkylamine to a vinyl arene [16].

$$\text{(11.5)}$$

Enhanced reactivity of strained polycyclic olefins such as norbornene has made them attractive model compounds for intermolecular hydroamination. In most cases, the reaction employs a late transition metal-based catalyst and requires elevated temperatures and extended reaction times. The first chiral iridium-based catalyst system was reported by Togni in 1997 (Eq. 11.6) [26]. The activity and enantioselectivity of this catalyst system was significantly enhanced by addition of Schwesinger's "naked" fluoride [N{P(NMe$_2$)$_3$}$_2$]F. In some cases, the addition of fluoride also resulted in a reversal of absolute product configuration. However, the reason for the strong fluoride effect remains unclear, but it has been suggested that hydrogen bridging of the fluoride could play a role.

Equation 11.6. Iridium-catalyzed asymmetric hydroamination of norbornene [26].

$$\text{(11.6)}$$

Earlier mechanistic studies by Milstein on a achiral Ir-catalyst system indicated that the iridium-catalyzed norbornene hydroamination involves amine activation as a key step in the catalytic cycle [27] rather than *alkene* activation, which is observed for most other late transition metal-catalyzed hydroamination reactions [28]. Thus, the iridium-catalyzed hydroamination of norbornene with aniline is initiated by an oxidative addition of aniline to the metal center, followed by insertion of the strained olefin into the iridium–amido bond (Scheme 11.4). Subsequent reductive elimination completes the catalytic cycle and gives the hydroamination product **11**. Unfortunately, this catalyst system seems to be limited to highly strained olefins.

This chemistry was extended to a number of bicyclic alkenes and dienes utilizing various chelating axially chiral bisphosphine iridium catalysts (Scheme 11.5) [29]. Further synthetic transformations of the chiral hydroamination product **13** provide access to functionally substituted chiral cyclopentylamines with multiple stereocenters, such as **14** and **15**. It should be noted that alkylamines, such as octylamine or *N*-methyl aniline, and sterically encumbered aniline derivatives, such as *o*-toluidine or *o*-anisidine did not undergo hydroamination reactions under these conditions.

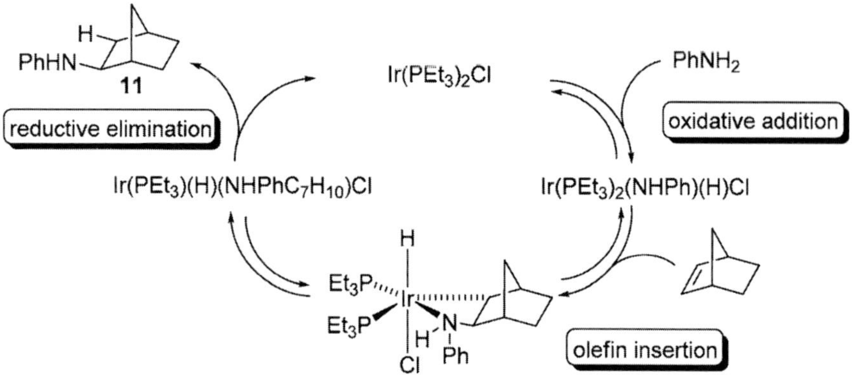

Scheme 11.4 Proposed mechanism for iridium-catalyzed hydroamination of norbornene via amine activation.

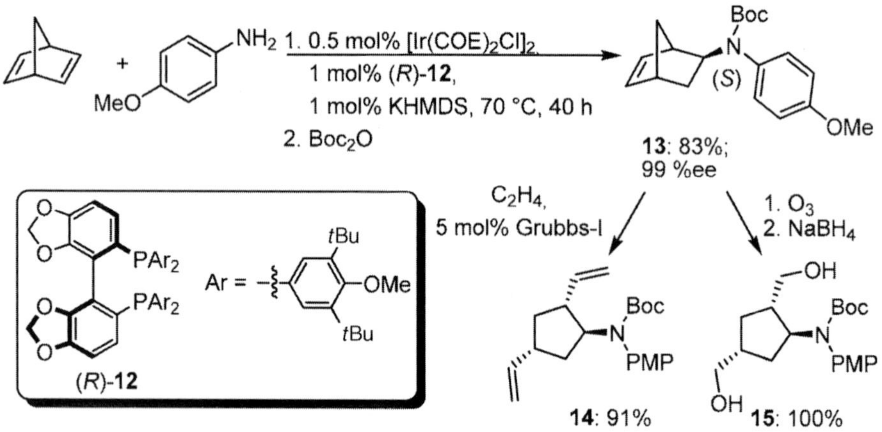

Scheme 11.5 Stereoselective synthesis of cyclopentylamines via asymmetric hydroamination of norbornadiene [29].

11.2.2
Intramolecular Asymmetric Hydroamination of Simple Aminoalkenes

11.2.2.1 Rare Earth Metal-Based Catalysts

Among hydroamination catalyst systems developed so far, organo rare earth metal complexes have been found to be the most versatile and most active catalysts for the hydroamination of nonactivated alkenes, allenes, 1,3-dienes, and alkynes [30]. However, intermolecular hydroamination reactions with these catalysts are problematic. Unfavorable factors are not only the negative reaction entropy but also the competition between strongly coordinating amines and weakly binding alkenes. Therefore, most rare earth metal-catalyzed reactions have been performed in an intramolecular fashion producing predominantly pyrrolidine and piperidine derivatives. Although rare earth metals have been shown to catalyze the intermolecular hydroamination of simple unactivated alkenes at elevated temperatures [31, 32], no example of an asymmetric intermolecular hydroamination has been reported to date.

The mechanism of the hydroamination/cyclization [33, 34] proceeds through a metal amido species, which is formed upon protonolysis of a metal amido, alkyl, or aryl bond in the precatalyst (Scheme 11.6), followed by insertion of the olefin into the metal amido bond with a seven-membered chair-like transition state (for $n=1$). The roughly thermoneutral insertion step [34] is usually rate determining, giving rise to a zero-order rate dependence on substrate concentration and first-order rate dependence on catalyst concentration. The resulting metal alkyl species undergoes fast protonolysis with a second amine molecule, regenerating the metal amido species and releasing the heterocyclic product. Although the mechanism was established for rare earth metal complexes, alkali and alkaline earth metal-based systems are apparently operating in agreement with this mechanism.

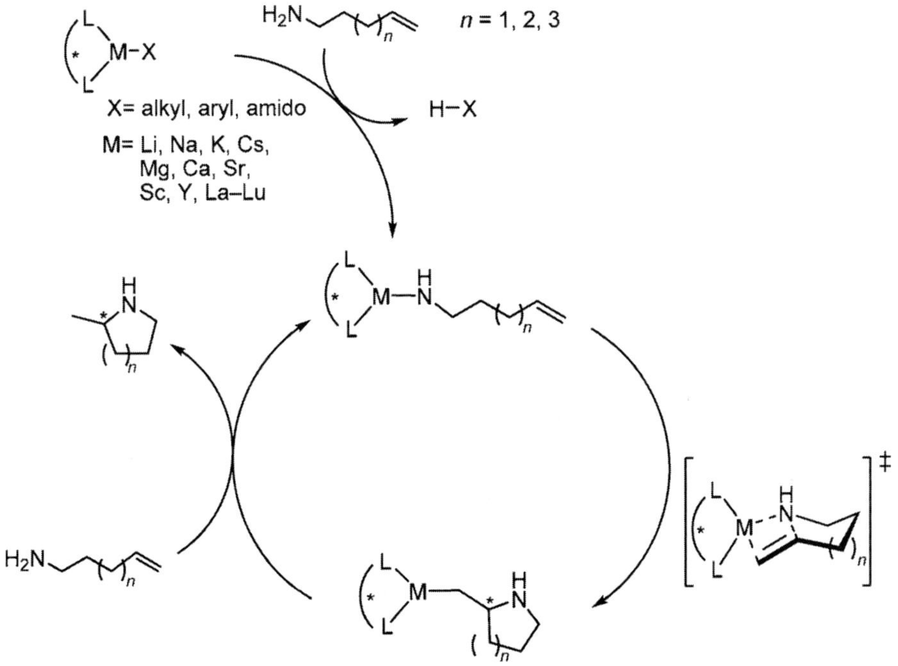

Scheme 11.6 Proposed mechanism for the hydroamination/cyclization of aminoalkenes using alkali, alkaline earth, and rare earth metal-based catalysts.

Chiral rare earth metal-based catalysts for the asymmetric hydroamination/ cyclization of aminoalkenes were first introduced in 1992 by Marks and co-workers [35, 36]. The C_1-symmetric chiral *ansa*-lanthanocene complexes **16–18** with (+)-neomenthyl, (−)-menthyl, or (−)-phenylmenthyl substituents attached to one of the cyclopentadienyl ligands (Figure 11.1) exist in two diastereomeric forms,

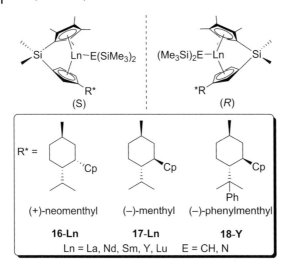

Figure 11.1 Chiral lanthanocene precatalysts for asymmetric hydroamination [35, 36].

depending on which diastereotopic face of the cyclopentadienyl ligand coordinates to the metal center. Generally, one of the two diastereomers predominates in solution and most of the complexes can be obtained diastereomerically pure by fractional crystallization.

The lanthanocene catalysts display high catalytic activity, which is proportional to the ionic radius of the rare earth metal (Table 11.1). The rate of cyclization depends on the ring size of the azacyclic product (5 > 6 ≫ 7) and the presence of rate-enhancing *gem*-dialkyl substituents [37]. The proposed stereomodel for aminopentene substrates (Figure 11.2) [36] predicts preferred formation of the (S)-pyrrolidine enantiomer due to unfavorable steric interactions between an axial substituent of the substrate with the substituents on the cyclopentadienyl ligands of the (S)-lanthanocene diastereomer in the transition state of the cyclization. Hence, dominance of one of the two catalyst epimers is essential to achieve high enantioselectivities.

Unfortunately, the complexes underwent facile epimerization under the conditions of catalytic hydroamination via reversible protolytic cleavage of the metal cyclopentadienyl bond (Scheme 11.7) [36, 38–40]. Thus, the product enantioselectivity was limited by the catalyst's epimeric ratio in solution and the absolute configuration of the hydroamination product was independent of the diastereomeric purity of the precatalyst. Complexes with a (+)-neomenthyl substituent on the cyclopentadienyl ligand generally produced the (R)-(−)-pyrrolidines, whereas (−)-menthyl and (−)-phenylmenthyl-substituted complexes yielded the (S)-(+)-pyrrolidines, which is in agreement with the proposed stereomodel and solution studies on the equilibrium epimer ratios in the presence of simple aliphatic amines.

The highest enantioselectivities of up to 74% ee were achieved with the (−)-menthyl-substituted samarium complex (S)-**17-Sm** in the formation of pyrrolidine products, whereas formation of piperidines suffered from significantly diminished

11.2 Hydroamination of Simple, Nonactivated Alkenes

Table 11.1 Lanthanocene-catalyzed aminoalkene hydroamination/cyclization [35, 36].

19 $n = 1$; R = Me
21 $n = 1$; R = H
23 $n = 2$; R = Me

20, 22, 24

Catalyst	Substrate	T (°C)	N_t,[a] h^{-1}	% ee (config)
(R)-16-Sm	19	−30		64 (R)
(R,S)-16-Y	19	25	38	36 (R)
(R)-16-Y	19	25	21	40 (R)
(S)-17-Sm	19	25	84	53 (S)
(S)-17-Sm	19	−30		74 (S)
(R)-18-Y	19	25	8	56 (S)
(60/40) (R,S)-18-Y	19	25		54 (S)
(S)-16-Sm	21	25	33	55 (R)
(R)-16-Sm	21	25	62	52 (R)
(S)-17-Sm	21	0		72 (S)
(R)-18-Y	21	25		64 (S)
(R)-16-Sm	23	25		17 (R)
(R)-17-Sm	23	25	2	15 (R)

a) N_t = turnover number per hour.

selectivities. The extended "wingspan" of the octahydrofluorenyl ligand in complex **25** was supposed to provide an increased enantiofacial discrimination for prochiral substrates, but only isolated examples of improved selectivities were observed (Scheme 11.8) [38].

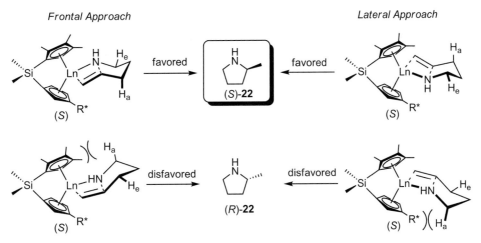

Figure 11.2 Proposed stereomodels for the lanthanocene-catalyzed hydroamination/cyclization of pent-4-enylamine (**21**) [36].

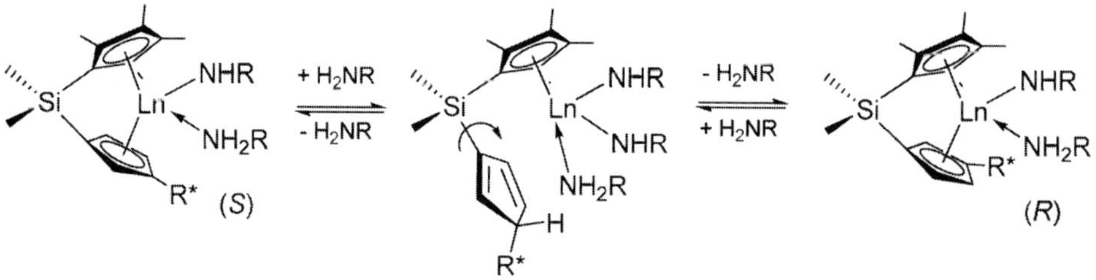

Scheme 11.7 Amine-induced epimerization of chiral lanthanocene complexes.

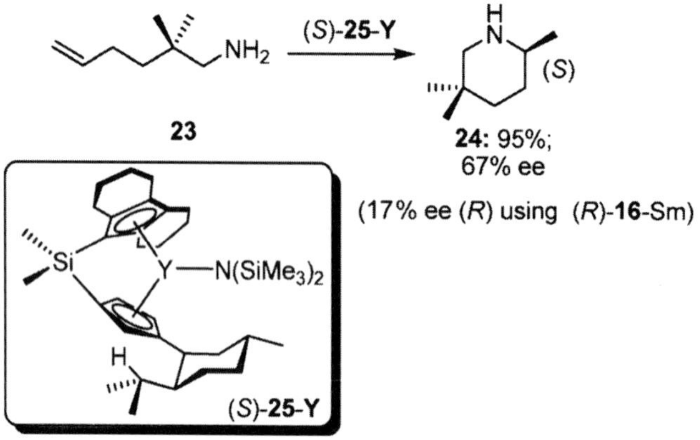

Scheme 11.8 Improved enantioselectivities in the hydroamination/cyclization of aminohexenes using a chiral octahydrofluorenyl yttrocene catalyst [38].

Internal 1,1- or 1,2-disubstituted olefins 26 and 28 are much less reactive for hydroamination and require significantly harsher reaction conditions [39, 41–44]. The formation of pyrrolidines and piperidines often proceeds at comparable rates (Eq. 11.7), contrasting the general trend of significantly faster five-membered ring formation observed with terminal aminoalkenes [39]. Despite these harsh reaction conditions, moderate enantioselectivities of up to 58% ee at 100 °C (up to 68% at 60 °C) were observed.

Equation 11.7. Enantioselective hydroamination/cyclization of internal aminoalkenes [39].

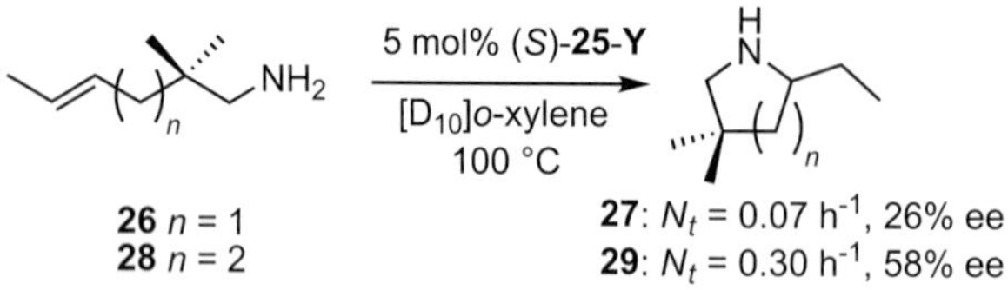

26 n = 1
28 n = 2

27: N_t = 0.07 h^{-1}, 26% ee
29: N_t = 0.30 h^{-1}, 58% ee

(11.7)

Significant progress in this area has been made since 2003 utilizing noncyclopentadienyl-based ligand sets [5–7, 45–68], thus avoiding configurational instability issues of the chiral lanthanocene complexes. However, while enantioselectivities

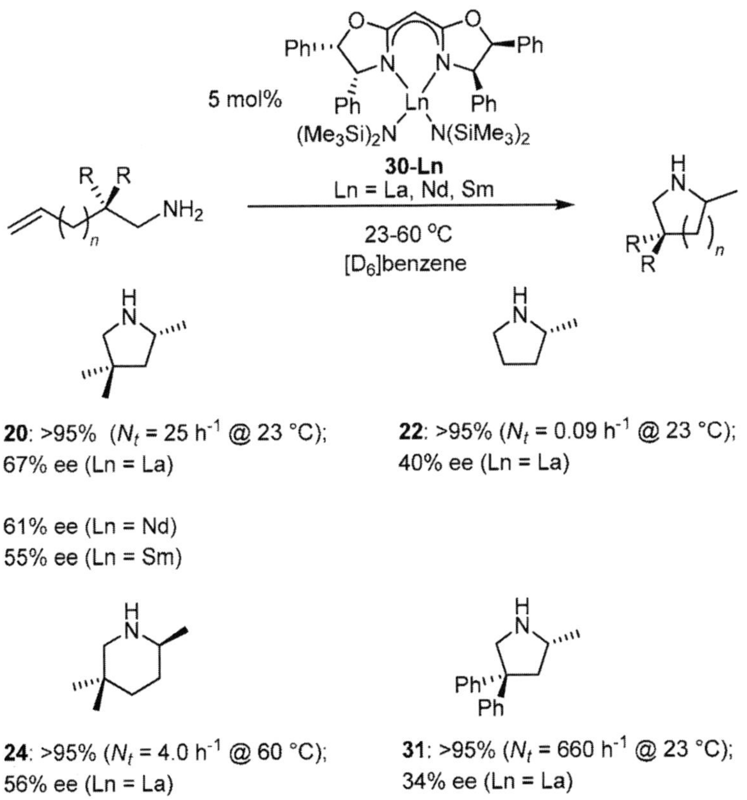

11.2 Hydroamination of Simple, Nonactivated Alkenes

Scheme 11.9 Bisoxazolinato rare earth metal catalysts in the hydroamination/cyclization of aminoalkenes [49].

20: >95% (N_t = 25 h^{-1} @ 23 °C);
67% ee (Ln = La)

61% ee (Ln = Nd)
55% ee (Ln = Sm)

22: >95% (N_t = 0.09 h^{-1} @ 23 °C);
40% ee (Ln = La)

24: >95% (N_t = 4.0 h^{-1} @ 60 °C);
56% ee (Ln = La)

31: >95% (N_t = 660 h^{-1} @ 23 °C);
34% ee (Ln = La)

have improved rather significantly, only few catalyst systems [52] display the same high catalytic activity as the lanthanocene catalysts.

A variety of bisoxazolinato rare earth metal complexes have been studied with regard to their hydroamination/cyclization catalytic activity (Scheme 11.9) [49]. The precatalysts could be conveniently generated *in situ* from [Ln{N(SiMe$_3$)$_2$}$_3$] (Ln = La, Nd, Sm, Y, Lu) and the corresponding bisoxazoline, resulting in minimal reactivity loss but with similar enantioslectivity, in comparison to the isolated precatalysts. The ligand-accelerated catalyst system showed the highest rates for a 1 : 1 metal to ligand ratio.

The ate complexes [Li(THF)$_4$][Ln{(R)-1,1'-{C$_{10}$H$_6$N(R)}$_2$}$_2$] (R)-**32**; Ln = Sm, Yb, Y; R = alkyl) [53–57, 60] are unusual hydroamination catalysts as they lack an obvious leaving amido or alkyl group that is replaced during the initiation step by the substrate. It is very likely that at least one of the amido groups is protonated during the catalytic cycle, analogous to the mechanism proposed for Michael additions and aldol reactions catalyzed by rare earth metal–alkali metal–BINOL heterobimetallic complexes [69, 70]. The observation of similar selectivities for rare earth metal

Table 11.2 Catalytic hydroamination/cyclization of an aminoalkene using diamidobinaphthyl ate complexes.

R	Ln	t (h)	% ee	Conversion (%)	Reference
iPr	Y	20	67	100	[56]
Cy	Yb	18	65	94	[55]
C_5H_9	Yb	20	87	90	[60]

complexes containing only one diamidobinaphthyl seem to support this proposal [7, 56, 59].

The best catalytic results were obtained using a small rare earth metal (Yb) and a large cyclopentyl substituent on the diamidobinaphthyl amido ligand (Table 11.2), though the moderate catalytic activity required the presence of *gem*-dialkyl substituents in the aminoalkene substrates. The low reactivity of **32** should mainly be attributed to the rather electron-rich diamidobinaphthyl ligand framework, as indicated by comparison to other diamidobinaphthyl [7, 56, 59] or diamidobiphenyl [46, 48], catalyst systems, rather than the unusual mode of catalyst activation.

Good to high enantioselectivities for a wide range of aminoalkene substrates, including internal olefins or secondary amines, were achieved using the aminothiophenolate catalyst system (*R*)-**35** (Scheme 11.10), which was also obtained *in situ* [61]. Variation of the steric demand of the silyl substituent attached to the thiophenolate moiety allowed facile fine-tuning of the enantiomeric excess, resulting in increasing selectivity with increasing steric hindrance. While the larger bite angle of the amino (thio)phenolate ligand is believed to improve enantiofacial differentiation through a better "reach around" of the chiral ligand around the metal center [47], the multidentate nature of the ligand also electronically saturates the metal center, resulting in diminished catalytic performance. Enantiomeric excess of up to 89% could be achieved at 30 °C, though reactions at this temperature required a long period of time to reach completion.

Significantly higher catalytic activities are achievable, if more electron-deficient ligand sets are employed. 3,3′-Disubstituted binaphtholate aryl complexes (*R*)-**38-Ln**, (Ln = Sc, Y, Lu) with sterically demanding tris(aryl)silyl substituents in the 3 and 3′ position showed not only superior catalytic activity at room temperature, comparable

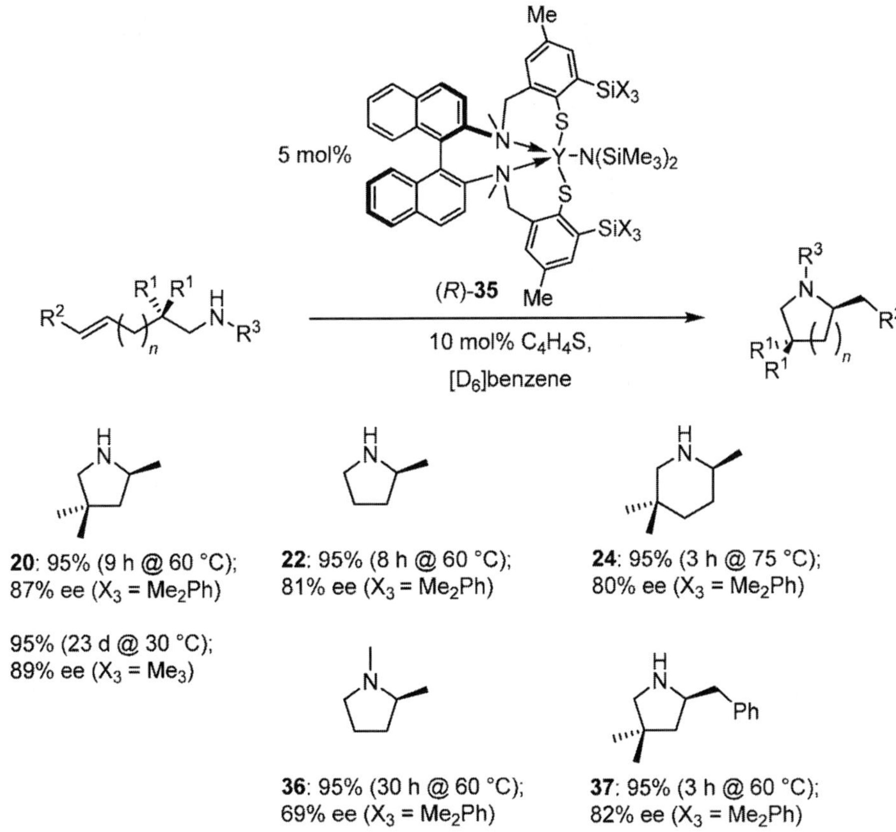

Scheme 11.10 Catalytic hydroamination/cyclization of aminoalkenes using chiral aminothiophenolate yttrium complexes [61].

in magnitude to lanthanocene catalysts, but also achieved up to 95% ee in hydroamination/cyclization reactions of aminoalkenes, among the highest enantioselectivities observed so far (Scheme 11.11) [52]. The sterically demanding tris(aryl)silyl substituents in the diolate complexes not only play a pivotal role in achieving high enantioselectivities but are also crucial in preventing undesired complex aggregation [50, 51] and reducing detrimental amine binding of the substrate and product to the catalytic active metal centers [52].

11.2.2.2 Alkali Metal-Based Catalysts

While hydroamination catalysts based on transition metals have been studied intensively over the past two decades, only a limited number of reports on alkali metal-based hydroamination catalysts have emerged, although the first reports date back 60 years [71]. In particular, the application of chiral alkali metal complexes in asymmetric hydroamination of nonactivated aminoalkenes has drawn little attention to date [72, 73]. Also, attempts to perform asymmetric hydroamination utilizing

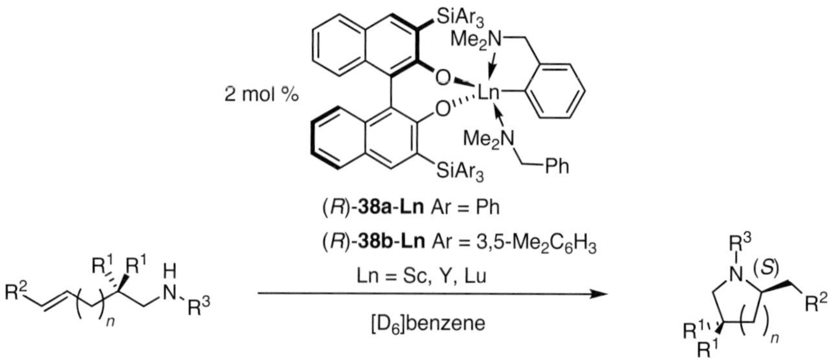

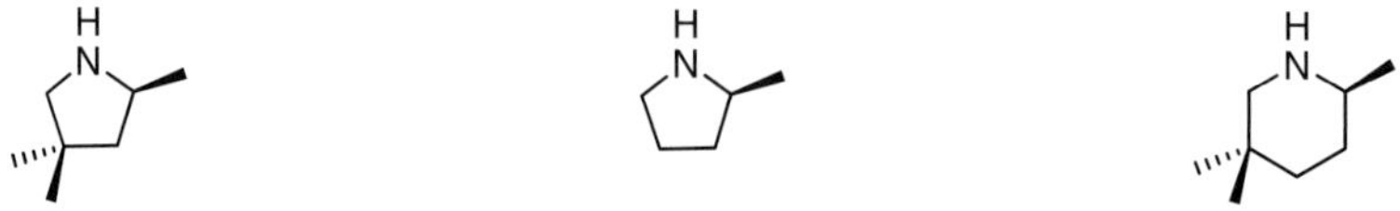

20: 93% (N_t = 13 h^{-1} @ 60 °C); 73% ee ((*R*)-**38a-Sc**)

22: 93% (N_t = 1.7 h^{-1} @ 22 °C); 90% ee ((*R*)-**38b-Lu**)

24: 97% (N_t = 0.9 h^{-1} @ 60 °C); 61% ee ((*R*)-**38b-Sc**)

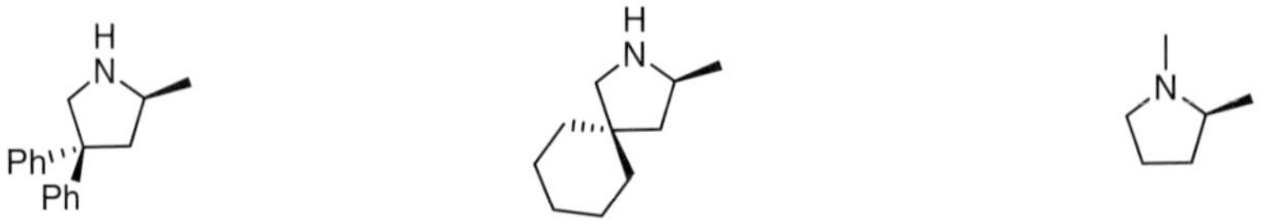

31: 94% (N_t = 110 h^{-1} @ 25 °C); 95% ee ((*R*)-**38a-Sc**)

34: 97% (N_t = 11 h^{-1} @ 25 °C); 85% ee ((*R*)-**38a-Sc**)

36: 93% (N_t = 0.9 h^{-1} @ 60 °C); 53% ee ((*R*)-**38b-Sc**)

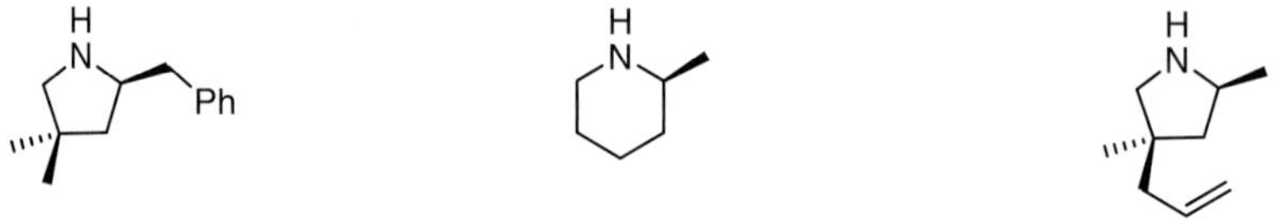

37: 91% (N_t = 0.6 h^{-1} @ 100 °C); 72% ee ((*R*)-**38a-Sc**)

39: 94% (N_t = 2.1 h^{-1} @ 80 °C); 55% ee ((*R*)-**38b-Lu**)

40: 93% (N_t = 3 h^{-1} @ 22 °C); dr 2.7:1 88:78% ee ((*R*)-**38a-Sc**)

Scheme 11.11 Catalytic hydroamination/cyclization of aminoalkenes using 3,3′-bis(trisarylsilyl)-substituted binaphtholate rare earth metal complexes [52].

chiral alkaline earth metal complexes has been thwarted so far by facile Schlenk equilibria of the metal species in solution [74, 75].

The proline-derived diamidobinaphthyl dilithium salt (*S,S,S*)-**41**, which is dimeric in the solid state and can be prepared via deprotonation of the corresponding tetraamine with *n*BuLi, represents the first example of a chiral main group metal-based catalyst for asymmetric intramolecular hydroamination reactions of aminoalkenes [72]. The unique reactivity of (*S,S,S*)-**41**, which allowed reactions at or below ambient temperatures with product enantioselectivities of up to 85% ee (Scheme 11.12) [76], is believed to derive from the proximity of the two lithium

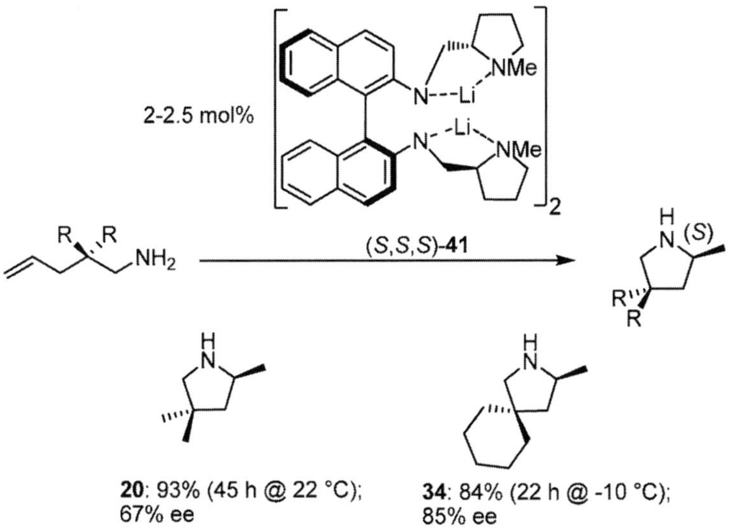

11.2 Hydroamination of Simple, Nonactivated Alkenes

Scheme 11.12 Lithium-catalyzed asymmetric hydroamination/cyclization of aminoalkenes [72, 76].

centers chelated by the proline substituents, because more simple lithium amides required significantly higher reaction temperatures and gave inferior selectivities.

More recently, the asymmetric hydroamination/cyclization of amino-substituted stilbenes was studied utilizing chiral bisoxazoline lithium catalysts [73]. Enantioselectivities reaching as high as 91% ee were achieved (Scheme 11.13). The reactions were performed in toluene at −60 °C to give the *exo*-cyclization product **43** under

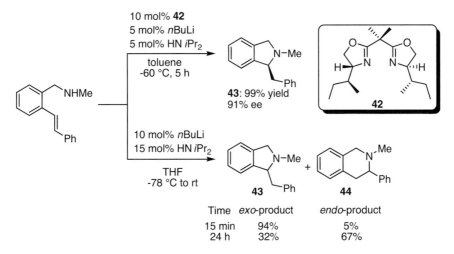

Scheme 11.13 Kinetic versus thermodynamic control in the lithium-catalyzed cyclization of aminostilbenes [73].

kinetic control. However, the hydroamination/cyclization reaction in THF solution is reversible, producing the thermodynamically favored *endo*-cyclization product **44**.

11.2.2.3 Group 4 Metal-Based Catalysts

The chemistry of organometallic group 4 metal compounds is well developed, thanks to their importance in polyolefin synthesis. Hence, their application in catalytic asymmetric hydroamination reactions is highly desirable. Group 4 metal complexes are commonly less sensitive and easier to prepare than rare earth metal complexes. Most important of all, many potential precatalysts or catalyst precursors are commercially available.

Group 4 metal-based catalysts have been studied intensively in hydroamination reactions involving alkynes and allenes [77–81], but (achiral) hydroamination reactions involving aminoalkenes were only recently reported [82–84]. The reactivity of these catalysts is significantly lower than that of rare earth, alkali, and alkaline earth metal-based catalysts. In most instances, *gem*-dialkyl activation [37] of the aminoalkene substrate is required for catalytic turnover.

The first chiral group 4 metal catalyst system for asymmetric hydroamination/ cyclization of aminoalkenes was based on the cationic aminophenolate complex (*S*)-**45** [85]. Secondary aminoalkenes reacted readily to yield hydroamination products with enantioselectivities of up to 82% ee (Scheme 11.14). For catalyst solubility reasons, reactions were commonly performed at 100 °C in bromobenzene using

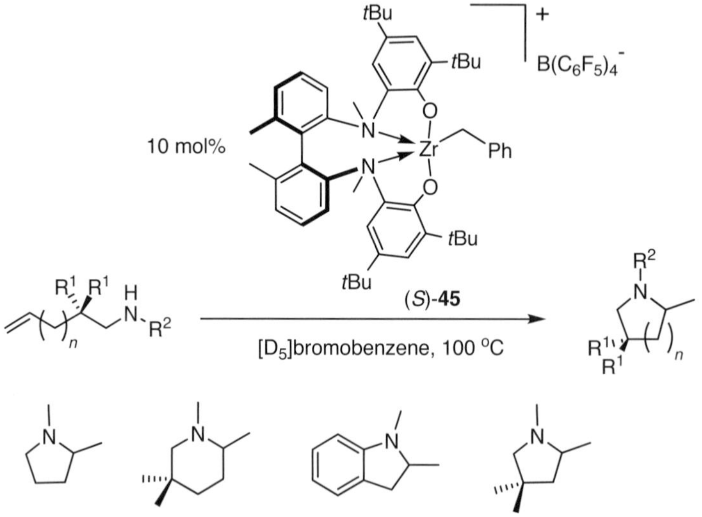

36: 100% (4 h); **46**: 100% (3 h); **47**: 100% (3 h); **48**: 70% (48 h @ 70 °C);
64% ee   82% ee   20% ee   14% ee
                                                       30% double bond
                                                       isomerization of substrate

Scheme 11.14 Hydroamination/cyclization of secondary aminoalkenes using a cationic chiral zirconium catalyst system [85].

10 mol% catalyst loading. The mechanism of this cationic system is thought to proceed similar to the σ-bond metathesis mechanism of rare earth metal-based catalyst systems (Scheme 11.6). Primary aminoalkenes did not react, which is thought to be caused by facile α-deptrotonation of the catalytic active cation metal-amido species leading to an unreactive metal-imido species [82, 85, 86]. The cationic catalyst systems are also prone to double bond isomerization via C−H activation that can significantly reduce product enantioselectivity and yield [82, 85, 87].

In contrast to the cationic group 4 metal hydroamination catalysts, their neutral counterparts will react only with primary aminoalkenes. For example, the chiral bis (phosphinic amido) zirconium complex (*R*,*R*)-**49** (Scheme 11.15) was found to be superior in reactivity and enantioselectivity for the cyclization of primary aminoalkenes in comparison to a wide range of diamido, diolate, and aminoalcoholate titanium, zirconium, and hafnium complexes [88]. Cyclization of aminopentenes proceeded with enantioselectivities as high as 80% ee, but formation of six-membered rings, for example, **52**, was somewhat less selective. Unfortunately, preliminary mechanistic studies indicate that this catalyst system undergoes slow ligand

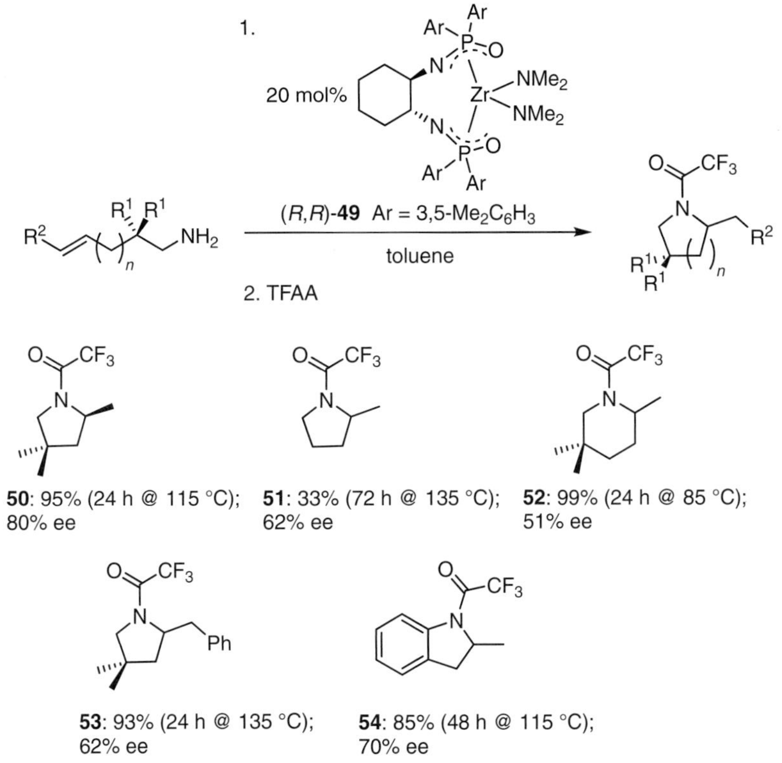
1.
20 mol%
Ar
P=O
N
Zr
NMe2
NMe2
(R,R)-49 Ar = 3,5-Me2C6H3
toluene
2. TFAA
R1 R1
R2
NH2
n
O
CF3
N
R2
R1
R1
n
50: 95% (24 h @ 115 °C); 80% ee
51: 33% (72 h @ 135 °C); 62% ee
52: 99% (24 h @ 85 °C); 51% ee
Ph
53: 93% (24 h @ 135 °C); 62% ee
54: 85% (48 h @ 115 °C); 70% ee

Scheme 11.15 Hydroamination/cyclization of primary aminoalkenes using a neutral bis (phosphinic amido) zirconium catalyst system [88].

redistribution reactions, leading to chiral catalytically inactive as well as achiral catalytically active species.

Enantioselectivities of up to 93% ee may be achieved using chiral bis(amidate) zirconium complex (*S*)-**55** (Scheme 11.16), first introduced by Schafer [89] and soon after by Scott [90].

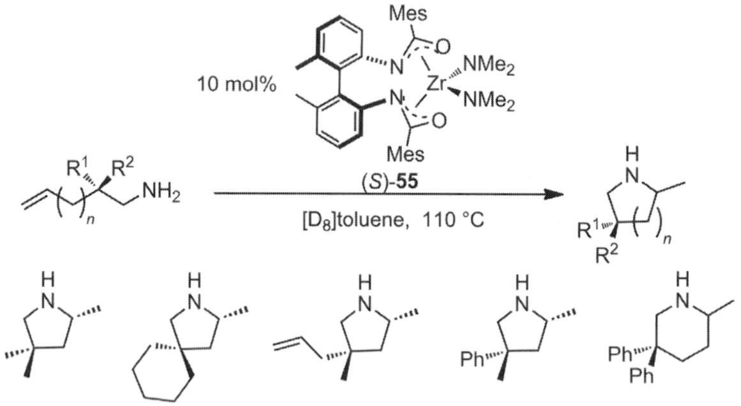

20: 80% (3 h); 93% ee
34: 96% (3 h); 82% ee
40: 82% (2 h); dr 2.3:1 92 : 88% ee
56: 94% (12 h); dr 1.9:1 62 : 86% ee
57: 91%; 29% ee

Scheme 11.16 Hydroamination/cyclization of primary aminoalkenes using a neutral bis(amidate) zirconium catalyst system [89].

The mechanism of alkene hydroamination is much less well understood than the mechanism of alkyne and allene hydroamination [91–98]. Based on the observation that most neutral group 4 metal alkene hydroamination catalysts are unreactive toward secondary aminoalkene substrates, a mechanism analogous to that for alkyne and allene hydroamination involving metal imido species as catalytically active species has been proposed (Scheme 11.17) [83, 84, 88, 89]. The metal imido species is generated via reversible α-elimination of an amine from a bis(amido) precursor. It can undergo a (most likely reversible) [93, 95] [2 + 2]-cycloaddition with the alkene moiety subsequently. The resulting azametallacyclobutane is protolytically cleaved to regenerate the metal imido species and release the hydroamination product. However, a few (achiral) neutral group 4 metal catalyst systems have been reported to catalyze cyclization of secondary aminoalkenes and it has been suggested that a lanthanide-like σ-bond metathesis mechanism (Scheme 11.6) is operational in these cases [99, 100].

11.2.2.4 **Organocatalytic Asymmetric Hydroamination**

In the last decade, the field of asymmetric organocatalysis has seen significant progress. In particular, the application of Brønsted acids in metal-free enantioselective catalysis is rapidly increasing [101–103]. Several research groups have recently demonstrated that strong Brønsted acids can be used in both *intra-* [104–106] and

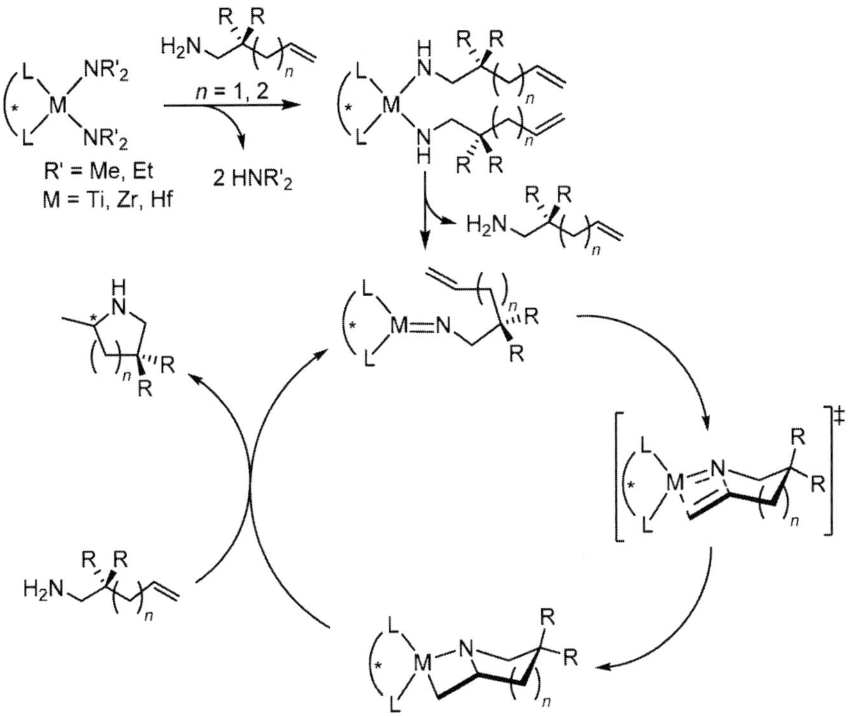
R' = Me, Et
M = Ti, Zr, Hf
n = 1, 2
2 HNR'2

Scheme 11.17 Proposed mechanism for the hydroamination/cyclization of primary aminoalkenes involving group 4 metal imido species.

*inter*molecular [107–110] hydroamination reactions, although protection of the amino group is often necessary.

The first and so far only example of catalytic asymmetric metal-free hydroamination using the chiral phosphoric acid diester (*R*)-**58** was recently disclosed (Eq. 11.8) [111]. Although both selectivity and catalytic activity for the hydroamination of the secondary aminopentene **59** are rather low, the method itself remains very promising.

Equation 11.8. Organocatalytic asymmetric hydroamination/cyclization of a secondary aminoalkene [111].

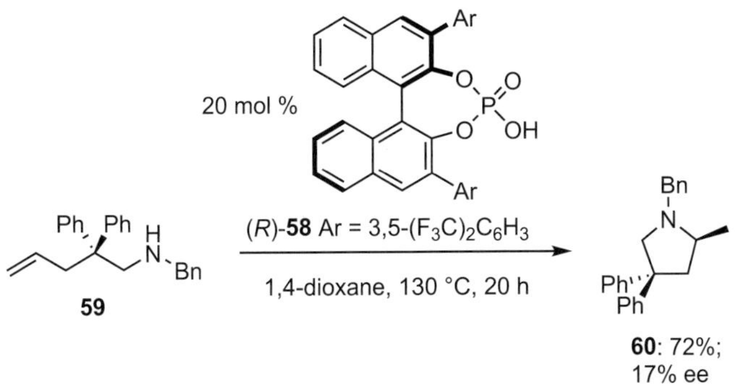
20 mol %
(R)-58 Ar = 3,5-(F3C)2C6H3
1,4-dioxane, 130 °C, 20 h
59
60: 72%;
17% ee

(11.8)

11.3
Hydroamination of Dienes, Allenes, and Alkynes

11.3.1
Intermolecular Hydroaminations

While intermolecular hydroamination of *simple* alkenes remains a great challenge as of now, addition of amines to alkynes, allenes, and dienes proceeds more easily. However, the addition of an amine to a polyene moiety may not necessary result in the formation of a new stereocenter, if an imine is generated (Scheme 11.18).

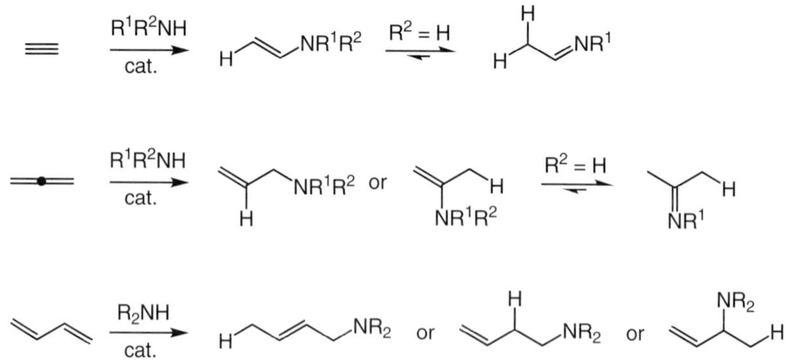

Scheme 11.18 Intermolecular hydroamination of alkynes, allenes, and dienes.

The palladium-catalyzed asymmetric hydroamination of cyclohexadiene with arylamines utilizing a variant of Trost's ligand (R,R)-**61** proceeds with high enantio-selectivities under mild conditions (Scheme 11.19) [21]. The mechanism is believed to follow a similar pathway as proposed for palladium-catalyzed hydroamination of vinyl arenes (Scheme 11.2) [17].

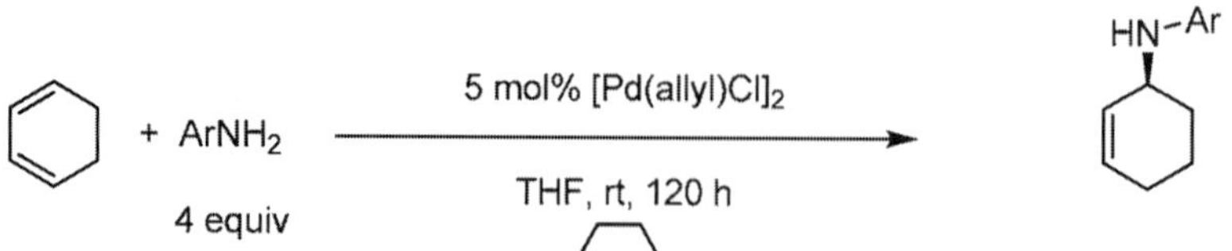

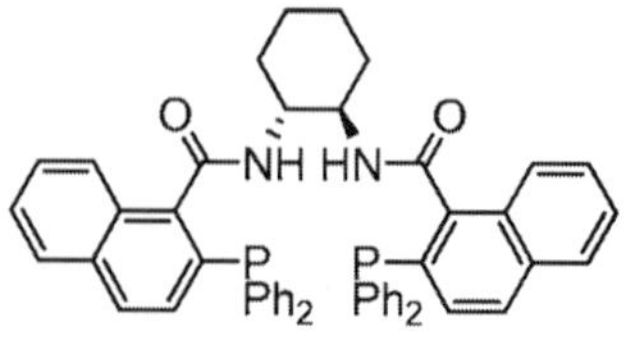

Ar	% yield	% ee
Ph	63	92
4-MeC$_6$H$_4$	78	86
2-MeC$_6$H$_4$	59	90
4-EtO$_2$CC$_6$H$_4$	83	95
4-CF$_3$C$_6$H$_4$	73	95

Scheme 11.19 Palladium-catalyzed asymmetric hydroamination of cyclohexadiene [21].

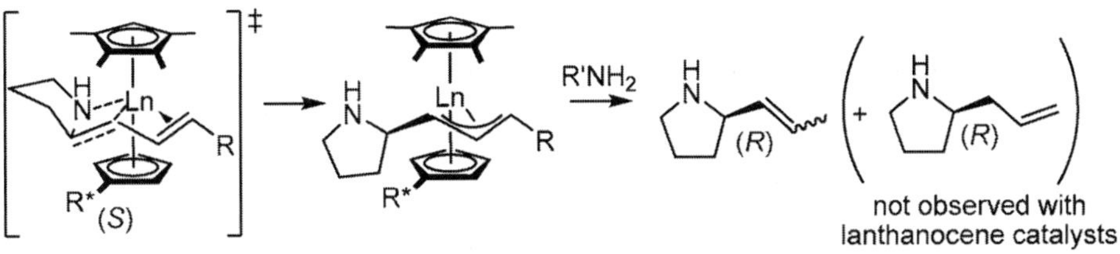

Scheme 11.20 Stereomodel for the lanthanocene-catalyzed hydroamination/cyclization of aminodienes. The silicon linker bridging the two cyclopentadienyl ligands was omitted for the sake of clarity.

11.3.2
Intramolecular Reactions

Hydroamination of 1,3-dienes is quite facile due to the transient formation of an η^3-allyl intermediate, which forms E/Z vinylpyrrolidines and vinylpiperidines upon protonation, and, under certain conditions, also allyl isomers (Scheme 11.20). Cyclizations with chiral lanthanocenes generally produced the E olefins with high E selectivity ($E/Z \geq 93:7$) [112, 113]. The reaction rates are higher for the aminodienes compared to the corresponding aminoalkenes, despite increased steric encumbrance of the cyclization transition state (Scheme 11.20). However, in most cases, the increased reactivity is at the expense of enantioselectivity. The aminooctadiene **62** is an exception with 63% ee observed in a benzene solution at 25 °C (71% ee in methylcyclohexane at 0 °C) using (S)-**25-Sm**, which gave facile access to (+)-coniine **64** after hydrogenolysis of the Cbz-protected vinylpiperidine **63** (Scheme 11.21) [113].

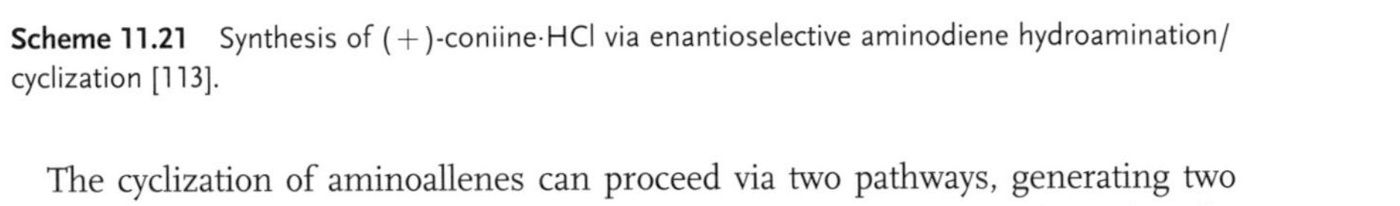

Scheme 11.21 Synthesis of (+)-coniine·HCl via enantioselective aminodiene hydroamination/cyclization [113].

The cyclization of aminoallenes can proceed via two pathways, generating two regioisomeric products (Scheme 11.22). Formation of an imine via the endocyclic pathway **a** usually predominates for monosubstituted aminoallenes, whereas vinylpyrroldines (pathway **b**) are generated preferentially when 1,3-disubstituted or trisubstituted aminoallenes are employed. Hence, application of the presumably dimeric chiral amino-alcoholate titanium complex **65** in the cyclization of aminoallene **66** produced the desired vinylpyrroldine **67**, although the enantioselectivities were disappointing [114].

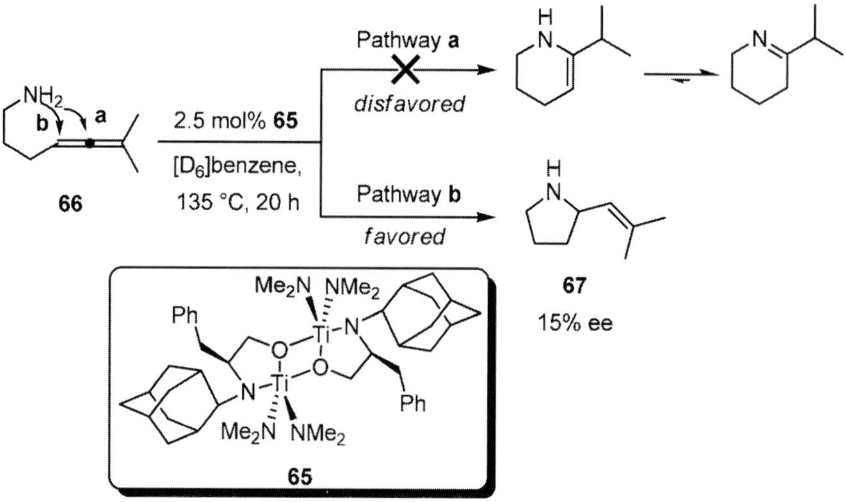

Scheme 11.22 Titanium-catalyzed asymmetric hydroamination/cyclization of an aminoallene.

Intermolecular hydroamination of alkynes, which is a process with a relatively low activation barrier, has not been used for the synthesis of chiral amines, since the achiral Schiff base is a major reaction product. However, protected aminoalkynes may undergo an interesting intramolecular allylic cyclization using a palladium catalyst with a chiral norbornene-based diphosphine ligand (Eq. 11.9) [115]. Unfortunately, significantly higher catalyst loadings were required to achieve better enantioselectivities of up to 91% ee.

Equation 11.9. Palladium-catalyzed asymmetric intramolecular hydroamination of aminoalkynes [115].

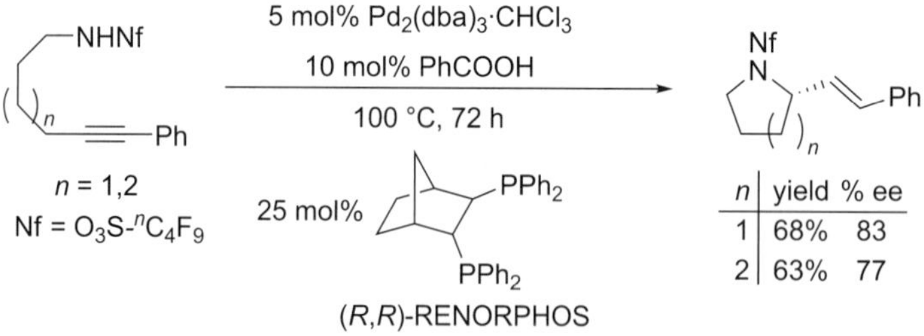

n	yield	% ee
1	68%	83
2	63%	77

(11.9)

The highly stereoselective cyclization of protected aminoallenes was recently demonstrated utilizing a dinuclear gold(I)–phosphine complex with a coordinating *p*-nitrobenzoate (OPNB) counteranion (Scheme 11.23) [116]. The most enantioselective catalytic active species is believed to be monocationic with one of the counteranions remaining coordinated to the dinuclear gold complex. Similarly, carbamate-protected aminoallenes have been cyclized using (S)-1/AgClO$_4$ with up to 81% ee [117].

The unique counteranion effect was also crucial in analogous catalytic reactions of allenylsulfonamides using an achiral gold(I) precatalyst in the presence of the

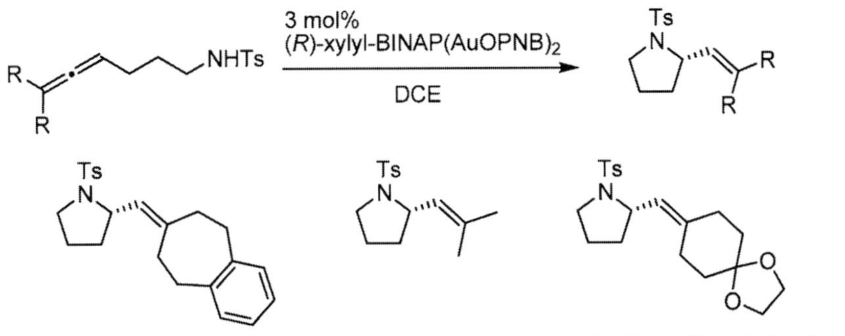

Scheme 11.23 Gold-catalyzed asymmetric hydroamination/cyclization of sulfonyl-protected aminoallenes [116].

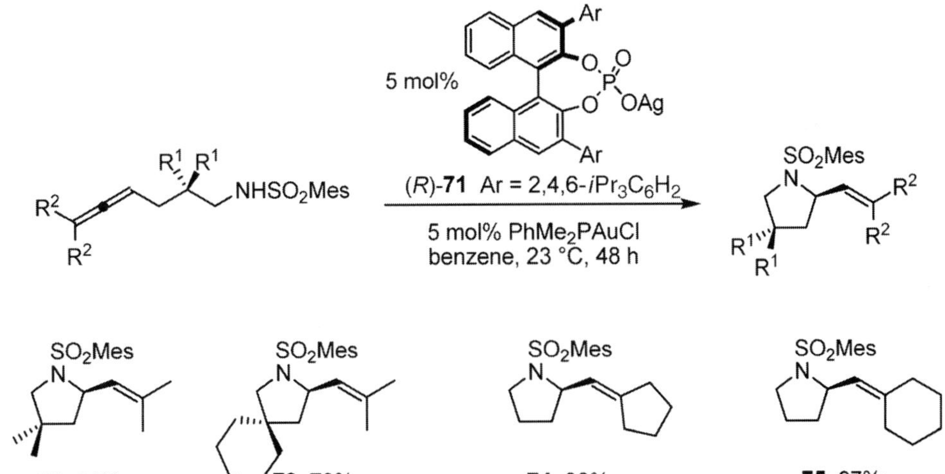

68: 80% (25 h @ 23 °C); 98% ee
69: 98% (15 h @ 23 °C); 99% ee
70: 79% (25 h @ 50 °C); 98% ee

72: 84%; 99% ee
73: 73%; 98% ee
74: 88%; 98% ee
75: 97%; 96% ee

Scheme 11.24 Chiral counteranion effect on the gold-catalyzed hydroamination/cyclization of aminoallenes [118].

chiral counteranion (R)-71, which proceeded with excellent enantioselectivities (Scheme 11.24) [118].

11.4
Hydroamination with Enantiomerical Pure Amines

11.4.1
Hydroaminations Using Achiral Catalysts

The diastereoselective cyclization of chiral aminoalkenes, such as α-substituted hex-5-enylamines, can be utilized in the synthesis of piperidine-based alkaloids, such as

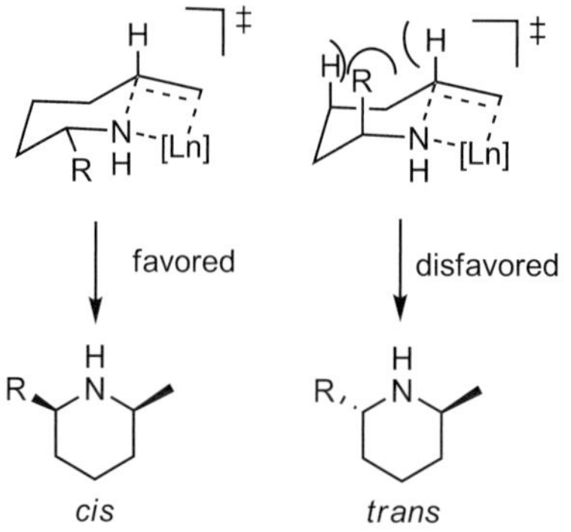

Figure 11.3 Proposed cyclization transition states for the preferred formation of cis-2,6-disubstituted piperidines.

(−)-pinidinol (**76**) (Eq. 11.10) [119]. Depending on the catalyst structure, the cyclization proceeds with excellent cis-diastereoselectivity, which can be explained with unfavorable 1,3-diaxial interactions in the chair-like cyclization transition state leading to the trans-product (Figure 11.3).

Equation 11.10. Diastereoselective synthesis of (−)-pinidinol [119].

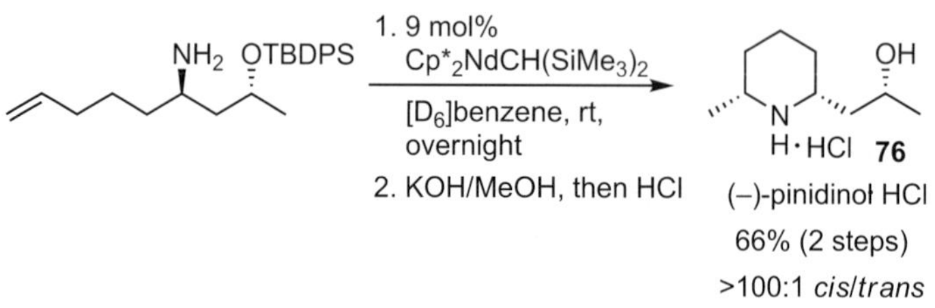

$$\text{1. 9 mol\% Cp*}_2\text{NdCH(SiMe}_3)_2, \text{[D}_6\text{]benzene, rt, overnight} \quad \text{2. KOH/MeOH, then HCl}$$

H·HCl **76**
(−)-pinidinol HCl
66% (2 steps)
>100:1 cis/trans

(11.10)

Also the hydroamination/cyclization of chiral aminoallenes has been utilized in the synthesis of various alkaloid skeletons [120]. The pyrrolidine alkaloid (+)-197B (Scheme 11.25), as well as the indolizidine alkaloid (+)-xenovenine (Scheme 11.26),

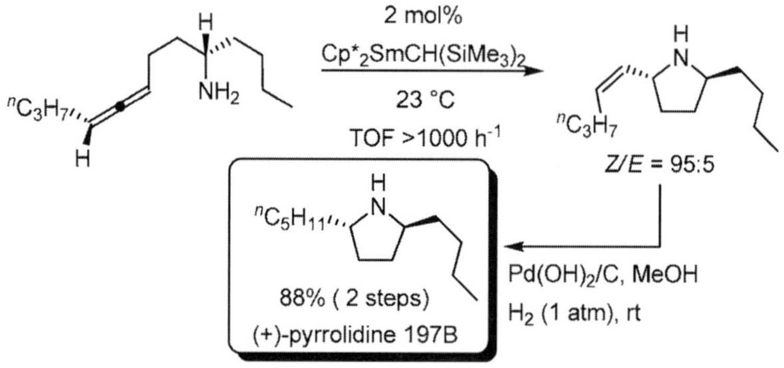

2 mol% Cp*$_2$SmCH(SiMe$_3$)$_2$
23 °C
TOF >1000 h^{-1}

Z/E = 95:5

Pd(OH)$_2$/C, MeOH
H$_2$ (1 atm), rt

88% (2 steps)
(+)-pyrrolidine 197B

Scheme 11.25 Diastereoselective synthesis of pyrrolidine (+)-197B via hydroamination/cyclization of an aminoallene [120].

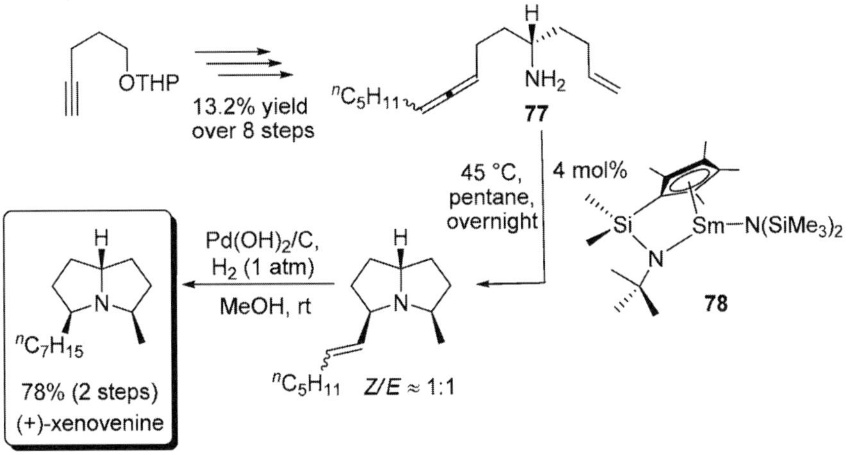

11.4 Hydroamination with Enantiomerical Pure Amines

Scheme 11.26 Diastereoselective synthesis of (+)-xenovenine via hydroamination/bicyclization of the aminoallene–alkene **77** [120].

was prepared using hydroamination/cyclization reaction as a key step. Reaction of the aminoallene–alkene **77** containing an allene and a terminal alkene moiety positioned at equal distance to the amino group requires the sterically open constrained geometry catalyst Me$_2$Si(C$_5$Me$_4$)(tBuN)SmN(SiMe$_3$)$_2$ (**78**) to achieve facile and stereospecific bicyclization, while sterically more encumbered lanthanocene catalysts react selectively at the allene moiety and leave the alkene moiety untouched.

Organolithium-mediated cycloisomerization of *chiral* aminoalkenes may proceed with high diastereoselectivities when favored by the substrate geometry. The last step of the enantioselective synthesis of (−)-codeine (**80**), reported by Trost and Tang [121], consisted of an intramolecular hydroamination. Treatment of amine **79** with LDA or nBuLi in refluxing THF left the starting material unreacted. However, irradiation of the amine/LDA solution with a 150 W tungsten lamp led to diastereoselective cycloisomerization to form (−)-codeine (**80**) (Eq. 11.11).

Equation 11.11. Synthesis of (−)-codeine (**80**) via base-catalyzed hydroamination/cyclization [121].

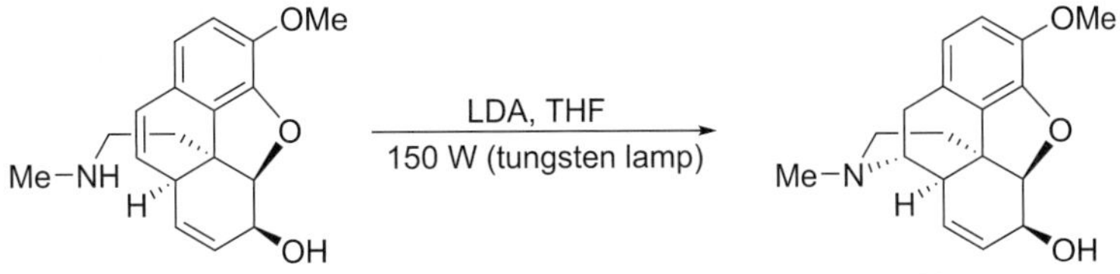

While no catalytic intermolecular asymmetric hydroamination of allenes has yet been reported, the addition of aniline to a chiral allene has been demonstrated to proceed with significant axial chirality transfer in the presence of gold salts (Eq. 11.12) [122].

Equation 11.12. Stereoselective gold-catalyzed intermolecular hydroamination of a chiral allene [122].

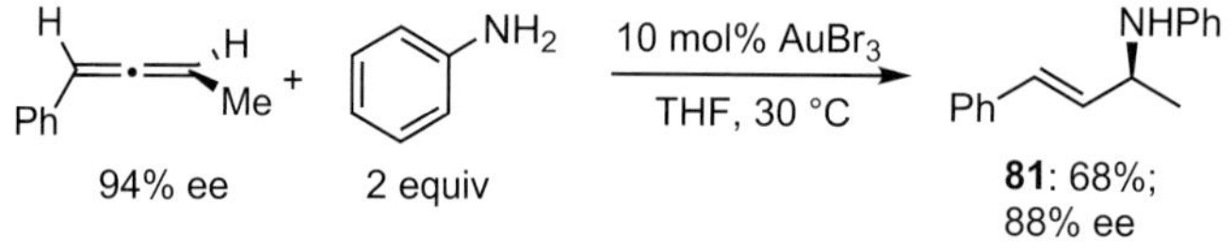

$$\underset{\substack{Ph \\ 94\% \text{ ee}}}{\overset{H}{\underset{Me}{\diagdown}}C=C=C\overset{H}{\diagdown}} + \underset{2 \text{ equiv}}{PhNH_2} \xrightarrow[\text{THF, 30 °C}]{10 \text{ mol\% AuBr}_3} \underset{\substack{81: 68\%; \\ 88\% \text{ ee}}}{Ph\diagdown\diagup\overset{NHPh}{\diagup}} \quad (11.12)$$

Hydroamination reactions involving alkynes and enantiomerically pure chiral amines can produce novel chiral amine moieties after single-pot reduction of the Schiff base intermediate **82** (Scheme 11.27) [123]. Unfortunately, partial racemization of the amine stereocenter was observed with many titanium-based hydroamination catalysts, even in the absence of an alkyne substrate. No racemization was observed when the sterically hindered Cp*$_2$TiMe$_2$ or the constrained geometry catalyst Me$_2$Si(C$_5$Me$_4$)(tBuN)Ti(NMe$_2$)$_2$ was used in the catalytic reaction. Also, the addition of pyridine suppressed the racemization mostly.

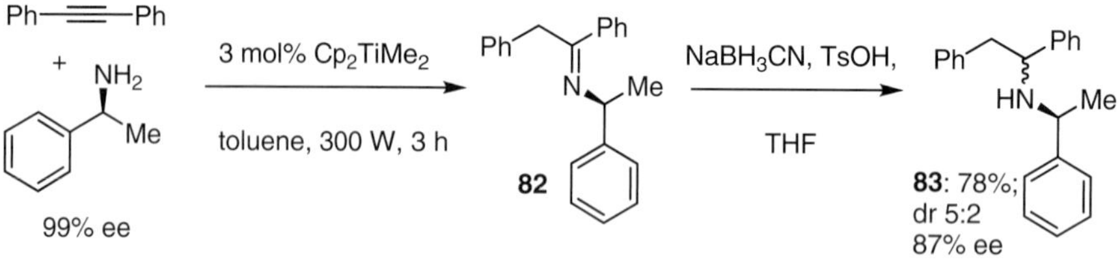

Scheme 11.27 Partial racemization observed in the titanium-catalyzed hydroamination of tolane with 1-phenylethylamine [123].

11.4.2
Kinetic Resolution of Chiral Aminoalkenes

The efficient kinetic resolution of chiral aminoalkenes has been achieved utilizing the binaphtholate complexes (*R*)-**38-Ln** (Table 11.3) [52, 124]. Various chiral aminopentenes were kinetically resolved with resolution factors f (defined as $f = K^{dias} \times k_{fast}/k_{slow}$, where K^{dias} is the Curtin–Hammett equilibrium constant between the two diastereomeric substrate/catalyst complexes and k_{fast}/k_{slow} being the ratio between the faster and the slower reaction rate constant) as high as 19 and enantiomeric excess for recovered starting material reaching ≥80% ee at conversions close to 50%. The 2,5-disubstituted pyrrolidines were obtained in good to excellent *trans*-diastereoselectivity, depending on the steric hindrance of the α-substituent. Kinetic resolution of **84e** using 1 mol% (*R*)-**38a-Lu** gave enantiopure (*S*)-**84e** (≥99% ee) in 33% reisolated yield at 64% conversion [52].

The preferred formation of (2*S*,5*S*)-**85a** using (*R*)-binaphtholate complexes can be attributed to impeded cyclization of (*R*)-**84a** due to unfavorable steric interactions of the vinylic methylene protons with a trisarylsilyl substituent in the chair-like transition state (Figure 11.4). Fast exchange between matching and mismatching aminoalkenes prior to cyclization is imperative for an effective kinetic resolution

Table 11.3 Catalytic kinetic resolution of chiral aminopentenes [52, 124].

84a R = Me   84e R = Ph
84b R = Et   84f R = 4-ClC$_6$H$_4$
84c R = iPr 84g R = 4-MeOC$_6$H$_4$
84d R = CH$_2$Ph

Substrate	Catalyst	t (h)	Conversion (%)	trans: cis	% ee of recovered starting material	f
84a	(R)-38b-Y	26	52	13: 1	80	16
84b	(R)-38b-Y	6	51	20: 1	57	5.9
84c	(R)-38a-Y	6	50.5	18: 1	37	3.0
84d	(R)-38a-Y	9	50	20: 1	42	3.6
84e	(R)-38a-Lu	15[a]	52	≥50: 1	83	19
84f	(R)-38a-Y	18[a]	50	≥50: 1	71	12
84g	(R)-38a-Y	8[a]	50	≥50: 1	78	19

a) At 40 °C.

process. Kinetic analysis of the kinetic resolution process has revealed that the Curtin–Hammett equilibrium favors the matching substrate/catalyst combination in aminopentene substrates **84e–84g** containing α-aryl substituents [52], whereas aliphatic substituents in **84a–84d** shift this equilibrium in favor of the mismatching substrate/catalyst combination [125].

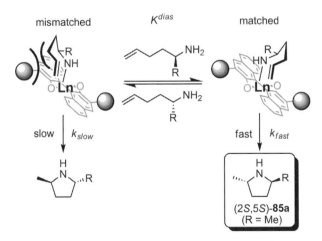

Figure 11.4 Proposed stereomodel for kinetic resolution of chiral aminopentenes with an equatorial approach of the olefin.

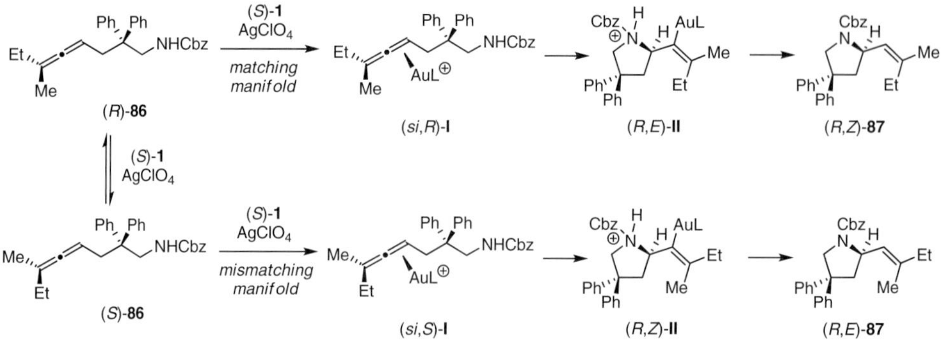

Scheme 11.28 Proposed mechanism for the gold-catalyzed dynamic kinetic resolution of aminoallenes.

Cationic gold(I) complexes are well known to racemize allenes [126], which can be exploited in the facile dynamic kinetic resolution of axially chiral N-(γ-allenyl) carbamates with trisubstituted allenyl groups [127]. A mixture of the dinuclear gold(I) phosphine complex (S)-1 and AgClO$_4$ catalyzed the cyclization of the Cbz-protected aminoallene **86** to yield predominantly the (Z)-vinylpyrrolidine (R,Z)-**87** with excellent enantioselectivity (Eq. 11.13). Initial mechanistic studies suggest that the (E)-vinylpyrrolidine (R,E)-**87** is formed in the mismatched reaction manifold (Scheme 11.28). In the matching manifold, the cationic gold(I) species is believed to coordinate preferentially to the *si* face of the allene substrate ((*si*,R)-**I**). Subsequent attack of the tethered nucleophilic carbamate nitrogen leads to the σ-alkenyl-gold complex (R,E)-**II**. Finally, protonolysis of the Au—C bond proceeds with retention of configuration, releasing the (R,Z)-vinylpyrrolidine product.

Equation 11.13. Gold-catalyzed dynamic kinetic resolution of aminoallenes [127].

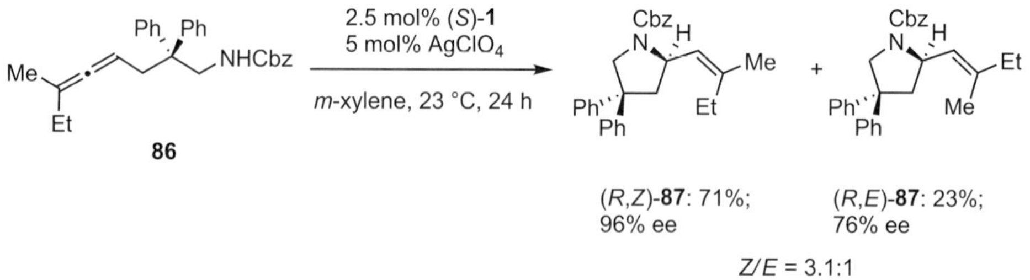

(R,Z)-**87**: 71%; 96% ee
(R,E)-**87**: 23%; 76% ee
Z/E = 3.1:1

(11.13)

11.5
Synthesis of Chiral Amines via Tandem Hydroamination/Hydrosilylation

One-pot multistep reaction sequences [128–132] are an efficient method to introduce additional complexity and versatility to the hydroamination reaction. Imines, the products of alkyne hydroamination, are a good target for further functionalization, in particular, as they tend to be prone to hydrolysis. While the hydroamination of alkynes does not generate a new chiral center, the hydrosilylation of the imine moiety can yield a chiral amine. An intramolecular hydroamination/asymmetric hydrosilylation sequence was accomplished using the chiral catalysts (S,S)-(ebthi)TiMe$_2$ (ebthi = ethylenebis(tetrahydroindenyl)). The aminoalkynes **88a** and **88b** underwent efficient hydroamination (>95% conv.), followed by enantioselective hydrosilylation in up to 66% ee and moderate overall yields (Eq. 11.14) [133].

Equation 11.14. Titanium-catalyzed tandem hydroamination/hydrosilylation of aminoalkynes [133].

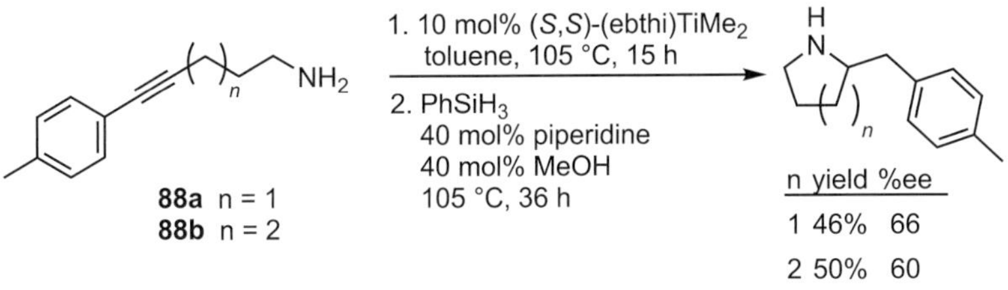

n	yield	%ee
1	46%	66
2	50%	60

(11.14)

11.6
Conclusions

As already mentioned, there has been significant progress in the development of chiral catalysts for asymmetric hydroamination reactions over the last decade. However, significant challenges remain, such as asymmetric intermolecular hydroaminations of simple nonactivated alkenes and the development of a chiral catalyst, which is applicable to a wide variety of substrates with consistent high stereochemical induction and tolerance of a multitude of functional groups as well as air and moisture. Certainly, late transition metal-based catalysts show promising leads that could fill this void, but to date, early transition metal-based catalysts (in particular, rare earth metals) remain the most active and most versatile catalyst systems.

11.7
Experimental Section

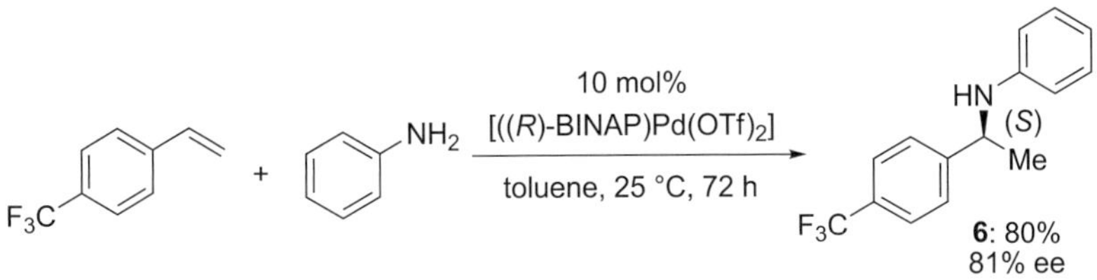

6: 80%
81% ee

(S)-N-phenyl-N-[1-{4-(trifluoromethyl)phenyl}ethyl]amine (6) [13]. [{(R)-BINAP}Pd(OTf)$_2$] (103 mg, 0.10 mmol) was suspended in toluene (0.5 ml) in the glove box. The suspension was placed into a vial, which was sealed with a cap containing a PTFE septum, and the vial was removed from the glove box. 4-(Trifluoromethyl)styrene (258 mg, 1.50 mmol) and aniline (93 mg, 1.00 mmol) were added to the reaction mixture by syringe. The reaction mixture was stirred at 25 °C for 3 days. The reaction mixture was adsorbed on silica gel and isolated by eluting with 10% EtOAc/hexanes to give **6** (212 mg, 80%). The enantiomeric purity was determined to be 81% ee (S) by capillary GLC analysis with a chiral stationary phase column (β-cyclodextrine permethylated). t_R, min 48.7 (S), 49.4 (R); ^{1}H NMR (400 MHz, CDCl$_3$): δ 7.49 (d, $J_{H,H}$ = 8.0 Hz, 2H), 7.40 (d, $J_{H,H}$ = 8.0 Hz, 2H), 7.05–6.99 (m, 2H), 6.59 (t, $J_{H,H}$ = 7.6 Hz, 1H), 6.39 (d, $J_{H,H}$ = 7.6 Hz, 2H), 4.44 (q, $J_{H,H}$ = 6.8 Hz, 1H), 4.05–3.90 (br s, 1H, NH), 1.44 (d, $J_{H,H}$ = 6.8 Hz, 3H); ^{13}C{^{1}H} NMR (100 MHz, CDCl$_3$): δ 148.7, 146.1, 128.5, 125.4, 124.9 (q, $J_{C,F}$ = 4.2 Hz), 124.8, 122.2, 116.9, 112.5, 52.6, 24.4.

S-(+)-1-Phenylpent-4-enylamine ((S)-84e) [52]. In the glove box, a 25 ml flask was charged with rac-1-phenylpent-4-enylamine (**84e**) (1.600 g, 9.92 mmol), benzene (7 ml), and (R)-**38b-Lu** (160 mg, 0.128 mmol, 1.3 mol%). The reaction mixture was kept at 45 °C. To monitor the conversion, small aliquots (50 μl) were taken and transferred into a NMR tube, diluted with CDCl$_3$ to 0.5 ml volume and a ^{1}H NMR spectrum was recorded. The conversion was 45% after 18.5 h and 64% after 27 h. The reaction flask was removed from the glove box and the reaction mixture was treated with acetic acid (320 μl, 330 mg, 5.50 mmol) followed by addition of benzaldehyde (640 mg, 6.03 mmol). The mixture was kept at room temperature for 2 h and was then transferred into a separatory funnel. Water (20 ml), benzene (10 ml), and hexanes (15 ml) were added. The funnel was shaken vigorously and the layers were separated. The organic layer was extracted with 1% (0.2 M) acetic acid (10 ml). The hexanes/benzene layer containing the benzaldimine was evaporated and the residue was treated with 2 M HCl (20 ml) and Et$_2$O (30 ml) and the mixture was stirred for 24 h at room temperature. The two layers were separated and the organic layer was extracted with water (10 ml). The combined aqueous layers were evaporated. The residue was treated with sodium hydroxide solution followed by extraction with Et$_2$O. The ether solution was dried (Na$_2$SO$_4$) and evaporated. The residue was distilled in vacuo to give the enantioenriched starting material (530 mg, 33%, 99%ee) in analytically pure form; ^{1}H NMR (400 MHz, CDCl$_3$): δ 7.31 (m, 4H, C$_6$H$_5$), 7.23 (m, 1H, C$_6$H$_5$), 5.80 (m, 1H, =CH), 5.03–4.93 (m, 2H, =CH$_2$), 3.89 (t, $J_{H,H}$ = 6.9 Hz, 1H, PhCHNH$_2$), 2.04 (m, 2H, CH_2CH=CH$_2$), 1.75 (m, 2H, CH_2CHNH$_2$), 1.45 (br s, 2H, NH$_2$); ^{13}C{^{1}H} NMR (100.6 MHz, CDCl$_3$): δ 146.4, 138.2, 128.4, 126.9, 126.3, 114.7, 55.6, 38.6, 30.7. (R)-Mosher adduct: ^{19}F NMR (CDCl$_3$, 60 °C) δ −69.3 (R), -69.4 (S).

2-(Cyclohexylidenemethyl)-1-[(4-methylphenyl)sulfonyl]pyrrolidine [116]. To a solution of N-(5-cyclohexylidenepent-4-enyl)-4-methylbenzenesulfonamide (50 mg, 0.157 mmol) in DCE (0.5 ml) was added (R)-xylyl-BINAP(AuOPNB)$_2$ [116] (6.9 mg, 4.7 μmol). The resulting homogeneous mixture was protected from ambient light and left to stir at 23 °C. Upon completion, as judged by TLC analysis of the reaction mixture, the solution was purified by flash column chromatography (hexanes/EtOAc 12: 1) to afford the desired pyrrolidine as a colorless oil (44 mg, 88%). R_f = 0.43 (3: 1 hexanes/EtOAc). ^{1}H NMR (400 MHz, CDCl$_3$): δ 7.68 (d, $J_{H,H}$ = 8.4 Hz, 2H), 7.26 (d, $J_{H,H}$ = 8.0 Hz, 2H), 4.99 (d, $J_{H,H}$ = 9.2 Hz, 1H), 4.39–4.34 (m, 1H), 3.40–3.30 (m, 2H), 2.40 (s, 3H), 2.26–2.20 (m, 1H), 2.15–2.09 (m, 1H), 2.01–1.96 (m, 2H), 1.85–1.78 (m, 2H), 1.65–1.43 (m, 8H); ^{13}C{^{1}H} NMR (100 MHz, CDCl$_3$): δ 143.0, 140.7, 136.0, 129.4, 127.5, 122.7, 57.2, 48.6, 36.9, 34.0, 29.1, 28.2, 27.5, 26.7, 24.2, 21.5. [α]$_D$ = −54° (c = 1.0, CHCl$_3$). HPLC (Chiralpak AD-H column (95: 5 hexanes:isopropanol, 1 ml/min) t_R 10.8 min (minor), 13.4 min (major): 98% ee.

Questions

11.1. The addition of water or alcohols to alkenes is readily catalyzed by Brønsted acids. The hydroamination reaction constitutes the analogous process using amines. Why do acids commonly not catalyze this process?

11.2. The addition of aniline to the enantioenriched allene (−)-**I** is known to proceed with excellent regio- and stereoselectivity to give **II** in good yield in presence of a robust gold catalyst:

The following mechanisms can be envisioned for this reaction

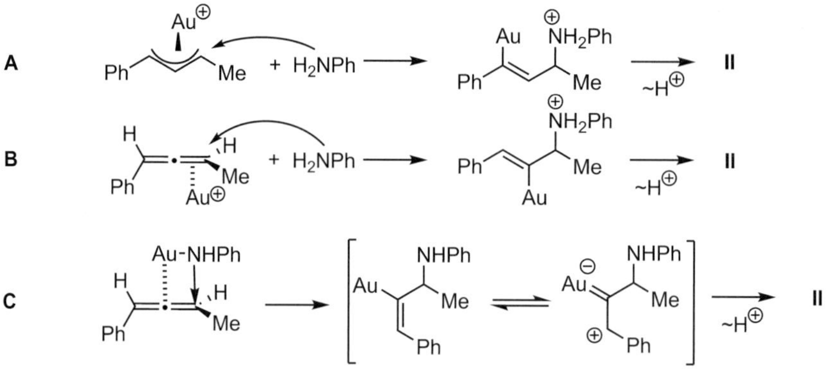

A [Au-activated alkene mechanism: Ph-CH=CH-Me with Au⁺ + H₂NPh → Ph-CH(NH₂Ph)-CH(Au)-Me → –H⁺ → **II**]

B [Au-vinyl cation / π-allyl mechanism with H₂NPh → intermediate → –H⁺ → **II**]

C [Au-amide insertion mechanism: Au–NHPh adds to alkene giving Au/NHPh intermediate in equilibrium with zwitterionic form → –H⁺ → **II**]

Select the mechanism most likely operating in this reaction.

11.3. The rare earth metal-catalyzed cyclization of the *primary* α-substituted aminopentene **III** proceeds with good to excellent *trans*-diastereoselectivity.

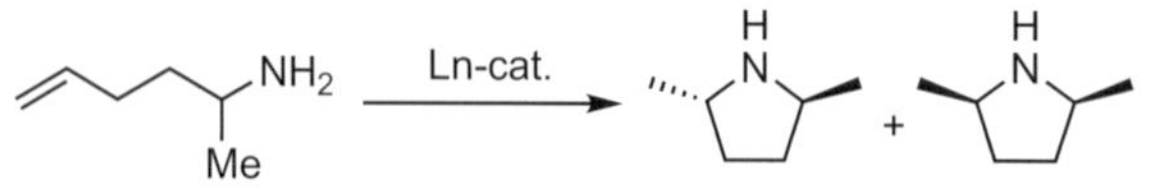

[Scheme: pentenylamine **III** (CH₂=CH-CH₂-CH₂-CH(Me)-NH₂) with Ln-cat. → *trans*-**IV** + *cis*-**IV** (2-methylpyrrolidines); *trans/cis* up to 20:1]

On the other hand, cyclization of the *secondary* α-substituted aminopentene **V** proceeds with moderate *cis*-diastereoselectivity. Provide stereomodels that accommodate the different stereoselectivities observed in the two cyclization reactions.

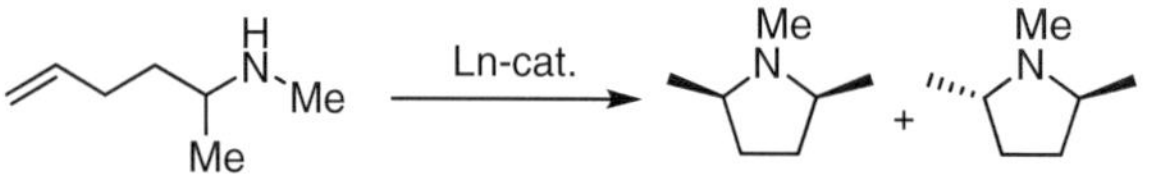

[Scheme: **V** (N-methyl analog) with Ln-cat. → *cis*-**VI** + *trans*-**VI** (N-methyl-2-methylpyrrolidines); *cis/trans* ≈ 3:1]

References

1 Müller, T.E. and Beller, M. (1998) *Chem. Rev.*, **98**, 675–703.
2 Müller, T.E., Hultzsch, K.C., Yus, M., Foubelo, F., and Tada, M. (2008) *Chem. Rev.*, **108**, 3795–3892.
3 Brunet, J.J. and Neibecker, D. (2001) in *Catalytic Heterofunctionalization from Hydroamination to Hydrozirconation* (eds A. Togni and H. Grützmacher), Wiley-VCH, Weinheim, pp. 91–141.
4 Roesky, P.W. and Müller, T.E. (2003) *Angew. Chem.*, **42**, 2708–2710.
5 Hultzsch, K.C. (2005) *Adv. Synth. Catal.*, **347**, 367–391.
6 Hultzsch, K.C. (2005) *Org. Biomol. Chem.*, **3**, 1819–1824.
7 Aillaud, I., Collin, J., Hannedouche, J., and Schulz, E. (2007) *Dalton Trans.*, 5105–5118.
8 Steinborn, D. and Taube, R. (1986) *Z. Chem.*, **26**, 349–359.
9 Taube, R. (2002) in *Applied Homogeneous Catalysis with Organometallic Compounds*, vol. 1, 2nd edn (eds B. Cornils and

W.A. Herrmann), Wiley-VCH, Weinheim, pp. 513–524.
10. Li-Wen Xu, C.-G.X. (2005) *Eur. J. Org. Chem.*, 633–639.
11. Hii, K.K. (2006) *Pure Appl. Chem.*, **78**, 341–349.
12. Zhang, Z., Lee, S.D., and Widenhoefer, R.A. (2009) *J. Am. Chem. Soc.*, **131**, 5372–5373.
13. Kawatsura, M. and Hartwig, J.F. (2000) *J. Am. Chem. Soc.*, **122**, 9546–9547.
14. Li, K., Horton, P.N., Hursthouse, M.B., and Hii, K.K. (2003) *J. Organomet. Chem.*, **665**, 250–257.
15. Nettekoven, U. and Hartwig, J.F. (2002) *J. Am. Chem. Soc.*, **124**, 1166–1167.
16. Utsunomiya, M. and Hartwig, J.F. (2003) *J. Am. Chem. Soc.*, **125**, 14286–14287.
17. Johns, A.M., Utsunomiya, M., Incarvito, C.D., and Hartwig, J.F. (2006) *J. Am. Chem. Soc.*, **128**, 1828–1839.
18. Besson, L., Goré, J., and Cazes, B. (1995) *Tetrahedron Lett.*, **36**, 3857–3860.
19. Al-Masum, M., Meguro, M., and Yamamoto, Y. (1997) *Tetrahedron Lett.*, **38**, 6071–6074.
20. Meguro, M. and Yamamoto, Y. (1998) *Tetrahedron Lett.*, **39**, 5421–5424.
21. Löber, O., Kawatsura, M., and Hartwig, J.F. (2001) *J. Am. Chem. Soc.*, **123**, 4366–4367.
22. Minami, T., Okamoto, H., Ikeda, S., Tanaka, R., Ozawa, F., and Yoshifuji, M. (2001) *Angew. Chem.*, **40**, 4501–4503.
23. Pawlas, J., Nakao, Y., Kawatsura, M., and Hartwig, J.F. (2002) *J. Am. Chem. Soc.*, **124**, 3669–3679.
24. Sakai, N., Ridder, A., and Hartwig, J.F. (2006) *J. Am. Chem. Soc.*, **128**, 8134–8135.
25. Landis, C.R. and Halpern, J. (1987) *J. Am. Chem. Soc.*, **109**, 1746–1754.
26. Dorta, R., Egli, P., Zürcher, F., and Togni, A. (1997) *J. Am. Chem. Soc.*, **119**, 10857–10858.
27. Casalnuovo, A.L., Calabrese, J.C., and Milstein, D. (1988) *J. Am. Chem. Soc.*, **110**, 6738–6744.
28. Beller, M., Breindl, C., Eichberger, M., Hartung, C.G., Seayad, J., Thiel, O.R., Tillack, A., and Trauthwein, H. (2002) *Synlett*, 1579–1594.
29. Zhou, J. and Hartwig, J.F. (2008) *J. Am. Chem. Soc.*, **130**, 12220–12221.
30. Hong, S. and Marks, T.J. (2004) *Acc. Chem. Res.*, **37**, 673–686.
31. Li, Y. and Marks, T.J. (1996) *Organometallics*, **15**, 3770–3772.
32. Ryu, J.-S., Li, G.Y., and Marks, T.J. (2003) *J. Am. Chem. Soc.*, **125**, 12584–12605.
33. Gagné, M.R., Stern, C.L., and Marks, T.J. (1992) *J. Am. Chem. Soc.*, **114**, 275–294.
34. Motta, A., Lanza, G., Fragalà, I.L., and Marks, T.J. (2004) *Organometallics*, **23**, 4097–4104.
35. Gagné, M.R., Brard, L., Conticello, V.P., Giardello, M.A., Stern, C.L., and Marks, T.J. (1992) *Organometallics*, **11**, 2003–2005.
36. Giardello, M.A., Conticello, V.P., Brard, L., Gagné, M.R., and Marks, T.J. (1994) *J. Am. Chem. Soc.*, **116**, 10241–10254.
37. Jung, M.E. and Piizzi, G. (2005) *Chem. Rev.*, **105**, 1735–1766.
38. Douglass, M.R., Ogasawara, M., Hong, S., Metz, M.V., and Marks, T.J. (2002) *Organometallics*, **21**, 283–292.
39. Ryu, J.-S., Marks, T.J., and McDonald, F.E. (2004) *J. Org. Chem.*, **69**, 1038–1052.
40. Vitanova, D.V., Hampel, F., and Hultzsch, K.C. (2007) *J. Organomet. Chem.*, **692**, 4690–4701.
41. Molander, G.A. and Dowdy, E.D. (1998) *J. Org. Chem.*, **63**, 8983–8988.
42. Molander, G.A. and Dowdy, E.D. (1999) *J. Org. Chem.*, **64**, 6515–6517.
43. Ryu, J.-S., Marks, T.J., and McDonald, F.E. (2001) *Org. Lett.*, **3**, 3091–3094.
44. Kim, Y.K. and Livinghouse, T. (2002) *Angew. Chem.*, **41**, 3645–3647.
45. Hultzsch, K.C., Gribkov, D.V., and Hampel, F. (2005) *J. Organomet. Chem.*, **690**, 4441–4452.
46. O'Shaughnessy, P.N. and Scott, P. (2003) *Tetrahedron: Asymmetry*, **14**, 1979–1983.
47. O'Shaughnessy, P.N., Knight, P.D., Morton, C., Gillespie, K.M., and Scott, P. (2003) *Chem. Commun.*, 1770–1771.
48. O'Shaughnessy, P.N., Gillespie, K.M., Knight, P.D., Munslow, I., and Scott, P. (2004) *Dalton Trans.*, 2251–2256.
49. Hong, S., Tian, S., Metz, M.V., and Marks, T.J. (2003) *J. Am. Chem. Soc.*, **125**, 14768–14783.
50. Gribkov, D.V., Hultzsch, K.C., and Hampel, F. (2003) *Chem. Eur. J.*, **9**, 4796–4810.
51. Gribkov, D.V., Hampel, F., and Hultzsch, K.C. (2004) *Eur. J. Inorg. Chem.*, 4091–4101.
52. Gribkov, D.V., Hultzsch, K.C., and Hampel, F. (2006) *J. Am. Chem. Soc.*, **128**, 3748–3759.

53 Collin, J., Daran, J.-D., Schulz, E., and Trifonov, A. (2003) *Chem. Commun.*, 3048–3049.
54 Collin, J., Daran, J.-D., Jacquet, O., Schulz, E., and Trifonov, A. (2005) *Chem. Eur. J.*, **11**, 3455–3462.
55 Riegert, D., Collin, J., Meddour, A., Schulz, E., and Trifonov, A. (2006) *J. Org. Chem.*, **71**, 2514–2517.
56 Riegert, D., Collin, J., Daran, J.-D., Fillebeen, T., Schulz, E., Lyubov, D., Fukin, G., and Trifonov, A. (2007) *Eur. J. Inorg. Chem.*, 1159–1168.
57 Aillaud, I., Wright, K., Collin, J., Schulz, E., and Mazaleyrat, J.-P. (2008) *Tetrahedron: Asymmetry*, **19**, 82–92.
58 Hannedouche, J., Aillaud, I., Collin, J., Schulz, E., and Trifonov, A. (2008) *Chem. Commun.*, 3552–3554.
59 Aillaud, I., Lyubov, D., Collin, J., Guillot, R., Hannedouche, J., Schulz, E., and Trifonov, A. (2008) *Organometallics*, **27**, 5929–5936.
60 Aillaud, I., Collin, J., Duhayon, C., Guillot, R., Lyubov, D., Schulz, E., and Trifonov, A. (2008) *Chem. Eur. J.*, **14**, 2189–2200.
61 Kim, J.Y. and Livinghouse, T. (2005) *Org. Lett.*, **7**, 1737–1739.
62 Kim, H., Kim, Y.K., Shim, J.H., Kim, M., Han, M., Livinghouse, T., and Lee, P.H. (2006) *Adv. Synth. Catal.*, **348**, 2609–2618.
63 Meyer, N., Zulys, A., and Roesky, P.W. (2006) *Organometallics*, **25**, 4179–4182.
64 Yu, X. and Marks, T.J. (2007) *Organometallics*, **26**, 365–376.
65 Heck, R., Schulz, E., Collin, J., and Carpentier, J.-F. (2007) *J. Mol. Cat. A*, **268**, 163–168.
66 Xiang, L., Wang, Q., Song, H., and Zi, G. (2007) *Organometallics*, **26**, 5323–5329.
67 Zi, G., Xiang, L., and Song, H. (2008) *Organometallics*, **27**, 1242–1246.
68 Wang, Q., Xiang, L., Song, H., and Zi, G. (2008) *Inorg. Chem.*, **47**, 4319–4328.
69 Shibasaki, M. and Yoshikawa, N. (2002) *Chem. Rev.*, **102**, 2187–2209.
70 Yamagiwa, N., Matsunaga, S., and Shibasaki, M. (2004) *Angew. Chem.*, **43**, 4493–4497.
71 Seayad, J., Tillack, A., Hartung, C.G., and Beller, M. (2002) *Adv. Synth. Catal.*, **344**, 795–813.
72 Horrillo Martínez, P., Hultzsch, K.C., and Hampel, F. (2006) *Chem. Commun.*, 2221–2223.
73 Ogata, T., Ujihara, A., Tsuchida, S., Shimizu, T., Kaneshige, A., and Tomioka, K. (2007) *Tetrahedron Lett.*, **48**, 6648–6650.
74 Buch, F. and Harder, S. (2008) *Z. Naturforsch.*, **63b**, 169–177.
75 Horrillo-Martínez, P. and Hultzsch, K.C. (2009) *Tetrahedron Lett.*, **50**, 2054–2056.
76 Horrillo-Martínez, P. (2008) New Catalysts for Base-Catalysed Hydroamination Reaction of Olefins, PhD thesis, Universität Erlangen-Nürnberg.
77 Pohlki, F. and Doye, S. (2003) *Chem. Soc. Rev.*, **32**, 104–114.
78 Bytschkov, I. and Doye, S. (2003) *Eur. J. Org. Chem.*, 935–946.
79 Doye, S. (2004) *Synlett*, 1653–1672.
80 Odom, A.L. (2005) *Dalton Trans.*, 225–233.
81 Severin, R. and Doye, S. (2007) *Chem. Soc. Rev.*, **36**, 1407–1420.
82 Gribkov, D.V. and Hultzsch, K.C. (2004) *Angew. Chem., Int. Ed.*, **44**, 5542–5546.
83 Bexrud, J.A., Beard, J.D., Leitch, D.C., and Schafer, L.L. (2005) *Org. Lett.*, **7**, 1959–1962.
84 Kim, H., Lee, P.H., and Livinghouse, T. (2005) *Chem. Commun.*, 5205–5207.
85 Knight, P.D., Munslow, I., O'Shaughnessy, P.N., and Scott, P. (2004) *Chem. Commun.*, 894–895.
86 Kissounko, D.A., Epshteyn, A., Fettinger, J.C., and Sita, L.R. (2006) *Organometallics*, **25**, 1076–1078.
87 Gribkov, D.V. (2005) Novel Catalysts for Stereoselective Hydroamination of Olefins Based on Rare Earth and Group IV Metals, PhD thesis, Universität Erlangen-Nürnberg.
88 Watson, D.A., Chiu, M., and Bergman, R.G. (2006) *Organometallics*, **25**, 4731–4733.
89 (a) Wood, M.C., Leitch, D.C., Yeung, C.S., Kozak, J.A., and Schafer, L.L. (2007) *Angew. Chem.*, **46**, 354–358; (b)(2009) *Angew. Chem.*, **48**, 6937.
90 Gott, A.L., Clarke, A.J., Clarkson, G.J., and Scott, P. (2007) *Organometallics*, **26**, 1729–1737.
91 Walsh, P.J., Baranger, A.M., and Bergman, R.G. (1992) *J. Am. Chem. Soc.*, **114**, 1708–1719.

92 Baranger, A.M., Walsh, P.J., and Bergman, R.G. (1993) *J. Am. Chem. Soc.*, **115**, 2753–2763.
93 Walsh, P.J., Hollander, F.J., and Bergman, R.G. (1993) *Organometallics*, **12**, 3705–3723.
94 Lee, S.Y. and Bergman, R.G. (1995) *Tetrahedron*, **51**, 4255–4276.
95 Polse, J.L., Andersen, R.A., and Bergman, R.G. (1998) *J. Am. Chem. Soc.*, **120**, 13405–13414.
96 Pohlki, F. and Doye, S. (2001) *Angew. Chem., Int. Ed.*, **40**, 2305–2308.
97 Straub, B.F. and Bergman, R.G. (2001) *Angew. Chem.*, **40**, 4632–4635.
98 Tobisch, S. (2007) *Chem. Eur. J.*, **13**, 4884–4894.
99 Stubbert, B.D. and Marks, T.J. (2007) *J. Am. Chem. Soc.*, **129**, 6149–6167.
100 Majumder, S. and Odom, A.L. (2008) *Organometallics*, **27**, 1174–1177.
101 Dalko, P.I. and Moisan, L. (2004) *Angew. Chem., Int. Ed.*, **43**, 5138–5175.
102 Seayad, J. and List, B. (2005) *Org. Biomol. Chem.*, **3**, 719–724.
103 Akiyama, T. (2007) *Chem. Rev.*, **107**, 5744–5758.
104 Schlummer, B. and Hartwig, J.F. (2002) *Org. Lett.*, **4**, 1471–1474.
105 Haskins, C.M. and Knight, D.W. (2002) *Chem. Commun.*, 2724–2725.
106 Ackermann, L., Kaspar, L.T., and Althammer, A. (2007) *Org. Biomol. Chem.*, **5**, 1975–1978.
107 Anderson, L.L., Arnold, J., and Bergman, R.G. (2005) *J. Am. Chem. Soc.*, **127**, 14542–14543.
108 Rosenfeld, D.C., Shekhar, S., Takemiya, A., Utsunomiya, M., and Hartwig, J.F. (2006) *Org. Lett.*, **8**, 4179–4182.
109 Marcseková, K. and Doye, S. (2007) *Synthesis*, 145–154.
110 Yang, L., Xu, L.-W., and Xia, C.-G. (2008) *Tetrahedron Lett.*, **49**, 2882–2885.
111 Ackermann, L. and Althammer, A. (2008) *Synlett*, 995–998.
112 Hong, S. and Marks, T.J. (2002) *J. Am. Chem. Soc.*, **124**, 7886–7887.
113 Hong, S., Kawaoka, A.M., and Marks, T.J. (2003) *J. Am. Chem. Soc.*, **125**, 15878–15892.
114 Hoover, J.M., Petersen, J.R., Pikul, J.H., and Johnson, A.R. (2004) *Organometallics*, **23**, 4614–4620.
115 Patil, N.T., Lutete, L.M., Wu, H., Pahadi, N.K., Gridnev, I.D., and Yamamoto, Y. (2006) *J. Org. Chem.*, **71**, 4270–4279.
116 LaLonde, R.L., Sherry, B.D., Kang, E.J., and Toste, F.D. (2007) *J. Am. Chem. Soc.*, **129**, 2452–2453.
117 Zhang, Z., Bender, C.F., and Widenhoefer, R.A. (2007) *Org. Lett.*, **9**, 2887–2889.
118 Hamilton, G.L., Kang, E.J., Mba, M., and Toste, F.D. (2007) *Science*, **317**, 496–499.
119 Molander, G.A., Dowdy, E.D., and Pack, S.K. (2001) *J. Org. Chem.*, **66**, 4344–4347.
120 Arredondo, V.M., Tian, S., McDonald, F.E., and Marks, T.J. (1999) *J. Am. Chem. Soc.*, **121**, 3633–3639.
121 Trost, B.M. and Tang, W. (2002) *J. Am. Chem. Soc.*, **124**, 14542–14543.
122 Nishina, N. and Yamamoto, Y. (2006) *Angew. Chem.*, **45**, 3314–3317.
123 Pohlki, F., Bytschkov, I., Siebeneicher, H., Heutling, A., König, W.A., and Doye, S. (2004) *Eur. J. Org. Chem.*, 1967–1972.
124 Gribkov, D.V. and Hultzsch, K.C. (2004) *Chem. Commun.*, 730–731.
125 Reznichenko, A.L., Hampel, F., and Hultzsch, K.C. (2009) *Chem. Eur. J.*, **15**, 12819–12827.
126 Sherry, B.D. and Toste, F.D. (2004) *J. Am. Chem. Soc.*, **126**, 15978–15979.
127 Zhang, Z., Bender, C.F., and Widenhoefer, R.A. (2007) *J. Am. Chem. Soc.*, **129**, 14148–14149.
128 Ajamian, A. and Gleason, J.L. (2004) *Angew. Chem.*, **43**, 3754–3760.
129 Fogg, D.E. and dos Santos, E.N. (2004) *Coord. Chem. Rev.*, **248**, 2365–2379.
130 Eilbracht, P., Barfacker, L., Buss, C., Hollmann, C., Kitsos-Rzychon, B.E., Kranemann, C.L., Rische, T., Roggenbuck, R., and Schmidt, A. (1999) *Chem. Rev.*, **99**, 3329–3366.
131 Eilbracht, P. and Schmidt, A.M. (2006) *Top. Organomet. Chem.*, **18**, 65–95.
132 Müller, T.E. (2006) *Top. Organomet. Chem.*, **19**, 149–205.
133 Heutling, A., Pohlki, F., Bytschkov, I., and Doye, S. (2005) *Angew. Chem.*, **44**, 2951–2954.

12
Enantioselective C−H Amination
Nadège Boudet and Simon B. Blakey

12.1
Introduction

In the last decade, C−H functionalization has emerged as a powerful tool for the synthesis of complex organic molecules [1, 2]. Given the ubiquity of the C−H bond and the fact that by definition, disconnections engaging the C−H bond require no preexisting functionality, this methodology can considerably abridge traditional synthetic sequences. The attractive synthetic efficiencies that can be obtained utilizing such a strategy have led to an explosion of research in this area. A variety of new catalysts have been designed promoting coordination-directed metallations as well as carbene/nitrene insertions [3].

In this chapter, we will focus on metallonitrene chemistry that allows the direct transformation of a C−H into a C−N bond. During this transformation, the substrate does not react directly with a transition metal but instead with a metallonitrene species (**1**) that inserts into a well-differentiated C−H bond (typically benzylic, allylic, 3° or α to a heteroatom) (Scheme 12.1). Independently, the groups of Du Bois and Che revolutionized this area by developing Rh- and Ru-complexes able to support metallonitrene species and perform direct regioselective aminations. While several methods have been successfully realized for intramolecular amination, intermolecular reactions remain challenging. Considering chiral amines are key building blocks for many pharmaceutical and crop protection agents, asymmetric C−H amination has arisen as an important goal in this field. In this chapter, we will outline the evolution of practical C−H amination protocols, highlighting important initial developments in racemic chemistry, that have led to exciting recent advances for both diastereoselective and enantioselective versions of this important reaction.

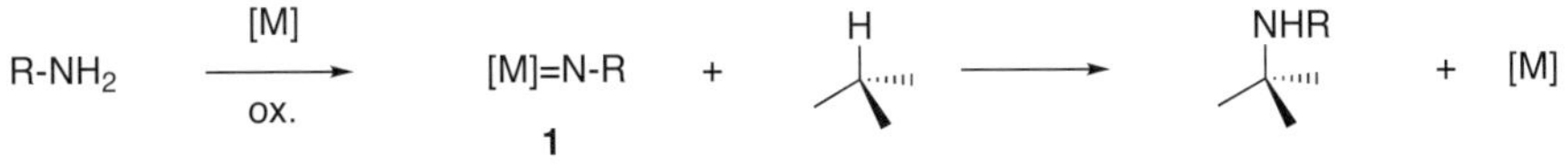

Scheme 12.1 Metal nitrenoid C−H functionalization.

Chiral Amine Synthesis: Methods, Developments and Applications. Edited by Thomas C. Nugent
Copyright © 2010 WILEY-VCH Verlag GmbH & Co. KGaA, Weinheim
ISBN: 978-3-527-32509-2

12.2
Background

Despite the fact that synthetically useful C–H amination reactions have only emerged in the last decade, the pioneering studies representing the bedrock of our present knowledge took place over 40 years ago. The first transition metal-catalyzed C–H amination involving a nitrene was reported by Kwart and Khan in 1967 [4]. Using copper as a catalyst, they described the decomposition of benzenesulfonylazide in the presence of cyclohexene furnishing a mixture of C–H insertion and aziridination products (Scheme 12.2). This transformation, which was postulated to occur through a copper nitrenoid intermediate, not only set the stage for the development of C–H amination technologies but also highlighted many of the current challenges related to C–H amination. Specifically, a large excess of substrate was required for the intermolecular reaction and competitive aziridination was observed in the case of olefinic substrates.

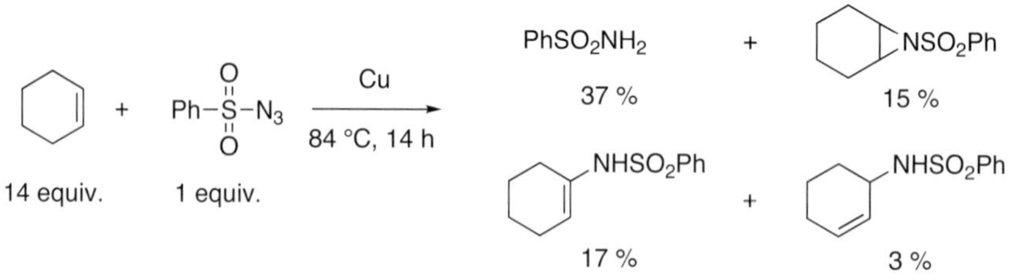

Scheme 12.2 Copper-catalyzed decomposition of benzenesulfonyl azide leads to a mixture of C–H amination and aziridination products.

The first breakthroughs in this chemistry appeared with the independent works of Breslow [5] and Mansuy [6] who reported the use of Fe(III)- and Mn(III)-porphyrins complexes with tosylimidoiodobenzene (PhI=NTs) [7] as an N-centered electrophilic oxidant (Scheme 12.3). Treatment of an alkene with a high-valent Mn-nitrene favored the C–H insertion product over the competing aziridination reaction. However, yields in these intermolecular reactions remained low. A mixture of N-tosylated allylic products was obtained with the product distribution best rationalized by a radical C–H abstraction/rebound process [6].

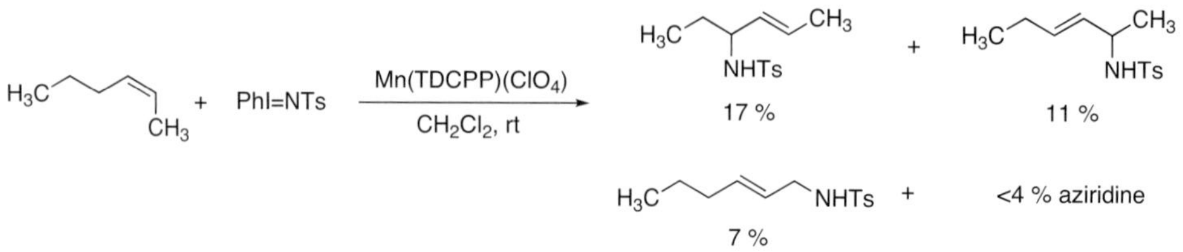

Scheme 12.3 Tosylamidation of cis-2-hexene (TDCPP = meso-tetra-(2,6-dichlorophenylporphyrin).

More impressively, Breslow and Gellman reported an intramolecular C–H amination involving an *ortho*-alkylated arylsulfonamide [8]. After preoxidation of the sulfonamide to the nitrene oxidation state, a number of catalysts were investigated for the amidation reaction (2 → 3, Scheme 12.4). $Rh_2(OAc)_4$, a well-established catalyst for metallocarbene insertions [9, 10], showed excellent reactivity providing the C–H amination product in 86% HPLC yield. In this case, the rhodium–nitrenoid and the reacting C–H bond are connected by a suitable tether, allowing proximity to induce a regioselective reaction. Although these observations lay dormant for more than a decade, this pioneering study set the stage for much of the subsequent work in this area.

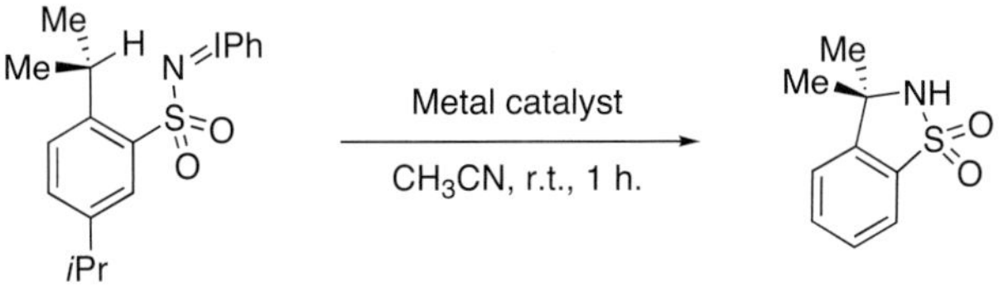

Catalyst	Yield
Mn(TPP)Cl	16 %
Fe(TPP)Cl	77 %
{Fe(cyclam)Cl$_2$}	42 %
FeCl$_3$	16 %
Rh$_2$(OAc)$_4$	86 %

Scheme 12.4 Intramolecular nitrene C–H insertion of iminoiodinane (TPP, tetraphenylphorpyrin; cyclam, 1,4,8,11-tetraazacyclotetradecane).

12.3
Racemic C–H Amination

12.3.1
Intramolecular C–H Amination

In the last decade, intramolecular C–H amination has evolved into a powerful tool for direct synthesis of functionalized N-containing heterocycles (Table 12.1). While many of these reactions are catalyzed by dirhodium tetracarboxylate salts, ruthenium and silver complexes are also effective. Others transition metals such as Mn [11] or Co [12] have been investigated, but, to date, efficient catalysts utilizing these metals have not been disclosed. Early studies by Du Bois and coworkers demonstrated the use of readily available carbamates as nitrogen sources and PhI(OAc)$_2$ as inexpensive oxidant allowing an *in situ* preparation of the iminoiodinane [13]. MgO was used as additive and is believed to neutralize 2 equiv of acetic acid formed during this oxidation. Under $Rh_2(OAc)_4$ catalysis, the carbamate tether controls the regioselectivity of the C–H amination, leading to five-membered ring insertion products (e.g., oxazolidinone 4), obtained in excellent yield under mild conditions (Table 12.1, entry 1). This one-pot procedure is applicable to 2°, 3°, and benzylic C–H bonds furnishing a wide variety of 1,2-aminooxygenated products. In a twist on this chemistry, Lebel and coworkers employed N-tosyloxycarbamates (e.g., 5) as a preoxidized source of nitrene, avoiding the use of the hypervalent iodine (entry 2) [14, 15]. In addition, a silver catalyst developed by He also provides excellent yields

Table 12.1 Catalytic intramolecular C–H amination.

Entry	Substrate	Catalyst	Conditions	Heterocyclic product yield %[a]
1		$Rh_2(OAc)_4$, 5 mol%	$PhI(OAc)_2$ (1.4 equiv), MgO (2.3 equiv), CH_2Cl_2, 40 °C	**4**: 74
2	**5**	$Rh_2(TPA)_4$, 6 mol%	K_2CO_3, CH_2Cl_2, 25 °C	84
3		$AgNO_3$, $tBu3tpy$, 4 mol%	$PhI(OAc)_2$ (2 equiv), CH_3CN, 82 °C	83
4		$Rh_2(OAc)_4$, 2 mol%	$PhI(OAc)_2$ (1.1 equiv), MgO (2.3 equiv), CH_2Cl_2, 40 °C	**7**: 84
5	**6**	[Ru(tpfpp)(CO)], 1.5 mol%	$PhI(OAc)_2$ (2.0 equiv), Al_2O_3 (2.5 equiv), CH_2Cl_2, 40 °C	77

#	Substrate	Conditions	Reagents	Product	Yield
6	sulfamate with Me,Me,OH chain	AgNO₃, tBu3tpy, 4 mol%	PhI(OAc)₂ (1.4 equiv), CH₃CN, 82 °C	cyclic sulfamate	87
7	H₂N-sulfamate Me,Me	Rh₂(esp)₂, 0.15 mol%	PhI(OAc)₂ (1.1 equiv), MgO (2.3 equiv), CH₂Cl₂, 40 °C	cyclic HN-sulfamate Me,Me	92
8	H₂N-C(O)NTces tryptamine-Boc	Rh₂(esp)₂, 1.0 mol%	PhI(OAc)₂ (1.5 equiv), MgO (2.5 equiv), toluene, 40 °C	indole-fused imidazolidinone	73
9	H₂N-C(NTces)-NH-CMe₂	Rh₂(esp)₂, 2.0 mol%	PhI(OAc)₂ (1.5 equiv), MgO (2.5 equiv), toluene, 40 °C	cyclic guanidine	74
10	H₂N-SO₂-NHBoc, EtO₂C, Me	Rh₂(esp)₂, 2.5 mol%	PhI(OAc)₂ (1.1 equiv), MgO (2.3 equiv), ⁱPrOAc	cyclic sulfamide	99

a) Yield of isolated product.

for the cyclization of carbamates, albeit with a requirement for elevated temperatures (entry 3) [16].

The Du Bois group recognized early on that the nature of the tether attaching the reactive nitrogen to the organic molecule could have a significant influence on the regioselectivity of the reaction. Thus, rhodium-catalyzed oxidative amination of sulfamate esters was developed, and proceeds with complete selectivity for γ-C—H insertion, leading to a six-membered ring product (e.g., 7, entry 4) [17]. These oxathiazinanes are synthetically useful intermediates and a simple protection/ring-opening sequence was developed to reveal 1,3-difunctionalized amines such as a 1,3 amino alcohol [17]. Ru–phorphyrin [18, 19] and Ag–pyridine complexes [16] were also proven to be successful catalysts for these one-pot C—H amination (entries 5 and 6). A significant advance in this field occurred with Espino and Du Bois' development of the strapped dicarboxylate catalyst $Rh_2(esp)_2$ [20]. The design of this new bridged Rh-dimer catalyst and its substantially improved stability under typical amination conditions facilitated a dramatic expansion in the scope of the C—H amination reaction. Using a low loading of this catalyst, C—H amination has been demonstrated utilizing sulfamate [20, 21], sulfamide [20, 22], urea [20, 23], and guanidine [23] substrates, all with predictable regioselectivity and routinely excellent yields (entries 7–10). In summary, intramolecular Rh-catalyzed C—H amination has been developed into a powerful tool for the generation of an impressive array intermediates with the predictability that has allowed many stunning applications in the total synthesis of complex nitrogen-containing natural products [24–28].

Despite these significant advances, allylic C—H insertion under these conditions remains problematic, with competing olefin aziridination often observed as the dominant reaction pathway (Scheme 12.5) [29].

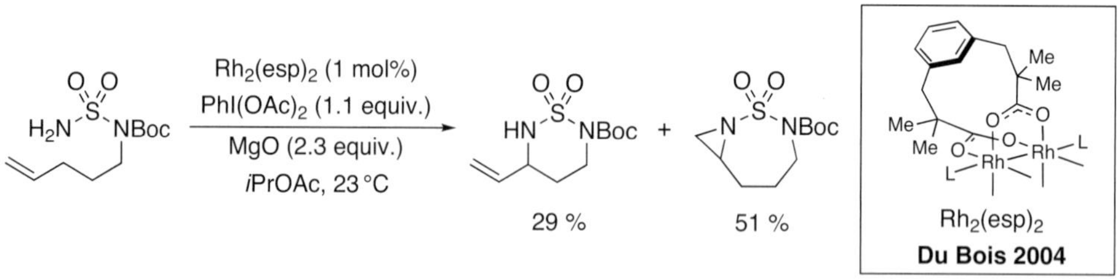

Scheme 12.5 Aziridination as a side reaction of Rh-catalyzed C—H amination of olefinic sulfamide.

12.3.2
Intermolecular C—H Amination

Catalytic intermolecular C—H amination has been extensively investigated but the development of broadly applicable conditions for this transformation remains a substantial challenge. Intramolecular C—H amination is often observed as a side reaction of Cu- [30] and Rh- [31, 32] catalyzed aziridination of olefins. Despite significant attempts to exploit this process further under Cu [33–36], Co [37], Pd [38],

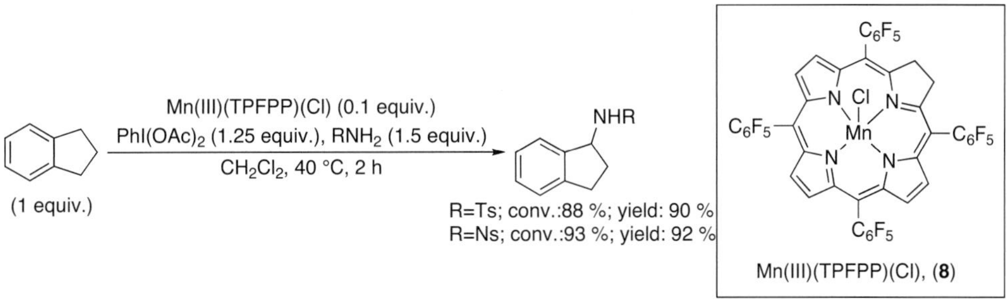

Scheme 12.6 Amidation of indane catalyzed by manganese(III)-porphyrins (TPFPP, tetrakis-(pentafluorophenyl) porphyrinato).

Ag [39], or Au catalysis [40], most protocols continue to call for significant excess of the starting alkane relative to oxidant ((5–100 equiv) and the reactions are often limited in scope to benzylic or allylic positions and cyclic ethers. Readily available iminoiodinanes of formula PhI=NR (R=Ts, Ns) are generally utilized as the source of nitrogen, but the employment of commercially available chloro- or bromamine-T [34, 37, 41–44] or arylazides [35, 45] has also been reported.

Despite the significant challenges involved, some striking advances have been reported. Specifically Che and coworkers employed a Mn complex **8** bearing a electron-deficient porphyrin macrocycle (TPFPP) [46] (Scheme 12.6). Noteworthy elements of this process included the use of the alkane as the limiting substrate, the *in situ* formation of the iminoiodinane, and a high turnover (up to 2600). Good to excellent yields were obtained with hydrocarbons such as indane, cyclohexene, or tetrahydrofuran. However, competing aziridination as side reaction continues to limit the scope of this methodology.

The bridged $Rh_2(esp)_2$ complex has proven to be a particularly robust and versatile catalyst for C—H amination and, indeed, Du Bois and coworkers have shown its efficiency for intermolecular C—H amination. Employing a 1 : 1 ratio of substrate and the commercially available trichloroethylsulfamate ($TcesNH_2$) nitrogen source, with $PhI(O_2CtBu)_2$ as the oxidant [47], a range of benzylic amines were generated in good yield (Scheme 12.7) [48]. The Tces group is easily removed under mild reductive conditions, enhancing the utility of this protocol.

With respect to the C—H functionalization/aziridination selectivity challenge, White and coworkers recently reported a significant advance [49]. Utilizing a heterobimetallic Pd/Cr(salen)Cl catalyst system, a wide variety of alpha olefins were

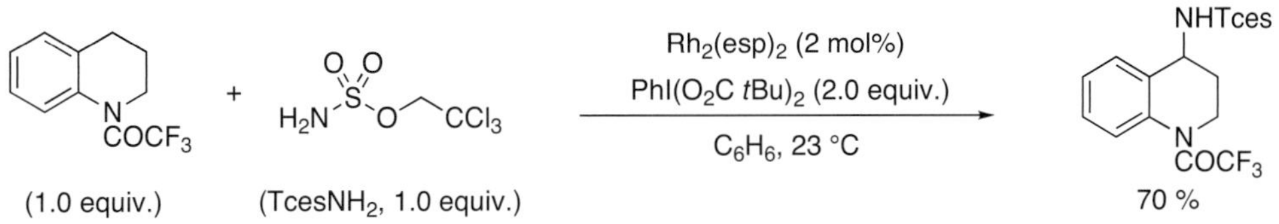

Scheme 12.7 Rhodium(II)-catalyzed intermolecular amination using $TcesNH_2$.

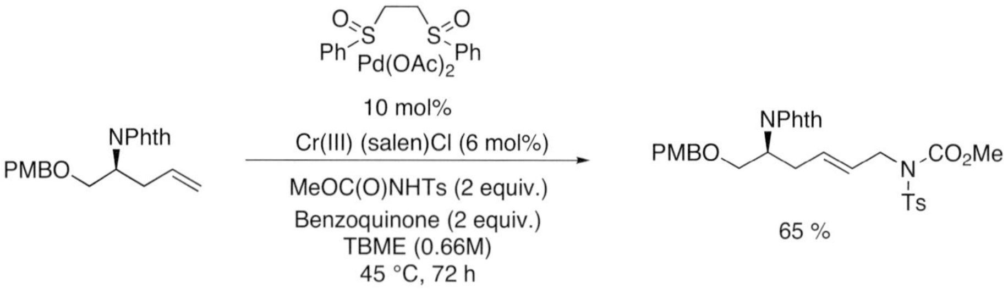

Scheme 12.8 Palladium-catalyzed intermolecular allylic amination.

transformed directly into the corresponding allylic carbamates with excellent selectivity for the *E*-linear product (Scheme 12.8). The White group has applied comparable catalytic systems to the synthesis of enantioenriched allylic alcohols setting the stage for the development of an enantioselective allylic amination in the near future [50].

12.4
Substrate-Controlled Chiral Amine Synthesis via C—H Amination

The use of substrate control in rhodium-catalyzed C—H aminations is covered in detail in Espino and Du Bois' recent review of rhodium-catalyzed oxidative amination [51]. A brief summary of relevant material is provided here, leading to a discussion of recent advances in the synthesis of chiral amines from achiral substrates. Rhodium-catalyzed C—H amination proceeds via a concerted insertion process rendering it a stereospecific transformation. Thus, the appropriate choice of an enantioenriched starting material can facilitate the synthesis of enantioenriched amines, which would often be particularly difficult to access in any other manner. As exemplified in Scheme 12.9, the C—H insertion reaction of enantiomerically pure carbamate **9** was accomplished with complete retention of configuration providing the chiral oxazolidinone **10** in greater than 98% ee [13].

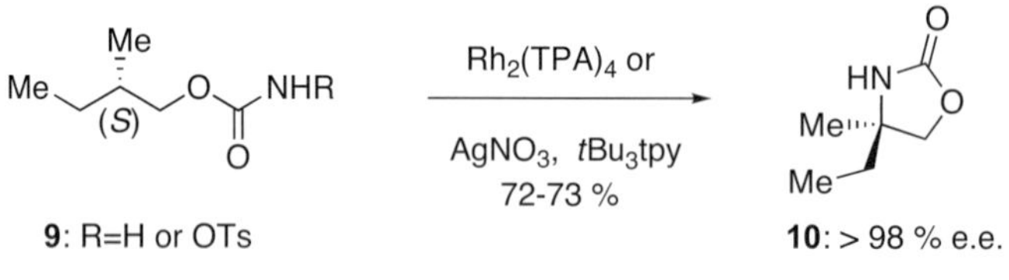

Scheme 12.9 Stereospecific C—H insertion of a carbamate substrate.

Du Bois and coworkers have demonstrated that this stereospecificity applies to the intramolecular cyclization of chiral sulfamate [17], urea [20, 23], and sulfamide [20, 22] substrates. Furthermore, an intermolecular variant was also confirmed to be stereospecific although a reduced yield was observed for this process (Scheme 12.10). The Rh(II)-catalyzed C—H amination can be conducted in the presence of a broad

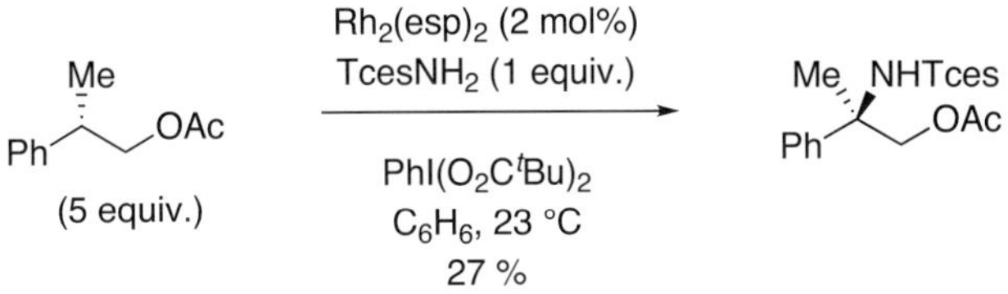

Scheme 12.10 Intermolecular stereospecific C–H amination.

range of functional groups making this transformation a tool of choice for the synthesis of functionalized chiral quaternary α-amine derivatives.

Highlighting the synthetic utility of this methodology, the group of Du Bois reported a three-step synthesis of the amino acid (R)-β-isoleucine starting from the chiral sulfamate ester **11**. Oxidative cyclization with retention of stereochemistry and a simple Cbz protection–hydrolysis–oxidation sequence provided the amino acid (Scheme 12.11) [17].

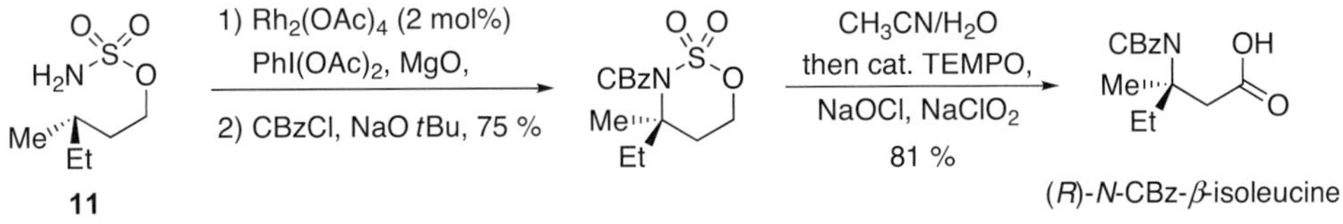

Scheme 12.11 Synthesis of a chiral amino acid using Rh-catalyzed C–H amination.

In addition to stereospecific insertion into methine C–H bonds, When et al. have demonstrated that excellent stereoinduction can be achieved in the cyclization of branched sulfamate esters. A chair-like transition state with an equatorial C–H insertion of the metallonitrene is proposed and this stereochemical model explains the *syn*-selectivity observed for the α-branched substrate and the *anti*-selectivity for the β-branched case (Scheme 12.12) [52]. In contrast, α- and β-branched chiral

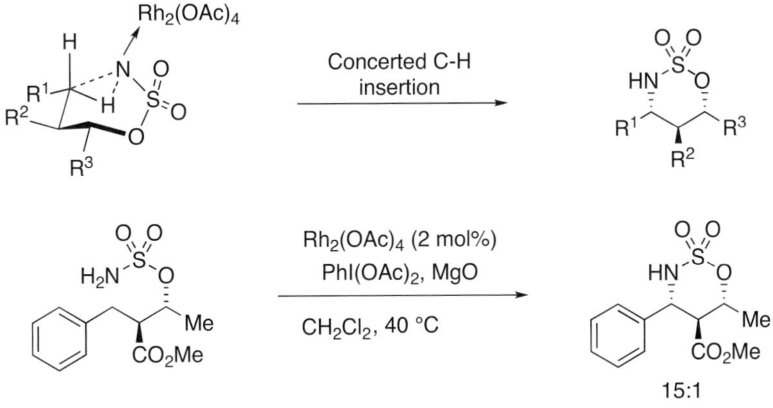

Scheme 12.12 Proposed stereochemical model for the C–H insertion of chiral sulfamate esters.

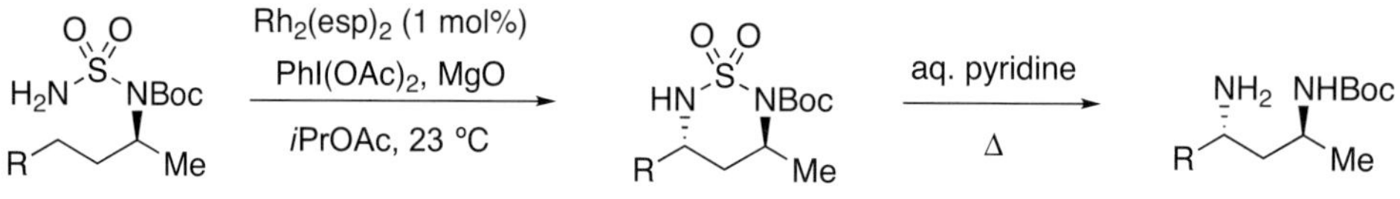

Scheme 12.13 Sulfamides provide the opposite diastereoselectivity to sulfamate esters.

sulfamides [20, 22] offered a high substrate-based diastereocontrol favoring the *anti*-diastereoisomer (Scheme 12.13) [53].

12.5
Enantioselective C—H Amination of Achiral Substrates

While high levels of enantiocontrol in aziridinations or carbene insertions have already been achieved using chiral Cu(I), Mn(III), or Rh(II) catalysts, enantioselective C—H amination has been a long-standing problem. A significant challenge exists in the design of chiral catalysts that are able to support a reactive oxidant and distinguish two hydrogen atoms on a prochiral CH_2 center. Recently, exciting advances have been reported for both intra- and intermolecular versions of this reaction opening a new chapter in the synthesis of chiral amines.

12.5.1
Enantioselective C—H Amination with Rhodium(II) Catalysts

Asymmetric induction for intermolecular C—H amination (up to 33% ee) was first demonstrated by Müller and coworkers using dirhodium(II) tetrakis [(R)-binaphthyl-phosphate] (**12**) (Rh$_2$(R-BNP)$_4$, Figure 12.1) in the reaction of PhI=NNs and indane [54]. Subsequently, Hashimoto developed new chiral dirhodiumtetracarboxylates with pendent phthalamide arms (**13**, Rh$_2$(S-TCPTTL)$_4$) [55] facilitating enantioselective aminations with enantioselectivities as high as 73% [56]. These promising catalysts inspired further ligand development in a number of laboratories. A major breakthrough was realized by Reddy and Davies who developed the chlorinated derivative Rh$_2$(S-TCPTAD)$_4$ (**14**) that was applied for the enantioselective C—H amination of benzylic substrates using NsNH$_2$/PhI(OAc)$_2$ as an *in situ* generating system of iminoiodinane (Scheme 12.14) [57]. High yields were obtained and in the case of indane, an astonishing enantioselectivity (**15**, 94% ee) was observed. However, in all cases, these early protocols required the use of excess substrate with respect to the nitrogen source, and the substrate scope remained slim.

The Davies catalyst **14** was also found to be useful for intermolecular cyclizations, and good enantioselectivities (up to 82% ee) were obtained in the synthesis of oxazolidinones (**16–18**) using Lebel's N-tosylcarbamates (Scheme 12.15).

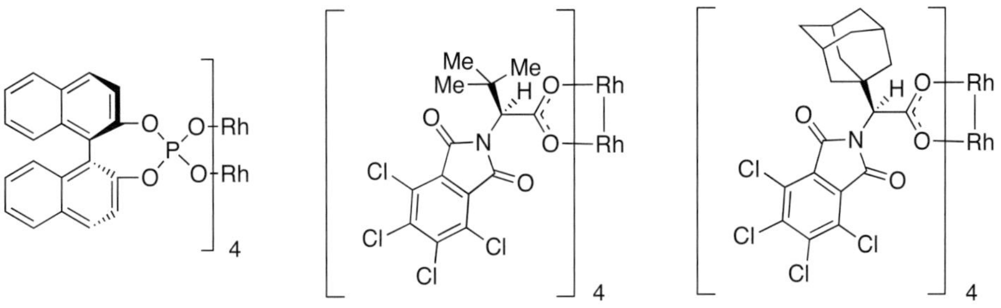

12.5 Enantioselective C—H Amination of Achiral Substrates | 387

(12) Rh$_2$(R-BNP)$_4$
Müller 1997

(13) Rh$_2$(S-TCPTTL)$_4$
Hashimoto 2002

(14) Rh$_2$(S-TCPTAD)$_4$
Davies 2006

Figure 12.1 Dirhodium catalysts (TCPTTL, N-tetrachlorophthaloyl-(S)-tert-leucinate; TCPTAD, N-tetrachloro-phthaloyl-(S)-(1-adamantyl)glycinate).

A significant breakthrough in the field of asymmetric C—H amination arrived with the introduction of enantioenriched sulfoxamines as amine sources, arising from a body of research in the laboratories of Müller and coworkers [58–60] (Scheme 12.16). This conceptually new approach offered high enantioselectivities based on the matching effect between a chiral sulfonimidamide **19** and an appropriately chosen chiral dirhodiumtetracarboxylate catalyst (**20** or **21**). The solvent system (CHCl$_2$)$_2$/MeOH was essential for optimum results. While protic solvents have been shown to hydrolyze iminoiodinanes [30], MeOH is believed in this case to better solubilize the sulfoniminamide and helps the formation of the iminoiodidane by complexing the hypervalent iodine. The substrate scope of this methodology is particularly impressive, and includes benzylic, allylic, and alkane substrates. Significantly, the

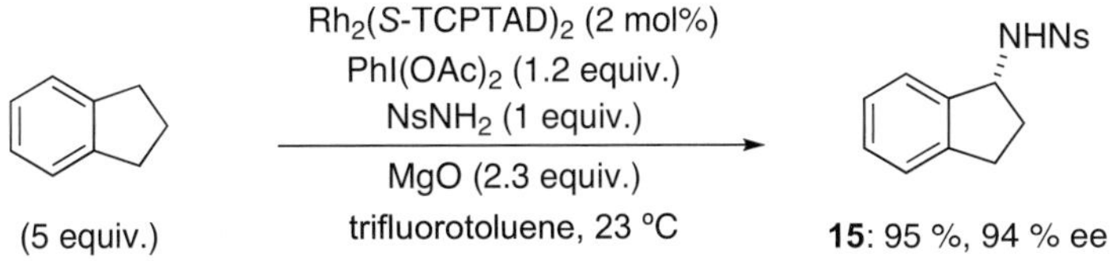

Scheme 12.14 Rh$_2$(S-TCPTAD)$_4$-catalyzed intermolecular C—H amination.

Conditions: Rh$_2$(S-TCPTAD)$_2$ (2 mol%), PhI(OAc)$_2$ (1.2 equiv.), NsNH$_2$ (1 equiv.), MgO (2.3 equiv.), trifluorotoluene, 23 °C. (5 equiv.) → **15**: 95 %, 94 % ee

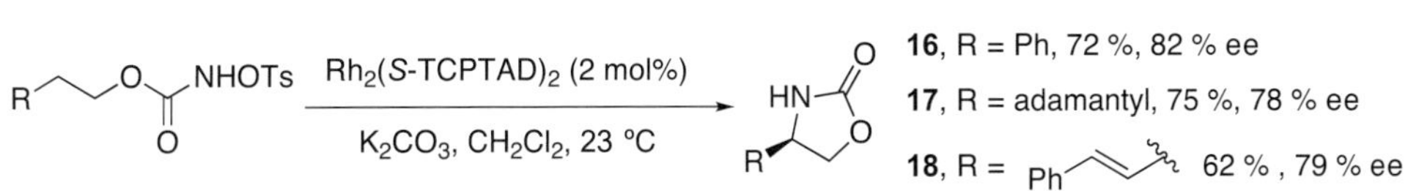

Scheme 12.15 Rh$_2$(S-TCPTAD)$_4$-catalyzed intramolecular C—H amination.

Conditions: Rh$_2$(S-TCPTAD)$_2$ (2 mol%), K$_2$CO$_3$, CH$_2$Cl$_2$, 23 °C.

16, R = Ph, 72 %, 82 % ee
17, R = adamantyl, 75 %, 78 % ee
18, R = Ph-CH=CH-, 62 %, 79 % ee

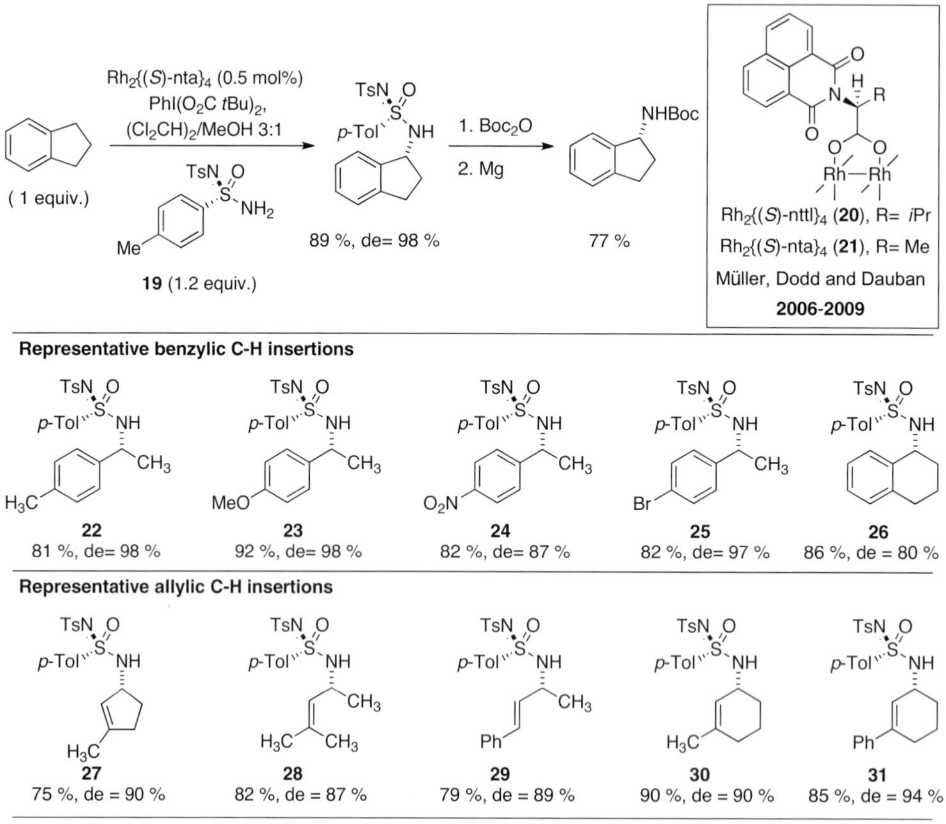

Scheme 12.16 Rh$_2${(S)-nta}$_4$-catalyzed enantioselective intermolecular C–H amination (nttl, N-1,8-naphthoyl-tert-leucine; nta, N-1,8-naphthoylalanine).

substrate can be used as limiting component. Cleavage of the sulfoniminamide products is achieved by Boc protection followed by reduction with magnesium in methanol, ensuring that the product amines are readily accessible.

Selectivity is observed for more substituted benzylic positions, allowing para-methylethylbenzene to be aminated in excellent yield and selectivity (**22**). Both electron-donating and electron-withdrawing substituents on the aromatic ring are tolerated (**23–25**). Both enantiomers of the sulfoxamine and catalyst are available, ensuring that both enantiomers of the product are accessible using this method [58]. Moreover, kinetic resolution of racemic sulfoxamine **19** has been observed, alleviating the requirement to synthesize this aminating reagent in an enantiocontrolled fashion [59].

Of particular note, enantioselective intermolecular allylic amination is also possible using this protocol, and no competing aziridination is observed. Cyclic and linear alkenes were selectively aminated with high selectivities (**27–31**).

Although the use of chiral aminating reagent in concert with chiral catalysts produces excellent selectivity in these reactions, the generation of the aminating reagent remains an impediment to the broad use of this methodology.

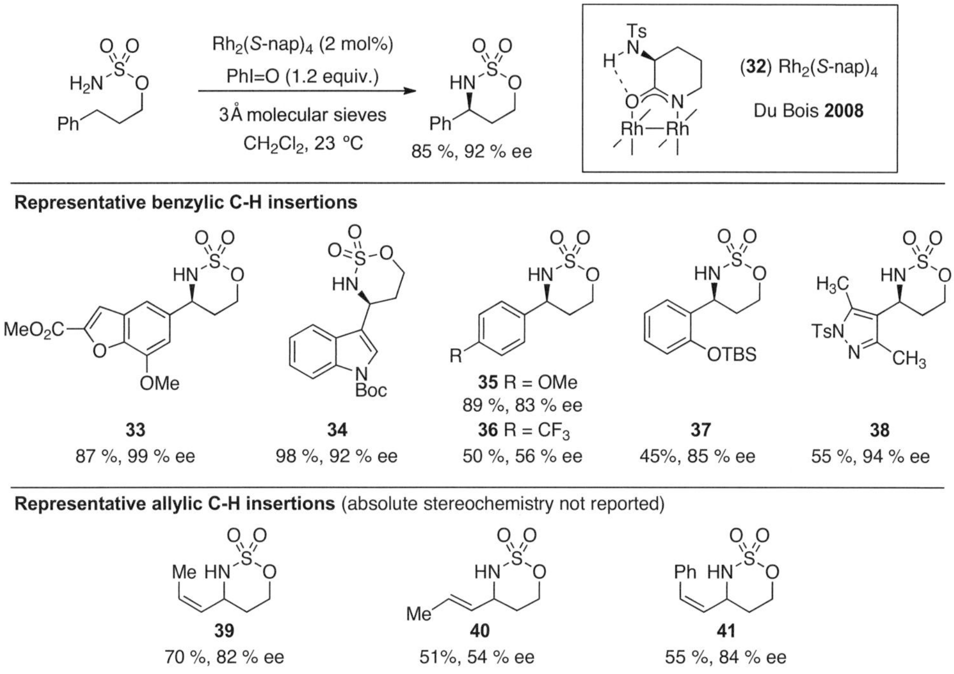

12.5 Enantioselective C—H Amination of Achiral Substrates

Scheme 12.17 Enantioselective C—H amination catalyzed by Rh$_2$(S-nap)$_4$ (**32**).

To promote the enantioselective amination of simple substrates, Zalatan and Du Bois designed the tetracarboxamidate catalyst Rh$_2$(S-nap)$_4$ (**32**, Scheme 12.17), a new valerolactam-derived catalyst that affords excellent enantiocontrol for a wide variety of benzylic and allylic C—H insertion reactions [61]. The critical design element in this catalyst system is the use of an electronically tuned carboxamide ligand that is proposed to reduce the electrophilicity of the reactive metallonitrene, thus promoting selectivity, but that is not so electron rich as to render the catalyst unstable to the hypervalent iodine oxidant. The enantioselective C—H amination was performed on sulfamate esters using 2 mol% of catalyst **32** and PhI=O as an oxidant in dichloromethane. C—H bonds adjacent to electron-rich aryl rings are particularly suitable and are aminated with excellent enantioselectivities (up to 99% ee) and yield. Common functionalities such as methoxy substituent, an ester, and a silyl alcohol are compatible with these conditions (**33–38**). However, electron-withdrawing groups are detrimental to both reaction efficiency and selectivity (e.g., **36**).

Remarkable enantiocontrol was obtained using N-heterocyclic substrates such as protected indole **34** and pyrazole **38**, showing the potential of this method in the synthesis of biologically active chiral amines. Another striking element of this catalyst is its reactivity toward alkene substrates. While rhodium tetracarboxylate catalysts tend to promote both C—H insertion and aziridination, the Rh$_2$(S-nap)$_4$ (**32**) is particularly selective for C—H insertion. *cis*-Olefins were well tolerated, providing the aminated product in good yield and enantioselectivity (**39, 41**). However, the use of *trans*-isomers resulted in reduced yield and selectivity (e.g., **40**).

12.5.2
Enantioselective C−H Amination with Ruthenium(II) Catalysts

Concurrent with the development of chiral dirhodium tetracarboxylate and tetracarboxamide catalysts for asymmetric C−H amination, significant advances have also been made in this field using chiral ruthenium complexes. Che and coworkers have demonstrated that the modified Ru–porphyrin complex **42** can promote the intramolecular C−H insertion of sulfamate esters (Scheme 12.18) [62, 63]. In these systems, there is a fine balance between achieving useful reaction progression and high enantioselectivities. Impressive levels of enantioselectivity (up to 86% ee) can be obtained at low temperature (5 °C), but reaction yields are reduced. The Ru(por)*(CO) **42** can catalyze the formation of five- as well as six-membered ring sulfamidates – a regioselectivity profile not usually observed for sulfamate esters. Later, chiral manganese(III) Schiff-base complexes were found particularly effective for the five-membered ring formation of electron-poor sulfamate esters [11]. Chiral ruthenium(II) porphyrin complexes have also been used to catalyze intermolecular C−H amination of benzylic substrates, but neither the reaction efficiency nor the selectivity are competitive with the previously described rhodium protocols [64].

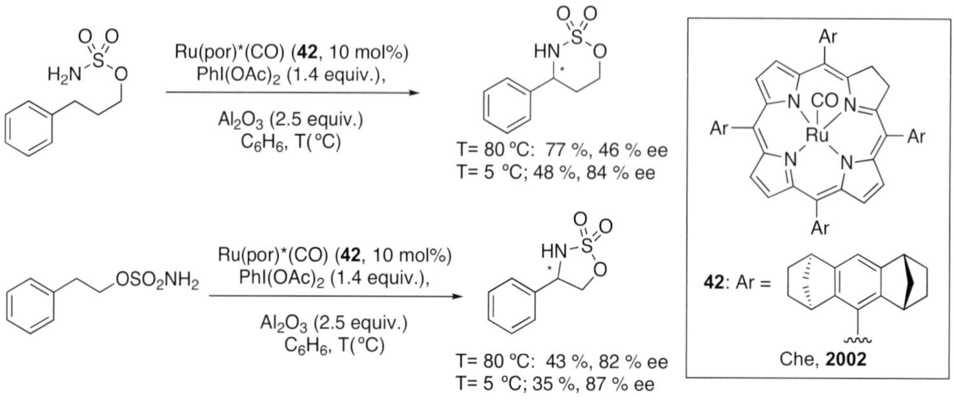

Scheme 12.18 Ru-catalyzed enantioselective C−H amination of sulfamate esters.

Cognizant of the reactivity differences between ruthenium nitrenoid systems and the highly active rhodium nitrenoid systems, Blakey and coworkers [65] have designed a new RuBr$_2$-pybox (pybox = pyridine bisoxazoline) catalyst **43** that offered high to excellent enantioselectivities under mild conditions for the C−H amination of sulfamate esters (Scheme 12.19). This Ru complex is readily obtained in one step from the commercially available pybox ligand. The key design principle in this protocol is the use of a silver salt to abstract bromide from the Ru(II) complex, generating the active cationic catalyst. The cationic catalyst is proposed to generate a more electrophilic Ru–imido complex, facilitating C−H amination at lower

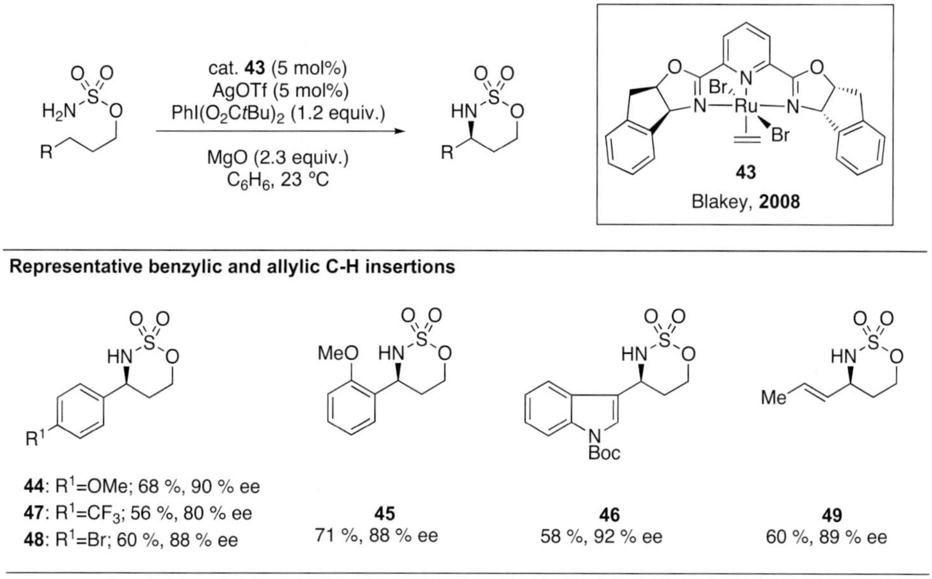

12.5 Enantioselective C–H Amination of Achiral Substrates

Scheme 12.19 Ru(II)-pybox catalyst and its application to enantioselective C–H amination.

temperatures allowing high enantioselectivities to be realized. Optimized conditions of this protocol included 1.1 equiv of PhI(O$_2$tBu)$_2$ in benzene at room temperature. This mild procedure is tolerant of both *ortho*- and *para*-substituted benzylic substrates. Although the best reactions occur with electron-donating substitutents on the aromatic ring (**44–46**), good enantioselectivities for substrates possessing electron-withdrawing groups are also possible (**47** and **48**).

Allylic C–H insertion proceeds with complete selectivity for the C–H bond, and competing aziridination is not observed. In contrast to the Rh$_2$(S-nap)$_4$ catalyst system, a preference for *trans*-olefins is observed under these conditions. Thus, *trans*-olefin is converted to oxathiazinane **49** in good yield (60%) and selectivity (89% ee). Conversely, *cis*-olefins are not good substrates for this catalyst.

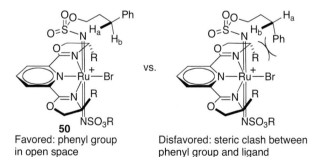

Figure 12.2 Stereochemical model for the cationic Ru(pybox)-catalyzed C–H amination.

Additional studies demonstrated that replacement of the ligand ethylene by CO or PPh$_3$ did not influence the enantioselectivity of the reaction, suggesting that the ancillary ligand dissociates before the C—H insertion. A bisimido-ruthenium(VI) complex **50** operating through an H$^\bullet$ abstraction/radical rebound mechanism, established in ruthenium–porphyrin systems by Che, is proposed to account for the observed stereochemistry (Figure 12.2).

12.6
Conclusion

Significant recent advances in catalyst design have led to the development of synthetically useful protocols for enantioselective intramolecular C—H aminations of sulfamate esters. The development of enantioenriched sulfoxamines as amine sources in conjunction with chiral catalysts has significantly advanced the science of intermolecular asymmetric C—H amination. However, many challenges remain. In particular, robust chiral catalysts capable of engaging a wide variety of nitrogen sources for both regio- and stereocontrolled C—H amination, mirroring the substrate diversity described for the achiral Rh$_2$(esp)$_2$ catalyst, have yet to be realized, and the development of such catalysts remains one of the prevalent challenges in this field today.

Representative Protocol

Rh$_2$(S-nap)- Catalyzed C—H Amination

A 10 ml round-bottom flask was charged with sulfamate **34s** (0.50 mmol), catalytic Rh$_2$(S-nap)$_4$ (0.01 mmol, 0.02 equiv), powdered 3 Å molecular sieves (265 mg), and 1.0 ml of CH$_2$Cl$_2$. A single portion of PhI=O (0.60 mmol, 1.2 equiv) was then added and the resulting pink suspension was stirred for 2 h. The reaction mixture was loaded directly onto silica gel and purified by chromatography (4 : 1 hexanes/EtOAc). The chiral oxathiazinane **34** was obtained as a white solid in 98% yield and 92% ee (Chiralcel-OD-H, 10% isopropanol/hexanes, 1.0 ml/min, 210 nm, tr(minor) 15.9 min, tr(major) = 22.1 min); $\alpha_D^{25} = -22.6$ (c = 1.0, CHCl$_3$). For characterization data, see the corresponding literature [17, 61].

Ru(Pybox)-Catalyzed C–H Amination

cat. 43 (5 mol%)
AgOTf (5 mol%)
PhI(O$_2$C tBu)$_2$ (1.2 equiv.)

MgO (2.3 equiv.)
C$_6$H$_6$, 22 °C

49s → **49**: 60 %, 89 % ee

A 25 ml round-bottom flask was charged with olefin **49s** (82 mg, 0.046 mmol), bis(*tert*-butylcarbonyloxy)-iodobenzene (0.206 g, 0.506 mmol, 1.1 equiv), magnesium oxide (0.039 g, 1.07 mmol, 2.3 equiv), catalyst **43** (0.016 g, 0.023 mmol, 5 mol%), and silver triflate (6 mg, 0.023 mmol, 5 mol%). Benzene (3 ml) was added, and the resulting suspension was stirred at 22 °C for 24 h. The reaction was then quenched by the addition of methanol (2.0 ml) and filtered over celite. The filter cake was washed with ethyl acetate (3 × 15 ml) and DCM (2 × 15 ml). The filtrate was concentrated under reduced pressure, and the crude oil purified by chromatography on silica gel (3 : 1 hexanes/EtOAc) afforded oxathiazinane **49** as a colorless oil (49 mg, 60%, 89% ee). GC (CHIRASIL-DEX CB column); tr(major) = 47.9 min, tr(minor) 49.2 min; $[\alpha]_D^{22} = +5°$ (c 0.25, CHCl$_3$). For characterization data, see the corresponding literature [65].

Questions

12.1. Suggest appropriate conditions for the following transformations.

12.2. Starting from chiral alcohol **51**, suggest synthesis of β-amino acid **52**.

12.3. Rationalize the difference in diastereocontrol observed in the cyclization of sulfamate **53** compared with sulfamine **54**.

Ph–CH₂CH₂–CH(Me)–O–S(=O)₂–NH₂ (**53**) →[Rh-catalyzed C–H amination] Ph–CH₂CH–CH(Me)–O–S(=O)₂–NH (cyclic)

Ph–CH₂CH₂–CH(Me)–N(Boc)–S(=O)₂–NH₂ (**54**) → Ph–CH₂CH–CH(Me)–N(Boc)–S(=O)₂–NH (cyclic)

References

1. Godula, K. and Sames, D. (2006) *Science*, **312**, 67.
2. Davies, H.M.L. and Manning, J.R. (2008) *Nature*, **451**, 417.
3. Dick, A.R. and Sanford, M.S. (2006) *Tetrahedron*, **62**, 2439.
4. Kwart, H. and Khan, A.A. (1967) *J. Am. Chem. Soc.*, **89**, 1951.
5. Breslow, R. and Gellman, S.H. (1982) *J. Chem. Soc. Chem. Comm.*, 1400.
6. Mahy, J.P., Bedi, G., Battioni, P., and Mansuy, D. (1988) *Tetrahedron Lett.*, **29**, 1927.
7. Yamada, Y., Yamamoto, T., and Okawara, M. (1975) *Chem. Lett.*, 361.
8. Breslow, R. and Gellman, S.H. (1983) *J. Am. Chem. Soc.*, **105**, 6728.
9. Hubert, A.J., Feron, A., Warin, R., and Teyssie, P. (1976) *Tetrahedron Lett.*, **17**, 1317.
10. Paulisse, R., Reimling, H., Hayez, E., Hubert, A.J., and Teyssie, P. (1973) *Tetrahedron Lett.*, **14**, 2233.
11. Zhang, J., Chan, P.W.H., and Che, C.M. (2005) *Tetrahedron Lett.*, **46**, 5403.
12. Ruppel, J.V., Kamble, R.M., and Zhang, X.P. (2007) *Org. Lett.*, **9**, 4889.
13. Espino, C.G. and Du Bois, J. (2001) *Angew. Chem., Int. Ed.*, **40**, 598.
14. Lebel, H., Huard, K., and Lectard, S. (2005) *J. Am. Chem. Soc.*, **127**, 14198.
15. Huard, K. and Lebel, H. (2008) *Chemistry*, **14**, 6222.
16. Cui, Y. and He, C. (2004) *Angew. Chem., Int. Ed.*, **43**, 4210.
17. Espino, C.G., Wehn, P.M., Chow, J., and Du Bois, J. (2001) *J. Am. Chem. Soc.*, **123**, 6935.
18. Liang, J.L., Yuan, S.X., Huang, J.S., Yu, W.Y., and Che, C.M. (2002) *Angew. Chem., Int. Ed.*, **41**, 3465.
19. Liang, J.L., Yuan, S.X., Huang, J.S., and Che, C.M. (2004) *J. Org. Chem.*, **69**, 3610.
20. Espino, C.G., Fiori, K.W., Kim, M., and Du Bois, J. (2004) *J. Am. Chem. Soc.*, **126**, 15378.
21. Olson, D.E. and Du Bois, J. (2008) *J. Am. Chem. Soc.*, **130**, 11248.
22. Kurokawa, T., Kim, M., and Du Bois, J. (2009) *Angew. Chem., Int. Ed.*, **48**, 2777.
23. Kim, M., Mulcahy, J.V., Espino, C.G., and Du Bois, J. (2006) *Org. Lett.*, **8**, 1073.
24. Wehn, P.M. and Du Bois, J. (2002) *J. Am. Chem. Soc.*, **124**, 12950.
25. Hinman, A. and Du Bois, J. (2003) *J. Am. Chem. Soc.*, **125**, 11510.
26. Fleming, J.J., McReynolds, M.D., and Du Bois, J. (2007) *J. Am. Chem. Soc.*, **129**, 9964.
27. Conrad, R.M. and Du Bois, J. (2007) *Org. Lett.*, **9**, 5465.
28. Fleming, J.J., McReynolds, M.D., and Du Bois, J. (2007) *J. Am. Chem. Soc.*, **129**, 9964.
29. Kurokawa, T., Kim, M., and Du Bois, J. (2009) *Angew. Chem., Int. Ed.*, **48**, 2777.
30. Evens, D.A., Faul, M.M., and Bilodeau, M.T. (1994) *J. Am. Chem. Soc.*, **116**, 2742.

31 Muller, P., Baud, C., Jacquier, Y., Moran, M., and Nageli, I. (1996) *J. Phys. Org. Chem.*, **9**, 341.
32 Muller, P., Baud, C., and Jacquier, Y. (1996) *Tetrahedron*, **52**, 1543.
33 Diaz-Requejo, M.M., Belderrain, T.R., Nicasio, M.C., Trofimenko, S., and Perez, P.J. (2003) *J. Am. Chem. Soc.*, **125**, 12078.
34 Fructos, M.R., Trofimenko, S., Diaz-Requejo, M.M., and Perez, P.J. (2006) *J. Am. Chem. Soc.*, **128**, 11784.
35 Badiei, Y.M., Dinescu, A., Dai, X., Palomino, R.M., Heinemann, F.W., Cundari, T.R., and Warren, T.H. (2008) *Angew. Chem., Int. Ed.*, **47**, 9961.
36 He, L., Yu, J., Zhang, J., and Yu, X.Q. (2007) *Org. Lett.*, **9**, 2277.
37 Harden, J.D., Ruppel, J.V., Gao, G.Y., and Zhang, X.P. (2007) *Chem. Commun.*, 4644.
38 Thu, H.Y., Yu, W.Y., and Che, C.M. (2006) *J. Am. Chem. Soc.*, **128**, 9048.
39 Lin, X., Zhao, C., Che, C.M., Ke, Z., and Phillips, D.L. (2007) *Chem. Asian J.*, **2**, 1101.
40 Li, Z., Capretto, D.A., Rahaman, R.O., and He, C. (2007) *J. Am. Chem. Soc.*, **129**, 12058.
41 Albone, D.P., Aujla, P.S., Taylor, P.C., Challenger, S., and Derrick, A.M. (1998) *J. Org. Chem.*, **63**, 9569.
42 Chanda, B.M., Vyas, R., and Bedekar, A.V. (2001) *J. Org. Chem.*, **66**, 30.
43 Albone, D.P., Challenger, S., Derrick, A.M., Fillery, S.M., Irwin, J.L., Parsons, C.M., Takada, H., Taylor, P.C., and Wilson, D.J. (2005) *Org. Biomol. Chem.*, **3**, 107.
44 Bhuyan, R. and Nicholas, K.M. (2007) *Org. Lett.*, **9**, 3957.
45 Ragaini, F., Penoni, A., Gallo, E., Tollari, S., Li Gotti, C., Lapadula, M., Mangioni, E., and Cenini, S. (2003) *Chemistry*, **9**, 249.
46 Yu, X.Q., Huang, J.S., Zhou, X.G., and Che, C.M. (2000) *Org. Lett.*, **2**, 2233.
47 This oxidant is generally more soluble than its acetate counterpart.
48 Fiori, K.W. and Du Bois, J. (2007) *J. Am. Chem. Soc.*, **129**, 562.
49 Reed, S.A. and White, M.C. (2008) *J. Am. Chem. Soc.*, **130**, 3316.
50 Covell, D.J. and White, M.C. (2008) *Angew. Chem., Int. Ed.*, **47**, 6448.
51 Espino, C.G. and Du Bois, J. (2005) in *Modern Rhodium-Catalyzed Organic Reaction* (ed. P.A. Evans), Wiley-VCH, Weinheim, Germany, pp. 379–416.
52 Wehn, P.M., Lee, J.H., and Du Bois, J. (2003) *Org. Lett.*, **5**, 4823.
53 The observed diastereoslectivity might be explained as a reduction of the 1,3 strain in the product.
54 Nageli, I., Baud, C., Bernardinelli, G., Jacquier, Y., Moran, M., and Muller, P. (1997) *Helv. Chim. Acta*, **80**, 1087.
55 Yamawaki, M., Tsutsui, H., Kitagaki, S., Anada, M., and Hashimoto, S. (2002) *Tetrahedron Lett.*, **43**, 9561.
56 Muller, P., Allenbach, Y., and Robert, E. (2003) *Tetrahedron:Asymmetry*, **14**, 779.
57 Reddy, R.P. and Davies, H.M.L. (2006) *Org. Lett.*, **8**, 5013.
58 Liang, C., Robert-Peillard, F., Fruit, C., Muller, P., Dodd, R.H., and Dauban, P. (2006) *Angew. Chem., Int. Ed.*, **45**, 4641.
59 Liang, C., Collet, F., Robert-Peillard, F., Muller, P., Dodd, R.H., and Dauban, P. (2008) *J. Am. Chem. Soc.*, **130**, 343.
60 Sulfonimidamides are prepared from the sulfinyl chloride in two steps.
61 Zalatan, D.N. and Du Bois, J. (2008) *J. Am. Chem. Soc.*, **130**, 9220.
62 Liang, J.L., Yuan, S.X., Huang, J.S., and Che, C.M. (2004) *J. Org. Chem.*, **69**, 3610.
63 Liang, J.L., Yuan, S.X., Huang, J.S., Yu, W.Y., and Che, C.M. (2002) *Angew. Chem., Int. Ed.*, **41**, 3465.
64 Zhou, X.G., Yu, X.Q., Huang, J.S., and Che, C.M. (1999) *Chem. Commun.*, 2377.
65 Milczek, E., Boudet, N., and Blakey, S. (2008) *Angew. Chem., Int. Ed.*, **47**, 6825.

13
Chiral Amines Derived from Asymmetric Aza-Morita–Baylis–Hillman Reaction
Lun-Zhi Dai and Min Shi

13.1
Introduction

Chiral amines are widely distributed in nature and extensively included in many biologically active molecules. In these compounds, the nitrogen-containing units are known to play important roles for their bioactivities [1]. For the synthesis of these chiral nitrogen-containing building blocks, specifically nonamino acid-derived chiral amines, the use of imines as electrophiles is the most promising and convenient route [2]. As one of the most important tools for converting simple starting materials to densely functionalized products in a catalytic and atom economic way, the Morita–Baylis–Hillman (MBH) reaction has made great progress since it was first reported by Morita in 1969 [3]. Use of an imine electrophile, instead of an aldehyde, is a fascinating alternative that would allow for even greater product diversity. The reaction with an imine electrophile came to be known as the aza-Morita–Baylis–Hillman (aza-MBH) and was first reported by Perlmutter and Teo in 1984 [4]. The model aza-MBH reaction is shown in Scheme 13.1. By reacting electron-deficient alkene with acylimine in the presence of nucleophilic catalysts such as tertiary amines and phosphines, these operationally simple and atom economic reactions afforded α-methylene-β-aminocarbonyl derivatives in good yields under mild conditions, which comprise a contiguous assembly of three different functionalities. Obviously, the highly efficient asymmetric aza-MBH reaction can be an effective synthetic method to afford nonamino acid-derived chiral amines.

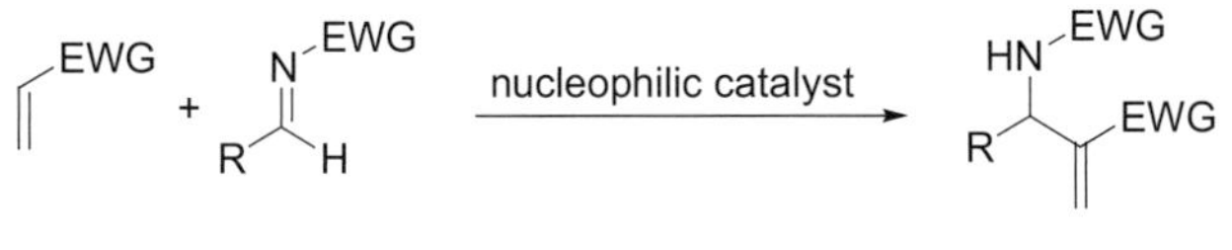

Scheme 13.1 A general aza-MBH reaction. EWG: electron-withdrawing group.

Chiral Amine Synthesis: Methods, Developments and Applications. Edited by Thomas C. Nugent
Copyright © 2010 WILEY-VCH Verlag GmbH & Co. KGaA, Weinheim
ISBN: 978-3-527-32509-2

13.2
Recent Mechanistic Insights

In accordance with the generally accepted mechanism of the MBH reaction, the aza-MBH reaction involves, formally, a sequence of Michael addition, Mannich-type reaction, and β-elimination. A commonly accepted mechanism is depicted in Scheme 13.2. A reversible conjugate addition of the nucleophilic catalyst to the Michael acceptor generates an enolate, which can intercept the acylimine to afford the second zwitterionic intermediate. A proton shift from the α-carbon atom to the β-amide followed by β-elimination then affords the aza-MBH adduct with concurrent regeneration of the catalyst [5].

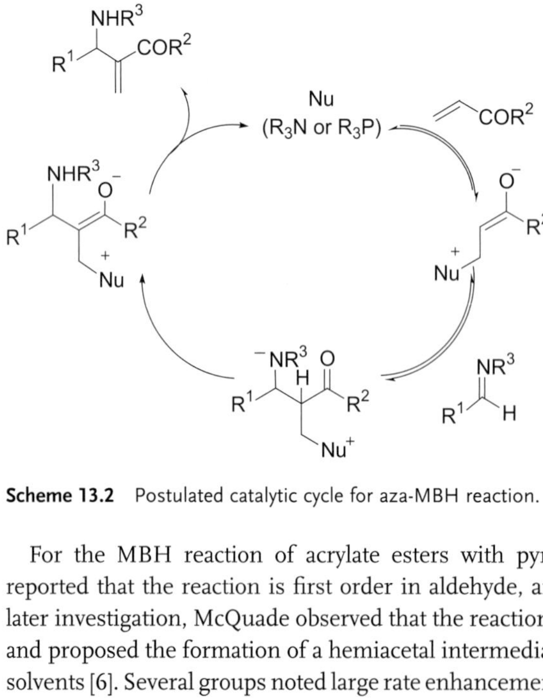

Scheme 13.2 Postulated catalytic cycle for aza-MBH reaction.

For the MBH reaction of acrylate esters with pyridinecarboxaldehydes, Kaye reported that the reaction is first order in aldehyde, amine, and acrylate [5b]. In a later investigation, McQuade observed that the reaction is second order in aldehyde and proposed the formation of a hemiacetal intermediate for the reaction in aprotic solvents [6]. Several groups noted large rate enhancements caused by water and other protic additives [7]. Aggarwal and Lloyd-Jones concluded that proton transfer is rate limiting in the initial stage of the reaction, but becomes increasingly efficient at higher conversion causing the aldol-type coupling to become rate limiting and making the reaction autocatalytic [8]. However, mechanistic studies on aza-MBH reaction were unavailable until 2005. In this regard, Jacobsen and Leitner have provided influential work that greatly promoted the development of the aza-MBH reaction and its enantioselective variant.

Jacobsen and Raheem chose the reaction between N-nosylbenzaldimine and methyl acrylate catalyzed by chiral thiourea derivatives **1a** and 1,4-diazabicyclo [2.2.2]octane (DABCO) as a representative system to study the mechanism of the thiourea-catalyzed aza-MBH reaction [9]. The physical properties of the reaction mixtures were observed to change within a few hours of initiation, with the system becoming more viscous and the appearance of a yellow precipitate. ESI mass spectral analysis of the precipitate isolated from reactions revealed a prominent peak with an m/z value of 489.2, consistent with a structure corresponding to zwitterionic **2** (Scheme 13.3). Further, addition of excess 4N HCl to the crude reaction mixture, followed by aqueous extraction, afforded a glassy, off-white solid isolated from the aqueous layer that was characterized by ESIMS, ^{1}H NMR, ^{13}C NMR, and IR as the dihydrochloride salt **3**. Salt **3** was isolated in high diastereomeric purity, with the relative stereochemistry of the major isomer assigned as *anti*.

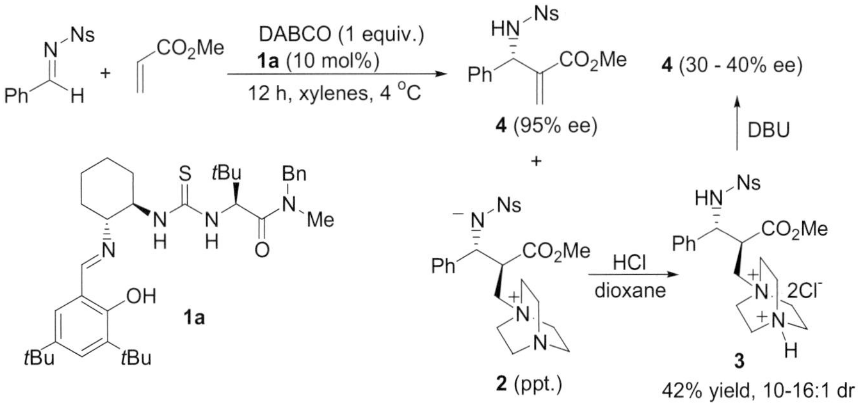

Scheme 13.3 Preparation of **3**.

Moreover, the reaction of methyl acrylate and N-nosylbenzaldimine, which was promoted by DABCO in the absence of **1a**, was homogeneous in CHCl$_3$ and proved amenable to kinetic analysis. The reaction was monitored by GC analysis, and was found to display a first-order kinetic dependence on both DABCO and methyl acrylate. In contrast to the MBH reaction, rate saturation with respect to the imine electrophile was observed. A prominent primary kinetic isotope effect was also observed ($k_H/k_D = 3.81$) by comparison of initial reaction rate of methyl acrylate with separate reactions of α-deuterio-methyl acrylate, strongly suggesting that deprotonation of the α-H(D) was rate limiting.

Meanwhile, Leitner and coworkers monitored the aza-MBH reaction of methyl vinyl ketone (MVK) with N-tosylated imine in the presence of PPh$_3$ in THF-d_8 at room temperature by ^{19}F NMR spectroscopy [10]. The rate law was derived by analyzing the initial rates as a function of concentration for the individual components. The broken order of 0.5 in imine indicated that the rate-determining step was partially influenced by proton transfer. A variety of Brønsted acidic additives with different pK_a values were examined with 3,5-bis(CF$_3$)phenol at p$K_a \approx 8$ corresponding to a 14-fold rate enhancement as compared to the reaction without additive. Examination of the kinetics in the presence of phenol as prototypical additive revealed that the rate law of the reaction changed in the presence of the Brønsted acid, showing first-order dependence on imine. Thus, the elimination step was not involved in the rate-determining step anymore, and that the proton transfer was accelerated by protic additive. The results obtained so far substantiate that bifunctional activation using a basic and a protic center is a viable strategy for catalyst design in the asymmetric aza-MBH reaction.

13.3
Asymmetric Aza-MBH Reaction

Very few efficient catalytic enantioselective versions of MBH reaction were known up to 1999 despite a considerable amount of efforts devoted to the field. A breakthrough came in 1999 when Hatakeyama and coworkers discovered that β-isocupreidine (β-ICD) is an efficient catalyst for the MBH reaction [11]. Meanwhile, the use of small organic molecules as catalysts to perform asymmetric transformations has received increasing attention over the past decade. Therefore, chiral multifunctional organocatalysts have also been developed rapidly to promote successful enantioselective MBH/aza-MBH processes. This chapter mainly summarizes recent advances in the design and synthesis of small organic molecules for the enantioselective aza-MBH reactions from 2000. On the basis of these enantioselective aza-MBH reactions, a variety of chiral amines can be easily prepared under mild conditions.

13.4
Chiral Auxiliary-Induced Diastereoselective Aza-MBH Reaction

In the past years, only a few reports have dealt with a chiral auxiliary-induced diastereoselective aza-MBH reaction. In 1994, Kündig *et al.* explored the reaction of methyl acrylate and acrylonitrile with enantiopure planar chiral *o*-substituted $Cr(CO)_3$. Under the catalysis of DABCO, the corresponding aza-MBH adducts were obtained in good yields. Removal of the metal provided chiral amines in high yields and enantiomeric excesses [12]. Later on, Aggarwal *et al.* used enantiopure *N*-sulfinimines in the aza-MBH reaction with methyl acrylate in the presence of 3-hydroxyquinuclidine (3-HQD) and Lewis acid. The desired adducts, functionalized β-sulfonated amino acid derivatives, were obtained with good diastereoselectivities (Scheme 13.4) [13].

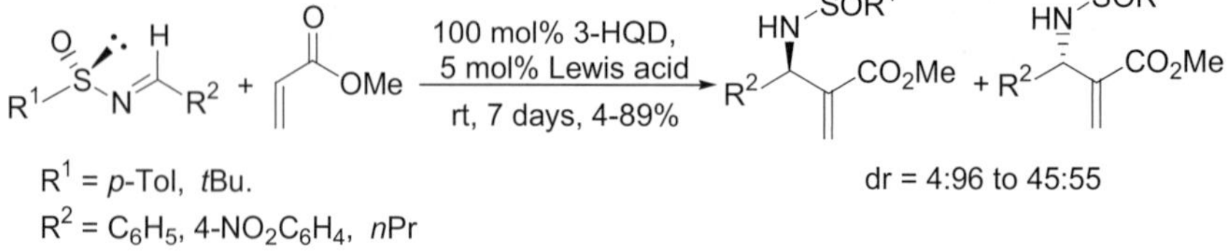
100 mol% 3-HQD,
5 mol% Lewis acid
rt, 7 days, 4-89%
OMe
CO_2Me
SOR^1
HN
dr = 4:96 to 45:55
R^1 = p-Tol, tBu.
R^2 = C_6H_5, 4-$NO_2C_6H_4$, nPr

Scheme 13.4 Aza-MBH reactions of *N*-sulfinimines with methyl acrylate.

Subsequently, enantiopure N-sulfinimines were also adopted by Shi and Xu in the aza-MBH reaction of N-sulfinimines with 2-cyclopenten-1-one in the presence of PPhMe$_2$ (10 mol%). Good yields and diastereoselectivities were achieved with the aza-MBH adducts (Scheme 13.5) [14].

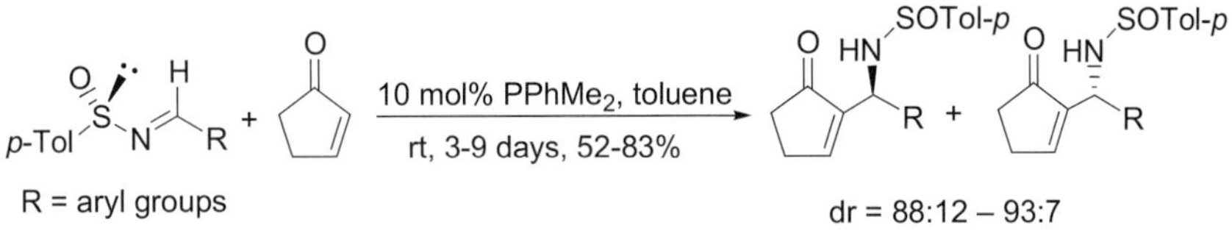

Scheme 13.5 Aza-BMH reactions of N-sulfinimines with 2-cyclopenten-1-one.

Recently, in the presence of an effective air-stable nucleophilic trialkylphosphine organocatalyst, 1,3,5-triaza-7-phosphaadamantane (PTA), a chiral N-thiophosphoryl imine bearing a (S)-binaphthalene skeleton induced a diastereoselective aza-MBH reaction that was carried out with fair chemical yields and moderate to excellent diastereoselectivities (up to >99% de) (Scheme 13.6) [15]. Removal of the chiral auxiliary provides the corresponding chiral amines in good yields.

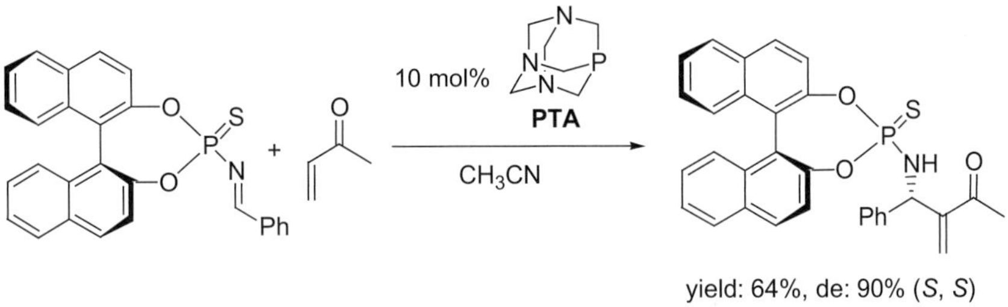

Scheme 13.6 PTA-catalyzed diastereoselective aza-MBH reaction.

13.5
Chiral Tertiary Amine Catalysts

13.5.1
Cinchona-Derived Bifunctional Catalysts

In 1999, Hatakeyama and coworkers [11a] discovered that β-ICD is an efficient catalyst for the MBH reaction. Both the rigid tricyclic structure and the phenolic OH group are indispensable for obtaining a high degree of asymmetric induction as well as rate acceleration. Thus, neither O-methyl-β-isocupreidine nor the O-demethylated hydroquinidine that lacks the tricyclic cage structure is effective for the asymmetric MBH reaction (Figure 13.1). Overall, the nucleophilic nitrogen atom in the quinuclidine moiety of β-ICD acts as a Lewis base (LB) to initiate the MBH reaction, whereas the phenolic OH group (Brønsted acid (BA)) acts as a Lewis acid to stabilize

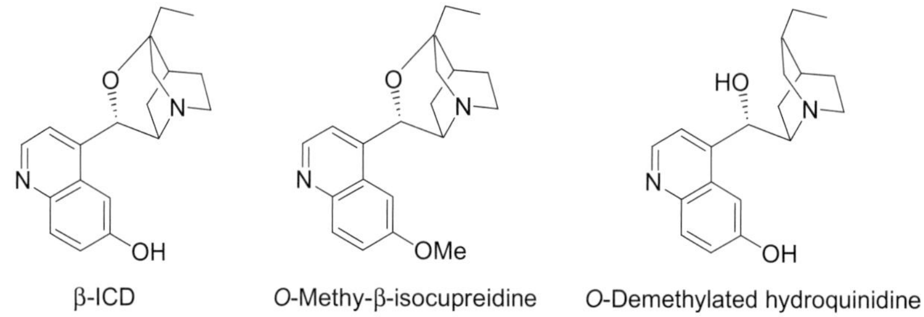

Figure 13.1 Several cinchona derivatives.

and organize the enolate intermediate and also to promote the subsequent aldol addition. Consequently, β-ICD is considered to be a typical bifunctional (LBBA) chiral organocatalyst.

Although the field of catalytic enantioselective MBH reactions has progressed greatly since 1999, the catalytic enantioselective aza-MBH reaction was notably missing until 2002 [16]. Shi and coworkers applied chiral amine quinidine derivatives of β-ICD to promote the aza-MBH reaction of N-sulfonyl imines with MVK. The reactions optimally performed in MeCN/DMF (1/1) at −30 °C gave the best results. The corresponding products were obtained in moderate to good yields with good to high enantiomeric excesses (Table 13.1). However, the reactions of N-tosylated imines with other activated olefins such as acrylonitrile, acrolein, methyl acrylate, phenyl acrylate, or naphthyl acrylate did not give good results. After careful examination of the solvent and temperature, it was found that the reaction could proceed smoothly to afford the products in moderate to good yields and ee's under specialized conditions (Scheme 13.7) [17]. Interestingly, the absolute configurations of the products derived from acrylonitrile, acrolein, methyl acrylate, phenyl acrylate, or naphthyl acrylate are opposite to those of the products derived from MVK and EVK. Removal of the N-tosyl group provides the corresponding chiral amines in good yields.

Table 13.1 The scope of asymmetric aza-MBH reaction of N-sulfonated imine with MVK catalyzed by β-ICD.

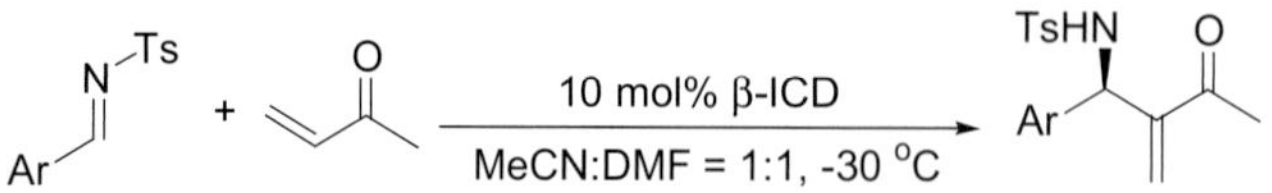

Entry	Ar	Yield (%)	ee (%)	Absolute configuration
1	C$_6$H$_5$	80	97	R
2	4-MeC$_6$H$_4$	76	96	R
3	4-MeOC$_6$H$_4$	64	99	R
4	3-FC$_6$H$_4$	55	90	R
5	2-Furyl	58	73	R
6	C$_6$H$_5$CH=CH	54	46	R

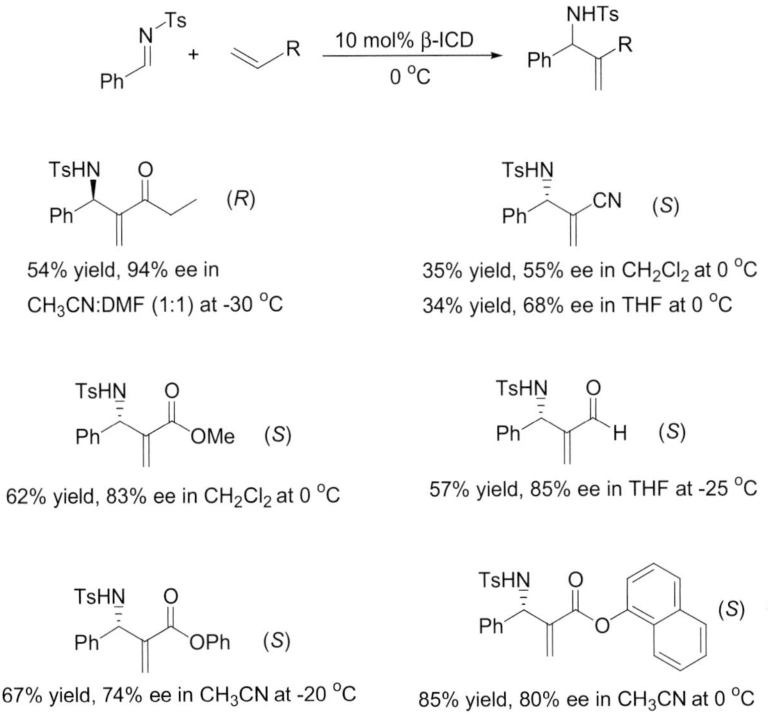

13.5 Chiral Tertiary Amine Catalysts

Ph—CH=N—Ts + CH₂=CH—R → (10 mol% β-ICD, 0 °C) → Ph—CH(NHTs)—C(=CH₂)—R

TsHN O
Ph—*—CH₂CH₃ (R)

54% yield, 94% ee in
CH₃CN:DMF (1:1) at -30 °C

TsHN
Ph—*—CN (S)

35% yield, 55% ee in CH₂Cl₂ at 0 °C
34% yield, 68% ee in THF at 0 °C

TsHN O
Ph—*—OMe (S)

62% yield, 83% ee in CH₂Cl₂ at 0 °C

TsHN O
Ph—*—H (S)

57% yield, 85% ee in THF at -25 °C

TsHN O
Ph—*—OPh (S)

67% yield, 74% ee in CH₃CN at -20 °C

TsHN O
Ph—*—O-(naphthyl) (S)

85% yield, 80% ee in CH₃CN at 0 °C

Scheme 13.7 Asymmetric aza-MBH reaction of N-sulfonated imine with various activated olefins.

Recently and significantly, Shi further discovered that the aza-MBH reaction between N-tosyl salicylaldehyde imines and α,β-unsaturated ketones catalyzed by β-ICD afforded the corresponding aza-MBH products in good to high yields (90% quant.) and excellent ee's (up to 99% ee) [18]. Interestingly, the products showed an opposite absolute configuration compared with the products resulting from the similar aza-MBH reaction of N-sulfonyl imines with α,β-unsaturated ketones, suggesting that the tethered phenolic hydroxyl group in the imine substrate can remarkably affect the reaction outcome. The representative results are shown in Scheme 13.8.

The function of the *ortho*-phenolic hydroxy group in N-tosyl salicylaldehyde imines was investigated through two control experiments. The reaction of N-tosylimines **5** and **6** with MVK in the presence of 10 mol% of β-ICD in THF at room temperature afforded the corresponding aza-MBH adducts in low yields and ee's even after 8 days, suggesting that the *ortho*-phenolic hydroxyl group in N-tosyl salicylaldehyde imines plays a key role in this reaction, providing the aza-MBH product in high yield and ee (Scheme 13.9).

In the case of using salicyl N-tosylimine in the reaction, two intermediates would be formed and intermediate **7** is more stable than intermediate **8**, because the *ortho*-phenol containing aromatic moiety in the chiral pocket can form a stronger net type or branched hydrogen bonding system, thereby affording the corresponding

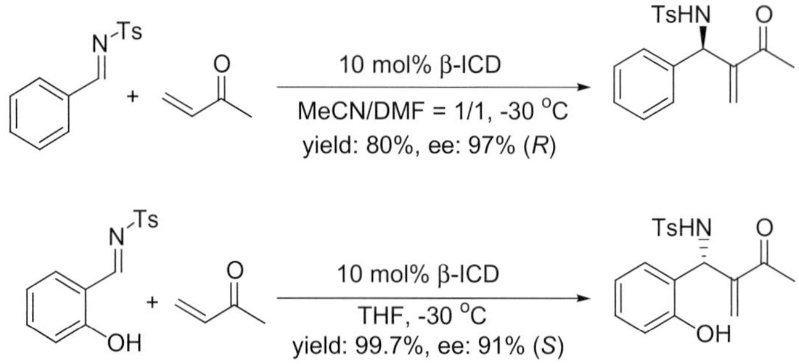

Scheme 13.8 Aza-MBH reactions with MVK and salicyl N-tosylimine or N-tosylimine of benzaldehyde.

aza-MBH product in the S-configuration. However, in intermediate **8**, the *ortho*-phenol group is unable to join in the transition state to influence the reaction outcome since it takes up a position outside of the chiral environment. As for N-tosylimines **5** or **6**, the phenolic hydroxy group is located at a position outside the hydrogen bonding network system and only acts as an electron-donating group on the benzene ring, retarding the reaction rate (Scheme 13.10).

Adolfsson and Balan [19] applied this catalyst to the three-component reaction between acrylates, aldehydes, and tosylamines with Ti(O^{i}Pr)$_4$ and molecular sieves as additives. Good yields and enantiomeric excesses were similarly achieved (Scheme 13.11).

The reactions of N-arylidenediphenylphosphinamides with hexafluoroisopropyl acrylate (HFIPA) to give the corresponding aza-MBH adducts in moderate to excellent yields with good enantiomeric excesses was also reported by Hatakeyama [20] (Scheme 13.11). In contrast to the corresponding aldehydes, imines show the opposite enantioselectivity, and Hatakeyama and coworkers indicated that H-bonding was responsible for the reversal in enantioselectivity. Michael addition

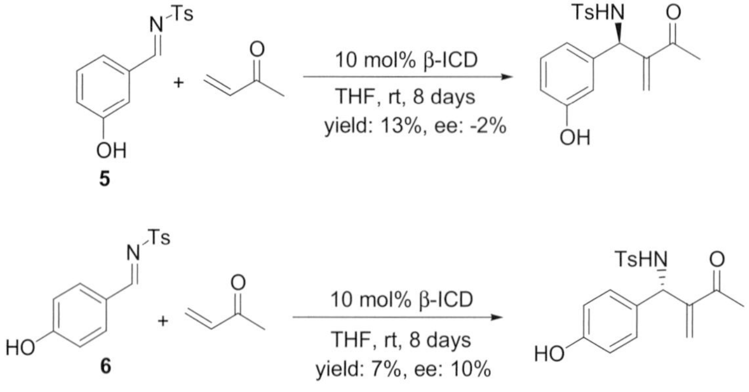

Scheme 13.9 Two controlled experiments.

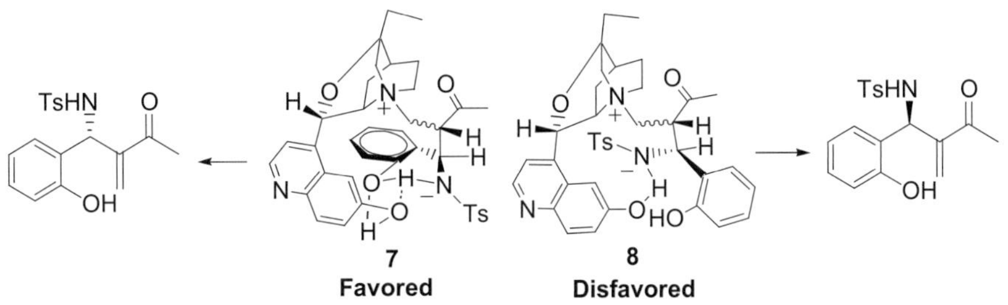

Scheme 13.10 Two competitive intermediates in the reaction.

of β-ICD to HFIPA forms enolate **9**, which in turn undergoes Mannich reaction with the imine to furnish an equilibrium mixture of several diastereomers. Among them, there would be two intermediates **10** and **11** that are stabilized through hydrogen bonding between the amidate ion and the phenolic OH. On taking the *anti*-periplanar arrangement of the ammonium portion and the α-hydrogen of the ester group in the subsequent E2 or E1cb reaction (see Newman projection **12**), intermediate **11** suffers from more severe steric interaction than intermediate **10**, as depicted in Scheme 13.12. Thus, the difference in the reaction rate of the elimination step of **10** and **11** would result in (S)-enriched enantioselectivity through equilibration. Aliphatic imines were poor substrates for the reaction and this is believed to be due to their extremely labile nature.

Very recently, Zhu and coworkers have developed a novel catalysis that was applicable to the reaction of acrylate with both aromatic and aliphatic imines [21]. The 6′-OH group in β-ICD provided a handle for the introduction of other H-bond donors that were capable of modulating both the steric and electronic parameters. Consequently, a series of amides and thioureas were synthesized from β-ICD and screened as catalysts using N-(p-methoxybenzenesulfonyl)imine and β-naphthyl acrylate as model substrates. As summarized in Table 13.2, after a survey of reaction conditions by varying the solvents, the temperatures, the stoichiometries, and the

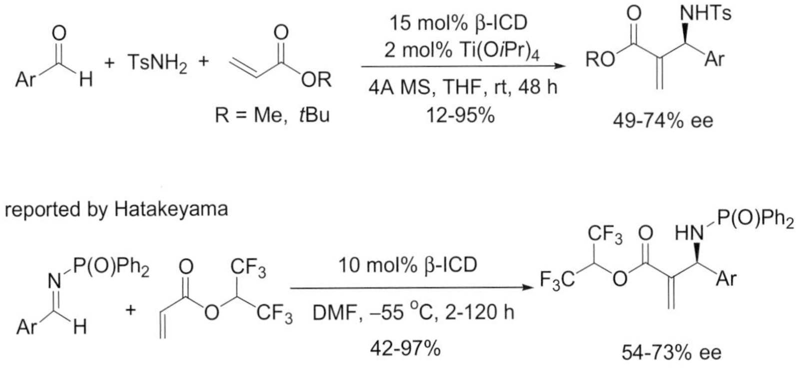

Scheme 13.11 Asymmetric aza-MBH reactions catalyzed by β-ICD.

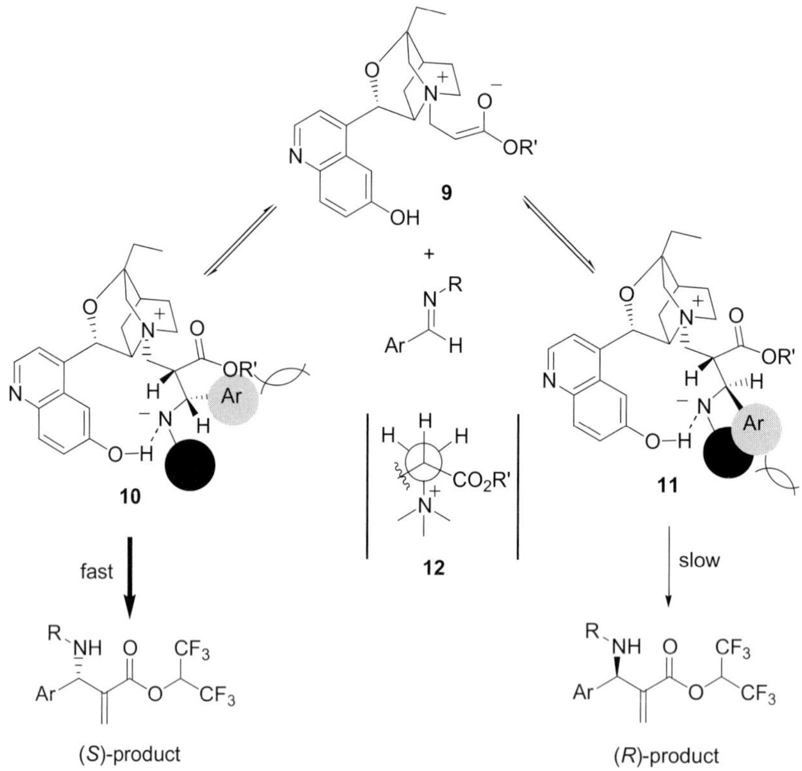

Scheme 13.12 Proposed reaction mechanism.

additives, the optimal conditions consisted of performing the reaction in CH_2Cl_2 at −30 °C in the presence of 0.1 equiv of catalyst **13c** and 0.1 equiv of β-naphthol as an additive. Too strong acids instead of β-naphthol as an additive either reduced the yield ($PhCO_2H$) or completely inhibited the reaction ($PhSO_3H$) probably due to the acid–base quench of active catalyst.

For aromatic imines, the yield and the ee of adducts were not sensitive to the electronic properties of the aromatic ring and the presence of strong electron withdrawing (NO_2, Table 13.3, entry 1) or donating group (MeO, Table 13.3, entry 2) was well tolerated. All previous efforts to include aliphatic imines as electrophilic partner in the aza-MBH reaction have met with failure. Importantly, under these reaction conditions, aliphatic imines, including (-unbranched imines, readily participated in the aza-MBH reaction to yield adducts in moderate yields and good enantioselectivities (Table 13.3, entries 4–7). Naphthol had a lesser impact on the enantioselectivity when aliphatic imines were used. However, higher yield was obtained in its presence (Table 13.3, entry 5).

The coexistence of H-bond pairings between two H-bond acceptors (enolate oxygen and imine nitrogen) and two H-bond donors (naphthol and amide NH) could explain the observed enantioselectivity (Scheme 13.13). Thus, the *E*-enolate

Table 13.2 Aza-MBH reaction: survey of catalysts.

13a (β-ICD) X = OH
13b X = NHC(S)NH-(3,5-di-CF₃)Ph

13c R¹ = R² = R³ = H, R⁴ = Boc
13d R¹ = R² = R³ = H, R⁴ = Tf
13e R¹ = R² = R³ = H, R⁴ = C(S)NH-(3,5-di-CF₃)Ph
13f R¹ = Bn, R² = R³ = H, R⁴ = Boc
13g R¹ = R³ = H, R² = Bn, R⁴ = Boc
13h R¹ = R² = H, R³ = Me, R⁴ = Boc

PMP = p-methoxyphenyl

Entry	Catalyst	Temperature (°C)	Yield (%)[a]	ee (%)[b]
1[c]	13b	rt	81	30[d]
2[c]	13c	rt	99	76[d]
3[c]	13d	rt	88	69[d]
4[c]	13e	rt	64	55[d]
5	13c	−30	87	91[d]
6	13a	−30	39	77[d]
7	13f	−30	75	78[d]
8	13g	−30	57	86[d]
9	13c	−30	80	79[e]
10	13c	−30	95	92[f,g]

a) Isolated yield after column chromatography.
b) Determined by chiral HPLC analysis.
c) Use of 20 mol% of catalyst **13**.
d) Use of commercially available β-naphthyl acrylate (2.0 equiv).
e) Use of freshly prepared β-naphthyl acrylate (2.0 equiv).
f) The reaction was carried out in the presence of **13c** (10 mol%), β-naphthol (10 mol%), and freshly prepared β-naphthyl acrylate (2.0 equiv).
g) The absolute configuration of the product was determined to be (S)-enriched.

formed upon addition of **13c** to acrylate, being more Lewis basic, would form a H-bond with the β-naphthol, which could in turn be stabilized by π–π interaction between two naphthyl groups. The less basic neutral imine would H-bond to the less acidic amide NH. These H-bonding pairs would be stronger and sterically less constrained than the alternative intramolecular H-bond. Among the two possible transition states, that of **14** was less crowded and should lead, after the C−C bond

Table 13.3 Selected examples for the enantioselective aza-MBH reaction of aromatic and aliphatic imines.[a]

Entry	R	Time (h)	Yield (%)[b]	ee (%)[c]
1	p-NO$_2$C$_6$H$_4$	47	67	97
2	p-MeOC$_6$H$_4$	80	84	93
3	3-Furyl	60	80	98
4	PhCH=CH	72	52	90
5	iPrCH$_2$	24	57 (43)[d]	87 (82)[d]
6	nPentyl	24	38	85
7	Ph(CH$_2$)$_2$	24	39	86

a) Reaction conditions: imine (1 mmol), β-naphthyl acrylate (2 mmol), β-naphthol (0.1 mmol), **13c** (0.1 mmol) in CH$_2$Cl$_2$ (0.35 ml) at −30 °C for aromatic imines and at 0 °C for aliphatic imines.
b) Isolated yield after column chromatography.
c) Determined by chiral HPLC analysis.
d) Use of β-naphthyl acrylate (2.0 equiv) freshly prepared without β-naphthol.

formation, to the adduct **15**. The β-naphthol-mediated proton transfer followed by β-elimination would then afford the observed (S)-adduct (Scheme 13.13).

13.5.2
Chiral Binol-Derived Bifunctional Amine Catalysts

Sasai and coworkers reported that a chiral BINOL-derived amine **16a** catalyzed asymmetric aza-MBH reaction of N-tosyl imines with acrolein and alkyl vinyl ketones [22]. The corresponding aza-MBH adducts were obtained in good to excellent yields with high enantiomeric excesses (Table 13.4). Replacing the iPr group with other substituents in amine **16a** provided less-effective catalysts regarding yield or enantioselectivity [23].

The catalyst **16a** possesses two phenolic hydroxy groups and two nitrogen atoms. To clarify the role of each unit on the reaction, the monoprotected catalyst **16b** with

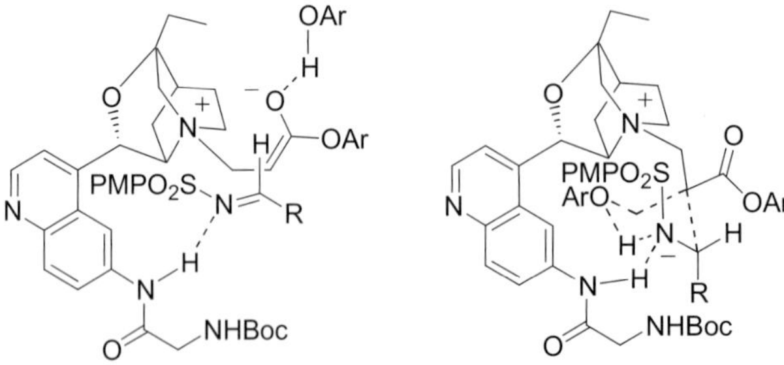

Scheme 13.13 Two possible transition states for the aza-MBH reaction catalyzed by **13c**.

Table 13.4 Enantioselective aza-MBH reaction.

Entry	Ar	R	Time (h)	Yield (%)	ee (%)
1	C_6H_5	Me	168	93	87
2	4-MeOC$_6$H$_4$	Me	132	93	94
3	4-ClC$_6$H$_4$	Me	60	96	95
4	2-Furyl	Me	48	100	88
5	2-Naphthyl	Me	108	94	91
6	4-NO$_2$C$_6$H$_4$	Me	12	91	91
7	4-NO$_2$C$_6$H$_4$	Et	96	88	88
8	4-NO$_2$C$_6$H$_4$	H	36	95	94

the 2'-OMe group, **16c** with the 2-OMe group and the aniline derivative **17**, the pyridine derivative **18a** and **18b** with varying chain length of the spacer, and **18c** with the oxygen atom linker were examined (Figure 13.2). Although the catalyst **16b** was not effective in promoting the reaction (5% yield, 24% ee), **16c** showed a slightly decreased activity (85% yield, 79% ee) compared to the parent catalyst **16a**. Interestingly, **17** and **18** did not promote the reaction. These results and the significant enantioselectivity of **16a** are consistent with the notion that the designed organic molecules can act as bifunctional catalysts utilizing both the phenolic 2'-hydroxy

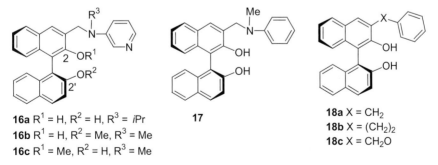

Figure 13.2 Various catalysts to this reaction.

13.5.3
Chiral Acid/Achiral Amine

Over the past several years, chiral thiourea derivatives have been extensively used as an efficient class of asymmetric catalysts. In 2005, Jacobsen and Raheem applied chiral thiourea derivatives to promote aza-MBH reaction of nosylimines with methyl acrylate [9]. A careful screening of thiourea catalysts led to the identification of **1a** and **1c** as optimal, with **1a** selected for further study due to its greater synthetic accessibility. By further selecting reaction solvent and nucleophilic additive, markedly higher reactivities and enantioselectivities were obtained. Addition of exactly 1 equiv of DABCO relative to methyl acrylate led to the highest enantioselectivity. Pronounced solvent and concentration effects were observed, with reactions carried out in nonpolar solvents (toluene, Et_2O, xylenes) affording the highest ee's, but producing highly heterogeneous mixtures and generally low yields. As can be seen from Table 13.5, while product yields still require improvement (<50%), the enantioselectivities for a variety of aromatic imines are high and unprecedented for aza-MBH reactions with acrylate derivatives.

Table 13.5 Thiourea-catalyzed aza-MBH reaction.

1a R^1 = Bn, R^2 = Me, R^3 = tBu, X = S
1b R^1 = R^2 = Me, R^3 = OCO tBu, X = O
1c R^1 = R^2 = Me, R^3 = OCO tBu, X = S

Ar-CH=NNs + CH$_2$=CHCO$_2$Me → (10 mol% **1a**, DABCO, MS 3 Å, xylenes, 4 °C) → Ar-CH(NHNs)-CH(=CH$_2$)-C(O)OMe

Entry	Ar	Time (h)	Yield (%)	ee (%)
1	C_6H_5	36	49	95
2	3-MeC$_6$H$_4$	24	40	93
3	3-MeOC$_6$H$_4$	24	42	96
4	3-ClC$_6$H$_4$	16	33	94
5	2-Thiophenyl	36	30	99
6	1-Naphthyl	24	27	91
7	3-Furyl	36	25	98

13.6
Chiral Phosphine Catalysts

As triaryl- or trialkylphosphines are effective catalysts for the MBH reaction, it appears logical that one expects to develop an enantioselective version by using chiral phosphines. Unfortunately, of the many chiral phosphines that have been screened, only a few of them displayed reasonable catalytic activity to afford, with limited substrate scope, the desired product with low to moderate enantioselectivity [24].

Shi and coworkers found that chiral phosphine **19a** was a successful catalyst for the asymmetric aza-MBH reaction of N-sulfonyl imines with acrolein, alkyl vinyl ketones, and acrylates (Table 13.6) [25]. In the presence of a catalytic amount of chiral phosphine and 4 Å molecular sieves, the corresponding aza-MBH adducts could be obtained in good yields with good to high ee's (70–95% ee) at low temperature (-30 to $-20\,^\circ$C) in THF or at room temperature in CH_2Cl_2. In CH_2Cl_2 upon heating at $40\,^\circ$C, the aza-MBH reaction of N-sulfonated imines with phenyl acrylate or naphthyl acrylate gave adducts in good to high yields (60–97%) with moderate ee's (52–77%). Both ^{31}P and ^{1}H NMR spectroscopic studies indicated the bifunctional role of the catalyst. The phosphine acts as a Lewis base to initiate the reaction sequence, whereas

Table 13.6 Selected examples for aza-MBH reaction catalyzed by chiral phosphine **19a**.

Entry	Ar	R^1	R^2	Conditions	Yield (%)	ee (%)
1	C$_6$H$_5$	Ts	Me	THF, $-30\,^\circ$C, 4A MS, 36 h	83	83 (S)
2	4-MeC$_6$H$_4$	Ts	Me	THF, $-30\,^\circ$C, 4A MS, 36 h	82	81 (S)
3	3-FC$_6$H$_4$	Ts	Me	THF, $-30\,^\circ$C, 4A MS, 24 h	96	85 (S)
4	C$_6$H$_5$CH=CH	Ts	Me	THF, $-30\,^\circ$C, 24 h	94	95 (S)
5	C$_6$H$_5$	Ts	Et	THF, $-20\,^\circ$C, 4A MS, 96 h	52	80 (S)
6	4-ClC$_6$H$_4$	Ts	OMe	CH_2Cl_2, rt	67	18 (S)
7	4-NO$_2$C$_6$H$_4$	Ts	OPh	CH_2Cl_2, $40\,^\circ$C, 12 h	97	75 (S)
8	C$_6$H$_5$	Ts	O-1-naphthyl	CH_2Cl_2, rt	49	58
9	C$_6$H$_5$	Ts	O-2-naphthyl	CH_2Cl_2, rt	57	53
10	4-MeOC$_6$H$_4$	Ts	H	THF, rt, 12 h	99	78 (S)
11	4-ClC$_6$H$_4$	Ts	Ms	THF, $-30\,^\circ$C, 24 h	94	82 (S)
12	4-ClC$_6$H$_4$	Ts	Ns	THF, $-30\,^\circ$C, 24 h	—	—
13	4-ClC$_6$H$_4$	Ts	4-ClC$_6$H$_4$SO$_2$	THF, $-30\,^\circ$C, 24 h	94	89 (S)
14	4-ClC$_6$H$_4$	Ts	SES	THF, $-30\,^\circ$C, 72 h	53	89 (S)

the phenolic OH group serves as a Lewis acid to activate the electrophile and to stabilize the enolate intermediate through hydrogen bonding.

However, for the reaction of N-arylidenediphenylphosphinamides with other ordinary activated alkenes such as MVK, acrylonitrile, or phenyl acrylate, neither the chiral amine β-ICD nor the chiral phosphine **19a** could provide good levels of enantioinduction (Scheme 13.14) [26].

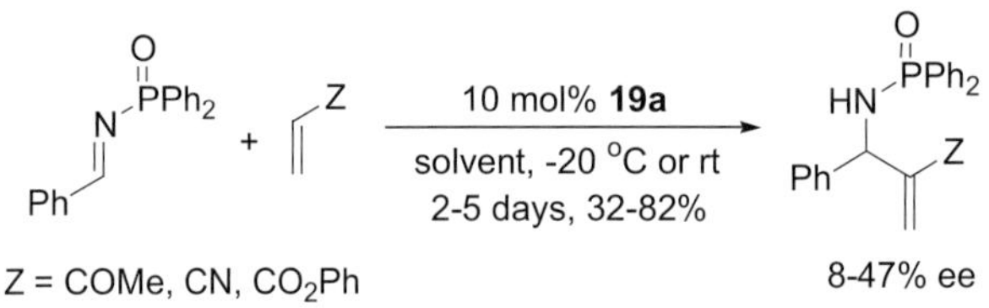

Scheme 13.14 Reaction of N-arylidenediphenylphosphinamides with activated alkenes.

Regarding the phosphine catalyst, the phenolic hydroxy group on the naphthyl ring is crucial for this reaction because the reaction becomes sluggish. For example, the aza-MBH product is found in 13% yield with only 20% ee (R-configuration) using the O-methylated ligand **19b** as a chiral Lewis base and in 17% yield with 22% ee (R-configuration) when using **19c** as a chiral catalyst under the same conditions. **19d** shows no catalytic activity for this reaction. **19e** that bears a thiophenol moiety on the naphthyl ring and **19f** that has two aliphatic hydroxy groups show low catalytic activities under the same conditions. These results suggest that the acidity of the hydroxy group plays a very important role for **19a** to be effective in the aza-MBH reaction. Chiral ferrocene-based phosphine ligands **19g** and **19h** also have low catalytic activities for this reaction under identical conditions (Scheme 13.15).

Leitner and coworkers [27] found that triphenylphosphine either alone or in combination with protonic additives could cause racemization of the aza-MBH product by proton exchange at the stereogenic center, but the chiral catalyst **19a** developed by Shi's group did not induce any racemization on a similar time scale.

Subsequently, Shi and Li synthesized a more nucleophilic chiral phosphine **19i** and subjected it to the aza-MBH reaction of N-sulfonyl imines with 2-cyclohexen-1-one or 2-cyclopenten-1-one [28]. The resulting adducts could be obtained in good yields, but with moderate enantiomeric excess (Scheme 13.16).

One phenyl group and an electron-donating alkyl group on the phosphorus atom was also designed and successfully synthesized (Scheme 13.17) [29]. Shi and coworkers found that these chiral bifunctional phosphane catalysts are very effective for this reaction under mild and concise conditions to produce the corresponding adducts in good to excellent yields and moderate to good enantioselectivities within 1–5 h. To the best of our knowledge, this is the fastest catalytic asymmetric aza-MBH reaction reported thus far.

Sasai and coworkers also functionalized the 3-position of BINOL with a series of aryl phosphines [30]. It was found that catalyst **20** could efficiently catalyze the asymmetric aza-MBH reaction of N-tosyl imines with vinyl ketones albeit slowly (Scheme 13.18).

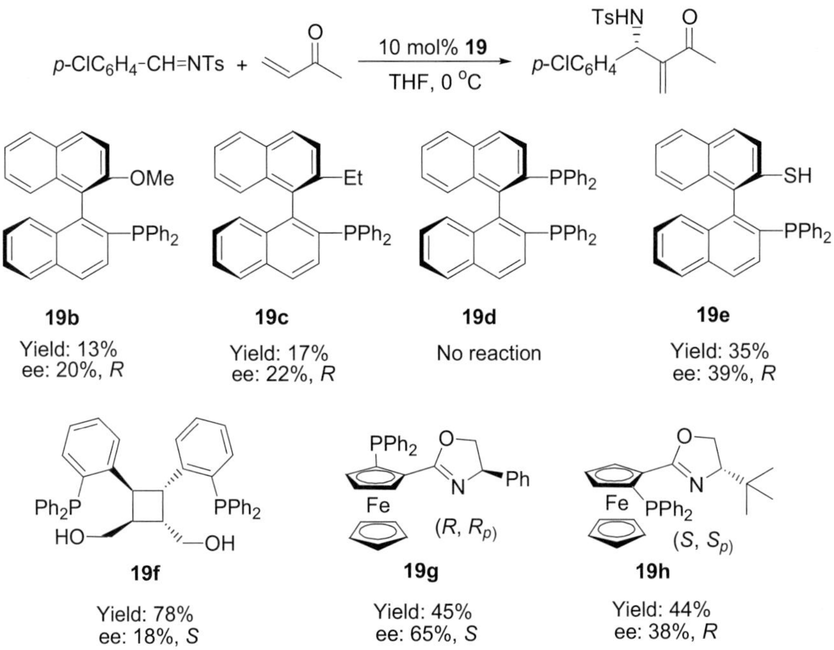

13.6 Chiral Phosphine Catalysts

$p\text{-ClC}_6\text{H}_4\text{-CH=NTs}$ + [MVK] $\xrightarrow[\text{THF, 0 °C}]{\text{10 mol\% 19}}$ $p\text{-ClC}_6\text{H}_4$—[product with TsHN]

19b
Yield: 13%
ee: 20%, R

19c
Yield: 17%
ee: 22%, R

19d
No reaction

19e
Yield: 35%
ee: 39%, R

19f
Yield: 78%
ee: 18%, S

19g (R, R$_p$)
Yield: 45%
ee: 65%, S

19h (S, S$_p$)
Yield: 44%
ee: 38%, R

Scheme 13.15 Survey of chiral phosphine catalysts.

To fine-tune the reactivity and enantioselectivity of the catalysts, chiral phosphine **21** with a "ponytail" was devised and synthesized, the corresponding adducts could be obtained in good yields with good to high ee's [31]. Catalyst **22** also turned out to be more effective in this reaction than the previously reported original chiral phosphine **19a** (Scheme 13.19).

Later on, Shi and coworkers further modified catalyst **19a**, incorporating multiple phenol moieties, and found that catalyst **23** gave better asymmetric induction [32]. The corresponding adducts could be obtained in >90% ee and good to high yields at −20 °C or room temperature (25 °C) in THF for most of the substrates using MVK, EVK, or acrolein as a Michael acceptor (Scheme 13.20).

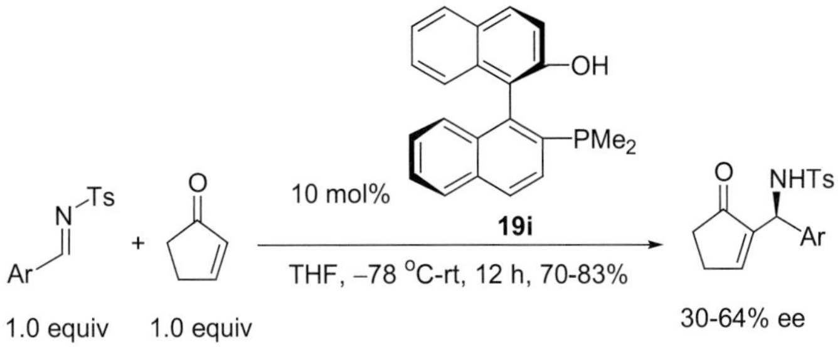

Scheme 13.16 Aza-MBH reactions using 2-cyclopenten-1-one.

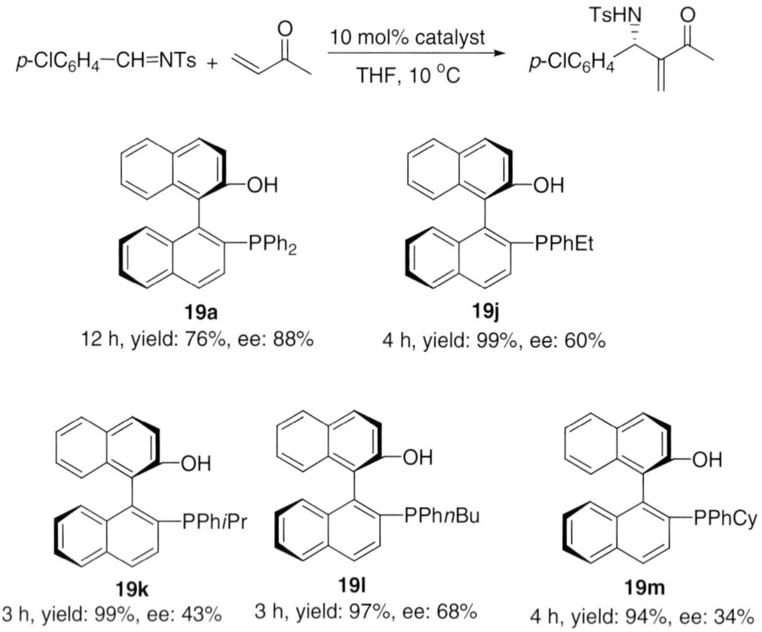

414 | *13 Chiral Amines Derived from Asymmetric Aza-Morita–Baylis–Hillman Reaction*

p-ClC$_6$H$_4$−CH=NTs + [MVK] → (10 mol% catalyst, THF, 10 °C) → p-ClC$_6$H$_4$-CH(NHTs)-C(=CH$_2$)-C(O)CH$_3$

19a
12 h, yield: 76%, ee: 88%
(BINOL-OH, PPh$_2$)

19j
4 h, yield: 99%, ee: 60%
(BINOL-OH, PPhEt)

19k
3 h, yield: 99%, ee: 43%
(BINOL-OH, PPh*i*Pr)

19l
3 h, yield: 97%, ee: 68%
(BINOL-OH, PPh*n*Bu)

19m
4 h, yield: 94%, ee: 34%
(BINOL-OH, PPhCy)

Scheme 13.17

Dendrimers are highly branched and well-defined macromolecules with controllable structure, which offer a unique tool for fine-tuning catalytic activity through the microenvironment. Subsequently, Shi and Liu have successfully anchored chiral phosphine catalysts onto a dendrimer support and investigated their catalytic activity in aza-MBH reactions. These polyether dendritic chiral phosphine catalysts [(R)-24a–24g] are simple to synthesize [33]. The initial examination was carried out using p-chlorobenzylidene-4-methylbenzenesulfonamide and MVK as the substrates in the presence of 10 mol% of these catalysts (R)-24 in THF at room temperature (25 °C). As can be seen from Table 13.7, regardless of the benzyl group or long carbon chain in the polyether dendritic catalysts, the asymmetric aza-MBH reaction

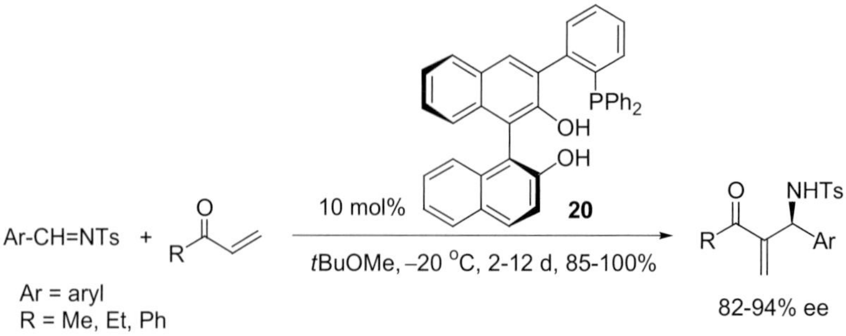

Ar-CH=NTs + [vinyl ketone with R] → (10 mol% **20**, tBuOMe, −20 °C, 2-12 d, 85-100%) → R-C(O)-C(=CH$_2$)-CH(NHTs)-Ar

Ar = aryl
R = Me, Et, Ph

82-94% ee

Scheme 13.18

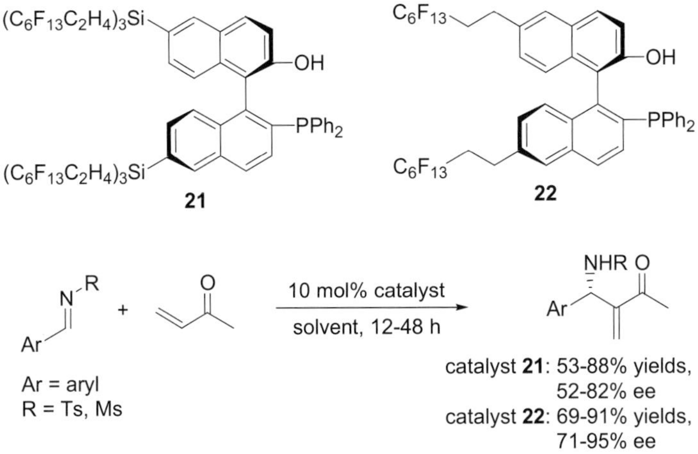

Scheme 13.19 The use of catalysts with "ponytails."

proceeded smoothly to give the adducts in excellent yield and ee, respectively, as high as 99 and 94%. Notably, the second generation of dendritic chiral phosphine Lewis base (R)-**24c** provided the best result in the catalytic activity (99% yield, 93% ee) under identical conditions. Although the third generation of dendritic chiral phosphine Lewis base (R)-**24d** showed slightly higher enantioselectivity (94% ee), a huge drop in reactivity was also observed (67% yield) under identical conditions.

It should be pointed out that the dendritic oganocatalyst (R)-**24c** can be recovered by filtration after the reaction was complete and the product was separated from the reaction mixtures by washing it with the solvent mixture hexane/ether (8/1) and can be reused in the same reaction to give the adducts with similar results.

Shi and Shi further demonstrated that replacing the phenol −OH group with a (thio)-urea group also gave similar catalytic activity and good asymmetric induction ability in aza-MBH reaction [34]. As is well known, the steric and electronic nature of the (thio)-urea group can be easily tuned by reacting the corresponding amine with different iso(thio)cyanates. The catalytic activity of **25a** was tested using the reaction of N-benzylidene-4-methylbenzenesulfonamide and MVK at room

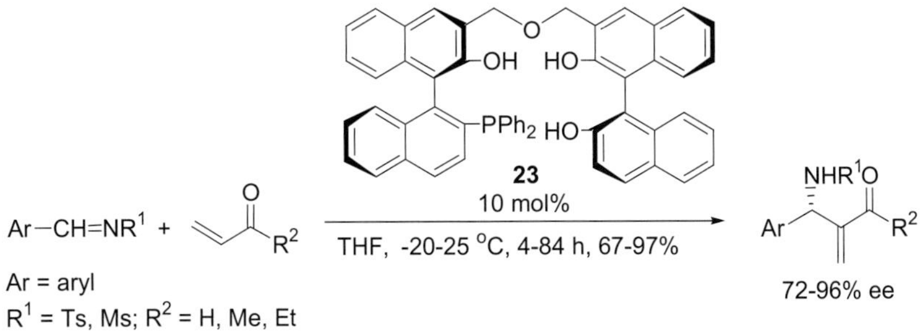

Scheme 13.20 Catalyst bearing multiple phenol groups.

Table 13.7 Screening of polyether dendrimer-supported chiral phosphine catalysts.

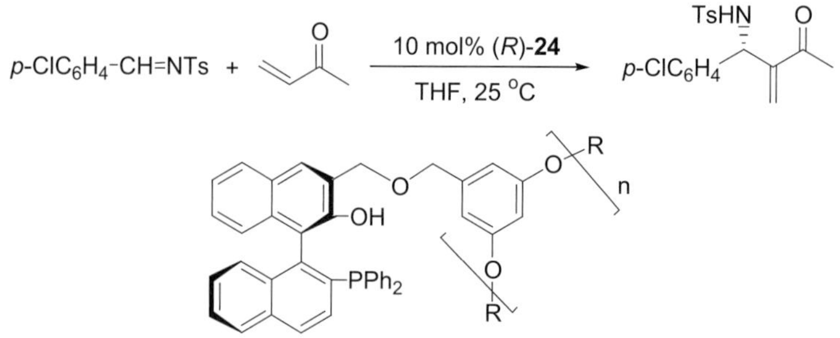

(*R*)-24	**Time (h)**	**Yield (%)**	**ee (%)**
(*R*)-**24a**, $n = 0$, R = Bn	24	90	93
(*R*)-**24b**, $n = 1$, R = Bn	24	99	89
(*R*)-**24c**, $n = 2$, R = Bn	24	99	93
(*R*)-**24d**, $n = 3$, R = Bn	48	67	94
(*R*)-**24e**, $n = 1$, R = C_6H_{13}	24	93	93
(*R*)-**24f**, $n = 2$, R = C_6H_{13}	24	99	53
(*R*)-**24g**, $n = 2$, R = $C_{13}H_{27}$	24	99	92

temperature as a model. Screening of solvent demonstrated dichloromethane was the best solvent in terms of both yield and ee of the corresponding product. Using dichloromethane as the solvent, the loading of benzoic acid was also examined from the range of 2.0–50 mol%. It was found that using 5.0 mol% of benzoic acid provided the highest ee (Scheme 13.21). Additives with weaker acidity or stronger acidity all resulted in lower yields and ee, and only additives with acidity similar to benzoic acid

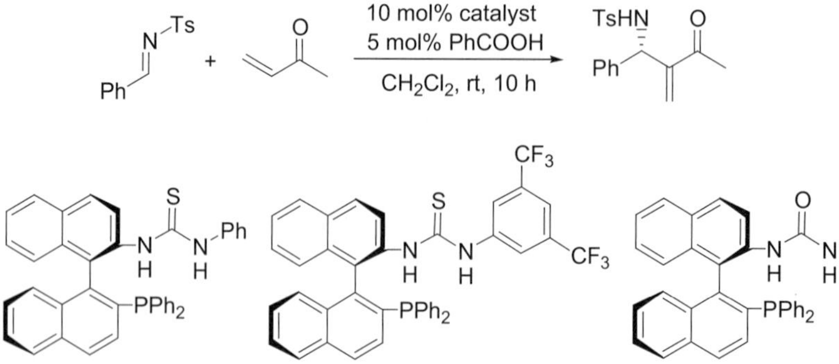

25a
yield: 96%, ee: 91%

25b
yield: 50%, ee: 70%

25c
yield: 63%, ee: 64%

Scheme 13.21 Asymmetric aza-MBH reaction catalyzed by **25a–25c**.

Table 13.8 Asymmetric aza-MBH reaction of N-sulfonated imines with MVK, phenyl vinyl ketone, EVK, and acrolein.

$$R^1\text{-CH=N-Ts} + \text{CH}_2\text{=CH-C(O)R}^2 \xrightarrow[\text{CH}_2\text{Cl}_2, \text{rt, 10 h}]{\substack{10 \text{ mol\% } \mathbf{25a} \\ 5 \text{ mol\% PhCOOH}}} R^1\text{-CH(NHTs)-CH(=CH}_2\text{)-C(O)R}^2$$

Entry	R^1	R^2	Time (h)	Yield (%)	ee (%)	Absolute configuration
1	4-MeC$_6$H$_4$	Me	10	98	90	S
2	4-ClC$_6$H$_4$	Me	10	98	90	S
3	4-FC$_6$H$_4$	Me	10	94	90	S
4	4-BrC$_6$H$_4$	Me	10	97	88	S
5	3-FC$_6$H$_4$	Me	10	96	91	S
6	C$_6$H$_5$CH=CH	Me	10	61	67	S
7	2-ClC$_6$H$_4$	Me	19	90	70	R
8	C$_6$H$_5$	H	3	81	67	S
9	C$_6$H$_5$	Et	56	69	77	S
10	C$_6$H$_5$	C$_6$H$_5$	5	80	73	S

(pK_a = 4.20) gave satisfactory results. The catalytic activity of **25b** and **25c** were also examined, but the results were unsatisfactory.

The scope of this reaction was also examined, and selected results were summarized in Table 13.8. Control experiments demonstrated that the ee of the product was independent of the reaction time and that racemization of the product or autocatalysis was not probable for this catalytic system. Moreover, when the thiourea group in **25a** was removed, both the catalytic activity and the level of asymmetric induction became significantly lower. Thus, introducing a thiourea group to the catalyst was pivotal to accelerate the reaction rate and improve the enantioselectivity of the product. Based on the observation and earlier reports, a rational mechanism was also proposed based upon ^{31}P NMR spectroscopic investigations.

Furthermore, other similar catalysts **26a–26j** have also been synthesized and applied to the aza-MBH reaction, the corresponding aza-MBH adducts were obtained in good yields (75–99%) and sometimes with very good enantiomeric excesses (51–95%) under mild conditions (Scheme 13.22) [35].

Recently, Liu has developed a Brønsted acid-activated trifunctional organocatalyst, based on the BINAP scaffold, that was used for the first time to catalyze aza-MBH reactions between N-tosylimines and MVK with fast reaction rates and good enantioselectivity at room temperature. This trifunctional catalyst containing a Lewis base, a Brønsted base and a Brønsted acid, required acid activation to confer its enantioselectivity and rate improvement for both electron-rich and electron-deficient imine substrates. The role of the amino Lewis base of **27** was investigated and found to be the activity switch in response to an acid additive. The counterion of the acid additive was found to influence not only the excess ratio but also the sense of asymmetric induction (Scheme 13.23) [36].

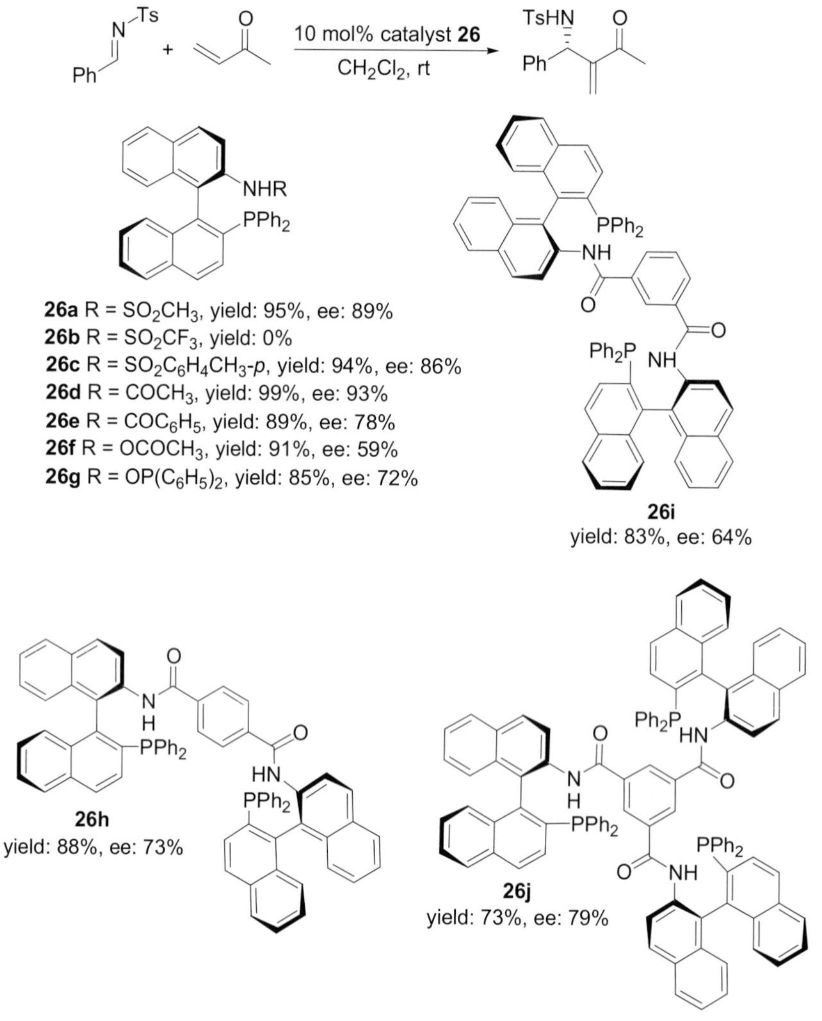
10 mol% catalyst 26
CH2Cl2, rt
TsHN
Ph
NHR
PPh2
26a R = SO2CH3, yield: 95%, ee: 89%
26b R = SO2CF3, yield: 0%
26c R = SO2C6H4CH3-p, yield: 94%, ee: 86%
26d R = COCH3, yield: 99%, ee: 93%
26e R = COC6H5, yield: 89%, ee: 78%
26f R = OCOCH3, yield: 91%, ee: 59%
26g R = OP(C6H5)2, yield: 85%, ee: 72%
26i
yield: 83%, ee: 64%
26h
yield: 88%, ee: 73%
26j
yield: 73%, ee: 79%

Scheme 13.22 Asymmetric aza-MBH reaction catalyzed by other chiral multifunctional phosphine catalysts.

13.7
Chiral Bifunctional *N*-Heterocyclic Carbenes

Chiral *N*-heterocyclic carbenes (NHCs), as Lewis basic organocatalysts, have been synthesized and applied to enantioselective organocatalytic reactions in recent years. Encouraged by Sheehan and Hunneman's first report of chiral thiazolium salts as NHC precursors for organocatalytic reactions [37], Leeper, Enders, Rovis, Glorius, Herrmann, and others have synthesized series of novel chiral NHCs with monocyclic, bicyclic, or tricyclic backbones [38]. Recently, a series of bifunctional NHCs were synthesized and applied to aza-BMH reaction of cyclopent-2-enone with

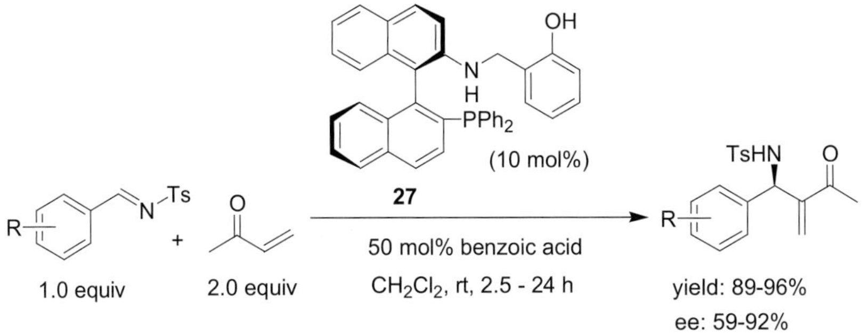

Scheme 13.23 Asymmetric aza-MBH reaction catalyzed by trifunctional phosphine catalyst.

N-tosylphenylmethanimine by Ye and coworkers. Unfortunately, poor results were obtained and the best result was shown in Scheme 13.24 [39].

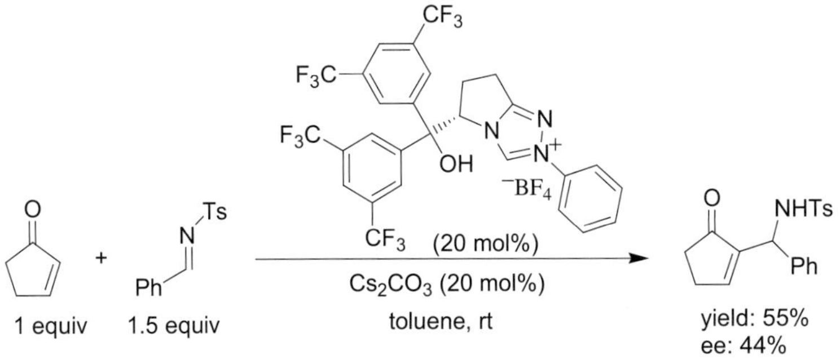

Scheme 13.24 Chiral NHC-catalyzed asymmetric aza-MBH reaction.

13.8
Chiral Ionic Liquids as Reaction Medium

Besides chiral catalysts, chiral reaction mediums were also explored and provided good asymmetric induction. Very recently, Leitner and coworkers performed the reactions in the chiral ionic liquid methyltrioctylammonium dimalatoborate **28** [40]. For the aza-MBH reaction of MVK with N-(4-bromobenzylidene)-4-toluenesulfonamide, up to 84% ee and 39% conversion were observed (Scheme 13.25).

13.9
Aza-MBH-Type Reaction to Obtain Chiral Amines

In addition to the imine as electrophilic reagent, N,O-acetals were also employed for the aza-MBH reaction. Using the camphor sulfonic acid derivative, sulfide catalyst

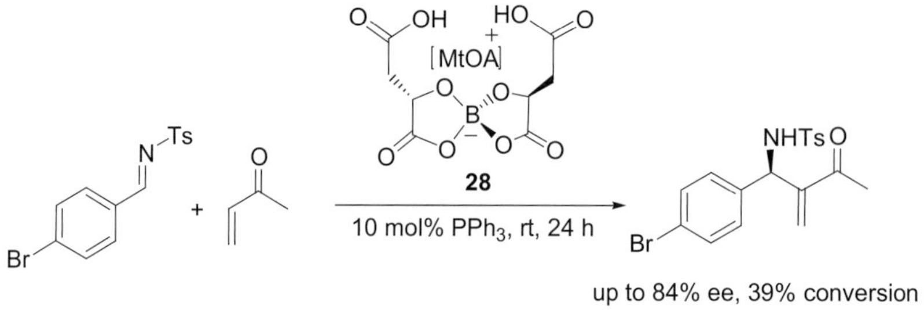

Scheme 13.25 Asymmetric aza-MBH reaction in chiral media.

29, allowed the reaction between N,O-acetals and cyclic enones in good yields and with high ee values (Table 13.9) [41]. Sulfide **29** was easily recovered (>90% yield) after column chromatography of the crude mixtures. When the acyclic enone MVK was employed, the corresponding adduct was obtained with a low ee value.

The reaction mechanism has been investigated on the basis of NMR spectroscopic data. Low-temperature NMR studies of a solution of cyclopentenone, sulfide **29**, and TMSOTf in CD_2Cl_2 at −90 °C revealed a mixture of diastereomeric β-sulfonium silyl enol ethers **30a** and **30b** in an unassigned ratio of 2 : 1 as the predominant species. On warming the mixture to −10 °C, in increments of 20 °C, the equilibrium shifted in favor of starting material, at −30 °C only starting material was detected. On recooling to −90 °C, the 2 : 1 mixture of β-sulfonium silyl enol ethers was formed again. This reveals that the silyl enol ethers **30a** and **30b** are formed reversibly, at the reaction temperature, and are in dynamic equilibrium with the starting material.

Table 13.9 Chiral sulfide catalyzed aza-MBH reactions.[a]

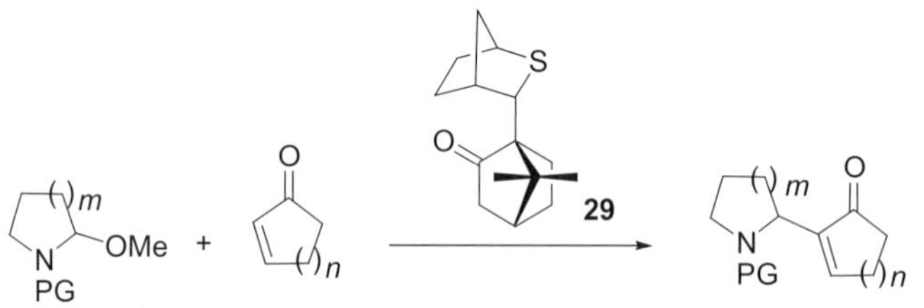

Entry	PG	m	n	Yield (%)	ee (%)	Product (R/S)
1	Cbz	1	1	69	82	S
2	Cbz	1	2	86	80	S
3	Boc	1	1	75	88	S
4	Boc	1	2	90	88	S
5	Cbz	2	1	88	94	S
6	Cbz	2	2	49	98	S

a) Alkene (2.0 equiv), sulfide **29** (1.5 equiv), TMSOTf (2.5 equiv), CH_2Cl_2, ≤60 °C, 5 h.

13.9 Aza-MBH-Type Reaction to Obtain Chiral Amines

Thus, enantioselectivity is not determined at the stage of formation of the β-sulfonium silyl enol ethers. The origin of the enantioselectivity must therefore result from a dynamic kinetic transformation of β-sulfonium silyl enol ether **30a** and **30b**, in which either the major or minor isomer reacts faster, thereby allowing the remaining diastereomer to revert to starting material for repartitioning. The synclinal approaches can be tentatively discounted since altering the size of the silyl moiety from trimethylsilyl to triisopropylsilyl had little impact on the ee value. Of the two remaining antiperiplanar approaches, attack on the *Si* face of the iminium ion would be disfavored because of nonbonding interactions between the two rings. This suggests that **30a** should favor attack on the *Re* face of the iminium ion, thereby leading to the *S* enantiomer of the product as observed (Scheme 13.26).

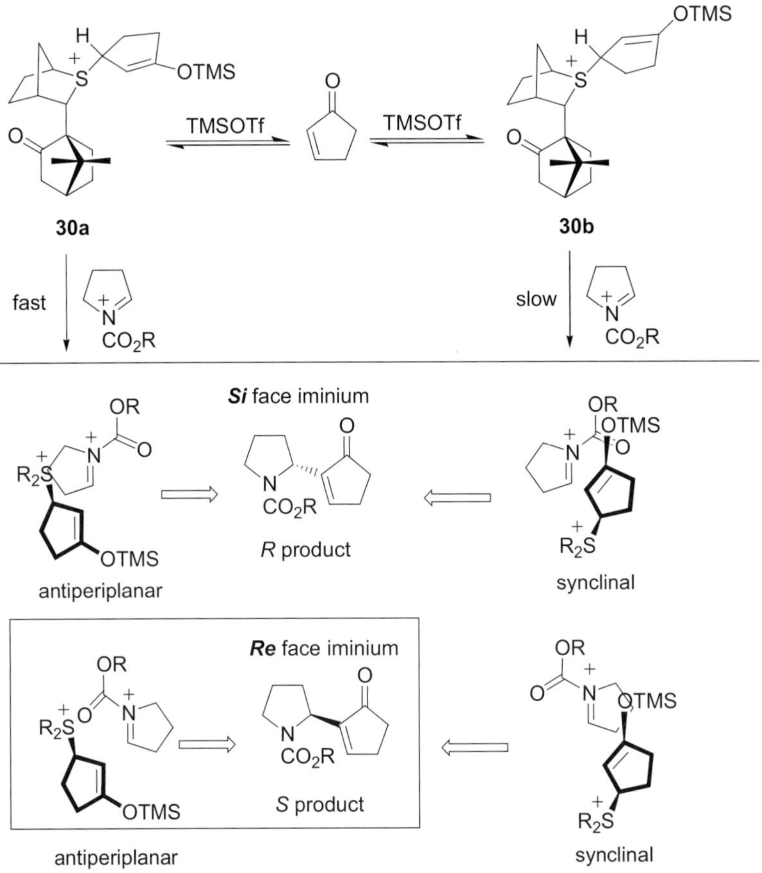

Scheme 13.26 Model proposed to explain the origin of the enantioselectivity. (*Source:* E.L. Myers, J.G. de Vries, V.K. Aggarwal, Reactions of iminium ions with Michael acceptors through a Morita–Baylis–Hillman-type reaction: enantiocontrol and applications in synthesis. *Angew. Chem., Int. Ed.* **2007**, *46*, 1893–1896. Wiley-VCH Verlag GmbH. Reproduced with permission.)

13.10
Strategies for the Removal of Protecting Groups

The N-P(O)Ph$_2$ group can be easily removed after treatment with 20% aqueous hydrochloric acid solution in excellent yield (Scheme 13.27) [20].

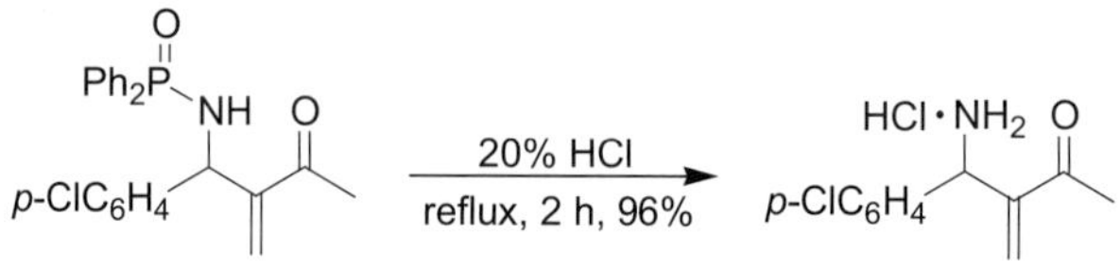

Scheme 13.27 Strategy for the removal of N-P(O)Ph$_2$ group.

However, the N-tosyl group is usually very difficult to be removed, and a general method was developed by Shi and Xu [42]. The adduct derived from the aza-MBH reaction of N-tosylated imine with acrolein was first reduced by LiAlH$_4$ to give the corresponding γ-amino alcohol in 89% yield. Then the amino alcohol was transformed to the Boc-protected amino alcohol by reaction with di-*tert*-butyl dicarbonate and DMAP in dichloromethane in 99% yield. Treatment of the obtained amino alcohol with Mg/MeOH resulted in the selective removal of the N-tosyl substituent [43], and the Boc-protected amino alcohol was isolated in 82% yield. Thus, the N-tosyl group has been changed to the synthetically more useful Boc group because the Boc protecting group can be easily removed by various methods (Scheme 13.28).

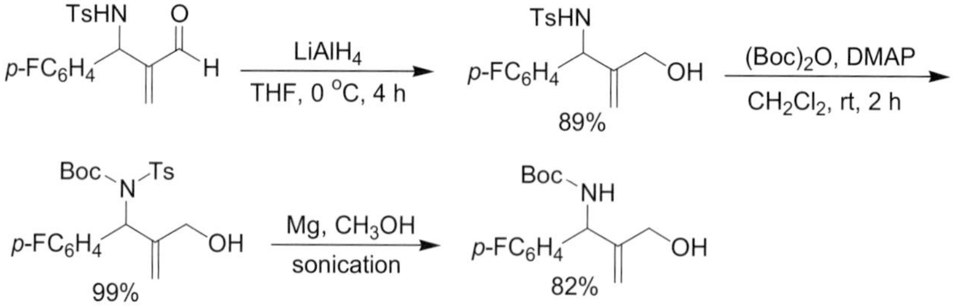

Scheme 13.28 Strategy for the removal of N-tosyl group.

An efficient deprotection of N-*p*-toluenesulfinyl group of (-branched aza-MBH adducts has been developed by using Amberlite IR-120 (plus) ion exchange resin in methanol solution at room temperature. No racemization was detected under the mild conditions. This new method provides a simultaneous deprotection/purification, which has the great advantage over the solution-phase deprotection technique of simplicity of workup (Scheme 13.29) [44].

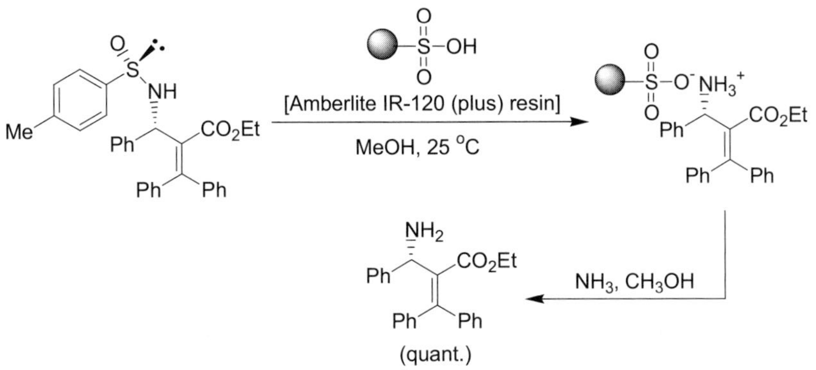

Scheme 13.29 Strategy for the removal of N-p-toluenesulfinyl group.

13.11
Selected Typical Experimental Procedures

13.11.1
Typical Procedures for 1a-Catalyzed Aza-MBH Reaction of Methyl Acrylate with N-Benzylidene-4-Nitrobenzenesulfonamide [9]

An oven-dried vial was charged with N-benzylidene-4-nitrobenzenesulfonamide (1 equiv), catalyst **1a** (10 mol%), DABCO (1 equiv), and activated 3 Å MS. The vial was evacuated and purged with N_2. Precooled, freshly distilled, anhydrous xylenes (0.15 M) and methyl acrylate (8 equiv) were added via syringe at 48 °C, and the mixture was stirred for 36 h. The mixture was diluted with anhydrous MeOH and then quenched immediately with 4N HCl in dioxane. The crude adduct was purified by flash chromatography (100% CH_2Cl_2) to afford the pure aza-MBH adduct as a white solid.

13.11.2
Typical Procedures for β-ICD-Catalyzed Aza-MBH Reaction of MVK with N-(p-Ethylbenzenesulfonyl)Benzaldimine [16]

MVK (0.5 mmol) was added to a solution of N-(p-ethylbenzenesulfonyl)benzaldimine (0.25 mmol) and β-ICD (0.025 mmol) in CH_3CN/DMF (1:1, 1.0 ml) at −30 °C. The reaction mixture was stirred at −30 °C for 24 h. After the reaction completed, the solvent was removed under reduced pressure, and the residue was purified by flash column chromatography (SiO_2, EtOAc/petroleum ether = 1 : 5) to yield the aza-MBH adduct as a colorless solid (74% yield, 96% ee).

13.11.3
Typical Procedures for Chiral Phosphine 23-Catalyzed Aza-MBH Reaction of MVK with N-(Benzylidene)-4-Chlorobenzenesulfonamide [32]

A 10 ml Schlenk tube containing N-(benzylidene)-4-chlorobenzenesulfonamide (0.5 mmol) and chiral phosphine **23** (0.05 mmol) was degassed and the reaction vessel was protected under an argon atmosphere. Then, THF (1.0 ml) was added. After the reaction mixture was cooled to −30 °C, MVK (1.5 mmol) was added into the Schlenk tube. The reaction mixture was stirred at −30 °C for 24 h. The solvent was removed under reduced pressure and the residue was purified by flash column chromatography (SiO$_2$, eluent: EtOAc/petroleum ether = 1/5) to yield the corresponding aza-MBH adduct as a colorless solid (94% yield, 96% ee).

13.11.4
General Procedures of Aza-MBH Reactions Involving Aliphatic Imines [21]

To a solution of corresponding imine freshly prepared (1.0 equiv) in dried dichloromethane at 0 °C, catalyst (**13c**) (0.1 equiv), β-naphthol (0.1 equiv), and β-naphthyl acrylate (2.0 equiv) were added. The reaction mixture was stirred under argon atmosphere at 0 °C for the required time indicated in the tables. The reaction was stopped by passing the mixture through a short pad of silica gel using ethyle acetate for the elution. Solvents were removed in vacuo and the residue was purified by flash chromatography on silica gel (eluent: n-Hept/EtOAc = 90/10) to afford the corresponding pure product.

13.11.5
Typical Procedures for 25a and Benzoic Acid-Catalyzed Aza-MBH Reaction of N-Sulfonated Imine with MVK [34]

To a solution of imine (65 mg, 0.25 mmol), **25a** (15 mg, 0.025 mmol), and benzoic acid (0.15 mg, 0.0125 mmol) in dichloromethane (1.0 ml) was added MVK (42 ml, 0.5 mmol) at room temperature. Then reaction mixture was stirred at room temperature for 10 h. After the reaction completed, the solvent was removed under reduced pressure and the residue was purified by a flash column chromatography (SiO$_2$, eluent: EtOAc/petroleum ether = 1/5) to afford 4-methyl-N-(2-methylene-3-oxo-1-phenylbutyl) benzenesulfonamide as a colorless solid.

13.11.6
Typical Procedures for Trifunctional Phosphine 27-Catalyzed Aza-MBH Reaction of N-Tosylimines with MVK [36]

N-Tosylimine (0.5 mmol), trifunctional phosphine catalyst **27** (10 mol%, 0.05 mmol), and benzoic acid (50 mol%, 0.25 mmol) were combined under N$_2$. CH$_2$Cl$_2$ (0.1 ml per mg of catalyst) was added dropwise followed by distilled MVK (1.0 mmol). The reaction mixture was stirred until completion. The solvent was evaporated and the crude mixture was subjected to chiral HPLC for the ee analysis and purified by silica gel flash column chromatography.

13.11.7
General Procedures for the Synthesis of Enantiomerically Enriched Aza-MBH-Type Adducts Catalyzed by Chiral Sulfide 29 [41]

To a solution of aminal (0.21 mmol), alkene (0.43 mmol), and chiral sulfide 29 (23 μl, 0.32 mmol) in CH_2Cl_2 (1 ml) at −78 °C was added TMSOTf (96 μl, 0.53 mmol). After stirring the solution for 5 h, maintaining the temperature below −60 °C, the reaction was quenched with saturated aq. $NaHCO_3$ (2 ml) and was allowed to warm to room temperature. Water (25 ml) was added and the resultant mixture was extracted with CH_2Cl_2 (2 × 25 ml). The combined organic layers were then dried with $MgSO_4$ and concentrated in vacuo. The resultant oil was taken up in CH_2Cl_2 (1 ml) and DBU (48 μl, 0.32 mmol) was added. After stirring for 30 min at room temperature, CH_2Cl_2 (25 ml) was added and the solution was washed with saturated aq. NH_4Cl (3 × 20 ml). The combined aqueous washes were then extracted with CH_2Cl_2 (25 ml) and the combined organic layers were dried with $MgSO_4$ and concentrated in vacuo. The resultant oil was purified by flash chromatography to give the required adduct.

13.11.8
General Procedures for the Removal of N-p-Toluenesulfinyl Group [44]

Into a clean dry vial was loaded the N-p-toluenesulfinyl-β, β-diphenyl-α-(aminoalkyl) acrylate (50 mg, 0.10 mmol) and methanol (5.0 ml). Into this solution was added Amberlite IR-120 (plus) ion exchange resin (0.50 g) in one portion. The resulting heterogeneous mixture was stirred in the capped vial at room temperature for 12 h without argon protection. TLC (EtOAc/hexane, 1/2, v/v) showed complete disappearance of the starting material in methanol solution. The Amberlite resin, which absorbed the deprotection product, was separated by suction filtration and then washed with 5.0 ml of methanol. No product was detected in the methanol. The free amino ester was released from the resin by immersing the loaded resin in 15% NH_4OH in methanol (10 ml) overnight. The resin was filtered off and further washed with dilute aqueous ammonia. The combined filtrates were concentrated to dryness to give product (36.0 mg, quant.) as a light yellow oil.

13.11.9
General Procedures for the Removal of N-Tosyl Group [42]

13.11.9.1 Reduction of the Aza-MBH Reaction Product with LiAlH$_4$
The aza-MBH product N-[1-(4-fluorophenyl)-2-formylallyl]-4-methylbenzenesulfonamide (166 mg, 0.5 mmol) was placed in a flask under argon atmosphere. THF (3.0 ml) was added into the solution, and then LiAlH$_4$ (38 mg, 1.0 mmol) was added

into the reaction mixture at 0 °C. The reaction mixture was further stirred at room temperature and the reaction was monitored by TLC plate. When the starting material was consumed completely, the reaction was quenched by addition of saturated NH$_4$Cl aqueous solution at 0 °C. The reaction mixture was extracted with CH$_2$Cl$_2$ for three times. The organic phases were combined and dried over anhydrous MgSO$_4$. The solvent was removed under reduced pressure, and the residue was subject to a flash column chromatography (SiO$_2$, EtOAc/petroleum ether = 1:2) to give the product (148 mg, 89%) as a colorless solid.

13.11.9.2 Boc-Protection

The obtained product from step 1 (68 mg, 0.20 mmol) was placed in a flask under argon atmosphere. CH$_2$Cl$_2$ (2.0 ml) was added into the reaction vessel, and then DMAP (4.5 mg, 0.04 mmol, 0.2 equiv) and di-*tert*-butyl dicarbonate (87 mg, 0.4 mmol, 2 equiv) were added into the reaction mixture at room temperature. The reaction mixture was further stirred at room temperature. The reaction was monitored by TLC plate. When the substrate was consumed completely, the solvent was removed under reduced pressure and the residue was subject to a flash column chromatography (SiO$_2$, EtOAc/petroleum ether = 1:10) to give the product (85 mg, 99%) as a viscous liquid.

13.11.9.3 Detosylation

The obtained product from step 2 (85 mg, 0.194 mmol) was transferred to an oven-dried flask under argon atmosphere. Then, this reaction vessel was charged with 1.5 ml of dry MeOH, followed by the addition of Mg (24 mg, 5 equiv). The reaction mixture was sonicated for 45 min and then was poured into 1.0 M HCl. The organic compound was extracted with ether (3 × 15 ml). The organic phase was sequentially washed with saturated NaHCO$_3$ solution and brine, and then dried over anhydrous MgSO$_4$. The solvent was removed under reduced pressure and the residue was subject to a flash column chromatography (SiO$_2$, EtOAc/petroleum ether = 1:6) to give the product (48 mg, 82%) as a colorless solid.

13.12
Summary and Outlook

Catalytic asymmetric aza-MBH reactions and related reactions provide a direct and efficient synthetic approach to a variety of functionalized nonamino acid-derived chiral amines under mild conditions. These novel chiral amines can be directly used to prepare many other useful chiral building blocks in organic synthesis. A variety of transformations of the chiral aza-MBH adducts were recently demonstrated by Raheem and Jacobsen (Scheme 13.30) [9].

Although excellent progress has been made in the area of aza-MBH reaction, many challenges and questions still remain:

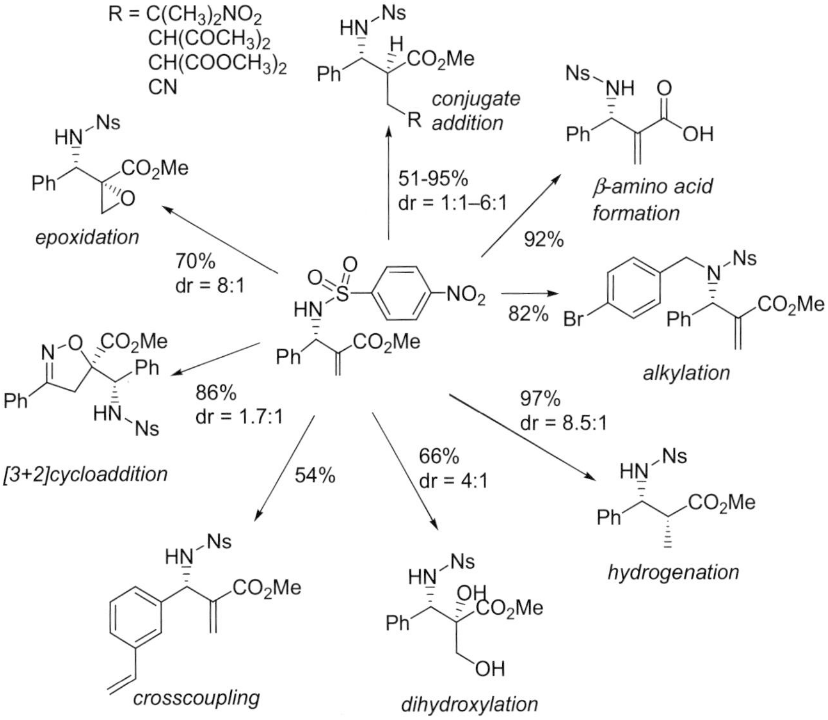

Scheme 13.30 Transformations of aza-MBH adduct. (Source: I.T. Raheem, E.N. Jacobsen, Highly enantioselective aza-Baylis–Hillman reactions catalyzed by chiral thiourea derivatives. *Adv. Synth. Catal.* **2005**, *347*, 1701–1708. Wiley-VCH Verlag GmbH. Reproduced with permission.)

1) The mechanism of the reaction largely depends on the reaction conditions and thus not as clear as desired.
2) Successful examples using aliphatic imines in the aza-MBH reaction is still limited.
3) Further synthetic applications of the corresponding aza-MBH adducts are still being optimized [45].
4) Previous reports usually referred to make small quantities of these aza-MBH products in high ee and yield. Methods for larger quantities were still rare. Supported chiral nitrogen or phosphine-containing catalysts [46], which could be used for many times, are probably a good choice for the above purpose.
5) Due to the low activity of β-substituted α,β-unsaturated ketone or ester, the asymmetric aza-MBH reaction using β-substituted α,β-unsaturated ketone or ester is still a challenge [47].
6) Because α,β-unsaturated ketone or ester can dimerize in the presence of phosphine or amine catalyst, more than 1 equiv of α,β-unsaturated ketone or ester usually need to be used. How to reduce the amount of α,β-unsaturated ketone or ester in the aza-MBH reaction is still a challenge for the future.

This field is developing at a rapid pace and further progress will surely be forthcoming. It can be envisioned that the catalytic asymmetric aza-MBH reaction will be used more often in the near future.

Questions

13.1. Synthesize the following products **a–c** starting from *p*-nitrobenzenesulfonamide, benzaldehyde, and methyl acrylate.

a

b

c

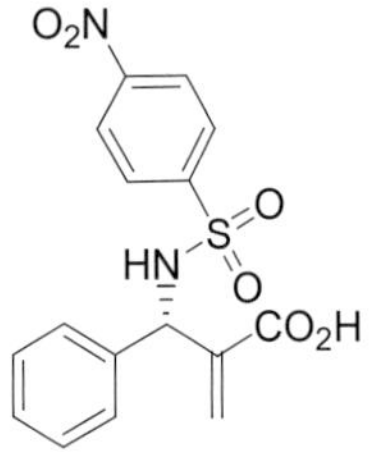

13.2. Explain why the following reactions give the products with opposite absolute configurations.

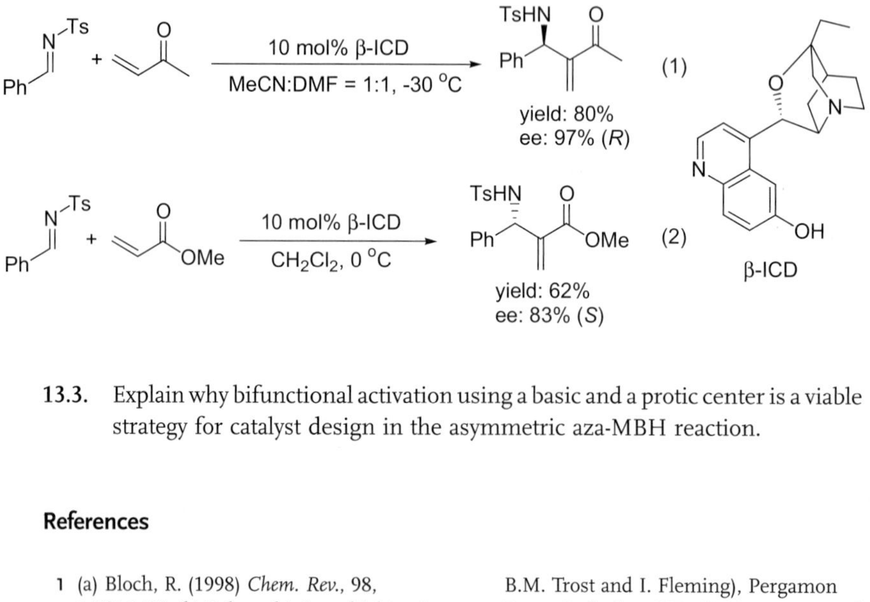

13.3. Explain why bifunctional activation using a basic and a protic center is a viable strategy for catalyst design in the asymmetric aza-MBH reaction.

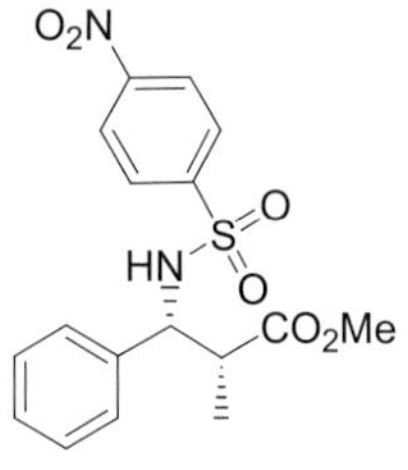

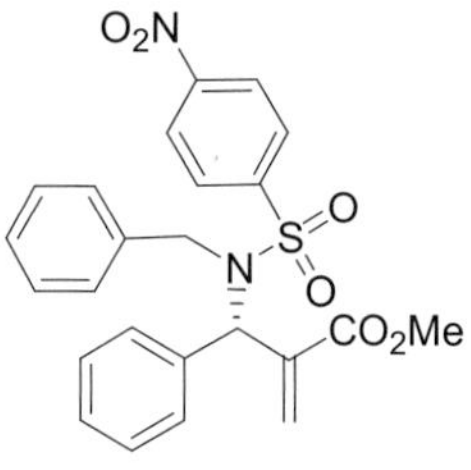

References

1 (a) Bloch, R. (1998) *Chem. Rev.*, 98, 1407–1438; (b) Kobayashi, S. and Ishitani, H. (1999) *Chem. Rev.*, 99, 1069–1094.

2 (a) Kleinman, E.F. (1991) Chapter 4, in *Comprehensive Organic Synthesis*, vol. 2 (eds B.M. Trost and I. Fleming), Pergamon Press, Oxford, p. 893; (b) Trost, B.M. and Fleming, I. (eds) (1991) Chapter 1, in *Comprehensive Organic Synthesis*, vol. 8, Pergamon Press, Oxford.

3 (a) Morita, K., Suzuki, Z., and Hirose, H. (1968) *Bull. Chem. Soc. Jpn.*, 41, 2815; (b) Baylis, A.B. and Hillman, M.E.D. (1972) Offenlegungsschrift 2155113. U.S. Patent 3,743,669;*Chem. Abstr.* 77 (1972) 34174q.

4 (a) Perlmutter, P. and Teo, C.C. (1984) *Tetrahedron Lett.*, 25, 5951–5952; (b) Basavaiah, D., Rao, A.J., and Satyanarayana, T. (2003) *Chem. Rev.*, 103, 811–891; (c) Masson, G., Housseman, C., and Zhu, J. (2007) *Angew. Chem., Int. Ed.*, 46, 4614–4628; (d) Basavaiah, D., Rao, K.V., and Reddy, R.J. (2007) *Chem. Soc. Rev.*, 36, 1581–1588; (e) Shi, Y.-L. and Shi, M. (2007) *Eur. J. Org. Chem.*, 2905–2916; (f) Singh, V. and Batra, S. (2008) *Tetrahedron*, 64, 4511–4574.

5 (a) Hill, J.S. and Isaacs, N.S. (1990) *J. Phys. Org. Chem.*, 3, 285–288; (b) Bode, M.L. and Kaye, P.T. (1991) *Tetrahedron Lett.*, 32, 5611–5614; (c) Santos, L.S., Pavam, C.H., Almeida, W.P., Coelho, F., and Eberlin, M.N. (2004) *Angew. Chem., Int. Ed.*, 43, 4330–4333.

6 (a) Price, K.E., Broadwater, S.J., Jung, H.M., and McQuade, D.T. (2005) *Org. Lett.*, 7, 147–150; (b) Price, K.E., Broadwater, S.J., Walker, B.J., and McQuade, D.T. (2005) *J. Org. Chem.*, 70, 3980–3987.

7 (a) Augé, J., Lubin, N., and Lubineau, A. (1994) *Tetrahedron Lett.*, 35, 7947–7948; (b) Aggarwal, V.K., Dean, D.K., Mereu, A., and Williams, R. (2002) *J. Org. Chem.*, 67, 510–514; (c) Cai, J., Zhou, Z., Zhao, G., and Tang, C. (2002) *Org. Lett.*, 4, 4723–4725; (d) Yu, C., Liu, B., and Hu, L. (2001) *J. Org. Chem.*, 66, 5413–5418; (e) Yamada, Y.M.A. and Ikegami, S. (2000) *Tetrahedron Lett.*, 41, 2165–2169; (f) McDougal, N.T. and Schaus, S.E. (2003) *J. Am. Chem. Soc.*, 125, 12094–12095.

8 (a) Aggarwal, V.K., Fulford, S.Y., and Lloyd-Jones, G.C. (2005) *Angew. Chem., Int. Ed.*, 44, 1706–1708; (b) Robiette, R., Aggarwal, V.K., and Harvey, J.N. (2007) *J. Am. Chem. Soc.*, 129, 15513–15525.

9 Raheem, I.T. and Jacobsen, E.N. (2005) *Adv. Synth. Catal.*, 347, 1701–1708.

10 Buskens, P., Klankermayer, J., and Leitner, W. (2005) *J. Am. Chem. Soc.*, 127, 16762–16763.

11 (a) Iwabuchi, Y., Nakatani, M., Yokoyama, N., and Hatakeyama, S. (1999) *J. Am. Chem. Soc.*, 121, 10219–10220; (b) Iwabuchi, Y. and Hatakeyama, S. (2002) *J. Synth. Org. Chem. Jpn.*, 60, 2–14.

12 Kündig, E.P., Xu, L.H., and Schnell, B. (1994) *Synlett*, 413–414.

13 Aggarwal, V.K., Castro, A.M.M., Mereu, A., and Adams, H. (2002) *Tetrahedron Lett.*, 43, 1577–1581.

14 Shi, M. and Xu, Y.-M. (2002) *Tetrahedron: Asymmetry*, 13, 1195–1200.

15 Lu, A., Xu, X., Gao, P., Zhou, Z., Song, H., and Tang, C. (2008) *Tetrahedron: Asymmetry*, 19, 1886–1890.

16 Shi, M. and Xu, Y.-M. (2002) *Angew. Chem., Int. Ed.*, 41, 4507–4510.

17 Shi, M., Xu, Y.-M., and Shi, Y.-L. (2005) *Chem. Eur. J.*, 11, 1794–1802.

18 Shi, M., Qi, M.-J., and Liu, X.-G. (2008) *Chem. Commun.*, 6025–6027.

19 Balan, D. and Adolfsson, H. (2003) *Tetrahedron Lett.*, 44, 2521–2524.

20 Kawahara, S., Nakano, A., Esumi, T., Iwabuchi, Y., and Hatakeyama, S. (2003) *Org. Lett.*, 5, 3103–3105.

21 Abermil, N., Masson, G., and Zhu, J. (2008) *J. Am. Chem. Soc.*, 130, 12596–12597.

22 Matsui, K., Takizawa, S., and Sasai, H. (2005) *J. Am. Chem. Soc.*, 127, 3680–3681.

23 Matsui, K., Tanaka, K., Horii, A., Takizawa, S., and Sasai, H. (2006) *Tetrahedron: Asymmetry*, 17, 578–583.

24 (a) Roth, F., Gygax, P., and Frator, G. (1992) *Tetrahedron Lett.*, 33, 1045–1048; (b) Hayase, T., Shibata, T., Soai, K., and Wakatsuki, Y. (1998) *Chem. Commun.*, 1271–1272; (c) Li, W., Zhang, Z., Xiao, D., and Zhang, X. (2000) *J. Org. Chem.*, 65, 3489–3496; (d) Pereira, S.I., Adrio, J., Silva, A.M.S., and Carretero, J.C. (2005) *J. Org. Chem.*, 70, 10175–10177.

25 (a) Shi, M. and Chen, L.-H. (2003) *Chem. Commun.*, 1310–1311; (b) Shi, M., Chen, L.-H., and Li, C.-Q. (2005) *J. Am. Chem. Soc.*, 127, 3790–3800; (c) Shi, M., Ma, G.-N., and Gao, J. (2007) *J. Org. Chem.*, 72, 9779–9781.

26 Shi, M. and Zhao, G.-L. (2004) *Adv. Synth. Catal.*, 346, 1205–1219.

27 Buskens, P., Klankermayer, J., and Leitner, W. (2005) *J. Am. Chem. Soc.*, 127, 16762–16763.
28 Shi, M. and Li, C.-Q. (2005) *Tetrahedron: Asymmetry*, 16, 1385–1391.
29 Lei, Z.-Y., Ma, G.-N., and Shi, M. (2008) *Eur. J. Org. Chem.*, 3817–3820.
30 Matsui, K., Takizawa, S., and Sasai, H. (2006) *Synlett*, 761–765.
31 Shi, M., Chen, L.-H., and Teng, W.-D. (2005) *Adv. Synth. Catal.*, 347, 1781–1789.
32 Liu, Y.-H., Chen, L.-H., and Shi, M. (2006) *Adv. Synth. Catal.*, 348, 973–979.
33 Liu, Y.-H. and Shi, M. (2008) *Adv. Synth. Catal.*, 350, 122–128.
34 Shi, Y.-L. and Shi, M. (2007) *Adv. Synth. Catal.*, 349, 2129–2135.
35 (a) Qi, M.-J., Ai, T., Shi, M., and Li, G. (2008) *Tetrahedron*, 64, 1181–1186; (b) Guan, X.-Y., Jiang, Y.-Q., and Shi, M. (2008) *Eur. J. Org. Chem.*, 2150–2155.
36 (a) Garnier, J.-M., Anstiss, C., and Liu, F. (2009) *Adv. Synth. Catal.*, 351, 331–338; (b) Garnier, J.-M. and Liu, F. (2009) *Org. Biomol. Chem.*, 7, 1272–1275.
37 Sheehan, J.C. and Hunneman, D.H. (1966) *J. Am. Chem. Soc.*, 88, 3666–3667.
38 (a) Knight, R.L. and Leeper, F.J. (1997) *Tetrahedron Lett.*, 38, 3611–3614; (b) Knight, R.L. and Leeper, F.J. (1998) *J. Chem. Soc., Perkin Trans. 1*, 1891–1894; (c) Enders, D., Breuer, K., Kallfass, U., and Balensiefer, T. (2003) *Synthesis*, 1292–1295; (d) Enders, D., Niemeier, O., and Balensiefer, T. (2006) *Angew. Chem., Int. Ed.*, 45, 1463–1467; (e) Kerr, M.S., de Alaniz, J.R., and Rovis, T. (2002) *J. Am. Chem. Soc.*, 124, 10298–10299; (f) Kerr, M.S. and Rovis, T. (2003) *Synlett*, 1934–1936; (g) Kerr, M.S., de Alaniz, J.R., and Rovis, T. (2005) *J. Org. Chem.*, 70, 5725–5728; (h) Glorius, F., Altenhoff, A.G., Goddard, R., and Lehmann, C. (2002) *Chem. Commun.*, 2704–2705; (i) Herrmann, W.A., Goossen, L.J., Köcher, C., and Artus, G.R.J. (1996) *Angew. Chem., Int. Ed.*, 35, 2805–2807; (j) Baskakov, D., Herrmann, W.A., Herdtweck, E., and Hoffmann, S.D. (2007) *Organometallics*, 26, 626–632; (k) Suzuki, Y., Yamauchi, K., Muramatsu, K., and Sato, M. (2004) *Chem. Commun.*, 2770–2771; (l) Li, G.-Q., Dai, L.-X., and You, S.-L. (2007) *Chem. Commun.*, 852–854; (m) Struble, J.R., Kaeobamrung, J., and Bode, J.W. (2008) *Org. Lett.*, 10, 957–960; (n) Takikawa, H. and Suzuki, K. (2007) *Org. Lett.*, 9, 2713–2716.
39 (a) Zhang, Y.-R., He, L., Wu, X., Shao, P.-L., and Ye, S. (2008) *Org. Lett.*, 10, 277–280; (b) He, L., Zhang, Y.-R., Huang, X.-L., and Ye, S. (2008) *Synthesis*, 2825–2829.
40 Gausepohl, R., Buskens, P., Kleinen, J., Bruckmann, A., Lehmann, C.W., Klankermayer, J., and Leitner, W. (2006) *Angew. Chem., Int. Ed.*, 45, 3689–3692.
41 Myers, E.L., de Vries, J.G., and Aggarwal, V.K. (2007) *Angew. Chem., Int. Ed.*, 46, 1893–1896.
42 Xu, Y.-M. and Shi, M. (2004) *J. Org. Chem.*, 69, 417–425.
43 Juhl, K., Gathergood, N., and Jøgenson, A.K. (2001) *Angew. Chem., Int. Ed.*, 40, 2995–2997.
44 Li, G., Kim, S.H., and Wei, H.-X. (2000) *Tetrahedron*, 56, 719–723.
45 Declerck, V., Martinez, J., and Lamaty, F. (2009) *Chem. Rev.*, 109, 1–48.
46 (a) Huang, J.-W. and Shi, M. (2003) *Adv. Synth. Catal.*, 345, 953–958; (b) Zhao, L.-J., Song He, H., Shi, M., and Toy, P.H. (2004) *J. Comb. Chem.*, 6, 680–683; (c) Zhao, L.-J., Kwong, C.K.-W., Shi, M., and Toy, P.H. (2005) *Tetrahedron*, 61, 12026–12032.
47 (a) Shi, Y.-L., Xu, Y.-M., and Shi, M. (2004) *Adv. Synth. Catal.*, 346, 1220–1230; (b) Shi, Y.-L. and Shi, M. (2006) *Tetrahedron*, 62, 461–475; (c) Back, T.G., Rankic, D.A., Sorbetti, J.M., and Wulff, J.E. (2005) *Org. Lett.*, 7, 2377–2379.

14
Biocatalytic Routes to Nonracemic Chiral Amines
Nicholas J. Turner and Matthew D. Truppo

14.1
Introduction

Enantiomerically pure chiral amines are highly valuable functionalized molecules with a wide range of applications including (i) intermediates for the synthesis of pharmaceutical and agrochemical active ingredients, (ii) resolving agents for the separation of enantiomers via diastereomeric salt formation, and (iii) ligands for asymmetric synthesis using either transition metal catalysis or organocatalysis [1]. The option of preparing chiral amines using biocatalytic approaches is now viewed as attractive as a result of recent developments in biocatalyst availability, methods for improving biocatalyst stability, and the inherent high selectivity and catalytic activity that can be obtained through enzyme catalysis. This chapter reviews the methods that are currently available for the preparation of chiral amines using biocatalysis. The review focuses on those biocatalytic methods in which the amine functionality is involved directly in the biocatalytic transformation (e.g., the synthesis of amino alcohols in which the alcohol functionality is manipulated by lipase-catalyzed acylation or hydrolysis are omitted). In addition, biocatalytic methods for the preparation of amino acids are also outside the scope of this review. The chapter is organized according to the three general approaches that have been developed, namely, kinetic resolution (KR), dynamic kinetic resolution (DKR) and deracemization, and finally asymmetric synthesis. A final section deals with emerging approaches including the use of imine reductases. As summarized in Figure 14.1, using α-methylbenzylamine **1** as a model compound, each of the methods discussed in this chapter possesses advantages and disadvantages in terms of (i) availability of starting material, (ii) efficiency of the process (i.e., kinetic resolution versus asymmetric synthesis), (iii) range of chiral amines that can be prepared (1°, 2°, 3°), (iv) availability of biocatalyst, and (v) possible requirement for cofactor recycling. Each of these issues is dealt with throughout the chapter. Table 14.1 provides details of commercial sources of the enzymes referred to in the text.

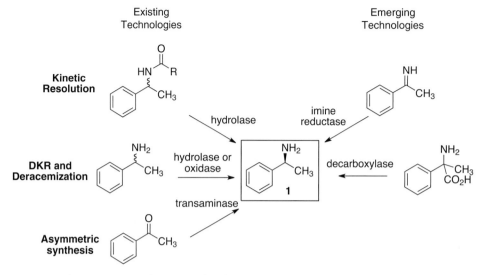

Figure 14.1 Biocatalytic approaches for the synthesis of chiral amines.

14.2
Kinetic Resolution of Racemic Amines

14.2.1
Hydrolytic Enzymes

Hydrolytic enzymes have long been used for the kinetic resolution of racemic alcohols and carboxylic acids. In recent years, the corresponding transformation on

Table 14.1 List of suppliers of enzymes referred in this chapter.

A. woodii (imine reductase)	DSM
Amino acid dehydrogenase	Codexis
Amino acid oxidase	Codexis
A. melleus (aminoacylase I)	Amano, Fluka, Lowry, Sigma
C. antarctica lipase B (Novozyme-435/CAL-B)	Novozymes, Sigma
C. parapsilosis ATCC 7330 (imine reductase)	American Type Culture Collection (ATCC)
C. rugosa lipase	Amano, Fluka
Cyclohexanone monooxygenase	Sigma
Glucose dehydrogenase	Amano, Codexis, Fisher, Fluka, Sigma
L-Tyrosine decarboxylase	Sigma
Lactate dehydrogenase	Amano, Codexis, Fisher, Fluka, Sigma
penicillin G acylase	Sigma
P. aeruginosa (LIP-300)	Toyobo
P. aeruginosa (LPL-311)	Toyobo
Subtilisin (protease)	Fisher, Sigma
Threonine aldolase	Sigma
Transaminases (ATAs)	Codexis
Vibrio fluviavialis (transaminase)	Fluka

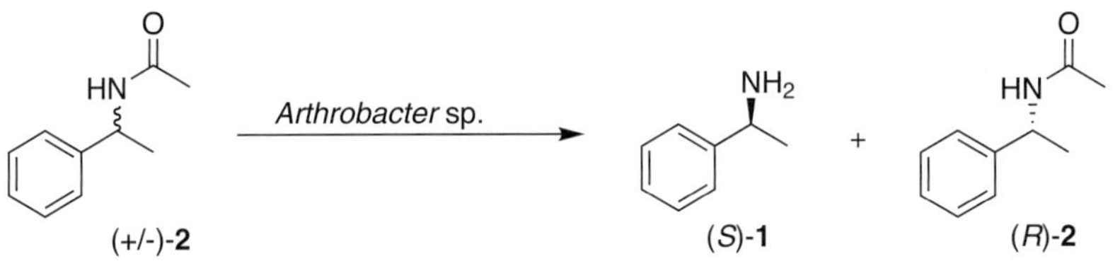

Figure 14.2 Hydrolysis of racemic N-1-phenylethylacetamide **2** using *Arthrobacter* sp.

racemic amines has received more attention [2, 3]. Most of the reported examples are run under low water conditions, with *Candida antarctica* lipase B (CAL-B) as the most commonly employed hydrolase. Amines by nature are more nucleophilic than alcohols and are therefore more prone to undergo background reaction with the acyl donor leading to lower overall enantiomeric excess (ee) of the product. This nonspecific acylation has resulted in a significant screening effort to identify less reactive acylating agents and appropriate solvent conditions that favor enzyme-mediated acylation over noncatalyzed acylation.

The first reported use of hydrolytic enzymes for the resolution of racemic amines was done in 1989 [4]. Kitaguchi *et al.* tested various lipases and proteases for their ability to catalyze the acylation of α-methylbenzylamine using trifluoroethyl butyrate as the acylating agent, finding a strong effect of solvent choice on the enantioselectivity of the reaction. Using subtilisin as a catalyst and 3-methyl-3-pentanol as solvent gave the best results, with ee values up to 99%. Subsequently, several groups investigated the scope of this methodology, examining different enzyme/solvent/acyl donor combinations to optimize activity, conversion, and enantioselectivity.

Scientists at Shell began work on the enantioselective hydrolysis of racemic amides in the early 1990s [5]. Enzyme-mediated hydrolysis of racemic N-1-phenylethylacetamide **2** using whole cells of *Arthrobacter* sp. enabled the production of enantiomerically pure (S)-**1** and (R)-**2** (Figure 14.2). The use of whole cells was not optimal and long reaction times were required to obtain pure (R)-**2**, as the selectivity of the catalyst was not very high.

Bayer improved upon this concept with the selective hydrolysis of racemic acetamide **3** using *C. antarctica* lipase B (Figure 14.3) [6]. In this case, the (R)-acetamide was deacylated, forming the (R)-amine **4**. Although the enantioselectivities obtained were excellent, the process was not used industrially due to low space time yields.

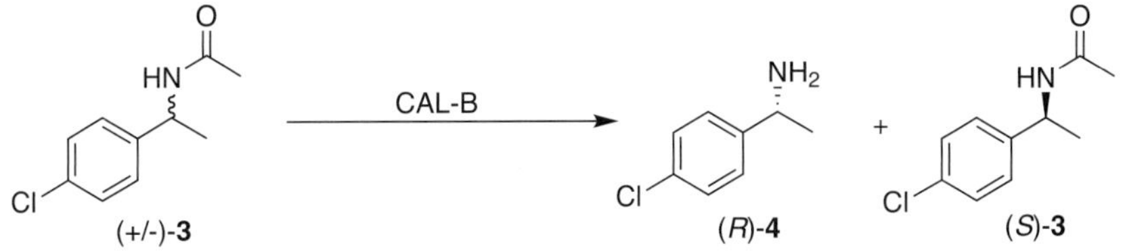

Figure 14.3 Hydrolysis of racemic acetamide **3** using *C. antarctica* lipase B.

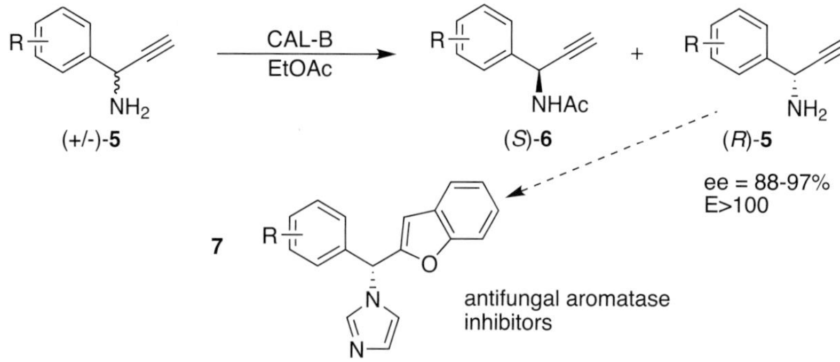

Figure 14.4 Synthesis of building blocks for aromatase inhibitors.

A wide range of biologically active compounds contain chiral amine functionalities, and several groups have reported on the kinetic resolution of amine building blocks that are useful in the synthesis of pharmaceutical intermediates. Immobilized C. antarctica lipase B (Novozyme-435 or CAL-B) has been used for the enantioselective acylation of 1-ethynylbenzylamine derivatives **5** (Figure 14.4) [7]. Ethyl acetate was used as the acyl donor, and the reactions were run in diethyl ether. Very high selectivities ($E > 100$) were obtained in many cases, yielding the corresponding (R)-amines **5** and (S)-amides **6**. The (R)-amine products are important building blocks in the synthesis of a range of antifungal aromatase inhibitors **7** (Note: E = enantiomeric ratio and is defined as the ratio of the second-order rate constants for the individual enantiomers S and R, hence $E = v_S/v_R$. E can be calculated from the following: $E = \ln[1 - c(1 + ee_p)]/\ln[1 - c(1 - ee_p)]$ where c = conversion and ee_p = ee of the product).

Pfizer demonstrated the resolution of a simple chiral building block **8** that was used as an intermediate in the synthesis of cyclin-dependent kinase (CDK2) inhibitors (Figure 14.5) [8]. Again, CAL-B was used with ethyl acetate as the acylating agent to give amide **9** in >95% ee and good yield (>45%).

A synthetic precursor to monoamine oxidase (MAO) inhibitors, (R)-1-aminoindan **10**, was produced using a continuous-flow column bioreactor containing subtilisin immobilized on glass beads [9]. The process was run in 3-methyl-3-pentanol and utilized the active ester 2,2,2-trifluoroethyl butyrate **11** as the acylating agent to give amide **12** (Figure 14.6).

The first chemoenzymatic synthesis of organoselenium containing amines was recently reported by Andrade and Silva (Figure 14.7) [10]. Compounds containing a selenium atom have important antioxidant and anti-inflammatory activities. Lipase-mediated acylation of amine **13** gave the corresponding chiral amides **14** and amines **13** with excellent enantioselectivity (up to 99% ee).

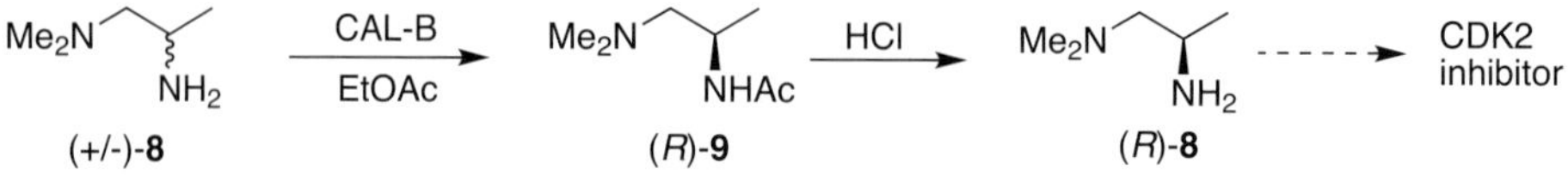

Figure 14.5 Kinetic resolution of chiral amines as intermediates for CDK2 inhibitors.

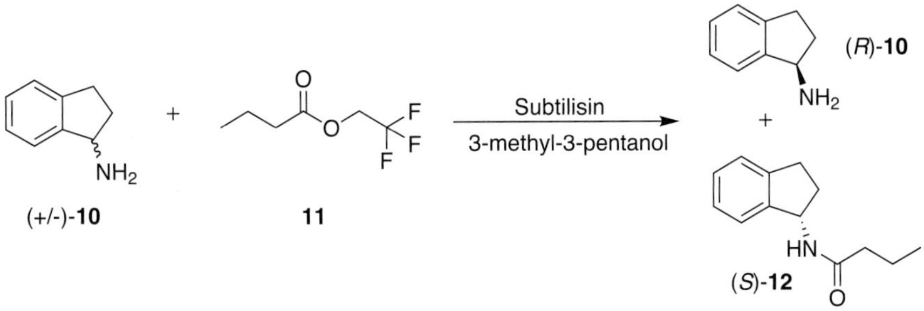

Figure 14.6 Kinetic resolution of 1-aminoindan **10**.

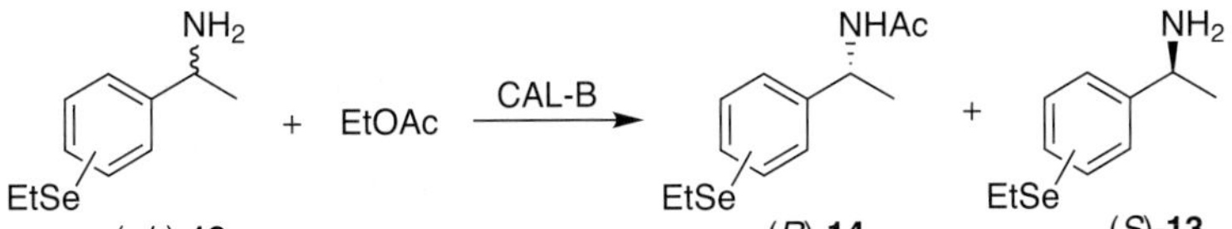

Figure 14.7 CAL-B catalyzed resolution of organoselenium containing amines.

Optically active 1,1′-binaphthylamine derivatives **15** are useful chiral ligands for various asymmetric reactions, such as the asymmetric hydrogenation of ketones [11]. Aoyagi and Izumi reported an effective kinetic resolution of 1,1′-binaphthylamines **15** via lipase-catalyzed amidation (Figure 14.8). LIP-300 (*Pseudomonas aeruginosa* lipase immobilized on Hyflo Super-Cel) and LPL-311 (*P. aeruginosa* lipase immobilized on Toyonite 200-M) were found to effectively catalyze the kinetic resolution yielding amides **16**. Increasing the alkyl chain length of the R substituent on the acyl donor improved the enantioselectivity of the reaction [12].

Some groups have reported on their search for less reactive acylating agents, to suppress noncatalyzed chemical acylation and increase product enantiomeric excess. Irimescu and Kato carried out an enantioselective lipase-catalyzed acylation of 1-phenylethylamine and 2-phenyl-1-propylamine by reacting the amines with carboxylic acids in a nonsolvent system or in ionic liquids (Figure 14.9). The reaction equilibrium was shifted toward amide synthesis by the continuous removal of the

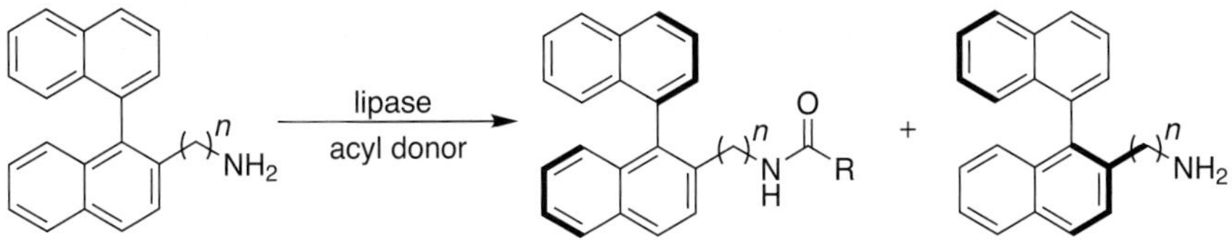

$n = 1,2$ or 3 R = CH_3, CH_2CH_3, $CH_2CH_2CH_3$, or CF_3

Figure 14.8 Synthesis of optically active 1,1′-binaphthylamines.

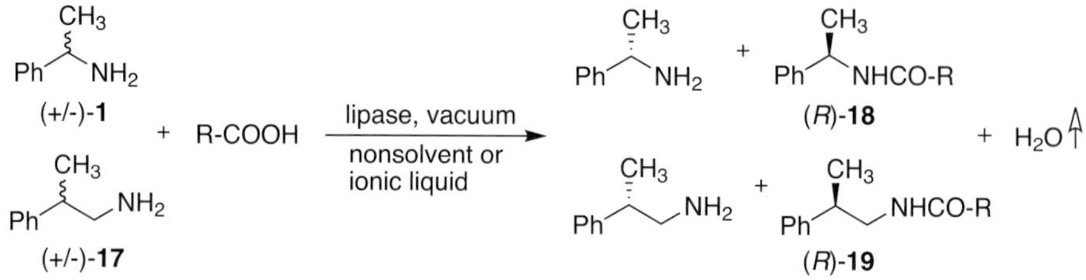

Figure 14.9 Lipase-catalyzed reactions in ionic liquids.

by-product water under reduced pressure. Ionic liquids were investigated because (i) they have no detectable vapor pressure and can be used under reduced pressure to drive off water, and (ii) they have the ability to dissolve a wide range of organic compounds with different polarities (both the substrates and the amide products). The initial reaction rates increased in the ionic liquid system relative to the nonsolvent system, and excellent enantioselectivities were obtained [13].

Goswami et al. resolved (±)-sec-butylamine **20** with CAL-B using ethyl esters of long-chain fatty acids as acylating agents to give **21** [14]. The enantioselectivity of the resolution was significantly improved when using the less reactive long-chain fatty acids (ee >99%) compared to ethyl or vinyl butyrate (ee <75%) (Figure 14.10).

Perhaps the best example of the large-scale application of lipases to the kinetic resolution of amines has been demonstrated by BASF. The BASF process is capable of multi-ton-scale production of chiral amines using lipase-catalyzed acylation in concert with base-catalyzed hydrolysis of the resulting amide [15]. Extensive screening of a range of acyl donors identified isopropyl methoxyacetate **22** as a promising candidate that led to a remarkable increase in selectivity and reactivity [16]. After the lipase-catalyzed formation of optically pure amine and amide, the products are separated and the amide is hydrolyzed under basic conditions to release the corresponding free amine without racemization and in quantitative yield. Recycling of the undesired enantiomer via racemization and recovery of the acylating help to lower the cost of this process. A wide variety of aryl and alkyl amines have been successfully resolved in this way. BASF has the production capacity to produce >3000 ton/year of chiral amines by lipase-catalyzed resolution (Figure 14.11).

Alternative hydrolytic enzymes, such as amidases and acylases, have also been employed for the resolution of racemic amines. Unlike lipases, these enzymes are typically not commercially available but they can be isolated from microorganisms.

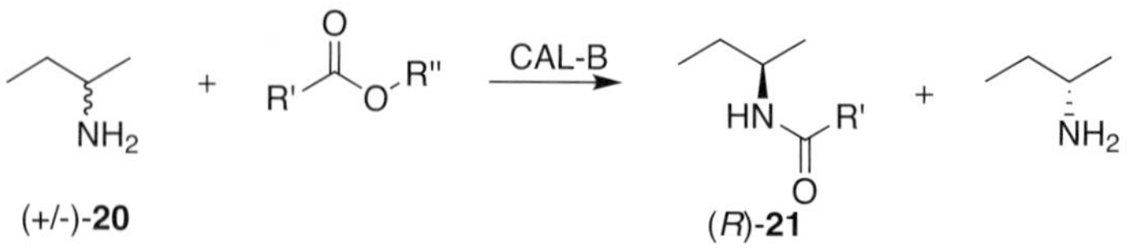

Figure 14.10 CAL-B-catalyzed resolution of (±)-sec-butylamine.

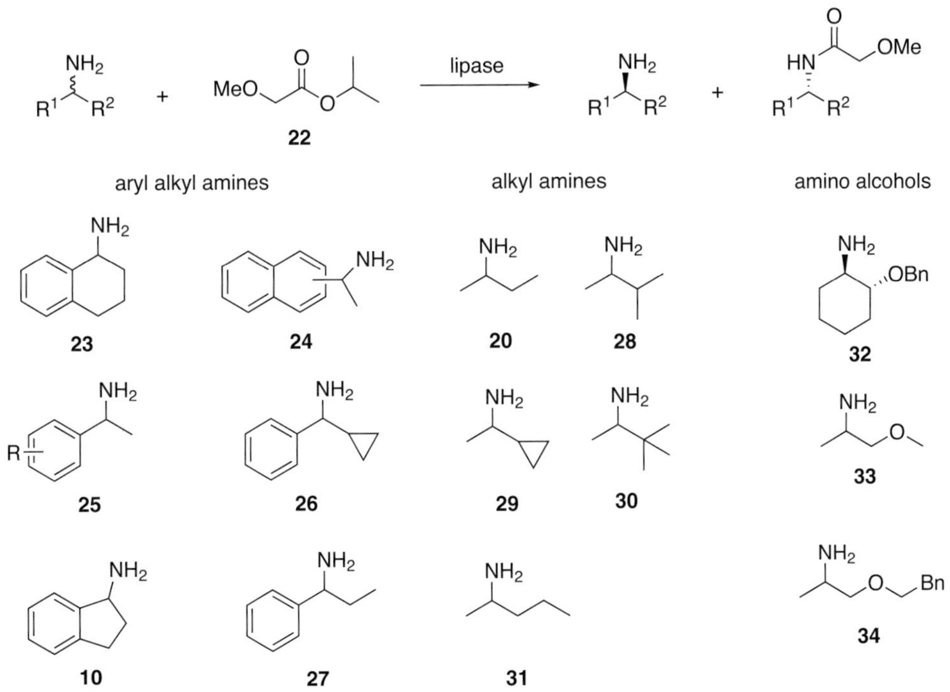

Figure 14.11 BASF process for the synthesis of a wide range of chiral amines.

Avecia identified approximately 60 microorganisms with amidase activity capable of resolving racemic amines [17, 18]. *Arthrobacter* species predominated in the list of microorganisms identified. The kinetic resolution of N-acetyl-1-aminoindanol **35** by a freeze-dried microbial sample (BH2-N1 amidase) allowed access to (1S,2R)-N-acetyl-1-aminoindanol **35** in high enantiomeric excess (96%). This compound is a key intermediate in the synthesis of Merck's HIV protease inhibitor Crixivan **37** (indinavir) (Figure 14.12).

Acylases have also been applied to the kinetic resolution of amines. Aminoacylase I from *Aspergillus melleus* was used for the resolution of a range of arylalkylamines and amino alcohols via acylation with methyl 2-methoxyacetate (Figure 14.13) [19]. Excellent chemoselectivity was also observed in all cases, as the amino group was preferentially acylated in the presence of a primary alcohol functionality. However, poor to moderate enantioselectivity was observed, with E values <10. The best result ($E = 9.3$) was obtained with 1-aminoindane **10** during the conversion to ester **38**.

An enzymatic procedure for amine resolution, employing acylation by *C. antarctica* lipase B and deacylation by penicillin G acylase, has been demonstrated by Ismail et al. (Figure 14.14) [20]. The acylase-catalyzed deacylation provides a "greener" process than the standard chemical deacylation as a result of the elimination of the salt waste stream typically generated by deacylating under strongly alkaline conditions. It is also more amenable to sensitive functional groups that are not stable under basic conditions. A drawback of this approach is that the amide hydrolysis step is

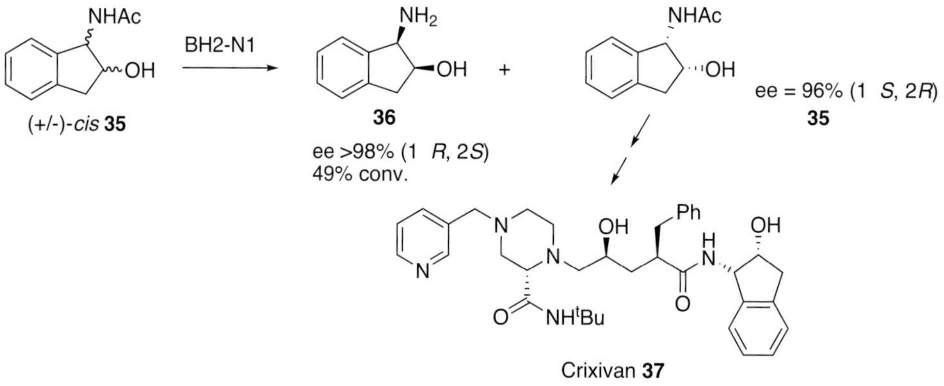

Figure 14.12 Resolution of a key intermediate for Crixivan **37**.

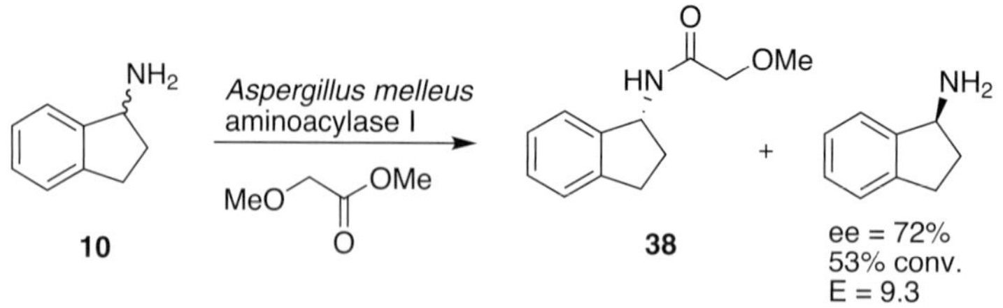

Figure 14.13 Amino acylase I-catalyzed amidation of 1-aminoindane **10**.

typically very slow. To overcome this challenge, more readily removable acylating groups, (*R*)-phenylglycine esters, were employed to increase the reaction rate. However, these acylating reagents provided low selectivities in the acylation of aliphatic amines compared to common acylating reagents.

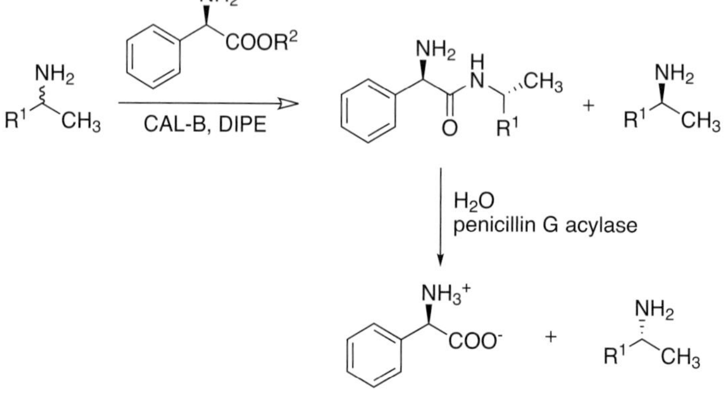

Figure 14.14 Combination of CAL-B and penicillin G acylase.

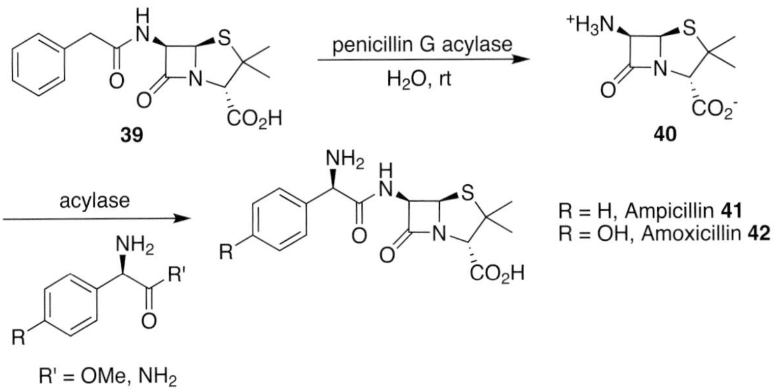

Figure 14.15 Synthesis of semisynthetic penicillins using penicillin G acylase.

Enzymatic deacylation of penicillin G **39** by penicillin G acylase can be run in water at room temperature and does not require protection or deprotection of other functional groups present in the molecule (Figure 14.15). Immobilized penicillin acylase was then used to catalyze acylation of 6-APA **40** and 7-ADCA to produce a range of semisynthetic β-lactam antibiotics such as ampicillin **41** and amoxicillin **42** [21].

All of the examples discussed so far focus on the resolution of primary amines; however, secondary amines are also important building blocks in the synthesis of biologically active compounds. The development of general, efficient methods for the production of optically pure secondary amines has remained a challenge. However, there are a few reported methods for enzyme-mediated kinetic resolution of secondary amines.

The preparation of optically pure secondary amines was accomplished via the resolution of oxalamic esters **43** to give acids **44** and ultimately amines **45** [22]. Excellent selectivities ($E > 200$) were obtained using a variety of proteases. Background chemical hydrolysis was suppressed by keeping the pH between 7.0 and 7.5 (Figure 14.16).

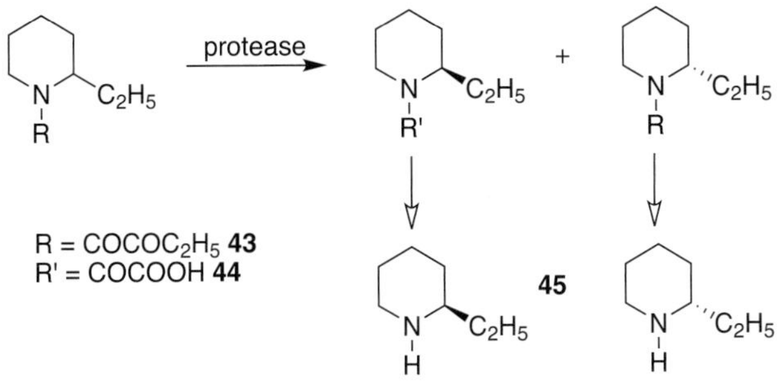

Figure 14.16 Protease-catalyzed resolution of secondary amines.

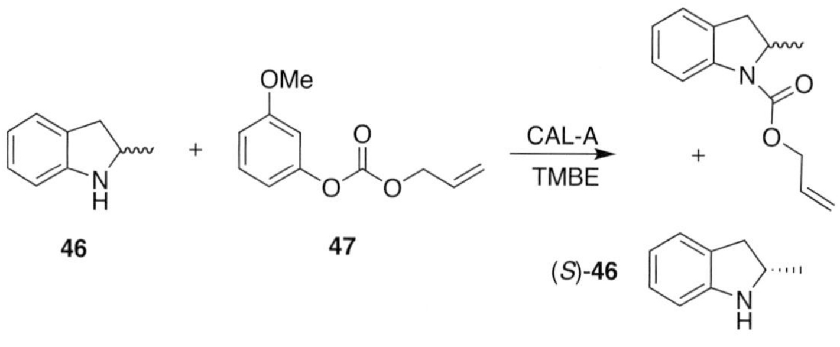

Figure 14.17 CAL-A-catalyzed resolution of racemic indoline.

The synthesis of optically active substituted indolines **46** has been achieved via *C. antarctica* lipase A-catalyzed resolution using 3-methoxyphenyl allyl carbonate **47** as the acylating agent (Figure 14.17) [23]. Indole and indoline cores are present in a variety of natural products, biologically active alkaloids, and pharmaceuticals. Excellent selectivity values ($E > 200$) were achieved.

GlaxoSmithKline also demonstrated the lipase-catalyzed resolution of a secondary amine. GSK required both enantiomers of 1-methyl-tetrahydroisoquinoline (MTQ) **48** as part of the development of YH1885 **49**, a potential treatment for gastroesophageal reflux disease (GERD) and duodenal ulcers [24]. They obtained good results using the lipase from *Candida rugosa* and substituted phenylallyl carbonate (e.g., **47**) acylating reagents. The (S)-1-MTQ amine product was obtained with >99% ee and good recovery (46% yield) (Figure 14.18).

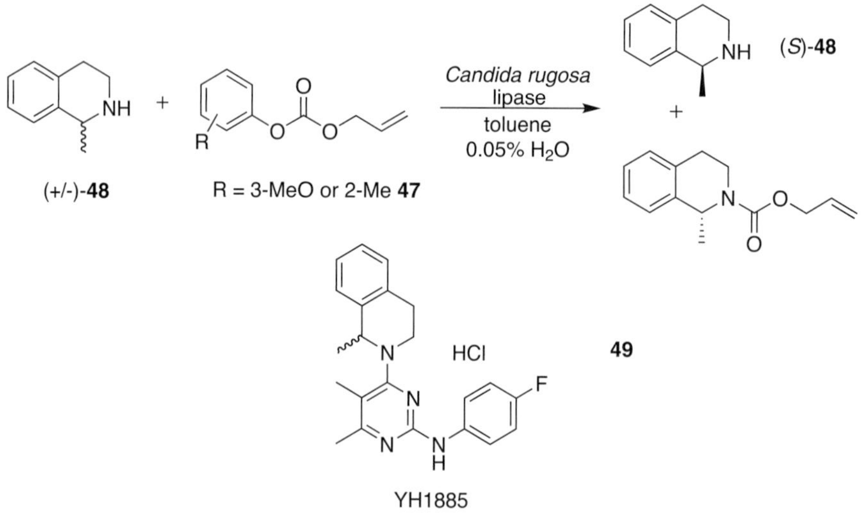

Figure 14.18 Resolution of an intermediate for YH1885.

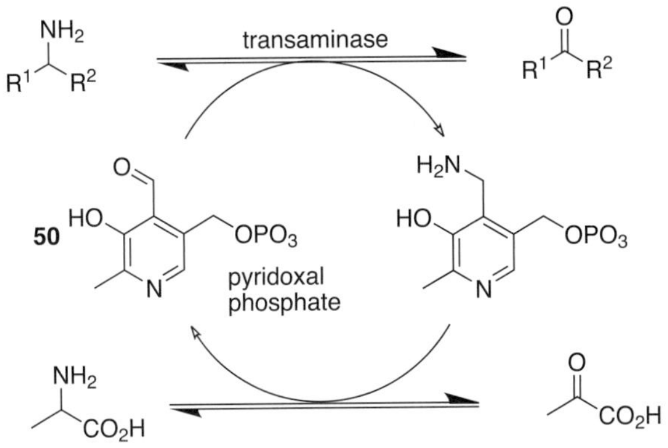

Figure 14.19 Kinetic resolution of racemic amines using transaminases.

14.2.2
Transaminases

The late 1990s saw the development of an alternative methodology for the enzymatic resolution of racemic amines using transaminases. Transaminases are pyridoxal phosphate **50** dependent enzymes that catalyze the transfer of an amine group to a carbonyl compound (amine group acceptor), such as a ketone, aldehyde, or keto acid (Figure 14.19).

Celgene developed transaminase technology for the enantioselective conversion of chiral amines to ketones [25]. Low-molecular-weight aldehydes, such as propionaldehyde, were used as the amine group acceptor. This process has been used on a 2.5 m^3 scale and has been demonstrated on several different aliphatic and aromatic amines. The disadvantage of this approach was the relatively low product concentrations attainable in an all-aqueous reaction system with hydrophobic compounds. Another disadvantage was the significant inhibition of the transaminases by both the amine substrates and the ketone products (Figure 14.20).

Shin and Kim developed various methods aimed at increasing the product concentrations of transaminase-catalyzed amine resolutions, through the continuous removal of product ketone from the reactions (Figure 14.21). The application of an aqueous/organic two-phase system to the ω-transaminase-catalyzed resolution of racemic (-methylbenzylamine **1** was found to be superior to an aqueous-only system in product concentration obtained [26, 27]. A drawback of the biphasic system was an increased enzyme deactivation rate compared to the aqueous-only system due to the aqueous/organic emulsion. Another disadvantage was the

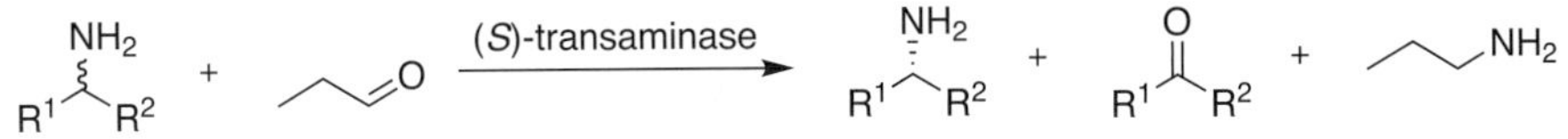

Figure 14.20 Use of propionaldehyde as amine acceptor.

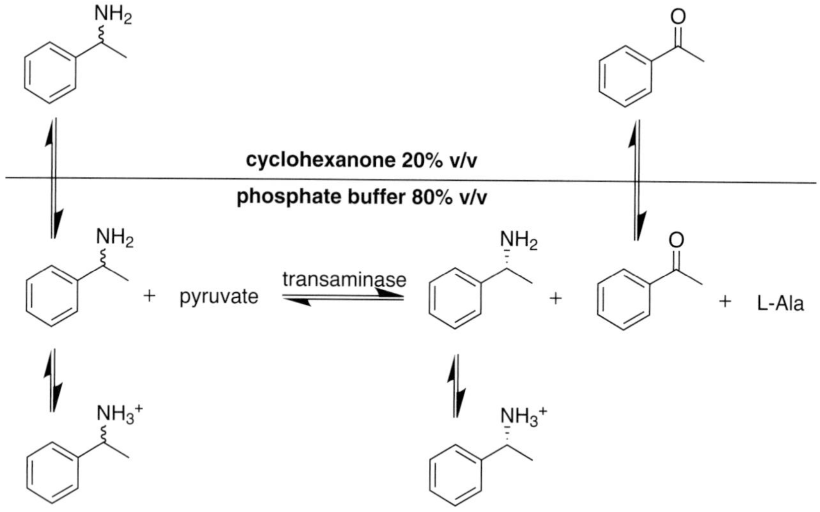

Figure 14.21 Removal of ketone product from transaminase reaction.

significant inhibition of the transaminase by the amine acceptor substrate (pyruvate) used.

In a further development, the kinetic resolution of (±)-**1** was demonstrated using an enzyme membrane reactor and a hollow fiber membrane contactor for the continuous removal of the inhibitory product acetophenone from the reaction system without incurring the penalty of enzyme deactivation associated with the biphasic emulsion system [28]. An alternative system utilizing a packed-bed reactor with a membrane contactor also showed promise [29]. Whole cells of *Vibrio fluvialis*, entrapped in Ca-alginate gel beads, were packed in a plug-flow column reactor. Fresh substrate solution was passed through the column reactor, and product was removed by flowing the column effluent through the membrane contactor. The range of substrates that could be prepared by these methods were enhanced via redesign of the substrate specificity of ω-transaminase to include the kinetic resolution of aliphatic chiral amines [30].

In 2008, Hanson *et al.* identified an (S)-amine transaminase from *Bacillus megaterium* that was used for the preparation of (R)-1-cyclopropylethylamine **51** and (R)-*sec*-butylamine **52** by resolution of the racemic amines [31]. Pyruvate was used as the amine acceptor (Figure 14.22).

Höhne *et al.* reported a substrate protection strategy that enhanced both the rate and the enantioselectivity of transaminase-catalyzed kinetic resolution reactions [32]. The ω-transaminase-catalyzed resolution of the pharmaceutically important synthons 3-aminopyrrolidine **53** and 3-aminopiperidine **54** was improved by the addition of protecting groups to the substrate amines. Reaction rates were improved by up to 50-fold, and product ee was improved from 86 to 99% (Figure 14.23).

A significant drawback of the transaminase approach is the need for a stoichiometric or superstoichiometric amount of amine acceptor to drive the reaction

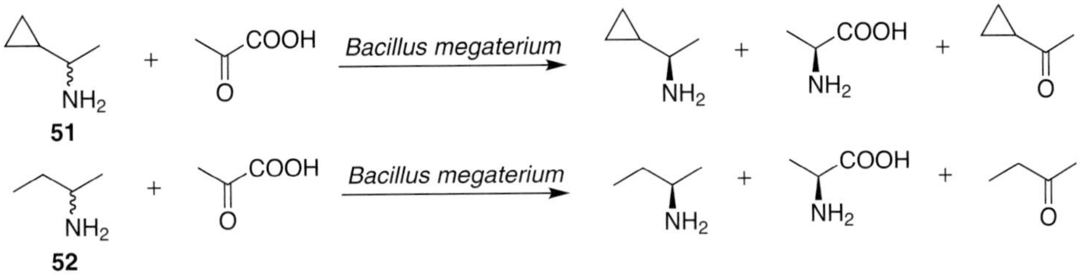

Figure 14.22 Resolution of racemic chiral amines using a transaminase from *B. megaterium*.

equilibrium toward the desired products. In addition, inhibition of the transaminase by the amine acceptor cosubstrate significantly reduces the productivity of transaminase mediated-amine resolutions. Truppo *et al.* have recently developed a strategy that enables the kinetic resolution of racemic amines using a transaminase in combination with an amino acid oxidase (AAO) to reduce both the quantities and the enzyme inhibition associated with the amine acceptor pyruvate [33]. A transaminase was used to selectively convert one enantiomer of a racemic amine substrate to the ketone product, resulting in concurrent amination of the amine acceptor (pyruvate) to produce the coproduct alanine. Addition of AAO regenerates the amine acceptor *in situ* using molecular oxygen, thereby requiring only a catalytic amount of amine acceptor to be used (amine to amine acceptor molar ratio = 1000 : 1). A significant reduction in the quantity and cost of the cosubstrate was realized with this methodology. In addition, this system drives the unfavorable reaction equilibrium and significantly reduces transaminase inhibition by employing a very low concentration of amine acceptor to run the reaction (Figure 14.24).

14.2.3
Amine Oxidases

The enantioselective oxidation of amphetamine **55** by copper containing amine oxidases from *Escherichia coli* and *Klebsiella oxytoca* was reported in 2000 by Hacisalihoglu *et al.* [34] Moderate *E* values of ~15 were obtained, opening up the possibility for future applications of amine oxidase-catalyzed resolutions of substituted amphetamines (Figure 14.25).

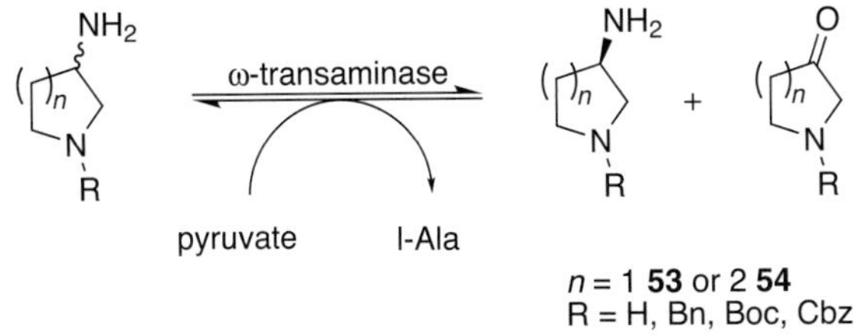

Figure 14.23 Use of protecting groups in transaminase-catalyzed amination reactions.

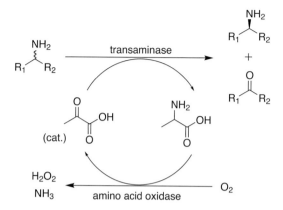

Figure 14.24 Dual transaminase/amino acid oxidase-mediated resolution of amines.

14.3
DKR and Deracemization of Amines

14.3.1
DKR Using Hydrolytic Enzymes and Racemization Catalysts

A number of different groups have recently investigated the dynamic kinetic resolution of racemic chiral amines [35, 36] using an enantioselective lipase (often CAL-B or Novozyme-435) in combination with a chemocatalyst that effects racemization of the unreactive amine enantiomer under the reaction conditions. A key issue with these types of DKR processes is finding conditions under which the bio- and chemocatalysts can function efficiently together. The catalytic cycle for a DKR is shown in Figure 14.26 in which it is essential to identify methods for selective racemization of the substrate but not the product.

The first reported example of DKR of an amine derives from Reetz et al. in which they employed palladium as a catalyst for the racemization of α-methylbenzylamine **1** in the presence of CAL-B [37]. The reaction was carried out in pyridine, with ethyl acetate as the acyl donor, at a temperature of 50–55 °C. In order to achieve total conversion to the *N*-acylated product, it was necessary to leave the reaction for 5 days giving the (*R*)-amide in 64% yield and 99% ee. Kim and coworkers extended this approach by employing the *in situ* reduction of the corresponding oxime **56** to the

(*R*)-(-)-amphetamine **55**   (*S*)-(+)-amphetamine **55**
                               (dexedrine)

Figure 14.25 Synthesis of (*S*)- and (*R*)-**55** using enantioselective monoamine oxidases.

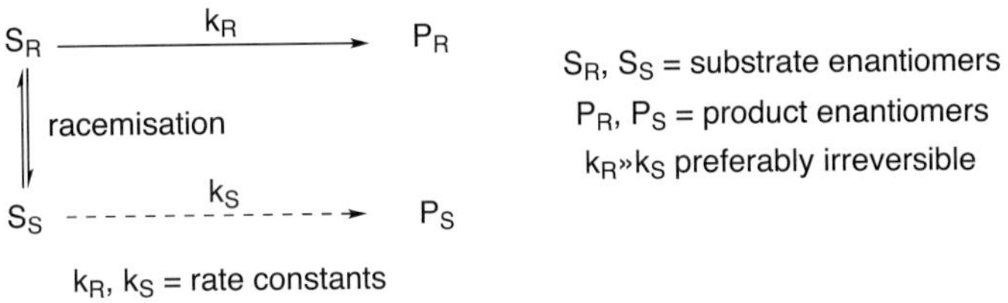

14.3 DKR and Deracemization of Amines

$$S_R \xrightarrow{k_R} P_R$$

racemisation

$$S_S \xdashrightarrow{k_S} P_S$$

S_R, S_S = substrate enantiomers
P_R, P_S = product enantiomers
$k_R \gg k_S$ preferably irreversible

k_R, k_S = rate constants

Figure 14.26 Reaction scheme for a dynamic kinetic resolution.

racemic amine, thereby avoiding the presence of high concentrations of the amine at the beginning of the reaction (Figure 14.27) [38].

Bäckvall and coworkers have worked extensively in this area and reported the DKR of a range of amines **57a–57k** by combining CAL-B with a ruthenium catalyst to promote racemization (Figure 14.28). This protocol allows unfunctionalized primary amines **57a–57j** to be converted to the corresponding amides **58a–58k** in good yields and high enantiomeric purities as shown. The addition of sodium carbonate increases the yield of the reaction and is believed to act by preventing traces of acid interfering with the activity of the metal catalyst [39].

Hult and coworkers have reported a modification to the Bäckvall system in which 1 equiv of isopropyl 2-methoxyacetate was used as the acyl donor in combination with

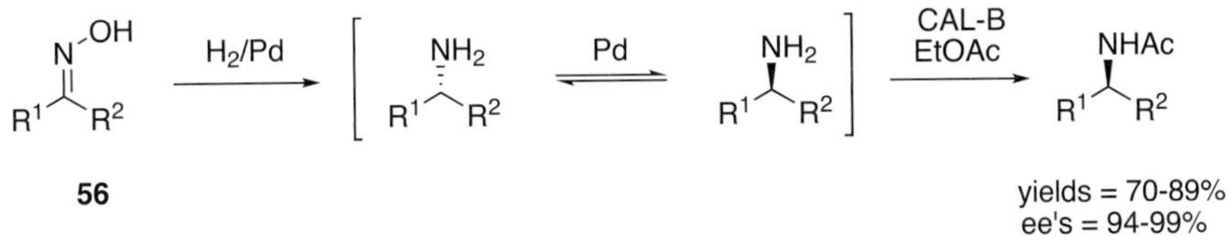

56 → yields = 70–89%
ee's = 94–99%

Figure 14.27 DKR of racemic chiral amines using oxime precursors.

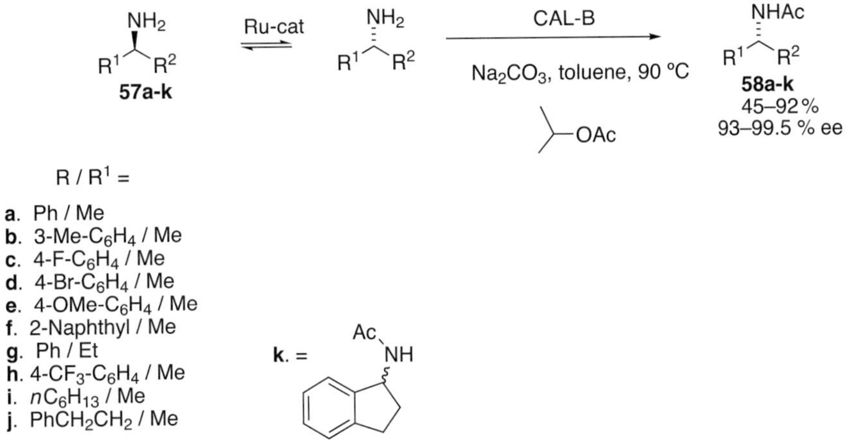

57a–k → **58a–k**
45–92%
93–99.5 % ee

R / R^1 =

a. Ph / Me
b. 3-Me-C_6H_4 / Me
c. 4-F-C_6H_4 / Me
d. 4-Br-C_6H_4 / Me
e. 4-OMe-C_6H_4 / Me
f. 2-Naphthyl / Me
g. Ph / Et
h. 4-CF_3-C_6H_4 / Me
i. nC_6H_{13} / Me
j. $PhCH_2CH_2$ / Me

k. = (2-acetamidoindane structure)

Figure 14.28 DKR of chiral amines using CAL-B and a ruthenium catalyst.

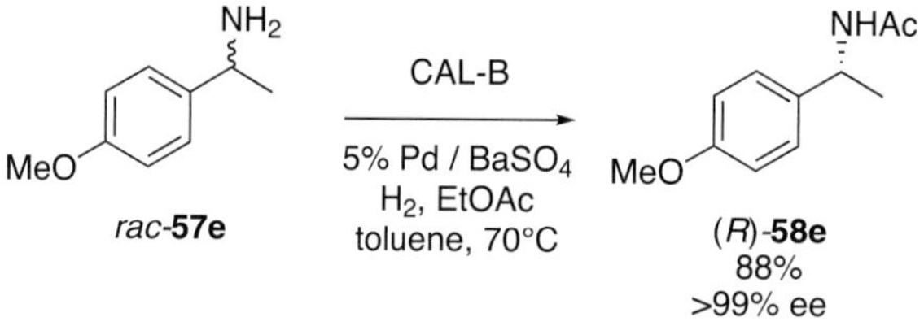

Figure 14.29 DKR of racemic **56** using CAL-B and a palladium racemization catalyst.

the modified Shvo complex and CAL-B. Carrying out the reaction at 100 °C ensured a high reaction rate. Under these conditions, complete conversion of most substrates was observed within 26 h compared to 72 h for the previous approach [40].

Similarly, CAL-B has been used in combination with a palladium/alkaline earth metal-based racemization catalyst to effect a dynamic kinetic resolution on the benzylic amine **57e** (Figure 14.29). The (R)-amide **58e** was obtained in very good yield and excellent optical purity. Several other substrates also underwent the reaction [41, 42].

Kim et al. have developed a practical procedure for the dynamic kinetic resolution of primary amines illustrated by substrate **57h** (Figure 14.30). They employed a supported palladium nanocatalyst as the racemization catalyst and the commercially available Novozym-435 as the enantioselective catalyst for acylation of the amine using ethyl methoxyacetate as the acyl donor. High yields and enantiomeric excesses were achieved [43].

Gastaldi et al. discovered that *in situ* racemization of a chiral amine **59** was mediated by the addition of thiyl radicals (Figure 14.31). Combination with CAL-B enabled the DKR of nonbenzylic amines **59** yielding amides (R)-**60** in good yield and high enantioselectivities [44].

Scientists at Huddersfield University in collaboration with Avecia have developed a DKR process involving the combination of immobilized *C. rugosa* lipase and an iridium-based racemization catalyst (Figure 14.32). By using carbonate **62** as the acyl

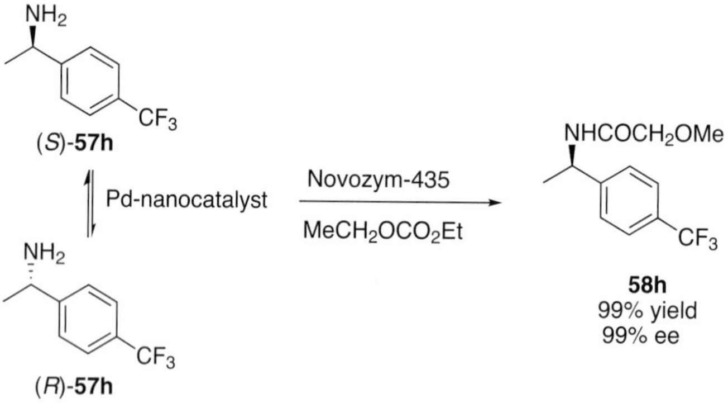

Figure 14.30 DKR of (±)-**56c** using CAL-B and a palladium nanocatalyst.

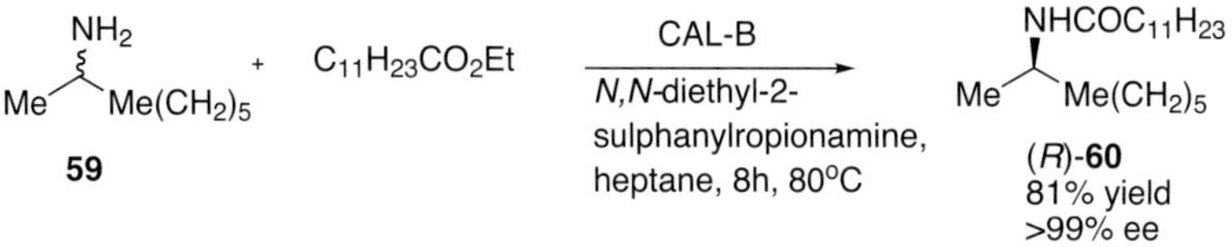

Figure 14.31 Racemization of amine **59** via addition of thiyl radicals.

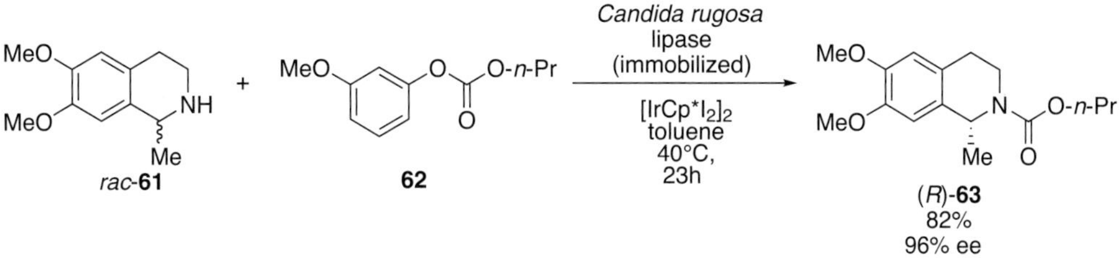

Figure 14.32 DKR of secondary amine **61** using a novel iridium catalyst.

donor, the racemic secondary amine **61** was converted to the corresponding carbamate (R)-**63** in high yield and ee [45].

A group at AnorMED in Canada have reported a highly unusual example of racemization in the DKR of the substrate 8-amino-5,6,7,8-tetrahydroquinoline **64**. Using CAL-B as enzyme and ethyl acetate as the acylating agent, they isolated the corresponding (R)-acetamide in >60% yield and 99% ee. They propose a mechanism involving spontaneous formation of a catalytic quantity of ketone **66** that reacts with the amine **64** to form imine **67** that then undergoes spontaneous racemization (Figure 14.33) [46].

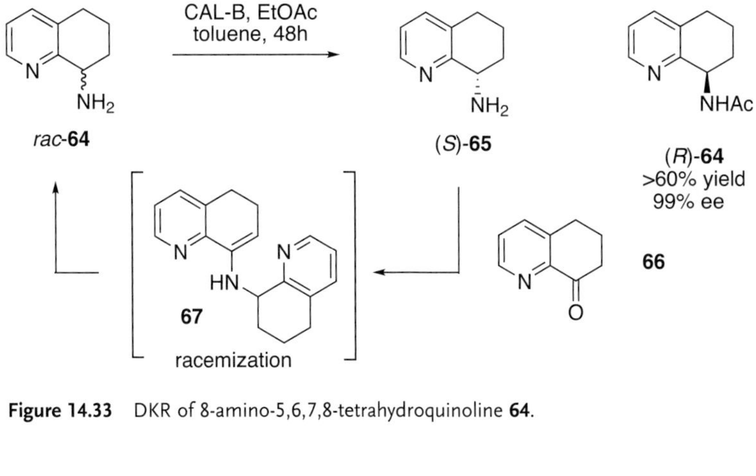

Figure 14.33 DKR of 8-amino-5,6,7,8-tetrahydroquinoline **64**.

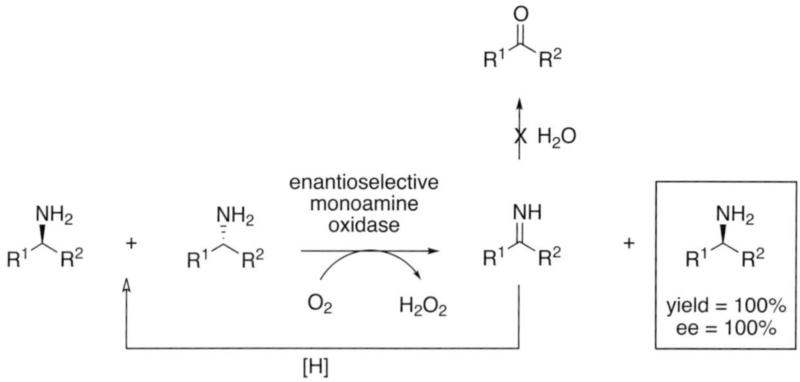

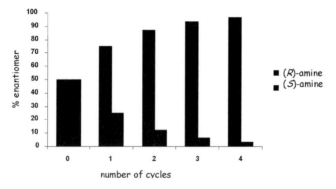

Figure 14.34 Deracemization of racemic amines by repeated cycles of enzyme-catalyzed enantioselective oxidation followed by nonselective chemical reduction.

14.3.2
Deracemization Reactions Using Amine Oxidases

Turner *et al.* have developed a general method for the deracemization of racemic 1°, 2°, and 3° amines as shown in Figure 14.34. In this catalytic cycle, monoamine oxidase N (MAO-N) from *Aspergillus niger* catalyzes enantioselective oxidation of one enantiomer of a racemic amine to the corresponding imine. Rather than allow the imine to simply hydrolyze to the corresponding ketone, a chemical reducing agent (e.g., ammonia borane) is added to the reaction resulting in *in situ* reduction of the imine back to the racemic amine that then undergoes another round of enantioselective oxidation to the imine. Provided that the enzyme is highly enantioselective, eventually complete "deracemization" of the racemate occurs yielding one enantiomer of the amine in high yield and ee (Figure 14.35).

In order to be able to apply this deracemization method to a wide range of chiral amines, it was essential to identify monoamine oxidase enzymes possessing both

Figure 14.35 Schematic illustration of deracemization.

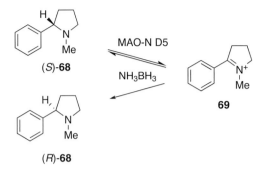

Figure 14.36 Deracemization of N-methylpyrrolidine **68**.

broad substrate specificity and high enantioselectivity. The wild-type enzyme was found to have high activity toward simple achiral amines (e.g., 1-butyl amine, benzylamine) but very poor activity toward chiral amines. Starting with the wild-type MAO-N, the Turner group subjected the enzyme to several rounds of random mutagenesis coupled with screening using (-methylbenzylamine **1** as the substrate. Initially, a variant (Asn336Ser) was identified that had *about* 50-fold improved in activity [47]. In addition, this variant has generally good activity toward a wide range of chiral 1° amines. Further rounds of random mutagenesis, coupled with screening, identified a MAO-N variant termed D5 with good activity and high enantioselectivity toward both 2° [48] and 3° amines [49]. For example, racemic N-methylpyrrolidine **68** was subjected to deracemization, *via* the intermediate iminium ion **69**, yielding (R)-**68** in 75% isolated yield and 99% ee within 24 h (Figure 14.36). The possibility of using the approach for enantioselective intramolecular reductive amination reactions was also investigated [49].

Interestingly, this D5 variant is also able to catalyze the enantioselective oxidation of racemic O-methyl-N-hydroxylamines **70** yielding the unreacted (R)-substrate (ee = 99%) and the E-oxime **71** product (Figure 14.37) [50].

In order to facilitate broader application of these MAO-N variants, the Turner group has developed a useful "desktop" model that can be used to predict the suitability of particular substrate for oxidation by MAO-N. This model was successfully used to predict the deracemization of racemic crispine A **72**, an alkaloid with potent biological activity (Figure 14.38) [51].

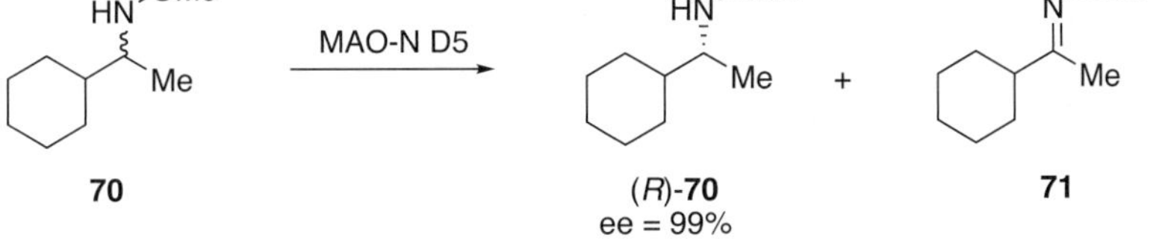

Figure 14.37 Enantioselective oxidation of racemic O-methyl-N-hydroxyamines.

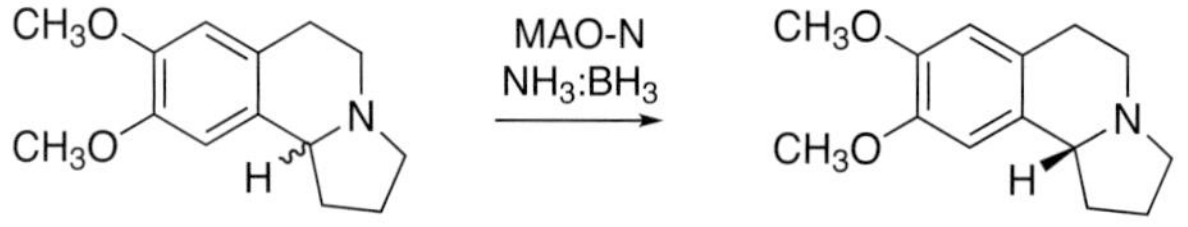

Figure 14.38 Deracemization of crispine A.

Recently, in collaboration with scientists at the University of York, the crystal structure of MAO-N D5 has been determined at 1.8 Å resolution (Figure 14.4) [52]. This new structure provides insights into the nature of the mutations and their possible effect in terms of determining the substrate specificity of the enzyme.

14.4
Asymmetric Synthesis of Amines Using Transaminases

Transaminases, which were discussed earlier, have long been used for the production of amino acids, and a primary goal has been to extend their use to the production of simple amines [53]. The first reported transaminase capable of transferring an amine to a simple ketone substrate without a carboxylic acid group was a diamine-ketoglutaric transaminase identified in 1964 [54]. However, it was not until the end of the 1980s that transaminase technology became an industrially viable option for the synthesis of chiral amines (Figure 14.39).

Researchers at Celgene developed both (R)- and (S)-selective transaminases that were active on a range of aliphatic and aromatic ketones and amines [25, 55–57]. Two approaches were employed based upon kinetic resolution, which has been discussed above, and asymmetric synthesis. The asymmetric synthesis approach starts with a

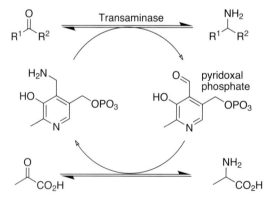

Figure 14.39 Asymmetric synthesis of chiral amines using transaminases.

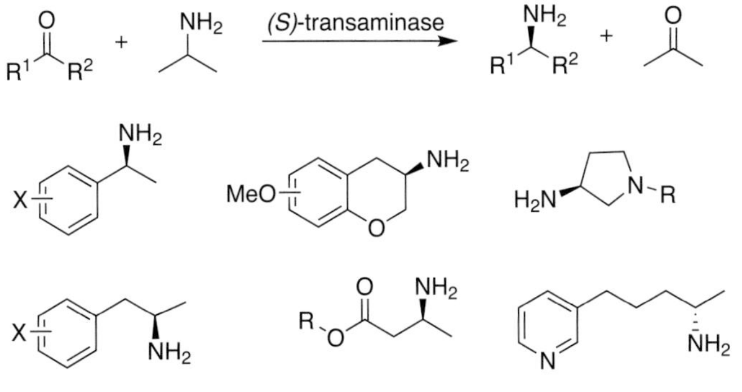

Figure 14.40 Celgene process for the conversion of ketones to amines using isopropylamine.

prochiral ketone substrate that served as an amine acceptor and isopropylamine as the amine group donor. The product of the reaction was the desired chiral amine and acetone. The Celgene processes have been run successfully at 2.5 m^3 scale; however, poor equilibrium necessitates the use of excess reagents to drive the reactions to completion, and low product concentrations were obtained with hydrophobic substrates (Figure 14.40).

Several investigations into the challenges facing transaminase technology were conducted by Shin and Kim in the late 1990s. Using an (S)-selective ω-transaminase isolated from *V. fluvialis*, Shin and Kim demonstrated the synthesis of aliphatic and aryl containing chiral amines using L-alanine as the amine donor [58, 59]. Severe inhibition of the transaminase by the amine and keto-acid reaction products, as well as extremely poor conversion, were observed. Even with a large excess of amine donor, reactions proceeded to <2% conversion. The continuous *in situ* removal of the inhibitory pyruvate product *via* reduction using lactate dehydrogenase (LDH) boosted reaction yields to >90%; however, inhibition by the desired amine product resulted in poor productivity (Figure 14.41).

Studies comparing ω-transaminases from different microorganisms further demonstrated the extreme inhibition of transaminases by reaction products [60] and also identified the rarity of (R)-selective transaminases [61]. Iwasaki *et al.*

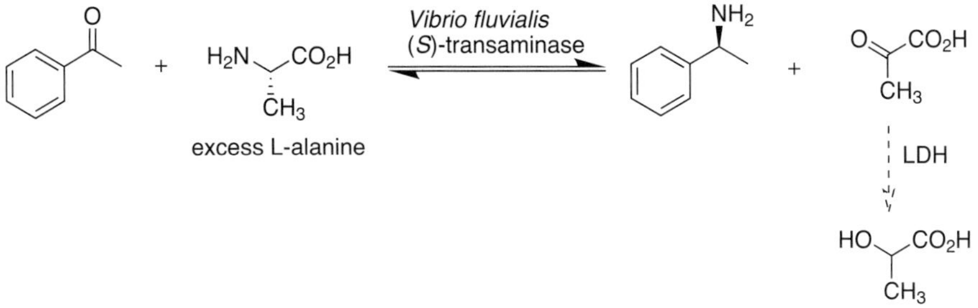

Figure 14.41 Removal of pyruvate by addition of LDH.

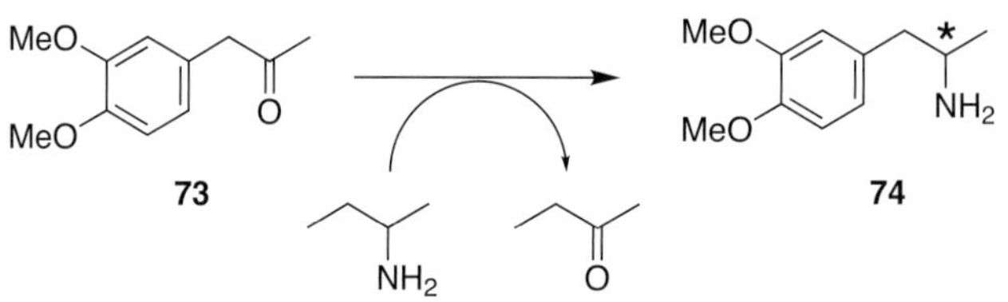

Figure 14.42 Synthesis of (R)-3,4-dimethoxyamphetamine **74**.

improved the substrate scope of transaminase technology through the identification of an (R)-selective transaminase in a soil isolate from *Arthrobacter sp.* that catalyzed amination of 3,4-dimethoxyphenylacetone **73** in the presence of *sec*-butylamine to give (R)-3,4-dimethoxyamphetamine (DMA) **74** [62, 63] The corresponding (S)-DMA could be obtained using an alternative transaminase from *Pseudomonas sp.* (Figure 14.42).

The evolution of a transaminase from *Arthrobacter citreus* to a thermostable transaminase with increased specific activity and decreased inhibition by the amine product was accomplished using error-prone polymerase chain reaction (PCR) [64] The reaction of substituted tetralone **75** and isopropylamine to produce substituted (S)-aminotetralin **76** was carried out at greater than 50 °C to facilitate the removal of the acetone by-product and drive reaction equilibrium (Figure 14.43).

An alternative method for driving the reaction equilibrium of transaminations was developed by Höhne *et al.* (Figure 14.44) [65]. Alanine served as the amine donor, and pyruvate decarboxylase was used to remove the pyruvate coproduct by decarboxylation to acetaldehyde. An advantage of pyruvate decarboxylase over lactate dehydrogenase is that it requires no cofactor recycling system, and the high volatility of the coproducts allows for the desired shift of equilibrium.

The synthesis of a range of optically pure pharmacologically relevant amines has been reported by Koszelewski *et al.* [66]. Various ω-transaminases were tested for the

Figure 14.43 Transaminase-catalyzed synthesis of (S)-aminotetralin **76**.

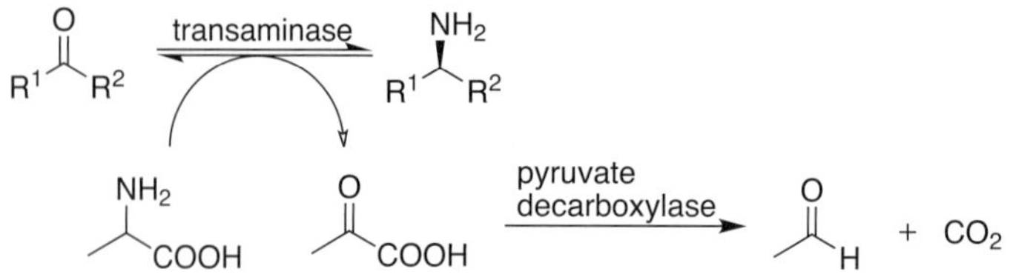

Figure 14.44 Removal of pyruvate using pyruvate decarboxylase.

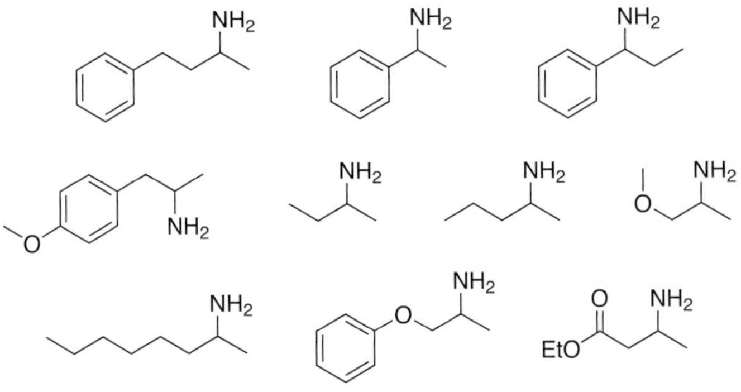

Figure 14.45 Synthesis of a range of chiral amines using transaminases.

synthesis of enantiomerically pure amines from their corresponding ketones using D- or L-alanine as the amine donor and lactate dehydrogenase to drive the reaction equilibrium via the removal of the coproduct pyruvate. Excellent enantioselectivities (>99% in many cases) were achieved for each enantiomer of amine product (Figure 14.45).

Most recently, a focus on the practical industrial use of transaminases including rapid screening and process development has been conducted by Truppo et al. A rapid, high-throughput screening methodology for the determination of transaminase activity has been developed that uses a pH-based colorimetric assay (Figure 14.46) [67]. As before, LDH is used to reduce the pyruvate product, eliminating inhibition by pyruvate and driving the reaction to completion. Recycling of the NADH cofactor is accomplished using glucose dehydrogenase (GDH) and glucose resulting in the formation of gluconic acid that causes a drop in pH that is detected spectrophotometrically through the addition of a pH indicator dye to the reaction system (Figure 14.46).

This system, which can be operated in 96-well plate format, was also used for process development studies including pH, temperature, and optimization of

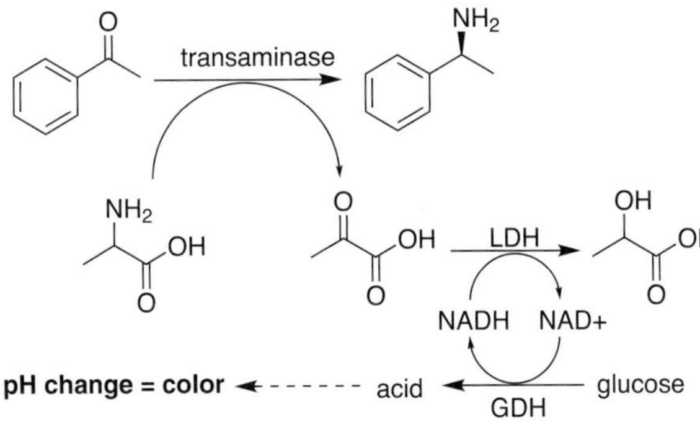

Figure 14.46 Monitoring transaminase-catalyzed reactions by use of a pH indicator.

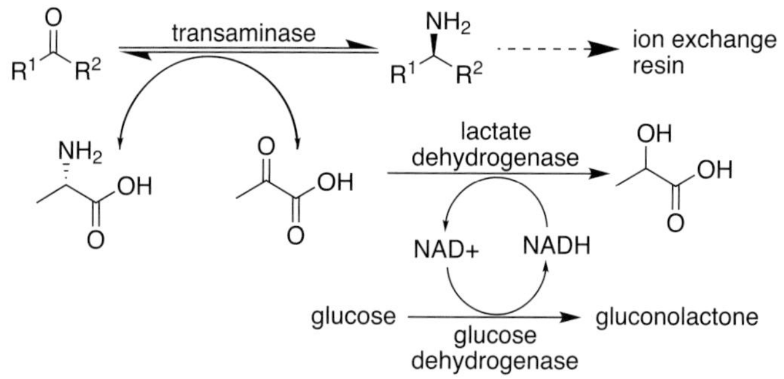

Figure 14.47 Use of ion-exchange resin to selectively remove the product amine from the reaction.

cosolvent concentration. Optimal conditions were successfully scaled directly from 100 μl scale to 50 ml demonstration scale. Inhibition of the transaminase by the enantiopure amine product was alleviated *via* continuous product removal using an ion exchange resin that selectively removed the desired amine without affecting the ketone substrate, allowing for the first time general, scaleable 50 g/l transamination processes for the production of chiral amines (Figure 14.47) (M. D. Truppo, J. D. Rozzell, and N. J. Turner, unpublished results).

An alternative, "reductive amination" approach was developed simultaneously by the Kroutil and Turner groups using an amino acid dehydrogenase (AADH) and a catalytic amount of alanine [67, 68]. Aqueous ammonia was used to regenerate alanine from pyruvate, and a glucose dehydrogenase system was employed for the recycling of the NADH cofactor (Figure 14.48).

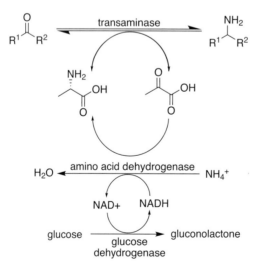

Figure 14.48 Asymmetric "reductive amination" procedure.

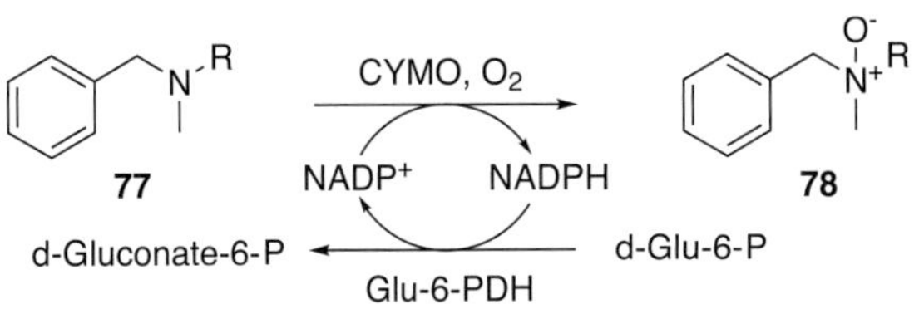

Figure 14.49 Enantioselective oxidation of tertiary amines.

14.5
Conclusions and Future Perspectives

Researchers in biocatalysis continue to explore novel routes to enantiomerically pure amines, with an emphasis on asymmetric methods or their equivalent. Cyclohexanone monooxygenase (CYMO) has been shown to catalyze the asymmetric oxidation of some tertiary amines **77** to the corresponding amine N-oxides **78** (Figure 14.49). It was found that the structure of the amine has a significant effect on the enantiomeric excess of the products obtained. For example, when R = CH_2CH_2OH, CH_2CCH or R = (R)-CHPh-CH_2OH, the ee's were in the range 24–32%. However, for R = CH_2Ph-pCl or R = CH_2-Ph-pCF3, no oxidation occurred [69].

The asymmetric reduction of imines represents a potentially very attractive route to chiral amines and has been studied by two groups (Figure 14.50). Vaijayanthi and Chadha examined the reduction of a series of aryl imines **79** using *Candida parapsilosis* ATCC 7330. A range of different aryl substituents were found to be tolerated yielding (R)-secondary amines **80** in good yields (55–80%) and high ee (95–99%) [70].

Stephens and coworkers have reported a novel method of screening for imine reductase activity in which a library of structurally different imines is assembled under dynamic combinatorial conditions. The library was screened against potential microorganisms for activity that led to the discovery that the anaerobic bacterium *Acetobacterium woodii* possessed imine reductase activity [71]. The same group has recently reported the reduction of amides to amines. Washed cells of *Clostridium sporogenes* reduced benzamide (up to 20 mM) to benzylamine in yields up to 73% using H_2 as electron donor with less than 10 g biocatalyst/l over 24 h. Product formation exhibited complex kinetics, with a lag before benzylamine production began. Very little substrate was hydrolyzed since the maximum yield of benzoic acid was only 9% of the substrate added. Boiled cells were inactivated, thus confirming that amide reduction was enzyme catalyzed [72].

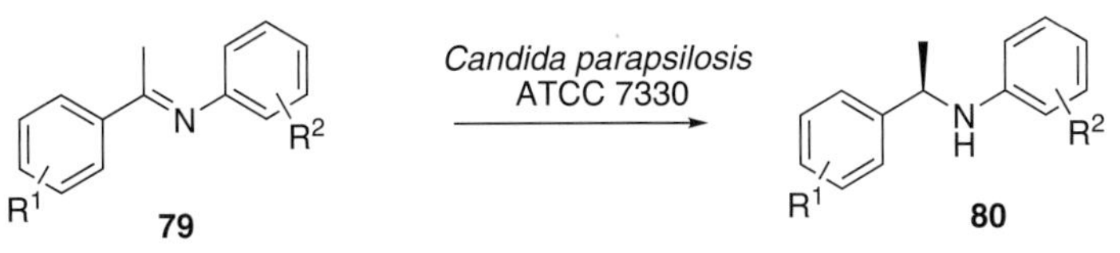

Figure 14.50 Asymmetric reduction of imines.

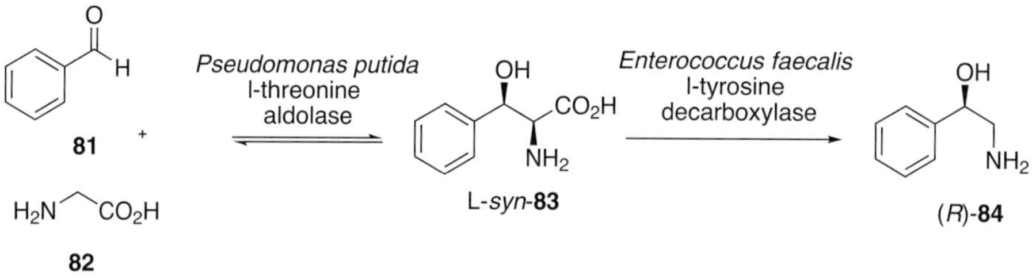

Figure 14.51 Combined use of threonine aldolase and L-tyrosine decarboxylase.

Researchers at the University of Graz, in collaboration with scientists from DSM, have developed an elegant and novel approach to the synthesis of β-amino alcohols using two different enzymes in one pot (Figure 14.51). For example, a threonine aldolase-catalyzed reaction was initially used to prepare L-**83** from glycine **82** and benzaldehyde **81**. L-**83** was then converted to (R)-**84** in high ee by an irreversible decarboxylation catalyzed by L-tyrosine decarboxylase [73].

The use of biocatalysts for the synthesis of enantiomerically pure chiral amines offers a highly attractive and competitive option compared to other approaches involving chemocatalysis, resolution of enantiomers, use of stoichiometric chiral auxiliaries, or transition metal catalysis. As shown above, different approaches can be used and the method of choice will depend upon a number of factors including the availability of the starting material and the scale of operation. A key factor driving the more widespread use of biocatalysis will be the increased availability of the biocatalysts themselves (in order to assist the reader, a list of enzymes mentioned in this chapter, together with commercial suppliers, is given in Table 14.1). Moreover, the techniques of directed evolution are being increasingly applied to develop biocatalysts with improved properties for practical applications, particularly in terms of broader substrate specificity, enhanced enantioselectivity, and most importantly, improved stability for operation under process conditions, particular in the presence of high concentrations of organic cosolvent.

Questions

14.1. Give an overview of the different biocatalytic methods that can be used to prepare enantiomerically pure chiral amines.

14.2. Compare and contrast the dynamic kinetic resolution of amines with deracemization. For each approach, outline the basic principle of the method and then suggest suitable biocatalyst/chemocatalyst combinations necessary for each process.

14.3. Outline the problems associated with transaminase-catalyzed amination of ketones to produce chiral amines, particularly with respect to product inhibition and equilibrium. Suggest two different methods that can be used to overcome these problems.

References

1. Turner, N.J. and Carr, R. (2007) in *Biocatalysis in the Pharmaceutical and Biotechnology Industries* (ed. R.N. Patel), CRC Press LLC, Boca Raton, FL, pp. 743–755.
2. van Rantwijk, F. and Sheldon, R.A. (2004) *Tetrahedron*, **60**, 501.
3. Garcia-Urdiales, E., Alfonso, I., and Gotor, V. (2005) *Chem. Rev.*, **105**, 313.
4. Kitaguchi, H., Fitzpatrick, P.A., Huber, J.E., and Klibanov, A.M. (1989) *J. Am. Chem. Soc.*, **111**, 3094.
5. Phillips, G.T. and Shears, J.H. (1990) Patent EP 399589.
6. Smidt, H., Fischer, A., Fischer, P., Schmid, R.D., and Stelzer, U. (1996) Germany Patent DE 19507217.
7. Messina, F., Botta, M., Corelli, F., Schneider, M.P., and Fazio, F. (1999) *J. Org. Chem.*, **64**, 3767.
8. Tao, J. (2003) *Chiral USA*, Scientific Update, Chicago.
9. Gutman, A.L., Meyer, E., Kalerin, E., Polyak, F., and Sterling, J. (1992) *Biotechnol. Bioeng.*, **40**, 760.
10. Andrade, L.H. and Silva, A.V. (2008) *Tetrahedron: Asymmetry*, **19**, 1175.
11. Noyori, R. (2003) *Adv. Synth. Catal.*, **345**, 15.
12. Aoyagi, N. and Izumi, T. (2002) *Tetrahedron Lett.*, **43**, 5529.
13. Irimescu, R. and Kato, K. (2004) *Tetrahedron Lett.*, **45**, 523.
14. Goswami, A., Guo, Z.W., Parker, W.L., and Patel, R.N. (2005) *Tetrahedron: Asymmetry*, **16**, 1715.
15. Breuer, M., Ditrich, K., Habicher, T., Hauer, B., Kesseler, M., Sturmer, R., and Zelinski, T. (2004) *Angew. Chem., Int. Ed.*, **43**, 788.
16. Ditrich, K. (2008) *Synthesis*, 2283.
17. Reeve, C.D. (1999) Patent WO 99/31264.
18. Reeve, C.D., Holt, R.A., Rigby, S.R., and Hazell, K. (2001) *Chim. Oggi*, **19**, 31.
19. Youshko, M.I., van Rantwijk, F., and Sheldon, R.A. (2001) *Tetrahedron: Asymmetry*, **12**, 3267.
20. Ismail, H., Lau, R.M., van Langen, L.M., van Rantwijk, F., Svedas, V.K., and Sheldon, R.A. (2008) *Green Chem.*, **10**, 415.
21. Bruggink, A., Roos, E.C., and de Vroom, E. (1998) *Org. Proc. Res. Dev.*, **2**, 128.
22. Hu, S.H., Tat, D., Martinez, C.A., Yazbeck, D.R., and Tao, J.H. (2005) *Org. Lett.*, **7**, 4329.
23. Gotor-Fernandez, V., Fernandez-Torres, P., and Gotor, V. (2006) *Tetrahedron: Asymmetry*, **17**, 2558.
24. Breen, G.F. (2004) *Tetrahedron: Asymmetry*, **15**, 1427.
25. Matcham, G.W. and Bowen, A.R.S. (1996) *Chim. Oggi*, **14**, 20.
26. Shin, J.S. and Kim, B.G. (1997) *Biotechnol. Bioeng.*, **55**, 348.
27. Shin, J.S. and Kim, B.G. (1998) *Biotechnol. Bioeng.*, **60**, 534.
28. Shin, J.S., Kim, B.G., Liese, A., and Wandrey, C. (2001) *Biotechnol. Bioeng.*, **73**, 179.
29. Shin, J.S., Kim, B.G., and Shin, D.H. (2001) *Enzyme Microb. Technol.*, **29**, 232.
30. Cho, B.K., Park, H.Y., Seo, J.H., Kim, J.H., Kang, T.J., Lee, B.S., and Kim, B.G. (2008) *Biotechnol. Bioeng.*, **99**, 275.
31. Hanson, R.L., Davis, B.L., Chen, Y.J., Goldberg, S.L., Parker, W.L., Tully, T.P., Montana, M.A., and Patel, R.N. (2008) *Adv. Synth. Catal.*, **350**, 1367.
32. Höhne, M., Robins, K., and Bornscheuer, U.T. (2008) *Adv. Synth. Catal.*, **350**, 807.
33. Truppo, M.D., Rozzell, J.D., and Turner, N.J. (2009) *Chem. Commun.*, 2127.
34. Hacisalihoglu, A., Jongejan, A., Jongejan, J.A., and Duine, J.A. (2000) *J. Mol. Catal. B: Enzym.*, **11**, 81.
35. For a review on DKR and deracemization reactions, see: Turner, N.J. (2004) *Curr. Opin. Chem. Biol.*, **8**, 114.
36. DKR reviews Pàmies, O. and Bäckvall, J.-E. (2003) *Curr. Opin. Biotechnol.*, **14**, 407; Ahn, Y., Ko, S.-B., Kim, M.-J., and Park, J. (2008) *Coord. Chem. Rev.*, **252**, 647; Martin-Matute, B. and Bäckvall, J.-E. (2007) *Curr. Opin. Chem. Biol.*, **11**, 226; Kamal, A., Ameruddin Azhar, M., Krishnaji, T., Shaheer Malik, M., and Azeeza, S. (2008) *Coord. Chem. Rev.*, **252**, 569.
37. Reetz, M.T. and Schimossek, K. (1996) *Chimia*, **50**, 668.
38. Choi, Y.K., Kim, M.J., Ahn, Y., and Kim, M.-J. (2001) *Org. Lett.*, **3**, 4099.

39 Pàmies, O., Éll, A.H., Samec, J.S.M., Hermanns, N., and Bäckvall, J.-E. (2002) *Tetrahedron Lett.*, **43**, 4699;Paetzold, J. and Bäckvall, J.-E. (2005) *J. Am. Chem. Soc.*, **127**, 17620;Hoben, C.E., Kanupp, L., and Bäckvall, J.-E. (2008) *Tetrahedron Lett.*, **49**, 977.

40 Veld, M.A.J., Hult, K., Palmans, A.R.A., and Meijer, E.W. (2007) *Eur. J. Org. Chem.*, 5416.

41 Parvulescu, A., De Vos, D., and Jacobs, P. (2005) *Chem. Commun.*, **42**, 5307; Parvulescu, A.N., Jacobs, P.A., and De Vos, D.E. (2007) *Chem. Eur. J.*, **13**, 2034; Parvulescu, A.N., Van der Eycken, E., Jacobs, P.A., and De Vos, D.E. (2008) *J. Catal.*, **255**, 206.

42 Parvulescu, A.N., Jacobs, P.A., and De Vos, D.E. (2008) *Adv. Synth. Catal.*, **350**, 113.

43 Kim, M.-J., Kim, W.-H., Han, K., Choi, Y.K., and Park, J. (2007) *Org. Lett.*, **9**, 1157.

44 Gastaldi, S., Escoubet, S., Vanthuyne, N., Gil, G., and Bertrand, M.P. (2007) *Org. Lett.*, **9**, 837.

45 Stirling, M., Blacker, J., and Page, M.I. (2007) *Tetrahedron Lett.*, **48**, 1247;Blacker, A.J., Stirling, M.J., and Page, M.I. (2007) *Org. Proc. Res. Dev.*, **11**, 642.

46 Crawford, J.B., Skerlj, R.T., and Bridger, G.J. (2007) *J. Org. Chem.*, **72**, 669.

47 Alexeeva, M., Enright, A., Dawson, M.J., Mahmoudian, M., and Turner, N.J. (2002) *Angew. Chem., Int. Ed.*, **41**, 3177;Carr, R., Alexeeva, M., Enright, A., Eve, T.S.C., Dawson, M.J., and Turner, N.J. (2003) *Angew. Chem., Int. Ed.*, **42**, 4807.

48 Carr, R., Alexeeva, M., Dawson, M.J., Gotor-Fernández, V., Humphrey, C.E., and Turner, N.J. (2005) *ChemBioChem.*, **6**, 637.

49 Dunsmore, C.J., Carr, R., Fleming, T., and Turner, N.J. (2006) *J. Am. Chem. Soc.*, **128**, 2224.

50 Eve, T.S.C., Wells, A.S., and Turner, N.J. (2007) *Chem. Commun.*, 1530.

51 Bailey, K.R., Ellis, A.J., Reiss, R., Snape, T.J., and Turner, N.J. (2007) *Chem. Commun.*, 3640.

52 Atkin, K.E., Reiss, R., Koehler, V., Bailey, K.R., Hart, S., Turkenburg, J.P., Turner, N.J., Brzozowski, A.M., and Grogan, G. (2008) *J. Mol. Biol.*, **384**, 1218–1231;Atkin, K.E., Reiss, R., Turner, N.J., Brzozowski, A.M., and Grogan, G. (2008) *Acta Crystallogr. F*, **64**, 182–185.

53 Stewart, J.D. (2001) *Curr. Opin. Chem. Biol.*, **5**, 120.

54 Kim, K.H. (1964) *J. Biol. Chem.*, **239**, 783.

55 Sterling, D.I., Zeitlin, A.L., and Matcham, G.W. (1989) U.S. Patent 4,950,606

56 Matcham, G.W., Rozzell, J.D., Sterling, D.I., and Zeitlin, A.L. (1990) U.S. Patent 5,169,780

57 Matcham, G.W., Sterling, D.I., and Zeitlin, A.L. (1992) U.S. Patent 5,300,437.

58 Shin, J.S. and Kim, B.G. (1999) *Biotechnol. Bioeng.*, **65**, 206.

59 Shin, J.S., Yun, H., Jang, J.W., Park, I., and Kim, B.G., (2003) *Appl. Microbiol. Biotechnol.*, **61**, 463.

60 Shin, J.S. and Kim, B.G. (2002) *Biotechnol. Bioeng.*, **77**, 832.

61 Shin, J.S. and Kim, B.G. (2001) *Biosc. Biotechnol. Biochem.*, **65**, 1782.

62 Iwasaki, A., Yamada, Y., Ikenaka, Y., and Hasegawa, J., (2003) *Biotechnol. Lett.*, **25**, 1843.

63 Iwasaki, A., Yamada, Y., Kizaki, N., Ikenaka, Y., and Hasegawa, J. (2006) *Appl. Microbiol. Biotechnol.*, **69**, 499.

64 Martin, A.R., DiSanto, R., Plotnikov, I., Kamat, S., Shonnard, D., and Pannuri, S. (2007) *Biochem. Eng. J.*, **37**, 246.

65 Höhne, M., Kühn, S., Robins, K., and Bornscheuer, U.T. (2008) *ChemBioChem*, **9**, 363.

66 Koszelewski, D., Lavandera, I., Clay, D., Rozzell, J.D., and Kroutil, W. (2008) *Adv. Synth. Catal.*, **350**, 2761.

67 Truppo, M.D., Rozzell, J.D., Moore, J.C., and Turner, N.J. (2009) *Org. Biomol. Chem.*, **7**, 395.

68 Koszelewski, D., Lavandera, I., Clay, D., Guebitz, G.M., Rozzell, D., and Kroutil, W. (2008) *Angew. Chem., Int. Ed.*, **47**, 9337.

69 Ottolina, G., Bianchi, S., Belloni, B., Carrea, G., and Danieli, B. (1999) *Tetrahedron Lett.*, **40**, 8483.

70 Vaijayanthi, T. and Chadha, A. (2008) *Tetrahedron: Asymmetry*, **19**, 93.

71 Dipeolu, O., Gardiner, J., and Stephens, G. (2005) *Biotechnol. Lett.*, **27**, 1803.

72 Li, H., Williams, P., Micklefield, J., Gardiner, J.M., and Stephens, G. (2004) *Tetrahedron*, **60**, 753.

73 Steinreiber, J., Schürmann, M., Wolberg, M., van Assema, F., Reisinger, C., Fesko, K., Mink, D., and Griengl, H. (2007) *Angew. Chem., Int. Ed.*, **46**, 1624; Steinreiber, J., Schürmann, M., van Assema, F., Wolberg, M., Fesko, K., Reisinger, C., Mink, D., and Griengl, H. (2007) *Adv. Synth. Catal.*, **349**, 1379.

Appendix: Solution

Chapter 1

1.1. A synthesis of (S)-(+)-cryptostyline II is presented in Miyaura's paper (K. Kurihara, Y. Yamamoto, `N. Miyaura, *Adv. Synth. Catal.* **2009**, *351*, 260):

Chiral Amine Synthesis: Methods, Developments and Applications. Edited by Thomas C. Nugent
Copyright © 2010 WILEY-VCH Verlag GmbH & Co. KGaA, Weinheim
ISBN: 978-3-527-32509-2

1.2. This reaction was developed by Scheidt (T.E. Reynolds, M.S. Binkley, K.A. Scheidt, *Org. Lett.* **2008**, *10*, 5227). The first key step is the generation of an α-acylvinyl anion by the Brook rearrangement followed by the House–Stork reaction:

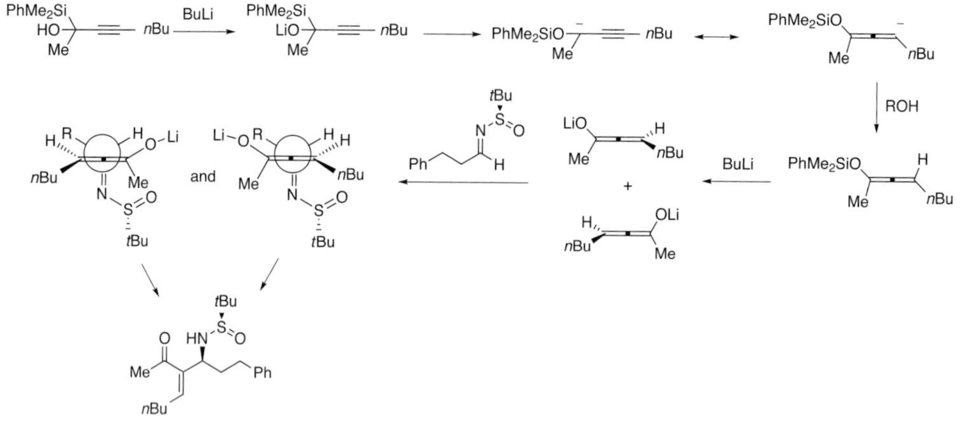

1.3. One potential approach is to use Knochel's catalytic asymmetric alkynylation reaction, the two precursors being readily available from the suggested starting materials. The product is then converted into *N*-acetylcolchinol (J.S. Sawyer, T. L. Macdonald, *Tetrahedron Lett.* **1988**, 4839; T.R. Wu, J.M. Chong, *Org. Lett.* **2006**, *8*, 15).

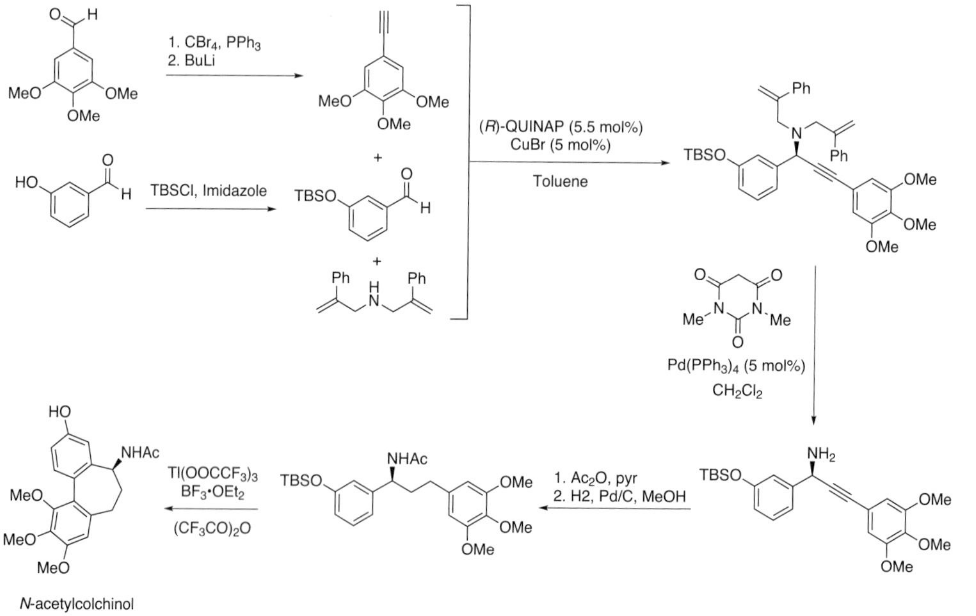

N-acetylcolchinol

Chapter 2

2.1. Only (b) would give synthetically useful yields. In (a) a primary alkyl radical is used, which under the conditions of Table 2.3 will suffer from competition with the ethyl radical generated from Et_3B; the Mn-mediated conditions (i.e., Table 2.6) would be a better choice for using primary halides. In (c) an aryl halide is used, and these are generally not useful in intermolecular radical addition reactions due to the very high reactivity of aryl radicals in comparison to alkyl radicals; relative reactivity may be gauged by examining C−H bond dissociation energies.

2.2. The solvent (CH_2Cl_2) likely is the source of the N−H hydrogen.

2.3. Following the examples of Schemes 2.5 and 2.7, the target may be synthesized from butyraldehyde as shown below. An alternative is radical addition of an enantiopure 2-hydroxy-1-pentyl group (perhaps from Jacobsen hydrolytic kinetic resolution of 1,2-epoxypentane,) to the N-acylhydrazone derived from 5-chloropentanal.

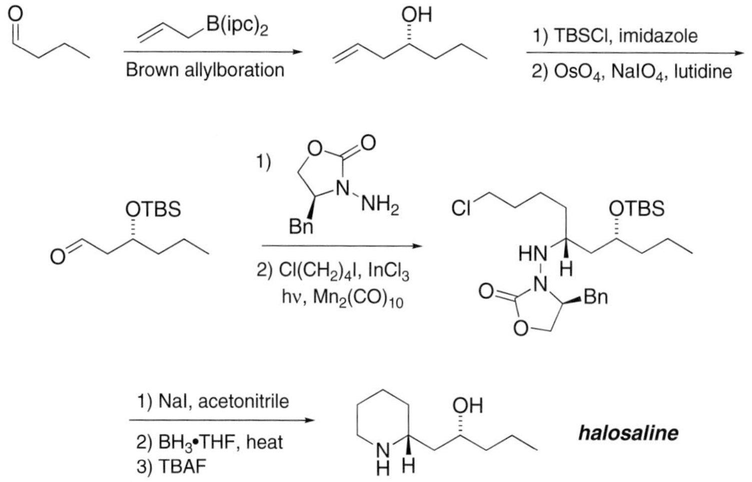

Chapter 3

3.1. In this transfer hydrogenation, aromatization of the dihydropyridine (Hantzsch ester) to form a pyridine derivative is essential for it to act as the hydride source.

3.2. The reversal of enantiofacial selectivity could be explained by the difference in the hydrogen bonding mode. In the reaction of vinyl ether, phosphoric acid would activate the imine only though a double hydrogen bonding interaction (see Scheme 3.3). The *re*-face attack by the vinyl ether on the imine results in the formation of the (2*R*,4*R*)-isomer via transition state **TS1**. On the other hand, in the reaction of the enecarbamate, phosphoric acid would act as an acid/base dual functional catalyst (see Figure 3.2), and both the imine and the enecarbamate would be activated by the catalyst to allow the *si*-face attack by the enecarbamate on the imine, giving the (2*S*,4*S*)-isomer via transition state **TS2**.

3.3. The diastereoselectivity difference between MeAl-BINOL catalysis (>95% *exo*) and Brønsted acid catalysis (97% *endo*) can be rationalized by the transition-state structures. Secondary π-orbital interactions, which are the dominant factor in the *endo*-selective Diels–Alder reactions, are weak in the 1,3-dipolar cycloaddition reaction. In the MeAl-BINOL catalyzed reaction, the *endo* approach (**TS3**) of ethyl vinyl ether to nitrone is disfavored because of the repulsive interaction between the ethoxy group and the bulky Lewis acid. Therefore, the reaction proceeds via the *exo*-orientation (**TS4**) predominantly. In contrast, in the Brønsted acid catalyzed reaction, the much smaller acidic proton allows ethyl vinyl ether to approach in an *endo*-orientation (**TS5**). The *exo*-orientation (**TS6**) might be unfavorable because of the repulsive interaction between the ethoxy group and the aromatic (Ar) group (K.B. Simonsen, P. Bayón, R.G. Hazell, K.V. Gothelf, K.A. Jørgensen, *J. Am. Chem. Soc.* **1999**, *121*, 3845–3853).

Chapter 4

4.1. *Advantages:* (a) No toxic heavy metals involved; (b) catalysts made from natural amino acids; (c) high enantioselectivities with quite a broad range of substrates; starting materials (imines) readily obtained from the commercially available ketones; (d) nonexpensive stoichiometric reagent (Cl_3SiH), which requires anhydrous (but not necessarily anaerobic) conditions; and (e) workup generating environmentally innocuous by-products (NaCl and SiO$_2$).

Limitations: Most of the catalysts currently require imines derived from aromatic amines. *N*-aryl amines are thus easily made (as in the synthesis of SCH48461) but if primary amines are the target, the *N*-aryl group has to be removed (as in the synthesis of colchinol).

4.2. The reagent must be sufficiently Lewis-acidic to effectively coordinate the catalysts. This condition is only met by Cl$_3$SiH. Other silanes are less Lewis-acidic and cannot be used.

4.3. The catalyst must act as a Lewis base. The structural features of Kočovský's catalysts (e.g., **35**) are detailed right below Table 4.1: (a) Valine *i*-Pr (optimized); (b) *N*-Me; (c) *N*–CH=O (formyl); and (d) another anilide group in the molecule. Other catalysts share similar features, in some cases the formamide group can be replaced by α-picolinyl (**41–43**) of sulfinamide group (**46**). Other functional groups that can be considered are, for example, imidazolyl (high affinity to Si), oxazolidinyl, *N*-oxide, phosphine oxides, and so on.

Chapter 5

5.1. Control experiments conducted by the Snapper and Hoveyda group show that electronically activated imines readily decompose in the reaction. Hence, highly electron-rich and as such deactivated imines that are less prone to decomposition but still capable of chelating the silver catalyst were used and furnished the products in high yields.

5.2. The *para*-position of the 3,3'-aryl groups appears only remote on paper. In fact, the 3,3'-aryl groups effectively shield one of the enantiotopic faces of the iminium ion that is bound to the chiral phosphate counterion in an enzyme-like pocket. The *para*-substituents actually play a very supportive role in this scenario.

5.3.

Chapter 6

6.1.

(a) A = acetone; B = sodium cyanoborohydride or Pd/C and H_2.

(b) [structure: 8-fluoro-5-methoxy-chroman with N=C(CH$_3$)$_2$ substituent at 3-position]

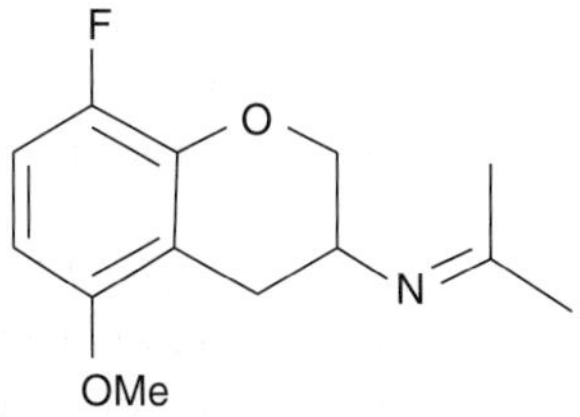

6.2.

(a) [structure: piperazine derivative with tBuO-C(=O)-N, N-C(=O)-OBn, and HN-tBu amide substituents]

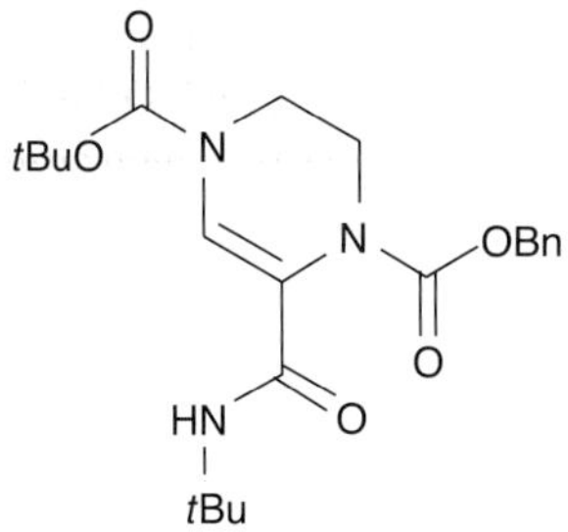

(b) 94% ee

Chapter 7

7.1. (a) See the text of any reliable undergraduate organic chemistry textbook for the mechanism of imine formation. For the hydrogen addition step, addition will be favored from the least hindered face of the *in situ* formed chiral imine. To better understand the low energy conformations of the *cis-* and *trans-N*-phenylethyl imines influencing hydrogen addition, see T.C. Nugent, M. El-Shazly, and V.N. Wakchaure, *J. Org. Chem.* **2008**, *73*, 1297–1305.

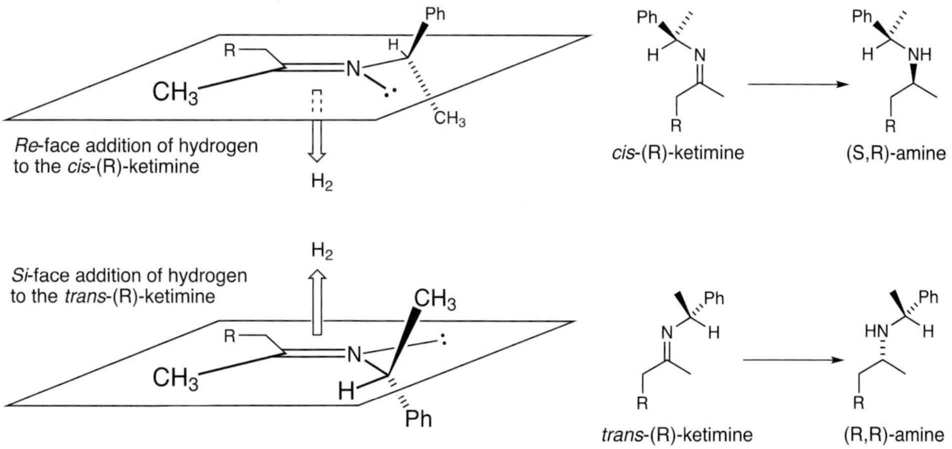

Re-face addition of hydrogen to the *cis*-(R)-ketimine → *cis*-(R)-ketimine → (S,R)-amine

Si-face addition of hydrogen to the *trans*-(R)-ketimine → *trans*-(R)-ketimine → (R,R)-amine

(b) The role of Ti(iOPr)$_4$ has not been clearly defined, but it likely facilitates imine formation via hemiaminal titanate intermediates, collapse of which provides the imine *in situ*. The most important role of Ti(iOPr)$_4$ is thought to be its ability to trap water (ultimately as TiO$_2$) and thereby shift the ketone–imine equilibrium in favor of the imine. For further insight, the reader is directed to (a) T.C. Nugent, A.K. Ghosh, V.N. Wakchaure, R.R. Mohanty, *Adv. Synth. & Catal.* **2006**, *348*, 1289–1299; (b) R.J. Mattson, K.M. Pham, D.J. Leuck, K.A. Cowen, *J. Org. Chem.* **1990**, *55*, 2552–2554; (c) M.T. Reetz, J. Westermann, R. Steinbach, B. Wenderoth, R. Peter, R. Ostarek, S. Maus, *Chem. Ber.* **1985**, *118*, 1421–1440; (d) M.T. Reetz, in *Organotitanium Reagents in Organic Synthesis* (ed. M.T. Reetz), Springer-Verlag, Berlin, 1986, specifically see p. 107.

7.2. The noncompatibility of transition metal hydrides with coexisting ketones result in alcohol by-product formation. The starting (and product) amines are limited to those that only weakly complex with the transition metal species in the catalytic cycle.

7.3. (a) Aniline, its substituted derivatives, and *N*-alkylated forms are well known to be genotoxic, see (a) T.F. Braish, F. De Knaep, K. Gadamasetti, Emerging trends in process chemistry, in *Process Chemistry in the Pharmaceutical Industry, Second Edition, Vol 2: Challenges in an Ever-Changing Climate*, T.F. Braish, K. Gadamasetti (eds), CRC Press-Taylor and Francis Group, New York, 2008, pp. 13–22, specifically see page 17; (b) H.-J. Federsel, A. Sveno, To overcome the hurdles: coping with the synthesis of robalzotan, a complex chroman antidepressant, in *Process Chemistry in the Pharmaceutical Industry, Second Edition, Vol 2: Challenges in an Ever-Changing Climate*, T.F. Braish and K. Gadamasetti (eds), CRC Press-Taylor and Francis Group, New York, 2008, pp. 111–136, see specifically page 121.

(b) *Nitrogen–aryl bond cleavage:* The deprotection conditions for aniline-based reductive amination products are 3.0–5.0 equiv of CAN (cerium (IV) ammonium nitrate), MeOH/H$_2$O, 0 °C, and 6–16 h; see (a) M.N. Cheemala, P. Knochel, *Org. Lett.* **2007**, *9*, 3089–3092; (b) S. Hoffmann, A.M. Seayad, B. List, *Angew. Chem., Int. Ed.* **2005**, *44*, 7424–7427; Y. Chi, Y.G. Zhou, X. Zhang, *J. Org. Chem.* **2003**, *68*, 4120–4122.

Alternatives to CAN exist, for example, use of periodic acid or trichloroisocyanuric acid in CH$_3$CN and H$_2$SO$_4$ (aqueous, 1.0 N) for 16 h; see J.M.M. Verkade, L.J.C. van Hemert, P.J.L.M. Quaedflieg, P.L. Alsters, F.L. van Delft, F.P.J.T. Rutjes, *Tetrahedron Lett.* **2006**, *47*, 8109–8113.

Nitrogen–benzylic carbon cleavage: The deprotection conditions for the reductive amination products of phenylethylamine, also known as α-methylbenzylamine, are Pd-C (0.1–0.5 mol% Pd), 8–12 bar (120–180 psi) H$_2$, MeOH, and 60 °C. Room temperature can be used, but then the Pd catalyst loading must be greatly increased: 1.0–10.0 mol% Pd.

Nitrogen–sulfur bond cleavage: The deprotection conditions for the amine products of *tert*-butylsulfinamide are HCl in MeOH or MeOH/dioxane, 25 °C.

Chapter 8

8.1. Although the polymerized or linked monodentate phosphoramidites are no longer monodentate phosphorus ligands, the nature of their coordination to the central metal of the catalysts resembles to that of the monodentated phosphoramidites. The catalysts are not chelate complexes like those with bidentate ligands. So it is suitable to discuss the polymerized or linked phosphoramidites as monodentate phosphorus ligands.

8.2. According to the mechanism of transition metal-catalyzed asymmetric hydrogenation of N-acetyl enamines, N-acetyl group is considered indispensable for the substrates to form a chelate complex with the metal of catalyst, which is important for the enantiocontrol of reaction. However, there is no N-acetyl group in N,N-dialkyl enamines to form such a chelate complex in the catalytic asymmetric hydrogenation, resulting in a low enantioselectivity.

8.3. In the Rh-catalyzed olefin hydrogenation using monodentate ligands (phosphoramidites, phosphites, or phosphonites), there are two monophosphorus ligands bound to the metal of catalyst in the transition state of reaction. The use of a mixture of two different chiral monodentate ligands L^a and L^b would lead to the formations of two homocombination catalysts [RhL^aL^a and RhL^bL^b] and one heterocombination catalyst [RhL^aL^b]. These three species can be formed in various ratios and are in equilibrium with one another. If the heterocombination catalyst [RhL^aL^b] dominates and/or shows higher activity and enantioselectivity, an improved result can be expected.

Chapter 9

9.1. Rh and Ru. Ir is seldom used in the enamide hydrogenation.
9.2. First example: Cyclic enamides derived from α-tetralone

C   D

There are few efficient catalysts for the hydrogenation of this type of substrates. Recently, Bruneau found that some Ru-catalysts showed good enantioselectivities (over 90% ee) for the hydrogenation of substrates A, but moderate enantioselectivities (up to 72% ee) for the hydrogenation of substrates B (see Refs [60] and [61])

Second example: *ortho*-Substituted α-arylenamides and α-(1-naphthyl)enamides

$$\text{4a: R = H} \quad \text{1a: R = Me} \xrightarrow[\text{H}_2,\ \text{solvent}]{\text{Rh-catalyst}} \text{15a: R = H} \quad \text{2a: R = Me}$$

(substrate: Ph-C(=CHR)-NHAc → product: Ph-CH*(R)-NHAc)

Zhang and Zhang found that the Rh-catalyst with a triphosphorus bidentate phosphine-phosphoramidite ligand can effectively catalyze the hydrogenation of this type of substrates (see Ref. [53]).

9.3. Possible reasons:

(a) Lack of efficient synthetic methods for making enamides.
(b) Difficulty in the removal of the protecting group (normally acetyl group).
(c) High cost of chiral catalysts, and relatively high catalyst loadings.

Chapter 10

10.1. For the hydrogenation of pyridine to piperidine, 3 equiv of molecular hydrogen are needed.

10.2. For the mechanism of Ir-catalyzed asymmetric hydrogenation of quinoline derivatives in the presence of iodine, the catalytic process is a cascade reaction involving a 1,4-hydride addition, isomerization, and 1,2-hydride addition. The mechanism was proposed as follows: The oxidative addition of I_2 to the Ir(I) species precursor **A** generates the Ir(III) species, and subsequent heterolytic cleavage of H_2 can occur to form the Ir(III)-H species **B** with hydrogen iodide elimination. The quinoline substrate could coordinate with Ir(III) species **B** (the I and Cl were omitted for clarity), and then 1,4-hydride transfer affords the intermediate **D**. Subsequent heterolytic cleavage of H_2 with intermediate **D** gives an enamine **F** and regenerates the Ir(III)-H species **B**. The enamine **F** isomerizes to yield imine **G**, which might be catalyzed by the generated HI that acts as strong Brønsted acid, which was also explained by Rueping. Imine intermediate **G** could coordinate with Ir(III)-H species **B** to form the intermediate **H**, followed by the insertion and sigma-bond metathesis to release the product 1,2,3,4-tetrahydroquinolines **P** to complete the catalytic cycle.

10.3. The hydrogenation mechanism of 2,3-disubstituted quinolines was somewhat different from that of 2-substituted quinolines. For the hydrogenation of 2-substituted quinoline, the hydrogenation of C=N bond is the enantioselectivity-controlled step, while the enantioselectivity-controlled step of 2,3-disubstituted quinolines is the isomerization of enamine to imine and the hydrogenation of C=N bond, which is in fact a dynamic kinetic resolution process. To obtain high enantioselectivity, it should meet the equation $K_{iso} \gg K_{hy}$. High temperature could accelerate the rate of isomerization (K_{iso}) of (S)-**G** and (R)-**G**, and low pressure of hydrogen gas can decrease the rate of hydrogenation (K_{hy}). Therefore, the asymmetric hydrogenation reactions should be performed in high reaction temperature and low hydrogen pressure.

Chapter 11

11.1. The acid-catalyzed addition of water or alcohols (ROH) to alkenes involves formation of a carbocation intermediate.

In the analogous reaction with amines, protonation occurs primarily at the basic amine nitrogen rather than the alkene moiety, forming a nonnucleophilic ammonium salt.

11.2. Mechanism **C**, since the addition proceeds via a *syn*-addition and the amine attacks from the less hindered π-face of the allene moiety, which is also more likely coordinated to gold. However, mechanism **B** cannot be ruled out completely as it involves attack *trans* to the coordinated gold, which has to occupy the more hindered π-face of the double bond. Mechanism **A** involving a η^3-allyl intermediate would result in racemization and can be excluded.

11.3. The preferred formation of *trans*-**IV** results from a preferred equatorial placement of the α-methyl substituent that minimizes 1,3-diaxial interactions in the chair-like cyclization transition state.

Preferred formation of *cis*-**VI** can be rationalized with unfavorable *gauche* interactions of the *N*-methyl group with the equatorial α-methyl group in the seven-membered chair-like transition state of the cyclization step leading to *trans*-**VI**. Avoidance of the *gauche* interaction prevails over the unfavorable 1,3-diaxial interactions present in the cyclization transition state leading to *cis*-**VI**.

Chapter 12

12.1. Rh$_2$(*S*-nap)$_2$ is an effective catalyst for the enantioselective C−H amination of the *cis*-olefin but gives lower yield and poor enantioselectivity in the case of the corresponding *trans*-olefin. The opposite behavior is observed with the RuBr$_2$-pybox catalyst **43**.

12.2. Proposed synthesis of the chiral amino acid **52**.

12.3. The proposed chair-like transition state for the α-branched sulfamate minimizes gauche interactions between substituents and lead to the *syn* diastereoisomer. In the case of the sulfamide oxidation, the preference for the *anti*-isomer may be rationalized as a minimization of allylic 1,3 strain.

Minimize allylic 1,3 strain

Chapter 13

13.1.

13.2. Because MVK has α-protons in its structure, severe steric interactions between the Ph and Me group exist in the intermediate **A**, which suffers from larger steric interaction than the intermediate **C**, which has only the steric interaction between the aromatic group and the *N*-sulfonated group. Therefore, intermediate **C** undergoes facile elimination to furnish *R*-enriched adducts and regenerates the nitrogen Lewis base. When methyl acrylate is used in the aza-MBH reaction, the reaction proceeds via intermediate **B** rather than **D** to give *S*-enriched adducts, because methyl acrylate is sterically smaller than MVK bearing α-protons in their structures, and steric interaction between the Ph and OMe group existing in the intermediate **B** is smaller than that between the aromatic group and the *N*-sulfonated group in the intermediate **D**.

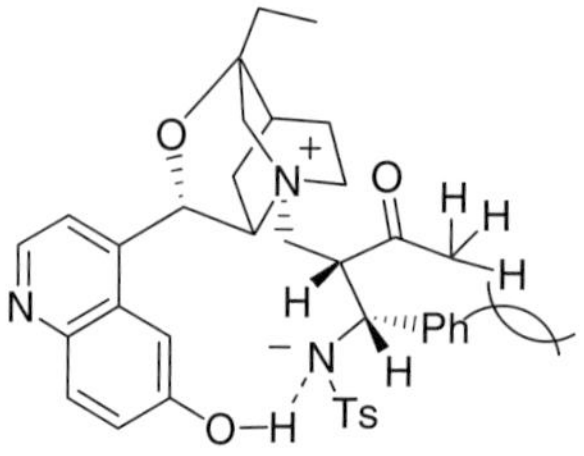

A (give *S*-adduct)

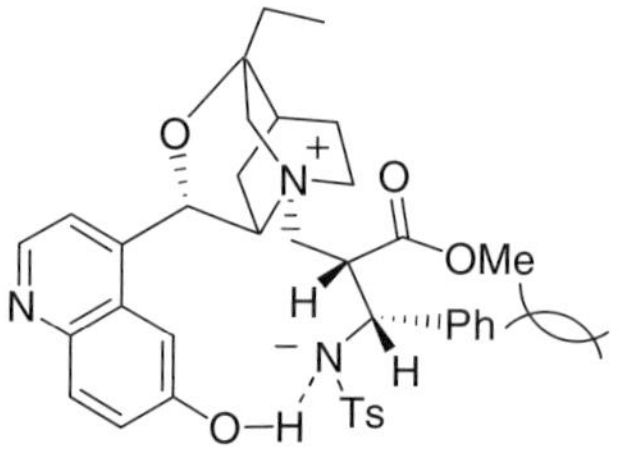

B (give *S*-adduct)

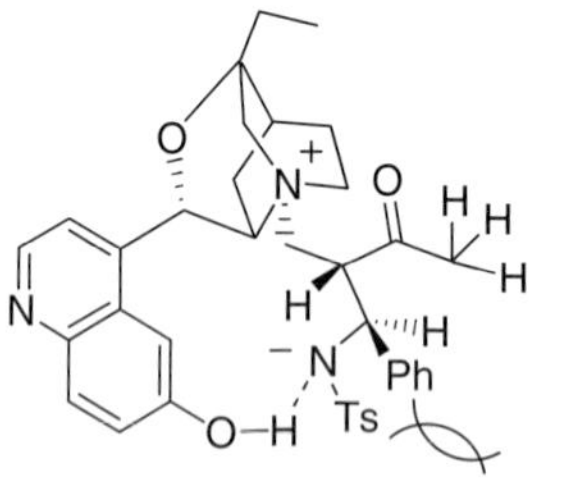

C (give *R*-adduct)

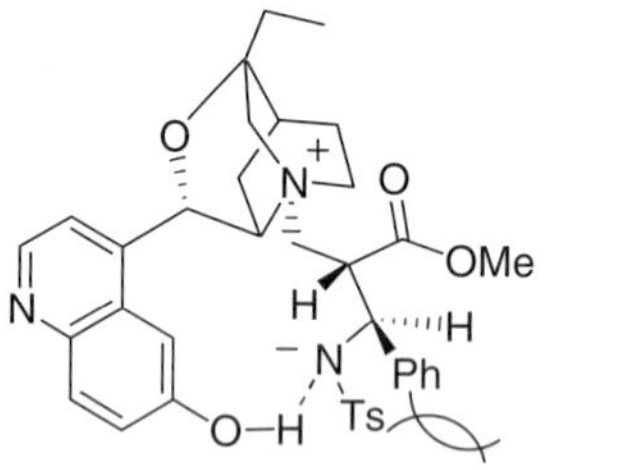

D (give *R*-adduct)

13.3. Kinetic studies on the aza-MBH reaction showed the broken order of 0.5 in imine, which indicates that the proton transfer is the rate-determining step. However, the rate law of the reaction changes in the presence of the Brønsted acid, showing first-order dependence on imine. This clearly demonstrates that the elimination step is not involved in the rate-determining step anymore, and that the proton transfer must be accelerated by these additives. The results obtained substantiate that bifunctional activation using a basic and a protic center is a viable strategy for catalyst design. Furthermore, it was found that phosphine catalysts either alone or in combination with protic additives can cause racemization of the aza-MBH product by proton exchange at the stereogenic center. In striking contrast, the bifunctional catalyst did not induce any racemization on a similar time scale.

Chapter 14

14.1. The principle methods used to prepare enantiomerically pure chiral amines are

(i) kinetic resolution of racemic mixtures using hydrolytic enzymes including lipases, acylases, and proteases. The enzymes can often be used either in deacylation mode under aqueous conditions or acylation mode under low water conditions. The addition of a suitable racemization catalyst allows these kinetic resolutions to be converted to dynamic kinetic resolutions.

(ii) transamination of ketones using transaminases and a suitable amine donor such as L-alanine. Transaminases can also be used for the kinetic resolution of racemic amines by employing a ketone acceptor such as acetone that is converted to isopropylamine in the reaction.

(iii) deracemization of racemic amines using an enantioselective amine oxidase in combination with a nonselective chemical reducing agent in a one-pot process.

(iv) new methods are emerging such as the use of imine reductases and amino acid decarboxylases.

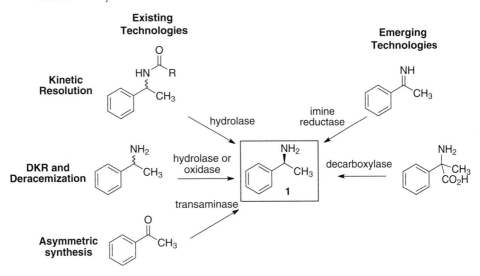

14.2. DKR involves the combination of an enantioselective enzyme (often a hydrolytic enzyme such as a lipase) with a racemization catalyst (often a transition metal catalyst such as Shvo's ruthenium-based catalyst and derivatives thereof) that is compatible with the enzyme. Under these conditions, both enantiomers of the starting material are converted to the product in yields and ee's that can approach 100%.

$$S_R \xrightarrow{k_R} P_R$$
$$\updownarrow \text{racemization}$$
$$S_S \xdashrightarrow{k_S} P_S$$

S_R, S_S = substrate enantiomers
P_R, P_S = product enantiomers
$k_R \gg k_S$ preferably irreversible

k_R, k_S = rate constants

Deracemization involves the combination of an enantioselective oxidase enzyme (e.g., amine oxidase, amino acid oxidase) with a nonselective chemical reducing agent (e.g., amine borane, cyanoborohydride). The key to the process is the generation of an imine intermediate by the oxidase enzyme that is intercepted by the chemical reducing agent *in situ*. As with DKR, yields and ee's in deracemization processes can approach 100%.

14.3. Three key problems are (i) inhibition of the transaminase enzyme by the product amine, (ii) low conversion of the ketone to the amine at equilibrium, and (iii) inhibition of the transaminase by the by-product from the amine donor (e.g., pyruvate if alanine is used as the amine donor). One way to overcome these problems is to use pyruvate decarboxylase to remove the pyruvate, which is converted to acetaldehyde.

Another approach involves the addition of lactate dehydrogenase to the reaction that reduces the pyruvate to lactate. In addition, the use of an ion-exchange resin allows *in situ* product recovery of the product amine that reduces product inhibition of the transaminase.

Index

a

Acetobacterium woodii 455
acetophenone derivatives 230
– Ir-based reductive amination 230
N-acetyl-2-butylindole 323
– asymmetric hydrogenation 323
N-acetyl indoles 324
– asymmetric hydrogenation 324
N-acetyl-1-methylene-1,2,3,4-
 tetrahydroisoquinoline 293, 295
– asymmetric hydrogenation 293
– Rh-catalyzed asymmetric
 hydrogenation 295
N-acetylcolchinol 292
– synthesis 292
achiral imines 15ff.
– catalytic asymmetric allylation of
 imines 24ff.
– sp hybridized carbanions 39
– sp^2 hybridized carbanions 29ff.
 – catalytic asymmetric arylation 32ff.
 – catalytic asymmetric vinylation 31
– sp^3 hybridized carbanions 16ff.
 – copper-catalyzed dialkylzinc
  additions 16ff.
 – early transition metal-catalyzed dialkylzinc
  additions 20ff.
 – rhodium-catalyzed dialkylzinc addition
  reactions 23
 – zinc alkoxide-catalyzed dialkylzinc
  additions 20
achiral iminium cations 231
achiral Ir-catalyst system 345
achiral substrates 386
– enantioselective C-H amination 386ff.
 – with rhodium(II) catalysts 386ff.
 – with ruthenium(II) catalysts 390
acid/base dual function 76

acid-catalyzed aza–ene-type reaction 99
acrylate esters, MBH reaction 398
activated imine 202f.
– asymmetric hydrogenation 203
– direct asymmetric hydrogenation 202ff.
activated pyridines 200
– asymmetric hydrogenation 200
acyclic enamides 276
– catalytic asymmetric hydrogenation
 276ff.
acyclic ketimines
– transfer hydrogenation 110
acyclic ketones
– Mannich reaction 80
acyclic olefins 207
– asymmetric transfer hydrogenation 207
– selectivity 207
acyclic silyl dienolates 165
(Z)-N-acyl-1-alkylidenetetra-
 hydroisoquinolines 293
– asymmetric hydrogenation 293
N-acyl aromatic moiety, *para*-substituents 81
N-acyl-α-arylenamides 274, 276
– synthesis methods 274
N-acyl imines, crotylation 27ff.
– Mannich reaction 79
– transition-state models 29
N-acylhydrazones 51f., 55
– intermolecular radical additions 51f., 59
– tin-mediated radical addition 55ff.
– two-point binding 67
aldol/reduction amination strategy,
 application 236
aliphatic amines 408, 438
– acylation 438
– enantioselective aza-MBH reaction 408
alkanone-based substrates 228
2-alkanones 239

Chiral Amine Synthesis: Methods, Developments and Applications. Edited by Thomas C. Nugent
Copyright © 2010 WILEY-VCH Verlag GmbH & Co. KGaA, Weinheim
ISBN: 978-3-527-32509-2

alkene hydroamination mechanism 358
α-alkoxy-γ-amino acid, synthesis 63
– β-elimination of alkoxy group 63f.
1-alkyl-5-aryl-2,3-dihydro-1H-pyrroles 268
– asymmetric hydrogenation 268
alkyl chain, extension 139
β-alkyl-β-aminocarbonyl derivatives 92
N-alkyl-substituted imines 2
– electrophilicity 2
– nucleophiles type 2
alkyl-substituted N-phosphinoyl imines, enantioselective addition 20
2-alkyl-substituted quinolines 300
alkyl vinyl ketones 411
alkylenamide 254
– asymmetric hydrogenation 254
allylic/benzylic amines, approaches to 31
aminals, preparation 120
amines 343
– deracemization 444ff.
– direct addition 342
– DKR 444ff.
  – using hydrolytic enzymes/racemization catalysts 444ff.
– nucleophilic attack 343
– synthesis 131f., 341f., 450
  – asymmetric, using transaminases 450ff.
  – enantioselective 75
– transaminase/amino acid oxidase-mediated resolution 444
– via hydroamination 341
amino acid dehydrogenase (AADH) 454
amino acid oxidase (AAO) 443
amino acids 431, 450
– α-amino acids 8
– β-amino acid
  – esters, formation 149
  – SCH48461, synthesis 149
  – α-substituted esters 139
– preparation, biocatalytic methods 431
aminoalcohol formation 235
α-amino aldehydes 7
– from glyoxal 7
β-amino aldehydes 85
– double bond isomerization/aza-Petasis–Ferrier rearrangement 85
β-amino esters 81
– single crystallization 140
β-amino imines 99
8-amino-5,6,7,8-tetrahydroquinoline 447
– DKR 447

β-amino-α,β-unsaturated esters 213
– rhodium-catalyzed asymmetric hydrogenations 213
aminoalkenes 347, 350ff.
– catalytic hydroamination/cyclization 347, 351ff.
  – bisoxazolinato rare earth metal catalysts 351
  – enantioselective 350
  – lithium-catalyzed asymmetric 354f.
  – using 3,3′-bis(trisarylsilyl)-substituted binaphtholate rare earth metal complexes 353f.
  – using chiral amino-thiophenolate yttrium complexes 353
  – using diamidobinaphthyl ate complexes 352
aminoalkynes 362, 369
– palladium-catalyzed asymmetric intramolecular hydroamination 362
– titanium-catalyzed tandem hydroamination/hydrosilylation 369
aminoallenes 362f., 368
– gold-catalyzed dynamic kinetic resolution mechanism 368
– gold-catalyzed hydroamination/cyclization 363
  – chiral counteranion effect 363
– titanium-catalyzed asymmetric hydroamination/cyclization 361f.
o-aminobenzamides
– direct transformation 121
δ-amino carbonyl compounds 83
aminodienes 361
– lanthanocene-catalyzed hydroamination/cyclization stereomodel 361
γ-aminoester 65
aminohexenes 350
1-aminoindane 435, 438
– amino acylase I-catalyzed amidation 438
– kinetic resolution 435
aminopentene substrates stereomodel 348
ortho-aminophenol 164
aminostilbenes 355
– lithium-catalyzed cyclization 355
  – kinetic vs. thermodynamic control 355
(S)-aminotetralin 452
– transaminase-catalyzed synthesis 452
aminothiophenolate catalyst system (R) 352
amphetamine, enantioselective oxidation 443
(R)-(−)-amphetamine synthesis 444
(S)-(+)-amphetamine synthesis 444
aniline nucleophile, external attack 344

p-anisidine reductive amination products   232
antifungal aromatase inhibitors, synthesis   434
antimalarial alkaloid quinine synthesis   63
aqueous/organic two-phase system   441
– application   441
N-arenesulfonyl imines   32, 34f., 39
– rhodium-catalyzed arylation   35, 39
– rhodium-catalyzed vinylation   32, 34
aromatic hydrazones   68
– electron-rich/electron-deficient   68
– isopropyl additions   68
– radical additions, development   68
aromatic imines   175, 406, 408
– enantioselective aza-MBH reaction   408
aromatase inhibitors   434
– building blocks synthesis   434
Arthrobacter citreus   433, 452
aryl-alkyl ketones   227, 229, 231, 238
– alkyl-methyl ketones   231
– reductive amination   227
arylenamide   258f., 262, 264
– aryl-substituted imines, direct asymmetric hydrogenation   179–190
– aryl-substituted phosphines   263
– 2-aryl-substituted quinolines   300
– α-arylenamides   280
– asymmetric hydrogenation   258f., 262
– rhodium-catalyzed asymmetric hydrogenation   264, 280
(*Z*)-3-arylidene-4-acyl-3,4-dihydro-2*H*-benzoxazines   295
– exocyclic double bond   295
 – asymmetric hydrogenation   295
N-arylidenediphenylphosphinamides   404, 412
– alkenes   412
N-aryl imines   20, 22f., 80
– copper-catalyzed alkynylation   39f.
– preparation   3
– zirconium-catalyzed alkynylation   42
– zirconium-catalyzed allylation   26
– zirconium-catalyzed dialkylzinc addition   20, 22f.
1-(*N*-arylimino)-1-(aryl)-ethanes   185
– asymmetric hydrogenations   185
Aspergillus melleus   437
Aspergillus niger   448
– monoamine oxidase N (MAO-N)   448
asymmetric hydroamination reactions   341, 348, 369
– chiral catalysts, development   369
– chiral lanthanocene precatalysts   348
– experimental section   369ff.
   – 2-(cyclohexylidenemethyl)-1-[(4-methylphenyl)sulfonyl]pyrrolidine   371

– (*S*)-*N*-phenyl-*N*-[1-{4-(trifluoromethyl)phenyl}ethyl]amine   370
– *S*-(+)-1-phenylpent-4-enylamine (*S*)   370
asymmetric hydrogenation   188, 250, 278, 283ff., 287, 289
– bidentate ligands   289
– catalyst   188
– chiral biphosphonamidite ligands   283
– chiral phosphine-aminophosphine ligands   285
– enamine hydrogenation   215
– imine hydrogenation   186, 202
– phosphine-phosphoramidite ligands   287
– phospholane ligands   278
– pyridine hydrogenation   201
– quinoline hydrogenation   194
– rhodium-catalyzed hydrogenation   344
– transfer hydrogenation   213
asymmetric imine reduction   150
atom transfer radical polymerization (ATRP)   147
axially chiral phosphoric acid   111
aza-Achmatowicz reaction   88
aza-Darzens reaction, *see* aziridine formation reactions
aza–ene-type mechanism   99, 102
– enantioselective, low catalyst load   100
aza-Morita–Baylis–Hillman (aza-MBH) mechanism   397ff., 406f., 416, 419f., 424
– aliphatic imines   424
– catalysts   407
– chiral amines   397, 419ff.
– chiral media   420
– postulated catalytic cycle   398
aza-Petasis–Ferrier rearrangement   84
azadecalinone derivatives   118
aziridine formation reactions   81
aziridines
– *meso*-aziridines, enantioselective desymmetrization   122
– formation reactions   81
– ring opening   121
meso-aziridinium ion
– enantioselective desymmetrization   123
– nucleophilic ring opening   123
azlactones, aldol-type reaction   105
azomethine, 1,3-dipolar cycloaddition reaction   98

b

Bacillus megaterium   442
– (*S*)-amine transaminase   442
Bäckvall system   445

(*S,S*)-BDPMI, Newman projection   282
benzene-sulfonyl groups   301
benzenesulfonyl azide   378
– copper-catalyzed decomposition   378
benzoic acid   416
benzoxathiazine dioxides   204
N-benzoyl imines   83, 120
– allylation   27f.
N-benzoyliminopyridinium ylide   333
N-benzyl 3,4-dihydroisoquinolinium ions   212
– ruthenium-catalyzed transfer hydrogenation   212
N-benzyl imines   26
– catalytic cycle reaction   26
– palladium-catalyzed allylation   26
N-benzyl ketimines, allylation   27, 29
N-benzyl-1-phenyl-6,7-dimethoxyisoquinolinium bromide   212
N-(benzylidene)-4-chlorobenzenesulfonamide   424
– chiral phosphine 23-catalyzed aza-MBH reaction   424
N-benzylidene-4-methylbenzenesulfonamide   415
– reaction   415
N-benzylidene-4-nitrobenzenesulfonamide   423
– methyl acrylate   423
– 1α-catalyzed aza-MBH reaction   423
2-benzylquinolines   300f.
– Ir-catalyzed asymmetric hydrogenation   301
benzyltributylammonium chloride   209
1,1′-bi-2-naphthol (BINOL)-derived monophosphoric acid   76
bicyclic alkenes   345
bidentate phosphorus ligands   266
– BINAP   266
– JosiPhos (1-[2-(diphenylphosphino)ferrocenyl] ethyldicyclohexylphosphine)   266
– SDP7,7′-bis(diphenylphosphino)-1,1′-spirobiindane   266
Biginelli reaction
– enantioselective, three-component coupling   81
– three-component coupling   80
BINAP-Ru(II) complex   295
binaphthyl backbone, 3,3′-positions   78
binaphthyl-carbohydrate monodentate phosphites   258
binaphthyl-type monophosphoramidite ligands   253
BINOL-based chiral phosphate counteranion   229

BINOL-based phosphoric acids   76, 164, 231
– catalysts   231
biocatalysis, use   456
bisimido-ruthenium(VI) complex   392
– H• abstraction/radical rebound mechanism   392
bisoxazolinato rare earth metal complexes   351
– enantioselective hydroamination/cyclization   350
– variety   351
N-Boc amine, arylation   36, 38
N-Boc aromatic imines   87
– enantioselective 1,2-aza-F–C reaction   88
N-Boc-indoles   325f.
– Ru-catalyzed asymmetric hydrogenation   326
Boc-protected amine   324
Boc-protected amino alcohol   422
N-Boc-protected arylimines
– catalytic reaction, *ortho/meta/para*-substituted   77
N-Boc-protected imine   77
N-Boc-pyrroles   328
– asymmetric hydrogenation   328
α-branched amines   1, 8, 19, 23
– *N-tert*-butanesulfinyl chiral auxiliaries   11
– chiral *N*-sulfinyl imines chiral auxiliaries   11
– generation   8
– importance   42
– one-pot preparation   19
– preparation   11
– stereoselective synthesis   1
  – by nucleophilic addition, unstabilized carbanions to imines   1
– three-component synthesis   23
β-branched amines   113
Brassard's diene
– hetero-D–A reaction   95
Brønsted acid-activated trifunctional organocatalyst   417
Brønsted acid binary system
– relay catalysis   108
Brønsted acid-catalyzed Friedel–Crafts-type reaction   170
Brønsted acid-catalyzed vinylogous Mannich reaction   165
Brønsted acids   93, 341, 399, 401
– catalyst   104
– catalyzed isomerization   149
Buchwald titanium catalysts   184
γ-butenolides   173
– γ-aminoalkyl-substituted   160

- furyl ring 88
- products 83
- vinylogous Mannich reaction 173
2,3-butanedione 229
tert-butanesulfinamide, preparation 12
N-tert-butanesulfinyl imines, diastereoselective addition 13
- derived from aldehydes 13
(+)-sec-butylamine 436
- CAL-B-catalyzed resolution 436
1-butyl-3-methylimidazolium (BMIM) 188

c
C–H amination reactions 378
C=N bonds 51, 53, 58, 67f.
- LUMO energy 54
- nucleophilic additions 51
- radical additions 58
  - asymmetric catalysis 67
  - development 68
  - hypothetical N-linked auxiliary approach design 53
  - stereocontrol approaches 51, 53
- reactivity 52
Candida antarctica lipase B (CAL-B) 433f., 437f.
- racemic acetamide hydrolysis 433
carbamate substrate 384
- stereospecific C–H insertion 384
carbanion reagents, nonstabilized 2
- catalytic asymmetric addition 2
α-carbon atom 105
carbon–carbon bond construction approach 51, 66
carbon-carbon bond forming reactions 85
- aza-Cope rearrangement 104
- aza–ene-type reactions
  - aldimines with enecarbamates 99
  - cascade transformations 99
  - enecarbamates, homocoupling reaction 102
  - hemiaminal ethers, two-carbon homologation 100ff.
- aza-Petasis–Ferrier rearrangement 84
- azlactones, aldol-type reaction 104ff.
- chiral phosphoric acids 106ff.
- cycloaddition reactions
  - aldimines, hetero-Diels–Alder reaction 94
  - aldimines with cyclohexenone 95
  - 1,3-dipolar cycloaddition reaction 97
  - Pavarov reaction 96
- diazoacetates to aldimine
  - nucleophilic addition 81ff.

- Friedel–Crafts reactions
  - aldimines, activation 87ff.
  - electron-rich alkenes, activation 91ff.
  - Pictet–Spengler reaction 93
- Mannich reaction 76ff.
- metal complexes, cooperative catalysis 106ff.
- one-carbon homologation reactions
  - aza-Henry reaction 86
  - imino-azaenamine reaction 87
  - Strecker reaction 85
- vinylogous Mannich reaction 83
carbon-heteroatom bond forming reactions
- aminals, formation 119ff.
- aziridines, nucleophilic ring opening 121ff.
- hydrophosphonylation 117ff.
carbon-hydrogen bond forming reactions
- acyclic/cyclic imines, transfer hydrogenation 109ff.
- cascade transformations 116
- ketimines, enantioselective reduction 108
- pyridine derivatives 113ff.
- quinoline, cascade transfer hydrogenation 113ff.
- transfer hydrogenation, application 116
carbonyl compounds, asymmetric α-alkylation 53
catalyst system 341, 345, 389
- design element 389
- development 341
- enantioselectivity 345
catalytic complexes 264
catalytic cycle 345
catalytic intramolecular C–H amination 380f.
cationic gold(I) complexes 368
cationic group 4 metal hydroamination catalysts 357
cationic system 357
Cbz-protected vinylpiperidine, hydrogenolysis 361
Cbz protection-hydrolysis-oxidation sequence 385
Celgene process 451
cetyltrimethylammonium bromide (CTAB) 207
chiral N-acylhydrazones 52ff.
- design 52ff.
- intermolecular radical addition 52ff.
- manganese-mediated coupling with multifunctional precursors 60ff.
  - ester-containing N-acylhydrazones 64
  - hybrid radical–ionic annulation 60

– ketone hydrazones, additions 65ff.
– precursors containing hydroxyl/protected hydroxyl groups 60ff.
– manganese-mediated radical addition 59
– preparation 54
– secondary/tertiary radicals, tin-mediated addition 55ff.
– tin-free radical addition 58
chiral allene 366
– stereoselective gold-catalyzed intermolecular hydroamination 366
chiral amides 179
chiral aminating reagent 388
chiral amines 1, 179ff., 247, 397, 431, 437, 445f.
– β-chiral amines 232
– from carbonyl derivatives 1
– disubstituted nitrogen atom 179ff.
– dynamic kinetic resolution (DKR) 445
    – for CDK2 inhibitors 434
    – using CAL-B 445
    – using ruthenium catalyst 445
– *in situ* racemization 446
– synthesis 368, 431f., 453
    – BASF process 437
    – biocatalytic approaches 432
    – using transaminases 453
    – via tandem hydroamination/ hydrosilylation 368
– trisubstituted nitrogen 211ff.
chiral amino acid synthesis 385
– using Rh-catalyzed C–H amination 385
chiral aminoalkenes 365f.
– kinetic resolution 366ff.
– organolithium-mediated cycloisomerization 365
chiral 1-aminoindanes 251
chiral aminopentenes 367
– catalytic kinetic resolution 366ff.
chiral auxiliary-based approaches 6–15
– imines bearing chiral protecting/activating group 8ff.
– imines derived from chiral aldehydes 7
– induced diastereoselective aza-MBH reaction 400
chiral bidentate phosphorus ligands 273
– BINAP 273
– DuPHOS 273
– PPFAPhos 273
– TangPhos 273
chiral bifunctional *N*-heterocyclic carbenes 418
chiral binaphthyl moiety 261
chiral BINOL-derived amine 408

chiral BINOL-derived diphosphonite ligand 304
chiral bisphosphane ligands 282
chiral Brønsted acid
– catalysis 78
– catalysts 75f.
chiral copper complexes 99
1,4 chiral diamine building block 236
chiral drugs 280
– β-arylisopropylamine units 280
chiral ferrocene-based phosphine-phosphoramidite ligands 286
chiral formamide ligand, *see* sigamide
chiral *N*-heterocyclic carbenes (NHCs) 418
– asymmetric aza-MBH reaction 419
chiral β-hydroxy ketones 235
chiral imines, derived from glyoxal/chiral α-hydroxyaldehydes 8
chiral indoline alkaloid 325
chiral ionic liquids 419
chiral iridium-diphosphine catalysts 182
chiral lanthanocene complex 350
– amine-induced epimerization 350
chiral lanthanum dienolate 174
chiral Lewis base 412
chiral ligand(s) 22, 39
chiral monodentate phosphite ligands 257ff.
chiral monodentate phosphoramidite ligands 249ff.
chiral monodentate phosphorus ligands 262ff.
chiral monophosphoramidites 249, 252
chiral-monosulfonated DPEN ligands 229
chiral multifunctional phosphine 418
– asymmetric aza-MBH reaction catalyzed 418
chiral nitriles, diastereoselective addition 9
chiral 1,3-oxazolidin-2-ones, use 199
chiral α-phenylethylamine-derived phosphine-phosphoramidite ligands 287
chiral phosphine
– aza-MBH reaction 411
– catalysts 411ff.
– ligand Pd complex 343
chiral phosphoric acids 76f.
– catalysts, novel structural motifs 79
– chiral Brønsted acid catalysts 76
– diester (*R*), catalytic asymmetric metal-free hydroamination 359
chiral primary amine synthesis, steps 237
chiral-protected amines 207
chiral *N*-protecting indolines, synthesis 326
chiral rhodium complexes 247

chiral secondary amines 76
chiral sulfamate esters 385
– C–H insertion 385
chiral sulfide 425
– catalyzed aza-MBH reactions 420
chiral tertiary amine catalysts 401
– chiral acid/achiral amine 410
– chiral binol-derived bifunctional amine catalysts 408ff.
– cinchona-derived bifunctional catalysts 401ff.
β-chloro tertiary amines 122
3-chlorobutyraldehyde hydrazone 60, 62
– ethyl addition 60, 62
chloromethyl imines 140
N-(1-(4-chlorophenyl)vinyl)acetamide, hydrogenation 251
cholesteryl ester transfer protein 97
cinacalcet hydrochloride 288f.
– asymmetric hydrogenation 289
cinchona derivatives 402
cinchona-modified heterogeneous Pd/TiO$_2$ 330
cinchona-modified Pd/TiO$_2$ catalyst 199
cis-2,3-dimethylindoline 326
cis-2,6-disubstituted piperidines formation 364
– cyclization transition states 364
cis-2-hexene, tosylamidation 378
Clostridium sporogenes 455
colchinol, synthesis 151
N-containing heterocycles 379
– backbones 255
– direct synthesis 379
– substrates, protected indole/pyrazole 389
Copper–Fesulphos-catalyzed vinylogous Mannich reaction 169
crispine A, deracemization 450
Crixivan, key intermediate, resolution 438
Curtin–Hammett equilibrium constant 366f.
cyano-substituted pyridine derivative 334
cyclic aminals
– benzo(thia)diazine class 120
– synthesis 120
cyclic amino acids, formamides 133
cyclic enamides 253, 289ff.
– asymmetric hydrogenation 253, 290, 292
– catalytic 289ff.
cyclic imines, transfer hydrogenation 111
cyclin-dependent kinase (CDK2) inhibitors, synthesis 434
cyclization cascade, one-pot entry to piperidine derivatives 101

cyclohexadiene 360
– palladium-catalyzed asymmetric hydroamination 360
cyclohexanone monooxygenase (CYMO) 455
cyclopent-2-enone 418
– aza-BMH reaction 418
2-cyclopenten-1-one 401, 413
– aza-MBH reactions 413
cyclopentylamines 346
– stereoselective synthesis 346
   – via asymmetric hydroamination of norbornadiene 346

d

Danishefsky's dienes 95
Dean–Stark apparatus 274
Dean–Stark trap synthesis 238
dehydrating agent 3
– magnesium sulfate/molecular sieves 3
dendrimers 414
– supported monophosphoramidites 257
dendritic phosphoramidite ligands 258
deracemization method 448
– schematic illustration 448
– using amine oxidases 448ff.
N,N-dialkyl enamines 264, 265
– asymmetric hydrogenation 264ff.
   – catalytic 265
diamidobinaphthyl amido ligand 352
diamidobinaphthylate complexes 352
1-(1,2-diarylvinyl)pyrrolidines 267
– asymmetric hydrogenation 267
syn-diastereoisomer 141
diastereoselective aza-MBH reaction 401
– PTA-catalyzed 401
diastereoselective reductive amination 234ff.
– tert-butylsulfinamide auxiliary 240ff.
– chiral ketones 234ff.
– phenylethylamine auxiliary 237ff.
α-diazocarbonyl compounds 81
DIBAL-H 243
α,α-dibranched propargylamines 14f.
α,α-dicyanoalkenes 171
– vinylogous Mannich reaction 171
dienes/allenes/alkynes hydroamination 360ff.
– intramolecular reactions 361ff.
dihydrochloride salt, preparation 399
4,7-dihydroindoles 90
– enantioselective 1,2-aza-F–C reaction 90
7,8-dihydroquinolin-5(6H)-ones 201
– asymmetric hydrogenation 201

(R)-3,4-dimethoxyamphetamine
  synthesis 452
1,2-dimethyl-1,2,3,4-tetrahydroisoquinoline,
  synthesis 197
diorganozinc reagents 18f.
– catalytic asymmetric addition to N-
  phosphinoyl imines 19
– copper-catalyzed catalytic asymmetric
  addition 18
1-(1,2-diphenylvinyl)pyrrolidine 266
– asymmetric hydrogenation 266
1,3-dipolar cycloaddition reaction
– of nitrones 97f.
direct hydrogenation reactions 185
dirhodium catalysts 386f.
N,N-disubstituted iminium 211
– transfer hydrogenation 211–213
2,3-disubstituted indoles 325f.
– asymmetric hydrogenation 326
2,3-disubstitued quinolines 313f.,
  319–321
– asymmetric hydrogenation 313
  – Ir-catalyzed 314
  – transfer hydrogenation 320
– asymmetric reduction 321
DMPEG-hexane biphasic system 306
dominant reaction pathway 382
DpenPhos ligand 255f.
– synthesis process 255
dynamic kinetic resolution (DKR) 314, 445
– reaction scheme 445

e
electron-deficient alkenes 342
electron-deficient porphyrin macrocycle
  383
electron-donating alkyl group 412
electron-donating BINAP chiral ligands 307
electron-poor imines 140, 204, 417
– asymmetric hydrogenation 204
electron-rich diamidobinaphthyl ligand
  framework 352
electron-withdrawing group (EWG) 397
electrophiles 2
– LUMO energy, *ab initio* calculations 2
Ellman's chiral auxiliary method 42
enamides 247ff., 254, 263, 265, 273f.,
  277
– asymmetric hydrogenation 249ff., 273ff.
  – Rh-DIOP catalyzed 247
  – Rh-DuPHOS/BPE-catalyzed 248
  – rhodium-catalyzed 254, 277
  – Ru-BINAP catalyzed 248
  – transfer hydrogenation 112

– homocoupling reaction 103
– reduction 112
– synthesis 274ff.
– tautomerization 112
enamines 213, 216, 247
– alkynylation 39f.
– asymmetric hydrogenation 216, 247
– catalytic hydrosilylation 152f.
– nitrogen analogue 87
– synthesis 313
– transfer hydrogenation 213ff.
β-enamino esters 149
β-enamino nitriles 149
enantioenriched isoindolines, one-pot
  synthesis 91
enantiomerically enriched aza-MBH adducts,
  synthesis 425
enantiomerically pure chiral amines 431
– applications 431
enantioselective aldol reaction 235
enantioselective aza-Cope
  rearrangement 104
enantioselective aza-Darzens reaction 83
enantioselective aza-MBH reaction 409
enantioselective C–H amination 377, 389
– background 378
– representative protocol 392
  – Rh$_2$(S-nap)-catalyzed C–H
    amination 389, 392
  – Ru(Pybox)-catalyzed C–H amination
    393
enantioselective direct Mannich reaction,
  mechanism 77
enantioselective imine reductions 181
enantioselective intermolecular C–H
  amination, Rh$_2${(S)-nta}$_4$-catalyzed 388
enantioselective organocatalytic reductive
  amination 231ff.
enantioselective Pictet–Spengler reaction
– of tryptamine derivatives 94
enantioselective reductive amination 227
enantioselective sigmatropic
  rearrangements 104
enantioselectivity 421
enolate, asymmetric Mannich reaction 157
1,3-enynes/1,3-diynes 13
– hydrogenation 13
ester group, α-hydrogen 405
ethyl nicotinate 199, 330f.
– hydrogenation 330f.
– net enantioselective hydrogenation 199
ethyl transfer process 16
N-(p-ethylbenzenesulfonyl)benzaldimine
  423

– MVK 423
  – β-ICD-catalyzed aza-MBH Reaction 423
exocyclic enamides, synthesis 275

f

fiber membrane contactor 442
formamides
– Lewis-basic organocatalysts 132ff.
N-formyl imines 21
– arylation 34
– copper-catalyzed diethylzinc addition 21
– diethylzinc addition 21
Friedel–Crafts (F–C) reaction 87
– N-Boc aldimines 87
– chiral phosphoric acid catalysts 87
– enantioselective 87
– homologation sequential transformation 102
– indole, mechanistic considerations 92f.
– nitrogen atom, enantioselective formation 93
Friestad's auxiliary 10
frontier orbital calculations 157
functionalized chiral amine synthesis 242
2-functionalized quinolines 302
– Ir-catalyzed hydrogenation 302
furan, organocatalytic aza-Friedel–Crafts reaction 170
furan-2-yl amine products, synthetic utility 88

g

Galipea officinalis 195
gastroesophageal reflux disease (GERD) 440
glucose dehydrogenase (GDH) 453
glyoxal, conversion 7
glyoxylate-derived aldimine 100
– catalyzed 99
glyoxylate hydrazone, reactions 59
gold complex/chiral phosphoric acid binary system
– enantioselective relay catalysis 109
gold nanoparticles 147
Grignard reagent(s) 2, 10f., 13, 18, 43
group 4 metal complexes 356

h

H_8-BINOL-based phosphine-phosphoramidite ligand 286
Hantzsch dihydropyridine ester 113, 318
Hantzsch esters 109f., 132, 316
Hantzsch pyridine 335
hemiaminal ethers
– methyl ether, acid catalyzed reaction 101

– preparation 122
– two-carbon homologation 100ff.
heteroaromatic aldimines 166
heteroaromatic compounds 190, 322
– asymmetric hydrogenation 322
– direct asymmetric hydrogenation 190ff.
hexafluoroisopropyl acrylate (HFIPA) 404
hexamethylphosphorus triamide (HMPT) 255
N-(1H-inden-1-yl)acetamide 261
– asymmetric hydrogenation 261
high-throughput screening methodology 453
HIV-protease inhibitors 236
– lopinavir 236
– ritonavir 236
homoallylamines
– Claisen rearrangement 104
human papillomavirus (HPV) 239
hydrazone 11, 25, 27, 65f.
– allylation 25, 27
– diastereoselective allylation 11
– enantioselective imino-azaenamine reaction 87
– preparation 65f.
hydride reagents 234
– $NaBH_3CN$ 234
– $NaBH_4$ 234
hydroamination 363
– catalyst systems 346, 351
– cyclization mechanism 347
– with enantiomerically pure amines 363ff.
– using achiral catalysts 363ff.
hydrolytic enzymes 432ff.
– amidases/acylases 436
– list of suppliers 432
– use 433
ortho-hydroxyl functionality 78
hydroxylamine hydrochloride 104

i

β-ICD, asymmetric aza-MBH reactions 405
imines 30, 179, 181, 184, 204ff., 208, 210f., 214, 455
– allylation, catalytic systems 30
  – asymmetric reduction 133, 142, 455
  – Brønsted acid-catalyzed isomerization 149
  – catalytic hydrosilylation 152
– asymmetric transfer hydrogenation 204ff.
  – silica-bound catalysts 210
– copper-catalyzed diorganozinc addition 18
  – catalytic cycle 18

- diazoacetate reactions
  - mechanistic proposal 82
  - α-substitution reaction 81f.
- direct cycloaddition reaction 96
- electrophilic activation 76
- enamine 148
  - equilibration 148
  - reduction 140
- enantioselective alkynylation 39
- hydrocyanation 85
- hydrophosphonylation 118
- hydrosilylation
  - catalyst, immobilization 145
  - fluorous silica gel 145
  - formamides, Lewis-basic organocatalysts 132ff.
  - formamide-type catalysts 144
  - mechanistic 147
  - organocatalysts amides 141ff.
  - organocatalysts, sulfinamides 143
  - resin-supported catalysts 147
  - synthetic applications 149f.
- iridium-catalyzed asymmetric hydrogenation 179
  - ligands 184
  - low-pressure hydrogenation 184
- metal-catalyzed hydrogenations, test substrates 181
- Mukaiyama-type Mannich reaction 79
- nitrogen nucleophiles 120
- nucleophilic addition 9
  - chiral auxiliaries 9
- organocatalysts, asymmetric reduction 153
  - reduction 131
- radical addition asymmetric methods 51
- reduction 225
- rhodium-catalyzed arylation mechanism 38
  - diene-catalyzed arylation 38
  - phosphine-catalyzed arylation 34ff.
- rhodium-catalyzed asymmetric transfer hydrogenation 206
- titanium-catalyzed hydrogenation 214
- transition metal-catalyzed asymmetric alkylation 15
- vinylation 31f.

imines preparation methods 3ff.
- N-acyl/N-carbamoyl imines, preparation 6
  - from nonenolizable aldehydes 6
- N-aryl/N-alkyl imines/hydrazones, preparation 3
  - Dean–Stark conditions 3
- N-phosphinoyl imines, preparation methods 5
  - Kresze reaction 6
- N-sulfinyl imines, preparation 3f.
- N-sulfonyl imines, preparation 4

α-imino ester
- enantioselective alkynylation 107
  - chiral phosphoric acid 107
  - silver complex binary catalytic system 107
- enantioselective aza-Henry reaction 86
- enantioselective reduction 111
- transfer hydrogenation 111f.

α-iminoglyoxylate derivatives 52
iminoiodinane 379, 383
- in situ formation 379, 383
- intramolecular nitrene C–H insertion 379

N-iminopyridinium ylides 334
- Charette's enantioselective hydrogenation 334
immobilized phosphoramidites 256
indane, amidation 383
- by manganese(III)-porphyrins 383
N-(1H-inden-1-yl)acetamide 261
- asymmetric hydrogenation 261
indoles 322, 324, 328
- asymmetric hydrogenation 322ff.
- 1,2-aza-F–C reaction 88
- enantioselective 1,2-aza-F–C reaction 89
- enantioselective F–C reaction 88
intermolecular stereospecific C–H amination 385
intramolecular aldol condensation 117
inverse electron-demand aza-Diels–Alder reaction 96f.
ion-exchange resin, use 454
ionic liquids (ILs) 282, 436
- lipase-catalyzed reactions 436
iridium-based racemization catalyst 446
iridium catalysts 187, 200, 266
iridium-catalyzed imine hydrogenation 183
- asymmetric 187
- stereoselectivity 183
iridium-catalyzed vinylation mechanism 33
iridium(III) complexes 305
isopropyl radical addition 58
- diastereoselectivity, stereocontrol elements effect 57f.
isoquinolines 196, 320, 322
- alkaloids, asymmetric synthesis 322
- asymmetric hydrogenation 320ff.
- hydrogenation of, chloroformates 322
- representative synthesis 196f.

k

Kabachnik–Fields reaction   119, *see also* hydrophosphonylation
- β-branched α-amino phosphonates
  - enantioselective synthesis   119
ketene silyl acetals   79
- Mannich reaction   78
ketimines
- reduction   134, 136ff.
  - trichlorosilane   142, 144ff.
ketone hydrazones, preparation   55f.
β-keto nitriles   140

l

lactate dehydrogenase (LDH)   451
lanthanocene catalysts   348
- catalyzed aminoalkene hydroamination/cyclization   349
lanthanum(III)-pybox catalyst   174
Leuckart–Wallach transfer hydrogenation conditions   228
Lewis acid   1, 4, 59, 412
- activation   53
- catalysis, *endo/exo*-isomers   97
- catalyzed vinylogous Mukaiyama–Mannich reactions   159
- heterogeneous hydrogenation catalyst   237
- magnesium atom   13
Lewis base   26, 401
- catalyzed imine reduction   147
- organocatalysts   418
- promoter   132
- sulfinamide   143
ligands
- design, conformational analysis   281
- effect   300
ligated iridium complexes   193
lipases, large-scale application   436
lopinavir, synthesis   236
low-molecular-weight aldehydes   441
- propionaldehyde   441
L-selectride   243
L-tyrosine decarboxylase   456
- use   456

m

Mannich reaction   76, 158, 398, 405
- enantioselective   78
- three-component coupling   80
Markovnikov addition   342ff.
N-Me-pyrroles   327
- asymmetric hydrogenation   327
MeOBIPHEP-ligated bis(gold(I))-catalyst system (S)   342

Merck's HIV protease inhibitor Crixivan synthesis   437
metal cyclopentadienyl bond   348
- catalytic hydroamination *via* reversible protolytic cleavage   348
metal dienolate   157
- ambident reaction profile   157
metal-DIOP complex   279
metal nitrenoid C–H functionalization   377
N-methanesulfonyl indole   325
N-(*p*-methoxybenzenesulfonyl)imine   405
p-methoxyphenyl-protected (PMP) imines   83
methyl acrylate   400
- aza-MBH reactions   400
N-methyl imine   2
- LUMO energy, calculations   2
methylbenzenesulfonamide   425
- Boc-protection   426
- detosylation   426
α-methylbenzylamine (α-MBA)   234, 240
- acylation   433
2-methyl-5-ethyl-aniline (MEA)   226
- imine   184
- iridium-catalyzed asymmetric hydrogenation   184
o-methyl-β-isocupreidine   401
1-methyl-5-phenyl-2,3-dihydro-1H-pyrrole   268
- asymmetric hydrogenation   268
N-methylpyrrolidine   449
- deracemization   449
2-methylnicotinic acid   332
- diastereoselective hydrogenation   332
2-methylquinoline   190ff., 194
- asymmetric hydrogenation   191f.
  - ligands   191
- ruthenium-catalyzed hydrogenation   194
2-methylquinoxaline   197f., 322, 329f.
- asymmetric hydrogenation   197f., 329f.
methyl-substituted silyloxyfurans, reaction   162
2-methyl-substituted vinylketene acetal   159
1-methyl-tetrahydroisoquinoline (MTQ)   440
- enantiomers   440
2-methyl-1,2,3,4-tetrahydroquinoline   193
methyl vinyl ketone (MVK)   399
- aza-MBH reaction   399
(*S*)-metolachlor   180
- industrial synthesis   180
- synthesis   226
Michael acceptor   398, 413
Michael addition   398, 404
model substrates   279, 283f.

– Rh-catalyzed asymmetric hydrogenation   279, 283f.
MOM-protected β-hydroxy-α-arylenamides   281
– Rh-catalyzed asymmetric hydrogenation   281
monoamine oxidase-N (MAO-N) variants
– application   449
– inhibitors   434
monodentate chiral ligands   248
– phosphites   248
– phosphonites   248
– phosphoramidites   248
monodentate phosphite ligands   257, 264
monodentate phosphoramidites   255
monodentate phosphorus ligands   249
monophosphite ligands   249, 257ff.
– application   257
– rhodium complex   258
monophosphonites ligands   262
monophosphoramidite ligands   251, 253ff.
Morita–Baylis-Hillman (MBH) reaction   397, 400
– (β-isocupreidine (β-ICD)   400
MorfPhos ligands   253
Mukaiyama-type Mannich reaction   76

n

N–N bond, cleavage   11
– hydrogenolysis   10
NADH cofactor, recycling   454
nicotinic derivatives   332
– diastereoselective hydrogenation   332
β-naphthol   406ff.
– mediated proton transfer   408
β-nitro amines, synthesis   86
β-nitro-α-amino acid esters   86
nitrogen-based heteroaromatic compounds   299
– asymmetric hydrogenation   299
– enantioselective reduction   299
nitrogen-containing compounds   76
nitrogen heterocycles   158
nitrogen nucleophile, azidotrimethylsylane   122
noncyclic amino acids, formamides   133
noncyclic N-methyl amino acids   135
nonracemic chiral amines   431
– biocatalytic routes   431
nonsolvent system   435
norbornene   345f.
– iridium-catalyzed asymmetric hydroamination   345f.
nosylimines   410
– aza-MBH reaction   410

Noyori catalyst   206
– enantiomeric rhodium analogues   206
Noyori's enantioselective transfer hydrogenation   239
Noyori's [trans-Cl$_2$Ru(diphosphine)(diamine)] motif   180
nucleophiles   1f., 8
– catalytic asymmetric addition   16
– organolithium reagents/cuprates   8
nucleophilic chiral phosphine   412
nucleophilic enamides   102
nucleophilic nitrogen atom   401

o

octahydrofluorenyl ligand, wingspan   349
olefinic sulfamide   382
– Rh-catalyzed C-H amination, side reaction aziridination   382
oligoamide dendrimer-bound ligand   208
optically active amines   299
optically active 1,1′-binaphthylamines synthesis   435
optically pure secondary amines, preparation   439
organoboron reagents, rhodium-catalyzed addition   35
organocatalytic chloroamine synthesis   233
organocatalytic reductive amination   233
organolithium reagent   10
– catalytic asymmetric addition to achiral aldimines   15
organometallic reagents   51
organoselenium containing amines   434f.
– CAL-B catalyzed   435
– chemoenzymatic synthesis   434
organozinc reagents, amino alcohol-catalyzed addition   32–34
oxazolidinones   55, 332f.
– amination   55
– condensation with aldehydes   55
– deprotonation   54f.
– substituted pyridine, asymmetric hydrogenation   332f.
oxime, in situ reduction   444
oxocarbenium ion   104

p

palladium/alkaline earth metal-based racemization catalyst   446
palladium-based reductive amination   229
palladium-catalyzed asymmetric Markovnikov addition   343
palladium-catalyzed intermolecular allylic amination   384
palladium/diphosphine catalysts   203

Pavarov reaction, *see* inverse electron-demand aza-Diels–Alder reaction
penicillin G acylase   437ff.
– enzymatic deacylation   439
– semisynthetic penicillin synthesis   439
pent-4-enylamine   349
– lanthanocene-catalyzed hydroamination/ cyclization stereomodels   349
peptides, amino acid mimics   242
Petasis reaction   32, 34
phase-transfer catalyst   172
phenolic hydroxy groups   408f.
(S)-phenylethylamine, *see* α-methylbenzylamine
phenylethylamine (PEA)   237
2-phenylindole   324
phenyl vinyl ketone   417
N-(1-phenylvinyl)acetamide   253
– asymmetric hydrogenation   253
pH indicator, monitoring transaminase-catalyzed reactions   453
phosphine-aminophosphine ligands, BoPhoz   285
phosphine-phosphinite ligands   182
phosphine-phosphoramidite ligands   288
phosphinite-oxazoline ligands   302
phosphino-oxazoline ligands   187
N-phosphinoyl imines   18, 20
– arylboronic acid addition   36f.
– diorganozinc reagents, nucleophilic attack   20
– *in situ* formation   18
phospholane ligands   279
phosphoramidite ligands   256
phosphoric acids
– catalyst, F–C reaction   92
– features   75
N-phthaloyl enamides, synthesis   275
Pictet–Spengler reaction   93
– aldehydes, tryptamine derivatives   94
(+)-pinidinol, diastereoselective synthesis   364
piperidine alkaloid synthesis   60, 62
piperidinium hydrochloride   333
PipPhos, *see* MorfPhos ligands
polycyclic olefins   345
polyether dendrimer-supported chiral phosphine catalysts, screening   416
polyether dendritic catalysts   414
polymer-based catalysts   257
polymer-supported monophosphite ligands   261
polystyrene-bound ligands   189
ponytails   415
– catalysts   415
N-P(O)Ph$_2$ group, removal strategies   422
practical C–H amination protocols, evolution   377

primary amines   446
– dynamic kinetic resolution   446
primary aminoalkenes   357f., 359
– hydroamination/cyclization mechanism   357ff.
 – using neutral bis(amidate) zirconium catalyst system   358
 – using neutral bis (phosphinic amido) zirconium catalyst system   357
 – group 4 metal imido species   359
prochiral ketones   131
propargylamines, three-component synthesis   40f.
propionaldehyde N-acylhydrazone   56, 58
– 3-chloro-1-iodopropane addition   60, 62
– metal-mediated radical addition   60f.
– tin-free radical addition   58
– use   441
PTA-catalyzed diastereoselective aza-MBH reaction   401
pyridines   198ff., 331, 335
– asymmetric hydrogenation   329ff.
 – auxiliary-directed   200
 – diastereoselective hydrogenation   331
 – homogeneous enantioselective hydrogenation   331
 – transfer hydrogenation   115
– Brønsted acid-catalyzed enantioselective reduction   335
– derivative   329, 331, 409
– nitrogen   138
pyrroles   327f.
– asymmetric hydrogenation   327
– enantioselective 1,2-aza-F–C reaction   90
– hydrogenation   328
pyrrolidine (+)-197B diastereoselective synthesis   364
– via hydroamination/ cyclization   364
pyrrolidinium salts   212
– ruthenium-catalyzed asymmetric hydrogenation   212
pyrroline-based alkaloids   327
pyrroline-based chiral amines synthesis   328
pyruvate, removal   451f.
– by addition of LDH   451
– using pyruvate decarboxylase   452
pyruvate decarboxylase, advantage   452

q

quinolines   190ff., 299, 301ff., 305, 316
– asymmetric hydrogenation   299, 303ff., 310, 312f., 317, 319
 – catalytic systems   309
 – chloroformates   316f.
 – enantioselective   194

– iridium-catalyzed 195, 299ff.
– organocatalyzed 318ff.
– ruthenium-catalyzed 299ff.
– transfer hydrogenation 115, 316
– asymmetric reduction 318
quinoxalines 197, 329
– asymmetric hydrogenation 329
quinuclidine intermediate synthesis 235

r

racemic amides, enantioselective hydrolysis 433
racemic amines 432, 441, 443, 445, 448
– deracemization 448
– DKR, using oxime precursors 445
– *in situ* reduction 448
– kinetic resolution 432ff.
– amine oxidases 443
– *B. megaterium* 443
– hydrolytic enzymes 432ff.
– using transaminases 441ff.
racemic C–H amination 379ff.
– intermolecular C–H amination 382ff.
– intramolecular C–H amination 379ff.
racemic indoline 440
– CAL-A-catalyzed resolution 440
racemic *O*-methyl-*N*-hydroxyamines 449
– enantioselective oxidation 449
racemic *N*-1-phenylethylacetamide, hydrolysis 433
– using *Arthrobacter* species 433
racemic sulfoxamine 388
– kinetic resolution 388
radical–polar crossover reactions 60
Raney-Ni substrate classes 238
rare earth metal-based catalyst systems 347, 357
– σ-bond metathesis mechanism 357
reactive alkenes 343
– sterically strained polycycles 343
– vinyl arenes 343
reductive alkylation, indirect 225
reductive amination strategy 230, 454
– Ir catalysts 230
– product, hydrogenolysis 238
resin-supported catalysts 147
rhodium catalyst 214, 255f., 260f., 285
– BICP catalyst 279, 291
– DIOP catalyst 278
– Me-DuPHOS catalytic system 296
rhodium-catalyzed asymmetric hydrogenations 186, 208, 256, 276, 278, 283f.
– bisaminophosphine ligands 283
– chiral 1,4-diphosphine ligands 278ff.

– chiral phospholane ligands 276ff.
– imine-hydrogenation catalysts 180
– unsymmetrical hybrid phosphorus-containing ligands 284ff.
rhodium(I)-catalyzed arylboronic acid, diastereoselective addition 13f.
rhodium(II)-catalyzed intermolecular amination 383
– using TcesNH$_2$ 383
Ru-porphyrin complex 390
Ru(pybox)-catalyzed C–H amination 391
– stereochemical model 391
ruthenium catalysts 180, 183, 292, 294
– (II)-pybox catalyst, application 390f.
ruthenium-catalyzed asymmetric hydrogenation 247
ruthenium complex
– PhTPAP catalytic system 328
– reductive amination 228
– relay catalysis 108

s

Schiff base 362
– ligand, dipeptide ligands 20
secondary amine(s) 439, 447
– DKR, using iridium catalyst 447
– protease-catalyzed resolution 439
secondary aminoalkene 356, 359
– organocatalytic asymmetric hydroamination/cyclization 359
– hydroamination/cyclization, using cationic chiral zirconium catalyst system 356
semisynthetic β-lactam antibiotics 439
– ampicillin/amoxicillin 439
sigamide 232
sigmatropic rearrangements 104
silica gel 153
– chromatography 241
siliceous mesoporous foam 209
siloxydienes
– hetero-Diels–Alder (D–A) reaction 94
– nitrogen heterocycles 95
2-siloxyfuran 83
– vinylogous Mannich reaction 84
silyl dienolates 159ff.
– Mannich reactions 159ff.
silyl dienol ether, vinylogous Mannich reaction 84
N-silyl imines, allylation 24
2-silyloxy furan 161, 163f.
– silver-catalyzed vinylogous Mannich reaction 161, 163f.
simple alkenes 342, 360
– gold-catalyzed asymmetric hydroamination, with cyclic ureas 342

– intermolecular hydroamination 342ff., 360
simple aminoalkenes 346
– intramolecular asymmetric
  hydroamination 346
  – alkali metal-based catalysts 353ff.
  – group 4 metal-based catalysts 356ff.
  – organocatalytic asymmetric
    hydroamination 358
  – rare earth metal-based catalysts 346ff.
solid-bound catalysts 211
solid-supported ligands 189
spiro monophosphoramidite ligands 250
– SIPHOS 250
spirobiindane phosphoramidite ligand 249
– SIPHOS 249
stereodivergent amine synthesis 241
stereoselective reductive amination 235
Strecker reaction, enantioselective 86
substance P receptor antagonists 234
ortho-substituted arylenamides 288
– Rh-catalyzed asymmetric
  hydrogenation 288
β-substituted enamides 254, 259
– asymmetric hydrogenation 254, 259
1-substituted isoquinolines, asymmetric
  hydrogenations 196
2-substituted N-alkyl-4,5-dihydropyrroles,
  asymmetric hydrogenation 216
3-substituted pyridine derivatives 334
– asymmetric hydrogenation process 334
N-substituted quaternary carbon centers 22
– enantioselective synthesis 22
2-substituted quinolines 196
– asymmetric reduction 321
3-substituted quinolines 320
– asymmetric reduction 320
– transfer hydrogenation 116
substituted pyridines 198
– asymmetric hydrogenations 198
substrate-controlled chiral amine
  synthesis 384
– via C–H amination 384ff.
substrate protection strategy 442
sulfamate esters 382, 386, 390
– diastereoselectivity 386
– enantioselective C–H amination 390
– rhodium-catalyzed oxidative amination 382
N-sulfamoyl imines, arylation 36, 38
sulfinamides 143
N-sulfinimines 400f.
– aza-BMH reactions 401
– aza-MBH reactions 400
N-sulfinyl imines 12
– allylation/benzylation/alkenylation 12
– methylmagnesium chloride, addition 11

N-sulfinyl ketimines, diastereoselective
  addition 14
sulfonamide 379
– preoxidation 379
N-sulfonated imine 402, 411, 417, 424
– asymmetric aza-MBH reaction 402f., 411, 417
– benzoic acid-catalyzed aza-MBH
  reaction 424
sulfoniminamide products 387
– Boc protection 387
– cleavage 387
β-sulfonium silyl enol ethers 421
N-sulfonyl group, cleavage 17
N-sulfonyl imines 18, 31, 33
– copper-catalyzed diorganozinc addition,
  chiral ligands test 18
– iridium-catalyzed vinylation 31, 33
sulfonyl-protected aminoallenes 363
– gold-catalyzed asymmetric hydroamination/
  cyclization 363
sulfur
– as heteroatom 139
– nitrogen heterocycles 139
supercritical fluid chromatography (SFC) 239

t
Takahashi's valinol/phenylglycinol 10
tandem aza–ene-type reaction
– one-pot entry to piperidine derivatives 101
tandem transfer hydrogenation cascade 117
terminal α-arylenamides 260
– asymmetric hydrogenation 260
terminal arylenamides 252, 260
– asymmetric hydrogenation 252
  – rhodium-catalyzed 260
tertiary amines 455
– enantioselective oxidation 455
tetrahydroisoquinoline alkaloid 320
tetrahydroquinoline alkaloids 312, 317, 319
– asymmetric synthesis 312, 317, 319
1,2,3,4-tetrahydroquinolines 190
4,4′,6,6′-tetrakis-trifluoromethyl- biphenyl-
  2,2′-diamine (TF-BIPHAM) 284
tetramethylethylenediamine (TMEDA) 174
tetra-substituted cyclic enamides 291
– asymmetric hydrogenation 291
thiourea-catalyzed aza-MBH reaction 410
three-component coupling reaction
– chiral phosphoric acid 107
– rhodium complex binary catalytic
  system 107
threonine aldolase, use 456
tolane 366
– titanium-catalyzed hydroamination 366

- with 1-phenylethylamine 366
N-p-toluenesulfinyl group 423, 425
- removal strategy 423, 425
N-tosyl-4-alkylidene-1,3-oxazolidin-2-ones 294
- di-substituted exocyclic double bond 294
- asymmetric hydrogenation 294
N-tosyl group 422, 425
- removal procedures 422, 425
- aza-MBH product N-[1-(4-fluorophenyl)-2-formylallyl]-4-
N-tosyl imines 16f., 203, 402, 412, 424
- arylation 35f.
- arylboronic acids addition 36f.
- asymmetric aza-MBH reaction 412
- asymmetric hydrogenation 203
- catalytic asymmetric addition 16f.
- generation by Hudson reaction 5
- reactions 402
- rhodium-catalyzed dimethylzinc addition 24
- synthesis 5
N-tosyl salicylaldehyde imines 403
- aza-MBH reaction 403
- ortho-phenolic hydroxy group 403
transaminase approach 442
- catalyzed amination reactions, protecting groups, use 443
- drawback 442
- ketone product, removal 442
transfer hydrogenation 204
transition metal 341, 377
- based catalysts 16
- mediated homogeneous reductive amination 226ff.
trichloroethoxycarbonyl (Troc) group 101
trichloroisocyanuric acid (TCCA) 149
trichlorosilane 151f.
tricyclic enamines 269
- asymmetric hydrogenation 269
trifluoromethyl-substituted ketimines preparation 6
- via ethanolate adduct 6
α,α,α-trifluoromethylamines 19
trifunctional phosphine catalyst 419, 424
- asymmetric aza-MBH reaction 419
trimethylsilyl (TMS) group 302
- N-trimethylsilyl imine 24
triphosphorus bidentate phosphine-phosphoramidite ligands 288
trisubstituted pyridine derivatives 335
- Ir-catalyzed hydrogenation 335
2,3,5-trisubstituted pyrroles 327
tubulysin γ-amino acids 63
- C–C bond disconnections 63

two-carbon homologation reaction, application 103

u

unmodified substrates 170
- direct vinylogous Mannich reactions 170ff.
unprotected propargylamines, preparation 41

v

valerolactam-derived achiral N-acylhydrazone acceptor 67
- isopropyl addition 67
- radical addition, scope 68
valerolactam-derived catalyst 389
Vibrio fluvialis 442
vinyl arenes 344, 360
- palladium-catalyzed hydroamination mechanism 344, 360
vinyl ethers
- activation 104
- protonation 105
vinylketene N,O-acetal 168
- Brønsted acid-catalyzed vinylogous Mannich reaction 168
vinylketene O,O-acetals 159, 166
- Brønsted acid-catalyzed vinylogous Mannich reaction 166
- TiCl$_4$-catalyzed vinylogous Mannich reaction 159
vinylketene silyl N,O-acetal 167
vinylketene silyl O,O-acetal 167
vinylogous Mannich reaction 158, 161, 167
- products 162, 172
- Ti(IV)-BINOLate-catalyzed 160
- transition states 161
- use 158
vinylorganometallic reagent 31
- catalytic asymmetric addition 31
vinyloxiranes 174f.
- Sc(OTf)$_3$-catalyzed vinylogous Mannich-type reaction 175

w

Wang resins 145
wild-type enzyme 449

x

(+)-xenovenine diastereoselective synthesis 365
- via hydroamination/bicyclization 365

z

Zhou's catalyst 215